Series: Marine Science

Series Editor: Andrés Arias

Volume 1

Marine Biology

A Functional Approach to the Oceans and their Organisms

Editors

Jerónimo Pan

National Scientific and Technical Research Council (CONICET)
Instituto de Geología de Costas y del Cuaternario (IGCyC)
Instituto de Investigaciones Marinas y Costeras (IIMyC)
Mar del Plata, Argentina

Paula D. Pratolongo

Centro de Recursos Naturales Renovables de la Zona Semiárida (CERZOS-CONICET-UNS)
Dto. Biología, Bioquímica y Farmacia-Universidad Nacional del Sur (BByF-UNS)
National Scientific and Technical Research Council (CONICET)
Bahía Blanca, Argentina

CRC Press
Taylor & Francis Group
Boca Raton London New York

CRC Press is an imprint of the
Taylor & Francis Group, an **informa** business
A SCIENCE PUBLISHERS BOOK

Front cover: Oral disc and tentacles of the sea anemone *Bunodosoma zamponii*, from an intertidal pool. The sea lettuce, *Ulva* (*Enteromorpha*) *compressa*, is seen in the upper background.
Photo credit: Jerónimo Pan

First edition published 2022
by CRC Press
6000 Broken Sound Parkway NW, Suite 300, Boca Raton, FL 33487-2742

and by CRC Press
2 Park Square, Milton Park, Abingdon, Oxon, OX14 4RN

CRC Press is an imprint of Taylor & Francis Group, LLC

Library of Congress Cataloging-in-Publication Data

Names: Pan, Jerónimo, editor. | Pratolongo, Paula D., editor.
Title: Marine biology : a functional approach to the oceans and their organisms / Jerónimo Pan, Paula D. Pratolongo.
Other titles: Marine science series.
Description: First edition. | Boca Raton, FL : CRC Press, 2021. | Series: Marine science | Includes bibliographical references and index.
Identifiers: LCCN 2021028238 | ISBN 9780367024987 (hardcover)
Subjects: LCSH: Marine biology.
Classification: LCC QH91 M25 2021 | DDC 578.77--dc23
LC record available at https://lccn.loc.gov/2021028238

ISBN: 978-0-367-02498-7 (hbk)
ISBN: 978-1-032-19390-8 (pbk)
ISBN: 978-0-429-39924-4 (ebk)

DOI: 10.1201/9780429399244

Typeset in Times New Roman
by Radiant Productions

Preface to the Series

Covering more than 70% of the Earth, the oceans regulate our environment including the climate, water purification, carbon storage and primary production. In addition, they also provide provisioning and cultural services, including harvestable goods (fish, shellfish and seaweed), mangroves timber, algae, minerals, recreation, beauty and plenty of cultural benefits. At present, the human population is reaching 8000 million while about half of the planet's suitable land has already been cultivated for crops, exacerbating climate change, biodiversity loss and compromising other ecosystems services. In this scenario, our oceans, which hold 97 per cent of the Earth's water, emerge as the immediate supply answer, not only in the food production sector, but also as a source of energy, biomaterials and even biochemicals/pharmaceuticals.

In these times, when global attention is focusing on oceans and their role in present and future human development, we are humbled to deliver this Marine Book Series through Taylor & Francis, aiming to cover all aspects of oceans and marine research. The present series will provide state-of-the-art coverage of important topics in marine biology and biodiversity, marine ecology, marine biotechnology, marine bioproducts, marine bioprocessing and marine chemistry. We believe this series will be of interest to specialists and nonspecialists alike as well as to scientists and managers engaged in the study of marine biosphere, addressing natural phenomena occurring at coasts, wetlands, estuaries, continental slopes, continental rises, abyssal regions, mid-ocean ridges, along with neritic and pelagic waters of the ocean.

We are pleased to deliver the Marine Series Collection starting point "Marine Biology: A Functional Approach to the Oceans and their Organisms", edited by Dr. Jerónimo Pan and Dr. Paula Pratolongo, which includes contributions from an impressive group of international scientists which will bring a comprehensive analysis ranging from the ocean fundamentals to the coupling and functionalities of different ecosystems—from primary producers to the oceanic megafauna. This first volume will provide a well-rounded perspective of the ocean essentials at a glance, including a reader-friendly style that does not lose any scientific accuracy.

Dr. Andrés H. Arias

Preface

Why a functional approach to marine biology?

Upon opening this book, you are about to embark on a trip into the world's oceans that may lead you from a familiar sandy beach to the far reaches of the open ocean and the deep sea, on a fascinating journey studying marine life and how living creatures, from giant whales to tiny microbes, interact with the environment in which they thrive. There are many possible approaches to the study of a certain subject in the life sciences, and when it comes to a complex, interdisciplinary subject such as Marine Biology, the approaches can be even more challenging. Biology has traditionally put an emphasis on structure over function. Taxonomy and systematics have played a significant role in the development of Marine Biology as a discipline, especially in the 19th and 20th century, following the Linnean tradition established amidst the Romantic Naturalism movement.[1]

Are Marine Biology and Biological Oceanography the same thing? No. Marine Biology is the study of the biological aspects of marine organisms, meaning that the focus is put in the biota itself rather than the marine environment which they inhabit. On the other hand, Biological Oceanography is the branch within the science of oceanography that focuses on the processes mediated by the biota, which have implications for the oceans. Traditionally, scientists who go on cruises look into distribution patterns of biota and relate them to environmental properties are considered biological oceanographers, and those who explore the biological functioning of organisms, e.g., conducting laboratory experiments, consider themselves marine biologists. *Are these two disciplines too far apart from each other?* No, not necessarily. In fact, they only differ in their focuses, while they have many points of contact and overlap. As it has been proposed, the combination of the two disciplinary approaches briefly sketched above leads to a better understanding of the ecology of the oceans.[2] In that sense, this book tries to "reconcile" (for lack of a better term) these two seemingly different disciplines, in the sense that we believe the study of marine organisms may be incomplete, and even limited, without considering them in the context of the environment that they thrive in, and how they interact and modify it through their metabolic, behavioral and reproductive processes.

Undoubtedly, the 20th century saw the advent and consolidation of Ecology as a new discipline in the biological sciences, drawing much of its conceptual framework from experimental developments in marine systems (e.g., island biogeography experiment by MacArthur and Wilson,[3] the concept of keystone species and bottom-up *vs.* top-down controls elegantly introduced by Robert Paine[4] through rocky-shore communities). *Can we separate the study of marine organisms from their environment?* No. The study of Marine Biology necessarily requires an ecological approach; it is even implicit in the very term (*Marine* refers to the environment, *Biology* refers to the life that thrives in and alters such environment). Remember that the term *ecology* was coined by the German zoologist Ernst Haeckel, after the Greek term οἶκος, meaning "house". That is enough reason for us to have primed an ecological approach to the treatment of each chapter in this book.

---

[1] Larson, J.L. 1967. Goethe and Linnæus. J. Hist. Ideas 28(4): 590–596.

[2] Kiørboe, T. 2008. A Mechanistic Approach to Plankton Ecology. Princeton University Press, Princeton, NJ. 209 pp.

[3] MacArthur, R.H. and E.O. Wilson. 1967. The Theory of Island Biogeography. Princeton University Press, Princeton, NJ. 203 pp.

[4] Payne, R.T. 1966. Food web complexity and species diversity. Am. Nat. 100: 65–75.

So why focus on function over structure? In its early stages, the blooming science of Ecology heavily relied on the characterization of communities from a structural point of view; the Braun-Blanquet cover-abundance scale[5] for the analysis of plant communities is perhaps the most remarkable example of such disciplinary approach, and this approach was imprinted into many ecological studies throughout the 20th century. More recently, the "ecosystem service" framework, introduced some 15 years ago,[6] emphasized the (economic) value that the natural functioning of ecosystems provides not only to the biosphere, but also to mankind. It has since been successfully adopted in academic and environmental management circles. What would be the economic costs of clearing a tidal prism of suspended particles in an enclosed embayment if we should need to resort to engineering devices rather than relying on benthic suspension feeders? On a closer look, the ecosystem service framework is inherently linked to ecosystem functioning rather than structure. The Anthropocene has witnessed many changes in the way biogeochemical processes and rates work at a planetary scale. As such, it is the functioning of marine ecosystems that is being compromised.[7] The case of an invading seagrass in the Caribbean Sea, mentioned in Chapter 8, is one example of how ecosystem functioning may be altered, while the structure might remain quite the same.

The field of Marine Biology certainly is not new; in fact, there are some excellent textbooks on the matter that have become classics over the decades since their publication. For our book, we have strived to seek the contributions of renowned and up-and-coming professionals in the different fields of Marine Biology, aiming to produce a fresh text for the new audiences of students and young academics. This was done in our hopes to deliver consolidated knowledge and contribute new perspectives to the young actors that will make a difference in the way we study, make use and preserve our oceans in the future. This book was conceived as an introduction to Marine Biology, for a new generation of scientists, teachers and academics interested in the world's oceans. Of course there are many remarkable Marine Biology textbooks out there, which paved the way for the enlargement of this discipline, and we hope this book measures up to the legacy of renowned authors who conceived those wonderful books.

The book is organized into three sections, namely, an introductory part separated into two chapters and providing a general background to the marine environment; a mid-section consisting of 10 Chapters, which delve into the functional biodiversity of marine organisms (this represents the core of the book); and finally two closing chapters that focus on current topics of special interest such as climate change.

We hope to have produced a high-quality text that will appeal both to upper-level undergraduate and first-year graduate students, professors teaching those courses, and members of the scientific community. For such endeavor, we have taken special care contacting renowned specialists on different topics, which is reflected in the diversity of institutions around the world that are represented among the authors included in this book. With contributions from authors from North and South America, Europe and Japan, we made a conscious effort as editors for the book to be appealing to a broad audience of readers from around the world. We are thankful to the authors who produced high-quality material, mostly based on primary literature sources (i.e., publications from specific scientific journals), covering examples from all major ocean basins and coastal seas on Earth. We are also thankful to those researchers who contributed original photographs and illustrations to enrich the text (credits are given where appropriate).

This is a peer-reviewed book, meaning that all chapters have been evaluated by experts in each field, on the basis of their scientific quality, innovative approaches and potential contributions to each sub-discipline within Marine Biology. Accordingly, this preface would not be complete without the acknowledgement of the superb job carried out by the many reviewers who contributed their expertise and insight to the improvement of each chapter and the book as a whole. They come from broad areas of academia, scientific research, conservation charities, NGOs, and environmental business and technology. They are:

N. Aberle-Malzahn: Department of Biology, Norwegian University of Science and Technology

M. Ahrens: Universidad de Bogotá Jorge Tadeo Lozano

---

[5] Braun-Blanquet, J. 1932. Plant Sociology (Transl. G.D. Fuller and H.S. Conrad). McGraw-Hill, New York. 539 pp.

[6] Millenium Ecosystem Assessment. 2005. Ecosystems and Human Well-Being: Biodiversity Synthesis. World Resources Institute, Washington DC.

[7] Margalef, R. 1963. On certain unifying principles in ecology. Am. Nat. 97: 357–374.

J.M. Bedmar: Argentinian Geological Service (SEGEMAR)
C.F. Berghoff: National Institute for Fisheries Research and Development - Argentina (INIDEP)
G.S. Blanco: Marine Organismal Biology Institute (IBIOMAR)
P.E. Bourdeau: Humboldt State University
G. Cavieres P.: Center of Applied Ecology and Sustainability, Pontificia Universidad Católica de Chile
G. Cepeda: National Institute for Fisheries Research and Development - Argentina (INIDEP) & IIMyC
R.R. Christian: East Carolina University
G. Giardino: Marine and Coastal Research Institute (IIMyC)
S. Gonçalves Leles: University of Southern California
A. Jordi: Jupiter Intelligence
R. McBride: NOAA Fisheries, Northeast Fisheries Science Center
A.C. Morandini: Center for Marine Biology (CEBIMar) & Zoology Dept., Universidade de São Paulo
V. Reyes: Whale and Dolphin Conservation UK & Fundación Cethus
A.F. Rosenthal: Asociación Naturalistas Geselinos-NGO
V. Segura: INIDEP
I. Sunesen: CONICET, Phycology Division at La Plata Museum

Some others have opted to remain anonymous. Through their contributions, they have made this book much more sound and robust; of course they are not liable for any inaccuracies and errors that may remain in the text.

Last but not the least, we would like to express our gratitude to our families and support systems, without whose help the road leading up to the completion of this project would have been even longer and more winding. JP would especially like to thank his twins, Lena and Braulio, for being understanding, encouraging and patient whenever their Daddy had to take time off to work on this book, and also for being the best help on field trips. PP wants to mention the inspiring questions of her son Ulises, a 10-year-old eager explorer of the beautiful oceans.

The final stages of writing and editing for this book took place during the COVID-19 pandemic, a time when most of humankind halted their daily lives and had to take a step aside. During this time, a deep collective conscience on the impacts we have created on our planet started to arise, and many people, perhaps millions who have never given a thought to such matters, noticed their magnitude and importance for the first time. Ours is a pivotal time for the care and preservation of the Earth's oceans, and we hope this is collectively embraced among other new paradigms; as such we hope to have made a small contribution with this book.

February 2021

Jerónimo Pan
Paula D. Pratolongo
Mar del Plata & Bahía Blanca

Contents

Section III: Topics of Current Interest

Section I

Introduction to the Marine Environment

CHAPTER 1

Introduction to the Marine Environment from a Geological Perspective

Paula D. Pratolongo,[1,*] *Jerónimo Pan*[2] and *Diana G. Cuadrado*[3]

If we are to study the biology of marine organisms, we first need to grasp a very important concept that stems from Ecology. According to this principle, the environment in which organisms thrive and develop their life histories has a profound effect on their biology, and the same applies the other way around, as we will learn through countless examples throughout this book. So naturally, the best way to start this book is to provide you with an introduction to the marine environment. This introduction will necessarily be concise and not at all encyclopedic in its scope, but it would rather cover those topics which we deem most relevant for the subject matter of this book. It may be worth mentioning here that there are entire disciplines such as marine geology, sedimentology, fluid and radiation physics, biogeochemistry, oceanography (most notable its geological, physical, and chemical branches) and such, that deal with these issues as their subject matter, and whole treaties have been written on these topics.

This is not a book on oceanography, but on marine biology. Marine Biology is a sub-set of disciplines within the biological sciences that draw mostly on—and deal with the physiology and ecology of marine organisms. Still, it is interesting, if not necessary, to get acquainted with a few facts of our subject matter, the oceans, in a few, concise statements. So, before we move on to our main topic, i.e., the geology of ocean basins and the geological processes that shape them, let us dive head on into some striking facts and figures, to appreciate the grandness of the world's oceans (Fig. 1.1).

Box 1: The oceans in numbers

The oceans cover 71% of the Earth's surface, and harbor 97.5% of its superficial water (much more is trapped within the planet's interior). This represents the largest habitat for the development of life (roughly 97% of the biosphere, Prager and Earle 2000; Fig. 1.1) and in fact, all the major phyla of prokaryotes (archaea, bacteria, cyanobacteria) and eukaryotes (protists, fungi, plants and animals) have marine representatives. On a planetary scale, oceans make up only 0.13% of Earth's volume (0.02% of its mass). The age of the ocean is about 4,000 million yr (MY), but the ocean and atmosphere have been in a steady state for around 800 MY.

Box 1 Contd. ...

[1] Centro de Recursos Naturales Renovables de la Zona Semiárida (CERZOS-CONICET-UNS); Dto. Biología, Bioquímica y Farmacia-Universidad Nacional del Sur (BByF-UNS); National Scientific and Technical Research Council, Argentina (CONICET).

[2] National Scientific and Technical Research Council, Argentina (CONICET); Instituto de Geología de Costas y del Cuaternario (IGCyC); Instituto de Investigaciones Marinas y Costeras (IIMyC).

[3] Instituto Argentino de Oceanografía (IADO); Dto. de Geología-Universidad Nacional del Sur; National Scientific and Technical Research Council, Argentina (CONICET).

* Corresponding author: ppratolongo@gmail.com

...Box 1 Contd.

The average temperature of the ocean is 3.9°C, and the average salinity is 34.48 g kg^{-1} (since salinity is a dimensionless ratio of masses, it does not have units; *this parameter will be expressed without units throughout this book*). The average depth of the ocean is 3,796 m (in contrast, the average land elevation is 840 m above sea level). Ocean's greatest depth is the Challenger Deep (11,022 m below sea level) located in the Mariana Trench, in the western Pacific. An estimate of 440,000 km of shorelines border the world's oceans (Garrison 2009, Seibold and Berger 2017).

The Pacific Ocean basin is the largest and most prominent feature visible from outer space. At 4,207 m above sea level, Mauna Kea is the highest peak, located in the Hawaiian volcanic island chain; with its base lying about 6,000 m below sea level, it has a total elevation of over 10,000 m, making it Earth's highest mountain (from its base, Mount Everest rises 8,848 m above sea level). The worldwide mid-ocean ridge, formed by seafloor spreading, is more than 60,000 km long (1.5 times the Earth's circumference). It takes up roughly one third of the ocean floor surface and represents 22% of the planet's solid surface (as a means of comparison, all continents account for 29%). The average elevation of the Mid-Atlantic Ridge is 2,351 m (Seibold and Berger 2017).

The oceans produce 50 Pg (1 petagram = 10^{15} g) of organic C yr^{-1}, roughly corresponding to ~ 40% of the annual global primary productivity (Libes 2009). This represents an annual uptake of 23% of the anthropogenic CO_2 emitted to the atmosphere. The primary oceanic carbon reservoir is in the form of biogenic calcite produced by marine organisms; these calcareous sedimentary oozes make up more than half (55%) of the oceanic and continental slope deposits.

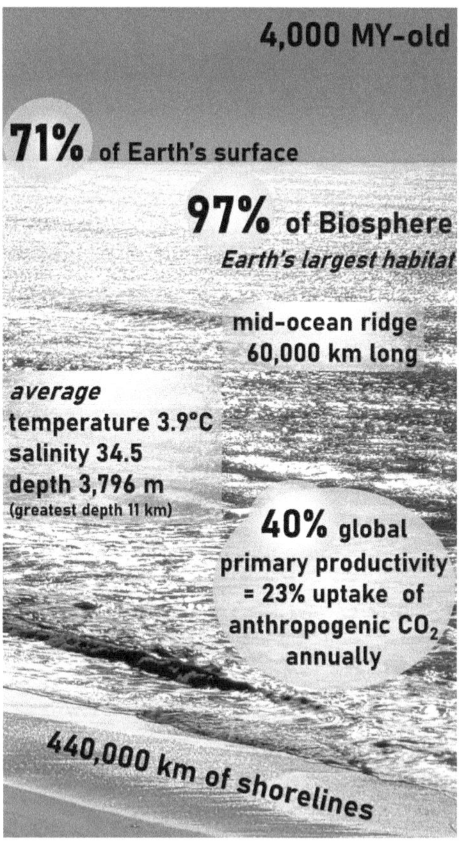

Figure 1.1. A quick reference to ocean basin facts, illustrating their extent and some of their most striking features.

1.1 The formation of ocean basins

In the first place, the most obvious question that someone interested in marine science might come up with is: why do oceans exist? How did they come to be on the surface of a planet that solidified as it cooled? The answer to that (as to many other phenomena in Earth Sciences) is a matter of density. Simply put, the ocean floor (better said, the oceanic crust) is made up of basaltic material, which is denser than the granitic rocks that continental crusts are made of. Therefore, continental crusts appear to be 'floating' on top of the denser material of the ocean floor, and thus ocean basins were and are constantly being formed. But let us first take a step backwards and introduce the concept of plate tectonics.

According to its chemical composition, the Earth's inner structure can be divided into the core, the mantle, and a crust (Fig. 1.2). This division results from density differences between layers as a function of variations in composition, temperature, and pressure (Boillot et al. 1980). The core makes up only 16% of Earth's volume and its density increases with depth. Indirect evidence indicates that the core is mostly composed of iron, making it distinctly different from the silicate material of the mantle. The mantle constitutes the greatest bulk of the Earth's volume (82%), and it is composed of silicate rocks enriched with iron and magnesium relative to the overlying crust. In this outermost layer, the continental crust is different from the crust beneath the ocean basins. The continental crust, much thicker and less dense than the oceanic crust, is composed of granitic, sedimentary, and metamorphic rocks, which form the continents and continental shelves. It is also strongly deformed by faulting and folding and includes the oldest rocks on Earth (in the order of billions of yr, BY). By contrast, the oceanic crust is only about 8 km-thick, composed of dense volcanic rocks like basalt, comparatively undeformed, and geologically younger (~ 200 MY or less).

Regarding mechanical properties, the lithosphere is the outermost shell of the planet and encompasses the crust and upper mantle. Although it is a rigid layer, the lithosphere deforms elastically in geological time scales and varies largely in thickness, from 10s of km beneath the mid-ocean ridges, to as much as 300 km in some continental areas. These differences in thickness relate to the hypotheses of isostasy (Watts 2001). Underlying the lithosphere is the asthenosphere, composed of a highly viscous and ductile material, and isostasy occurs because the lighter lithosphere must be floating on this denser underlying material in the mantle. Isostatic equilibrium occurs where the crust and mantle settle into place in absence of disturbing forces (Fig. 1.3). Thus, the concept of isostasy is used to explain the regional elevations of continents and ocean floors in accordance with the densities of their underlying rocks.

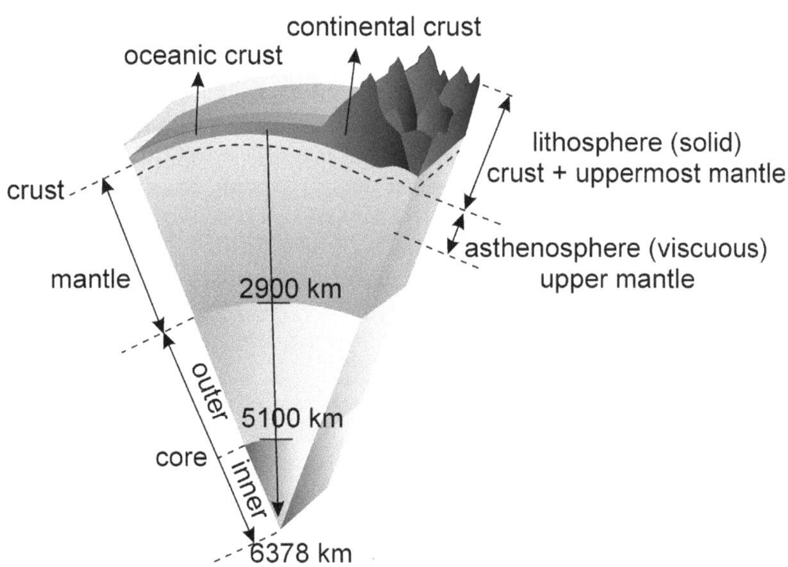

Figure 1.2. Earth's inner structure and its division into three different layers (based on composition and density): the core, mantle, and crust (not to scale).

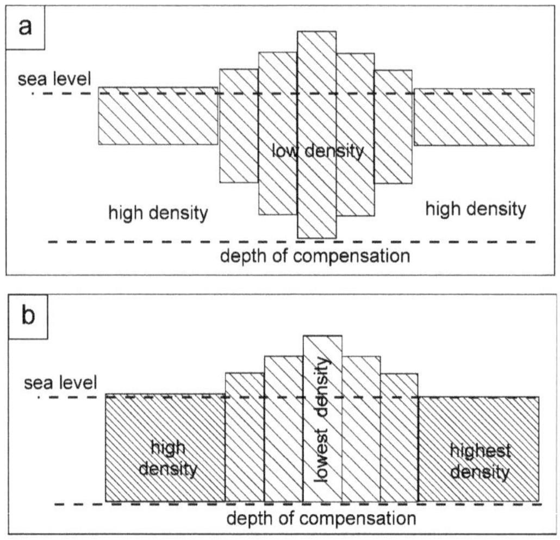

Figure 1.3. Schematic representation of the two main hypotheses explaining how isostatic adjustment is attained. (a) Airy's, and (b) Pratt's hypotheses.

There are two main hypotheses, each named after the scientist that formulated them in the mid-19th century, explaining how isostatic adjustment supports mountain masses. Both of them assume a denser fluid or plastic layer that buoys up the rocky lithosphere. On the one hand, Airy's hypothesis (Fig. 1.3a) assumes that rock density is nearly the same across the lithosphere. As crustal blocks have different thicknesses, thicker blocks of the crust would sink deeper into the asthenosphere, while thinner blocks would be 'buoyed up' by the opposing force created by the asthenosphere. On the contrary, Pratt's model (Fig. 1.3b) assumes that there are lateral changes in rock density across the lithosphere. Thus, blocks would have a uniform thickness below sea level, with its base supporting an equal weight per unit area everywhere, at the depth of compensation. In other words, those areas of the Earth with lower density, such as mountain ranges, project higher above sea level than do those of greater density. Thus, Airy's model provides a better explanation for mountains within continental regions, but Pratt's theory fits better to explain the differences between the granitic continental crust and the denser basaltic ocean basin.

Throughout the Earth's history, there have been several supercontinents (e.g., Kenorland, Columbia, Rodinia, Pannotia, Pangaea, among others) that formed by the grouping of formerly dispersed continents. Pangaea was the most recent supercontinent. It formed during the late Paleozoic and early Mesozoic eras (some 250 MY ago) and was surrounded by the Panthalassic Ocean, the Paleo-Pacific Ocean. Pangaea began to break apart about 175 MY ago, and since then, the modern continents 'drifted' to their present locations. Continental drift describes one of the earliest ways in which geologists thought continents moved over time, as if they had 'drifted' across the ocean bed. In 1912, Alfred Wegener postulated his hypothesis of continental drift to the German Geological Society (Wegener 1912), but it was not accepted for many years by the scientific community, as it was lacking the mechanism for why and how the continents drift. Later on, the British geologist Arthur Holmes proposed that the Earth's mantle contains convection cells that move the crust at the surface (Fig. 1.4). The hypothesis of continental drift has since been incorporated by the theory of plate tectonics, which explains that the continents move as part of plates of Earth's lithosphere.

According to plate tectonics, the asthenosphere and the lower mantle have the same composition but the asthenosphere behaves plastically and flows more slowly. The decay of radioactive elements in the mantle is a heat source that partially melts the asthenosphere, creating a less dense material and forming convention currents (Fig. 1.4). Thus, the solid upper mantle is broken into seven massive and many smaller individual lithosphere plates that move on top of the asthenosphere as a result of the underlying convection cells. Where two plates meet, there might be different types of boundaries (Meissner 2002). Divergent or constructive boundaries occur where two plates are pulled apart from each other. At mid-ocean ridges, new oceanic crust and lithosphere is created by seafloor spreading. As the ocean plate splits,

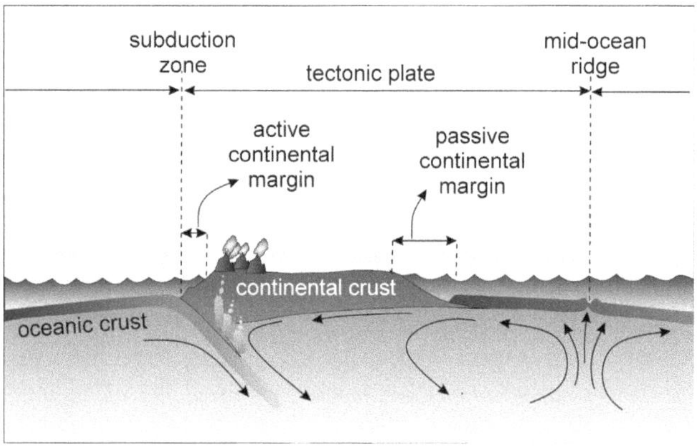

Figure 1.4. Schematic representation of the main mechanisms involved in the theory of plate tectonics (not to scale). Magma convective currents in the asthenosphere are responsible for the circulation of fluidized material. The collision and subduction of oceanic crust under continental crustal plates, and the expansion and formation of new oceanic crust at mid-ocean ridges are illustrated.

the ridge forms at the spreading center, and the ocean basin expands. At zones of continent-to-continent rifting, divergent boundaries may cause new ocean basin to form as the continent splits, spreads, the central rift collapses, and seawater fills the basin. Active zones of mid-ocean ridges (like the Mid-Atlantic Ridge; Fig. 1.5), and continent-to-continent rifting (such as the East African Rift) are examples of divergent boundaries.

Convergent or destructive boundaries, on the other hand, occur where two plates slide toward each other to form either a subduction zone or a continental collision. At zones of ocean-to-continent subduction (e.g., the Andes mountain range in South America), the dense oceanic lithosphere moves underneath the less dense continent. Earthquakes occur as the moving plate sinks into the asthenosphere. As the subducted plate is heated, its materials melt in the upper mantle. The ocean water contained in the seafloor sediments of the subducting plate lowers the melting temperature of the asthenosphere materials, and the resulting magma typically leads to volcanism. There are also convergent boundaries of ocean-to-ocean subduction where older and denser oceanic lithosphere slips underneath a less dense ocean floor. This type of subduction also causes earthquakes and volcanism, and is typically associated with curving chains of volcanic islands (e.g., the Scotia Arc). In continent-to-continent convergent boundaries (e.g., the Himalayas range), there is a collision between masses of granitic continental lithosphere, without subduction. Subduction occurs in a previous stage, as continents approach, but it stops once continents collide. After collision, plate boundaries are compressed, folded, and uplifted and these processes might lead to the closure of ocean basins. Finally, in transform or conservative boundaries two lithospheric plates slide along transform faults, where plates are neither created nor destroyed (e.g., the San Andreas Fault in western North America).

The Wilson cycle is a simple model that explains the cyclical opening and closing of ocean basins. A continent-to-continent divergent boundary leads to the formation of an ocean basin between two lithospheric plates, followed by seafloor spreading, and widening of the basin. Subsequently, subduction is initiated at one of the ocean basin's margins, leading to the closure of the basin, and eventually to the collision of the two continental blocks. Accordingly, continental margins (e.g., the submarine edges of the continental crust) can be classified as active or passive. Active continental margins are typically associated with convergent or transform plate boundaries (e.g., the west coast of South and North America). Because of the mountainous terrain, most of the rivers are short, and the narrow continental shelf quickly drops off into the depths of the subduction trench. Passive continental margins, on the other hand, occur along the remaining coastlines, away from plate boundaries (e.g., the east coast of South and North America). Because there is no collision or subduction, tectonic activity is minimal, and the gentle-sloped passive margins allow the development of long river systems and sediment accumulation on wider continental shelves (Fig. 1.4).

1.2 Glaciations and sea level change

Over the past 3,000 MY there have been several major ice ages in the history of Earth, and the Huronian is the oldest known glacial period. Glacial deposits around Lake Huron (Ontario, Canada) showed that the Huronian glaciation lasted from approximately 2.4 to 2.1 BY ago, in coincidence with the GOE (see Box 2). The intensity or global extent of the Huronian could not be established directly, but rather through indirect evidences that relate this glaciation event with the breakup of the Kenorland supercontinent, along with the significant atmospheric changes introduced by oxygenic photosynthesis (Tang and Cheng 2013).

Late in the Precambrian, for reasons that are not fully understood, the global climate cooled again, and the Earth underwent its most intense glaciation: the Cryogenian period. It is also known as the Snowball Earth glaciation because it is hypothesized that the entire planet was covered by a thick mantle of ice. But nonetheless life survived in the oceans, and the end of the Cryogenian glaciation coincides with the evolution of relatively large and complex life forms that evolved during the Precambrian-to-Cambrian transition, and then continued with the explosion of life forms in the Cambrian (see Box 2). It is currently believed the changing environmental conditions of the Cryogenian may have had a significant impact on the evolution of large and complex marine life (Long et al. 2019). Global sea level fluctuations throughout the Phanerozoic Eon (542 MY to present) are illustrated in Fig. 1.6.

The most recent period of repeated glaciations occurred during the Pleistocene Epoch, beginning about 2.6 MY ago and lasting until about 11,700 yr ago. The end of the last glacial period marks the start of the Holocene Epoch that is not significantly different from previous interglacial intervals within the Pleistocene (De Blij 2012). In the 1920s, the Serbian astrophysicist Milutin Milankovitch proposed a hypothesis to explain the episodic sequence of glacial and interglacial periods within the Pleistocene. According to Milankovitch, this sequence was caused by cyclical changes in the Earth's circumnavigation of the Sun. Variations in Earth's eccentricity (i.e., the deviation of the orbit from a perfect circle), obliquity or axial tilt (i.e., the inclination of Earth's axis with respect to the orbital plane), and precession (i.e., the wobble of Earth's axis) comprise the three dominant parameters of what are collectively known as Milankovitch Cycles (further explained in Chapter 14). The combination of these three parameters defines the amount of solar radiation reaching the Earth's surface at any season and latitude (Seibold and Berger 2017). In turn, solar radiation has a profound effect on the global climate and relates cyclical changes in these parameters to periods of advance and retreat of glaciers (i.e., the extent of the cryosphere; see Chapter 14).

During glacial periods, large amounts of water are stored in ice sheets on land, lowering the mean global sea level, also known as eustatic sea level. Eustatic sea level changes occur when there is a global change in the amount of water stored in ocean basins, or a change in their geometry which alters the volume of water they can hold. When the sea level changes in response to the amount of water stored in ice sheets, it is called glacial-eustasy. During interglacial periods, sea level rises due to land-based ice melting, but also because ocean waters become warmer and therefore expand. Thus, a change in the temperature of seawater also contributes to a change in sea level because seawater expands at higher temperatures.

Figure 1.5. The emerged Mid-Atlantic Ridge at Thingvellir National Park, Island (64°15′ N). Visitors literally walk through a rift between the North American and Eurasian plates. Photo credit: J. Pan.

Box 2: The early Earth, young oceans and the beginning of life

The age of the Earth is approximately 4.5 billion yr (BY). According to the Nebular theory, our planet formed from a disk of dust and gas orbiting the newly-formed Sun *via* accretion, a process where small rocky bodies called planetesimals collided, gradually growing into a hot and completely molten protoplanet. As the rate of planetary accretion slowed down, the Earth began to cool, forming its primordial crust, a thin shell on its outside. At this early stage, the planet was waterless, and most scientists agree that the atmosphere and the oceans formed by the continual 'degassing' of the Earth's interior. Volcanic activity allowed gases formed within the Earth to escape, forming a primitive atmosphere mostly composed of water vapor, CO_2, and dinitrogen (N_2). Once the surface temperature was cool enough, the water condensed into torrential rains which filled the basins of the young oceans about 4,000 MY ago. CO_2 dissolved easily into water, and then created vast carbonatic deposits and surface rocks such as limestone, leaving an atmosphere mostly composed of N_2.

Relatively early in Earth's history, a variety of different metabolic pathways would have evolved and allowed various ways of life (Oschmann et al. 2002); the general consensus favors a heterotrophic metabolism for early life forms over an autotrophic one. Thus, before photosynthesis evolved, the Earth's atmosphere had no free oxygen (O_2), but sometime between 2.4 and 2.3 BY ago, during the Great Oxidation Event (GOE), the Earth's atmosphere and the shallow ocean experienced a seemingly enormous increment in O_2 content, with atmospheric O_2 reaching $> 1\%$ of modern levels (Hazen et al. 2008, Pufahl and Hiatt 2012). The driver of the GOE was the gradual accumulation of photosynthetic residual O_2 from cyanobacteria, which also presented a peak in their diversity (Noffke and Awramik 2013, Lyons et al. 2014). The transition from a non-oxidizing atmosphere to one characterized by free O_2 has been well recorded in sedimentary sequences (Eriksson and Cheney 1992), by the disappearance of banded-iron formations (BIFs) which had presented a continuous record dating back to at least 3.9 BY ago (Koehler et al. 2010). BIFs are controlled by the availability of ferrous Fe^{+2} in seawater, and from 1.9 BY ago onwards, they became discontinuous and were replaced by red beds (oxidized Fe^{+3}) in sediments.

Recent studies showed that fully oxygenated water columns existed as early as 2.5 BY ago, thus O_2 may have accumulated in Earth's oceans well before than in the atmosphere (Ostrander et al. 2019). Photosynthetic prokaryotic organisms that produced O_2 as a waste product lived long before the first build-up of free O_2 in the atmosphere, perhaps as early as 3.5 BY ago in shallow lagoonal and peritidal environments of an ancient ocean (Awramik et al. 1983). This represents the earliest undisputed evidence of life on Earth, although life may have evolved even earlier (see Mojzsis et al. 1996). The structure of these ancient microbes was noted to be similar to bacteria found in modern hydrothermal vents, supporting a hypothesis that life may have begun at deep marine hydrothermal settings (Nisbet and Sleep 2001, Dodd et al. 2017). Other hypotheses favor the existence of microbial mats of bacteria and archaea coexisting in simple biofilms in which microbes with different metabolisms cooperated in the exploitation of diversifying niches (Nisbet and Fowler 1999).

Stromatolites are conspicuous fossils left by some of the first life forms throughout the Precambrian (Grotzinger and Knoll 1999, Walter and Allwood 2005). Some modern stromatolites occur in shallow seas at very few spots on Earth (such as Hamelin Pool in Shark Bay, Western Australia, or in The Bahamas), and have recently been found on high-altitude volcanic lakes (Farías et al. 2013). The prevalence of well-illuminated, warm, alkaline (carbonatic), hypersaline waters seem to be a condition for their occurrence. Stromatolites consist of three-dimensional microbial mat laminae that are sequentially stacked-up in layers (*stroma* = Greek term for mattress) and grow in height by accretion and *in situ* microbial aragonite precipitation in the form of thin laminations. The thin limestone laminations in stromatolites can accrete over time, resulting in the banded pattern common to stromatolites. These layered carbonatic biosedimentary structures may have been laid down by the same cyanobacteria that were largely responsible for increasing O_2 through photosynthesis. Stromatolites peaked about 1.25 BY ago, when a wide range of prokaryotes existed (Noffke and Awramik 2013), and subsequently declined in abundance and diversity, so that by the start of the Cambrian they had fallen to 20% of their peak. The most widely supported explanation for their demise is that stromatolite builders fell victims to animal grazers.

Box 2 Contd. ...

...Box 2 Contd.

The so-called Cambrian substrate revolution occurred at different rates in different locations throughout much of the Cambrian period (542–488 MY ago). The first animals (metazoans) began to burrow vertically, forming a great diversity of burrows preserved in fossil records. These burrowing animals were able to break down microbial mats that extensively colonized shallow marine sediments up to that point, and thus allowed water and O_2 to penetrate down to considerable depths below the sediment surface (a process known as bioturbation and further explained in Chapter 8). Before the Cambrian, widespread microbial mats formed a 'blanket', cutting off the underlying sediments from the ocean water above. This meant that the sediments were anoxic and hydrogen sulfide (H_2S) was abundant. The exchange of sedimentary pore water with oxygenated ocean water enhanced by bioturbation was essential to make the sediments habitable for a much wider range of organisms. The upper layers of seafloor sediments became wetter and softer as they were constantly being bioturbated by burrowers, and the microbial mats that had covered the Precambrian sea floor became increasingly restricted to a limited range of environments such as peritidal hypersaline basins (Seilacher et al. 2005).

Sea level changes may be described as relative or absolute. Even though glacier melting produces a higher absolute height of the oceans, the relative height of the sea with respect to land (the relative sea level) may be quite different for a given coastal region. While the eustatic sea level has been rising from the Last Glacial Maximum (LGM) to the present, i.e., during the past 20,000 yr, the relative sea level can vary from place to place due to the effect of eustatic, isostatic, ocean dynamics, tectonic, and local factors. After the LGM, the extensive melting of large ice sheets produced a transfer of mass from the continents to the ocean, which increased the ocean volume and uplifted the Earth's crust in a process of glacio-isostatic adjustment (Khan et al. 2015).

Intermediate-field regions, like the temperate Atlantic and Pacific coasts on North America, extended for several thousand km beyond the ice sheet margins. In these regions, the mantle material flowed off the forebulge (bulge of the lithosphere peripheral to glaciated areas), in the direction of the area formerly depressed by the ice sheet. This adjustment of the lithosphere (called glacio-isostatic subsidence; Fig. 1.7a, b, c) combined with the eustatic effects to produce a continuously rising relative sea level

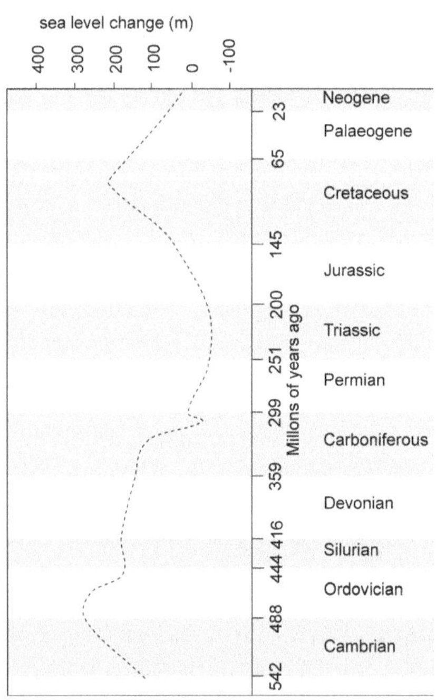

Figure 1.6. Global sea level fluctuations throughout the Phanerozoic Eon (542 MY to present).

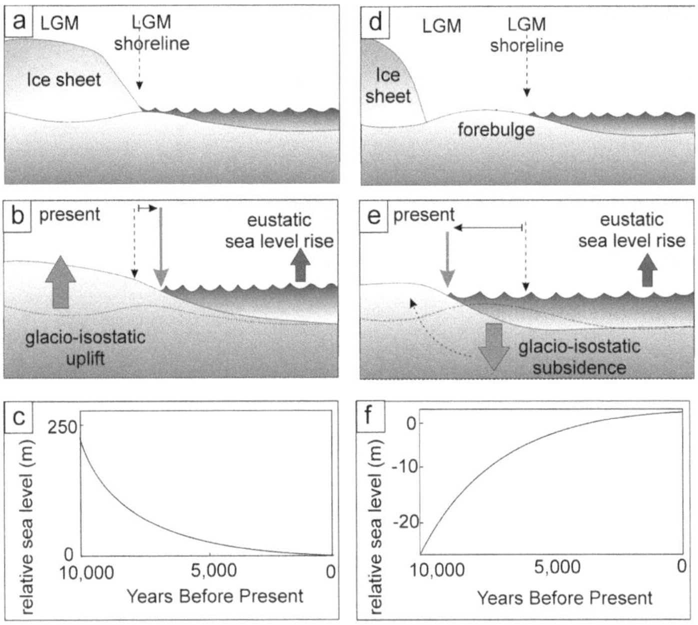

Figure 1.7. Schematic representation of relative sea level changes in regions experiencing glacial isostatic adjustment. (a–b–c) Holocene sea level changes for intermediate-field locations. (d–e–f) Holocene sea level changes for near-field locations. The sea level curves in (c) and (f) do not represent specific locations; the vertical scale represents relative sea level change in m (negative values are below present sea level; positive values above).

since the LGM. In coastal settings located near the edges of large ice sheets (near-field locations), the continental crust rose after the removal of the huge weight of ice (a process known as glacio-isostatic uplift; Fig. 1.7d, e, f). In those regions, the rate of crustal rebound was faster than the eustatic sea level rise, thus the relative sea level fell continuously. Thus, in these typically high-latitude coastal regions (e.g., Fennoscandia, Finland, Labrador), the shoreline prograded, and new land emerged.

Away from major ice sheets (far-field locations), eustatic contributions exceeded glacio-isostatic adjustments and relative sea levels commonly presented a mid-Holocene highstand (i.e., a sea level maximum) above the elevation of present shorelines (Fig. 1.8). In low latitude regions, the subsequent relative sea level fall to current position was mainly due to ocean siphoning (i.e., the migration of water away from the tropics, into subsiding forebulges at higher latitudes). Near continental margins at higher latitudes, a crustal tilting upwards towards the continent, due to continental shelves loading by meltwater, produced the observed relative sea level fall during the late Holocene (Kopp et al. 2015). In far-field locations (e.g., the East coast of South America, west Australia, east China, South Africa), the coast initially receded from a rising sea until the relative sea level reached a transgressive maximum. After the highstand the shoreline prograded, as the relative sea level decreased to its present elevation.

1.3 Estuaries and coastal zones

Estuaries are major components of coastal zones; the global estuarine surface area has been estimated at 1.1×10^6 km^2 (Dürr et al. 2011). There are many definitions of estuaries, and most of them follow the disciplinary background or research aims of those who study them and the processes of interest. From a classical perspective, estuaries may be defined as semi-enclosed coastal bodies of water, with an open connection to the ocean, and within which seawater is diluted with freshwater derived from land drainage (Pritchard 1967).

Worldwide, estuaries adapted to the recent rapid rise in sea level that took place during the Holocene (after the LGM ~ 20,000 yr ago to the end of deglaciation melting ~ 6,000 yr ago), along with the large environmental changes brought about by new climatic conditions. At any time, the balance between the

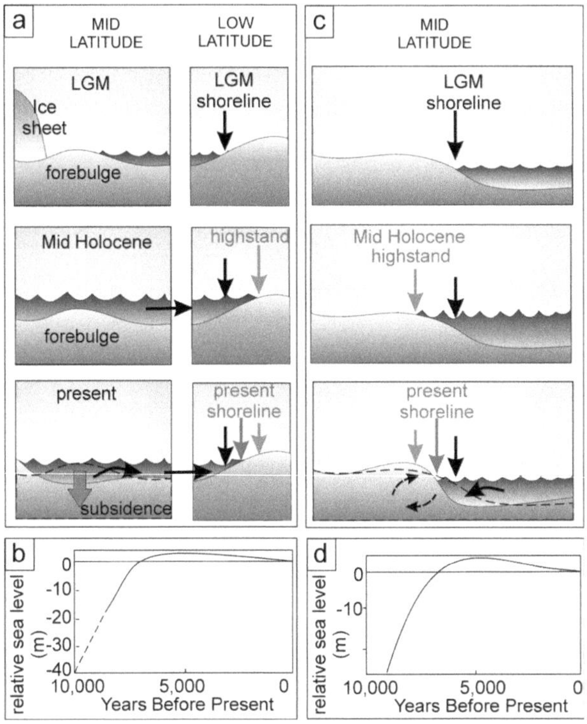

Figure 1.8. Schematic representation of relative sea level changes for far-field locations at low latitude (a–b) and mid-latitude (c–d) regions. The sea level curves in (b) and (d), showing a mid-Holocene highstand, do not represent specific locations; the vertical scale represents relative sea level change in m (negative values are below present sea level; positive values above).

regional rates of relative sea level change and the local sediment budget determined the evolution of an estuary. For instance, the evolution of estuaries under a rising relative sea level is mainly controlled by deposition of allochthonous mineral sediments or *in situ* accumulation of organic matter, which allow vertical accretion at rates that keep pace with the rise in sea level. In Chesapeake Bay, wide, shallow estuaries fringed by vast wetlands formed at locations with abundant sediment inputs from rivers and the sea, while experiencing rates of relative sea level rise of less than a few mm yr^{-1} throughout the late Holocene. However, these pre-historical rates contrast sharply with modern estimations based on Chesapeake Bay tide-gauge records (3.1–4.5 mm yr^{-1}, according to Cronin et al. 2019), raising concern on the wetland ability to keep pace with sea-level rise. Conversely, in more open systems with smaller sediment inputs, sea level rise ended up 'drowning' estuaries. On gently sloping coastal settings, sediment-starved estuaries moved upland along with coastal wetlands, creating a varied succession of wetland habitats, including the die-off of mangroves and salt marshes, their capping by inorganic sediment and later colonization by seagrasses. Where the relative sea level fell and new land emerged constantly, estuaries migrated seaward and the old coastal features became part of the terrestrial landscape. Estuaries trap sediments brought by rivers, thus silting occurs as shorelines prograde. Under these circumstances, organic estuarine soils that formed at or near sea level have been raised to higher elevations and effectively perched above their formation zone (Stevenson et al. 1986).

In this chapter, we first introduce the so-called geomorphologic classification of estuaries, which considers four different types. (1) Coastal plain estuaries, also termed drowned river valleys, occur in coastal settings under rising relative sea level. In these estuaries, seawater progressively inundated river valleys preserving their original topography (Fig. 1.9a). (2) Bar-built or lagoon-type estuaries typically develop on gently sloping plains, enclosed by sand spits or barrier islands, with narrow inlets allowing water exchange with the open ocean (Fig. 1.9b). (3) Fjords (or fiords, a word derived from the Norse *fjǫrðr*) were formed by Pleistocene glaciers that deepened and widened existing river valleys that

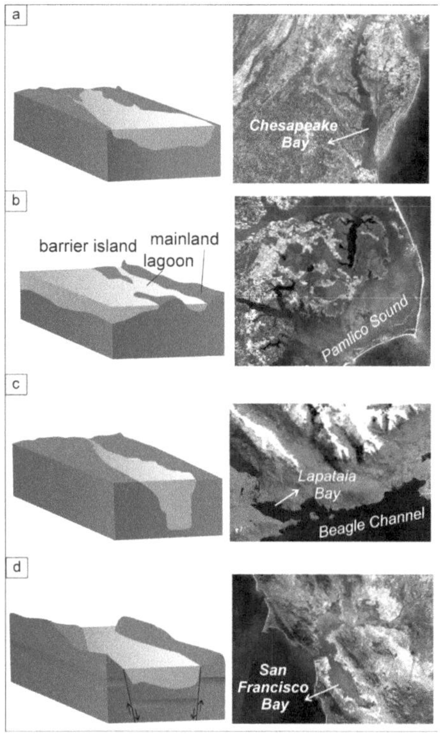

Figure 1.9. A geomorphological classification of estuaries. (a) Coastal plain estuary, or drowned river valley, exemplified by Chesapeake Bay in the East coast of the USA (37°27′ N); (b) bar-built or lagoon-type estuary, illustrated by Pamlico Sound in the state of North Carolina, USA (35°24′ N); (c) fjords, as the one forming Lapataia Bay (54°51′ S), emptying into the Beagle Channel in Argentina; and (d) a tectonic estuary, exemplified by San Francisco Bay, on the West coast of the USA (37°41′ N). Satellite images credits: USGS Landsat 8.

became U-shaped in cross-section. At fjord mouths, there are glacial deposits that form bars and sills, which limit the estuarine circulation and exchange with the adjacent coastal ocean (Fig. 1.9c). Finally, (4) tectonically-produced estuaries have been formed by subsidence or land cut off from the ocean by block slides associated with faulting, volcanoes, and landslides (Fig. 1.9d).

Apart from their geomorphology, estuaries may be classified according to their circulation patterns. Basically, estuarine circulation is created due to density differences between seawater and freshwater from rivers. Thus, where seawater and riverine water meet, seawater tends to move up into the estuary beneath the outward river flow of less dense water. Vertical mixing by turbulence, which is driven by tides, wind, and river flow, allows denser seawater to move landward in the lower water layers, while the less dense riverine water flows outwards in the upper layer. This pattern of inland-directed seawater circulation along the bottom and riverine water moving seaward along the surface is known as estuarine circulation. In general, the amount of mixing depends on local factors such as wind, tidal range, geomorphology, and river discharge, and often change seasonally within the same estuary (Sumich 1996).

Regarding circulation, four major types of estuaries are recognized: salt wedge, partially stratified, vertically homogeneous, and fjord estuaries. (1) Salt-wedge estuaries are the least mixed of all estuaries and occur when river discharge is large and tidal currents are weak. A sharp boundary is created between the upper freshwater layer and the bottom saline seawater, and the location of the wedge varies with the river flow and tidal conditions. (2) Partially-stratified estuaries occur when tidal flow intensifies and turbulence mixes the water column, creating a gradual increase in salinity from surface to bottom. In (3) vertically-homogeneous estuaries, strong tidal currents produce an intense turbulent mixing that eliminates the vertical salinity gradient. Salinity increases through the ocean and changes with tidal stage. In the last type, (4) fjord estuaries, the steep U-shaped basin and the presence of narrow glacial sills at their mouths that restrict water exchange with the open ocean, combine to create a pattern in which the denser seawater rarely overflows the sill and becomes stagnant within the basin, while the surface riverine outflow is relatively more intense. Thus, fjords have little tidal mixing, and the water remains highly stratified.

The sediment that characteristically abounds in estuaries is mud, which is defined as a mixture of fine-grained sediments (clays, silt and sometimes a minor fraction of fine sand), with relatively high content of organic matter and water. Muds are cohesive, a key feature distinguishing them from non-cohesive sand particles. The cohesive properties in muds come mostly from water tension and the charged surfaces of clay minerals, enhanced by the chemical properties of organic matter. Fine sediment in estuaries is transported mainly as suspended load and is deposited in the intertidal zone under calm conditions, during slack tides.

Hydrodynamics fosters sediment retention in estuarine systems (Gao et al. 2011). The estuarine circulation, along with stratification and tidal asymmetry, can promote the formation of the so-called estuarine turbidity maximum (ETM). This is a zone commonly located around the maximum seawater intrusion, most commonly found in coastal plain, salt wedge, and river-dominated estuaries (Jay et al. 2015), where the concentration of fine suspended sediments is much higher than upstream or in the adjacent sea. The high turbidity results from turbulent resuspension of sediment and flocculation of particulate matter (see below), and plays an important role in the sedimentation process, enhancing the efficiency of an estuary to trap fine sediments.

Flocculation is an important biophysical process that is essential for trapping fine sediments in estuaries (Jay et al. 2015). The interface at which fresh- and saltwater enter into contact at the heads of estuaries, experiences significant changes in water biogeochemistry (e.g., steep gradients in dissolved ions and pH) over short horizontal scales, where flocculation of suspended matter is largest, coinciding with the ETM. Flocculation is mainly influenced by the chemical nature of suspended sediments (clays

Box 3: Sediment and marine depositional environments

Depositional environments are those areas in which sediment is deposited. If the sediment is preserved in the rock record, the rock types that form after diagenesis and lithification represent the combination of physical, chemical, and biological processes associated with the depositional environment. Far from the continents, in flat areas on the deep ocean floor much of the sediment comes from hard body parts of dead pelagic organisms (either autotrophic or heterotrophic). Deep-sea pelagic sediments (collectively called oozes) are mostly composed of fine-textured small shells of calcium carbonate ($CaCO_3$) or silica (SiO_2). As oozes are *biogenous* (i.e., formed by what used to be living organisms), and even when a large proportion of the organic matter of dead marine organisms gets consumed by scavengers and decomposers in its vertical transit to the sea floor, pulses of intense pelagic production are reflected in- and incorporated as high contents of total organic carbon in oozes (Yoon et al. 2001). According to the composition of the tests, shells or other body parts of the pelagic organisms that formed them, oozes can be classified as calcareous (derived from coccolithophores, Foraminifera and pteropods-pelagic snails) or siliceous (formed mostly by diatoms, but also by radiolarians and silicoflagellates to a lesser extent). Deep waters are more corrosive to $CaCO_3$ than surface waters, so at a certain depth carbonatic shell dissolution becomes detectable (this is called the lysocline); the depth at which the *dissolution rate* of shells equals (or exceeds) the *rain rate* is termed the calcite compensation depth (CCD; Libes 2009). Below this depth, the corrosiveness of seawater seldom allows the deposition of calcareous material. The vertical location of the CCD varies between different ocean basins. Accordingly, the purity of siliceous oozes is greatest for depths below the CCD (Knauth 2003).

On the other hand, a large proportion (~ 38%) of the world's deep ocean floor is covered by terrigenous (i.e., land-derived) red clays. Of course deep ocean sediments are not entirely 'pure' in their composition, but rather a mixture of different sediment types including volcanic ashes, meteorite fragments, and other rare materials produced by chemical reactions and biological processes. A practical

Box 3 Contd. ...

...Box 3 Contd.

rule to distinguish between an ooze and a red clay is to determine the proportion of its biogenous component; oozes have 30% or more biogenous material.

Organosedimentary structures are sedimentary features built by the interaction of organisms and their environment. Coral reefs, for instance, are composed of a thin layer of living polyps, with a macroscopic exoskeletal framework of $CaCO_3$ underneath. Corals and calcareous algae (further discussed in Chapter 10) grow accretionally and/or on top of one another and form a three-dimensional framework that is modified in various ways by other organisms and inorganic processes. Another example of biogenic sedimentary structures are microbial mounds built by microorganisms like cyanobacteria. Filamentous cyanobacteria do not have skeletal parts, but their metabolism can facilitate the precipitation or accumulation of $CaCO_3$ to produce stromatolites with a laminated internal structure in shallow, well-illuminated seas (see Box 2). Bryozoans and crinoids are another example of mound that can persist over time and produce organo-sedimentary deposits.

Coastal environments are transitional zones between the open ocean and terrestrial watersheds. Thus, coastal depositional environments largely result from the redistribution of continental sediment inputs along shorelines by waves, tides and currents. The Earth's crust is made up of different types of rocks that break down into clastic sediments (a process known as weathering). The products of terrestrial rock weathering can be transported by rivers, wind, and ice-rafts through the ocean, while individual sediment grains get progressively rounded and reduced in size by abrasion. During their journey, grains are sorted according to density, size and shape, and some of their integral minerals may be dissolved.

Coastal depositional environments can be classified according to the relative importance of the sediment supply from riverine, tidal, and wave sources. Many coastal systems belong to an intermediate class, where all three source and transport agents (tides, waves, and river discharge) interact. Tide-dominated systems occur on coastal settings where the action of tides is much greater than the action of waves. An example of such environments are estuaries with a wide mouth and a main channel that narrows inland and connects to a river. A mouth barrier is absent or reworked into a submarine ebb-tidal delta and tidal flats are present along the channels. These tidal flats are commonly fed by marine sediment inputs, with grain sizes decreasing inland. On the other hand, in wave-dominated systems, a coastal barrier, built by wave-induced cross-shore sediment transport and longshore currents, typically exists. Breaching of a coastal barrier may occur during storm surges or periods of extreme river discharge. If the coastal barrier is backed by a low-lying wide coastal plain, the sea will scour a large inlet in the coastal barrier and transform the coastal plain into a lagoon. Tidal currents may also scour a channel system inside the lagoon and part of the sediment may be deposited in the lagoon to form extensive tidal flats.

that tend to aggregate due to van der Waals forces) and the concentration of particulate organic matter (Verney et al. 2009). Flocs are mainly formed by a combination of two groups of processes: those bringing particles together, and those keeping them together. The first group is related to the flow conditions, while the second is related to the formation of sticky organic compounds by a variety of organisms (e.g., bacteria, phytoplankton, Eisma 1986; more on these substances will be presented in Chapter 3). There are two main types of flocs (Eisma 1986): microflocs with a diameter up to 125 μm, and macroflocs with a maximum size of 3–4 mm. The microflocs are the basic building units of macroflocs, which are unstable and form under conditions of viscous flow. Flocs may represent a significant input of small sediment particles and organic matter, from the water column to the sediments. Yet, the basic problem in floc vertical transport and settling relates to the variable degree of flocculation (Manning et al. 2010), the turbulence-induced aggregation and floc breakup processes (Winterwerp 2002).

1.4 Divisions of the world's oceans

To close this chapter, we will introduce the different divisions of the world's oceans. Throughout the development of marine science, scholars and academics have divided the oceans into distinct divisions, as a means of better understanding their characteristics, and conceptualizing them. Some criteria for such division of the oceans reflect physical and/or ecological parameters determining actual patterns in the natural world, while in some other cases, the divisions have been traced along more or less arbitrary boundaries. Figure 1.10 provides a schematic representation of the ocean environment and how it is divided; it is important to note that this figure is not drawn to scale. The most basic division considers, on the one hand, all habitats in which bottom-dwellers are found, mostly a two-dimensional environment characterized by solid substrates (i.e., the benthic environment, from the Greek term *benthos*, 'depth of the sea'), and on the other hand, the three-dimensional pelagic environment (Greek *pélagos*, 'surface of the sea') that comprises habitats relating to fluid seawater.

Benthic environments run from the water line at high tide and the area immediately adjacent to it (the supralittoral, which is influenced by seawater through spraying or flooding during extreme events), to depths > 10,000 m. The littoral (or intertidal) zone is delimited by the extent of high and low water, determined by the tides, and obviously has local and regional variations in its extent. This zone includes coastal ecosystems such as mudflats, salt marshes and mangroves (which are treated at length in Chapter 9). The sublittoral or shelf zone extends from the low water level to depths of 200 m where the continental shelf break is located; it comprises important ecosystems such as the submerged areas of estuaries, seagrass beds and coral reefs, and the seafloor from where most ocean resources (trawling fisheries, mining) are extracted. From 200 m down to 3,000 m, the bathyal zone runs along continental slopes where the accumulation and mass-transport of terrigenous sediments takes place. From 3,000 to 6,000 m, the abyssal plains concentrate biogenic oozes produced by planktonic organisms; the basaltic seafloor of ocean basins here may be interrupted by trenches running down from 6,000 m to depths of 10,000 (the hadal zone), and other geomorphologic features such as mid-ocean ridges and seamounts.

Pelagic environments have both horizontal and vertical divisions. As such, the *neritic zone* comprises the water masses over the continental shelf up to the shelf break that provides most of fisheries' catches. The neritic zone may be narrow or comprise several hundred km in extension as exemplified by the Pacific

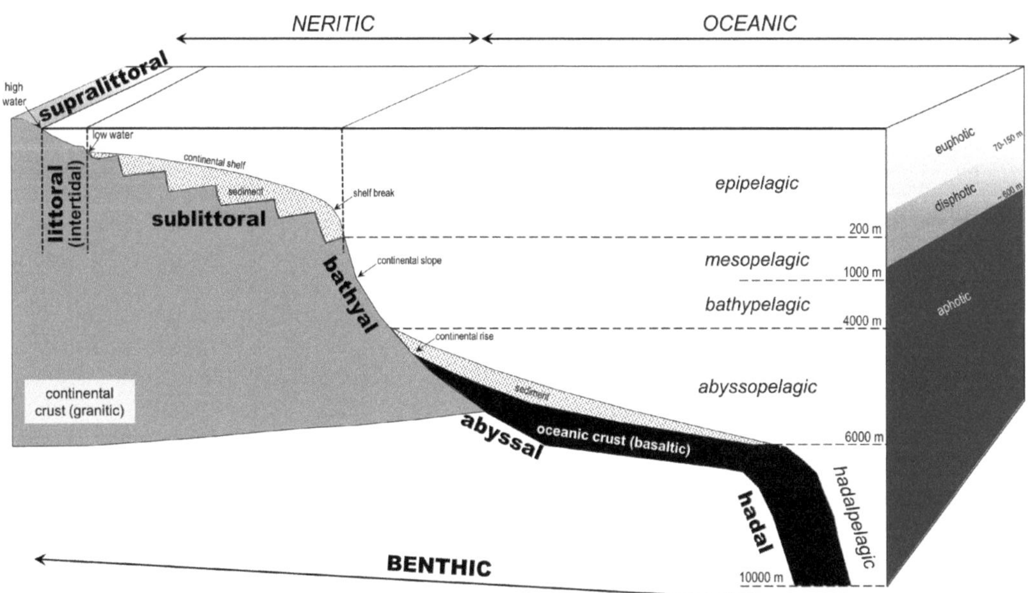

Figure 1.10. Schematic representation in cross section of the physical and ecological divisions of the ocean and its main habitats (not to scale). Benthic habitats are in bold; pelagic divisions in italics (modified from Marshall 1954, Garrison 2009).

and Atlantic coasts of southern South America, respectively. The *oceanic zone* is subdivided into different depth-layers. Perhaps the most significant of these is the *epipelagic zone* (0–200 m), which is the only one that receives abundant solar radiation as to concentrate oceanic primary production and support many oceanic fisheries (e.g., long-line). The water masses that lie beneath the epipelagic zone are permanently in darkness (collectively called the *aphotic zone*) or may at most be located in a transitional area which receives some solar radiation from the surface but not enough for photosynthesis (*disphotic zone*).

The divisions of the world's oceans respond to geological properties (e.g., plate tectonics, substrate type, restrictions to water circulation), but also physical (e.g., water mass movements in the form of wind-generated currents, tides and waves, differences in water density, solar radiation at the surface and penetration into the water column) and chemical attributes (temperature, salinity, nutrient concentration) that determine the characteristics of a certain environment and will be treated in the next chapter. All these properties of the abiotic environment have marked influences on the distribution and biological parameters of living organisms, and, as we will learn throughout this book, the organisms themselves contribute to various degrees to the alteration and modulation of the physical, geological and chemical environment in which they live.

Summary

1. The age of the ocean is about 4,000 MY. Considered the largest habitat for life on Earth (97% of the biosphere by volume), oceans cover 71% of its surface and harbor 97.5% of its superficial water. The oceans produce about 40% of the annual global primary productivity, and their average salinity is 34.48.
2. The oceanic crust is ~ 8 km-thick, and geologically younger ($\leq$ 200 MY) in relation to continental crust. Ocean basins are formed by virtue of the oceanic crust being made up of basaltic material, which is denser than the granitic rocks that continental crusts are made of.
3. The hypotheses of isostasy explains differences in thickness in Earth's lithosphere. The lighter lithosphere floats over the denser underlying material of the asthenosphere. Isostatic equilibrium occurs where the crust and mantle settle into place in absence of disturbing forces. There are two main hypotheses (Airy's and Pratt's) for isostatic adjustment.
4. The hypothesis of continental drift has been incorporated in the *Theory of Plate Tectonics*, which explains that continents move as part of lithospheric plates on top of the asthenosphere, as a result of the underlying convection cells in the upper mantle.
5. Where two plates meet, there might be divergent boundaries in the case of two lithospheric plates being pulled apart from each other. For instance, at mid-ocean ridges, new oceanic crust and lithosphere is created by seafloor spreading; at zones of continent-to-continent rifting, divergent boundaries may cause new ocean basin to form. On the other hand, convergent boundaries occur where two plates slide toward each other to either form a subduction zone or a continental collision. Finally, in transform or conservative boundaries, two lithospheric plates slide along transform faults, where plates are neither created nor destroyed.
6. The *Wilson cycle* explains the cyclical opening and closing of ocean basins. Active continental margins are associated with convergent or transform plate boundaries; a narrow continental shelf quickly drops off into the depths of the subduction trench. On the other hand, passive margins occur away from plate boundaries, tectonic activity is minimal, and there is sediment accumulation on the wider continental shelves.
7. The evolution of life on Earth (~ 3.5 BY ago) and the invention of oxygenic photosynthesis by cyanobacteria led to an unprecedented increase in O_2 ~ 2.4 BY ago (known as the *Great Oxidation Event*, GOE). This transition from a non-oxidizing atmosphere to one with free O_2 has been well recorded in sedimentary sequences by the disappearance of banded-iron formations (controlled by the availability of ferrous Fe^{+2} in seawater), and their replacement by red beds (oxidized Fe^{+3}). Some of the earlier life forms left conspicuous fossils known as stromatolites, which declined in abundance

with the appearance of animal grazers, later followed by bioturbation and the *Cambrian substrate revolution*.

8. The most recent period of repeated glaciations occurred during the Pleistocene Epoch, beginning about 2.6 MY ago and lasting until about 11,700 yr ago, coinciding with the start of the Holocene Epoch. The episodic sequence of glacial and interglacial periods throughout the Pleistocene is explained by *Milankovitch's hypothesis* for cyclical variations in Earth's eccentricity, obliquity and precession.
9. During glacial periods, large amounts of water are stored in ice sheets on land, lowering the mean global sea level; the sea level changes are called glacial-eustatic. After the *Last Glacial Maximum* (LGM), the melting of large ice sheets produced a net transfer of mass to the oceans, but also Earth's crust was uplifted ("bounced") in a process of glacio-isostatic adjustment.
10. Estuaries are semi-enclosed coastal bodies of water with an open connection to the ocean, and within which seawater is diluted by land freshwater inputs. Estuaries adapted rapidly to the Holocene sea level rise.
11. The geomorphologic classification of estuaries considers four different types: (1) coastal plain, (2) bar-built, (3) fjords, and (4) tectonically-produced estuaries. Additionally, estuaries may be classified according to the estuarine circulation created due to density differences between seawater and freshwater, into four major types: (1) salt-wedge, (2) partially-stratified, (3) vertically-homogeneous, and (4) fjord estuaries.
12. Deep-sea pelagic sediments, collectively called *oozes*, are mostly composed of fine-textured small shells of $CaCO_3$ or SiO_2 from once living organisms (hence, biogenous). At a certain depth (the lysocline), carbonatic shell dissolution becomes detectable. On the other hand, ~ 38% of the deep ocean floor is covered by terrigenous red clays. An ooze and a red clay are distinguished by the proportion of its biogenous component (30% or more for oozes).
13. Scholars have divided the oceans into distinct divisions, as a means of better understanding their characteristics, and conceptualizing them. These divisions respond to geological, physical and chemical attributes.
14. The most basic division considers benthic and pelagic environments. The former is sub-divided into the supralittoral, intertidal and sublittoral zones down to the shelf-break located at 200 m-depths; and the bathyal zone (200–3,000 m-depths), the abyssal plains' (3,000–6,000 m-depths) and the hadal zone (6,000–10,000 m-depths). Pelagic environments have both horizontal and vertical divisions. The neritic zone comprises the water masses over the continental shelf, while the oceanic zone is subdivided into different depth-layers (the most significant being the epipelagic zone from 0–200 m-depths).

References

Awramik, S.M., J.W. Schopf and M.R. Walter. 1983. Filamentous fossil bacteria from the Archaean of Western Australia. Precambrian Res. 20: 357–374.

Boillot, G., S. Grimaud, A. Mauffret, D. Mougenot, J. Kornprobst, J. Mergoil-Daniel and G. Torrent. 1980. Ocean-continent boundary off the Iberian margin: a serpentinite diapir west of the Galicia Bank. Earth. Planet. Sci. Lett. 48: 23–34.

Cronin, M.F., C.L. Gentemann, J.B. Edson, I. Ueki, M. Bourassa, S. Brown, C.A. Clayson et al. 2019. Air-sea fluxes with a focus on heat and momentum. Front. Mar. Sci. 6: 430.

De Blij, H. 2012. Why Geography Matters: More Than Ever. Oxford University Press, Oxford, UK.

Dodd, M.S., D. Papineau, T. Grenne, J.F. Slack, M. Rittner, F. Pirajno, J. O'Neil et al. 2017. Evidence for early life in Earth's oldest hydrothermal vent precipitates. Nature 543: 60–64.

Dürr, H.H., G.G. Laruelle, C.M. van Kempen, C.P. Slomp, M. Meybeck and H. Middelkoop. 2011. Worldwide typology of nearshore coastal systems: defining the estuarine filter of river inputs to the oceans. Estuaries Coast. 34: 441–458.

Eisma, D. 1986. Flocculation and de-flocculation of suspended matter in estuaries. Neth. J. Sea Res. 20: 183–199.

Eriksson, P.G. and E.S. Cheney. 1992. Evidence for the transition to an oxygen-rich atmosphere during the evolution of red beds in the lower Proterozoic sequences of southern Africa. Precambrian Res. 54: 257–269.

Farías, M.E., N. Rascovan, D.M. Toneatti, V.H. Albarracín, M.R. Flores, D.G. Poiré, M.M. Collavino, O.M. Aguilar, M.P. Vazquez and L. Polerecky. 2013. The discovery of stromatolites developing at 3570 m above sea level in a high-altitude volcanic lake Socompa, Argentinean Andes. PLoS ONE 8(1): e53497; doi:10.1371/journal.pone.0053497.

Gao, S., Y.P. Wang and J. Gao. 2011. Sediment retention at the Changjiang sub-aqueous delta over a 57 year period, in response to catchment changes. Estuar. Coast. Shelf Sci. 95: 29–38.

Garrison, T. 2009. Essentials of Oceanography, 5th edition. Brooks/Cole, Belmont, CA. USA.

Grotzinger, J.P. and A.H. Knoll. 1999. Stromatolites in Precambrian carbonates: evolutionary mileposts or environmental dipsticks? Annu. Rev. Earth Planetary Sci. 27: 313–358.

Hazen, R.M., D. Papineau, W. Bleeker, R.T. Downs, J.M. Ferry, T.J. McCoy, D.A. Sverjensky and H. Yang. 2008. Mineral evolution. Am. Mineral. 93: 1693–1720.

Jay, D.A., S.A. Talke, A. Hudson and M. Twardowski. 2015. Estuarine turbidity maxima revisited: instrumental approaches, remote sensing, modeling studies, and new directions. pp. 49–109. *In*: Ashworth, P.J., J.L. Best and D.R. Parsons [eds.]. Fluvial-Tidal Sedimentology. Developments in Sedimentology Vol. 68. Elsevier, Amsterdam. The Netherlands.

Khan, N.S., C.H. Vane and B.P. Horton. 2015. Stable carbon isotope and C/N geochemistry of coastal wetland sediments as a sea-level indicator. pp. 295–311. *In*: Shennan, I., A.J. Long and B.P. Horton [eds.]. Handbook of Sea-Level Research. John Wiley & Sons, Oxford. UK.

Knauth, L.P. 2003. Siliceous sediments. pp. 1095–1104. *In*: Middleton, V. [ed.]. Encyclopedia of Sediments and Sedimentary Rocks. Encyclopedia of Earth Sciences Series. Springer, Netherlands.

Koehler, I., K. Konhauser and A. Kappler. 2010. Role of microorganisms in banded iron formations. pp. 309–324. *In*: Barton, L.L., M. Mandl and A. Loy [eds.]. Geomicrobiology: Molecular and Environmental Perspective. Springer, Berlin, Heidelberg.

Kopp, R.E., C.C. Hay, C.M. Little and J.X. Mitrovica. 2015. Geographic variability of sea-level change. Curr. Clim. Chang. Reports 1: 192–204.

Libes, S.M. 2009. Introduction to Marine Biogeochemistry, 2nd edition. Elsevier/Academic Press, Burlington, MA. USA.

Long, J., S. Zhang and K. Luo. 2019. Cryogenian magmatic activity and early life evolution. Sci. Rep. 9: 6586.

Lyons, T.W., C.T. Reinhard and N.J. Planavsky. 2014. The rise of oxygen in Earth's early ocean and atmosphere. Nature 506: 307–315.

Mann, K.H. and J.R.N. Lazier. 2006. Dynamics of Marine Ecosystems: Biological-Physical Interactions in the Oceans. Blackwell Publishing, Malden, MA. USA.

Manning, A.J., W.J. Langston and P.J.C. Jonas. 2010. A review of sediment dynamics in the Severn Estuary: influence of flocculation. Mar. Pollut. Bull. 61: 37–51.

Marshall, N.B. 1954. Aspects of deep sea biology. Hutchinson, London. 380 pp.

Meissner, R. 2002. The Little Book of Planet Earth. Springer Science & Business Media, Berlin, Heidelberg.

Mojzsis, S.J., G. Arrhenius, K.D. McKeegan, T.M. Harrison, A.P. Nutman and C.R.L. Friend. 1996. Evidence for life on Earth before 3,800 million years ago. Nature 384: 55–59.

Nisbet, E.G. and C.M.R. Fowler. 1999. Archaean metabolic evolution of microbial mats. Proc. R. Soc. Lond. B 266: 2375–2382.

Nisbet, E.G. and N.H. Sleep. 2001. The habitat and nature of early life. Nature 409: 1083–1091.

Noffke, N. and S.M. Awramik. 2013. Stromatolites and MISS-Differences between relatives. GSA Today 23: 4–9.

Oschmann, W., M. Grasshof and M. Gudo. 2002. The early evolution of the planet earth and the origin of life. Senck. Leth. 82: 285–294.

Ostrander, C.M., S.G. Nielsen, J.D. Owens, B. Kendall, G.W. Gordon, S.J. Romaniello and A.D. Anbar. 2019. Fully oxygenated water columns over continental shelves before the Great Oxidation Event. Nat. Geosci. 12: 186–191.

Prager, E.J. and S.A. Earle. 2000. The Oceans. McGraw-Hill, New York, USA.

Pritchard, D. and A.J. Hogg. 2003. Cross-shore sediment transport and the equilibrium morphology of mudflats under tidal currents. J. Geophys. Res. 108(C10): 3313, doi:10.1029/2002JC001570.

Pritchard, D.W. 1967. What is an estuary: physical viewpoint. pp. 3–5. *In*: Lauf, G.H. [ed.]. Estuaries. Amer. Assoc. Adv. Sci. Vol. 83, Washington, DC. USA.

Pufahl, P.K. and E.E. Hiatt. 2012. Oxygenation of the Earth's atmosphere-ocean system: a review of physical and chemical sedimentologic responses. Mar. Pet. Geol. 32: 1–20.

Seibold, E. and W. Berger. 2017. The Sea Floor: An Introduction to Marine Geology, 4th edition. Springer, Cham. Switzerland.

Seilacher, A., L.A. Buatois and M.G. Mángano. 2005. Trace fossils in the Ediacaran–Cambrian transition: behavioral diversification, ecological turnover and environmental shift. Palaeogeogr. Palaeoclimatol. Palaeoecol. 227: 323–356.

Stevenson, J.C., L.G. Ward and M.S. Kearney. 1986. Vertical accretion in marshes with varying rates of sea level rise. pp. 241–259. *In*: Wolfe, D. [ed.]. Estuarine Variability. Proceedings of the Eighth Biennal International Estuarine Research Conference. Elsevier.

Tang, H. and Y. Chen. 2013. Global glaciations and atmospheric change at ca. 2.3 Ga. Geosci. Front. 4: 583–596.

Verney, R., R. Lafite and J.-C. Brun-Cottan. 2009. Flocculation potential of estuarine particles: the importance of environmental factors and of the spatial and seasonal variability of suspended particulate matter. Estuaries Coast. 32: 678–693.

Walter, M.R. and A.C. Allwood. 2005. Biosediments and biofilms. pp. 279–294. *In*: Selly, R.C., L.R.M. Cocks and I.R. Plimerm [eds.]. Encyclopedia of Geology. Elsevier, Amsterdam. Netherlands.
Watts, A.B. 2001. Isostasy and Flexure of the Lithosphere. Cambridge University Press, Cambridge. UK.
Wegener, A. 1912. Die entstehung der kontinente. Geol. Rundschau. 3: 276–292.
Winterwerp, J.C. 2002. On the flocculation and settling velocity of estuarine mud. Cont. Shelf Res. 22: 1339–1360.
Yoon, H.I., B.K. Park, M.W. Han, B.K. Khim, Y. Kim and C.Y. Kang. 2001. Origin of laminated diatom ooze in King George Basin sediments from Bransfield Strait, Antarctica. Geosci. J. 5: 225–231.

CHAPTER 2

Introduction to the Marine Environment from Physical and Chemical Perspectives

Paula D. Pratolongo[1,*] and *Jerónimo Pan*[2]

The oceans and the atmosphere exchange heat, salt, water, and momentum, and most physical and chemical processes in the oceans result from this interaction. Winds blowing over the ocean transfer energy to the surface layers allowing for the development of ocean currents. Solar radiation can evaporate water, removing heat from the ocean that is transferred to the atmosphere when water vapor is condensed into clouds. Evaporation and freezing, both resulting from the ocean-atmosphere interaction, turn liquid water saltier, but rainfall dilutes the salt concentration of the surface ocean. As mentioned, there are many feedback mechanisms between the oceans and the atmosphere. In the next sections, we will introduce some of the physical and chemical properties of the marine environment that result from the ocean-atmosphere coupling and are relevant for marine life.

Our coverage of these topics is by no means exhaustive, nor at length, but rather aiming to provide the basics anyone interested in marine biology should be familiar with. Those readers interested in more thorough and quantitative treatments of these topics should refer to specific oceanography and biogeochemistry textbooks.

2.1 The energy budget of the Earth

The oceans are fast, dynamic systems, with a fundamental role in regulating the Earth's climate. From global to regional scales, oceans store, transport, and exchange a huge amount of heat and solar radiation, which is the ultimate source of energy driving the motion of the atmosphere and the ocean. Let us first examine the ways in which the oceans and atmosphere interact with solar radiation to influence wind and ocean circulation.

The energy budget of the Earth is the balance of the various kinds and amounts of energy incoming and outgoing the Earth system (land, ocean, and atmosphere). The incoming solar radiation that reaches the Earth includes wavelengths of the electromagnetic spectrum ranging from infrared (heat) to ultraviolet radiation, but most of the energy reaches the Earth in the form of visible light (Fig. 2.1a). On average (over a full year and the entire planet), about 340 W m^{-2} of energy reaches the planet. About 23% of this energy is reflected back to space by clouds, dust particles, and other aerosols in the atmosphere (Fig. 2.1b). The Earth's surface also reflects a proportion of the incoming radiation (about 7%), especially

[1] Centro de Recursos Naturales Renovables de la Zona Semiárida (CERZOS-CONICET-UNS); Dto. Biología, Bioquímica y Farmacia-Universidad Nacional del Sur (BByF-UNS); National Scientific and Technical Research Council, Argentina (CONICET).

[2] National Scientific and Technical Research Council, Argentina (CONICET); Instituto de Geología de Costas y del Cuaternario (IGCyC); Instituto de Investigaciones Marinas y Costeras (IIMyC).

* Corresponding author: ppratolongo@gmail.com

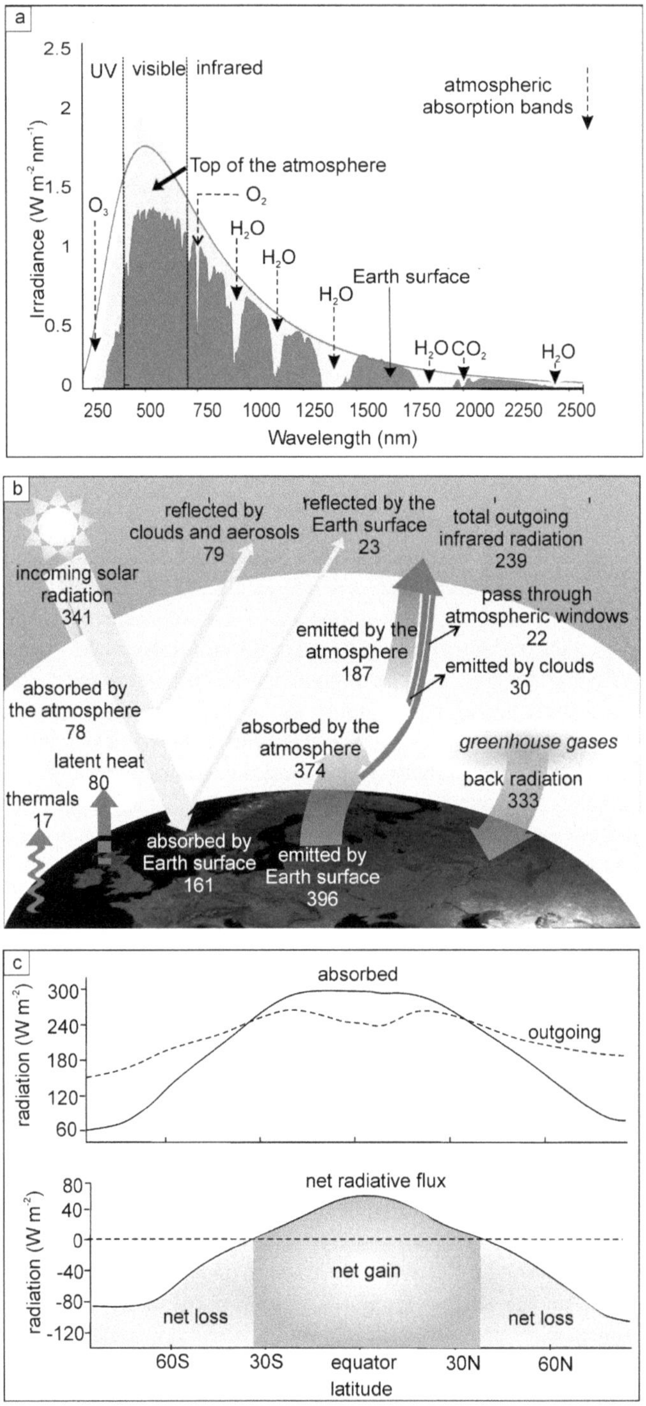

Figure 2.1. (a) Solar radiation spectrum that reaches the top of the atmosphere and the Earth surface. As light passes through the atmosphere, some wavelengths are absorbed by gases with specific absorption bands. Curves are based on the American Society for Testing and Materials (ASTM) Terrestrial Reference Spectra; (b) The energy budget of the Earth system. Energy fluxes (W m^{-2}) are based on recent satellite measurements from the Clouds and the Earth's Radiant Energy System (CERES; http://science.larc.nasa.gov/ceres/) along with information from other data sources (data sources are described in Loeb et al. (2009) and Trenberth et al. (2009)). According to these data, the radiation budget at the top of the atmosphere is not balanced for five years, from 2000 to 2005, and 0.85 W m^{-2} were added to the Earth system.

on bright surfaces, like snow-covered areas, continental ice sheets and glaciers. Adding up reflection in clouds, aerosols, and the Earth's surface, about 30% of the incoming solar radiation is directly reflected out of the Earth's system, and this proportion is called albedo. The albedo refers to reflectivity of a surface. Lighter surfaces are more reflective than darker surfaces (which absorb more energy), and therefore have a higher albedo.

The incoming sunlight that is not reflected (about 70%) is absorbed and heats up the atmosphere and Earth's surface. All the elements in the Earth's system emit infrared radiation (heat) according to their temperature. The colder the object, the lower the emission. Assuming an equilibrium state of the Earth, the infrared radiation that is emitted from the Earth system must be equivalent to that 70% of the incoming solar radiation that is absorbed. In other words, the Earth system tends to an equilibrium between the energy that it receives from the Sun and the energy that is emitted back to space. This represents the balanced radiation budget for the equilibrium state. However, there are situations that may throw the system out of the equilibrium. Examples of such situations are the ice ages during the Pleistocene (as we have seen in Chapter 1) triggered by changes in the orbit of the Earth that reduced the incoming solar radiation. Over time, the Earth surface cooled and heat emissions changed to balance the energy budget (Bielefeld 1997).

Therefore, in the overall energy balance at the top of the atmosphere, the incoming solar radiation equals light reflection plus heat emission, and there are different mechanisms by which the energy is transferred within the Earth system. About 22% of the solar energy that reaches the Earth system is absorbed by water vapor and other gases in the atmosphere, where it is converted into heat. Similarly, about 48% of the incoming solar radiation is absorbed by the land and ocean, and part of this energy returns to the atmosphere through thermal emission from the elements on the surface, according to their temperatures.

Apart from emission, heat is transferred from the surface to the atmosphere by latent heat, convection, and conduction. Latent heat is released or absorbed when water changes between solid, liquid, and vapor phases. For example, heat is removed from the ocean as liquid water evaporates into the atmosphere. When water vapor condenses into rain, that heat is then transferred to the atmosphere. Likewise, heat is absorbed by ice when it melts and released when ice forms. These phase changes account for an energy exchange between the Earth's surface (land or oceans) and the atmosphere equivalent to 25% of the incoming solar energy. Conduction is the transfer of heat through direct contact between the surface and the atmosphere, and convection is the transfer of heat through the movement of air. Both create thermals (columns of rising air in the lower atmosphere) that represent a small part of the energy transfer between the Earth and the atmosphere, equal to about 6% of the incoming solar energy (Loeb et al. 2009, Trenberth et al. 2009).

To complete the energy budget, part of the heat that is absorbed by the atmosphere is eventually radiated back to space. However, there is an interesting twist on the infrared emission because part of the heat is radiated back from the atmosphere to the Earth's surface. This is called the greenhouse effect and is mainly due to the presence of water vapor, CO_2, methane, and other infrared-absorbing gases in the atmosphere. In the greenhouse effect, incoming shortwave radiation from the sun passes through the atmosphere and is absorbed by the Earth's surface, warming it in the process. For certain wavelengths (called atmospheric windows), the infrared radiation that is re-emitted from the Earth's surface passes through the atmosphere to space, but much of the energy is absorbed by the greenhouse gases and the atmosphere heats up (Trenberth and Fasullo 2012). Without the greenhouse effect, the average temperature on Earth would be too cold to sustain present life forms, but since the Industrial Revolution the atmospheric concentrations of the major greenhouse gases have increased dramatically, and there is a great concern because it is causing warming at a global scale. The causes and consequences of climate change will be treated in detail in Chapter 14.

So far, we have explained the energy budget based on averaged fluxes at the planetary scale. However, because the Earth is an approximate sphere, the incoming solar radiation is not equally distributed and different regions receive and emit energy differentially (Fig. 2.1c). At any latitude, in any given season, there is an angle between the Earth's surface and the incoming solar radiation that may range from perpendicular between the tropics to 0° at latitudes higher that the polar circles. Thus, the same amount of incoming solar radiation will be concentrated in a smaller area close to the equator but will be spread over a much larger area close to the poles. Moreover, smaller incident angles at lower latitudes result in more of the energy being absorbed rather than reflected, as opposed to a larger proportion of the radiation being reflected close to the poles. Finally, the poles reflect more solar energy because the ice cover and higher albedo. Through all these mechanisms, the poles absorb much less solar radiation than equatorial regions, thus the poles are cold, and the tropics are warm. The differential heating of the Earth's surface leads to atmospheric convection, creates winds, waves, and surface currents. As we will see throughout the chapters of this book, these currents influence nutrient distribution and primary productivity, which support the rest of the living organisms in the ocean.

2.2 Seawater temperature, salinity, and heat transport

Water temperature is one of the most important physical properties of the marine environment, as it controls the rates at which chemical reactions and biological processes like metabolism and growth occur. Water temperature also has an effect on the concentration of dissolved gases in seawater, including oxygen and CO_2, which are relevant to biological processes. There is a latitudinal pattern of sea surface temperature produced because seawater is heated by solar radiation. The most obvious feature of the long-term average sea surface temperature is the latitudinal variation, from the warm regions along the equator to the cold regions at higher latitudes (Fig. 2.2a). The poles receive only 40% of the radiation that the equator does and these variations in solar energy translate into an ocean surface temperature variation from 30°C in the tropics to –2°C near the poles. At the ocean surface, solar radiation also produces evaporation, and water vapor rises into the atmosphere leaving dissolved salts and increasing seawater salinity. Thus, the surface salinity shows a latitudinal pattern mainly driven by the sea surface temperature but modified by freshwater inputs from major rivers and heavy rainfall in the intertropical region (Fig. 2.2b).

In addition to this latitudinal pattern, temperatures in the oceans decrease with increasing depth and water stratification is the process by which water is segregated into layers based on differences in

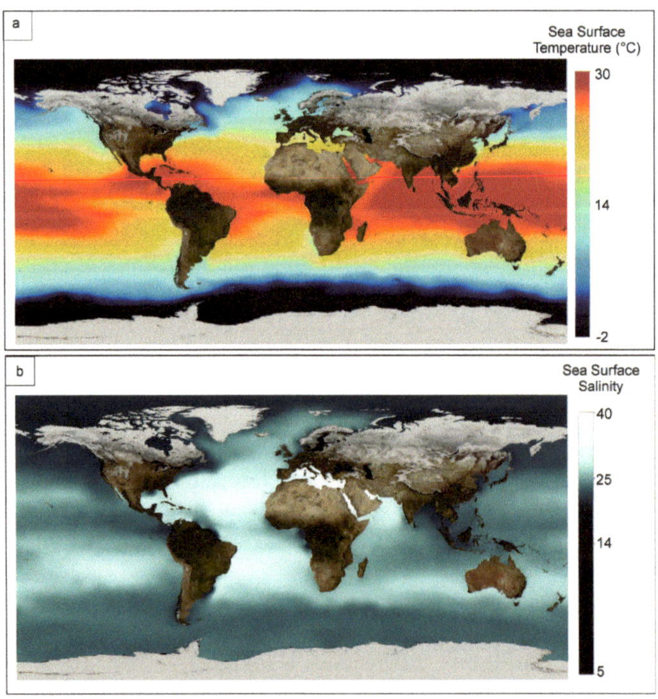

Figure 2.2. (a) Global map of average Sea Surface Temperature, and (b) Global map of average Sea Surface Salinity. Available through the Web Map Service, NASA/Goddard Space Flight Center Scientific Visualization Studio (at http://svs.gsfc.nasa.gov).

Box 1: Why oceans are salty?

Most of the salts and dissolved elements in seawater were probably outgassed along with the water vapor (see Box 1 in Chapter 1), so the primitive ocean might have been salty already. But runoff from the continents is a constant input of dissolved substances into the ocean contributing to its salinity. However, the oceans are in a steady state regarding salinity (e.g., the ocean's salt composition remains largely constant through time) because the rate of input of new material is balanced by the rate of removal. Dissolved ions enter the ocean from rivers, volcanic activity, hydrothermal vents, dissolution of particulate materials within the ocean, and groundwater input. Ions leave seawater when they are incorporated into living organisms, like calcareous algae and corals, by sea spray formation, or percolation of water into the crust. At this point, let us introduce the concept of residence time, which is the average time that a single atom of an element remains in the ocean before being removed. As a rule, substances that are readily used in biological processes have short residence times, while substances with longer residence times are less reactive, and may be a part of long-scale geological cycles.

On average, about 35 g of salts are present in 1 kg of seawater, thus the average ocean salinity is 35. It is most important to note here that, as salinity is a ratio of two masses, its units are dimensionless; for many years physical oceanographers used the practical salinity unit (PSU), but the use of this unit has been discontinued by convention. Many different substances contribute to this salinity, but six ions (chloride, sodium, sulfate, magnesium, calcium, and potassium) comprise about 99.4% of all the dissolved ions in seawater, and only two, chloride and sodium, make up over 85%. In addition to the major constituents, there are numerous minor constituents, like radionucleotides, organic compounds, and metals, but together they make up less than 1% of the ions in the seawater. Some of these trace elements may be vital to living organisms, but they have little effect on the overall salinity. Thus, even though the absolute salinity of seawater changes from place to place, the relative proportions of the six major ions are always constant. This is known as the rule of constant proportions, and we call these major elements conservative ions. Because the concentrations of the minor ions are not constant, these are called non-conservative ions, and as we will see, they have major impacts on the distribution of certain organisms.

temperature and salinity. In polar regions, the formation of dense water at the surface that sinks produces nearly isothermal conditions throughout the water column. In the low latitudes, on the other hand, the temperature gradient from top to bottom is large. In those areas where there is an annual variation in surface heating (such as temperate latitudes), there is a shallow isothermal layer of elevated temperature in summer. Below this wind-mixed layer that is 10 to 20 meters deep the temperature decreases rapidly with depth, forming a shallow seasonal thermocline (i.e., a layer of sharp vertical temperature change). During winter cooling, increased wind mixing at the ocean surface and convective overturning deepen the isothermal layer, and the shallow thermocline disappears. The seasonal thermocline forms again when summer returns. At greater depth, there are no seasonal changes but a non-seasonal thermocline below which temperatures decrease slowly.

Temperature and salinity variations combine to determine the density of seawater. As temperature decreases and salinity increases, seawater becomes denser. Thus, there is a buoyant layer below which there lies the denser, deep ocean, and the density gradient between them is called the pycnocline. Density, in turn, governs the vertical movements of water, and density-driven circulation has a fundamental role in redistributing solar heat from the tropical regions towards high latitudes. The Meridional Overturning Circulation is a system of surface and deep currents encompassing all ocean basins. It transports large amounts of heat around the globe and connects the deep sea, surface ocean, and atmosphere. At some locations, the surface waters of the Earth's oceans are dense enough to sink to depths of up to several thousand meters. These sites of deep-water formation are located at high latitudes, as in the North Atlantic. Subduction in the Labrador and Nordic Seas is fed by northward flow at the surface, transporting tropical and subtropical warm water masses into the subpolar and polar North Atlantic. As warm surface currents move northwards, they lose heat to the atmosphere and it has a large effect on the regional climate. Surface water in the North Atlantic is denser because it is very cold and because of ice formation, that turns seawater saltier and thus even denser. Newly formed deep water in the North Atlantic flows southward along the eastern margin of the Americas, crosses the equator, and eventually enters the Antarctic Circumpolar Current of the Southern Ocean, that connects all major ocean basins, allowing water, heat, and other properties to be carried from one basin to another.

2.3 Wind-driven circulation

The long-term averaged surface winds include the trade winds and westerlies. Warm air around the equator rises and flows toward the poles as it cools and descends near 30 degrees latitude (North and South). The descending air forms the trade winds that blow back toward the equator and meet at the Intertropical Convergence Zone. The Coriolis effect arises from the fact that the rotation speed of the Earth is faster at lower latitudes, and the apparent path taken by an object on the surface is deflected as it moves between areas of different rotation speeds. As the air moves away from the equator, the Coriolis force deflects it toward the right in the Northern Hemisphere, and towards the left in the Southern Hemisphere. Thus, in the Northern Hemisphere trade winds blow from the northeast to the southwest, and from the southeast toward the northwest in the Southern Hemisphere (see Chapter 14). At higher latitudes are the westerlies that flow towards the east, with a poleward component. The westerlies are associated with the strongest windstorms and atmospheric fronts.

The drag of trade winds on the ocean surface along the equator creates the North Equatorial and South Equatorial currents that flow east to west. When these currents meet the east coast of continents, they are deflected away from the equator by the Coriolis effect and become western boundary currents (that is, currents that run along the western side of an ocean basin, and the eastern coasts of the continents). At higher latitudes, the westerlies push surface water from west to east and the presence of the continents deflect the currents towards the equator, creating eastern boundary currents. These currents combine to produce major spirals of ocean surface currents called gyres that rotate clockwise in the Northern Hemisphere and counterclockwise in the Southern Hemisphere. There are five major ocean-wide gyres named after the ocean basins they occupy (e.g., the North Atlantic, South Atlantic, North Pacific, South Pacific, and Indian Ocean gyres). In the North Atlantic, for instance, the North Equatorial Current flows into the Gulf Stream, the western boundary current along the east coast of North America. The Gulf Stream blends into the North Atlantic Current that moves towards Europe, which then becomes the

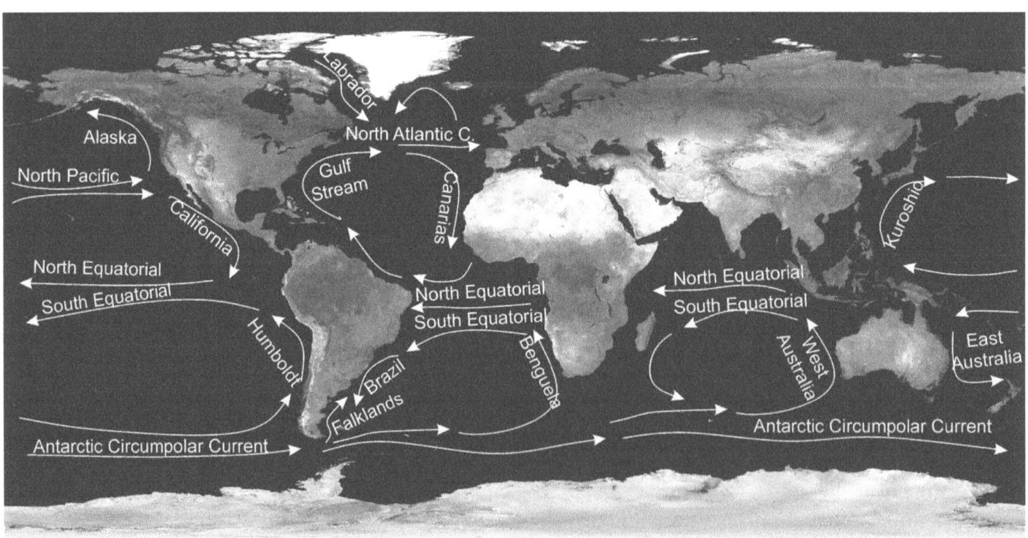

Figure 2.3. Major global surface currents. Base map (The Blue Marble Next Generation data) is courtesy of Reto Stockli (NASA/GSFC) and NASA's Earth Observatory.

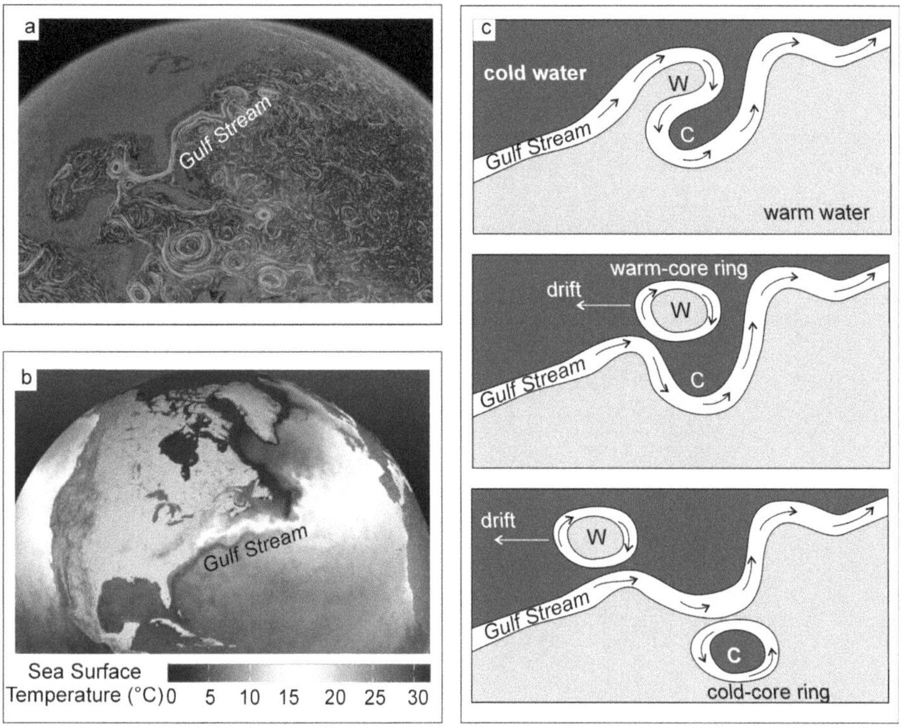

Figure 2.4. (a) Ocean currents off the coast of the Eastern United States from a computational model called ECCO-2. The image shows the turbulent nature of the Gulf Stream, one of the most powerful ocean currents; (b) Sea surface temperature of the same region on May 16, 2016. The data set used is from the Jet Propulsion Laboratory (JPL) Multi-scale Ultra-high Resolution (MUR) Sea Surface Temperature Analysis. Available through NASA's Scientific Visualization Studio; (c) Schematic representation of cold-core and warm-core ring formation along the Gulf Stream.

Canary Current. This is the eastern boundary current of the gyre, which flows south to join back the North Equatorial Current. In addition to these large ocean gyres, prevailing winds generate the Antarctic Circumpolar Current, that partially makes up the southern edges of the South Atlantic, South Pacific, and Indian Ocean gyres. The Antarctic Circumpolar Current flows east uninterrupted by any land mass across the Southern Ocean at 60° latitude and is the only globe-circling surface current (Fig. 2.3).

Major currents traced on Fig. 2.3 represent ocean conditions averaged over decades, but the ocean surface is much more dynamic due to its turbulent nature (Fig. 2.4a). Satellite images show that large turbulent rotating warm-core and cold-core eddies (also known as rings) detach from major currents. A ring forms when a meander in one of these currents forms a loop that separates from the main current and moves independently as an eddy. Rings are tridimensional features that extend to some depth and are bordered by fast rotating currents that isolate water and organisms within rings, sometimes playing an important role in the transport of larvae between regions. Rings form in all ocean basins, yet, perhaps the most remarkable examples of ring formation and dissipation are found on either side of the Gulf Stream (Fig. 2.4b). On the north side, warm-core rings detach from the current and transport southern warm waters towards the North Atlantic. These warm-core rings rotate in a clockwise direction (Fig. 2.4c) and typically have relatively low levels of biological production, but they can modify coastal currents and bring organisms that live in much warmer water onto normally cool coastal waters. On the south side of the Gulf Stream, cold-core rings entrain relatively cold and productive waters and rotate in a counterclockwise direction. Cold-core rings typically contain more nutrients than the surrounding waters, and unusually abundant marine life (Wiebe et al. 1976).

Winds drag on the ocean's surface and, because of the Coriolis force, water movement is not in the same direction of the wind. Instead, the surface layers of the ocean move in a spiral across a surface

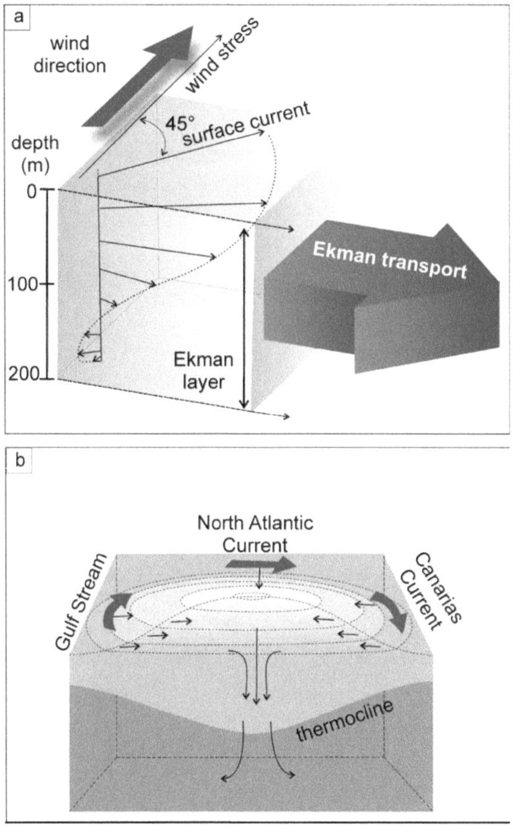

Figure 2.5. (a) Schematic representation of the Ekman spiral for the Northern Hemisphere and the net Ekman transport; (b) Water hill created by Ekman transport through the center of the North Atlantic gyre.

layer known as the Ekman spiral (Fig. 2.5a). The Ekman spiral forms because water molecules on the surface move by the drag of the wind and they, in turn, drag water molecules below them. Each layer of water molecules is slowed by friction with respect to the shallower layer. Thus, a given layer moves more slowly than the layer above it, until the movement stops at a depth of about 100 meters. Due to the Coriolis effect, water movement is deflected to the right in the Northern Hemisphere and to the left in the Southern Hemisphere. As a result, each successively deeper layer of water moves more slowly to the right or left, creating a spiral effect and the resultant water transport within this surface layer (called the Ekman layer) is 90° towards the right of the wind in the Northern Hemisphere and 90° to the left in the Southern Hemisphere (Mann and Lazier 2013).

Surface currents that form ocean gyres arise from a combination of global wind patterns and the Coriolis force, while the Ekman transport governs water movement below the sea surface. In the Northern Hemisphere, gyres rotate clockwise, and the net movement of the Ekman transport is 90° towards the right of the wind (or ocean current) direction. In the Southern Hemisphere, gyres rotate anticlockwise, and the Ekman transport is 90° towards the left. In both cases, the mean Ekman transport is towards the center of the gyre, where water piles up, and the sea surface has a higher level than on the edges (Fig. 2.5b). This pile of water then flows back by gravity towards the edges forming a current called geostrophic flow, that rotates clockwise in the Northern Hemisphere, and counterclockwise in the Southern Hemisphere.

While the Ekman transport produces convergence of surface waters in the center of gyres, in other regions it results in the divergence of surface waters. As the surface waters diverge, deeper water must move up to the surface to replace it, creating upwelling zones. Upwelling has a major consequence on marine ecosystems, as it brings cold nutrient-rich deep water to the surface, where light is available for phytoplankton growth, thus stimulating primary production (Fig. 2.6a), and many of the most productive

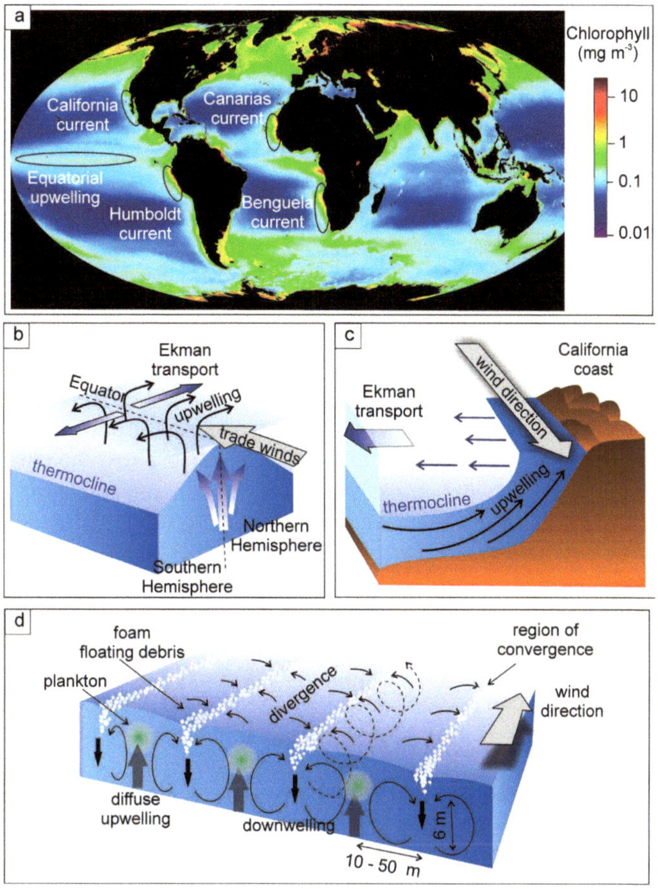

Figure 2.6. (a) Ocean chlorophyll concentration averaged over the full SeaWiFS mission (1997 to 2010) and major upwelling areas. Images are courtesy of NASA/Goddard Space Flight Center, The SeaWiFS Project and GeoEye; (b) Schematic representation of the equatorial upwelling; (c) Schematic representation of the coastal upwelling off the California coast; (d) Schematic representation of the Langmiur circulation.

regions on the Earth's oceans overlap with upwelling zones. Ekman divergence occurs along the equator. The North and South Equatorial Currents flow towards the west, and the net Ekman transport to the north and south in their respective hemispheres creates a zone of divergence, the equatorial upwelling (Fig. 2.6b). Analogously, coastal upwelling is created by the Ekman transport and occurs along the west coasts of the continents, associated with the eastern boundary currents. In the Northern Hemisphere (i.e., the coast of California), the Ekman transport moves the surface layer 90° to the right of the direction

Box 2: El Niño-Southern oscillation

Coastal upwelling off the coast of Peru makes that region one of the most productive fishing grounds in the world. Peru ranks second among the top seven countries of global capture fisheries which together account for ~ 50% of total captures worldwide; Peru alone produces 7% of the total, largely attributed to the Peruvian anchoveta, *Engraulis ringens* (Bertrand et al. 2020). In 2018, it yielded catches of 7 million tons. However, every three to seven years, the commonly cold and nutrient-rich upwelled water is replaced by an unusually warm and nutrient-poor water mass, leading to a decline in fish landings. The term "El Niño" (the child in Spanish) refers to the Child Jesus, because the phenomenon was described by fishermen along the Pacific coast of South America, who found that a significant warming of surface waters occurs in some years close to Christmas. More formally, the event is

Box 2 Contd. ...

...Box 2 Contd.

referred to as El Niño-Southern Oscillation (ENSO). There is an ocean component, characterized by the first observed anomalous warming of the sea surface in the Tropical Pacific (El Niño) that is linked to a perturbation in the atmospheric circulation patterns, which is known as the Southern Oscillation. Thus, ENSO is a coupled ocean-atmosphere mode of variability in the Earth's climate system and it has a periodicity of three to seven years.

Under normal conditions, the trade winds blow towards the west across the equatorial Pacific, moving warm surface water towards Southeast Asia. As the surface water moves west, it is replaced by upwelled cold, nutrient-rich water near the coast of South America. Regarding atmospheric circulation, the trade winds associate to a convection cell that forms because warm moist air rises over the western Pacific, leading to low atmospheric pressure and significant precipitation in the region. In the eastern Pacific, there is high pressure, leading to drier conditions. El Niño and La Niña events are cyclical departures from these conditions. During El Niño years, the high-pressure system over the eastern Pacific diminishes, the trade winds weaken, and warm surface water begins to flow eastwards across the Pacific, warming coastal waters in South America. In the atmosphere, the low-pressure system in the western Pacific is replaced by high pressure and the low-pressure system moves east across the Pacific. La Niña events, on the other hand, are characterized by surface waters colder than normal over the eastern Pacific, and stronger trade winds that increase coastal upwelling near the coast of South America. Because of the cyclic nature of ENSO, El Niño and La Niña events alternate nearly every three to seven years, and each event may last from a few months to a year or more.

These coupled ocean-atmosphere events impact global weather and climate through atmospheric teleconnections. For example, during El Niño events Australia and Indonesia receive a reduced amount of rainfall in response to the higher pressure over the western Pacific, but also India and the equatorial band in Africa experience weather conditions drier than normal. On the other hand, Peru and Chile have exceptional rains and floods, but excess rainfall was also described for central-east South America and southern North America. In recent years, noncanonical or "nontraditional" form of ENSO events have been documented and named central Pacific or Modoki type. In Modoki-El Niño events, the warmest sea surface temperature anomaly does not occur near the coast of South America but in the central equatorial Pacific, often accompanied by anomalously cold temperatures to the east and west. The Canonical and Modoki types of ENSO relate to different climatic impacts; for instance, Modoki-El Niño events are associated with a greater frequency of tropical cyclones and increasing landfall potential along the Gulf of Mexico and Central America (Kim et al. 2009). Thus, ENSO is regarded as one of the most important sources of global climate variability (Philander 1990).

ENSO effects on the eastern equatorial Pacific are variable among events, regions, species, and communities. ENSO impacts nutrient availability and productivity of primary producers and the effects propagate toward higher trophic levels. These effects include anomalous migration events and changes in population survival and recruitment, thus affecting the local abundance, geographical distribution, and gene flow structure of populations of higher organisms such as fish, seabirds, and marine mammals. During strong El Niño events, native fish populations like the Peruvian anchoveta may collapse. Many seabirds and mammals along the Peruvian coast and the nearby Galapagos Islands largely depend on anchoveta for their food, and populations decline dramatically. Although the strongest effects of ENSO are reported in the eastern Pacific, coral bleaching events are common on the western Pacific during extreme El Niño events, along with diminished landings of giant clams and reef fishes. ENSO teleconnections also alter the oceanic conditions and climate in the North Pacific, affecting the biologically productive region of the Pacific Arctic. El Niño events in this area are associated with dramatic changes in distribution and range expansions of several fish and invertebrate species of high relevance for small-scale fisheries supporting local communities. In temperate South America, there is a strong impact of ENSO in the most commercially-important marine species within the large estuarine complexes of Los Patos Lagoon (Brazil) and the Rio de la Plata Estuary (Argentina-Uruguay). It results from the negative relationship between freshwater discharge and fish production associated with heavy rainfall during El Niño events. In this area, low shrimp catches are related, with a one year-lag to El Niño floods (Bertrand et al. 2020).

of the eastern boundary current (Fig. 2.6c). In the southern counterpart, the coast of Peru, the Ekman transport is 90° to the left, meaning that in both cases the net Ekman transport is offshore. The water displaced near the coast is replaced by upwelled nutrient-rich deeper waters that support important fisheries.

Langmuir circulations, or Langmuir cells, are a type of wind-driven water circulation in a series of horizontal, rotating spirals (or cells), named after Irving Langmuir, the American scientist that first described them in 1938. These shallow-water (~ 6 m) circulation patterns can be readily appreciated on the ocean surface as parallel foam, floating debris, or seaweed accumulation lines roughly oriented in the prevailing direction the wind is blowing.

The direction of rotation alternates between spirals; there is a net water movement towards regions of convergence, where water from contiguous spirals meets at the surface and flows downwards. These areas of convergence is where the streaks or surface lines are formed. Each convection cell may be 10–50 m broad and run for hundreds of m. As the counter-rotating spirals occur in sets, there are typically multiple parallel streaks. The downwelling of water is balanced by intercalated diffuse flows of water back to the sea surface in broader areas between the streaks (Fig. 2.6d); downwelling velocities are roughly 1% of the wind speed (Knauss 1996). The maximum rate of water flow takes place at the surface, along the streaks, following a somehow horizontally-spiraling cone of water (much like a corkscrew).

Langmuir cells may be a coastal or open-ocean features. They are transient circulation features that form as a short-term response to wind drag, in a matter of minutes. As these are shallow-water circulation patterns, they have particular interest in the accumulation of phytoplankton cells or concentration of harmful algal blooms, and the formation of small-scale surface patterns (Franks 1997).

2.4 The coastal zone: tides, waves, and coastal currents

Tides are the rise and fall of sea levels that result from the gravitational attraction of the sun and moon on Earth's oceans. The relationship between the masses of the Earth, moon, and sun and their distances to each other play critical roles in affecting tides. As the tides rise and fall, the water moves horizontally, creating currents. Tidal currents that move toward the land, producing rising water levels, are called floods and currents moving toward the sea are called ebbs. There is a period of no velocity called slack water between floods and ebbs after which the current switches direction and increases in velocity. Slack water usually occurs near high tide (the highest water level) and low tide (the lowest water level) but there are locations where the moments of slack tide differ significantly from those of high and low water. As tidal currents move in opposite directions, they are called "rectilinear" or "reversing" currents.

Tidal currents are affected by the interactions of the Earth, sun, and moon, but the moon's force is much greater because it is 389 times closer to the Earth than the sun is. This gravitational force between the Earth and moon pulls fluids creating a "bulge" of water on the side of the Earth facing the moon. This bulge always faces the moon while the Earth rotates, and the regions moving through the bulge experience a high tide, while those regions opposed to the bulge experience a low tide. Thus, for any given coastal location there would be always one high tide per day, as that region passes through the bulge of water on the side closest to the moon.

However, in most places there are semi-diurnal tides (two high waters and two low waters each day). It happens because of the presence of a second bulge of water, formed by inertial force. The inward gravitational force between the Earth and moon is counteracted by an outward inertial force that keeps the Earth and moon apart. This inertial force, created by the rotation of the Earth and moon, has the same magnitude everywhere on Earth, and is always directed away from the moon. The gravitational force, on the other hand, is always directed towards the moon, and is stronger on the side of the Earth closest to the moon. Thus, on the side of Earth closest to the moon, the inward gravitational force is greater than the outward inertial force and the net resulting force creates a bulge of water facing the moon. On the opposite side of Earth, the outward inertial force is greater than the inward gravitational force and the net result creates a water bulge directed away from the moon (Fig. 2.7).

In addition to the moon, the sun creates its own, independent water bulges that contribute to tides, but the sun's gravitational forces are much smaller. During new and full moons, the sun, Earth, and moon

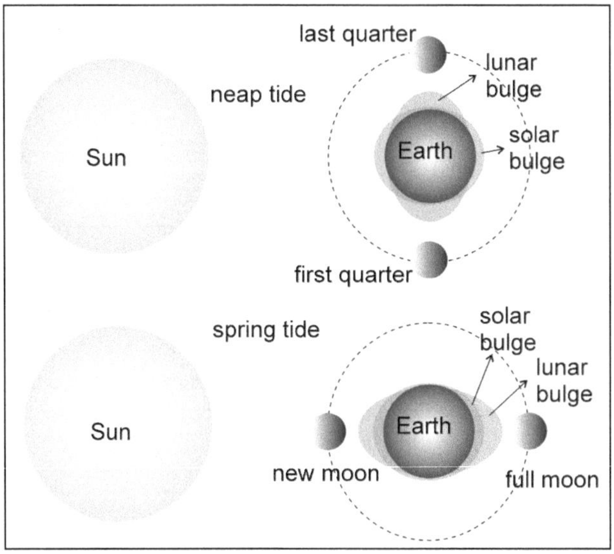

Figure 2.7. Ocean bulges created by the gravitational pull of the Moon and the Sun.

are aligned. The solar and lunar bulges are also aligned and add to each other to produce especially higher high tides and lower low tides. These situations of maximum tidal range are called spring tides and occur every two weeks. On the other extreme, when the moon is at first or third quarter, the solar and lunar bulges are out of phase and cancel each other out. The tides on these lunar phases are called neap tides, and tidal ranges are the smallest, with low high tides and high low tides. Accordingly, when the moon is at full or new phases, tidal current velocities are strong and are called spring currents. When the moon is at first or third quarter phases, tidal current velocities are weak and are called neap currents.

The shape of bays and estuaries can amplify the tidal range and intensify tidal currents. Funnel-shaped estuaries and bays are particularly effective in changing tidal dynamics, as the strength of a tidal current depends on the volume of water that enters and exits with each tidal cycle and the area through which the water flows. A large tidal volume moving through a large area, at the mouth of an estuary, may create only a weak tidal current. As the channel narrows through the head, it may produce a strong tidal current, as all the water is forced through a smaller area (van Rijn 2011). The Bay of Fundy in Nova Scotia (Canada) is a typical example of tidal amplification.

Waves are another important component of the near-shore circulation. Most ocean waves are created by wind blowing across the surface. Light winds in open water create capillary waves (or ripples) that have a V-shaped trough, rounded crest, and short wavelengths (less than 1.7 cm). Ripples are the first waves produced by light breezes on a completely flat surface and represent the initial stage of wave generation. For these short-wavelength capillary waves, the restoring force is surface tension. Simply put, the surface acts like a stretched membrane that provides a force that attempts to return the water to its equilibrium condition. As the wind energy increases, wavelengths get larger, and waves switch from capillary to wind waves. For waves with wavelengths greater than approximately 10 cm, gravity is the restoring force that causes any displacements of the surface to be accelerated back toward the rest position. The fluid overruns because of the kinetic energy that has been gained, resulting in the oscillating wave motion.

When wind waves travel from deep to shallow water, they break near the shoreline and generate currents. As waves get closer to the shore, friction with the sea floor increases, water piles up and the wave becomes steeper and less stable until they form breakers (or breaking waves) at a certain depth where waves become too steep and unstable (Fig. 2.8a). Oblique incoming wind squeezes water along the coast, and breaking waves release a burst of energy that creates a longshore current, which runs parallel to the shoreline. Longshore currents are affected by wave velocity and angle. A steeper breaking angle, a steeper shore slope, and a higher wave produce faster longshore currents. Conversely, a lower

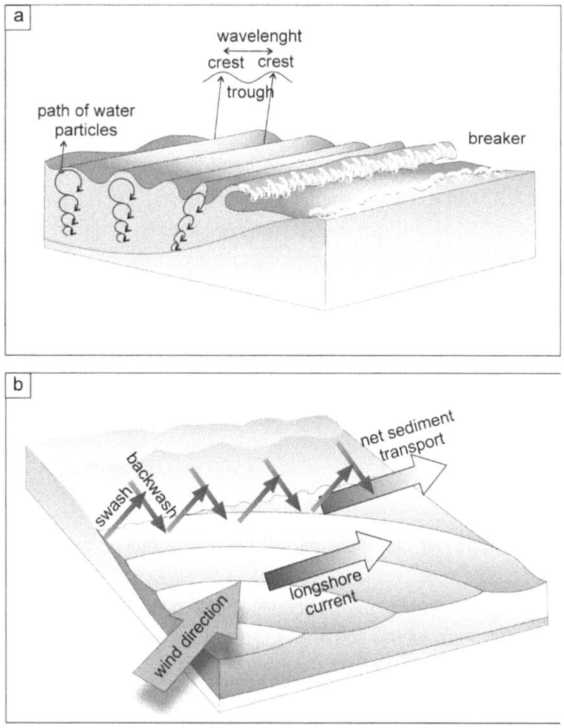

Figure 2.8. (a) Wind waves travelling from deep to shallow water and the formation of brakers; (b) Longshore current created by an oblique incoming wind.

breaking angle, gentler shore slope, and lower wave height slow down the longshore current's velocity. Breaking surf sends water up the beach (swash) at an oblique angle and gravity then drains the water straight downslope (backwash) perpendicular to the shoreline. As water moves on and off the beach, it can transport sediment with an overall direction parallel to the shoreline, in a process known as longshore drift (Fig. 2.8b).

Coastal waters, from estuaries and bays to the edge of the continental shelf, are the most biologically productive areas of the world's oceans (Mann 2009). Phytoplankton forms the base of most food webs in the ocean (see Chapters 3 and 4); for its growth both nutrients and light must be available to allow photosynthesis. Light usually is not a limiting factor, but the concentration of nutrients might be. Since light typically penetrates up to 50 to 100 meters from the ocean surface, there must be an efficient mechanism to supply nutrients to the upper layers of the water. In the open ocean, phytoplankton blooms at the surface but dead and decaying cells often sink to a depth where there is no light, and nutrients contained in their biomass are either remineralized in the vertical transit or accumulate in deeper layers. Conversely, in coastal zones there are specific mechanisms for bringing nutrient-rich deep-water masses to the surface layer.

The wind driven coastal upwelling described in Section 2.3 is one of these mechanisms. Strong winds blowing parallel to the coast push water downwind, while the Ekman transport deflects water to the right or left of the winds. In those coastal settings where waters in the upper layers are blown offshore, colder, nutrient-rich waters are upwelled from the depths. There are five major coastal currents associated with upwelling areas: the Canary Current (off northwest Africa), the Benguela Current (off southern Africa), the California Current (off California and Oregon), the Humboldt Current (off Peru and Chile), and the Somali Current (off Somalia and Oman). Up to 23% of the world's commercial fish catches come from these upwelling areas that comprise just 1% of the global ocean (FAO 2020).

Another mechanism is river runoff. Natural and anthropic nutrient loads are carried by river outflow, and the estuarine circulation also promotes the upwelling of nutrient rich bottom waters at the head of an estuary. River runoff is responsible for the high productivity of estuarine regions, but excess nutrients released into coastal seas, bays, and estuaries can create eutrophication. Anthropic eutrophication may lead to harmful algal blooms, increased growth of macroalgae, increased siltation, and increased oxygen demand and even oxygen depletion in bottom waters, often leading to mortality of benthic animals and fish (Diaz 2001).

Finally, tides and tidal currents cause turbulence and mixing, that may break down stratification and bring nutrient-rich bottom waters to the surface. The stronger the tidal current and the shallower the environment, the more efficient the tidal mixing. In most situations, tidal mixing all year round produces higher primary production and phytoplankton growth close to the shoreline. Depending on the current velocity (and thus, the tidal range), there will be a certain distance from the shoreline (or depth) at which the water column changes from being tidally mixed to being stratified. These transitions are called tidal fronts and there are often large concentrations of phyto- and zooplankton, fishes, and marine megafauna close to these fronts (Derisio et al. 2014).

2.5 The carbon cycle

Ocean's primary productivity will be treated in detail in Chapter 4; it broadly refers to the production of organic matter by phytoplankton. By means of photosynthesis, autotrophic phytoplankton fixes inorganic CO_2 into organic carbon, and makes this organic carbon available to heterotrophs, including bacteria as well as more complex organisms of the zooplankton (floating animals), nekton (swimming organisms), and benthos (organisms of the seafloor). There are many nested cycles of carbon associated with ocean productivity (Sigman and Hain 2012) (Fig. 2.9a). Gross primary productivity is the total rate (hence, expressed per units of time) of organic carbon production by photosynthesis, and respiration is the biological oxidation of organic carbon to CO_2. Net primary productivity is the gross primary productivity minus respiration by the autotrophs themselves (thus, it is the rate of phytoplankton biomass production). Secondary productivity is the growth rate or production of biomass by heterotrophs, and is much smaller than the net primary productivity, because the largest proportion of the organic carbon ingested by heterotrophs is respired. Net ecosystem productivity is the gross primary productivity minus the respiration of all organisms combined, usually expressed by unit area to which the ecosystem extends.

The euphotic zone has been traditionally assumed to be the depth to which 1% of surface radiation remains and defines the zone where autotrophs can catch enough light for primary production by photosynthesis to exceed the loss of carbon by autotrophic respiration (Garrison 2012). In this zone, there is an annual surplus of organic carbon that is exported out of the surface layer either as Particulate Organic Carbon (POC) or Dissolved Organic Carbon (DOC). The difference between POC and DOC is a matter of size and methodological tradition; POC is the organic carbon contained in suspended organic particles retained by a 0.45 μm filter and DOC is the organic carbon in the water solution that passes through the filter. Even if living biomass does not accumulate in the oceans as it does in, for example, terrestrial ecosystems, marine organisms are responsible for the largest carbon sequestration on Earth (Broecker 1982). This is achieved by the sinking of organic matter into the deep ocean before it turns into CO_2 and dissolved nutrients by bacterial decomposition. Oceanographers often refer to this process as the biological pump (this process will be treated in detail in Chapter 5), as large amounts of CO_2 are pumped out of the atmosphere and into the ocean. Yet, most of the organic carbon contained in phytoplankton biomass is respired back to the atmosphere within the surface layers (Eppley and Peterson 1979).

However, a small fraction of the organic carbon sinks to the deep ocean as aggregates of phytoplankton cells, fecal pellets, and remains of zooplankton, forming large particle aggregates known as "marine snow" (Shanks and Trent 1980). As the organic matter settles onto the seafloor, it is nearly entirely decomposed back to dissolved inorganic elements and typically much less than 1% of the organic carbon from photosynthesis in the euphotic zone gets buried in deep sea sediments (Emerson and Hedges 2003). A process that is also part of the biologically-mediated carbon cycle is the formation and sinking of calcareous skeletal material. Some marine organisms, like coccolithophores and foraminifera, combine calcium with carbonate ions to form hard body parts of calcium carbonate. Calcium carbonate is denser

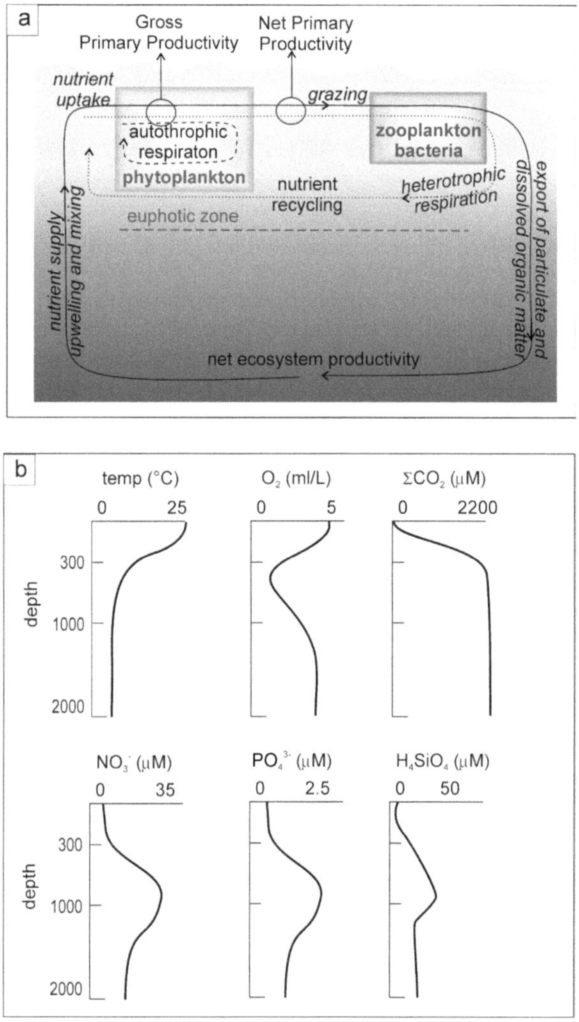

Figure 2.9. (a) Nested carbon cycles associated with ocean productivity (modified from Sigman and Hain 2012); (b) Schematic depth profiles of dissolved O_2, total dissolved inorganic carbon and macronutrients, showing the vertical segregation of biolimiting elements throughout the water column. These are schematic profiles which do not account for the geographic variation of different water masses; the lower limit of the thermocline is assumed to be at ~ 300 m [replotted and modified from Libes 1992].

than organic matter and sinks out of the surface water more rapidly. Globally, the CO_2 that effectively sinks into sedimentary rocks is four times greater than the sink in organic sediments (Carlson et al. 2008).

Ocean acidification is the ongoing decrease in the pH of seawater, caused by the increasing dissolution of atmospheric CO_2 into the oceans (this process will be treated in detail in Chapter 14). Dissolved CO_2 can react with seawater to form carbonic acid and some of the resulting carbonic acid molecules dissociate into a bicarbonate ion and a hydrogen ion (H^+), thus increasing ocean acidity (lowering pH). This process has led to a decrease in pH by 0.1 units since the industrial revolution, and increasing amounts of atmospheric CO_2 are projected to further decrease ocean pH by another 0.3–0.4 pH units by the end of the century (Hennige et al. 2014). This decrease in pH may have a wide range of impacts upon marine life, especially on calcifying organisms. By increasing the presence of free hydrogen ions, the additional carbonic acid that forms in the oceans ultimately results in the conversion of carbonate ions into bicarbonate ions. This net decrease in the amount of carbonate ions reduces the formation of biogenic calcium carbonate structures and makes them more vulnerable to dissolution. Ocean acidification, also referred to as the "evil twin of global warming", has many potential ecological, economic, and societal consequences, and most ocean and climate scientists view ocean acidification as a critical global-scale problem that must be addressed along with climate change (Cooke and Kim 2019).

2.6 Nutrient requirements and the distribution of nutrients

For calcifying algae, calcium is an element that is rapidly assimilated to produce calcium carbonate hard parts, but it has a uniformly high concentration in surface waters; so is dissolved CO_2 which is readily taken up by phytoplankton during photosynthesis. Those elements required for phytoplankton growth that may be scarce, and even limit growth and reproduction, are typically identified as nutrients. Elements universally accepted as nutrients include nitrogen (N) and phosphorus (P), and silicon (Si) for certain phytoplankton. These are required by phytoplankton in high concentrations in seawater (in the order of 10^{-6} M), to build up biomass, and are thus considered macronutrients. Other elements are needed in much less quantities but are nonetheless equally necessary for phytoplankton growth. Examples of micronutrients (required at concentrations in the order of 10^{-9} M) are iron (Fe), zinc (Zn), copper (Cu), manganese (Mn), molybdenum (Mo), and cobalt (Co). In general, micronutrients act as enzyme co-factors. More recently, organic molecules such as vitamins, which typically are at very low concentrations in seawater, have been identified as limiting factors for phytoplankton growth, and thus may be considered essential micronutrients too (Helliwell 2017).

The vertical segregation of macronutrients and other elements across the water column provides valuable clues about the processes (either physicochemical or biological) that determine their distributions. The schematic depth profiles in Fig. 2.9b illustrate this for a number of elements. For example, dissolved O_2 and total dissolved inorganic carbon (ΣCO_2) profiles give us an idea of the relative rates of photosynthesis and respiration. In surface waters, O_2 concentrations are high due to sea-atmosphere exchange and O_2 production by phytoplankton, whereas nutrients (NO_3^-, PO_4^{3-}, H_4SiO_4) and total dissolved inorganic carbon (ΣCO_2) are at low concentrations due to phytoplankton uptake. Below the mixed layer (300 m), respiration draws CO_2 levels to a maximum, and they remain constant with increasing depths. Conversely, O_2 saturation drops steadily at the thermocline and is drawn to a minimum below the mixed layer, only to reach maximum concentrations below 2000 m where the demand for aerobic respiration is less. Remineralization of the POM produced above the thermocline and the lack of nutrient uptake out of the euphotic zone create concentration maxima for PO_4^{3-} and NO_3^- by accumulation below the mixed layer. The minimum O_2 concentration coincides with this peak in nutrients, suggesting that the waters directly above the thermocline have received the greatest amount of POM and are stagnant (Libes 1992). Dissolved H_4SiO_4 is one of the biolimiting elements that does not have a gradient which coincides with the thermocline, presenting a maximum concentration at greater depth than N and P because silicon remineralization is not biologically-controlled (i.e., by bacterial respiration), but rather by the chemically-controlled dissolution of siliceous shells and frustules.

The American oceanographer Alfred C. Redfield proposed in 1958 that phytoplankton biomass has C:N:P stoichiometric ratios of approximately 106:16:1 and the dissolved N:P ratio in the deep ocean is close to the 16:1 ratio. On the other hand, only trace amounts of iron make up phytoplankton biomass, but iron's shortage can limit primary productivity in the open ocean, especially in those regions where nitrogen and phosphorus are in excess. Silicon is considered a nutrient only for some phytoplankton groups that use it to make opal hard body parts. Among these organisms are diatoms that thrive in silicon-rich surface waters (see Chapter 3). Mixing across the pycnocline requires energy, and so the transport of water with its dissolved elements. In the absence of stratification, mixing would carry phytoplankton away from sunlight for long periods of time. Thus, most open ocean biomass, including phytoplankton and their consumers, is found within the ocean surface, but dead organisms sink, and export their organic matter below the pycnocline. It depletes the surface ocean of nutrients, causing them to accumulate in deep waters where there is no light available for photosynthesis.

2.7 Corollary

As a corollary to these introductory chapters, we may well sum up the concept of both-ways interaction between the physical environment and its biotic components, which lies at the root of ecological theory. The geological features always are subjacent to everything else, and that applies either to the tectonics of whole ocean basins or the hardness of the bottom and the type of sedimentary processes that prevail in coastal areas. Superimposed on the geological imprint, physical processes that determine the nature and circulation of water masses are of utmost importance in shaping up the oceanic environment. Seawater is the inherent medium in which all marine organisms spend their whole or most of their life history; hence, seawater chemistry and the chemical processes that take part in the aqueous medium of seawater are of major importance.

The proximity to the continent of the different water masses relates to continental inputs of inorganic and organic nutrients (through continental rock weathering, river runoff and anthropogenic sources). Water circulation redistributes these nutrients into phytoplankton that lie at the bottom of marine food webs and fuel higher trophic levels. The distribution of nutrients is not uniform in either geographic scales or with depth; for instance, consider how nutrients entrained into large oceanic gyres are used up over long time scales, versus the upwelling of nutrient-rich water masses from below the thermocline that takes place seasonally in coastal areas representing only 1% of the world's oceans.

Wind-generated oceanic waves are responsible for air-sea exchange of gases (Libes 2011). On the other hand, the movement of water masses through tides and waves becomes a feature of great importance in the littoral zone, as wave energy represents a major physical force that shapes up benthic communities and contributes to sediment transport. In the same way, tides may not have marked effects on the organisms of the open ocean, but they do limit the distribution and shape up intertidal communities in coastal areas.

Rivers not only provide nutrients to the oceans, but also large amounts of sediments. The quantity of sedimentary particles in suspension on the water column determines the degree of light penetration in the ocean and as we mentioned, light penetration is of utmost importance for photosynthesis. In coastal areas, light penetration determines the depth to which submerged aquatic vegetation such as seagrasses occurs, and the characteristic zonation of seaweeds with depth. In the open ocean, it limits primary productivity down to the depth of the euphotic zone. On the other hand, freshwater inputs of continental origin create estuaries and river plumes. Estuaries are geomorphological features of major importance from an ecological point of view; as many characteristic communities develop within their reaches, they harbor oceanic life forms such as marine mammals throughout some parts of their life history, and most often act as rearing grounds for pelagic fish.

Summary

1. Most of the solar radiation reaches the Earth in the form of visible light; about 70% is absorbed and heats up the atmosphere and Earth's surface, producing a latitudinal pattern of the long-term average sea surface temperature. In addition to this latitudinal pattern, water stratification is the process by which water is segregated into vertical layers based on differences in temperature and salinity.
2. The drag of trade winds on the ocean surface along the equator creates the Equatorial currents that flow east to west. When these currents meet the east coast of continents, they are deflected away from the equator and become western boundary currents. At higher latitudes, the westerlies push surface water from west to east and continents deflect the currents creating eastern boundary currents. These currents combine to produce gyres that rotate clockwise in the Northern Hemisphere and counterclockwise in the Southern Hemisphere.
3. Tides are the rise and fall of sea levels that result from the gravitational attraction of the sun and moon. During new and full moons, the sun, Earth, and moon are aligned producing higher tidal ranges called spring tides. Waves are another important component of the near-shore circulation.

Most ocean waves are created by wind blowing across the surface. When wind waves travel from deep to shallow water, they break near the shoreline and generate currents.

4. Ocean primary productivity refers to the production of organic matter by photosynthesis. Phytoplankton fixes inorganic CO_2 into organic carbon and makes it available to heterotrophs. Within the euphotic zone, primary production by photosynthesis exceeds the loss of carbon by autotrophic respiration. Thus, there is an annual surplus of organic carbon that is exported out of the surface layer. By the sinking of organic matter, marine organisms are responsible for the largest carbon sequestration on Earth, and the process is often referred to as the biological pump.
5. Ocean acidification is the ongoing decrease in the pH of seawater, caused by the increasing dissolution of atmospheric CO_2 into the oceans. Dissolved CO_2 can react with seawater to form carbonic acid and some of the resulting carbonic acid molecules dissociate into a bicarbonate ion and a hydrogen ion (H^+), thus lowering pH. This decrease in pH may have a wide range of impact upon marine life, especially on calcifying organisms.
6. Those elements required for phytoplankton growth, that may be scarce, and even limit growth and reproduction are typically identified as nutrients. Those elements required by phytoplankton in high concentrations are considered macronutrients. Other elements called micronutrients are needed in much minor quantities. In general, micronutrients act as enzyme co-factors. More recently, organic molecules such as vitamins have been identified as limiting factors for phytoplankton growth, and thus may be considered essential micronutrients too.

References

Bertrand, A., K. Lengaigne, M. Takahashi, A. Avadí, F. Poulain and C. Harrod. 2020. El Niño Southern Oscillation (ENSO) effects on fisheries and aquaculture. FAO Fisheries and Aquaculture Technical Paper No. 660. Rome, Italy.

Bielefeld, B. 1997. Investigation into albedo-controlled energy loss during the last glaciation. GeoJournal 42: 329–336.

Broecker, W.S. 1982. Ocean chemistry during glacial time. Geochim. Cosmochim. Acta 46: 1689–1705.

Carlson, C.A., N.R. Bates, D.A. Hansell and D.K. Steinberg. 2008. Carbon cycle. pp. 477–486. *In*: Encycl. Ocean Sci. Second Ed. Elsevier Inc.

Cooke, S.L. and S.C. Kim. 2019. Exploring the "evil twin of global warming": Public understanding of ocean acidification in the United States. Sci. Commun. 41: 66–89.

Derisio, C., D. Alemany, E.M. Acha and H. Mianzan. 2014. Influence of a tidal front on zooplankton abundance, assemblages and life histories in Península Valdés, Argentina. J. Mar. Syst. 139: 475–482.

Diaz, R.J. 2001. Overview of hypoxia around the world. J. Environ. Qual. 30: 275–281.

Emerson, S. and J. Hedges. 2003. Sediment diagenesis and benthic flux. Tr. Geo. 6: 625.

Eppley, R.W. and B.J. Peterson. 1979. Particulate organic matter flux and planktonic new production in the deep ocean. Nature 282: 677–680.

FAO. 2020. The State of World Fisheries and Aquaculture 2020. Sustainability in action. Rome; doi.org/10.4060/ca9229en.

Franks, P.J.S. 1997. Models of harmful algal blooms. Limnol. Oceanogr. 45: 1273–1282.

Garrison, T.S. 2012. Oceanography: an invitation to marine science. Cengage Learning.

Helliwell, K.E. 2017. The roles of B vitamins in phytoplankton nutrition: new perspectives and prospects. New Phytol. 216: 62–68.

Hennige, S.J., L.C. Wicks, N.A. Kamenos, D.C.E. Bakker, H.S. Findlay, C. Dumousseaud and J.M. Roberts. 2014. Short-term metabolic and growth responses of the cold-water coral *Lophelia pertusa* to ocean acidification. Deep Sea Res. Part II Top Stud. Oceanogr. 99: 27–35.

Kim, H.-M., P.J. Webster and J.A. Curry. 2009. Impact of shifting patterns of Pacific Ocean warming on North Atlantic tropical cyclones. Science (80-) 325: 77–80.

Knauss, J.A. 1996. Introduction to Physical Oceanography, Second edition. Prentice Hall, Upper Saddle River, NJ. 309 pp.

Libes, S. 2011. Introduction to Marine Biogeochemistry. Academic Press.

Libes, S.M. 1992. An Introduction to Marine Biogeochemistry. John Willey & Sons. Inc, New York.

Loeb, N.G., B.A. Wielicki, D.R. Doelling, G.L. Smith, D.F. Keyes, S. Kato, N. Manalo-Smith et al. 2009. Toward optimal closure of the Earth's top-of-atmosphere radiation budget. J. Clim. 22: 748–766.

Mann, K.H. 2009. Ecology of Coastal Waters: With Implications for Management. John Wiley & Sons.

Mann, K.H. and J.R.N. Lazier. 2013. Dynamics of Marine Ecosystems: Biological-Physical Interactions in the Oceans. John Wiley & Sons.

Philander, S.G. 1990. El Niño, La Niña, and the southern oscillation, 293 pp. San Diego, CA Acad.

Shanks, A.L. and J.D. Trent. 1980. Marine snow: sinking rates and potential role in vertical flux. Deep Sea Res. Part A Oceanogr. Res. Pap. 27: 137–143.
Sigman, D.M. and M.P. Hain. 2012. The biological productivity of the ocean. Nat. Educ. Knowl. 3: 1–16.
Trenberth, K.E., J.T. Fasullo and J. Kiehl. 2009. Earth's global energy budget. Bull. Am. Meteorol. Soc. 90: 311–324.
Trenberth, K.E. and J.T. Fasullo. 2012. Tracking Earth's energy: From El Niño to global warming. Surv. Geophys. 33: 413–426.
van Rijn, L.C. 2011. Analytical and numerical analysis of tides and salinities in estuaries; part I: tidal wave propagation in convergent estuaries. Ocean Dyn. 61: 1719–1741.
Wiebe, P.H., E.M. Hulburt, E.J. Carpenter, A.E. Jahn, G.P. Knapp, S.H. Boyd, P.B. Ortner et al. 1976. Gulf stream cold core rings: large-scale interaction sites for open ocean plankton communities. Deep Sea Res. Oceanogr. Abstr. 23: 695–710.

Section II

Functional Biodiversity of Marine Organisms

CHAPTER 3
Marine Phytoplankton Diversity and Ecology

Jerónimo Pan

3.1 Introduction: What is phytoplankton?

Photosynthesis in the ocean is largely carried out by microscopic, single-celled photoautotrophs living at the surface and collectively known as phytoplankton. Phytoplankton is a functional term, which refers to the community of planktonic microscopic algae (prokaryotic and eukaryotic) that are responsible for primary production in aquatic environments. While individual phytoplankton cells may present motility structures and mechanisms to individually displace through a fluid medium, their collective movements and position in the water column are largely determined by water motion; the term plankton comes from the Greek word *planktos*, meaning 'drifter' or 'wanderer'. The term phytoplankton makes reference to a community or assemblage of cells, while each individual organism in that community is called a phytoplankter (a much less used term).

It is worth noting that the term *algae does not* a refer to any taxonomic category, but it is a functional term, by which almost all non-vascular, oxygen-producing photoautotrophs which do not develop complex, protected reproductive structures, are functionally grouped. Being both a polyphyletic (i.e., derived from more than one common evolutionary ancestor) and paraphyletic (i.e., not including all descendant clades from a common evolutionary ancestor) assemblage of photosynthetic organisms, phycologists have resorted to common features to group several unrelated clades into the "algae". Such features include photosynthetic pigments (most notably chlorophyll *a*, common to all algae), and reproduction through mitotic division or involving spores and gametes produced in morphologically-simple sporangia and gametangia, respectively, that are never encased by sterile, protective cells. Moreover, the somatic cells of many single-celled algae may act as gametes at a stage of their life history.

Through photosynthesis, marine phytoplankton yields globally some 50 billion tons of organic C, roughly corresponding to ~ 40% of the global primary production yr^{-1} (Longhurst et al. 1995), and representing one of the largest contributions to the 'active' carbon pools in the biosphere. Yet, as these are microscopic organisms with rapid turnover times, on a global scale, phytoplankton cells only account for ~ 1% of total plant biomass (compare to 'Blue Carbon' ecosystems, in Chapter 9). The entire phytoplankton biomass of the global oceans is consumed every 2 days to 2 weeks (Libes 2009). In other terms, according to the latest published global carbon budget, oceanic phytoplankton contributes to an annual uptake of ~ 23% of the CO_2 emitted to the atmosphere through fossil CO_2 and land-use change emissions (Friedlingstein et al. 2019). As it will be presented in Chapter 4, the rates of primary production are not uniformly distributed across oceans in either space, depth or time.

Let me continue by presenting some emergent characteristics and functional adaptations that are broadly shared by phytoplankton, to later introduce the diversity of the dominant groups in modern

National Scientific and Technical Research Council, Argentina (CONICET); Instituto de Geología de Costas y del Cuaternario (IGCyC); Instituto de Investigaciones Marinas y Costeras (IIMyC).
Email: jeronimopan@gmail.com

oceans. On parallel, as I present different groups of marine phytoplankton, some notable ecosystemic processes that are driven by their biology will be discussed.

3.2 Emergent properties and functional adaptations in marine phytoplankton

3.2.1 Phytoplankton size and allometric relationships

Phytoplankton life forms include individual cells, colonies, and pseudocolonies (resulting from genetically-determined incomplete cell division). This array of life forms also spans a huge range in size, with the smallest cells < 1 μm^3 (e.g., picocyanobacteria), and on the other end of the size spectrum, large unicellular diatoms attaining 10^9 μm^3, some even visible to the naked eye (Beardall et al. 2009). Regardless of this ~ 9 orders of magnitude variation in biovolume, most phytoplankton are typically small (i.e., large individual phytoplankton cells are rare, Beardall et al. 2009). Being small in a fluid medium translates to biophysical jargon as living at low Reynolds numbers,[a] in which viscous forces are much more important than inertial ones and the organisms are largely governed by Stokes' laws of viscous drag (Naselli-Flores and Barone 2011). In the case of marine phytoplankton, add to that the increased density of seawater due to salinity.

Plankton has traditionally been classified into size categories. This classification has responded to operational and methodological constraints and sampling habits, rather than ecological criteria. Thus, in relation to its length/diameter phytoplankton is categorized into picoplankton (0.2–2.0 μm in diameter), nanoplankton (2.0–20 μm), microplankton (20–200 μm), and mesoplankton (0.2–20 mm, comprising a few individual diatoms, chains and tufts of the cyanobacterium *Trichodesmium*). Such variation in cell size has to be accompanied by the up- or downscaling in physiological processes, even for single-celled organisms. The biological discipline studying such relationships between body size, shape and physiology is called allometry. Much research has been devoted to allometric relationships in phytoplankton, relating cell size (biovolume), surface area, and nutrient uptake.

It has long been established that small size plays a significant role in nutrient uptake (see below). Verdy et al. (2009) simulated the variation in phytoplankton growth and evolutionarily-attained stable cell size, with a steady input of a single nutrient. They posit that, while smaller cells have an advantage under uptake-limited conditions due to their large surface to volume ratio, the size-dependency of cellular quotas gives an advantage to large cells, which theoretically can evolve from small cells. In other words, larger cells would produce biomass faster due to their larger nutrient storage capacity, thus having an advantage under assimilation-limited conditions. Yet, it is worth pointing that this simulation focuses on a single ecological mechanism (i.e., cell size determination by nutrient uptake, storage and metabolism), leaving aside others, e.g., top-down controls.

The way in which cell-part proportions depart from an archetypical scheme within a phytoplankton taxon may also be regarded from the point of view of allometric relationships. As much as cell size plays a role in nutrient uptake, cell architecture plays its part in light harvesting (Naselli-Flores and Barone 2011). With regards to the latter, one evolutionary strategy seems to have favored large vacuoles, which often concentrate cytoplasm and plastids in that part of the cell not occupied by the vacuole, thus indirectly helping to increase the surface area for interception of PAR and for nutrient influx per unit metabolizing volume (Beardall et al. 2009).

While unicellular representatives are found in all phytoplankton taxa, colonies are rather less common; very large unicells are rare, while multicellular organisms only occur in two taxa and pseudocolonies are restricted to dinoflagellates (Beardall et al. 2009). The effective size of a phytoplankton colony may be increased by the formation of chains of cells. Chain formation is a common feature across various groups of phytoplankton, ranging from cyanobacteria filaments and tufts (e.g., *Trichodesmium* up to 0.5 mm), to dinoflagellates (e.g., *Margalefidinium polykrikoides*, up to ~ 240 μm) and diatoms (e.g., *Chaetoceros* sp. > 100 μm). As grazing modulates the size and the shape structure of phytoplankton assemblages

[a] The Reynolds number is a ratio obtained by comparing the relative importance of inertial to viscous forces within a fluid, and is used to describe the way in which fluids will behave at different scales (Vogel 1994, Naselli-Flores and Barone 2011).

(Naselli-Flores and Barone 2011), so it has been hypothesized that the formation of colonies may either be a deterrent to predation by zooplankton or at least modulate it. For example, as a response to the presence of cues from three species of copepods whose grazing rates on diatom chains are several times higher compared to that of single cells, the diatom *Skeletonema marinoi* suppresses chain formation, thus reducing grazing (Bergkvist et al. 2012). On the other hand, the formation of elongated chains enhances vertical migration of phytoplankton much more effectively than single cells (Lovecchio et al. 2019). Other benefits arise from the formation of colonies, such as some protection against viruses and parasites through the production of mucilage (Beardall et al. 2009). Furthermore, predation upon colonial phytoplankton not necessarily leads to death, as when a predator feeds on a colony the colony can fragment into smaller individuals that may be too small for the predator, and become propagules for new colonies themselves.

One notable effect arising from the formation of colonies and grouping of phytoplankton cells is known as the 'package effect'. In relation to spectral absorption, the package effect is responsible for decrease in the specific absorption coefficient of a pigment when it is aggregated *vs.* homogeneously distributed (Lohrenz et al. 2003, Beardall et al. 2009). Further, the package effect is responsible for the apparent restriction on nutrient uptake as a result of the greater thickness of the diffusion boundary layer (reviewed by Beardall et al. 2009). While the latter may be regarded as a disadvantage, colonial aggregation may, on the one hand, make the transport of nutrients among cells more efficient compared to a large single cell, and on the other hand, provide structural benefits for N_2 fixation, such as in *Trichodesmium*.

3.2.2 Phytoplankton cells as particles, and their behavior in a fluid medium

Most particles in suspension in the upper layers of the ocean are either organisms or produced by them. In fact, it is not uncommon for oceanographers to refer to phytoplankton cells indistinctively as 'particles' (Jackson 2012). This might be convenient, since inert particles interact with each other in analogous ways to organisms (Jackson 2012). Yet, a number of emergent properties make phytoplankton cells stand out. For once (i) phytoplankton have distinct pigments that determine their bio-optical properties in relation to light absorption; on the other hand, (ii) phytoplankters have distinctive and diverse cell 'architectural' types that distinguish them from an idealized spherical particle from the point of view of light-scattering and their behavior in suspension or sinking in a fluid medium (the latter is known as 'form resistance', Naselli-Flores and Barone 2011); and further (iii) most phytoplankton cells have motility organelles and mechanisms to modulate their position while in suspension.

Now let me discuss further the latter point. Particles (including organisms and biogenous particles) account for much of the spatial and temporal variation in optical properties of seawater (Jumars 1993). For instance, Kitchen and Zaneveld (1992) proposed a model consisting of three concentric spheres to quantify refraction of different parts of a phytoplankton cell. The model discerned between an outer sphere that represented the cell wall, the middle layer represented the chloroplasts, and the core sphere represented the cytoplasm. They found that altering cell-wall thickness produced much variation in refraction and scattering. Green et al. (2003) studied the variability in upper ocean optical properties for New England shelf waters, and found that larger eukaryotic phytoplankton were responsible for the majority of total particle absorption and scattering, while picoplanktonic *Synechococcus* and bacterioplankton did not contribute much to either parameter.

Cells (or "biological particles", *sensu* Jumars 1993) contain more organic mass in their fluid portions than they do in their solid components. For instance, up to 85% of the cell volume in marine diatoms is occupied by a central vacuole (Strathmann 1967). Maintaining a position in the water column in which the acquisition of nutrients and the harvesting of light are optimal for photosynthesis is of paramount importance for phytoplankton cells; yet, the organic or biomineral buildup of individual cells makes them denser than the surrounding seawater and prone to sink. Thus, attaining neutral buoyancy often is an ecophysiological target for phytoplankton, but neutral or positive buoyancy cannot be achieved solely by regulating concentrations of common inorganic ions. A means of compensating for the dense cellular materials is the incorporation of compounds that are less dense than seawater such as ammonium derivatives (Boyd and Gradmann 2002). In the case of planktonic cyanobacteria (which do not have intracellular membrane systems), gas vesicles aid in the regulation of vertical position in the water column (Walsby et al. 2006). Furthermore, siliceous or organic spines and bristles protruding from the cell margins, and chitin fibrils which may exceed in length the diameter of the cell, represent a kind of 'form resistance' with a role in floatation and antigrazing (Naselli-Flores and Barone 2011).

The capacity to move upwards relative to the surrounding water can be attained and controlled by flagellar beating (Beardall et al. 2009), with most flagellated phytoplankton being quite efficient in vertical positioning (Naselli-Flores and Barone 2011). The *stimulus* for motility relates to inverse gradients of PAR (highest at the surface) and nutrients (higher concentrations at depth), combined with a small vertical component of water movement in the upper mixed layer.

3.2.3 Pigments and phytoplankton

The functional role of phytoplankton in aquatic systems is in primary production. By this, I mean the harvesting and conversion of radiative energy from the sun into chemical energy stored (or fixed) in a stable molecule. This is achieved through the process of photosynthesis, and the initial photosynthate typically is sugar, which may then enter other metabolic pathways and be converted into other storage molecules or leave the cell as an exudate (see Chapter 4). The first step (harvesting of light) is achieved through specific pigments, molecules that are excited by certain wavelengths in the 400–700 nm range. In this process, pigments quench the radiative energy which is then transferred and used in the chemical reactions of photosynthesis. This may seem a very uniform and widespread process among phytoplankton, but as we will see, the metabolism, cell architecture and behavior of different phytoplankton groups is enormously diverse.

Phytoplankton present a wide array of pigments which allow them to fully exploit the PAR radiation spectrum. These are presented in Chapter 4 of this volume, so I will not discuss them here, but rather present some relationships of phytoplankton and their pigments. Ciotti et al. (2002) studied the relationship between cell size distribution and pigment absorption across several natural plankton assemblages. They found that as phytoplankton abundance increases, larger size classes are added incrementally to a background of smaller cells, and that the dominant size class (pico-, nano-, or microplankton) explains the variability in absorption coefficients. In other words, the dominant cell size determines the optical properties of surface waters to a large extent. As above mentioned, the 'package effect' resulting from cell aggregation reduces absorption in natural communities (Lohrenz et al. 2003). An adaptation to low light ($< 1\%$ surface light) in the picophytoplankter *Prochlorococcus* has been described for the Azores Front in the Atlantic Ocean (Stambler 2014). At the deep chlorophyll maximum (DCM) where *Prochlorococcus* is dominant, its absorption coefficient for chlorophyll is lowest due to a package effect, but this is compensated by its content of chlorophyll cell^{-1} being highest, which allows them to bloom at the DCM.

Furthermore, pigments have been used to classify phytoplankton into different taxonomic groups and establish their phylogenies before the advent of genomics. From an ecological perspective, pigments are indicative of (or "proxies", as ecologists say) the phytoplanktonic biomass at a certain geographic point and time of day. Remote sensing and satellite imaging makes use of ocean color and pigment composition to map surface ocean production over planetary scales (more on this in Chapter 4), but still some methodological constraints remain for determining phytoplankton diversity and community composition from space (Bracher et al. 2017). Regardless, this is an active field of study, and phytoplankton functional types (PFTs) based on ocean color are increasingly being incorporated into marine ecosystem and climate models (Ciavatta et al. 2018). PFTs should incorporate taxonomic, bio-optical, physiological (Segura et al. 2013) and cell-size information in order to better contribute to an integration of the effects of plankton community structure into biogeochemical simulations. For instance, Le Quéré et al. (2005) defined PFTs for phyto- and zooplankton, on the premise that biogeochemical processes are closely linked to functional types. Their criteria for defining PFTs used a combination of properties including (1) an explicit biogeochemical role (e.g., calcifiers, silicifiers, N_2 fixers, DMS producers); (2) distinct physiological, environmental or nutritional requirements controlling their biomass and productivity; and (3) the behavior of each group in relation to others.

Box 1: Light in marine environments

There is a little great book by Marston Bates (1960) called *The Forest and the Sea*, in which the author traces brilliant parallelisms between terrestrial and marine ecosystems. In one of its opening essays, Bates compares a South American rainforest with the marine environment. First, he ascertains that both share a truly three-dimensional nature, and that the vertical gradient in light that exists in both environments is a key, structuring feature. So let us take a brief look at light in the oceans.

The solar radiation spectrum is made up of an array of wavelengths which are grouped into three fractions (i) short wave radiation (ultraviolet, UV ~ 280–400 nm); (ii) the visible fraction also known as photosynthetically-active radiation (PAR, 400–700 nm); and (iii) long-wave infrared radiation (IR, 700–2,500 nm) [see Figs. 2.1a and 2.1b for further reference]. Despite being obvious, let me briefly mention that the amount of incoming solar radiation changes with time of day, season, latitude, and cloud cover; these factors combine to yield a net flux of incoming solar radiation for a certain area and moment. Sunlight is either reflected back to space by clouds, or eventually reaches the sea surface. The atmosphere is mostly transparent to incoming sunlight, yet gas molecules (cloud droplets, O_3, CO_2, etc.) absorb radiation in some bands, and aerosols and dust scatter it (Stewart 2008).

Now, any beam of light (actually a photon, if you want to get technical) passing through a medium may be *transmitted* (no change in its direction), *absorbed* (pigments usually play a significant role on this matter) or *scattered* (Jumars 1993). Water is a poor absorber of visible radiation for wavelengths < 560 nm, and the 475 nm wavelengths are the ones that penetrate deeper (Fig. 3.1b) rendering the characteristic blue-green color of the open ocean. The chlorophyll *a* molecule, on the other hand, presents two absorption peaks (at 435 and 675 nm; Fig. 3.1a), while other substances in seawater, collectively known as dissolved (DOM) and particulate (POM) organic matter (see Chapters 4 and 5), increase their absorption exponentially with decreasing wavelength in the visible and UV ranges.

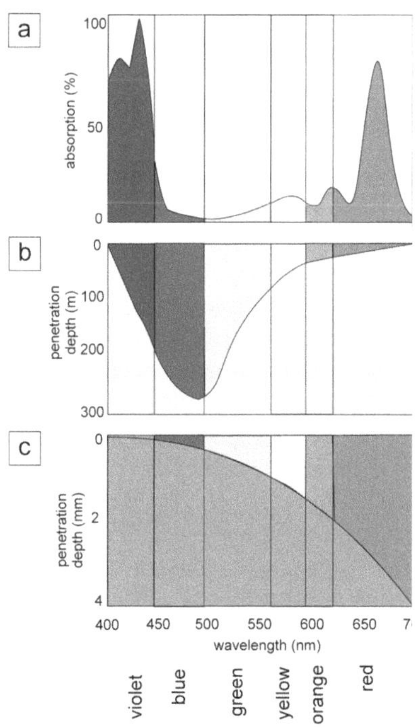

Figure 3.1. (a) The absorption spectrum for chlorophyll *a* in the range of photosynthetically-active radiation (PAR, 400–700 nm). (b) Schematic representation of the relative depth of penetration of different wavelengths from the solar radiation spectrum in open ocean water; the exponential decline in light penetration is much shallower for a coastal water column (modified from Garrison 2009). (c) Schematic representation of the relative depths of penetration of different wavelengths of PAR in microphytobenthos-colonized marine sediments (modified from Des Marais 2003).

Box 1 Contd. ...

...Box 1 Contd.

Particles in suspension in the uppermost layers of the water column (living and dead phytoplankton cells, phytoplankton exudates and broken parts, bacterioplankton, colloidal POM that is < 0.45 μm, and inert sedimentary particles), and colored DOM are largely responsible for the absorption of light throughout the water column. This results in a gradual attenuation of light penetration, which is by no means uniform across different ocean systems (e.g., neritic *vs.* oceanic zones). As a result, biological oceanographers have established different 'bands' or zones, in relation to the amount of light which is available to regulate biological activities. These are the *euphotic zone* (receiving enough solar radiation to support photosynthesis, and running from the surface down to 70 to 150 m depths); a transitional *disphotic zone* (receiving some radiation from the surface but not enough for photosynthesis; living phytoplankton cells may be entrained into this zone and fall into deeper layers or be resuspended into the euphotic zone); and the *aphotic zone* (in permanent darkness and physically detached from where primary production occurs) [see Fig. 1.10 for reference]. At the bottom of the euphotic zone, corresponding to depths where light ranges between 0.1 and 1% of its surface values, there is a deep chlorophyll maximum (DCM), characterized by the presence of low light-adapted species or ecotypes distributed in a distinct pattern (Latasa et al. 2017), that profit from the nutrients that are entrained up from below the nutricline.

In shallow, coastal areas where resuspension of particulate material is a recurring process, light is highly scattered and absorbed by particles and phytoplankton in suspension. Upon reaching the seafloor, the penetration of the different visible light wavelengths presents a pattern that is the opposite of light penetration in the water column (compare Figs. 3.1b and 3.1c). Accordingly, blue and green wavelengths become extinct within the uppermost mm, while red wavelengths are the ones that penetrate deeper (down to ~ 4 mm where filamentous cyanobacteria or diatoms may migrate down temporarily; Fig. 3.1c). Scattering by sediment particles plays a major role in the hyperbolic decline in light penetration, while absorption by microphytobenthos contributes a minor part [see Chapter 8 for more on this topic].

3.2.4 Phytoplankton nutrition

This section discusses different aspects of phytoplankton nutrition. Nutrients are essential for all living organisms, and in the case of phytoplankton they contribute significantly to the shaping of planktonic communities (i.e., through bottom-up controls, see below). Nutrients may be available in dissolved, colloidal or particulate forms to different phytoplankton (more on this will be discussed in the 'types of nutrition' topic). Traditionally, the research focus has been on dissolved nutrients, since most phytoplankton were regarded as osmotrophs (but see Worden et al. 2015). Within the dissolved fraction, nutrients may be inorganic or organic. Inorganic forms of N and P (and Si in the case of diatoms and other siliceous phytoplankton), which are required for the production of organic matter, have traditionally been regarded as the principal source of phytoplankton nutrition. In the past few decades, much research has been devoted to understanding how certain phytoplankton metabolize organic forms of nutrients, such as aminoacids and urea. For instance, *Aureococcus anophagefferens*, a ~ 2–3 μm coccoid pelagophyte that causes blooms in coastal bays of the USA mid-Atlantic coast, is capable of using a wide variety of nitrogen and carbon compounds (Mulholland et al. 2002), but its advantage over other phytoplankton lies in its ability to use organic nitrogen (Nuzzi and Waters 2004).

Besides their chemical composition, dissolved nutrients may be classified according to their concentration into macro- and micronutrients. Dissolved macronutrients such as NO_3^-, NO_2^-, NH_4^+ and PO_4^{3-} are found at micromolar concentrations in seawater (μM, 10^{-6} M), while micronutrients are found at nanomolar concentrations (nM, 10^{-9} M) or less, i.e., at least three orders of magnitude more dilute than macronutrients. Fe and other trace metals (Zn, Cu, Mn, Co, and Mo) dominated much of the research devoted to micronutrients during the 1990s, while the attention then shifted to vitamins (e.g., thiamin, biotin and cobalamin/B_{12}), at the turn of this century. Let us review some of the findings.

Dissolved Fe is a micronutrient almost exclusively (99%) found bound to organic ligands in seawater (Hutchins et al. 1999), and phytoplankton growth is limited by Fe availability in ~ 30–40% of the world ocean (Moore et al. 2001). The uptake of Fe is fundamentally different between prokaryote and eukaryotic phytoplankton. On the one hand, cyanobacteria release substances that strongly chelate iron (siderophores) and make it relatively more available to prokaryotes than to eukaryotes. Castruita et al. (2006) described the first Fe-binding protein for a marine organism, in the N_2-fixing marine cyanobacterium *Trichodesmium erythraeum*. Eukaryotic phytoplankton, on the other hand, efficiently assimilate porphyrin-complexed iron (formed by zooplankton-mediated degradation).

It has recently been established that vitamin auxotrophy (i.e., a requirement for these organic compounds) is widespread in the marine environment and occurs in both autotrophs and heterotrophs residing in oligotrophic and eutrophic waters (Sañudo-Wilhelmy et al. 2014). As in the case of Fe, large areas of the world ocean are devoid of B vitamins, suggesting that vitamin limitation could be impacting carbon and nitrogen fixation. For example, through several incubation experiments of surface water in the Ross Sea, Bertrand et al. (2007) demonstrated the importance of vitamin B_{12} to phytoplankton growth and its effects on community composition (inducing diatom population growth). Sañudo-Wilhelmy et al. (2006) found similar results in coastal environments, performing nutrient amendment experiments, in which B_{12} additions stimulated growth of large (> 5 µm) phytoplankton over unamended controls, while small phytoplankton (< 5 µm) densities remained unaffected.

I will continue describing the modes of nutrient acquisition found among phytoplankton, and discuss the mechanisms of nutrient uptake and how cell size influences this process, resulting in a characteristic elemental stoichiometry in phytoplankton biomass.

(i) Types of nutrition

In planktonic food webs, the acquisition of organic compounds necessary to sustain the metabolism of eukaryotic protists can take place through different processes. Nutrient acquisition strategies vary among different phytoplankton and their growth conditions, that contributes to generating the unique ecological properties of different phytoplankton. It may also help support high phytoplankton diversity by effectively subdividing the habitat with regard to different nutrients. This led mid-20th century plankton ecologists to propose the "paradox of the plankton" (Hutchinson 1961), which delved around the apparent paradox of many different microbial life forms having evolved in a seemingly uniform, resource-limited habitat such as the water column. Actually, we now know phytoplankton communities are plastic enough to exploit different resources (Record et al. 2014).

Regarding the modes of nutrient and organic carbon fixation/acquisition, I have already presented photoautotrophy when discussing phytoplankton pigments. On the other hand, heterotrophy comprises either the assimilation of particles (phagotrophy) or liquids (osmotrophy). In turn, omnivory takes place when a consumer feeds on two or more trophic levels (Ptacnik 2003) and mixotrophy may be regarded as a case of omnivory in which a phototrophic protist is additionally ingesting particles (usually prokaryotes or other protists) by phagotrophy. Furthermore, there are some variations, e.g., kleptochloroplastidy, the retention of a prey's functional chloroplasts by a phagotroph. The occurrence of phagotrophy has been a crucial element in the evolution of photosynthetic eukaryotes (Raven 1997) involved in the many endosymbiotic events that led to protistan clades (see Box 2).

Mixotrophic organisms combine light, mineral nutrients and prey as supplementary resources (Katechakis and Stibor 2006) and this allows some phytoplankton to function at multiple trophic levels (Cloern and Dufford 2005). Phagotrophy by phototrophic eukaryotes is mainly a response to a relative shortage of N, P, Fe, etc. (Raven 1997) with the heterotrophic nutrition of phytoplankton being a source of mineral nutrients (Stibor and Sommer 2003). Mixotrophs are found among virtually all flagellated taxa (Ptacnik 2003) and mixotrophy requires an investment in both photosynthetic and heterotrophic cellular apparati (Hammer and Pitchford 2005, Tittel et al. 2003) which implies a greater physiological cost for a cell to have both photosynthetic and phagotrophic machinery, than having either apparatus alone.

A semi-extracellular phagocytosis of *Prochlorococcus* by the smallest eukaryotic algae was recently described by Kamennaya et al. (2018). The tiny haptophyte, *Braarudosphaera bigelowii,* which is 1.3-µm in diameter and harbors a nitrogen-fixing endosymbiont in its cytoplasm, were photographed with SEM and TEM techniques holding individual *Prochlorococcus marinus* cells in their open hemispheric cytostomes despite being only 1.6 times bigger than their prey. This semi-extracellular mode

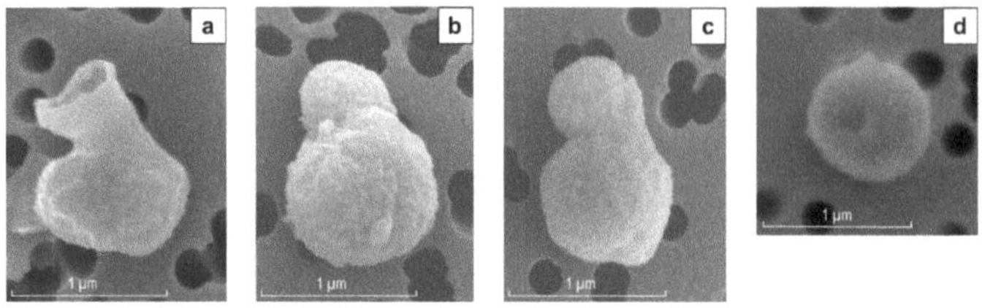

Figure 3.2. Semi-extracellular phagocytosis by the picoplanktonic haptophyte *Braarudosphaera bigelowii* on *Prochlorococcus marinus*. (a–d) SEM photomicrographs of the sequence involving protrusion of a cytostome, engulfing a *Prochlorococcus* cell and the donut-shape predated *Prochlorococcus* cell ends up with. Photo credits courtesy of Drs. M. Zubkov and N. Kamennaya, previously published in PLoS Biol. 16: e2003502.

of phagocytosis (obviously limited by the cell size of the predatory alga; Fig. 3.2) has been termed 'pomacytosis' (from a Greek word meaning 'plug') (Kamennaya et al. 2018).

Other strategies displayed by phytoplankton to optimize their nutrient uptake include nitrogen fixation (exclusive of prokaryotes that carry *nifH* genes encoding the enzyme nitrogenase, most remarkably cyanobacteria) and mucilage production which can help nutrient sequestration (Naselli-Flores and Barone 2011). I will discuss N_2 fixation when presenting colonial cyanobacteria below.

The interactions between bacteria and phytoplankton have been increasingly recognized as regulators of important biogeochemical reactions (Rolland et al. 2016, Seymour et al. 2017), and they relate to the phycosphere concept. The phycosphere is defined as the immediate surrounding environment extending beyond the plasma membrane (or other external cell coating), often maintained by exudates or secretions from an autotrophic phytoplankton cell, and harbors a distinct symbiotic microbial community with which it interacts through the exchange of metabolites and infochemicals that regulate biochemical reactions and ecological relationships ranging from mutualism to antagonism, parasitism and competition. For instance, marine diatoms harbor bacterial communities, and bacterial abundance and community structure greatly depended on the host diatom species (Grossart et al. 2005). Furthermore, it has been established that soluble B vitamins originate from the activities of some bacteria and archaea (Bertrand et al. 2007, Sañudo-Wilhelmy et al. 2014). While the interactions taking place within the phycosphere occur at the microscopic scales of individual microorganisms, they exert an ecosystem-scale influence on fundamental processes (Seymour et al. 2017). Bacterial cells in the phycosphere are organized in a biofilm, and exchange information through the diffusion of molecules; their physiological responses to the medium is coordinated through density-dependent *quorum sensing* and its antagonist *quorum quenching* (Rolland et al. 2016).

(ii) Cell size and nutrient uptake

Phytoplankton nutrient uptake follows Michaelis-Menten kinetics, as it is mediated by specific enzymes (Zevenboom 1986). Yet, nutrient acquisition is limited by their diffusion rate into individual cells, so the evolutionary strategy followed by phytoplankton is to maximize surface area, without becoming too bulky (as a larger volume would increase nutrient requirements). Thus, the only way of gaining surface area for diffusion while limiting volume is to be small. In other words, the smaller a cell, the greater is its surface-to-volume ratio. Furthermore, a large surface-to-volume ratio, as in prokaryotic cells, means that nutrients can easily and rapidly reach any part of the cell interior. In addition, small cells have a smaller diffusion boundary layer that limits nutrient transport (Naselli-Flores and Barone 2011). Conversely, in the larger eukaryotic cells, the limited surface area compared to its volume means nutrients cannot rapidly diffuse to all interior parts of the cell. That is why eukaryotic cells require a variety of specialized internal organelles to carry out metabolism, provide energy, and transport chemicals throughout the cell.

Even for the average phytoplankter living in a fluid at low Reynolds numbers, in which high viscous forces prevent sustained turbulence (Vogel 1994), the uptake of nutrients and food and the release of metabolic products creates localized patchiness that diffusion homogenizes. As organisms sit or move through the environment, they leave trails-regions depleted in some substances and enhanced in others (Jackson 2012). At an ecosystemic scale, phytoplankton cells and other particles in suspension take up or adsorb chemical compounds, depleting them in surface waters from which they are removed by gravitational settling (Jumars 1993).

(iii) Phytoplankton elemental stoichiometry

After pioneering empirical research on the elemental composition of various species of phytoplankton, Alfred C. Redfield (1958) established a well-known and fairly constant average atomic molar ratio for C:N:P of 106:16:1 (see also Chapter 4). Since then, empirical evidence has accumulated documenting deviations from the proportions of the Redfield ratio. The ratio varies in response to changes in algal nutrient status and among taxa (Klausmeier et al. 2008). A review of published data done by Geider and La Roche (2002) reported that under optimal nutrient-replete growth conditions, the cellular N:P ratio ranges from 5–19 N:1 P, with most observations below the Redfield ratio of 16:1. Apparently, the lowest values of N:P are associated with nitrate- and phosphate-replete conditions, while the highest values are observed in oligotrophic waters. Further, Sañudo-Wilhelmy et al. (2004) reported that Redfield ratios are also strongly affected by partitioning into surface-adsorbed and intracellular P pools; they found that if surface-adsorbed P in the cyanobacterium *Trichodesmium* is considered for the ratio, the values are close to the Redfield ratio. Conversely, intracellular ratios consistently exceeded the Redfield ratio (316–434 C:59–83 N:1 *intracellular* P), suggesting P deficiency. A considerable amount of research has been carried out trying to relate this internal ratio to the ratio of nutrients present in the water and identifying those nutrients that potentially limit phytoplankton growth by being present in quantities lower than predicted by the Redfield ratio (Naselli-Flores and Barone 2011).

3.3 Determinants of phytoplankton abundance and distribution

From what has been discussed so far, it should be clear that phytoplankton need nutrients and light in order to grow and reproduce in pelagic (Smetacek and Pollehne 1986) and coastal systems (Cloern and Dufford 2005). These two main factors combined largely determine phytoplankton abundance in the oceans in a most peculiar way, as they present opposing distribution patterns throughout the water column. Light availability is maximum at the surface where nutrients are more rapidly depleted, and conversely, maximum nutrient concentrations are found at depths where light becomes limiting. It is also important to note that CO_2, the prime matter that is fixed through the process of photosynthesis, is not a limiting factor. Other (secondary) factors also tune up phytoplankton distribution and abundance in the oceans such as temperature, salinity, organic micronutrients such as vitamins, trace metals and other enzyme co-factors.

Population dynamics within phytoplankton assemblages have traditionally been understood on the basis of differential growth rates among species, primarily determined by the availability of nutrients (Cloern and Dufford 2005), that is, if you will, the view from the *bottom-up*, considering the distribution of phytoplankton as being determined by that of the elements needed to photosynthesize and grow. There is, however, another view that is equally important in explaining the abundance of phytoplankton from the *top-down*, and has to do with phytoplankton losses due to predation and/or disease (e.g., Tillmann 2004, see also Chapters 5 and 6). There are also many biological interactions that take place within a phytoplankton community that need to be taken into account.

Chapter 4 will further present and discuss the patterns of variation of phytoplankton in space and time. Phytoplankton ecology is about understanding what processes drive these composition, abundance and distribution shifts at various scales. Depending on the scale, these processes have to do with physical and chemical environmental conditions and how these interact with the biological traits that predominate in different ecological and/or taxonomic groups of phytoplankton. The preeminent role of physical processes over biological controls (e.g., predator-prey interactions) has long been recognized in determining phytoplankton abundance and distribution. For example, physical processes such as vertical mixing, upwelling and turbulent advection have long been recognized as strong regulators of phytoplankton patchiness in shelf systems, at scales ranging from a few cm (microscale), to a few m (finescale; Owen 1989), to ~ 100s m (mesoscale) and larger (~ 1,000 km; Mackas et al. 1985). In estuarine and coastal systems, climate forces (e.g., temperature, solar radiation, precipitation and drought, freshwater runoff), that in turn modulate other physical parameters such as salinity (Nche-Fambo et al. 2015), and seasonality are superimposed to physical processes in determining phytoplankton distribution (Cloern and Jassby 2010).

In sum, the success of a certain phytoplankton population in exploiting an ecological niche created by, say, a combination of spatial gradients in light, temperature, water column mixing and nutrient concentration, eventually translates into its abundance and distribution in the plankton. Kiørboe (2008) refers to this as a "mechanistic" approach to the study of plankton, having to do with how individual organisms function and respond differently to their immediate surroundings. This mechanistic nature is limited (or enhanced) by the many morphological and physiological traits acquired over the course of evolution.

From a population ecology perspective, the logistic equation provides a conceptual framework for understanding the variables that affect population growth of a given phytoplankton species. Let us take a look at it:

$$\frac{dN}{dt} = rN\frac{(K - N)}{K}$$

N is the density of phytoplankton cells, and dN/dt represents the change in density over a certain period of time (t). The parameter r results from integrating the reproductive rate of a phytoplankter (b, usually expressed as doublings day^{-1}) minus d, the rate of death of such individual [$r = b - d$]. In the case of single-celled organisms such as phytoplankton, the value for this parameter is of course averaged for a species in the field (or in a monospecific culture). Allometric relationships among different phytoplankton species show that specific growth rate (b) decreases as a function of cell size (Beardall et al. 2009). It is generally assumed that r is directly proportional to the amount of resource in the system (be it light, nutrients, etc.), and this is expressed as K, the carrying capacity. K represents the capacity the environment has for sustaining or "carrying" phytoplankton at a maximum density so as to make use of resources to the fullest. In the logistic equation, the term $(K - N)/K$ is the fraction of total available resource that remains for use (i.e., the fraction of the carrying capacity that has not yet been taken up, Vandermeer and Goldberg 2003). It is important to bear in mind that density-dependence is implicit in the concept of carrying capacity; the mathematical expression implies that the population will grow slowly if it is nearing its carrying capacity (i.e., using up its resources), and will grow more rapidly if it is far below its carrying capacity. In general, density-dependent and density-independent forces may act together to regulate populations in nature, but the rate of growth or decline of a population does not necessarily suggest either one of these mechanisms (Vandermeer and Goldberg 2003).

3.4 Phytoplankton biodiversity and ecology

This section will discuss the diversity of phytoplankton in modern oceans. I will provide a general exposition of the biological traits of the most successful marine phytoplankton groups in modern oceans (e.g., cell coatings, modes of nutrition, motility structures and other essential biological attributes and/or strategies) that have contributed to their evolutionary radiation and ecological success. In other words, I will present those characteristics that determine an intrinsic ability to exploit resources, photosynthesize and divide for each group of phytoplankton, that are reflected in the b term of the r parameter described in the logistic equation above [$r = b - d$]. The following chapters in this book are devoted to micro- (Chapter 5) and mesozooplankton grazers (Chapter 6), and metazoan benthic suspension feeders (Chapters 8 and 10), and discuss at large their grazing mechanisms upon phytoplankton, largely responsible for the d term, in the r parameter.

A discussion of the four dominant phytoplankton groups in the modern oceans will follow: (i) the cyanobacteria, pigmented prokaryotes regarded as the inventors of oxygenic photosynthesis, a physiological trait shared by all phytoplankton; (ii) the dinoflagellates, a primitive polyphyletic group of protists whose representatives are thought to have acquired different plastids and pigments *via* secondary and tertiary symbioses (see Box 2); (iii) the coccolithophores within the Haptophytes (a sister clade to that of dinoflagellates and apicomplexans); and (iv) the diatoms, a group of siliceous Stramenopiles [from the Latin *stramen* (straw) + *pilus* (hair), referring to an anterior flagellum with short tripartite hairlike extensions = mastigonemes, present at some stage of the life history] which evolved and radiated in the Cenozoic era to occupy a prominent role in modern marine biogeochemical cycles.

3.4.1 Cyanobacteria (Phylum Cyanobacteria)

The cyanobacteria are an ancient group of metabolically-diverse photosynthetic prokaryotes found in most illuminated environment on Earth. Cyanobacteria are thought to have invented oxygenic photosynthesis, perhaps the most preeminent bioenergetic process on Earth. The date is uncertain, but biosignatures of planktonic picocyanobacteria lipids found in oils and bitumens from 2,500 MY-old Proterozoic shales indicate that they have dominated primary production in the oceans for more than half of Earth's history (Summons et al. 1999). A conservative estimate of their global collective biomass is 3×10^{14} g C or a thousand million tons (10^{15} g) wet biomass (Garcia-Pichel et al. 2003). In terms of individual cells, the picoplanktonic cyanobacterium *Prochlorococcus* (Fig. 3.3) is the most abundant organism on Earth.

Cyanobacteria are true Bacteria that synthesize chlorophyll *a*. *Prochlorococcus* has a unique antenna complex, in which divinyl chlorophyll a_2 and Chl b_2 replace the characteristic phycobilisomes of all other cyanobacteria (Barrera-Rojas et al. 2018, Scanlan 2012). Additionally, most cyanobacteria produce phycocyanin (a phycobilin pigment that gives the cells a bluish color and was the origin of the popular name 'blue-green algae' for this group); in some cases, the red accessory pigment phycoerythrin is also synthesized (Whitton and Potts 2012). Glycogen is usually the storage photosynthate. The cell structure of cyanobacteria is relatively simple, as these prokaryotes lack internal membrane systems such as a membrane-bound nucleus, plastids and mitochondria. In fact, even the genome of cyanobacteria is simple; *Prochlorococcus* possesses one of the smallest known genomes (1.66 Mbp) (Ting et al. 2007). Perhaps the most remarkable ultrastructural feature is the intracytoplasmic thylakoid membrane where the components of the electron transport chain (both respiratory and photosynthetic) are located. The possession of closely appressed photosynthetic thylakoid laminae is the primary ultrastructural feature that distinguishes *Prochlorococcus* from *Synechococcus* (Scanlan 2012); thylakoid membranes form complete concentric rings around the cell periphery in *Prochlorococcus* (Ting et al. 2007, Fig. 3.3b). The cell wall of cyanobacteria has basically the same composition and structure of that of Gram-negative bacteria (Lee 2008), consisting of a peptidoglycan layer.

A number of physiological and ecological features contribute to the success of cyanobacteria across a wide range of aquatic environments (Whitton and Potts 2012). For instance, the temperature optimum for most cyanobacteria is higher than for most eukaryotic algae, and most species are adapted to sub-optimal light regimes. Planktonic cyanobacteria are usually more tolerant to hypersalinity than eukaryotic phytoplankton (Nche-Fambo et al. 2015). Despite their morphological and genomic

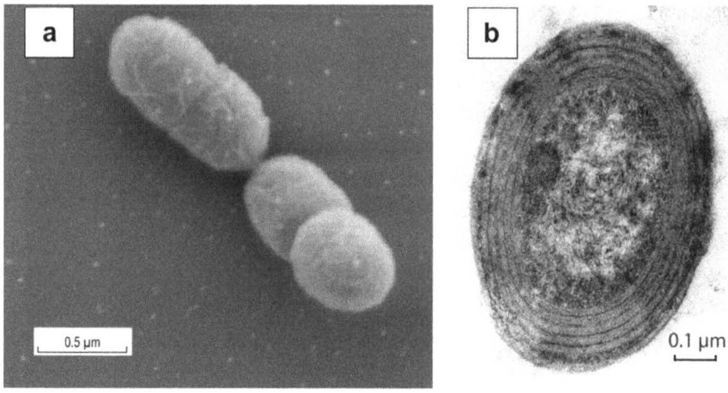

Figure 3.3. Representative marine planktonic cyanobacteria: *Prochlorococcus marinus*. (a) SEM photomicrograph of clone MIT9313. (b) TEM photomicrograph of clone SS120. Photo credits: (a) courtesy of Dr. E. Zinser, previously published in PLoS ONE 6(2): e16805; (b) courtesy of W.K.W. Li & F. Partensky.

simplicity, cyanobacteria have a circadian clock encoded by a gene cluster (Ishiura et al. 1998) which, for example, is reflected in highly synchronized diel patterns in photosynthesis, cell growth and division in the picocyanobacterium *Synechococcus* (Seymour et al. 2005).

In planktonic species (e.g., *Trichodesmium*), the ability to form gas vacuoles and increase cell buoyancy is an asset for maintaining their position in the water column at low rates of vertical mixing. Further, some cyanobacteria have evolved ways of overcoming oligotrophy (i.e., nutrient limitation/ depletion in surface waters), through fixation of atmospheric N_2, and efficient nutrient sequestration and storage (Paerl 2012). While all major cyanobacterial groups are present in marine plankton, specific environmental cues strongly select for certain species or ecotypes to dominate in geographically and climatically distinct regions of the oceans (Paerl 2012).

The ability of many species to fix N_2 provides a competitive advantage when environmental N concentrations are low. N_2 fixation requires anoxic conditions, so in most species this takes place inside a thick-walled cell that prevents contact with O_2, the heterocyst. However, heterocysts are absent from phytoplanktonic cyanobacteria, so the role of cyanobacteria as the main N_2 fixers in the oceans has only been realized a few decades ago (Díez et al. 2008). The pelagic cyanobacterium *Trichodesmium* forms very intense, transient blooms, estimated to fix ~ 42 Tg N yr^{-1} which is of biogeochemical significance on regional and global scales (Westberry and Siegel 2006). An estimated additional ~ 20 Tg N yr^{-1} are fixed under non-bloom conditions. On the other hand, Zehr et al. (2001) reported N_2 fixation in the North Pacific Ocean by unicellular cyanobacteria, shifting the paradigm that *Trichodesmium* was the only N_2 fixer in open waters.

Picocyanobacteria (*Prochlorococcus*, *Synechococcus*) and eukaryotic picoautotrophs are the dominant autotrophs in the oligotrophic subtropical gyres, the largest biome on Earth, contributing a significant proportion of primary production. *Synechococcus* strains have a diameter ~ 0.9 μm, while *Prochlorococcus* strains have a diameter ~ 0.6 μm, and are the smallest of all the phytoplankton (Fuhrman 2003). The ecological success of picocyanobacteria lies in the existence of specific ecotypes or lineages occupying different niches (reviewed by Scanlan 2012). In the case of *Prochlorococcus,* the physiological and genomic data compiled over the past two decades has evidenced the niche partitioning that the tiniest phytoplankter is capable of, particularly with respect to its vertical distribution in the water column (exploiting light and nutrient gradients) in the stratified warm oceans where it abounds (Fuhrman 2003, Rocap et al. 2003). There are even significant differences in the cell size (diameter and biovolume) and geometry of the different strains of *Prochlorococcus* (Ting et al. 2007). The recent finding of predation by a picoplanktonic haptophyte upon *Prochlorococcus marinus* (Kamennaya et al. 2018) suggests that this mechanism may account for significant losses of picocyanobacteria.

Filamentous and colonial cyanobacteria periodically accumulate as dense "blooms" that may discolor oceanic and coastal waters. Blooms are of considerable biogeochemical and ecological significance because they represent large concentrations of phytoplankton biomass that impact carbon, nutrient (N, P, Fe and micronutrients), and O_2 cycling (Paerl 2012). Some cyanobacteria increase their densities in response to environmental cues to form Harmful Algal Blooms (CyanoHABs). These can negatively affect water quality and disrupt habitat function, e.g., by producing toxins, exacerbating hypoxia, and altering food webs (Paerl 2012).

Box 2: Endosymbiosis and the origin of eukaryotic plastids

Eukaryotic phytoplankton are key drivers of ecological function and determinants of community structure in modern marine ecosystems (Falkowski et al. 2004). So, before moving on to the discussion of eukaryotic phytoplankton, let me introduce briefly some key concepts on the endosymbiotic acquisition of plastids and the diversification of metabolic pathways.

Oxygenic photosynthesis presumably evolved only once in the cyanobacteria and subsequently spread *via* endosymbiosis to a wide variety of eukaryotic clades (Falkowski et al. 2004). It appears that in the course of evolution, a single primary endosymbiosis (McFadden 2001) consisting in the engulfment of a coccoid cyanobacterium by a eukaryotic host cell that already contained a mitochondrion (Keeling

Box 2 Contd. ...

...Box 2 Contd.

2010) led to the formation of a eukaryotic type of cell with primary plastids. Modern representatives of these lineages (the so-called 'Archaeplastida') are the Glaucophytes, the red algae (Rhodophytes), and the clade of green algae (Chlorophytes) and higher vascular plants (Keeling 2010, Larkum 2020) (Fig. 3.4). Later on in protistan evolution, the incorporation of the photosynthetic machinery and genetic material from an ancestral red algae (in what may have implied a lateral transfer of plastids or secondary endosymbioses) determined the evolution of new cellular types with secondary plastids. The resultant cellular types are the so-called 'chromalveolate' ancestors that radiated into several lineages, some of which retained their plastids, some of which putatively lost them in the course of evolution (Keeling 2010, Bhattacharya and Price 2020). Representatives of chromalveolate group with secondary plastids include dinoflagellates (with a record dating back to the early Mesozoic era ~ 240 MY ago) and their related groups [some dinoflagellates acquired their plastids either by tertiary endosymbiosis with other chromalveolates or by serial secondary endosymbiosis with a green alga; Morden and Sherwood 2002, Keeling 2010]; coccolithophores (starting their record by the late Triassic period ~ 220 MY ago); and a clade including Cryptophytes and diatoms (the latter, with an obscure origin which according to fossil evidence dates back to the Jurassic period ~ 190 MY ago; Sims et al. 2006) (Fig. 3.4). In contrast to the ancestral photosynthetic eukaryotes which evolved through primary endosymbiosis ~ 1,500 MY ago, dinoflagellates, coccolithophores, and diatoms, the three principal phytoplankton clades that would dominate the modern seas, contain plastids derived from an ancestral red alga by secondary symbiosis (Falkowski et al. 2004). Biochemical evidence suggests that after the End-Permian mass extinction which marked a major transition in ocean ecosystem structure, the acquisition of a red alga secondary plastid would have allowed these heterotrophic eukaryotes to retain fixed nitrogen within the cell, thus favoring their radiation in early Mesozoic oceans, while ecologically displacing other eukaryotic algae (Grzebyk et al. 2003, Falkowski et al. 2004).

Finally, a word of caution: there is no such thing as a definitive cladogram (i.e., a phylogenetic tree) tracing the evolution of protists (be them plastid-harboring algae or not). Different methodological approaches (e.g., morphological, biochemical, genomic) contribute different results which, when combined, produce different trees, and these are bound to change, as new methodologies are applied. The evolution of plastid-protists is without doubt a fascinating field, to which genomic sequencing has contributed significantly in the past few decades (Fuhrman 2003); if you are interested in learning more, I suggest you consult the excellent revision by Bhattacharya and Price (2020). But, as this chapter is not about protistan molecular phylogenies, we will now move on to the dominant groups of eukaryotic marine phytoplankton, and present them mostly from an ecological perspective.

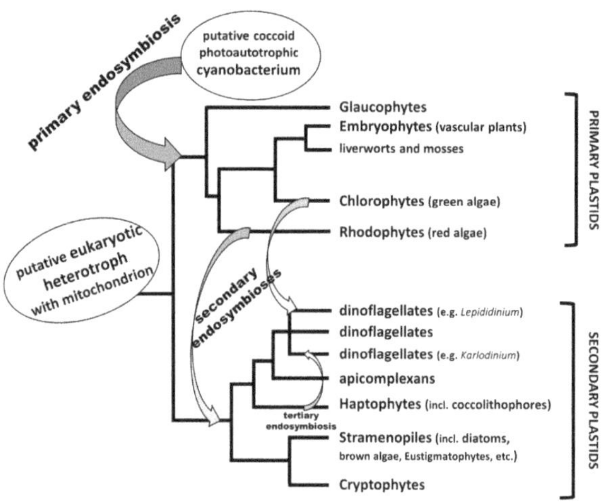

Figure 3.4. A simplified schematic representation of the evolutionary events (primary and secondary endosymbioses) leading to the formation of eukaryotic plastids, and some major evolutionary clades within photoautotrophic eukaryotes. See text in Box 2 for further reference. Modified and adapted from Larkum (2020).

3.4.2 Dinoflagellates (Phylum Myzozoa, Subphylum Dinozoa or Dinophyta; Muñoz-Gómez and Slamovits 2018)

The dinoflagellates (from the Greek *díni*, to whirl) are a major group of aquatic protists containing approximately 4,500 species in more than 550 genera, with many fossil representatives. Dinoflagellates play diverse trophic roles in aquatic food webs; furthermore, some of their representatives have notably established symbioses with corals (*Symbiodinium*, known as zooxanthellae), and some others produce bioluminescence (e.g., *Noctiluca, Pyrocistis*). A small sample of their astounding morphological diversity is presented in Fig. 3.5.

Dinoflagellates are derived from photosynthetic ancestors (the ancestral plastid type is the peridinin-containing plastid) and early in their evolutionary history exchanged most of the histones in their nuclei for dinoflagellate/viral nucleoproteins (DVNPs) which, as the term says, are proteins of viral origin that caused a complete reorganization of chromosomes that involved the loss of the typical eukaryotic nucleosomes and a very marked increase in total amounts of DNA per nucleus. Later on, they acquired other types of DNA-binding proteins, probably through lateral transfers from bacteria (Saldarriaga and Taylor 2017). The nucleus in a dinoflagellate cell is usually referred to as the dinokaryon.

The basic dinoflagellate cell is divided in halves by a median girdle trough called the *cingulum*. In most dinoflagellates, two flagella arise from the side (designated as ventral) and run along surface grooves: the transverse flagellum in the cingulum (responsible for the whirling motion of most dinoflagellates) and the longitudinal flagellum in the sulcus. The flagella have independent beating patterns.

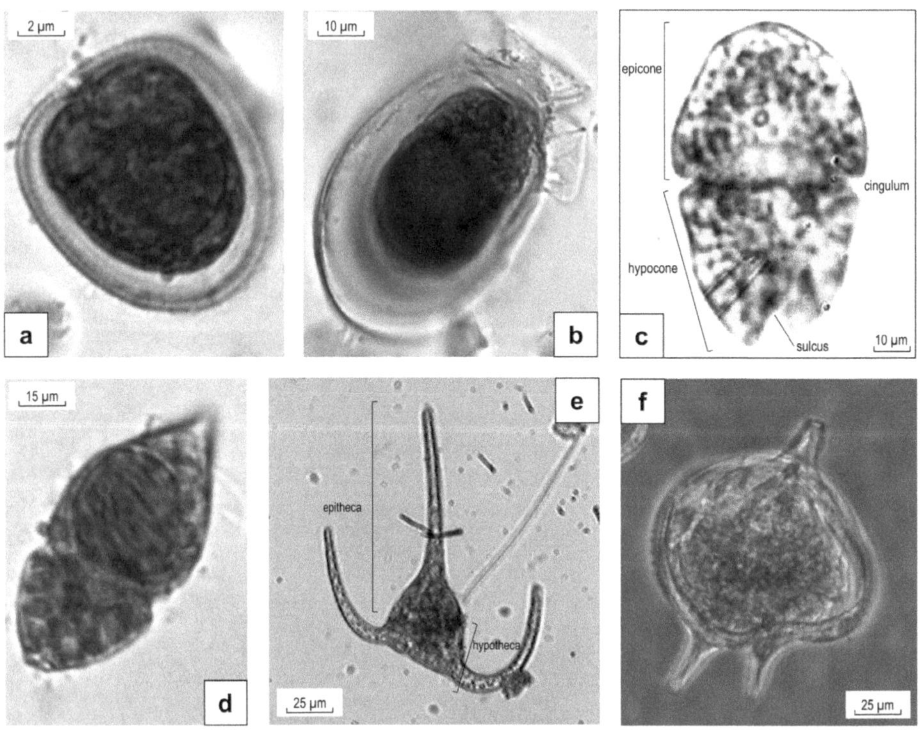

Figure 3.5. Representative marine planktonic dinoflagellates. (a) *Prorocentrum minimum*, a bloom-forming, toxic species in estuarine and neritic waters. (b) *Dinophysis acuminata*, an armored marine toxic species common in coastal waters; note the much larger hypotheca, in relation to the cap-like epitheca. (c) *Akashiwo sanguinea*, an athecate, coastal, red-tide producing species. (d) *Gyrodinium spirale*, an athecate species with a characteristic spindle-shaped asymmetric cell lacking chloroplasts. (e) *Tripos muelleri* (= *Ceratium tripos*), a coastal and oceanic armored species with apical- and hypothecal horns. (f) *Protoperidinium depressum*, a coastal and oceanic armored, phagotrophic species with a prominent apical horn and two divergent antapical horns. Morpho-ecological information from Steidinger and Tangen (1997), Faust and Gulledge (2002). Photo credits: (a–d, f) J. Pan; (e) courtesy of Dr. M.A. Marcoval.

Dinoflagellates possess quite a complex cell covering, the amphiesma (Kalinina et al. 2018). Dinoflagellates, ciliates and apicomplexans belong to the clade Alveolata and are characterized by a similar structure of the cell covering, consisting of a continuous plasma membrane and flattened single-membrane underlying vesicles (alveoli). In armored (thecate) dinoflagellates, amphiesmal vesicles may contain cellulosic material and form thecal plates (Fig. 3.6, *right scheme*); these dinoflagellates comprise a monophyletic group (Figs. 3.5a, b, e, f) and the two halves of the cell are called the *epitheca* and the *hypotheca*. In naked (athecate) species, amphiesmal vesicles lack cellulosic material and the cell is divided into an *epicone* and a *hypocone* (Fig. 3.5c, d). The amphiesma undergoes rearrangements at different stages of the life history, some of them implying ecdysis (Pozdnyakov and Skarlato 2012). Both thecate and athecate dinoflagellates are known to produce resistance stages (cysts) with thickened, dinosporin-containing cell walls. Cysts are usually resting benthic stages, while the motile, vegetative cell remains in motion in the water column. In one special case, the pelagic dinoflagellate *Pyrocystis noctiluca* alternates between a large, highly vacuolated positively buoyant vegetative cyst, and a smaller, compact and negatively buoyant cell. This strategy in *Pyrocystis* promotes vertical migration for nutrient uptake, as the cyst carries out photosynthesis, while the thecate form uptakes nutrients in deeper waters (Seo and Fritz 2000).

The amphiesmal vesicles complicate transport into the dinoflagellate cell, yet there exist permanent cytostomes and/or fixed sites for endocytosis (e.g., collared pits), and for exocytosis (e.g., trichocysts, a type of ejectile body). The pusule, a system of membrane vesicles, tubules and sacs connected with the flagellar canal by a permanent opening, is thought to play a role in osmoregulation and nutrient uptake (Kalinina et al. 2018).

Roughly half of the dinoflagellates are non-photosynthetic (Morden and Sherwood 2002), and the majority of the photosynthetic forms have plastids that contain the xanthophyll peridinin (which is exclusive of dinoflagellates), and chlorophylls *a* and *c*. Over the course of evolution, the peridinin plastid of dinoflagellates has dramatically reduced its own genome, a remarkable feature, considering it is still capable of encoding for an array of photosynthetic pigments (Muñoz-Gómez and Slamovits 2018). Other pigments found across plastids of various dinoflagellate species are Chl *b* (also in green algae), fucoxanthin, Chl c_1 and c_2 (also in stramenopiles and haptophytes) and Chl c_1 and phycobilins (also in cryptophytes). This remarkable diversity of light-harvesting pigments attests to the tertiary or serial secondary endosymbiotic events that took place throughout the evolution of different dinoflagellate lineages (Morden and Sherwood 2002, Keeling 2010, Muñoz-Gómez and Slamovits 2018; also see Box 2). Loss of photosynthesis has occurred repeatedly, but all free-living non-photosynthetic forms remain metabolically dependent on cryptic plastids; complete loss of plastid metabolic activity has only been shown in a few parasitic forms. Several lineages show a tendency for reacquisition of photosynthesis,

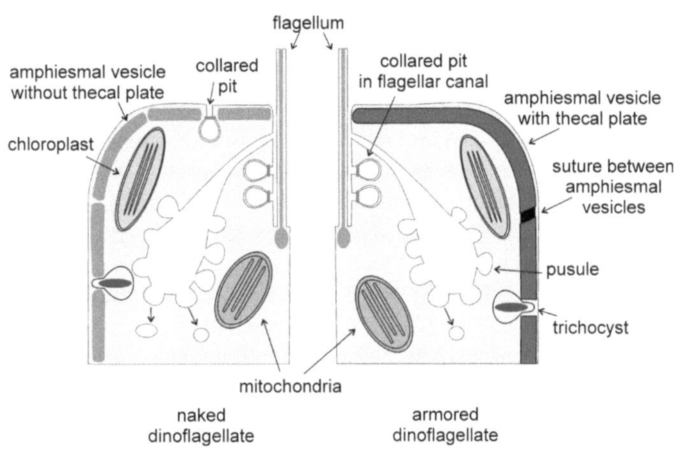

Figure 3.6. Schematic representation of the cell wall (amphiesma) of unarmored (naked, left sketch) and armored dinoflagellates (right sketch), and the cell structures involved in the acquisition of nutrients. Redrawn and modified from Kalinina et al. (2018).

often through permanent photosynthetic endosymbionts known as kleptochloroplasts (Saldarriaga and Taylor 2017).

The complicated phylogenetic relationships among dinoflagellates are starting to untangle in the light of new genomic approaches (e.g., Zhang et al. 2005, Hoppenrath and Leander 2010, Bachvaroff et al. 2014). Dinoflagellates have a continuous record (particularly of fossilized cysts) since the early Mesozoic era (Falkowski et al. 2004), although some indirect evidence from molecular fossils (i.e., dinosterols found in Precambrian to Cenozoic marine rocks) propose an earlier origin than the cyst record currently supports (Moldowan et al. 2000).

In ecological terms, dinoflagellates are responsible for a large proportion of marine primary production (Saldarriaga and Taylor 2017), and represent a significant component of protistan nano- and microplankton, with studies demonstrating complex trophic modes among their representatives (e.g., facultative heterotrophy, mixotrophy) (Jeong et al. 2005, Seong et al. 2006). They behave as ecophysiologically diverse annual species, whose blooms tend to be monospecific (Smayda and Reynolds 2003) as the most common form of reproduction in dinoflagellates is by mitotic division. Due to their motility and relatively larger size, heterotrophic dinoflagellates such as *Gyrodinium* sp. (Fig. 3.5d) often present biomass peaks at the subsurface chlorophyll maximum that forms seasonally above the thermocline in density-stratified shelf seas (Barnett et al. 2019).

Dinoflagellate harmful algal blooms (HABs, formerly known as "red tides") occur when cell densities of (generally) a single species are significantly elevated in relation to background concentrations in a short period of time, reaching concentrations in the order of 10^6 cells ml^{-1}, causing discoloration of surface waters, and leading to a disruption of aquatic ecosystem structure and functioning (Sunda et al. 2006). In the case of toxic dinoflagellates, toxins induce adverse effects on the physiology of other organisms, and through trophic transfer and biomagnification can impact all levels of marine food webs, including humans (e.g., the most common impact of HABs on humans is paralytic shellfish poisoning-PSP and ciguatera from toxins concentrated in the tissues of edible shellfish and finfish, respectively; Landsberg 2002, Sunda et al. 2006). Toxic dinoflagellates not necessarily need to reach high cell densities for their toxins to bioaccumulate in consumer tissues; for instance, *Alexandrium* PSP toxin becomes detectable at densities in the order of 10^3 cells l^{-1}, and the diarrhetic shellfish poisoning-DSP toxin produced by *Dinophysis* spp. is detectable at densities in the order of 10^2 cells l^{-1}. HABs occur in marine, estuarine, and freshwater systems and their frequency has increased globally due to anthropogenic influences (Hallegraeff 1993, Anderson 1997, Sunda et al. 2006). Recently, an experimental study by Costa et al. (2020) found that okadaic acid, a lipophilic toxin, produced by dinoflagellates of the genus *Dinophysis* and *Prorocentrum* was adsorbed to the surface of plastics, pointing to an overlooked mechanism of toxin dispersal and accumulation in waters where plastic pollution is an issue. Specific features of dinoflagellate biology (e.g., mixotrophy, vertical migration, and typical swimming and aggregation patterns) have been linked to their supremacy among HAB species (Smayda 1997).

3.4.3 Coccolithophores (Phylum Haptophyta)

As representatives of the Haptophytes, coccolithophores possess two similar flagella and the haptonema, an appendage used for attachment and food handling at some stage of their life history (Eikrem et al. 2017). Their chloroplasts contain a remarkable diversity of pigments, with chlorophylls *a* and c_2; Chl c_1 is synthesized as an accessory pigment in littoral, benthic or brackish water species, while the carotenoid hexanoyloxyfucoxanthin (HFx) is found in open-ocean species (Van Lenning et al. 2004). Coccolithophore cells range in size from 2–75 µm in diameter, and the majority of species are strictly oceanic, with a minority of neritic and brackish-water representatives.

The scales of coccolithophores consist in calcite platelets individually known as coccoliths. The term coccolithophore derives from combining the Greek words *kókkos* (grain, seed) + *líthos* (stone) + *phóros* (to carry). Coccoliths are produced internally in a Golgi-derived vesicle and then extruded (Fig. 3.7). Calcite precipitation in the coccolith vesicle is modulated by coccolith-associated polysaccharides; this organic coating is retained in coccoliths once they are extruded, and the extracellular polysaccharide-rich

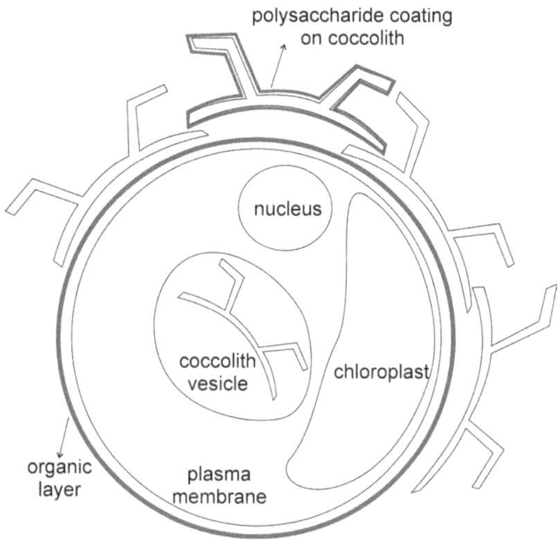

Figure 3.7. Schematic representation of the ultrastructure of a coccolithophore cell. Note that each coccolith has a polysaccharide coating (drawn only for one). Only one chloroplast has been schematized, for simplicity. Redrawn and modified from Walker et al. (2018).

organic layer in external coccoliths differs in structure and composition between species (Walker et al. 2018) and plays a role in coccolith adhesion to the cell surface.

The calcareous, protective exoskeleton made up of multiple coccoliths is known as the coccosphere. Coccospheres may be multilayered and monomorphic as in *Emiliania huxleyi* (Fig. 3.8a), have two layers of morphologically different coccoliths (dithecate), or have two types of coccoliths within one layer (dimorphic and monothecate) as in *Rhabdosphaera clavigera* (Fig. 3.8b) (Eikrem et al. 2017). Besides, not all stages of the life history of a coccolithophore present coccospheres; for example, *E. huxleyi* presents three cell types (one with a coccosphere, a completely naked type, and a motile cell with scales), each capable of vegetative reproduction. Entire coccospheres or individual fossil coccoliths are of stratigraphic value and, due to their role in biogeochemical cycles and climate modulation, they are used as proxies for paleoceanographic and paleoclimatic conditions.

Despite low seasonal and interannual variability, a study from the Central North Pacific Gyre (ALOHA Station, off Hawaii) found that the density of oceanic coccolithophores is strongly correlated with temperature, presenting spring and fall seasonal blooms. In the non-bloom months, coccolithophore densities are extremely low (Cortés et al. 2001). Of the ecological factors that regulate phytoplankton production (light, macronutrients, trace metals and CO_2), light intensity and day-length seem to play a major role in the initiation of coccolithophore blooms (Balch 2004, Zondervan 2007). This is usually studied in the field by estimating the ratio of calcium carbonate to organic carbon, as coccolithophore blooms are known to produce shifts through their relative contributions to calcification and photosynthetic carbon fixation in the surface ocean.

In terms of organic carbon fixation, coccolithophores are a minor player, representing only 6 to 8% of global marine primary production. But their detrital remains contribute disproportionately to the burial of carbon in marine sediments (Libes 2009). It is estimated that they are responsible for about half of all modern precipitation of $CaCO_3$, thus playing a significant role in ocean biogeochemistry (De Vargas et al. 2007). Coccolithophore biomineralization plays a major role in controlling the alkalinity of the ocean's euphotic zone by increasing pCO_2 and promoting its outgassing (i.e., the "alkalinity pump"; Cermeño et al. 2008), and their dense exoskeletons contribute to the rain of POM out of it, into the deep ocean.

Coccolithophores are usually cited as one of the main phytoplankton groups responsible for the synthesis of large quantities of dimethyl sulfide [DMS, $(CH_3)_2S$], in turn synthesized from the precursor dimethylsulfoniopropionate (DMSP). DMS and DMSP are ubiquitous in the euphotic zone of all oceans, and sea-to-air emission of these compounds accounts for about 23% of the sulfur entering the global atmosphere (Malin and Steinke 2004). Yet, the quantity of DMS that is emitted to the atmosphere is

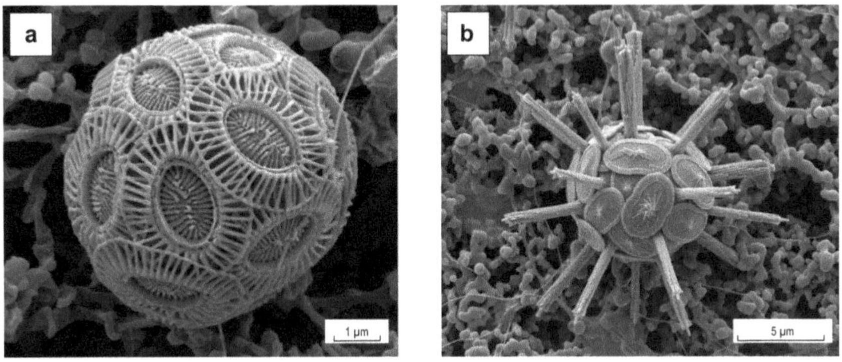

Figure 3.8. Representative coccolithophores. (a) SEM photomicrograph of the coccosphere of *Emiliania huxleyi*, a cosmopolitan oceanic coccolithophore. (b) SEM photomicrograph of *Rhabdosphaera clavigera*, another marine species with a club-shaped protrusion in some of its coccoliths. Photo credits: (a, b) courtesy of Dr. P. Ziveri.

a small fraction of the marine DMS + DMSP pools. DMS oxidation in the atmosphere leads to the formation of a number of sulfur compounds; these sulfur aerosols influence climate in two ways: directly, since aerosol particles scatter or reflect some of the incoming radiation back into space (Stewart 2008); and indirectly, because the aerosols act as cloud condensation nuclei, therefore influencing cloud formation and albedo, reflecting radiation from the sun and cooling the Earth. This was found in the mid-1980s and since then much research has been carried out on this topic for its obvious links to climate change. Haptophytes in general, and blooms of the cosmopolitan coccolithophore *E. huxleyi* in particular (Fig. 3.8a), are areas of high DMS production. *E. huxleyi* and its DMS production are usually considered as the 'model' coccolithophore species for biogeochemical and atmospheric models, but in ecosystems where coccolithophores form stable populations, other species than *E. huxleyi* can make significant contributions to the DMSP pool (Franklin et al. 2010).

Cell lysis, grazing, and viral attack have been pointed as processes that influence the production of DMS from haptophyte DMSP (Malin and Steinke 2004). Likewise, bacterial activity also influences the size of the DMS pool in seawater; these processes are particularly relevant due to their potential indirect impacts on world climate.

3.4.4 Diatoms (Phylum Ochrophyta, Class Bacillariophyceae)

Comprising ~ 100,000 species, diatoms are found in marine, brackish, and freshwaters worldwide. They are the archetypical representatives of modern marine phytoplankton assemblages with a remarkable morphological diversity (Fig. 3.9). Diatoms have traditionally been split into two clades, the Centrales and Pennales, presenting radial and bilateral symmetry, respectively. In present-day oceans, diatoms are responsible for about half of the global ocean net primary production (Libes 2009), or about 20–25% of global net primary production (Mann 1999). Due to their high affinity for nutrients, they dominate the early stages of the spring bloom in coastal seas (Barlow et al. 1993), with their abundances often peaking at subsurface and deep chlorophyll maxima (Gomi et al. 2010, Latasa et al. 2017, Barnett et al. 2019). Biogeochemically, diatoms are the most important silicifiers in modern marine ecosystems (Hendry et al. 2018); they are also the main Si exporters to the deep sea (Renaudie 2016), thus balancing global chemical weathering of Si (Milligan and Morel 2002).

Diatom chloroplasts occupy much of the cell biovolume (although a central vacuole may occupy a significant proportion; Strathmann 1967), and they possess a similar photosynthetic apparatus to that of other stramenopiles, with chlorophyll *a*, and fucoxanthin and Chl *c* as the principal accessory pigments (Mann et al. 2017). The storage photosynthate is chrysolaminarin.

Perhaps the single most distinctive feature of diatoms is their siliceous cell wall, termed the frustule. It consists of two overlapping halves or thecae (in fact, the term diatom derives from the Greek *diátomos*, meaning 'cut in two'). The thecae, in turn, consist of a larger end piece (valve) and a series of narrow

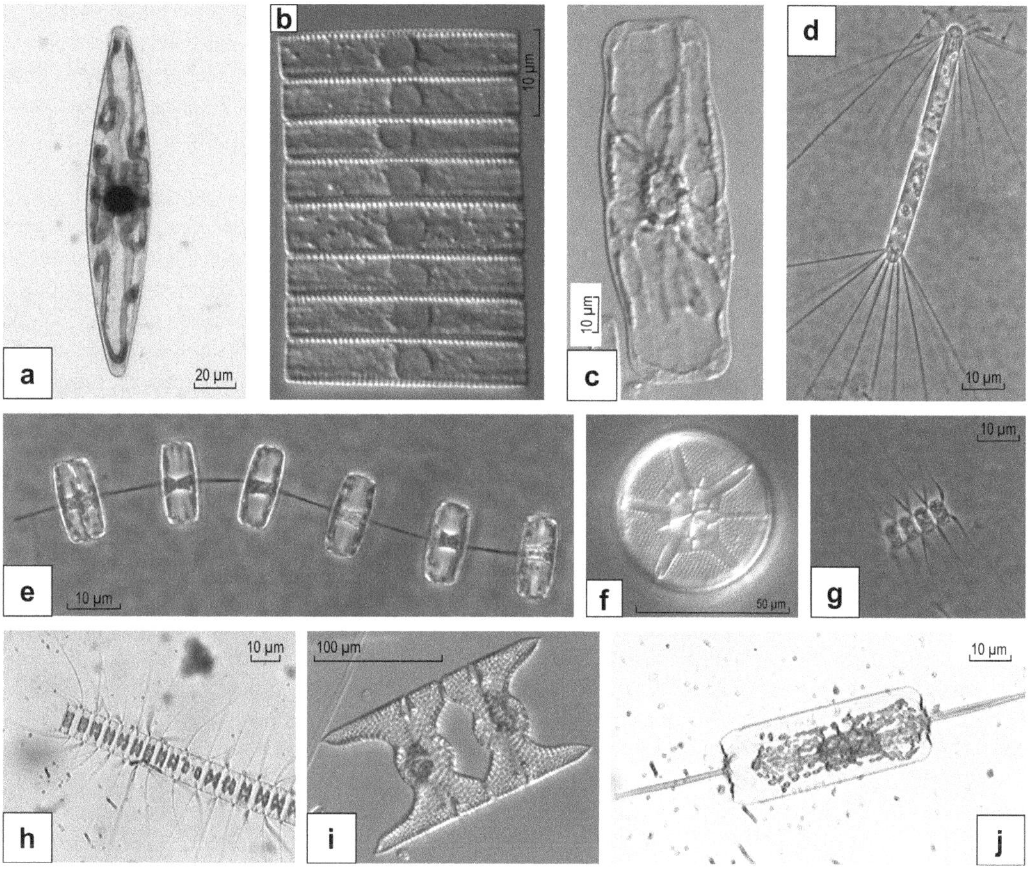

Figure 3.9. Representative marine planktonic diatoms. Pennate: (a) *Pleurosigma* sp., a common genus in coastal phytoplankton. (b) *Fragilariopsis kerguelensis*, an Antarctic species forming chains of cells, here in girdle view. (c) *Banquisia belgicae*, also from Antarctic waters. Centric: (d) *Corethron* sp., a planktonic genus with dome-shaped valves and spines; seen here in girdle view. (e) *Thalassiosira* sp., chain of cells bound by mucilage threads. (f) *Asteromphalus* sp., a genus with distinctive rays formed by siliceous thickenings. (g) *Chaetoceros* sp., a diverse genus of chain-forming centric diatoms; cells are united by interlocking of setae produced from the siliceous valve. (h) *Chaetoceros diadema*. (i) *Eucampia antarctica* var. *recta*, an ice-edge diatom with cells interlocking by two apical elevations to form chains, often curved in spirals. (j) *Ditylum brightwellii* in girdle view; cells with a characteristic spine at the center of each valve; cells are cylindric or triangular in valve view. Morpho-ecological information from Fryxell and Prasad (1990), Round et al. (1990), Hasle and Syvertsen (1997). Photo credits: (a–g, i) J. Pan; (h, j) courtesy of Dr. M.A. Marcoval.

strips (girdle bands). The addition of new bands to the inner, overlapped theca permits the expansion of the cell. The intricate architecture of frustules and the material properties of amorphous Si make them remarkably strong (Hamm et al. 2003), leading researchers to posit that the acquisition of this mechanical protection conferred a competitive advantage in diatom evolution. The geometries and patterns of diatom frustules are species-specific, and they are precisely reproduced in each generation, since the biomineralization process is under genetic control (Sumper and Brunner 2006). Furthermore, the porous structure of frustules increases their surface area reaching ~ 200 m^2g^{-1} (Wang et al. 2013). On the other hand, it has been demonstrated that the diatom frustule is an excellent optical structure for focusing incident light into the chloroplasts (De Stefano et al. 2007). Biophysical studies proved that strong interactions between blue light and the inner Si materials of the frustule enhance photosynthesis (Yamanaka et al. 2008). For more on the structure and functional properties of the diatom frustule, I suggest you read the excellent review by Gordon et al. (2008).

The Si content of diatoms affects their palatability and digestibility and consequently, the selectivity of protozoan grazers (Zhang et al. 2017). Yet, diatoms possess an organic coating (the perifrustular envelope) that renders the cells more palatable for consumers, independently of the carbon content of the cell (Beninger and Descottignies 2005). Due to the siliceous frustule, Si is a nutrient requirement for diatom growth. Diatoms take up dissolved silicic acid from the water and precipitate opaline Si to form their frustules, that constitutes an effective pH buffer (Milligan and Morel 2002). Experimental evidence indicates that diatoms compete for Si uptake with flagellated phytoplankton, and when grown at varying silicate:nitrate ratios, they become dominant at Si:N ratios > 25:1 (Sommer 1994).

The diatom life history alternates between several cycles of vegetative (asexual) and sexual reproduction (Fig. 3.10). At every asexual cell division, each daughter cell inherits one of the thecae of the parent and forms a new theca internally. Because of the silicified wall being rigid and inelastic (Hamm et al. 2003), the population cell size declines through successive vegetative divisions until reaching a critical size threshold, upon which cell size has to be restored through sexual reproduction. Meiotic gametes are implied in this process; the sperm cells of some centrics are the only known flagellate stage among diatoms. Fecundation leads to the formation of a special cell unique to the diatoms, the auxospore (Mann et al. 2017). After expansion of the auxospore, it buds into a new, larger vegetative cell which 'restores' the maximum cell size of the population, and further divisions are vegetative. Some planktonic species produce benthic resting stages, which constitute an important source of diaspores for the initiation of planktonic blooms (McQuoid and Godhe 2004). Likewise, in Polar ecosystems, sea ice diatoms that survive the summer melt and remain in the uppermost 20–80 cm of the ice column during winter, function as an algal seed repository, as they migrate toward the bottom in the subsequent thaw and initiate the ice algae spring bloom (Olsen et al. 2017).

The paramount role of diatoms in pelagic food webs has long been recognized, regarding them as valuable components of the planktonic community in terms of their nutritional quality. Various diatom species synthesize valuable lipids such as fatty acids, polar lipids and steroids (Yi et al. 2017). On the other hand, some diatoms are known for their toxicity. Some species, especially those in the bloom-forming genus *Pseudo-nitzschia,* produce domoic acid, a potent neurotoxin, with the potential of becoming a risk to human health when bioaccumulated by suspension-feeding benthos (Bates 2000). Domoic acid is a functional component of the unusual Fe-quenching system of *Pseudo-nitzschia* species found in oceanic HNLC (High Nutrient, Low Chlorophyll) regions (Wells et al. 2005). Also, diatoms are

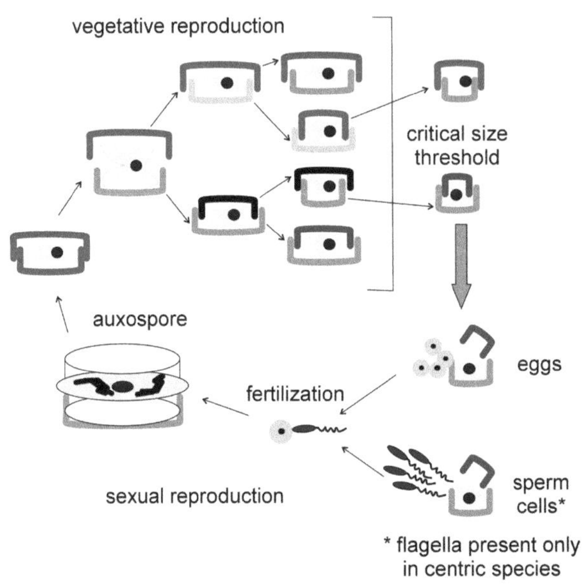

Figure 3.10. Schematic representation of reproduction in diatoms. Successive mitotic divisions in vegetative (asexual) reproduction, and the conservation of one half of the parental frustule valve in the daughter cells imply a progressive diminution of cell size. When the vegetative cell size reaches a critical size threshold for a certain population, diatoms resort to sexual reproduction, implying the production of gametes (oosfere and sperm in centrics or ameboid gametes in pennates) and an auxospore after fertilization. Redrawn and modified from Kale and Barthick (2015).

known to produce cytotoxic compounds (secondary metabolites released immediately after wounding, e.g., short-chain polyunsaturated aldehydes and oxylipins) that are responsible for growth inhibition and teratogenic activity upon their grazers (mainly copepods), ultimately impacting zooplankton recruitment (Ianora et al. 2003, Pohnert 2005, Ianora and Miralto 2010). These findings have led to a revision over the past two decades of the established paradigm that planktonic diatoms represent the prime food source for herbivorous copepods.

Diatoms have a rich fossil record dating back to the Jurassic period (~ 190 MY ago; Sims et al. 2006, Kooistra et al. 2007, Medlin 2016), and notably peaking in abundance during the mid-Cenozoic era (Renaudie 2016). Their origin, however, remains obscure. Medlin (2016) points to the major changes in ocean trace metal chemistry, and the widespread anoxic conditions at the Permo-Triassic boundary (~ 250 MY ago) as key environmental factors that gave putative diatom ancestors with an acquired red algal plastid an adaptive advantage over other phytoplankton, allowing their subsequent radiation. Even if the monophily of the different clades of diatoms is still under debate (Medlin 2016), we know from paleontological evidences that radial centrics first appeared in the Jurassic, multipolar centrics were present in the early Cretaceous, and pennates appeared in the late Cretaceous, and raphid diatoms in the Palaeocene (Sims et al. 2006, Kooistra et al. 2007), illustrating an evolutionary shift from centric to pennate symmetry (Medlin 2016). Marine planktonic diatoms rose to ecological prominence during the mid-Miocene, coinciding with global sea-level rise and increased flooded continental areas (Falkowski et al. 2004), and since then they have had a significant influence on Si cycles and the Si repository to the deep ocean (Milligan and Morel 2002). Since Si in the oceans is derived primarily from weathering of continental rocks, and knowing that grasses accelerate the weathering of Si, there appears to be a remarkable concordance in timing of the radiation of diatoms and the evolution of modern terrestrial ecosystems (Falkowski et al. 2004). Further, the diatom peak in abundance from the mid-Miocene onwards and its inferred acceleration of the "biological carbon pump" promoted a mid-Cenozoic drop in atmospheric CO_2 concentration and related climate changes (Milligan and Morel 2002, Cermeño et al. 2008, Renaudie 2016).

Marine diatoms, as any other phytoplankton, produce a wide variety of exudates that are released into the surrounding aquatic environment and contribute to the formation of marine snow and Transparent Exopolymer Particles (TEP). TEP are ubiquitous and abundant gel-particles with peak concentrations during phytoplankton blooms (Passow 2002, Decho and Gutierrez 2017). Chemically, TEP are acidic mucopolysaccharides that are formed abiotically from dissolved precursors exuded by senescent phytoplankton cells; these abundant gel particles (10^6 l^{-1} on average) are > 0.4 μm. There are known links between TEP concentration and diatom blooms, and most importantly, TEP from non-diatom origin may not be as sticky (reviewed by Passow 2002). For example, Engel (2000) described an increase in TEP concentration during the decline of a spring diatom bloom that translated into increased particle stickiness; and Verney et al. (2009) experimentally found that diatom blooms strongly enhanced flocculation in estuaries. Furthermore, diatom aggregates are very rich in TEP (Passow et al. 1994). It has been suggested that diatom TEP links the carbon flux between primary producers and heterotrophic bacteria, and its downward transport plays a significant role in the transfer of organic carbon produced in the surface to deeper waters below the surface mixed layer (Bar-Zeev et al. 2009). Thus, marine diatoms are not only important players in the "biological carbon pump", but through their exudates and the formation of large particle aggregates (marine snow and TEP) also promote vertical carbon flux.

Corollary

If you made it this far, by now you have quite a comprehensive first-hand knowledge of marine phytoplankton. I have strived to present the information so as to show you how significant photoautotrophic prokaryotes and protists are in biogeochemical cycles and ecological interactions in the current ocean; how the major lineages were established through some of the most intimate ecological relationships (endosymbioses); and how these photosynthetic microbes at the base of marine food webs have regulated and altered the structure and functioning of past oceans through their activities. New exciting avenues are just around the corner for the field of phytoplankton ecology (Chust et al. 2017). The wealth of high-throughput "OMICs" methods (i.e., genomics, transcriptomics, epigenomics, proteomics and metabolomics) are shedding new light onto previously overlooked components of phytoplankton communities (e.g., see Moon-van der Staay et al. 2001) and their ecological roles, which are constantly being revisited and reinterpreted (e.g., Worden et al. 2015, Kamennaya et al. 2018). One of the greatest challenges for the next generation of phytoplankton ecologists will be to integrate this new knowledge and ecological paradigms into a robust understanding of the links between photoautotrophic plankton diversity and marine ecosystem function.

Summary

1. Phytoplankton comprises planktonic prokaryotic and eukaryotic microalgae that are responsible for primary production in aquatic environments; their collective movement and position in the water column are largely determined by water motion. Marine phytoplankton are responsible for 40% of the global primary production yr^{-1}; yet they only account for 1% of total plant biomass.
2. Phytoplankton are typically small, responding to the evolutionary strategy of maximizing surface area:volume, as a means of optimizing nutrient uptake (governed by Michaelis-Menten kinetics). As small cells in a fluid medium, they live at low Reynolds numbers influenced by viscous forces.
3. Colony formation is not uncommon, with apparent benefits against predation, viruses and parasites, and modulation of pigment properties through the 'package effect'. The latter is notable in deep chlorophyll maximum (DCM) communities.
4. Individual phytoplankton cells are routinely considered 'particles'; yet, they differ from non-living/ inert particles in that (i) phytoplankton have distinct pigments that determine their bio-optical properties; (ii) they possess a distinctive cell 'architecture' that differs from an idealized spherical particle ('form resistance'); (iii) some phytoplankton have flagella and mechanisms to modulate their position in suspension.
5. Phytoplankton have traditionally been regarded as osmotrophs, basing their nutrition on the uptake of dissolved nutrients (inorganic forms of N, P, Si). Some species uptake organic forms of nutrients (aminoacids, urea). Macronutrients are found in seawater at μM concentrations, while micronutrients (Fe, trace metals and vitamins) exist at nM concentrations.
6. Besides photoautotrophy, other forms of nutrition include heterotrophy, mixotrophy, kleptochloroplastidy, as well as N_2 fixation (exclusive of prokaryotes). Phytoplankton present a fairly constant average atomic molar ratio of 106 C:16 N:1 P (known as *Redfield ratio*), although deviations from the ratio have been documented.
7. Phytoplankton abundance and distribution is jointly determined by bottom-up (nutrients, light) and top-down (predation, disease) controls. From a population ecology perspective, this balance is expressed by the logistic equation. Physical factors play a preeminent role in determining phytoplankton distribution.
8. The four dominant phytoplankton groups in the modern oceans are: (i) the prokaryotic Cyanobacteria which invented oxygenic photosynthesis; and three eukaryotic clades which originated from secondary and/or tertiary endosymbioses in the Mesozoic, after an initial primary endosymbiosis with a cyanobacterium plastid. These are (ii) the dinoflagellates—a primitive polyphyletic group; (iii) the coccolithophores within the Haptophytes; and (iv) the diatoms, which evolved and radiated in the Cenozoic.
9. Picoplanktonic cyanobacteria (*Prochlorococcus*, *Synechococcus*) are the most numerically-abundant phytoplankton with ecotypes that occupy segregated niches in the pelagic environment. The colonial cyanobacterium *Trichodesmium* forms blooms responsible for pelagic N_2 fixation.
10. Dinoflagellates have a complex cell cover (the amphiesma) and plastids containing chlorophyll *a* and peridinin, and a suite of accessory pigments reflecting their complex phylogeny. Most forms

exhibit facultative heterotrophy or mixotrophy and are known for other symbioses (e.g., with corals), and the formation of toxic harmful algal blooms (HABs).

11. Coccolithophore cells have a protective exoskeleton (the coccosphere) made up calcite platelets (coccoliths). Their biomineralizations are of stratigraphic importance in paleoceanographic reconstructions, and currently play a major role in controlling ocean alkalinity. Coccolithophores are one of the main phytoplankton responsible for the production of dimethyl sulfide (DMS), a volatile compound implied in the formation of aerosols and cloud condensation, with global impacts on climate.
12. Biogeochemically, diatoms are responsible for half of the global ocean net primary production and promote vertical carbon flux through transparent exopolymer (TEP) exudates and the formation of large particle aggregates; they are also the main Si exporters to the deep sea. The single most distinctive feature of diatoms is their siliceous cell wall (the frustule) which follows species-specific geometries and patterns. Their life history alternates between several cycles of vegetative divisions with sexual reproduction. Diatoms play a paramount role in pelagic food webs in terms of their nutritional quality, while some species produce neurotoxins (domoic acid). It is inferred that diatoms accelerated the 'biological carbon pump' with their peak in abundance from the mid-Miocene onwards.

Acknowledgments

Some sections of this chapter were drafted after a *Phytoplankton Ecology* course I took with Prof. C.J. Gobler more than a decade ago at Stony Brook University. Chris's class was thought-provoking, and largely based on first-hand literature sources. His teaching style fueled stimulating scientific discussions and critical thought among students; for this, I remain grateful to this day. In preparing this chapter, I have of course attempted to update its contents with the advancements that phytoplankton research has produced in the past decade or so, and imprint my personal view on many of the topics. As such, any inaccuracies, misconceptions or statements that rely on personal opinions, remain my sole responsibility. I owe my colleague P.D. Pratolongo for the digital design of most of the figures.

Further reading

The literature on Cyanobacteria and protists updates at a very dynamic pace. The introduction of molecular biology and genomic techniques to prokaryote and protistan phylogenies has revolutionized the systematics of these microbes. Accordingly, this is a very active field of scientific research, as well as phytoplankton ecology. If you are interested in keeping up to date with such issues, there are some excellent websites and reference publications not to be missed. I list a few of them below:

- AlgaeBase (www.algaebase.org): this website is a terrific, up-to-date database compiling systematic and morpho-ecological information for all groups of algae. The information is supported by photos and thorough bibliographic references. It is run and updated by M.D. Guiry and G.M. Guiry from the National University of Ireland (Galway), and sponsored by many scientific societies and academic institutions.
- Phyto'pedia - The Phytoplankton Encyclopaedia (www.eoas.ubc.ca/research/phytoplankton): this very resourceful website, similar to the AlgaeBase, is coordinated by Dr. D. Cassis, from the Department of Earth, Ocean and Atmospheric Sciences, University of British Columbia.
- The World Register of Marine Species (WoRMS, www.marinespecies.org) is also a very helpful source of updated information on marine species in general, compiled and curated by an editorial board of international specialists.
- The Tree of Life Web Project (ToL, www.tolweb.org) is a collaborative effort between experts and amateur naturalist aiming to build an ever more accurate phylogeny of life. It is not only graphically-appealing, but also provides a comprehensive bibliographic review for most taxonomic groups.

- B.A. Whitton's *Ecology of Cyanobacteria: Their diversity in space and time* (2nd Edition, 2012, Springer) is the ultimate reference for all aspects of the tremendously diverse and ubiquitous cyanobacteria.
- S.M. Adl, from the Department of Soil Sciences, University of Saskatchewan, and a string of collaborators periodically publish detailed revisions to the classification, nomenclature, and diversity of eukaryotic groups, in the *Journal of Eukaryotic Microbiology*. These papers are the accepted standard classification reference, and undoubtedly the most authoritative reviews on the subject. The latest one is Adl et al. (2019).
- The *Handbook of the Protists* (2nd Edition, 2017, Springer), edited by J.M. Archibald, A.G.B. Simpson, and C.H. Slamovits, is a terrific reference on all aspects of protistan biology, written by renowned specialists for each systematic group.

Besides, there are numerous classic phycology textbooks for reference, such as H.C. Bold and M.J. Wynne's *Introduction to the Algae* (2nd Edition, 1985, Prentice-Hall); C. van den Hoek, D.G. Mann and H.M. Jahns's *Algae: An Introduction to Phycology* (1995, Cambridge University Press); L.E. Graham and L.W. Wilcox's *Algae* (2nd Edition, 2008, Benjamin Cummings); R.E. Lee's *Phycology* (4th Edition, 2008, Cambridge University Press). Do not be discouraged by the publication dates of some, as these are oldies but goodies.

References

Adl, S.M., D. Bass, C.E. Lane, J. Lukeš, C.L. Schoch et al. 2019. Revisions to the classification, nomenclature, and diversity of eukaryotes. J. Eukaryot. Microbiol. 66: 4–119.

Anderson, D.M. 1997. Turning back the harmful red tide. Nature 388: 513–514.

Bachvaroff, T.R., S.G. Gornik, G.T. Concepcion, R.F. Waller, G.S. Mendez, J.C. Lippmeier and C.F. Delwiche. 2014. Dinoflagellate phylogeny revisited: using ribosomal proteins to resolve deep branching dinoflagellate clades. Mol. Phylogenetics Evol. 70: 314–322.

Balch, W.M. 2004. Re-evaluation of the physiological ecology of coccolithophores. *In*: Thierstein, H.R. and J.R. Young [eds.]. Coccolithophores, from Molecular Processes to Global Impact. Springer, Berlin. Germany.

Barlow, R.G., R.F.C. Mantoura, M.A. Gough and W.T. Fileman. 1993. Pigment signatures of the phytoplankton composition in the northeastern Atlantic during the 1990 spring bloom. Deep-Sea Res. II 40: 459–477.

Barnett, M.L., A.E.S. Kemp, A.E. Hickman and D.A. Purdie. 2019. Shelf sea subsurface chlorophyll maximum thin layers have a distinct phytoplankton community structure. Cont. Shelf Res. 174: 140–157.

Barrera-Rojas, J., L. González de la Vara, E. Ríos-Castro, L.E. Leyva-Castillo and C. Gómez-Lojero. 2018. The distribution of divinyl chlorophylls *a* and *b* and the presence of ferredoxin-NADPþ reductase in *Prochlorococcus marinus* MIT9313 thylakoid membranes. Heliyon 4: e01100; doi: 10.1016/j.heliyon.2018.e01100.

Bar-Zeev, E., I. Berman-Frank, N. Stambler, E. Vázquez Domínguez, T. Zohary et al. 2009. Transparent exopolymer particles (TEP) link phytoplankton and bacterial production in the Gulf of Aqaba. Aquat. Microb. Ecol. 56: 217–225.

Bates, M. 1960. The Forest and the Sea: A Look at the Economy of Nature and the Ecology of Man. Vintage/Random House, New York, NY. USA.

Bates, S.S. 2000. Domoic-acid-producing diatoms: another genus added! J. Phycol. 36: 978–983.

Beardall, J., D. Allen, J. Bragg, Z.V. Finkel, K.J. Flynn, A. Quigg, T.A.V. Rees, A. Richardson and J.A. Raven. 2009. Allometry and stoichiometry of unicellular, colonial and multicellular phytoplankton. New Phytol. 181: 295–309.

Beninger, P.G. and P. Decottignies. 2005. What makes diatoms attractive for suspensivores? The organic casing and associated organic molecules of *Coscinodiscus perforatus* are quality cues for the bivalve *Pecten maximus*. J. Plankton Res. 27: 11–17.

Bergkvist, J., P. Thor, H.H. Jakobsen, S.-Å. Wängberg and E. Selander. 2012. Grazer-induced chain length plasticity reduces grazing risk in a marine diatom. Limnol. Oceanogr. 57: 318–324.

Bertrand, E.M., M.A. Saito, J.M. Rose, C.R. Riesselman, M.C. Lohan, A.E. Noble, P.A. Lee and G.R. DiTullio. 2007. Vitamin B_{12} and iron colimitation of phytoplankton growth in the Ross Sea. Limnol. Oceanogr. 52: 1079–1093.

Bhattacharya, D. and D.C. Price. 2020. The algal tree of life from a genomics perspective. *In*: Larkum, A., A. Grossman and J. Raven [eds.]. Photosynthesis in Algae: Biochemical and Physiological Mechanisms, Advances in Photosynthesis and Respiration 45: 11–24. Springer, Cham. Switzerland.

Boyd, C.M. and D. Gradmann. 2002. Impact of osmolytes on buoyancy of marine phytoplankton. Mar. Biol. 141: 605–618.

Bracher, A., H.A. Bouman, R.J.W. Brewin, A. Bricaud, V. Brotas et al. 2017. Obtaining phytoplankton diversity from ocean color: a scientific roadmap for future development. Front. Mar. Sci. 4: 55; doi: 10.3389/fmars.2017.00055.

Castruita, M., M. Saito, P.C. Schottel, L.A. Elmegreen, S. Myneni, E.I. Stiefel and F.M.M. Morel. 2006. Overexpression and characterization of an iron storage and DNA-binding Dps protein from *Trichodesmium erythraeum*. Appl. Environ. Microbiol. 72: 2918–2924.

Cermeño, P., S. Dutkiewicz, R.P. Harris, M. Follows, O. Schofield and P.G. Falkowski. 2008. The role of nutricline depth in regulating the ocean carbon cycle. Proc. Natl. Acad. Sci. USA 105: 20344–20349.

Chust, G., M. Vogt, F. Benedetti, T. Nakov, S. Villéger et al. 2017. *Mare incognitum*: a glimpse into future plankton diversity and ecology research. Front. Mar. Sci. 4: 68; doi: 10.3389/fmars.2017.00068.

Ciavatta, S., R.J.W. Brewin, J. Skákala, L. Polimene, L. de Mora, Y. Artioli and J.I. Allen. 2018. Assimilation of ocean color plankton functional types to improve marine ecosystem simulations. J. Geophys. Res. Oceans 123: 834–854.

Ciotti, A.M., M.R. Lewis and J.J. Cullen. 2002. Assessment of the relationships between dominant cell size in natural phytoplankton communities and the spectral shape of the absorption coefficient. Limnol. Oceanogr. 47: 404–417.

Cloern, J.E. and R. Dufford. 2005. Phytoplankton community ecology: principles applied in San Francisco Bay. Mar. Ecol. Prog. Ser. 285: 11–28.

Cloern, J.E. and A.D. Jassby. 2010. Patterns and scales of phytoplankton variability in estuarine-coastal ecosystems. Estuar. Coast. 33: 230–241.

Cortés, M.Y., J. Bollmann and H.R. Thierstein. 2001. Coccolithophore ecology at the HOT station ALOHA, Hawaii. Deep-Sea Res. II 48: 1957–1981.

Costa, S.T., A. Rudnitskaya, C. Vale, L. Guilhermino and M.J. Botelho. 2020. Sorption of okadaic acid lipophilic toxin onto plastics in seawater. Mar. Pollut. Bull. 157: 111322.

De Stefano, L., I. Rea, I. Rendina, M. De Stefano and L. Moretti. 2007. Lensless light focusing with the centric marine diatom *Coscinodiscus walesii*. Opt. Express 15: 18083–18088.

De Vargas, C., M.-P. Aubry, I. Probert and J. Young. 2007. Origin and evolution of coccolithophores: from coastal hunters to oceanic farmers. pp. 251–285. *In*: Falkowski, P.G. and A.H. Knoll [eds.]. Evolution of Primary Producers in the Sea. Academic Press, New York, NY. USA.

Decho, A.W. and T. Gutierrez. 2017. Microbial Extracellular Polymeric Substances (EPSs) in ocean systems. Front. Microbiol. 8: 922; doi: 10.3389/fmicb.2017.00922.

Des Marais, D.J. 2003. Biogeochemistry of hypersaline microbial mats illustrates the dynamics of modern microbial ecosystems and the early evolution of the biosphere. Biol. Bull. 204: 160–167.

Díez, B., B. Bergman and R. El-Shehawy. 2008. Marine diazotrophic cyanobacteria: out of the blue. Plant Biotechnol. 25: 221–225.

Eikrem, W., L.K. Medlin, J. Henderiks, S. Rokitta, B. Rost, I. Probert, J. Throndsen and B. Edvardsen. 2017. Chapter 25: Haptophyta. pp. 893–953. *In*: Archibald, J.M., A.G.B. Simpson and C.H. Slamovits [eds.]. Handbook of the Protists, Second edition. Springer, Cham. Switzerland.

Engel, A. 2000. The role of transparent exopolymer particles (TEP) in the increase in apparent particle stickiness (α) during the decline of a diatom bloom. J. Plankton Res. 22: 485–497.

Falkowski, P.G., M.E. Katz, A.H. Knoll, A. Quigg, J.A. Raven, O. Schofield and F.J.R. Taylor. 2004. The evolution of modern eukaryotic phytoplankton. Science 305: 354–360.

Faust, M.A. and R.A. Gulledge. 2002. Identifying harmful marine dinoflagellates. Contr. US Natl. Herb. 42: 1–144.

Franklin, D.J., M. Steinke, J. Young, I. Probert and G. Malin. 2010. Dimethylsulphoniopropionate (DMSP), DMSPlyase activity (DLA) and dimethylsulphide (DMS) in 10 species of coccolithophore. Mar. Ecol. Prog. Ser. 410: 13–23.

Friedlingstein, P., M.W. Jones, M. O'Sullivan, R.M. Andrew, J. Hauck et al. 2019. Global carbon budget 2019. Earth Syst. Sci. Data 11: 1783–1838; doi.org/10.5194/essd-11-1783-2019.

Fryxell, G.A. and A.K.S.K. Prasad. 1990. *Eucampia antarctica* var. *recta* (Mangin) stat. nov. (Biddulphiaceae, Bacillariophyceae): life stages at the Weddell Sea ice edge. Phycologia 29: 27–38.

Fuhrman, J. 2003. Genome sequences from the sea. Nature 424: 1001–1002.

Garcia-Pichel, F., J. Belnap, S. Neuer and F. Schanz. 2003. Estimates of global cyanobacterial biomass and its distribution. Algol. Stud. 109: 213–227.

Garrison, T. 2009. Essentials of Oceanography, Fifth edition. Brooks/Cole Cengage Learning, Belmont, CA. USA.

Geider, R.J. and J. La Roche. 2002. Redfield revisited: variability of C:N:P in marine microalgae and its biochemical basis. Eur. J. Phycol. 37: 1–17.

Gomi, Y., M. Fukuchi and A. Taniguchi. 2010. Diatom assemblages at subsurface chlorophyll maximum layer in the eastern Indian sector of the Southern Ocean in summer. J. Plankton Res. 32: 1039–1050.

Gordon, R., D. Losic, M.A. Tiffany, S.S. Nagy and F.A.S. Sterrenburg. 2008. The Glass Menagerie: diatoms for novel applications in nanotechnology. Trends Biotechnol. 27: 116–127.

Green, R.E., H.M. Sosik and R.J. Olson. 2003. Contributions of phytoplankton and other particles to inherent optical properties in New England continental shelf waters. Limnol. Oceanogr. 48: 2377–2391.

Grossart, H.-P., F. Levold, M. Allgaier, M. Simon and T. Brinkhoff. 2005. Marine diatom species harbour distinct bacterial communities. Environ. Microbiol. 7: 860–873.

Grzebyk, D., O. Schofield, C. Vetriani and P. Falkowski. 2003. The Mesozoic radiation of eukaryotic algae: the portable plastid hypothesis. J. Phycol. 39: 259–267.

Hallegraeff, G.M. 1993. A review of harmful algal blooms and their apparent global increase. Phycologia 32: 79–99.
Hamm, C.E., R. Merkel, O. Springer, P. Jurkojc, C. Maier, K. Prechtel and V. Smetacek. 2003. Architecture and material properties of diatom shells provide effective mechanical protection. Nature 421: 841–843.
Hammer, A.C. and J.W. Pitchford. 2005. The role of mixotrophy in plankton bloom dynamics, and the consequences for productivity. ICES J. Mar. Sci. 62: 833–840.
Hasle, G.R. and E.E. Syvertsen. 1997. Marine diatoms. pp. 5–385. *In*: Tomas, C.R. [ed.]. Identifying Marine Phytoplankton. Academic Press, New York, NY. USA.
Hendry, K.R., A.O. Marron, F. Vincent, D.J. Conley, M. Gehlen, F.M. Ibarbalz, B. Quéguiner and C. Bowler. 2018. Competition between silicifiers and non-silicifiers in the past and present ocean and its evolutionary impacts. Front. Mar. Sci. 5: 22; doi: 10.3389/fmars.2018.00022.
Hoppenrath, M. and B.S. Leander. 2010. Dinoflagellate phylogeny as inferred from heat shock protein 90 and ribosomal gene sequences. PLoS ONE 5: e13220; doi:10.1371/journal.pone.0013220.
Hutchins, D.A., A.E. Witter, A. Butler and G.W. Luther. 1999. Competition among marine phytoplankton for different chelated iron species. Nature 400: 858–861.
Hutchinson, G.E. 1961. The paradox of the plankton. Am. Nat. 95: 137–145.
Ianora, A., S.A. Poulet and A. Miralto. 2003. The effects of diatoms on copepod reproduction: a review. Phycologia 42: 351–363.
Ianora, A. and A. Miralto. 2010. Toxigenic effects of diatoms on grazers, phytoplankton and other microbes: a review. Ecotoxicology 19: 493–511.
Ishiura, M., S. Kutsuna, S. Aoki, H. Iwasaki, C.R. Andersson et al. 1998. Expression of a gene cluster *kaiABC* as a circadian feedback process in cyanobacteria. Science 281: 1519–1523.
Jackson, G.A. 2012. Seascapes: the world of aquatic organisms as determined by their particulate natures. J. Experim. Biol. 215: 1017–1030.
Jeong, H.J., J.Y. Park, J.H. Nho, M.O. Park, J.H. Ha, K.A. Seong, C. Jeng, C.N. Seong, K.Y. Lee and W.H. Yih. 2005. Feeding by red-tide dinoflagellates on the cyanobacterium *Synechococcus*. Aquat. Microb. Ecol. 41: 131–143.
Jumars, P.A. 1993. Concepts in Biological Oceanography, An Interdisciplinary Primer. Oxford University Press, New York, NY. USA.
Kale, A. and B. Karthick. 2015. The diatoms, big significance of tiny glass houses. Resonance 20: 919–930.
Kalinina, V., O. Matantseva, M. Berdieva and S. Skarlato. 2018. Trophic strategies in dinoflagellates: how nutrients pass through the amphiesma. Protistology 12: 3–11.
Kamennaya, N.A., G. Kennaway, B.M. Fuchs and M.V. Zubkov. 2018. "Pomacytosis"-semi-extracellular phagocytosis of cyanobacteria by the smallest marine algae. PLoS Biol. 16: e2003502; doi.org/10.1371/journal.pbio.2003502.
Katechakis, A. and H. Stibor. 2006. The mixotroph *Ochromonas tuberculata* may invade and supress specialist phago- and phototroph plankton communities depending on nutrient conditions. Oecologia 148: 692–701.
Keeling, P.J. 2010. The endosymbiotic origin, diversification and fate of plastids. Phil. Trans. R. Soc. B 365: 729–748.
Kiørboe, T. 2008. A Mechanistic Approach to Plankton Ecology. Princeton University Press, Princeton, NJ. USA.
Kitchen, J.C. and J.R.V. Zaneveld. 1992. A three-layered sphere model of the optical properties of phytoplankton. Limnol. Oceanogr. 37: 1680–1690.
Klausmeier, C.A., E. Litchman, T. Daufresne and S.A. Levin. 2008. Phytoplankton stoichiometry. Ecol. Res. 23: 479–485.
Kooistra, W.H.C.F., R. Gersonde, L.K. Medlin and D.G. Mann. 2007. Chapter 11: The origin and evolution of the diatoms: their adaptation to a planktonic existence. pp. 207–249. *In*: Falkowski, P.G. and A.H. Knoll [eds.]. Evolution of Primary Producers in the Sea. Elsevier Academic Press, Burlington, MA. USA.
Landsberg, J.H. 2002. The effects of harmful algal blooms on aquatic organisms. Rev. Fish. Sci. 10: 113–390.
Larkum, A.W.D. 2020. Light-harvesting in cyanobacteria and eukaryotic algae: an overview. *In*: Larkum, A., A. Grossman and J. Raven [eds.]. Photosynthesis in Algae: Biochemical and Physiological Mechanisms, Advances in Photosynthesis and Respiration 45: 207–260. Springer, Cham. Switzerland.
Latasa, M., A.M. Cabello, X.A.G. Morán, R. Massana and R. Scharek. 2017. Distribution of phytoplankton groups within the deep chlorophyll maximum. Limnol. Oceanogr. 62: 665–685.
Le Quéré, C., S.P. Harrison, I.C. Prentice, E.T. Buitenhuis, O. Aumont et al. 2005. Ecosystem dynamics based on plankton functional types for global ocean biogeochemistry models. Glob. Change. Biol. 11: 2016–2140.
Lee, R.E. 2008. Phycology, 4th Edition. Cambridge University Press, Cambridge. UK.
Libes, S.M. 2009. Introduction to Marine Biogeochemistry, Second edition. Elsevier/Academic Press, Burlington, MA. USA.
Lohrenz, S.E., A.D. Weidemann and M. Tuel. 2003. Phytoplankton spectral absorption as influenced by community size structure and pigment composition. J. Plankton Res. 25: 35–61.
Longhurst, A., S. Sathyendranath, T. Platt and C. Caverhill. 1995. An estimate of global primary production in the ocean from satellite radiometer data. J. Plankton Res. 17: 1245–1271.
Lovecchio, S., E. Climent, R. Stocker and W.M. Durham. 2019. Chain formation can enhance the vertical migration of phytoplankton through turbulence. Sci. Adv. 5: eaaw7879; doi: 10.1126/sciadv.aaw7879.
Mackas, D.L., K.L. Denman and M.R. Abbott. 1985. Plankton patchiness: biology in the physical vernacular. Bull. Mar. Sci. 37: 652–674.

Malin, G. and M. Steinke. 2004. Dimethyl sulfide production: what is the contribution of the coccolithophores? *In*: Thierstein, H.R. and J.R. Young [eds.]. Coccolithophores, from Molecular Processes to Global Impact. Springer, Berlin. Germany.
Mann, D.G. 1999. The species concept in diatoms. Phycologia 38: 437–495.
Mann, D.G., R.M. Crawford and F.E. Round. 2017. Chapter 7: Bacillariophyta. pp. 205–266. *In*: Archibald, J.M., A.G.B. Simpson and C.H. Slamovits [eds.]. Handbook of the Protists, Second edition. Springer, Cham. Switzerland.
McFadden, G.I. 2001. Primary and secondary endosymbiosis and the origin of plastids. J. Phycol. 37: 951–959.
McQuoid, M.R. and A. Godhe. 2004. Recruitment of coastal planktonic diatoms from benthic versus pelagic cells: variations in bloom development and species composition. Limnol. Oceanogr. 49: 1123–1133.
Medlin, L.K. 2016. Evolution of the diatoms: major steps in their evolution and a review of the supporting molecular and morphological evidence. Phycologia 55: 79–103.
Milligan, A.J. and F.M.M. Morel. 2002. A proton buffering role for silica in diatoms. Science 297: 1848–1850.
Moldowan, J.M., S.R. Jacobson, J. Dahl, A. Al-Hajji, B.J. Huizinga and F.J. Fago. 2000. Molecular fossils demonstrate Precambrian origin of dinoflagellates. pp. 474–493. *In*: Zhuravlev, A.Y. and R. Riding [eds.]. The Ecology of the Cambrian Radiation. Columbia University Press, New York, NY. USA.
Moon-van der Staay, S.Y., R. De Wachter and D. Vaulot. 2001. Oceanic 18S rDNA sequences from picoplankton reveal unsuspected eukaryotic diversity. Nature 409: 607–610.
Moore, J.K., S.C. Doney, D.M. Glover and I.Y. Fung. 2001. Iron cycling and nutrient-limitation patterns in surface waters of the World Ocean. Deep. Res. II 49: 463–507.
Morden, C.W. and A.R. Sherwood. 2002. Continued evolutionary surprises among dinoflagellates. Proc. Natl. Acad. Sci. USA 99: 11558–11560.
Mulholland, M.R., C.J. Gobler and C. Lee. 2002. Peptide hydrolysis, amino acid oxidation and N uptake in communities seasonally dominated by *Aureococcus anophagefferens*. Limnol. Oceanogr. 47: 1094–1108.
Muñoz-Gómez, S.A. and C.H. Slamovits. 2018. Plastid genomes in the Myzozoa. Adv. Bot. Res. 85: 55–94.
Naselli-Flores, L. and R. Barone. 2011. Fight on plankton! Or, phytoplankton shape and size as adaptive tools to get ahead in the struggle for life. Cryptogam. Algol. 32: 157–204.
Nche-Fambo, F.A., U.M. Scharler and K. Tirok. 2015. Resilience of estuarine phytoplankton and their temporal variability along salinity gradients during drought and hypersalinity. Estuar. Coast. Shelf Sci. 158: 40–52.
Nuzzi, R. and R.M. Waters. 2004. Long-term perspective on the dynamics of brown tide blooms in Long Island coastal bays. Harmful Algae 3: 279–293.
Olsen, L.M., S.R. Laney, P. Duarte, H.M. Kauko, M. Fernández-Méndez et al. 2017. The seeding of ice algal blooms in Arctic pack ice: the multiyear ice seed repository hypothesis. J. Geophys. Res. Biogeosci. 122: 1529–1548; doi:10.1002/2016JG003668.
Owen, R.W. 1989. Microscale and finescale variations of small plankton in coastal and pelagic environments. J. Mar. Res. 47: 197–240.
Paerl, H.W. 2012. Chapter 5: Marine plankton. pp. 127–153. *In*: Whitton, B.A. [ed.]. Ecology of Cyanobacteria II: their Diversity in Space and Time, Second edition. Springer, Dordrecht. Netherlands.
Passow, U., A.L. Alldredge and B.E. Logan. 1994. The role of particulate carbohydrate exudates in the flocculation of diatom blooms. Deep-Sea Res. I 41: 335–357.
Passow, U. 2002. Transparent exopolymer particles (TEP) in aquatic environments. Prog. Oceanogr. 55: 287–333.
Pohnert, G. 2005. Diatom/copepod interactions in plankton: the indirect chemical defense of unicellular algae. ChemBioChem. 6: 946–959.
Pozdnyakov, I. and S. Skarlato. 2012. Dinoflagellate amphiesma at different stages of the life cycle. Protistology 7: 108–115.
Ptacnik, R. 2003. Omnivory in planktonic food webs: a study on the impact of mixotrophic flagellates and microzooplankton on food web dynamics and productivity. Ber. Inst. Meereskd. Christian-Albrechts-Univ. Kiel 328: 1–101.
Raven, J.A. 1997. Phagotrophy in phototrophs. Limnol. Oceanogr. 42: 198–205.
Record, N.R., A.J. Pershing and F. Maps. 2014. The paradox of the "paradox of the plankton". ICES J. Mar. Sci. 71: 236–240.
Redfield, A.C. 1958. The biological control of chemical factors in the environment. Am. Sci. 46: 205–221.
Renaudie, J. 2016. Quantifying the Cenozoic marine diatom deposition history: links to the C and Si cycles. Biogeosciences 13: 6003–6014.
Rocap, G., F.W. Larimer, J. Lamerdin, S. Malfatti, P. Chain et al. 2003. Genome divergence in two *Prochlorococcus* ecotypes reflects oceanic niche differentiation. Nature 424: 1042–1047.
Rolland, J.L., D. Stien, S. Sanchez-Ferandin and R. Lami. 2016. Quorum sensing and quorum quenching in the phycosphere of phytoplankton: a case of chemical interactions in ecology. J. Chem. Ecol. 42: 1201–1211.
Round, F.E., R.M. Crawford and D.G. Mann. 1990. The Diatoms, Biology and Morphology of the Genera. Cambridge University Press, Cambridge. UK.
Saldarriaga, J.F. and F.J.R. Taylor. 2017. Chapter 17: Dinoflagellata. *In*: Archibald, J.M., A.G.B. Simpson and C.H. Slamovits [eds.]. Handbook of the Protists, Second edition. Springer, Cham. Switzerland.
Sañudo-Wilhelmy, S.A., A. Tovar-Sanchez, F.-X. Fu, D.G. Capone, E.J. Carpenter and D.A. Hutchins. 2004. The impact of surface-adsorbed phosphorus on phytoplankton Redfield stoichiometry. Nature 432: 897–901.

Sañudo-Wilhelmy, S.A., C.J. Gobler, M. Okbamichael and G.T. Taylor. 2006. Regulation of phytoplankton dynamics by vitamin B_{12}. Geophys. Res. Lett. 33: L04604; doi:10.1029/2005GL025046.

Sañudo-Wilhelmy, S.A., L. Gómez-Consarnau, C. Suffridge and E.A. Webb. 2014. The role of B vitamins in marine biogeochemistry. Annu. Rev. Mar. Sci. 6: 339–367.

Scanlan, D.J. 2012. Chapter 20: Marine picocyanobacteria. pp. 503–533. *In*: Whitton, B.A. [ed.]. Ecology of Cyanobacteria II: their Diversity in Space and Time, Second edition. Springer, Dordrecht. Netherlands.

Segura, V., V.A. Lutz, A. Dogliotti, R.I. Silva, R.M. Negri, R. Akselman and H. Benavides. 2013. Phytoplankton types and primary production in the Argentine Sea. Mar. Ecol. Prog. Ser. 491: 15–31.

Seo, K.S. and L. Fritz. 2000. Cell-wall morphology correlated with vertical migration in the non-motile marine dinoflagellate *Pyrocystis noctiluca*. Mar. Biol. 137: 589–594.

Seong, K.A., H.J. Jeong, S. Kim, G.H. Kim and J.H. Kang. 2006. Bacterivory by cooccurring red-tide algae, heterotrophic nanoflagellates, and ciliates. Mar. Ecol. Prog. Ser. 322: 85–97.

Seymour, J.R., L. Seuront and J.G. Mitchell. 2005. Microscale and small-scale temporal dynamics of a coastal planktonic microbial community. Mar. Ecol. Prog. Ser. 300: 21–37.

Seymour, J.R., S.A. Amin, J.-B. Raina and R. Stocker. 2017. Zooming in on the phycosphere: the ecological interface for phytoplankton-bacteria relationships. Nat. Microbiol. 2: 17065; doi: 10.1038/nmicrobiol.2017.65.

Sims, P.A., D.G. Mann and L.K. Medlin. 2006. Evolution of the diatoms: insights from fossil, biological and molecular data. Phycologia 45: 361–402.

Smayda, T.J. 1997. Harmful algal blooms: their ecophysiology and general relevance to phytoplankton blooms in the sea. Limnol. Oceanogr. 42: 1137–1153.

Smayda, T.J. and C.S. Reynolds. 2003. Strategies of marine dinoflagellate survival and some rules of assembly. J. Sea Res. 49: 95–106.

Smetacek, V. and F. Pollehne. 1986. Nutrient cycling in pelagic systems: a reappraisal of the conceptual framework. Ophelia 26: 401–428.

Sommer, U. 1994. Are marine diatoms favoured by high Si:N ratios? Mar. Ecol. Prog. Ser. 115: 309–315.

Stambler, N. 2014. Relationships between picophytoplankton and optical properties in the Azores Front region in the Atlantic Ocean. J. Sea Res. 85: 144–154.

Steidinger, K.A. and K. Tangen. 1997. Dinoflagellates. pp. 387–584. *In*: Tomas, C.R. [ed.]. Identifying Marine Phytoplankton. Academic Press, New York, NY. USA.

Stewart, R.H. 2008. Introduction to Physical Oceanography. Texas A&M University, College Station, TX. USA.

Stibor, H. and U. Sommer. 2003. Mixotrophy of a photosynthetic flagellate viewed from an optimal foraging perspective. Protist 154: 91–98.

Strathmann, R.R. 1967. Estimating the organic carbon content of phytoplankton from cell volume or plasma volume. Limnol. Oceanogr. 12: 411–418.

Summons, R.E., L.L. Jahnke, J.M. Hope and G.A. Logan. 1999. 2-Methylhopanoids as biomarkers for cyanobacterial oxygenic photosynthesis. Nature 400: 554–557.

Sumper, M. and E. Brunner. 2006. Learning from diatoms: Nature's tools for the production of nanostructured silica. Adv. Funct. Mater. 16: 17–26.

Sunda, W.G., E. Granéli and C.J. Gobler. 2006. Positive feedback and the development and persistence of ecosystem disruptive algal blooms. J. Phycol. 42: 963–974.

Tillmann, U. 2004. Interactions between planktonic microalgae and protozoan grazers. J. Eukaryot. Microbiol. 51: 156–168.

Ting, C.S., C. Hsieh, S. Sundararaman, C. Mannella and M. Marko. 2007. Cryo-electron tomography reveals the comparative three-dimensional architecture of *Prochlorococcus*, a globally important marine cyanobacterium. J. Bacteriol. 189: 4485–4493.

Tittel, J., V. Bissinger and B. Zippel. 2003. Mixotrophs combine resource use to outcompete specialists: implications for aquatic food webs. Proc. Natl. Acad. Sci. USA 100: 12776–12781.

Van Lenning, K., I. Probert, M. Latasa, M. Estrada and J.R. Young. 2004. Pigment diversity of coccolithophores in relation to taxonomy, phylogeny and ecological preferences. pp. 51–73. *In*: Thierstein, H.R. and J.R. Young [eds.]. Coccolithophores, from Molecular Processes to Global Impact. Springer, Berlin. Germany.

Vandermeer, J.H. and D.E. Goldberg. 2003. Population Ecology, First Principles. Princeton University Press, Princeton, NJ. USA.

Verdy, A., M. Follows and G. Flierl. 2009. Optimal phytoplankton cell size in an allometric model. Mar. Ecol. Prog. Ser. 379: 1–12.

Verney, R., R. Lafite and J.-C. Brun-Cottan. 2009. Flocculation potential of estuarine particles: the importance of environmental factors and of the spatial and seasonal variability of suspended particulate matter. Estuar. Coast. 32: 678–693.

Vogel, S. 1994. Life in Moving Fluids: The Physical Biology of Flow, Second edition. Princeton University Press, Princeton, NJ. USA.

Walker, C.E., S. Heath, D.L. Salmon, N. Smirnoff, G. Langer, A.R. Taylor, C. Brownlee and G.L. Wheeler. 2018. An extracellular polysaccharide-rich organic layer contributes to organization of the coccosphere in coccolithophores. Front. Mar. Sci. 5: 306; doi: 10.3389/fmars.2018.00306.

Walsby, A.E., F. Schanz and M. Schmid. 2006. The Burgundy-blood phenomenon: a model of buoyancy change explains autumnal waterblooms by *Planktothrix rubescens* in Lake Zürich. New Phytol. 169: 109–122.

Wang, Y., J. Cai, Y. Jiang, X. Jiang and D. Zhang. 2013. Preparation of biosilica structures from frustules of diatoms and their applications: current state and perspectives. Appl. Microbiol. Biotechnol. 97: 453–460.

Wells, M.L., C.G. Trick, W.P. Cochlan, M.P. Hughes and V.L. Trainer. 2005. Domoic acid: the synergy of iron, copper, and the toxicity of diatoms. Limnol. Oceanogr. 50: 1908–1917.

Westberry, T.K. and D.A. Siegel. 2006. Spatial and temporal distribution of *Trichodesmium* blooms in the world's oceans. Global Biogeochem. Cy. 20: GB4016; doi:10.1029/2005GB002673.

Whitton, B.A. and M. Potts. 2012. Chapter 1: Introduction to the cyanobacteria. pp. 1–13. *In*: Whitton, B.A. [ed.]. Ecology of Cyanobacteria II: their Diversity in Space and Time, Second edition. Springer, Dordrecht. Netherlands.

Worden, A.Z., M.J. Follows, S.J. Giovannoni, S. Wilken, A.E. Zimmerman and P.J. Keeling. 2015. Rethinking the marine carbon cycle: Factoring in the multifarious lifestyles of microbes. Science 347: 1257594.

Yamanaka, S., R. Yano, H. Usami, N. Hayashida, M. Ohguchi, H. Takeda and K. Yoshino. 2008. Optical properties of diatom silica frustule with special reference to blue light. J. Appl. Phys. 103: 074701; doi:10.1063/1.2903342.

Yi, Z., M. Xu, X. Di, S. Brynjolfsson and W. Fu. 2017. Exploring valuable lipids in diatoms. Front. Mar. Sci. 4: 17; doi:10.3389/fmars.2017.00017.

Zehr, J.P., J.B. Waterbury, P.J. Turner, J.P. Montoya, E. Omoregie et al. 2001. Unicellular cyanobacteria fix N_2 in the subtropical North Pacific Ocean. Nature 412: 635–638.

Zevenboom, W. 1986. Ecophysiology of nutrient uptake, photosynthesis and growth. Can. Bull. Fish. Aquat. Sci. 214: 391–422.

Zhang, H., D. Bhattacharya and S. Lin. 2005. Phylogeny of dinoflagellates based on mitochondrial cytochrome *b* and nuclear small subunit rDNA sequence comparisons. J. Phycol. 41: 411–420.

Zhang, S., H. Liu, Y. Ke and B. Li. 2017. Effect of the silica content of diatoms on protozoan grazing. Front. Mar. Sci. 4: 202; doi: 10.3389/fmars.2017.00202.

Zondervan, I. 2007. The effects of light, macronutrients, trace metals and CO_2 on the production of calcium carbonate and organic carbon in coccolithophores—A review. Deep-Sea Res. II 54: 521–537.

CHAPTER 4

Planktonic Metabolism and Primary Production in the Oceans

Aurore Regaudie-de-Gioux[1,*] and *Aurea M. Ciotti*

4.1 Introduction

The metabolism in the ocean is controlled by planktonic microorganisms that play a major biogeochemical role on Earth. The photosynthetic portion of these communities contributes to roughly 50% of the organic carbon and oxygen (Antoine and Morel 1996, Behrenfeld and Falkowski 1997, Falkowski and Raven 1997, Field et al. 1998). Because of that, the rates of oceanic primary production alter global processes of air-sea exchange, particularly the atmospheric carbon dioxide (CO_2) partial pressure. Thus, planktonic primary production is a piece of essential knowledge for understanding the ocean ecosystem.

For simplicity, the pelagic primary producers can be denominated phytoplankton, which are a collection of very diverse taxonomic groups of microscopic algal organisms characterized by a great diversity of shapes and sizes with different nutritional requirements and with a diversity growth strategy (Worden et al. 2015, and also Chapter 3 in this book). Besides carbon, phytoplankton cells are composed also by phosphorous (P) and nitrogen (N) both, once mineralized in the water column, are defined as macronutrients. Thus, primary production rates affect the cycles of these elements in the ocean as well, and phytoplankton optimal growth depends on their availability. Phytoplankton cells assimilate phosphate to satisfy their requirement for P and nitrate and ammonium as sources of N. Nonetheless, all phytoplankton groups require many complementary chemical elements and while some of them are plentiful in the ocean and rarely limit phytoplankton growth (e.g., sodium, calcium, sulfur, carbon), others such as N, P, silicon (Si), iron (Fe) and zinc (Zn) are usually present in the sea, especially away from continental and deep water inputs, at very low concentrations, and these are called micro-nutrients. Furthermore, phytoplankton growth requires vitamin B_{12} that they either synthesize or assimilate from the environment. The proportion among major elements was theoretically described by the Redfield's ratio (Redfield 1958: C_{106}:N_{16}:P_1) but their internal compositions are functions of taxonomy and physiological state (Geider et al. 2002).

When phytoplankton growth discontinues through depletion of nutrients, phytoplankton cells are lysed by organisms such as viruses and zooplankton. During this lyse, part of the organic material previously constituting phytoplankton cells returns to the seawater, is recycled by bacteria and,thus, becomes available for phytoplankton growth. Because of that, one important concept in the ecology of phytoplankton attempts to delimit primary production into two categories. Production is indicated as "new" when it is fueled by nutrients supplied from outside the productive layer and as "regenerated"

[1] Ifremer, DYNECO PELAGOS, F-29280 Plouzané, France.
[2] Laboratório Aquarela, Centro de Biologia Marinha (CEBIMar), Universidade de São Paulo, Rod. Manoel Hipólito do Rego Km 131.5, São Sebastião-SP, 11600-000, Brazil.
* Corresponding author: Aurore.Regaudie.De.Gioux@ifremer.fr

when nutrient sources derive from recycled materials inside this layer (Dugdale and Goering 1967). The mass proportion between new and regenerated primary production rates is conceptually represented by the "*f*-ratio" that reveals values from around 0.05 to 0.5, approximately. These extreme values are found, respectively, in oligotrophic deep oceanic waters located away from continental influences that are also vertically stratified (i.e., the productive layers are physically isolated from the deep layers) and in coastal upwelling systems where nutrients are regularly supplied (Miller and Wheeler 2010).

The carbon originated from phytoplanktonic primary production fuels marine food webs with organic matter (e.g., Duarte et al. 1999) but many logistics problems exist for measuring primary production in the ocean (Regaudie-de-Gioux et al. 2014a), so studies remain scarce and limited to few regions of the global ocean. Indeed, only a few studies presented *in situ* primary production rates at the global scale (e.g., Robinson and Williams 2005, Regaudie-de-Gioux and Duarte 2013, Bouman et al. 2018, Kulk et al. 2020) and much work is still necessary to properly quantify primary production in the ocean (del Giorgio and Williams 2005). In this chapter, our goal is to describe different techniques and methods used to estimate planktonic metabolism and primary production highlighting their differences.

Planktonic Metabolism. Rigorously speaking, planktonic metabolism means the biochemical modifications of chemical components by planktonic organisms through the biosynthesis of complex organic molecules (anabolism) and their breakdown (catabolism). The metabolic rates of planktonic communities are composed of two principal processes: (1) the *production* and (2) the *respiration* of the organic matter.

1) Production of organic matter: The planktonic photosynthesis (or primary production) is the main pathway to produce organic matter. It is represented by the capture of the energy from the sunlight from electromagnetic radiation and turns it into chemical energy in the process called photosynthesis, resumed by the following equation:

$$CO_2 + H_2O + \text{light} \rightarrow CH_2O + O_2 \qquad \text{(Eq. 1)}$$

The planktonic primary production refers to the rate of carbon fixation by the photosynthetic processes (Falkowski and Raven 1997) that require the Rubisco (ribulose bisphosphate carboxylase-oxygenase), the enzyme responsible to metabolize CO_2 and O_2. Both the chemical activities of CO_2 and O_2 in solution and the kinetic properties of Rubisco determine the rate of CO_2 fixation, expressed as mol or g of C per unit of area or volume (m^2 or m^3) per unit of time (e.g., second, day, year). The energy produced is used to synthesize complex organic molecules (e.g., glucose) from inorganic compounds such as CO_2, water (H_2O), and inorganic nutrients (Field et al. 1998, Behrenfeld et al. 2001). Besides, microorganisms can also produce organic matter through chemosynthesis, mainly conducted by bacteria and archaea in ecosystems such as deep ocean hydrothermal vents. To date, chemosynthesis has been shown to play a minor role in the formation of organic matter at the biosphere scale.

2) Respiration of organic matter: The respiration planktonic organism represents the oxidative breakdown of organic molecules into its inorganic constituents (CO_2 and inorganic nutrients) to free energy to support various cellular processes and can be resumed by the following equation:

$$CH_2O + O_2 \rightarrow CO_2 + H_2O + \text{energy} \qquad \text{(Eq. 2)}$$

As a rule, respiration consumes O_2, although there are alternative respiratory pathways, confined to hypoxic and anoxic environments that use oxidized molecules other than oxygen as the electron donor.

Equations 1 and 2 show promptly that respiration is the opposing process to photosynthesis; while photosynthesis uses CO_2 to produce organic matter releasing O_2, respiration consumes O_2 while consuming organic matter and releasing CO_2. In addition to the fluxes of O_2 and CO_2, photosynthesis and respiration also support opposite nutrient fluxes such as nitrogen (N), phosphorus (P) and iron (Fe). Indeed, they are incorporated into the organic matter through photosynthesis and they are remineralized into inorganic forms (e.g., ammonium and phosphate) during the respiration process. They also can be considered as complementary processes considering that natural ecosystems show a close balance between photosynthesis and respiration, particularly in large spatial and temporal scales. The metabolic balance of a pelagic marine ecosystem, or a planktonic community, refers to the subtraction between photosynthesis (gross primary production, GPP) and respiration (community respiration, CR) assigned as the net community production (NCP = GPP – CR) (Duarte et al. 2004). Depending on the dominance of each of these processes, the ecosystems or communities may be classified as: (1) autotrophic, indicating that these communities produce organic matter in excess of its respiratory requirements or (2) heterotrophic, which consume more organic matter through respiratory processes that they produce, and the metabolic balance between production and respiration are of equal magnitudes (Duarte and Prairie 2005). For this reason, heterotrophic ecosystems require allochthonous inputs of organic matter to meet their requirements, allowing for links between autotrophic and heterotrophic ecosystems. There is evidence that low productive ecosystems, such as the open ocean, tend to be heterotrophic (Duarte et al. 1998, Duarte and Prairie 2005), whereas high productive ecosystems, such as vegetated coastal habitats, tends to be autotrophic (Duarte et al. 2005).

Several external factors affect directly and indirectly both the primary production and the metabolism of planktonic communities in the ocean, highlighting temperature, nutrient concentrations, and solar irradiation. Indeed, all metabolic processes are affected by the temperature of the environment, which is the foundation of the metabolic theory of ecology (Gillooly et al. 2001, Brown et al. 2004) that predicts that both respiration and photosynthesis increase with increasing temperature albeit at a different rate (Brown et al. 2004, Harris et al. 2006, López-Urrutia et al. 2006, Regaudie-de-Gioux and Duarte 2012). Additionally, lack of sufficient light is one of the main factors that lead to heterotrophic community metabolism in the ocean (Sverdrup 1953), responsible for the heterotrophic nature over most of the 95% of the ocean volume corresponding to the dark ocean. Although photosynthesis needs light as an energy source, solar irradiance can positively or negatively affect this process, depending on its amount and wavelength. The visible light (400–700 nm) is mainly used for photosynthesis and is so-called photosynthetically active radiation (PAR) (Fig. 4.1). While infrared light (700 nm–1 mm) does not have enough energy to sustain photosynthesis for most phytoplanktonic organisms, UV radiation (UVR; 10–400 nm) has too much energy. UVR is known to reduce plankton photosynthesis rates (e.g., Cullen and Neale 1994, Llabrés et al. 2013) and some studies showed that UVR enhances heterotrophic community metabolism (e.g., Regaudie-de-Gioux et al. 2014b, Agustí et al. 2014).

As for observed by temperature and light, nutrient availability is one of the major factors regulating marine productivity and phytoplankton community structure. The planktonic photosynthetic performance can reflect the nutritional status of phytoplankton (Cullen et al. 1992). However, the consequences of nutrient limitation differ remarkably among ecosystems and phytoplankton species. Indeed, some studies

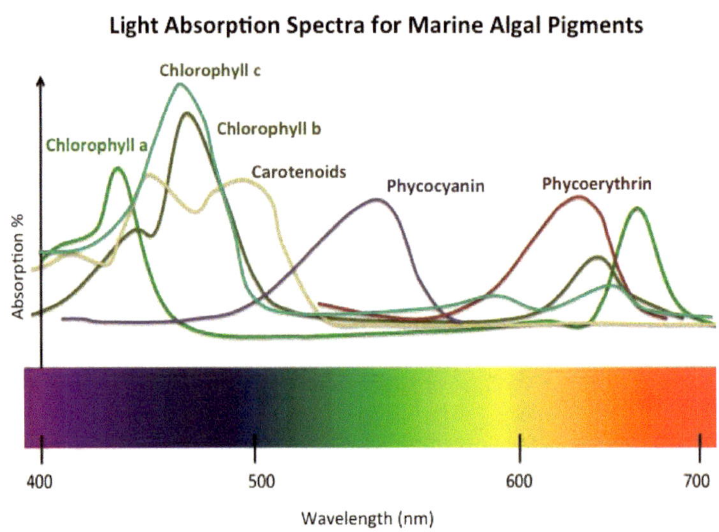

Figure 4.1. Light absorption spectra for photosynthetic pigments (Yarish et al. 2012).

showed maximum photosynthetic rates during growth rates experiments limited severely by N (Laws and Bannister 1980, Herzig and Falkowski 1989).

The prediction of temporal changes in phytoplankton in the ocean with emphasis on spring blooms was based on observations in coastal temperate environments in the early 1950s (Sverdrup 1953). In the classic model described as the critical depth model, light penetration and the depth of the mixed layer act together, regulating the balance between photosynthesis and respiration rates vertically. This model has been revisited many times, and the reader can find a complete discussion about it in Fischer et al. (2014).

4.2 Methods of measurement

The first photosynthetic rates of marine phytoplankton were measured using the oxygen evolution method in phytoplankton assemblages in the Oslo Fjörd by Gaarder and Gran (1927). This method originally presented a poor resolution but was enhanced with the development of the high-precision Winkler analyses using automatic titrators and end-point detection of the Winkler reaction (Carpenter 1965, Carrit and Carpenter 1966), which allow low primary production rates to be resolved. Before the improvement of the oxygen-based method, the ^{14}C technique was developed (Steemann Nielsen 1952) and was the only method that showed, at that time, the capacity to estimate low photosynthetic rates with high precision. For that reason, it rapidly became the golden standard for the oceanographic community, and it is the one used to calibrate remote sensing algorithms nowadays. However, this method has disadvantages like the confinement effect due to incubation (as for the O_2 evolution method) or the use of radioactive hazards (e.g., some countries prohibit its use). Other approaches have been proposed, such as the use of tracer methods based on stable isotope additions, such as ^{13}C (Slawyk et al. 1977) and ^{18}O (Bender et al. 1987). Several incubation-free techniques have been used including the use of active fluorescence, as the fast repetition rate fluorometry (FRRf) developed by Kolber and Falkowski (1993), the non-intrusive bio-optical (OPT) method (Loisel and Morel 1998, Claustre et al. 1999), the bio-optical model proposed by Marra et al. (2007), the triple oxygen isotopes method (Luz and Barkan 2000) or the analysis of oxygen records from buoys and gliders (Nicholson et al. 2008). Because of that, when different methods are blended or compared, care must be taken for comparing values in time or space.

4.3 Standard methods

Standard methods to estimate the metabolic balance of the planktonic community and primary production are *in vitro* oxygen-based and *in vitro* carbon-based methods. Each of these methods represents a particular process involved during the primary production process and tend to yield different results when applied to the same planktonic community (Marra 2002). Indeed, comparing estimates of primary production provided from different methods can be a challenging task and their varying results continue to fuel discussions (Quay et al. 2010, Marra 2012, Regaudie-de-Gioux et al. 2014a). The *in vitro* methods, as they require incubation, are prone to errors due to the confinement which may affect the organisms involved (e.g., excluding grazers) or the environmental conditions (e.g., temperature, nutrient concentration or light) (Maske and Garcia-Mendoza 1994, Karl et al. 1998, Robinson and Williams 2005, Duarte et al. 2013, Williams et al. 2013). The light field can be modified (Kirk 1994) as the incubations have been conducted with borosilicate or polycarbonate bottles, a material that partly excludes UV radiation. The variability in the time of the incubations is a major challenge when comparing measurements and techniques, as different fractions of the produced carbon may be respired by the algae or by the heterotrophic community during the experiments.

a. Oxygen-based methods

a.1 In vitro *oxygen methods*

In vitro oxygen-based method estimate changes in oxygen concentration during the photosynthesis process by high-precision Winkler titration (Carpenter 1965, Carrit and Carpenter 1966), triple O_2 isotope fields (Luz and Barkan 2000), or ^{18}O-labeled H_2O (Grande et al. 1982, Bender et al. 1987).

The dark-light method, or bulk oxygen evolution, consists in quantifying changes in oxygen concentration, utilizing an adaptation of the classic Winkler method that allows for precisions around 0.1% (Carpenter 1965, Carrit and Carpenter 1966). Changes in dissolved oxygen concentration are measured after incubating natural plankton communities in both clear (light) and dark bottles for about 24 h. Primary production is calculated using the changes in oxygen concentration in clear bottles, where photosynthesis and respiration are occurring, thus representing the net community production (NCP), and those in the dark bottles expressing only the community respiration (CR). The sum of NCP and CR over the 24 hours yields the gross primary production rates (GPP), assuming that the respiration rates in the clear and dark bottles are the same, which is one limitation. For instance, while some studies on natural communities suggested that respiratory losses are enhanced in the light (Harris and Lott 1973), others showed similar respiration rates during the day and night (e.g., Grande et al. 1989, Marra and Barber 2004), and the required incubations make the results prone to bottle effects. On the other hand, this method has several advantages, including being easy and reproducible, with enough sensitivity for use.

The ^{18}O-labelled H_2O method (Grande et al. 1982, Bender et al. 1987) measures GPP by using the stable isotope ^{18}O to trace oxygen production through photosynthesis. Water samples are enriched in ^{18}O derived by the photosynthetic release of ^{18}O-labeled from $H_2{}^{18}O$ during the daytime. Although this method does not assume the constancy of dark and light respiration as the ^{14}C does, it remains subject to bottle effects. GPP uncertainty evaluated by this method is considered lower than 2% (Bender et al. 1999). One advantage of this method regards the daily turnover time of phytoplankton, and the recycling of labeled O_2 will be weak (2%) during 24 h in comparison to that by labeled-^{14}C that should be faster (Bender et al. 1999). A second advantage of the ^{18}O-labeled method is that it can measure directly the gross photosynthetic O_2 production in comparison with the dark-light or ^{14}C-labeled method. On the other hand, one disadvantage of this method is that it includes gross oxygen production not related to organic carbon uptake (Bender et al. 1999, Robinson et al. 2009).

a.2 In situ *oxygen methods*

The O_2/Ar ratio method evaluates the GPP integrated throughout the entire mixed layer depth by using the similarity observed between the solubility of argon (Ar) and O_2 (Emerson 1987, Spitzer and Jenkins 1989, Emerson et al. 1993). Excesses of O_2 concentrations, or deficits in saturation, after removing the effects of physical processes, indicate either net autotrophy or net heterotrophy (Cassar et al. 2011). This is possible due to the similar solubilities of Ar and O_2 in seawater and that while Ar saturation is only affected by physical mechanisms, O_2 saturation changes with both biological and physical processes. Thus, a given value of O_2 supersaturation within the mixed layer derived from physical processes can be estimated from Ar supersaturation, yielding the biological O_2 supersaturation. The O_2/Ar ratio method requires mass spectrometers and the interpretation of O_2 concentration is sometimes challenged by chances of bubble injection during analyses (Hendricks et al. 2004). However, this method has the advantage to allow the direct evaluation of the primary production rates without the need of incubation.

The triple isotope method estimates time-integrated rates of photosynthesis from the difference among the uptake of ^{18}O, ^{17}O, and ^{16}O (Luz et al. 2000). One of the advantages of this method is the absence of incubations and thus bottle effects. Considering solely biological production and consumption, there is a relative excess of ^{17}O in the air O_2. This method allows evaluating photosynthesis rates *in situ* integrated over a week to month periods, which is impossible to measure directly with standard incubations. The rates computed are independent of respiration, but the isotope effects on respiration are not well known (Quay et al. 1993, Hendricks et al. 2004, Venkiteswaran et al. 2008). This method may involve multiple assumptions on the state of water column mixing and air-sea O_2 exchange rates (Robinson and Williams 2005). Although a complete error propagation analysis is still needed, reported uncertainties in GPP retrieved with the triple isotope method are between 15% (Stanley et al. 2010) and 40% (Reuer et al. 2007, Quay et al. 2010), thus results should be used with care when deriving carbon budgets (Juranek et al. 2005, Robinson and Williams 2005). Also, Kaiser (2011) highlighted that different studies have used different definitions of 17Δ (the ^{17}O excess) and so, the same measurements will give different results of production, causing systematic uncertainty. As for the O_2/Ar ratio method, the triple isotope method requires the injection of bubbles to induce O_2 supersaturation.

b. Carbon-based methods

In vitro carbon-based method estimates the assimilation of carbon (C) during the photosynthesis process using sodium bicarbonate marked with the radioactive ^{14}C (Steemann Nielsen 1952, Platt and Jassby 1976) or with stable mass isotope ^{13}C (Slawyk et al. 1977, Hama et al. 1983).

The ^{14}C-labeled method (Steemann Nielsen 1952) consists in measuring the uptake of ^{14}C-labelled inorganic carbon, added as a $NaH^{14}CO_3$ solution, into particulate and dissolved organic carbon during photosynthesis. This method provides estimates of the carbon incorporated into particles, that is, particulate organic carbon, POC (the material retained on filters) but also, if a high activity is provided for the $NaH^{14}CO_3$ added to the sample, estimates the contribution of dissolved organic carbon (DOC). The sum of ^{14}C incorporated into DOC and POC yields the total organic carbon (TOC) production. Differentiating between the photosynthetic carbon retained into dissolved and particulate fractions is important considering that DOC can represent 0 to 80% of the TOC production (Malinsky-Rushansky and Legrand 1996) and thus, underestimate primary production if DOC is not considered. This technique shows several advantages such as its high sensitivity in low productive oceans (Ichimura et al. 1962) as well as the estimation of DOC and POC incorporation over short time intervals. The use of ^{14}C due to international regulation and the potential of contamination on ships can be a problem for the ongoing use of this method. Indeed, radioisotope legislation in some countries restricts (e.g., Spain) or prohibits altogether (e.g., Japan) the use of ^{14}C on research vessels.

The ^{13}C-method can be also used to assess particulate organic carbon production (Slawyk et al. 1977, Hama et al. 1983) and is similar to the ^{14}C-labeled method except that the sodium bicarbonate solution is labeled with ^{13}C ($NaH^{13}CO_3$). The main advantages of this method compared to the ^{14}C are that the ^{13}C allows primary production incorporated by different components of the food web to be resolved using compound-specific analyses (Boschker and Middelburg 2002) and also that ^{13}C is a natural stable mass isotope, which does not involve any risks to the operator and can avoid any environmental contamination. However, this method is less sensitive than the ^{14}C method and the operator has to filtrate more volume of seawater in order to measure accurately the photosynthetic rates in low productive ecosystems.

c. Variable fluorescence

As mentioned previously, *in situ* data for ocean primary production rates has a poor temporal and spatial coverage globally (e.g., Regaudie-de-Gioux and Duarte 2013, Bouman et al. 2018), due to the logistic constraints and costs of ship operations, and time required for sampling and performing incubation procedures in the field. Because of that, new technologies aim to probe photosynthesis in the ocean in real-time, and currently, they are based on chlorophyll fluorescence transients, such as the FRRf (Fast Repetition Rate fluorometer) method (Kolber and Falkowski 1993), widely applied for measurements in the oceans (see discussion by Hughes et al. 2018). However, issues with the FRRf technique, as reported by Suggett et al. (2001, 2004), include uncertainties regarding the photosynthetic unit size that is a necessary knowledge for quantifying the excitation energy between the reaction centers between photosystems one (PSI) and two (PSII), the evaluation of the fraction of photochemically active reaction centers, as well as the use of the assumed values of the ratio of number of reaction centers of photosystem to the Chl-*a* concentration for different taxonomic groups. The FRRf method may also be prone to errors when the measured sample contains Coloured Dissolved Organic Matter (CDOM) that can affect the spectral absorption and so the accurate *in situ* measurement of active fluorescence. The FRRf method also provides a better signal-to-noise ratio and allows more robust measurements in oligotrophic ecosystems. To avoid this error, routine blanks must be incorporated in the observations. The advantages of this method are the potential to evaluate rapidly the depth and time-dependent value of PP (Sakshaug et al. 1997). Indeed, it resolves PP at spatial (< 1 m) and temporal (~ 1 s) resolutions that cannot be achieved by *in vitro* approaches (Robinson et al. 2009). However, proper guidelines for *in situ* measurements are under investigation.

d. Bio-optical method

Sosik (1996) proposed that the increasing measurements of light in the ocean could allow scientists to develop models to estimate PP rates using ocean color remote sensing data, but also pointed out that a major challenge in using this approach was quantifying the extent to which the photons absorbed by phytoplankton in the upper ocean could efficiently contribute to photosynthesis. Indeed, Chl-*a* concentration is, at first order, a good index of phytoplankton biomass and their activities (see Huot et al. 2008), and can be successfully retrieved by satellite data (Eppley et al. 1985), and thus used to estimate PP globally (Campbell et al. 2002, Saba et al. 2011). However, in nature, especially at lower spatial and temporal scales, the relationships between Chl-*a* concentration and phytoplanktonic biomass and growth are regulated by the physiological status and the taxonomic composition of communities (MacIntyre et al. 2002). Several phytoplankton characterizations have been proposed, using traits as dominant cell size and pigment composition, as well as phytoplankton functional types (see Chapter 3 in this book). A final consensus for a metric of phytoplankton communities is likely to be a combination of molecular and optical (i.e., spectral absorption and fluorescence) characteristics (Prézelin and Boczar 1986). The spectral absorption of light by phytoplankton ($a_{ph}(\lambda)$) is the most described bio-optical property in the oceans in time and space. As it contains information on pigment composition and cell size (Ciotti et al. 2002, Bricaud et al. 2004), it may be a better predictor of PP in models than Chl-*a*, as proposed by Marra et al. (2007) and more recently applied by Barnes et al. (2014) in coastal areas. More studies are still necessary but this method can also be implemented for remote sensing satellite data.

4.4 Quantitative overview of plankton metabolism and primary production in the ocean

The global annual PP in the ocean is about 50.1 Pg C yr^{-1}, ranging from 42.5 to 67 Pg C yr^{-1} (e.g., Falkowski et al. 1998, Gregg et al. 2002). Early surveys showed that *in situ* PP followed geographical trends, with higher values located near the coast and upwelling sites, and lower values over the Subtropical Gyres (e.g., Carr et al. 2006, Westberry et al. 2008, Regaudie-de-Gioux and Duarte 2013). PP in the ocean has been widely studied using remote sensing data derived by the sensors such as CZCS (Coastal Zone Color Scanner), SeaWiFS (Sea-Viewing Wide Field-of-view Sensor) and MODIS (Moderate Resolution Imaging Spectroradiometer) (e.g., Gregg et al. 2002, Carr et al. 2006, Westberry et al. 2008, Milutinovic et al. 2011, Kurk et al. 2020).

As observed for in situ PP estimations, modeled PP using satellite data showed regional variability of PP (Table 4.1). Indeed, subtropical oceans show in general higher PP (~ 46% of the global annual PP

Table 4.1. Area of each ocean (%), average annual integrated primary production (Mean PP; Gt C y^{-1} and %). Data compiled from Carr et al. 2006 (a, b), Kurk et al. 2020 (b), and Westberry et al. 2008 (c). n.d.: not determined.

		Area (%)	Mean PP (Pg C y^{-1})	Mean PP (%)
Oceans	Pacific	45[a]	19.47[b]	44[a]
	Atlantic	23[a]	12.14[b]	27[a]
	Indian	17[a]	8.305[b]	21[a]
	Southern	13[a]	3.656[b]	5.5[a]
	Artic	1.2[a]	0.33[a]	0.7[a]
	Mediterranean Sea	0.8[a]	0.45[a]	0.95[a]
Regions	Gyres	n.d.	12[c]	20[c]
	High latitudes	n.d.	14.3[c]	28[c]
	Subtropics	n.d.	25.3[c]	46[c]
	Southern Ocean (> 50°S)	n.d.	2.3[c]	4[c]

mean) than in the subtropical gyres and high latitudes regions (~ 20% and ~ 28% of the global annual PP mean, respectively; Westberry et al. 2008) (Table 4.1). The Pacific and the Arctic basins show the highest (19.5 Pg C yr^{-1}) and the lowest (0.3 Pg C yr^{-1}) annual PP mean in proportion to their area, respectively (Table 4.1; Carr et al. 2006, Kurk et al. 2020). While the Atlantic and Indian Oceans tend to be slightly more productive, consistently with their basin average latitudes, the Southern and Arctic Oceans tend to be less productive as a consequence of their respective incident solar radiation, and most of the global PP occurs in mesotrophic waters (Table 4.1; Carr et al. 2006, Kurk et al. 2020).

From large datasets (> 3000 paired GPP–CR of volumetric data) encompassing the five oceans and the Mediterranean Sea, the relationship between production and respiration shows that the CR of plankton communities are not proportional to their GPP (Duarte and Agustí 1998, Robinson and Williams 2005, Regaudie-de-Gioux and Duarte 2013; Fig. 4.2). Indeed, the more recent and expanded GPP–CR paired dataset (Regaudie-de-Gioux and Duarte 2013) shows that GPP ~ $CR^{0.79}$ indicating that CR increases more slowly with increasing GPP than GPP does, i.e., unproductive communities tend to have low P:R ratios (Duarte and Agustí 1998). Differences in the scaling of GPP to CR may occur, depending on the allochthonous inputs of organic carbon to the communities, which may elevate CR relative to GPP (i.e., higher intercept), particularly at low GPP (i.e., shallow slopes), as well as water temperature, which results in elevated CR for a given GPP in warm surface waters (Regaudie-de-Gioux and Duarte 2012). Heterotrophic ecosystems require allochthonous inputs of organic matter to meet their requirements which, for a long time, has been accepted by the scientific community to come from river inputs. However, additional organic carbon (~ 1 Pg C yr^{-1}; Siegenthaler and Sarmiento 1993) from atmospheric inputs to the ocean have been estimated to be about 11 Tg C yr^{-1} for dry deposition (i.e., dust) and between 47 and 187 Tg C yr^{-1} for wet (i.e., rainfalls) and gaseous (i.e., aerosols) depositions, respectively (Jurado et al. 2008). Furthermore, the relationship between production and respiration implies that GPP = R at an average GPP of 1.26 mmol O_2 $m^{-3}d^{-1}$ and those marine planktonic communities with a lower GPP are, therefore, likely to be heterotrophic (Regaudie-de-Gioux and Duarte 2013).

Examination of CO_2 fluxes in the ocean indicates that the equatorial areas and the unproductive subtropical gyres in the center of oceanic basins tend to be net sources of CO_2 to the atmosphere (Fig. 4.3), consistent with the net heterotrophy expected for communities in these ultraoligotrophic ecosystems (Prairie and Duarte 2007). Some coastal habitats composed of macrophytes, seagrass meadows, salt marshes, and mangrove forests show high production of organic matter over their respiration rates and thus, are strongly autotrophic (Duarte and Cebrián 1996, Gattuso et al. 1998; see also Chapter 8 in this book). Indeed, the excess production by these communities is so high that vegetated coastal habitats are important sinks for CO_2, accumulating organic carbon in their sediments (Duarte et al. 2005).

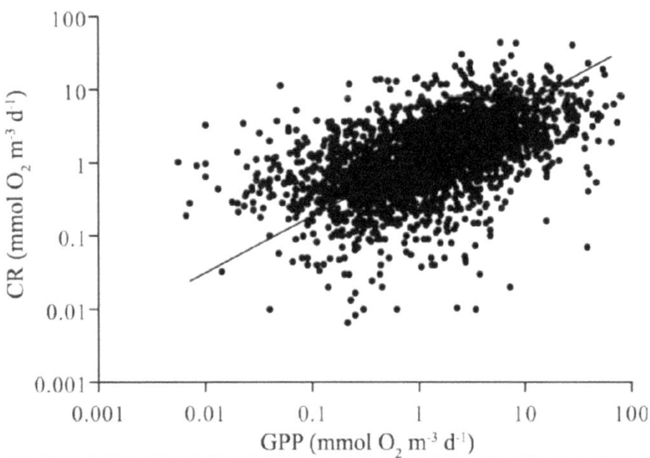

Figure 4.2. Log–log plot of GPP (mmol O_2 m^{-3} d^{-1}) and CR (mmol O_2 m^{-3} d^{-1}) from the dark-light method data set. The fitted model II regressions have the following equations: log CR = 0.79 (± 0.01) log GPP + 0.01 (± 0.01), $R^2 = 0.30$, $p < 0.05$, n = 3149 (Source: Regaudie-de-Gioux and Duarte 2013).

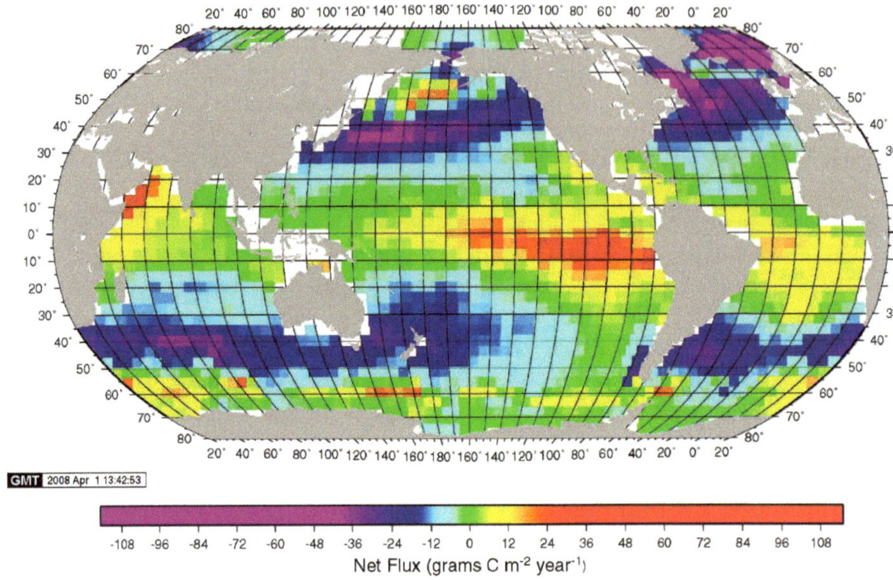

Figure 4.3. Climatological mean annual sea–air CO_2 flux (g C m^{-2} yr^{-1}) for the reference year 2000 (non-El Niño conditions) (Source: Takahashi et al. 2009). Intense blue-purple color indicates strong sinks whereas strong red color indicates strong emissions of CO_2 to the atmosphere.

4.5 Models and parametrization of primary production models

When oceanic primary production needs to be estimated for specific areas or time intervals, *in situ* approaches are not sufficient, and additional modeling and the use of satellite data become necessary. Although satellite observations register only the upper portion of the euphotic layer of the ocean, and that chlorophyll and light are not uniformly distributed in the water columns, several types of models were developed having distinct complexities (Carr et al. 2006), but the analyses of their performances, as well as improvements, remain challenged by the lack of *in situ* information (Saba et al. 2011). The application of remote sensing in PP models allows for estimations over daily, monthly, and annual time scales at regional-, basin- and global scales (Falkowski et al. 1998, Lee et al. 2015). There are simple to complex models that require depth-integrated and/or wavelength-integrated inputs and depth-resolved and/or wavelength-resolved inputs (e.g., Behrenfeld and Falkowski 1997, Carr et al. 2006, Saba et al. 2011). To estimate PP, main remote sensing models require the information of (1) phytoplankton biomass as indexed by carbon content or Chl-*a* concentration (necessary to use a conversion ratio to carbon), (2) light field that needs to be estimated vertically or assumed homogenous with depth, and (3) photosynthetic parameters (e.g., Behrenfeld et al. 1997). A major constraint of remote sensing models results from the fact that the available information is restricted to the sea surface; thus, premises about the vertical distribution of photosynthesis need to be made (Smith et al. 1982, Eppley et al. 1985), and although both Chl-*a* and solar irradiance can be easily modeled as functions of depth (e.g., Platt 1986), information on the quantum yield of photosynthesis as a function of time, space and depth is lacking.

Over the past decades, PP models improved mainly by combining a suite of remote sensing data: the commonly used data of Chl-*a* concentration, sea-surface temperature (SST), photosynthetically active radiation (PAR), and the coefficient of attenuation for PAR (k_{PAR}) or added by complementary data such as the particulate backscattering coefficient (b_{bp}), which can be a proxy for phytoplankton carbon (Behrenfeld et al. 2005). One of the most widely used remote sensing-derived models is the Vertically Generalized Production Model (VGPM) developed by Behrenfeld and Falkowski (1997). The maximum observed photosynthetic rate within the water column, P^B_{opt}, is obtained as a seventh-order polynomial of SST. Another commonly used PP model is the standard Howard, Yoder, Ryan (HYR) model with maximum growth rate parameterized as a function of SST according to Eppley (1972) and primary production integrated to the mixed-layer depth (Howard and Yoder 1997). The HYR model was used for many years as a standard MODIS algorithm. Other PP model was described by Marra et al. (2003) and is based on chlorophyll-specific absorption parametrized by SST and maximum quantum yield. The Carbon-based Production Model (CbPM) is estimated from the phytoplankton carbon content using the $b_{bp}(443)$ rather than from Chl-*a* concentration (Westberry et al. 2008).

The evolution of the models motivated a series of exercises supported by NASA, entitled Primary Production Algorithm Round Robin (PPARR). The distinct models that computed vertical profiles of PP into the euphotic zone were initially classified according to model complexity (Carr et al. 2006). In the following year, the performances of these same models were verified with *in situ* PP data available in the Equatorial Pacific (Friedrichs et al. 2009) and validated further with two time-series data obtained at two sites: the Bermuda Atlantic Time-series Study (BATS) and the Hawaii Ocean Time-series (HOT). As Chl-*a* concentration measurements were also available at both sites, it was possible to evaluate the influence of the main input variables of the models as well (Saba et al. 2010). During the last PPARR exercise (Saba et al. 2011), both performances and bias analyses were presented for the different models, using *in situ* data with greater spatial coverage on the globe, including observations in coastal regions for the first time. The main conclusion of the PPARR exercises highlighted the need for regional remote sensing algorithms for Chl-*a*, especially in coastal waters, emphasizing the need for high quantity and quality of *in situ* observations such as photosynthetic parameters.

4.6 Regulating factors and response to climate change

Since the industrial revolution, the ocean became a sink for atmospheric CO_2, removing about 2 Gt C yr^{-1} (Siegenthaler and Sarmiento 1993, Sarmiento and Gruber 2002) due in part to the major role of planktonic communities in the CO_2 air-sea exchange. Although several works (e.g., Behrenfeld et al. 2006, Chavez et al. 2011) using remote sensing and models observed an increase in global PP in the last 20 years, regional differences are apparent (Chavez et al. 2011, Kurk et al. 2020). Indeed, while low and mid-latitudes showed a decrease (up to –3%), polar and coastal regions showed an increase (up to 4.5%) in the last 20 years (Kurk et al. 2020), as planktonic communities may respond differently to climate change and in particular, to ocean warming. Increases in temperatures at the sea surface may shift its metabolic balance to a heterotrophic ecosystem, where the respiration could tend to increase faster than the production in the planktonic communities (e.g., Harris et al. 2006, Lopez-Urrutia et al. 2006, Regaudie-de-Gioux and Duarte 2012). Considering that CR intensifies faster than GPP at the same temperature increase, NCP (i.e., GPP – CR) should decrease with ocean warming (Regaudie-de-Gioux et al. 2012). Besides, the current P/R ratio of planktonic communities in the ocean is close to 1 (Duarte and Agustí 1998, Williams 1998, Robinson and Williams 2005, Regaudie-de-Gioux and Duarte 2013), but some studies estimated a decrease between 16 and 25% in this ratio with a 4°C of warming, predicted for the 21st century (Harris et al. 2006, Regaudie-de-Gioux and Duarte 2012). A decline of P/R ratios in portions of the oceans could favor the establishment of heterotrophic planktonic communities, changing locations normally acting as a sink to a source of CO_2 to the atmosphere.

Climate models predict that ocean warming would lead to significant changes in water stratification, thus affecting nutrient fluxes for an increase of vertical stability, reducing vertical mixing and, consequently, diminishing primary production (Cox et al. 2000, IPCC 2001, Boyd and Doney 2002, Sarmiento et al. 2004). On the other hand, experiments showed that ocean acidification led by atmospheric CO_2 increase may improve phytoplankton photosynthetic performance (Riebesell et al. 2007, Hendriks et al. 2010). These observations may lead to the assumption that an increase in CO_2 in ocean water would stimulate phytoplankton photosynthetic performances and thus would attenuate the decline in net primary production with ocean warming. Although the response of planktonic respiration to ocean warming is expected to be a steep and sustained increased response, the stocks of dissolved and particulate (i.e., sediment) organic carbon in the ocean, as well as the significant inputs of organic depositions of carbon in dry, wet (Jurado et al. 2008), and volatile components (Dachs et al. 2005, Ruiz-Halpern et al. 2010), are large enough to support the excess of respiration over production for some decades.

4.7 Acknowledgments

We would like to thank Dr. Pratolongo and Dr. Pan for their invitation to write this chapter. On Prof. Ciotti's part, this research was funded by the Brazilian National Council of Research and Development (CNPq, fellowship PQ 311362/2016-8). On Dr. Regaudie-de-Gioux's part, this work was funded by the INGENIO 2010 CONSOLIDER program (ref. CDS2008-00077) of the Spanish Ministry of Economy and Competitiveness as a contribution to the Malaspina Circumnavigation Expedition 2010, and also funded by Ifremer. This research is a contribution of the NP-BioMar (Center for Marine Biology-CEBIMar/USP), and INCT Mar-Coi (FURG).

Summary

The metabolism in the ocean is controlled by planktonic microorganisms that play a major biogeochemical role on Earth mainly through their metabolism. The metabolic rates of planktonic communities are composed of the production and the respiration of the organic matter. Depending on the dominance of each of these processes, the ecosystems or communities may be classified as: autotrophic, indicating that these communities produce organic matter in excess of its respiratory requirements, or heterotrophic, which consume more organic matter through respiratory processes that they produce, or in balance when production and respiration are of equal magnitudes. Several external factors affect directly and indirectly both the primary production and the metabolism of planktonic communities in the ocean such as temperature, nutrient concentrations, and solar irradiation. In this chapter, we briefly present the different methods allowing to measure planktonic primary production (PP), the general overview of planktonic metabolism in the ocean, the parameterization of PP through models and remote sensing and finally, the different responses of planktonic metabolism to climate change.

References

Agustí, S., A. Regaudie-de-Gioux, J.M. Arrieta and C.M. Duarte. 2014. Consequences of UV-enhanced community respiration for plankton metabolic balance. Limnol. Oceanogr. 59: 223–232. doi:10.4319/lo.2014.59. 1.0223.

Antoine, D. and A. Morel. 1996. Oceanic primary production: I. Adaptation of a spectral light-photosynthesis model in view of application to satellite chlorophyll observations. Global Biogeochem. Cycles. 10: 43–55. doi:10.1029/95GB02831.

Barnes, M.K., G.H. Tilstone, T.J. Smyth, D.J. Suggett, R. Astorenca, C. Lancelot and J.C. Kromkamp. 2014. Absorption-based algorithm of primary production for total and size-fractionated phytoplankton in coastal waters. Mar. Ecol. Progr. Ser. 504: 1–12. doi:10.3354/meps10751.

Behrenfeld, M.J. and P.G. Falkowski. 1997. Photosynthetic rates derived from satellite-based chlorophyll concentration. Limnol. Oceanogr. 42: 1–20. doi:10.4319/lo.1997.42.1.0001.

Behrenfeld, M.J., J.T. Randerson, C.R. McClain, G.C. Feldman, S.O. Los, C.J. Tucker, P.G. Falkowski, C.B. Field, R. Frouin, W.E. Esaias, D.D. Kolber and N.H. Pollack. 2001. Biospheric primary production during an ENSO transition. Science 291: 2594–2597. doi:10.1126/science.1055071.

Behrenfeld, M.J., E. Boss, D.A. Siegel and D.M. Shea. 2005. Carbon-based ocean productivity and phytoplankton physiology from space. Glob. Biogeochem. Cycles. 19: GB1006. doi:10.1029/2004GB002299.

Behrenfeld, M.J., R.T. O'Malley, D.A. Siegel, C.R. McClain, J.L. Sarmiento, G.C. Feldman, A.J. Milligan, P.G. Falkowski, R.M. Letelier and E.S. Boss. 2006. Climate-driven trends in contemporary ocean productivity. Nature 444: 752–755. doi:10.1038/nature05317.

Bender, M.L., K. Grande, K. Johnson, J. Marra, P.J.LeB. Williams, J. Sieburth, M. Pilson, C. Langdon, G. Hitchcock, J. Orchardo, C. Hunt, P. Donaghay and K. Heinemann. 1987. A comparison off our methods for determining. Planktonic community production. Limnol. Oceanogr. 32: 1085–1098. doi:10.4319/lo.1987.32.5.1085.

Bender, M., J. Orchardo, M.-L. Dickson, R. Barber and S. Lindley. 1999. *In vitro* O_2 fluxes compared with ^{14}C production and other rate terms during the JGOFS Equatorial Pacific experiment. Deep Sea Res. 46: 637–654. doi:10.1016/S0967-0637(98)00080-6.

Boschker, H.T.S. and J.J. Middelburg. 2002. Stable isotopes and biomarkers in microbial ecology. FEMS Microbiol. Ecol. 40: 85–95. doi:10.1111/j.1574-6941.2002.tb00940.x.

Bouman, H.A., T. Platt, M. Doblin, F.G. Figueiras, K. Gudmudsson, H.G. Gudfinnsson, B. Huang, A. Hickman, M. Hiscock, T. Jackson et al. 2018. Photosynthesis-irradiance parameters of marine phytoplankton: Synthesis of a global data set. Earth Syst. Sci. Data 10: 251–266.

Boyd, P.W. and S.C. Doney. 2002. Modeling regional responses by marine pelagic ecosystems to global climate change, Geophys. Res. Lett. 29: 1806. doi:10.1029/2001GL014130.

Bricaud, A., H. Claustre, J. Ras and K. Oubelkheir. 2004. Natural variability of Phytoplanktonic absorption in oceanic waters: influence of the size structure of algal populations. J. Geophys. Res. 109: C11010. doi:10.1029/2004JC002419.

Brown, J.H., J.F. Gillooly, A.P. Allen, V.M. Savage and G.B. West. 2004. Toward a metabolic theory of ecology. Ecology 85(7): 1771–1789. doi:10.1890/03-9000.

Campbell, J., D. Antoine, R. Armstrong, K. Arrigo, W. Balch, R. Balber, M. Behrenfeld, R. Bidigare, J. Bishop, M.-E. Carr, W. Esaias, P. Falkowski, N. Hoepffner, R. Iverson, D. Kiefer, S. Lohrenz, J. Marra, A. Morel, J. Ryan, V. Vedernikov, K. Waters, C. Yentsch and J. Yoder. 2002. Comparison of algorithms for estimating ocean primary production from surface chlorophyll, temperature, and irradiance. Glob. Biogeochem. Cycles. doi:10.1029/2001GB001444.

Carpenter, J.H. 1965. The accuracy of the Winkler method for dissolved oxygen analysis. Limnol. Oceanogr. 10: 135–140. doi:10.4319/lo.1965.10.1.0135.

Carr, M.E., M.A.M. Friedrichs, M. Schmeltz, M.N. Aita, D. Antoine, K.R. Arrigo, I. Asanuma, O. Aumont, R. Barber, M. Behrenfeld, R. Bidigare, E.T. Buitenhuis, J. Campbell, A. Ciotti, H. Dierssen, M. Dowell, J. Dunne, W. Esaias, B. Gentili, W. Gregg, S. Groom, N. Hoepffner, J. Ishizaka, T. Kameda, C. Le Quéré, S. Lohrenz, J. Marra, F. Mélin, K. Moore, A. Morel, T.E. Reddy, J. Ryan, M. Scardi, T. Smyth, K. Turpie, G. Tilstone, K. Waters and Y. Yamanaka. 2006. A comparison of global estimates of marine primary production from ocean color. Deep-Sea Res. Pt. II 53: 741–770. doi:10.1016/j.dsr2.2006.01.028.

Carrit, D.E. and J.H. Carpenter. 1966. Comparison and evaluation of currently employed modifications of the Winkler method for determining dissolved oxygen in sea-water. J. Mar. Res. 24: 286–318.

Cassar, N., P.J. Difiore, B.A. Barnett, M.L. Bender, A.R. Bowie, B. Tilbrook, K. Petrou, K.J. Westwood, S.W. Wright and D. Lefèvre. 2011. The influence of iron and light on net community production in the Subantarctic and Polar Frontal Zones. Biogeosciences 8: 227–237. doi:10.5194/bg-8-227-2011.

Chavez, F.P., M. Messié and J.T. Pennington. 2011. Marine primary production in relation to climate variability and change. Annu. Rev. Mar. Sci. 3: 227–260. doi:10.1146/annurev.marine.010908.163917.

Ciotti, A.M., M.R. Lewis and J.J. Cullen. 2002. Assessment of the relationships between dominant cell size in natural phytoplankton communities and the spectral shape of the absorption coefficient. Limnol. Oceanogr. 47: 404–417. doi:10.2307/3068986.

Claustre, H., A. Morel, M. Babin, C. Cailliau, D. Marie, J.C. Marty, D. Tailliez and D. Vaulot. 1999. Variability in particle attenuation and chlorophyll fluorescence in the Tropical Pacific: scales, patterns, and biogeochemical implications. J. Geophys. Res. 104: 3401–3422. doi:10.1029/98JC01334.

Cox, P.M., R.A. Betts, C.D. Jones, S.A. Spall and I.J. Totterdell. 2000. Acceleration of global warming due to carbon-cycle feedbacks in a coupled climate model. Nature 408: 184–187. doi:oi.org/10.1038/35041539.

Cullen, J.J., X. Yang and H.L. MacIntyre. 1992. Nutrient limitation of marine photosynthesis. pp. 69–88. *In*: Falkowski, P.G. and A.D. Woodhead [eds.]. Primary Productivity and Biogeochemical Cycles in the Sea. Plenum Press, New York.

Cullen, J.J. and P.J. Neale. 1994. Ultraviolet radiation, ozone depletion, and marine photosynthesis. Photosynth. Res. 39: 303–320.

Dachs, J., M.Ll. Calleja, C.M. Duarte, S. del Vento, B. Turpin, A. Polidori, G. Herndl and S. Agustí. 2005. High atmosphere-ocean exchange of organic carbon in the NE Subtropical Atlantic. Geophys. Res. Lett. 32. doi:10.1029/2005GL023799.

del Giorgio, P.A. and P.J.LeB. Williams. 2005. Respiration in aquatic ecosystems. Oxford Univ. Press.315 pp.ISBN 0-19-852708-X.

Duarte, C.M. and J. Cebrián. 1996. The fate of marine autotrophic production. Limnol. Oceanogr. 41: 1758–1766. doi:rg/10.4319/lo.1996.41.8.1758.

Duarte, C.M. and S. Agustí. 1998. The CO_2 balance of unproductive aquatic ecosystems. Science 281: 234–236. doi:10.1126/science.281.5374.234.

Duarte, C.M., S. Agustí, H. Kennedy and D. Vaqué. 1999. The Mediterranean climate as a template for Mediterranean marine ecosystems: the example of the northeast Spanish littoral. Prog. Oceanogr. 44: 245–270. doi:10.1016/S0079-6611(99)00028-2.

Duarte, C.M., S. Agustí and D. Vaqué. 2004. Controls on planktonic metabolism in the Bay of Blanes, north-western Mediterranean litoral. Limnol. Oceanogr. 49: 2162–2170. doi:10.4319/lo.2004.49.6.2162.

Duarte, C.M. and Y.M. Prairie. 2005. Prevalence of heterotrophy and atmospheric CO_2 emissions from aquatic ecosystems. Ecosystems 8: 862–870. doi:10.1007/s10021-005-0177-4.

Duarte, C.M., J.J. Middelburg and N. Caraco. 2005. Major role of marine vegetation on the oceanic carbon cycle. Biogeosciences 2: 1–8. doi:10.5194/bg-2-1-2005.

Duarte, C.M., A. Regaudie-de-Gioux, J.M. Arrieta, A. Delgado-Huertas and S. Agustí. 2013. The oligotrophic ocean is heterotrophic. Ann. Rev. Mar. Sci. 5: 551–569. doi: 10.1146/annurev-marine-121211-172337.

Dugdale, R.C. and J.J. Goering. 1967. Uptake of new and regenerated forms of nitrogen in primary productivity. Limnol. Oceanogr. 12(2): 196–206.

Emerson, S. 1987. Seasonal oxygen cycles and biological new production in surface waters of the sub-Arctic Pacific Ocean. J. Geophys. Res.: Oceans 92: 6535–6544. doi:10.1029/JC092iC06p06535.

Emerson, P., P.D. Quay and P.A. Wheeler. 1993. Biological productivity determined from oxygen mass-balance and incubation experiments. Deep-Sea Res. I 40: 2351–2358. doi:10.1016/0967-0637(93)90109-G.
Eppley, R.W. 1972. Temperature and phytoplankton growth in the sea. Fish. Bull. 70: 1063–1085.
Eppley, R.W., E. Stewart, M.R. Abbott and U. Heyman. 1985. Estimating ocean primary production from satellite chlorophyll. Introduction to regional differences and statistics for the Southern California Bight. J. Plankton Res. 7: 57–70. doi:110.1093/plankt/7.1.57.
Falkowski, P.G. and J.A. Raven. 1997. Aquatic Photosynthesis. Blackwell Science.
Falkowski, P.G., R.T. Barber and V. Smetacek. 1998. Biogeochemical controls and feedbacks on ocean primary production. Science 281: 200–206.
Field, C.B., M.J. Behrenfeld, J.T. Randerson and P.G. Falwoski. 1998. Primary production of the biosphere: integrating terrestrial and oceanic components. Science 281: 237–240. doi:10.1126/science.281.5374.237.
Fischer, A.D., E.A. Moberg, H. Alexander, E.F. Brownlee, K.R. Hunter-Cevera et al. 2014. Sixty years of Sverdrup: a retrospective of progress in the study of phytoplankton blooms. Oceanography 27(1): 222–235.
Friedrichs, M.A.M. et al. 2009. Assessing the uncertainties of model estimates of primary productivity in the tropical Pacific Ocean. J. Mar. Syst. doi: 10.1016/j.jmarsys.2008.05.010.
Gaarder, T. and H.H. Gran. 1927. Investigations of the production of plankton in the Oslo Fjörd. Rapp. P. v. Reun. Cons. Perm. Int. Explor. Mer. 42: 1–48.
Gattuso, J.-P., M. Franjignoulle and R. Wollast. 1998. Carbon and carbonate metabolism in coastal aquatic ecosystems. Annu. Rev. Ecol. Syst. 29: 405–433. doi:10.1146/annurev.ecolsys.29.1.405.
Geider, R.J. and J. La Roche. 2002. Redfield revisited: variability of C:N:P in marine microalgae and its biochemical basis. European Journal of Phycology 37: 1–17.
Gillooly, J.F., J.H. Brown, G.B. West, V.M. Savage and E.L. Charnov. 2001. Effects on size and temperature on metabolic rate. Science 293: 2248–2251. doi:10.1126/science.1061967.
Grande, K.D., P. Kroopnick, D. Burns and M.L. Bender. 1982. ^{18}O as a tracer for measuring gross primary productivity in bottle experiments. [Abstr.] Eos 63: 107.
Gregg, W.W., M.E. Conkright, J.E. O'Reilly, F.S. Patt, M.H. Wang, J.A. Yoder and N.W. Casey. 2002. NOAA-NASA coastal zone color scanner reanalysis effort. Appl. Opt. 41: 1615–1628. doi:10.1364/AO.41.001615.
Hama, T., T. Miyazaki, Y. Ogawa, T. Iwakumi, M. Takahashi, A. Otsuki and S. Ichimura. 1983. Measurement of photosynthetic production of a marine phytoplankton population using a stable ^{13}C isotope. Marine Biology 73: 31–36.
Harris, L.A., C.M. Duarte and S.W. Nixon. 2006. Allometric laws and prediction in estuarine and coastal ecology. Estuar. Coasts 29: 340–344. doi:10.1007/BF02782002.
Hendricks, M.B., M.L. Bender and B.A. Barnett. 2004. Net and gross O_2 production in the Southern Ocean from measurements of biological O_2 supersaturation and its triple isotope composition. Deep Sea Res. I 51: 1541–1561. doi:10.1016/j.dsr.2004.06.006.
Hendriks, I.E., C.M. Duarte and M. Alvarez. 2010. Vulnerability of marine biodiversity to ocean acidification: a meta-analysis. Estuar. Coast. Shelf. Sci. 86: 157–164. doi:https://doi.org/10.1016/j.ecss.2009.11.022.
Herzig, R. and P.G. Falkowski. 1989. Nitrogen limitation in *Isochrysis galbana*. I. Photosynthetic energy conversion and growth efficiencies. J. Phycol. 25: 462–471.
Howard, K.L., J.A. Yoder and C. Liu. 1997. Contribution of the subtropical oceans to global primary production. pp. 157–168. *In*: Pergamon, T. [ed.]. Proceedings of COSPAR Colloquium on Space Remote Sensing of Subtropical Oceans. New York.
Hughes, D.J., D.A. Campbell, M.A. Doblin, J.C. Kromkamp, E. Lawrenz, C.M. Moore, K. Oxborough, O. Prášil, P.J. Ralph, M.F. Alvarez and D.J. Suggett. 2018. Roadmaps and de-tours: Active chlorophyll-assessments of primary productivity across marine and freshwater systems. Environ. Sci. Technol. 52: 12039–12054. doi:10.1021/acs.est.8b03488.
Huot, Y., A. Morel, M.S. Twardowski, D. Stramski and R.A. Reynolds. 2008. Particle optical backscattering along a chlorophyll gradient in the upper layer of the eastern South Pacific Ocean. Biogeosciences 5: 495–507. doi:10.5194/bgd-4-4571-2007.
Ichimura, S., Y. Saijo and Y. Aruga. 1962. Photosynthetic characteristics of marine phytoplankton and their ecological meaning in the chlorophyll method. Bot. Mag. 75: 212–220.
Kaiser, J. 2011. Consistent calculation of aquatic gross production from oxygen triple isotope measurements. Biogeosciences 8: 1793–811. Doi:10.5194/bg-8-1793-2011.
Karl, D.M., D.V. Hebel, K. Björkman and R.M. Letelier. 1998. The role of dissolved organic matter release in the productivity of the oligotrophic North Pacific Ocean. Limnol. Oceanogr. 43: 1270–1286. doi:10.4319/lo.1998.43.6.1270.
Kirk, J.T.O. 1994. Light and Photosynthesis in Aquatic Ecosystems, 2nd Edn. Cambridge: Univ. Press. doi:10.1017/CBO9780511623370.
Kolber, Z. and P.G. Falkowski. 1993. Use of active fluorescence to estimate phytoplankton photosynthesis *in situ*. Limnol. Oceanogr. 38: 1646–1665. doi:10.4319/lo.1993.38.8.1646.
Kulk, G., T. Platt, J. Dingle, T. Jackson, B.F. Jönsson, H.A. Bouman, M. Babin, R.J.W. Brewin, M. Doblin, M. Estrada, F.G. Figueiras et al. 2020. Primary production, an index of climate change in the ocean: satellite-based estimates over two decades. Remote Sens. 12(5): 826. doi: 10.3390/rs12050826.

IPCC Climate Change. 2001. The Scientific Basis. Contribution of Working Group I to the Third Assessment Report of the Intergovernmental Panel on Climate Change. Cambridge University Press, Cambridge. doi:10.1093/ije/dyg059.
Jurado, E., J. Dachs, C.M. Duarte and R. Simó. 2008. Atmospheric deposition of organic and black carbon to the global ocean. Atmospheric Environment 42: 7931–7939. doi: 10.1016/j.atmosenv.2008.07.029.
Juranek, L.W. and P.D. Quay. 2005. *In vitro* and *in situ* gross primary and net community production in the North Pacific Subtropical Gyre using labeled and natural abundance isotopes of dissolved O_2. Global Biogeochem. Cycles 19: GB3009. doi:10.1029/2004GB002384.
Laws, E.A. and T.T. Bannister. 1980. Nutrient- and light-limited growth of *Thalassiosira fluviatilis* in continuous culture, with implications for phytoplankton growth in the ocean. Limnol. Oceanogr. 25: 457–473.
Lee, Z., J. Marra, M. Jane and M. Kahru. 2015. Estimating oceanic primary productivity from ocean color remote sensing: A strategic assessment. J. Mar. Syst. 149: 50–59. doi:10.1016/j.jmarsys.2014.11.015.
Llabrés, M., S. Agustí, M. Fernández, A. Canepa, F. Maurin, F. Vidal and C.M. Duarte. 2013. Impact of elevated UVB radiation on marine biota: A meta-analysis. Global Ecol. Biogeogr. 22: 131–144.
Loisel, H. and A. Morel. 1998. Lights cattering and chlorophyll concentration in case 1 waters: a reexamination. Limnol. Oceanogr. 43: 847–858. doi:10.4319/lo.1998.43.5.0847.
López-Urrutía, A., E. San Martin, R.P. Harris and X. Irigoyen. 2006. Scaling the metabolic balance of the oceans. Proc. Natl. Acad. Sci. USA 103: 8739–8744. doi:10.1073/pnas.0601137103.
Luz, B. and E. Barkan. 2000. Assessment of oceanic productivity with the triple-isotope composition of dissolved oxygen. Science 288: 2028–2031. doi:10.1126/science.288.5473.2028.
MacIntyre, H.L., T.M. Kana, T. Anning and R.J. Geider. 2002. Photoacclimation of photosynthesis irradiance response curves and photosynthetic pigments in microalgae and cyanobacteria. J. Phycol. 38: 17–38. doi:10.1046/j.1529-8817.2002.00094.x.
Malinsky-Rushansky, N.Z. and C. Legrand. 1996. Excretion of dissolved organic carbon by phytoplankton of different sizes and subsequent bacterial uptake. Mar. Ecol. Progr. Ser. 132: 249–255. doi:10.3354/meps132249.
Marra, J. 2002. Approaches to the measurement of plankton production. pp. 78–108. *In*: Williams, P.J.LeB., D.N. Thomas and C.S. Reynolds [eds.]. Phytoplankton Productivity: Carbon Assimilation in Marine and Fresh water Ecosystems. Blackwell Science, Oxford, UK.
Marra, J., C. Ho and C. Trees. 2003. An alternative algorithm for the calculation of primary productivity from remote sensing data. *In*: LDEO Technical Report, #LDEO-2003-1.
Marra, J. and R.T. Barber. 2004. Phytoplankton and heterotrophic respiration in the surface layer of the ocean. Geophys. Res. Lett. 31: 19–24. doi: 10.1029/2004GL019664.
Marra, J., C.C. Trees and J.E. O'Reilly. 2007. Phytoplankton pigment absorption: a strong predictor of primary productivity in the surface ocean. Deep-Sea Res. I 54: 155–163. doi:10.1016/j.dsr.2006.12.001.
Marra, J. 2012. Comment on "Measuring primary production rates in the ocean: enigmatic results between incubation and non-incubation methods at Station ALOHA" by Quay, P.D. et al. Glob. Biogeochem. Cycles 26: GB2031. doi:10.1029/2011GB004087.
Maske, H. and E. Garcia-Mendoza. 1994. Adsorption of dissolved organic matter to the inorganic filter substrate and its implications for ^{14}C uptake measurements. Appl. Environ. Microb. 60: 3887–3889.
Miller, C.B. and P.A. Wheeler. 2010. Biological Oceanography, 2nd edn. John Wiley & Sons, Ltd., Hoboken, NJ, 464 pp.
Milutinović, S. and L. Bertino. 2011. Assessment and propagation of uncertainties in input terms through an ocean-color-based model of primary productivity. Remote Sens. Environ. 115: 1906–1917. doi:10.1016/j.rse.2011.03.013.
Nicholson, D., S. Emerson and C.C. Erisken. 2008. Net community production in the deep euphotic zone of the subtropical North Pacific gyre from glider surveys. Limnol. Oceanogr. 53: 2226–2236. doi:10.4319/lo.2008.53.5_part_2.2226.
Platt, T. and A.D. Jassby. 1976. The relationship between photosynthesis and light for natural assemblages of coastal marine phytoplankton. J. Phycol. 12: 421–430.
Platt, T. 1986. Primary production of the ocean water column as a function of surface light intensity: algorithms for remote sensing. Deep Sea Res. 33: 149–163. doi:10.1016/0198-0149(86)90115-9.
Prairie, Y.T. and C.M. Duarte. 2007. Direct and Indirect Metabolic CO_2 release by Humanity. Biogeosciences 4: 215–217. doi:10.5194/bg-4-215-2007.
Prézelin, B.B. and B.A. Bozcar. 1986. Molecular bases of cell absorption and fluorescence in phytoplankton: potential applications to studies in optical oceanography. pp. 349–354. *In*: Round, F.E. and D.J. Chapman [eds.]. Progress in Phycological Research. Volume 4, Elsevier Biomedical Press, Uni. Michigan.
Quay, P.D., S. Emerson, D.O. Wilbur, C. Stump and M. Knox. 1993. The delta-^{18}O of dissolved O_2 in the surface waters of the Sub-Arctic Pacific: a tracer of biological productivity. J. Geophys. Res.: Oc. 98: 8447–8458. doi:10.1029/2007GB003162.
Quay, P.D., C. Peacock, K. Björkman and D.M. Karl. 2010. Measuring primary production rates in the ocean: Enigmatic results between incubation and non-incubation methods at Station ALOHA. Glob. Biogeochem. Cycles. 24: GB3014. doi:10.1029/2009GB003665.
Redfield, A.C. 1958. The biological control of chemical factors in the environment. Am. Sci. 46: 205–221.
Regaudie-de-Gioux, A. and C.M. Duarte. 2012. Temperature dependence of planktonic metabolism in the ocean. Global Biogeochem. Cycles. 26: GB1015. doi:10.1029/2010GB003907.

Regaudie-de-Gioux, A. and C.M. Duarte. 2013. Global patterns in oceanic planktonic metabolism. Limnol. Oceanogr. 58: 977–986. doi:10.4319/lo.2013.58.3.0977.

Regaudie-de-Gioux, A., S. Lasternas, S. Agustí and C.M. Duarte. 2014a. Comparing marine primary production estimates through different methods and development of conversion equations. Front. Mar. Sci. 1: 19. doi:10.3389/fmars.2014.00019.

Regaudie-de-Gioux, A., S. Agustí and C.M. Duarte. 2014b. UV sensitivity of planktonic net community production in ocean surface waters. J. Geophys. Res. 119: 929–936. doi:10.1002/2013JG002566.

Reuer, M., B. Barnett, M.L. Bender, P.G. Falkowski and M.B. Hendricks. 2007. New estimates of Southern Ocean biological production rates from O_2/Ar ratios and the triple isotope composition of O_2. Deep Sea Res. Part I 54: 951–974. doi:10.1016/j.dsr.2007.02.007.

Riebesell, U., K.G. Schulz, R.G.J. Bellerby, M. Botros, P. Fritsche, M. Meyerhöfer, C. Neill, G. Nondal, A. Oschlies, J. Wohlers and E. Zöllner. 2007. Enhanced biological carbon consumption in a high CO_2 ocean. Nature 450: 545–548. doi:10.1038/nature06267.

Robinson, C. and P.J.LeB. Williams. 2005. Respiration and its measurement in surface marine waters. pp. 147–181. *In*: del Giorgio, P.A. and P.J.LeB. Williams [eds.]. Respiration in Aquatic Ecosystems. Oxford University Press, Oxford, UK. doi:10.1093/acprof:oso/9780198527084.003.0009.

Robinson, C., G.H. Tilstone, A.P. Rees, T.J. Smyth, J.R. Fishwick, G.A. Tarran, B. Luz, E. Barkan and E. David. 2009. Comparison of *in vitro* and *in situ* plankton production determinations. Aquat. Microb. Ecol. 54: 13–34. doi:10.3354/ame01250.

Ruíz-Halpern, S., M.K. Sejr, C.M. Duarte, D. Krause-Jensen, T. Dalsgaard, J. Dachs and S. Rysgaard. 2010. Air-water exchange and vertical profiles of organic carbon in a subarctic fjord. Limnol. Oceanogr. 55: 1733–40. doi:10.4319/lo.2010.55.4.1733.

Saba, V.S., M.A.M. Friedrichs, M.-E. Carr, D. Antoine, R.A. Armstrong, I. Asanuma, O. Aumont, N.R. Bates, M.J. Behrenfeld, V. Bennington, L. Bopp, J. Bruggeman, E.T. Buitenhuis, M.J. Church, A.M. Ciotti, S.C. Doney, M. Dowell, J. Dunne, S. Dutkiewicz, W. Gregg, N. Hoepffner, K.J.W. Hyde, J. Ishizaka, T. Kameda, D.M. Karl, I. Lima, M.W. Lomas, J. Marra, G.A. McKinley, F. Mélin, J.K. Moore, A. Morel, J. O'Reilly, B. Salihoglu, M. Scardi, T.J. Smyth, S. Tang, J. Tjiputra, J. Uitz, M. Vichi, K. Waters, T.K. Westberry and A. Yool. 2010. Challenges of modeling depth integrated marine primary productivity over multiple decades: A case study at BATS and HOT. Global Biogeochem. Cycles. 24: GB3020. doi:10.1029/2009GB003655.

Saba, V.S., M.A.M. Friedrichs, D. Antoine, R.A. Armstrong, I. Asanuma, M.J. Behrenfeld, A.M. Ciotti, M. Dowell, N. Hoepffner, K.J.W. Hyde, J. Ishizaka, T. Kameda, J. Marra, F. Mélin, A. Morel, J. O'Reilly, M. Scardi, W.O. Smith Jr., T.J. Smyth, S. Tang, J. Uitz, K. Waters and T.K. Westberry. 2011. An evaluation of ocean color model estimates of marine primary productivity in coastal and pelagic regions across the globe. Biogeosciences 8: 489–503. doi:10.5194/bg-8-489-2011.

Sakshaug, E., A. Bricaud, Y. Dandonneau, P.G. Falkowski, D.A. Kiefer, L. Legendre, A. Morel, J. Parslow and M. Takahashi. 1997. Parameters of photosynthesis: definitions, theory and interpretation of results. J. Plankton Res. 19: 1637–1670. doi:10.1093/plankt/19.11.1637.

Sarmiento, J.L. and N. Gruber. 2002. Sinks for anthropogenic carbon. Phys. Today 55: 30–36. doi:10.1063/1.1510279.

Sarmiento, J.L., R. Slater, R. Barber, L. Bopp, S.C. Doney, A.C. Hirst, J. Kleypas, R. Matear, U. Mikolajewicz, P. Monfray, V. Soldatov, S.A. Spall and R. Stouffer. 2004. Response of ocean ecosystems to climate warming. Glob. Biogeochem. Cy. 18: GB3003. doi:10.1029/2003GB002134.

Siegenthaler, U. and J.L. Sarmiento. 1993. Atmospheric carbon dioxide and the ocean. Nature 365: 119–125. doi:10.1038/365119a0.

Slawyk, G., Y. Collos and J.-C. Auclair. 1977. The use of the ^{13}C and ^{15}N isotopes for the simultaneous measurement of carbon and nitrogen turnover rates in marine phytoplankton. Limnol. Oceanogr. 22: 925–932. doi:10.4319/lo.1977.22.5.0925.

Smith, R.C., R.W. Eppley and K.S. Baker. 1982. Correlation of primary production as measured aboard ship in southern California coastal waters and as estimated from satellite chlorophyll images. Mar. Biol. 88. doi:10.1007/BF00397033.

Sosik, H.M. 1996. Bio-optical modeling of primary production: consequences of variability in quantum yield and specific absorption. Mar. Ecol. Prog. Ser. 143: 225–238. doi:10.3354/meps143225.

Spitzer, W.S. and W.J. Jenkins. 1989. Rates of vertical mixing, gas-exchange and new production estimates from seasonal gas cycles in the upper ocean near Bermuda. J. Mar. Res. 47: 169–196. doi:10.1357/002224089785076370.

Stanley, R.H.R., J.B. Kirkpatrick, N. Cassar, B.A. Barnett and M.L. Bender. 2010. Net community production and gross primary production rates in the western equatorial Pacific. Glob. Biogeochem. Cycles 24: GB4001. doi:10.1029/2009GB003651.

Steeman Nielsen, E. 1952. The use of radioactive carbon (^{14}C) for measuring production in the sea. J. Cons. Perm. Int. Explor. Mer. 18: 17–140. doi:10.1093/icesjms/18.2.117.

Suggett, D., G. Kraay, P. Holligan, M. Davey, J. Aiken and R. Geider. 2001. Assessment of photosynthesis in a spring cyanobacterial bloom by use of a fast repetition rate fluorometer. Limnol. Oceanogr. 46: 802–810. doi: 10.4319/lo.2001.46.4.0802.

Suggett, D.J., H.L. MacIntyre and R.J. Geider. 2004. Evaluation of biophysical and optical determinations of light absorption by photosystem II in phytoplankton. Limnol. Oceanogr.: Methods 2: 316–332. doi:10.4319/lom.2004.2.316.

Svendrup, H.W. 1953. On conditions for the vernal blooming of phytoplankton. J. Cons. Int. Explor. Mer. 18: 28–295.

Takahashi, T., S.C. Sutherland, R. Wanninkhof, C. Sweeney, R.A. Feely, D.W. Chipman, B. Hales, G. Friederich, F. Chavez, C. Sabine, A. Watson, D.C.E. Bakker, U. Schuster, N. Metzl, H. Yoshikawa-Inoue, M. Ishii, T. Midorikawa, Y. Nojiri, A. Körtzinger, T. Steinhoff, M. Hoppema, J. Olafsson, T.S. Arnarson, B. Tilbrook, T. Johannessen, A. Olsen, R. Bellerby, C.S. Wong, B. Delille, N.R. Bates and H.J.W. de Baar. 2009. Climatological mean and decadal change in surface ocean pCO_2, and net sea-air CO_2 flux over the global oceans. Deep-Sea Res. Pt II 56: 554–577. doi:10.1016/j.dsr2.2008.12.009.

Venkiteswaran, J.J., S.L. Schiff and L.I. Wassenaar. 2008. Aquatic metabolism and ecosystem health assessment using dissolved O_2 stable isotope diel curves. Ecol. Appl. 18: 965–982. doi:10.1890/07-0491.1.

Westberry, T., M.J. Behrenfled, D.A. Siegel and E. Boss. 2008. Carbon-based primary productivity modeling with vertically resolved photophysiology. Glob. Biogeochem. Cycles. 22. doi:10.1029/2007GB003078.

Williams, P.J.LeB., P.D. Quay, T.K. Westberry and M.J. Behrenfeld. 2013. The oligotrophic ocean is autotrophic. Annu. Rev. Mar. Sci. 5: 535–549. doi:10.1146/annurev-marine-121211-172335.

Worden, A.Z., M.J. Follows, S.J. Giovannoni, S. Wilken, A.E. Zimmerman and P.J. Keeling. 2015. Rethinking the marine carbon cycle: Factoring in the multifarious lifestyles of microbes. Science 347: 1257594. doi:10.1126/science.1257594.

Yarish, C., S. Redmond and K.K. Jang. 2012. Gracilaria Culture Handbook for New England. Wrack Lines. pp 72.

CHAPTER 5

Microzooplankton

Functionally Diverse Protists in the Pelagic Realm

María Celeste López-Abbate

5.1 Introduction

In the last four decades, several scientific discoveries significantly improved the understanding of the ecology of ocean microbes. These new insights challenged the traditional view of pelagic marine food webs characterized as the linear sequence of nutrients, phytoplankton, zooplankton and fish. In 1974, Lawrence Pomeroy presented new hypotheses about the operation of pelagic trophic networks and stressed that the respiration of organisms smaller than 30 μm exceeds that of microplankton (20–200 μm) by up to ten times. Likewise, the disclosure of an extraordinary diversity of prokaryotic and eukaryotic microorganisms that inhabit marine environments in low concentrations ("the rare biosphere", Sogin et al. 2006) demonstrated the enormous adaptive potential of microbial communities against different environmental scenarios. In 1983, Farooq Azam et al. proposed the term "microbial loop" to define the microbial food web as a parallel energy conduit to the classical phytoplankton-zooplankton-fish pathway, where microzooplankton act as a trophic link between the pico- and nano-sized communities and larger zooplankton. Today, we know that microzooplankton are ubiquitous in marine environments and comprise an enormous functional diversity (Caron et al. 2016). Their feeding plasticity allows them to prey on bacteria, picoplankton and all kinds of phytoplankton and other microzooplankton species, with many species able to combine both phototrophy and heterotrophy. We also know that microzooplankton are key components within the microbial food web, and that this diverse network determines the fate of carbon and provides nutrition, along with large phytoplankton, to the rest of the pelagic food web (Worden et al. 2015).

This chapter focuses on microzooplankton, a complex and phylogenetic diverse group. Accordingly, while going through the different sections, the reader will note that such complexity prevents to state mechanistic and functional generalizations. As an attempt to illustrate traits, functional responses and ecological strategies, this chapter will show a bias towards ciliates (including the subclasses Oligotrichia and Choreotrichia) and dinoflagellates, whose numerical dominance in a wide range of marine environments has motivated a vast amount of research (Sherr and Sherr 2002). In the following sections, the main aspects that shape the life of microzooplankton are discussed, starting from individual- (feeding and locomotion) and community-levels processes (spatial distribution and predation), until reaching the group-specific responses that may influence whole ecosystems (feedbacks on carbon cycles and climate change). But first, a brief review concerning the sometimes problematic delimitation of this group is provided.

Instituto Argentino de Oceanografía (CONICET-UNS), Camino La Carrindanga km 7.5, 8000 Bahía Blanca, Argentina.
Email: clopezabbate@gmail.com

5.2 Classification

According to the classic plankton classification established by Sieburth et al. (1978), most heterotrophic protists are included within the microzooplankton fraction, which is defined by the size range between 20 and 200 μm. However, this fraction of marine microorganisms encompasses a heterogeneous group of heterotrophic protists, including mixotrophs, represented by ciliates, tintinnids, dinoflagellates, foraminiferans, radiolarians and euglenophytes, as well as some small metazoans such as rotifers and small larval stages of mero- and holoplankton. The first taxonomists created a classification of protists that was based on morphological characters (such as the presence of flagella or cilia, skeletal structures or the mode of swimming) and on the mode of nutrition (phototrophic vs. heterotrophic). However, after the 1980s, with the emergence of molecular techniques, it was recognized that this group encompasses different phylogenetic groups that are interspersed with multicellular forms of the domain "Eucaria" (Woese et al. 1990). Recent findings show that protists include organisms with a great variety of sizes, morphotypes and trophic habits, and that they represent most of the eukaryotic diversity in the pelagic ocean (Caron et al. 2012, Adl et al. 2019). Protists evolved from an osmotrophic or saprotrophic form, and extant forms radiated into morphologically and functionally diverse groups such as flagellates, dinoflagellates, ciliates, radiolarians and foraminiferans. The complex genetic and functional variation of protist has complicated the definition of the term *microzooplankton* in either ecological or operative terms. In an attempt to solve this problem, Mitra et al. (2016) introduced a non-phylogenetic classification consisting of "functional types". These functional types are groups of organisms of different lineages that respond similarly to environmental factors and have comparable ecological functions. Given the significance of protists in marine biogeochemistry and trophic dynamics, the authors focused on the trophic modes instead of morphological or phylogenetic traits that would group organisms in categories that do not have a homogeneous impact among relevant ecosystem-level processes. This function-based classification includes protists with the ability to consume prey by different mechanisms, although this may not be the unique energy acquisition method. That is, in addition to prey consumption, mixotrophs may acquire energy by photosynthesis. Mixotrophs are grouped into two main classes: those with the innate capacity to photosynthesize (constitutive mixotrophs) and those in which the photosynthetic ability is derived from their prey (non-constitutive mixotrophs). Accordingly, Mitra et al. (2016) defined the following categories: phago-heterotrophs, constitutive mixotrophs, generalist non-constitutive mixotrophs, plastidic specialist-non-constitutive mixotrophs and endosymbiotic specialist-non-constitutive mixotrophs. To date, this classification offers a close representation of natural microzooplankton communities, thus providing a conceptual framework that serves to optimize biogeochemical models (Mitra et al. 2014). Although the traditional use of the term *microzooplankton* refers to heterotrophic protists and small metazoans within 20–20 μm, here, the term *microzooplankton* is used to enclose protists within the functional types listed above and with no size delimitation. Parasitic forms and small metazoans are not included.

5.3 Small-scale driving forces

In the terrestrial environment, the resistance exerted by air is negligible and the main force an object or an organism must undertake in order to move is the resistance to change in velocity, or inertia. Unlike terrestrial animals, aquatic organisms move into a much denser environment, which sets the physical boundaries of movement. In addition, aquatic organisms are faced with the vertical dimension of the medium, and gravitational forces in water are counterbalanced by friction resistance, buoyancy devices and viscous forces. At the microscopic scale, inertial forces are almost insignificant and microorganisms must overcome the viscosity exerted by seawater, while they become less dependent on gravity and more dependent on currents (Kiørboe 2008). The significance of inertial *versus* viscous forces is defined by the Reynolds's number (Re). Re varies directly with the dimension of the moving object and their velocity, and is inversely proportional to the kinematic viscosity of seawater. Below Re = 1, the main resistance to movement is produced by viscosity, while inertia may be considered as negligible. For example, a ciliate swimming at 50 μm s^{-1} and with a mean diameter of 30 μm has a Re of approximately 10^{-3}, which is largely lower than the Re threshold of 1. This means that a ciliate would have to expend less energy to avoid sinking than, for example, a copepod (Re $\geq$ 1), but they will also be forced to constant propelling in order to overcome the diffusivity of momentum. That is, the dependency of micro- and nano-sized organisms on ambient viscosity will allow them to start movement fast, but to stop almost immediately after propelling ceases (Kiørboe 2008).

In view of this trade-off, one may answer why so many protists are motile. One of the most important advantages of motility is that it increases the chances to encounter prey and mates. In fact, the encounter rate depends on motility, including both the speed and the trajectory of encountering particles (e.g., ciliate/prey, bacteria/solute), and the motion of the ambient fluid (Visser and Kiørboe 2006). Interception of prey by motile protists is dominated by advection, i.e., constant movement of water around the cell body, and although diffusion may act together with advection in some cases, it can be largely ignored. Diffusion feeding is performed by non-motile heterotrophs such as foraminiferans and heliozoans that depend on prey, typically bacteria, approaching them. Advective encounter, in contrast, implies a net movement of water relative to the organism, and can be produced by the swimming or sinking of organisms or by turbulence of ambient seawater (Kiørboe 2011).

Sinking and positive buoyancy, as that performed by the dinoflagellate *Noctiluca scintillans*, produces a constant flow around the cell body. For organisms with low Re, sinking of quasi-spherical cells produces a regular and symmetrical laminar flow, the so called Stokes' flow (Fig. 5.1), whose velocities increase from the centre of the body (where velocity equals zero) towards the sides away from the body, until reaching ambient flow velocity. A similar flow around the cell body is created by the swimming of flagellates. This flow allows the capture of prey traveling in streamlines within one prey radius from the central axis of the predator cell. In each case, encounter rate is proportional to the prey size, while independent of the predator size.

Swimming produces a feeding current around the cell body, thus accelerating the clearance of new water parcels. It also creates a specific flow field, thus exposing the cell body to different concentrations of prey compared with a settling organism (Fig. 5.1). The enhancement of prey uptake due to flow can be estimated by the Sherwood number (Sh), defined as the ratio of uptake in the presence of flow to uptake solely due to diffusion. The Sherwood number is often expressed as a function of the dimensionless Péclet number (Pe), which estimates the relative importance of advection *versus* diffusion of solutes. Model comparisons on the efficiency of prey uptake of sinking *versus* swimming consumers show that

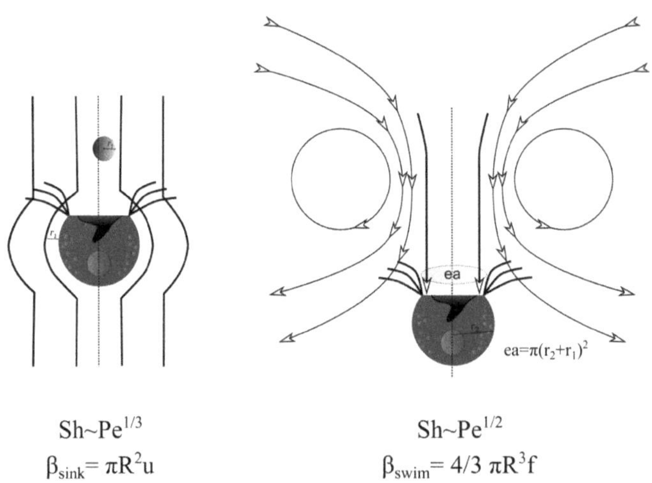

Figure 5.1. Idealized models of water currents around the cell body generated by sinking and positive buoyancy (left) and by swimming (right) ciliates. The scaling of the Sherwood number (Sh) with the Péclet number (Pe) denotes the enhancement of prey uptake due to movement behaviour. Encounter rate kernel or clearance rate (β, $mm^3 s^{-1}$) also varies among behavioural prey encountering. Clearance rate on sinking organisms (β_{sink}) varies with R, which is the separation distance between the predator and their prey below which an encounter is deemed to occur (R), and the velocity difference between predator and prey (u). Note that R is dependent on the prey radius. The frequency of stops in random walkers (f) comes into play for defining the clearance rate on swimming organisms (β_{swim}). Modified from Kiørboe (2008) and Montagnes (2012).

mass transport of a sinking quasi-spherical protist of low Re, scales as Sh ~ $Pe^{1/3}$. In contrast, a similar protist generating propulsion force will experience a significant enhancement of prey uptake that scales as Sh ~ $Pe^{1/2}$ (Guasto et al. 2010). This enhancement is also reflected by comparing the encounter rate kernel or clearance rate (β) of sinking *versus* swimming organisms (Fig. 5.1).

Swimming has additional advantages since it can be strategically modified according to organism's needs. Most motile protists do not perform guided swimming but instead, they swim along straight lines interrupted by sudden turns or tumbles. This mode of locomotion resembles that of the random walk of colliding molecules driven by Brownian diffusion, and hence the term "random walk" has been frequently coined to describe the swimming of motile protists. Conversely, ciliates and dinoflagellates may eventually change their random walk behaviour by changing their speed and the frequency of turnings when a food replete parcel is encountered (Buskey and Stoecker 1989, Fenchel 2001). The functional response is given by the kinetic recognition of a pattern rather than by the chemical detection of clues.

In additions, some naked oligotrichs and flagellates may implement sudden jumps in response to an approaching predator (Broglio et al. 2001, Jakobsen 2002). However, swimming also creates fluid disturbances that increases the probability of being detected by predators, and low-Re organisms produce larger fluid signals relative to organisms moving at higher Re values. Tintinnids are more vulnerable of being detected due to the presence of a bulky lorica that increases the intensity of fluid signals. Also, in most protists, locomotion and grazing are necessary linked, and are thereby more vulnerable of being detected by predators in comparison with protists that only swim for locomotion (Kiørboe et al. 2014). Organisms that need to swim in order to feed (propulsion feeders) generate constant flows with high attenuation timescales, thus deforming the far field spatial flow and increasing the detection distance of predators sensible to hydromechanical signals. Conversely, non-motile predators such as foraminiferans, heliozoans and radiolarians, create less fluid deformations but rely on diffusion feeding and the approaching of prey towards the cell body. Hence, their capture efficiency is proportional to predator and prey cell radius. Cell size is optimized when the net energetic gain and metabolic expenditure are balanced, and once the optimal size is reached, the cell continues to grow and divide. Optimum cell size will increase with the increase in solute and prey concentration. At a certain size threshold (bacteria), Brownian motion becomes the main force driving the movement of organisms.

Microscale turbulence produces eddies of different sizes and may reduce the sinking velocity of suspended organisms while enhancing prey capture and nutrient uptake (Pécseli et al. 2014). Turbulence also affects the sinking rate of protists (Martínez et al. 2017). Fast-sinking tintinnids with agglomerated lorica tend to be more dependent on turbulence than slower sinking oligotrichs to retain suspended populations. Given the viscosity of seawater, the smallest eddies created by turbulence are in the order of 1 mm, meaning that unicellular protists (≤ 500 μm) will essentially experience laminar shear characterized by the absence of disruptions among parallel layers with smooth velocity gradients. Hence, turbulence will renew the water around the cell, thus increasing the probability of predator/prey encounter. Experimental data revealed that the velocity difference created by turbulence is similar to that of swimming and sinking (Kiørboe 2008). As a general rule, the scale of fluid motion that influences protists is below the Kolmogorov length scale (range from 0.1 to 10 mm). As mentioned before, below this scale, turbulence is perceived as laminar shear and its effect is weak relative to viscosity. Hence, downscaling from the Kolmogorov length scale, behaviour becomes increasingly important for prey encounter. Only predators close to the Kolmogorov length scale (e.g., copepods) are considered to benefit from water renewal produced by turbulence (Kiørboe and Saiz 1995).

5.4 Feeding strategies

The functional plasticity of marine protists partly resides on their ability to obtain energy from widely diverse sources and to adjust their nutritional mode according to the environmental conditions. They can photosynthesize, graze on prokaryotes and eukaryotes, parasite, and absorb extracellular organic matter (Caron et al. 2012). Among microzooplankton, their feeding plasticity allows them to prey on a wide range of trophic items from bacteria to large chain-forming phytoplankton or other heterotrophic protists, toxic dinoflagellates and even copepods (Steinberg and Landry 2017, Berge et al. 2012). However, given the vast amount of phototrophs both in the open and the coastal ocean (~ 70% of total producer's biomass), microzooplankton have preferentially evolved to prey on phytoplankton (Sherr and Sherr 2002). The feeding plasticity of microzooplankton is based on their fast generation times and their ability to thrive among prey items with different intracellular composition and palatability (Ger et al. 2014). The uptake of chemically dissimilar resources gives microzooplankton an important advantage over mesozooplankton, whose diet tends to be more specific in terms of taxonomic groups and size structure and whose responses to the pulses of primary producers is much slower due to their lower growth rates and complex life cycle (Calbet 2001). This does not necessarily imply that microzooplankton act as generalist consumers. In fact, microzooplankton are able to structure phytoplankton and flagellate communities by active selective feeding (Löder et al. 2011, Rose et al. 2009, Aberle et al. 2007).

The review paper from Calbet and Landry (2004) summarized the magnitude of primary productivity grazed by microzooplankton and concluded that this group is one of the main factors driving phytoplankton mortality in a wide variety of environments. On average, microzooplankton consume between 49 and 77% of the daily primary productivity in coastal and open ocean environments, respectively, while the average consumption of the mesozooplankton is 10% (Schmoker et al. 2013). In addition, microzooplankton represent the main mortality source of bacterioplankton in pelagic environments, after viral lysis (Pernthaler 2005). In spite of their relevance as the primary mortality source for phytoplankton, microzooplankton are generally unable to prevent bloom formation in meso- and eutrophic ecosystems. This is attributable to several factors. In the first place, microzooplankton attain maximum growth rates under relatively high prey concentrations, meaning that they can match prey division rate only under near bloom conditions. Under pre-bloom conditions, microzooplankton standing stock is usually low as well as cell division rate (Sherr and Sherr 2009). The time taken to match prey division rate allows phytoplankton blooming species to temporarily escape predation and form blooms. In addition, booming species are generally provided with efficient defensive skills such as spines and grazing deterrents, while the formation of colonies deters predation by protists and small metazoans (Irigoien et al. 2005). Top-down control exerted by mesozooplankton under low prey concentration may additionally reduce the ability of microzooplankton to rapidly respond to bloom formation (Saiz and Calbet 2011).

Microzooplankton are thereby forced to unfold different strategies to survive the in-between bloom periods. Ciliates may exhibit different strategies to starvation according to the nature of food patchiness. If food is spatially irregular, ciliates may increase their division rate to form swarmer cells to increase the likelihood of encountering a food patch. In contrast, other ciliate species may form cysts to improve longevity during starving conditions, although this strategy reduces motility and is thus better suited to cope with seasonal pulses rather than to resist spatial patchiness (Fenchel 1990). Ciliates can also modify their behaviour to persist under starvation. For instance, when a food patch is met, ciliates can reduce their motility and accumulate at prey aggregations (Fenchel and Blackburn 1999). In contrast, the low prey specificity of dinoflagellates determines that these consumers have a higher ability to survive under low food conditions. Still, dinoflagellates are able to form cysts under a threshold food concentration (Sherr and Sherr 2007), and many species are able to substantially reduce their metabolism and survive for long periods of time without encystment by utilizing their storage reserves. The ability of dinoflagellates to relocate energy into storage reserves and to prey among a wide size spectrum of prey allows a longer survival time under starving conditions compared to ciliate's swarmer cells (Menden-Deuer et al. 2005). Moreover, when grown in the light, starved *Oxyrrhis marina* showed a higher rhodopsin gene expression compared to dark conditions, suggesting that some dinoflagellates may sustain growth from harvesting light, independent of the acquirement of plastids from prey (Guo et al. 2014).

The encounter between predators and their prey depends on the strategy of pursue and capture. Flagellates directly intercept prey that arrives within the capture area via their feeding currents and can passively (contact probability) or actively (item avoidance and selection) take up food particles (Fenchel 1982). In diffusion feeders, prey capture depends on the approaching of prey towards the cell body. When prey approaches the cell body, non-motile organisms collect and ingest them by the extension of pseudopodia. Most ciliates are filter feeders, meaning that they generate a swimming current by the movement of oral membranelles that forces water into their cytostome cavity. Particles within the capture volume larger than the distance between membranelles are effectively retained and ingested (Fenchel 1980a). Dinoflagellates, in turn, perform different feeding strategies; for example, species of the genera *Gyrodinium, Gymnodinium* and *Noctiluca*, gobble up their prey directly and digest it within the cell. The digestion of prey can also be extracellular by means of a cellular extension or pallium (e.g., *Protoperidinium, Diplopsalis, Diplopelta* and *Oblea*) or by means of tubes or peduncles (e.g., *Amphidinium, Gyrodinium, Gymnodinium* and *Dinophysis*), which in turn allow to take a whole prey or fractions of them (Hansen and Calado 1999). Species of the genus *Alexandrium* produce a toxic mucus trap to capture and immobilize their prey (Blossom et al. 2012). Moreover, many ciliates and dinoflagellates contain functional chloroplasts (Stoecker et al. 2009), which gives them an additional advantage under certain environmental conditions (see Section 5.5 "Mixotrophy: A pervasive mode of nutrition in protists"). The beating of flagella in dinoflagellates creates a feeding current that enhances prey encounter, and this behaviour has the additional advantage of increasing the renewal of nutrient-depleted water around the cell body. This constant renewal may increase nutrient uptake by 75%, giving plenty of nourishment to the mixotrophic way of life (Nielsen and Kiørboe 2015).

Prey abundance is a critical factor that determines the encounter rate between predators and their prey. Hence, in addition to modifications to their swimming behaviour, microzooplankton regulate their ingestion rate (I) in response to prey concentration (P) in order to balance the energetic cost of swimming and food acquisition. The functional response of zooplankton, i.e., the dependence of prey ingestion rate on prey concentration, has implications on carbon budgets and has been theoretically and experimentally described (Kiørboe 2008, Gallegos 1989, Fenchel 1980b). Typically, the ingestion rate of heterotrophic protists increases linearly with prey concentration until a saturating threshold, in which the consumption, handling and/or processing becomes saturated. After this threshold, I reaches a maximum and constant value (I_{max}). Since the clearance rate (C) is defined as $C = I/P$, further increases in P will result in a decrease in the clearance rate. This type of functional response resembles that of the Michaelis-Menten model kinetics, and is termed Type II functional response (Fig. 5.2), according to the classification of Holling (1959). Deviations from this functional response may appear when the concentration of prey is below the consumption threshold of heterotrophs (Moigis 2006). In fact, a certain prey concentration is required to activate feeding in many oligotrichs (Dolan et al. 2000) and dinoflagellates (Davidson et al. 1995) and hence, once this critical concentration is attained, the ingestion rate starts to respond linearly to the concentration of the food. This type of functional response fits well with the Type III curve (Fig. 5.2).

Prey concentration can also stimulate another type of grazing behaviour based on active prey selection. Intuitively, the primary metrics defining prey selection in ciliates is size because ingestion of particles is constrained by the oral diameter. The prey:predator size ratio in ciliates is approximately 1:8, while the prey size spectrum (ranging from 2.5:1 to 30:1) is centred at smaller ratios than that of flagellates (range from 2:1 to 8:1, mean of 1:3) and dinoflagellates (range from 0.4:1 to 7:1, mean of 1:1)

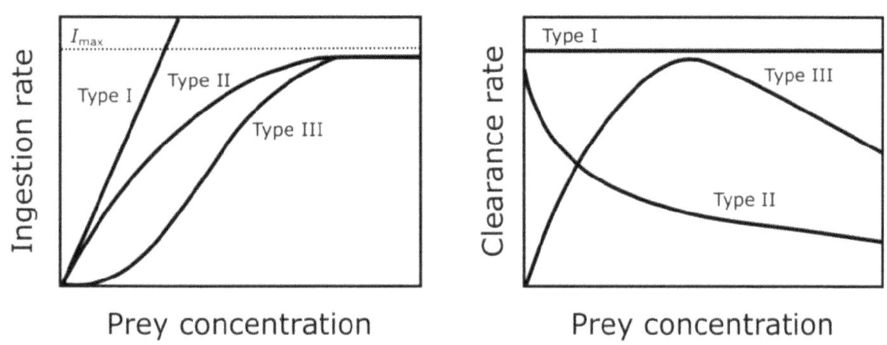

Figure 5.2. Functional responses defined by Holling (1959) denoting the evolution of the ingestion (expressed as the number of prey ingested per unit time) and clearance rates (expressed as the volume of water cleared of prey per unit time). Type I functional response assumes a linear increase in the ingestion rate with food density, denoting no handling and consumption time limitation. Types II and III functional responses are characterized by a threshold prey concentration in which I asymptotically stabilizes (I_{max}) due to predator-specific handling time.

(Hansen et al. 1994). Dinoflagellates are considered as relatively non-selective feeders; however, group-specific feeding strategies determine the scope of targeted prey. Species that directly engulf food particles (e.g., *Gyrodinium* spp.) select their prey according to their own body size. Both pallium and tube feeders are able to feed on prey 5–8 times larger than themselves, while the lower limit of prey size (2–4 μm) is set by their detection limit and is similar in all feeding types (Hansen and Calado 1999). Some pallium feeders such as *Protoperidinium* spp., *Oblea rotunda*, *Diplopsalis lenticula* and *Oxyrrhis marina* actively select their prey, evidencing the implementation of recognition mechanisms. These diverse feeding strategies imply that dinoflagellates may access a wider prey size spectrum and in fact constitute one of the most important grazers of large, bloom-forming diatoms in meso- and eutrophic areas of the ocean such as upwelling and coastal ecosystems (Sherr and Sherr 2007, Calbet 2008, Hansen et al. 1997). These species, as well as some ciliates, are able to respond to chemical cues or attractors that bind to specific cell surface receptors (Montagnes et al. 2008). In addition, dinoflagellates may alternate between raptorial and interception feeding according to prey motility. When offered non-motile bacteria, dinoflagellates generate feeding currents with the beating of their transverse flagellum, and directly engulf bacteria approaching them. In contrast, when available prey are motile protists, the beating of their flagellum produces a helical swimming to actively search and capture prey (Jeong et al. 2008).

5.5 Mixotrophy: A pervasive mode of nutrition in protists

The growing awareness of the capacity of marine protists in the photic zone to combine different trophic strategies has led many researchers to abandon the traditional dichotomy between phototrophy and heterotrophy. Organisms with this capacity are included in the category "mixotrophs", and hence they should be considered as both primary and secondary producers (Flynn et al. 2012). Although the relevance of mixotrophs was first stressed back in the 1980s (Stoecker et al. 1987, Bird and Kalff 1986), the ecological role and ubiquity of these organisms in aquatic sciences remained overlooked until recently. Perhaps, the fact that these studies appeared simultaneously with other "microbial revolutions" contributed to the underestimation of mixotrophy in marine and freshwater ecosystems. As a result, sampling procedures remained suboptimal for the detection of mixotrophs for a long time (Leles et al. 2019), but recent global data compilations have confirmed the widespread distribution of mixotrophs across biogeographic areas (Leles et al. 2019, 2017, Faure et al. 2019, Edwards 2019, Hartmann et al. 2012, Unrein et al. 2014). These compilations also revealed that mixotrophs have specific distributions implying that particular functional types tend to predominate under specific oceanographic regimes (Leles et al. 2019, 2017, Faure et al. 2019).

Mixotrophy has evolved from a heterotrophic ancestor which gained the ability to photosynthesise by hosting endosymbiotic algae or by retaining plastids through the phagocytosis of phototrophic prey (Raven et al. 2009). Ancestral heterotrophs appear to have lost and regained the ability to photosynthesize several times in their evolutionary history. Hence, in spite of the common misconception that mixotrophs evolved by combining ancestral heterotrophic and phototrophic forms, they truly embody functional forms that lost their exclusiveness for heterotrophy (Flynn et al. 2019).

This long evolutionary history has shaped a wide spectrum of mixotrophic strategies for energy acquisition, and recognizing different mixotrophic functional types among heterotrophic protists is fundamental because they occupy different ecological niches and have different impacts on element budgets (Leles et al. 2018). Accordingly, Mitra et al. (2016) proposed a classification of mixotrophs based on their physiological ability to combine trophic modes (Fig. 5.3). Constitutive mixotrophs (CM) are considered as phytoplankton with the ability to consume other organisms since they are fully able to synthetize their plastids and have the innate capacity to photosynthesise. In contrast, non-constitutive mixotrophs (NCM) are regarded as zooplankton that acquired the photosynthetic ability from the consumption of phototrophic prey. Between 40–60% of protists traditionally viewed as heterotrophic are actually NCM (Leles et al. 2017). Generalist NCM (GNCM) retain chloroplasts from diverse prey types, such as the majority of plastid-retaining oligotrich ciliates (Dolan and Pérez 2000). This group is tightly dependent on the availability of prey to engage with photosynthesis and to accomplish stable energy acquisition from phototrophy since functional plastids are retained for up 2 weeks (Schoener and

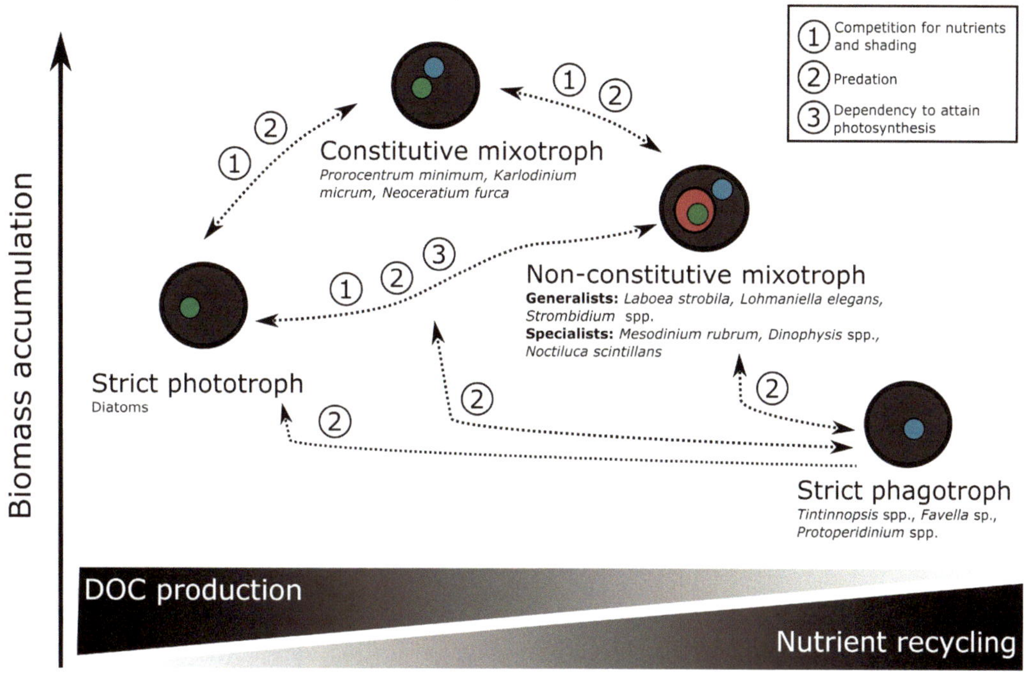

Figure 5.3. Conceptual model of ecological interactions and environmental feedbacks associated with different types of protists according to the classification proposed by Mitra et al. (2016).

McManus 2012). The acquisition of photosynthetic ability in specialists NCM (SNCM) can occur either by the retention of chloroplasts from their prey or by the retention of phototrophic prey as symbionts. Plastidic SNCM steal chloroplasts from specific prey, and typically have the capacity to retain functional plastids for longer time periods than GNCM. For instance, *Mesodinium rubrum* may retain functional plastids for up to 80 days in the absence of prey (Myung et al. 2013). Endosymbiotic SNCM harbour dinoflagellates, haptophytes, or green algae as symbionts during most of their life cycle and important representatives of this group are Foraminifera and Radiolaria (Mitra et al. 2016).

Although mixotrophy is generally regarded as a strategy to improve nutrient uptake in oligotrophic systems (Unrein et al. 2014), mixotrophic plankton are ubiquitous in aquatic systems (Faure et al. 2019) and can also prevail under highly variable resources supply rates (Selosse et al. 2017). This is based partly on the fact that the degree of phototrophic and heterotrophic activities is not predetermined but it is rather a direct response to the environmental conditions, such as light, nutrients and prey availability. CM and endosymbiotic SNCM are the most abundant among mixotrophic types, while GNCM revealed an almost ubiquitous distribution (Faure et al. 2019). In the Atlantic Ocean, it is presumed that mixotrophic protists can overcome the biomass of heterotrophic protists, and therefore can exert a greater control over the productivity of their prey (Hartmann et al. 2012), while in polar seas, mixotrophic ciliates usually dominate total oligotrichs biomass (Stoecker and Lavrentyev 2018).

The ability of mixotrophs to alternate among the utilization of CO_2 and particulate organic carbon (POC) from suspended prey adds a complexity level that needs to be adequately identified when modelling carbon pathways (Mitra et al. 2014). For instance, while phototrophic activities are associated with the production of dissolved organic carbon (DOC), heterotrophy enhances inorganic nutrient regeneration; hence, the extant degree of phototrophic/heterotrophy will imply important consequences on carbon and nutrient balance (Fig. 5.3). Moreover, CM exhibit a high potential of bloom formation and biomass accumulation since they can offset nutrient exhaustion after seasonal bloom events by the ingestion of prey. Modelling approaches assuming no nutrient limitation have revealed that CM can enhance primary productivity by 50% (Mitra et al. 2014). However, since many CM can form harmful algal bloom and produce toxic substances, or even outcompete heterotrophs by the rapid uptake of prey, carbon transfer toward higher trophic levels can, in some cases, be reduced with mixotrophy (Burkholder et al. 2008). On the other hand, NCM have higher gross growth efficiencies than strict heterotrophs and can double their biomass in natural conditions, specially under low food supply (Stoecker et al. 2009). This means that, when present, NCM can significantly increase the carbon transfer toward higher trophic levels (Stoecker et al. 2017). Eventually, NCM can also cause harm to environments by outcompeting resident phytoplankton. For instance, the influx of hypoxic waters into the Arabian Sea led to unprecedented blooms of green *Noctiluca scintillans,* facilitated by higher carbon fixation efficiency of endosymbionts under low oxygen compared to that of diatoms (do Rosário Gomes et al. 2014).

Attempts to model planktonic food webs have revealed that the competitive advantage of dual mode of nutrition produces a shift toward larger plankton (Ward and Follows 2016). The traditional dichotomy between phototrophy and heterotrophy states that large phytoplankton is outcompeted by small, high nutrient-affinity species. However, with the presence of mixotrophs, nutrient-limited photosynthesis of large cells is supplemented by the uptake of prey. The sustained production of large, fast-sinking cells, even after nutrient exhaustion, increases the export of carbon toward the bottom, and the inclusion of mixotrophs in model simulations has revealed that this group may be responsible of ~ 35% increase in global carbon export relative to the phototrophic/heterotrophy model (Ward and Follows 2016). It is worth noticing that the above model assumed all plankton to be mixotrophic, with no delimitation among functional types, thus introducing an overestimation error of the potential impact of mixotrophs. Also, functional types are expected to cause contrasting impacts on carbon budgets according to the balance among trophic strategy, the innate vs. the acquired capacity to photosynthesize, the specificity of prey from which plastids are acquired and the nature of the limiting resource (Leles et al. 2018). The incorporation of traits-based models is therefore fundamental to optimise the projection of carbon export in the ocean.

In spite of improvements made in the recognition of the ubiquity of mixotrophy among aquatic protists and in the understanding of their impacts in biogeochemical cycles, many challenges remain to be solved. For instance, how mixotrophs can coexist with their strict phototrophic and heterotrophic competitors and to what extent their ability to shift between phototrophic vs. heterotrophic nutrition will allow them to cope with growing environmental variability, are open fields of intense research (Edwards 2019, Leles et al. 2018).

5.6 Biogeography

The delimitation of geographic boundaries of plankton distribution is often problematic since although no significant dispersion limits exist in the ocean, truly cosmopolitan species are actually rare (Foissner 2006). At the same time, a review analysis revealed that only 8.5% of ciliates' morphospecies exhibit restricted distribution or endemism (Agatha 2011). Further difficulties arise from the fact that biogeography of microzooplankton is frequently based on taxa defined by morphological traits that may not fully match genetically-defined species (Dolan 2005). In fact, protists encompass such a wide phylogenetic diversity that common biogeographic patterns are exceptional since each taxon exhibits its unique dependence on environmental conditions (Caron 2009). Biogeography studies made so far, however, revealed that identifiable patterns emerge from the relative abundance of microbes rather than from the presence or absence of taxa (Dolan 2005).

In spite of the scarcity of sustained and spatially-wide observations on microzooplankton distribution, some global patterns have been reported in the literature. The latitudinal distribution of tintinnids diversity, for instance, is characterized by minimum species richness at high latitudes, which gradually increases until maximum values at 15–20° and decreases to some extent at the equator (Dolan and Pierce 2012). A similar distribution is found in dinoflagellates (Taylor et al. 2008), foraminiferans (Be 1977) and radiolarians (Boltovskoy et al. 2010), and the pattern is partly attributed to the temperature range effect on metabolic rates. Clear distribution patterns do not hold true for all clades. For instance, the Tara Oceans expedition which uncovered unprecedented taxonomic discrimination of the majority of ciliate's clades, including tintinnids, reported the absence of a clear latitudinal diversity gradient of planktonic ciliates (Gimmler et al. 2016). The analysis of DNA sequences in nearly all oceanic regions showed a close correlation between ciliate's diversity and environmental factors of local relevance (temperature, chlorophyll, salinity, and inorganic nutrients) but no clear large-scale latitudinal gradients. The authors found that only 17% of taxa were distributed among all oceanic regions, and that cosmopolitan ciliates had higher local abundances than ciliates with a restricted distribution, although in general, open-water species showed low diversity. Similarly, a lack of global distribution patterns was revealed for mixotrophic ciliates, even though mixotrophic forms revealed a higher dominance towards high latitudes than their strict heterotrophic counterparts (Leles et al. 2017). The lack of phylum-based gradients, however, does not imply that taxon-specific patterns, such as that of tintinnids, actually exists (Gimmler et al. 2016). A global compilation of *in situ* imaging revealed that Rhizaria constitutes a significantly higher proportion of planktonic carbon in the upper 200 m than previously thought (Biard et al. 2016). This compilation also revealed a latitudinal gradient, with maximal values observed near the Equator. In addition, Biard et al. (2016) observed that the contribution of mixotrophic Rhizaria to the total Rhizaria biomass decreased towards the poles, in contrast with observations made by Leles et al. (2017), who described a higher relative contribution of mixotrophic ciliates to total ciliate biomass towards the tropics.

It has been argued that the global diversity gradient of tintinnids results from the functional redundancy of species at low and mid-latitudes (i.e., high number of species with similar preferred prey size) that decreases toward higher latitudes (Dolan et al. 2016). Among open ocean waters, the contribution of tintinnids to total ciliate abundance is higher in productive areas including the tropical and subarctic regions (Sohrin et al. 2010). In addition to species diversity, functional traits may also define geographic pattern. As a general trend, most tintinnid species that show a quasi-cosmopolitan distribution form a hyaline lorica, while species with agglutinated lorica dominate in coastal environments and warm and temperate climate regimes (Dolan and Pierce 2012). Likewise, warm-water tintinnid species generally display ornate and complex morphology in order to avoid sinking in a less viscous medium. Dinoflagellates show remarkable similarity of species diversity among the southern and northern hemispheres. However, differences arise between neritic and oceanic regions due to the dominance of cyst-forming species in shallow and continental shelf areas (Taylor et al. 2008). Collodarians (Radiolaria) also display large-scale distribution patters. These organisms follow the general latitudinal distribution of radiolarians, with maximum species richness at 10–30°. The highest diversity occurs in the open ocean, with dominance of skeleton-bearing forms, while neritic areas are characterized by lower species diversity and the dominance of un-silicified forms (Biard et al. 2017).

Although mixotrophic ciliates are not regularly distributed at a global-scale, GNCM and plastidic SNCM showed a comparable distribution restricted to productive, coastal waters, while CM and endosymbiotic SNCM are ubiquitously distributed (Leles et al. 2019, 2017, Faure et al. 2019). However, the review made by Edwards (2019) revealed that CM increase in abundance at lower latitudes compared to their strict photo- or heterotrophic competitors and are more frequently encountered in eutrophic, shallow habitats and subtropical open-waters. Among CM, specific traits other than trophic strategies come into play. For example, larger CM are common in coastal waters, particularly in eutrophic and turbid ecosystems, while smaller CM can dominate in the open oceans, within clear and nutrient-poor waters (Leles et al. 2019). On the other hand, endosymbiotic SNCM are largely dominant in oligotrophic, open-waters (Leles et al. 2017).

Distribution patterns also vary at the temporal scale. Tintinnids show a succession pattern that is characterized by the dominance of species with agglutinated lorica in summer that are gradually replaced with species with hyaline lorica in winter. Maximum species richness occurs at transitional periods (spring and autumn), presumably as a result of more diverse ecological niches. Primary productivity is another factor that contributes to increase the richness of tintinnids species; however, many regional-scale exceptions to this rule are reported and the correlation is only occasionally valid for open waters (Dolan and Pierce 2012). Among coastal waters, GNCM are also associated with summer, post-bloom conditions, while plastidic SNCM are dominant during spring, when nutrients are close to saturation (Leles et al. 2017). This is thought to be related with the specificity of prey from which mixotrophs acquire their phototrophic potential as well as the lifetime of functional plastids (Leles et al. 2018). Microzooplankton also display distributional patterns along the water column. The vertical distribution of ciliates and phagotrophic nanoflagellates follows a power function with maximum values at surface layers that decrease until 1000 m and remain variable or constant afterwards (Nagata et al. 2010). Significant abundances of both ciliates and phagotrophic nanoflagellates have been found in the bathypelagic zone (2500 m) in close association with high concentrations of bacteria (Sohrin et al. 2010).

5.7 Importance of microzooplankton as prey

Microzooplankton represent an essential food resource for mesozooplankton in spite of their lower concentration compared to that of phytoplankton in natural environments (Calbet and Saiz 2005). Ciliates are typically the preferred prey of copepods among other microzooplankton species and they represent between 3 and 55% of mesozooplankton diet, while the contribution of dinoflagellates varies between 4% and 19% (Saiz and Calbet 2011, Calbet and Saiz 2005, Broglio et al. 2004). In general, the contribution of ciliates to the diets of mesozooplankton is greater in oligotrophic environments (~ 50%), while in productive environments (2.8–18%), diatoms represent the main trophic item (Saiz and Calbet 2011). Certain specific characteristics of prey such as body size, swimming mode and nutritional quality have been recognised as crucial factors determining the active grazing selectivity of mesozooplankton on microzooplankton (Stoecker and Capuzzo 1990).

Prey size is perhaps the most important factor that determines prey selection in plankton, since most consumers are size-limited by their feeding and handling apparatus (Hansen et al. 1994). The size relationship between planktonic prey and predator usually ranges around 0.1; however, this ratio varies among wide boundaries depending on the type of predator. For instance, copepods may feed on particles that are three to four orders of magnitude smaller than themselves (Wirtz 2012). Among the appropriate size spectrum of prey, large motile organisms are usually the preferred items since they move faster and create larger fluid deformations and a higher probability of being detected by predators. Feeding mode (i.e., suspension vs. ambush feeding) also affects the allometric relationship between predators and prey. Larger moving prey are usually actively grazed by ambush feeding copepods, while high concentrations of smaller, non-motile prey are preferred by suspension feeders (Saiz and Kiørboe 1995, Jonsson and Tiselius 1990).

Motile prey creates far field distortions that are perceived as hydromechanical cues by ambush feeders such as many copepods and chaetognaths (Jakobsen et al. 2005). In turn, various protists can avoid predators by detecting fluid deformations created by them (Jakobsen 2002). To minimize the escape response of protists, ambush feeders reduce fluid deformation in comparison to predators that create constant feeding currents by passively waiting for cues and then jumping to attack their prey. Simulations have demonstrated that fluid deformations created by these sudden jumps is only seldom detected by protistan prey and that jump velocity is approximately 10 times faster than protists' escape movement (Kiørboe et al. 2009). The fastest the jump, the thinner the viscous boundary layer that forms around the jumping predator and the lower the likelihood of being detected by prey beforehand. Since jump velocities scale with body size, the thickness of the boundary layer increases dramatically in small copepods (Kiørboe et al. 2009). In contrast, copepods that generate feeding currents produce large fluid deformations that prompt motile prey to undertake its escape. Hanging predators typically do not react to moving prey and the velocity of feeding currents is in most cases well below protists' escape velocities (Jakobsen et al. 2005). Overall, the active predation on motile protists depends on the feeding mode of predators, the fluid deformation created by both prey and predator, the escape velocity of protists, and the hydromechanical sensitivity of predators.

The third aspect that may influence positive selection upon microzooplankton is their nutritional quality. The ability of mesozooplankton to select nutritionally rich prey is difficult to demonstrate in field and experimental studies (Isari et al. 2013). Although no clear evidence exists on sensory and behavioural responses to the biochemical composition of prey, experimental data demonstrate that the nutritional value of prey does impact egg production and fecundity of copepods (Tang and Taal 2005). For instance, observations made under culture conditions have demonstrated that many crustaceans are not able to survive or reproduce by feeding only upon cyanobacteria or certain algal species. This implies that a trophic intermediary is needed to fulfil metabolic requirements of these animals (e.g., Koski et al. 1998, Martin-Creuzburg et al. 2005b). In simple chemical terms, the quality of prey can be defined by the intracellular ratio of C:N:P, which depends on the concentration of nutrients in the environment. Major disparities on internal element concentration exist between producers and consumers, and such disparities are predominantly evident for P and N (Hessen et al. 2013). The smaller the difference in the elemental stoichiometry of prey and consumers, the higher the quality of food resources (Sterner and Elser 2002). Microzooplankton are able to regulate their internal element stoichiometry among narrower boundaries than primary producers (Marki and Pahlow 2016) and usually are stoichiometrically closer to higher predators (Stoecker and Capuzzo 1990). Mixotrophs were also observed to buffer environmental stoichiometric unbalances by a higher capacity to constrain cellular nutrient ratios compared to primary producers (Moorthi et al. 2017). Experimental data revealed that different life stages of copepods are able to select prey according to its stoichiometry (Meunier et al. 2016). While rapidly growing nauplii tend to select P-rich prey, copepodites, that redirect relatively less nutrients to somatic growth, tend to select N-rich prey. Video-imaging trials revealed that *Acartia tonsa* can switch its swimming behaviour according to the N:P ratio of their prey, and that the responsiveness to prey stoichiometry improves with age (Herstoff et al. 2019).

In natural environments, feeding selectivity towards nutritionally rich prey has been observed after nutrient pulses. An external input of limiting nutrients can generate an exaggerated element uptake known as luxury consumption, in which prey undergoes an internal stoichiometric shift. This shift translates into an almost immediate stimulation in the production of their predators in a process known as trophic tunnelling (Thingstad 2005). But nutrient stoichiometry is not the only aspect relevant to mesozooplankton nutrition, and the presence of complex molecules such as lipids must be obtained from their prey. Some species of heterotrophic dinoflagellates (Klein Breteler et al. 1999) and ciliates (Martin-Creuzburg et al. 2005b) can produce specific essential fatty acids and sterols absent in phytoplankton, thus upgrading the food composition for mesozooplankton. Although many microzooplankton species, specially ciliates, may not be able to *de novo* synthetize complex fatty acids and will end up mimicking the composition of their prey (Klein Breteler et al. 2004), others can upgrade lipid molecules acquired from phytoplankton (Roohani et al. 2019). Consequently, as a strategy to cope with low nutritional quality of prey, many planktonic consumers adopt an omnivorous behaviour (Siuda and Dam 2010). In turn, some protists may reduce the performance of copepods through the effect of toxins, as seen by the reduction of gross growth efficiency of copepods fed with the mixotrophic *Karlodinium veneficum* (Traboni et al. 2020). Considering microzooplankton as prey, their key properties including body size, swimming behaviour, a closer cellular stoichiometry to that of higher consumers and the ability to upgrade the food quality for metazoans, partly offset their low numerical abundance in natural environments compared to phytoplankton of similar size. Some dinoflagellates are not only responsible for repackaging food into more suitable prey in terms of size and motility, but they may also produce specific fatty acids and sterols absent in phytoplankton, thus upgrading the food composition for mesozooplankton.

5.8 The role of microzooplankton in the ocean's carbon pump

When considering ocean's carbon cycle in which biological activity takes fundamental relevance, the classical view that compartmentalizes plankton as phytoplankton and microzooplankton results inadequate. In fact, unicellular plankton are functionally entrenched within the so called "ocean microbes", which includes archaea, bacteria, cyanobacteria and protists (Caron et al. 2009). Recognizing that unicellular eukaryotes are components of an interacting network with diverse microbial lifestyles, offers a wider perspective that helped researchers to create a more holistic conceptual model of ocean's carbon cycle (Worden et al. 2015). Ocean microbes take part essentially in every POC/DOC transformation process by unfolding a wide spectrum of functional traits. The emerging roles of microbes in carbon cycles, however, not only depends on their individual metabolic rates and trophic modes, but also, and perhaps most importantly, on microbe's interactions.

Carbon enters the ocean mainly by the assimilation of atmospheric CO_2 by primary producers, although assimilation of sedimentary dissolved inorganic carbon (DIC) and continental transport are also important. Hence, at this point, carbon transfer pathways depend largely on metabolic rates of phototrophs which, in turn, respond to the availability of nutrients. During phytoplankton growth, the exudation of labile and semilabile DOC provides an important source of reduced carbon that supports the growth of heterotrophic bacteria in the euphotic zone (Seymour et al. 2017). This means that under certain circumstances, phytoplankton and bacteria may collaborate (by the bacterial use of DOC exudates) or compete (by the joint utilization of shared resources, i.e., N and P). In general terms, the increase in non-shared resources (DOC and Si) will stimulate the development of its specific consumers (heterotrophic bacteria and diatoms, respectively), while the increase in shared resources (inorganic nutrients) will stimulate the development of all its consumers and give place to various interspecific relationships such as competition and differential predation by microzooplankton (Havskum et al. 2003).

These interactions are mechanistically illustrated by the model proposed by Thingstad et al. (1997). The model is based on the availability of inorganic nutrients, and has been widely supported by field and experimental observations. Starting from the oligotrophic extreme of an environmental trophic gradient, the model states that bacteria overcome the biomass of primary producers as they are more efficient than diatoms and other large phototrophs in the use of nutrients at low concentrations (left side of Fig. 5.4). As a result, much of the organic matter from the photic zone is remineralized and retained in surface waters (Legendre and Rassoulzadegan 1995). As the concentration of nutrients and the degree of eutrophication of the system increases, phototrophs respond proportionally by generating significant amounts of DOC through the exudation of organic molecules. This additional pool of carbon stimulates bacterial growth that, in turn, provides nutrients by remineralization processes, resulting in a relationship of commensalism and coupling. Simultaneously, an increase in system's productivity is accompanied by a proportional

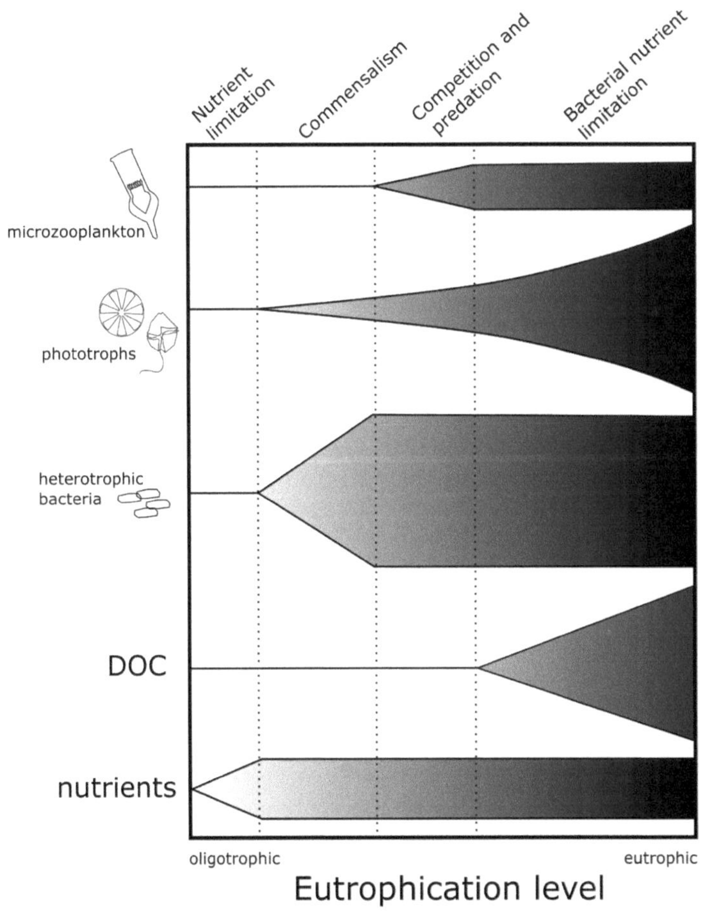

Figure 5.4. Simplified scheme of the model proposed by Thingstad et al. (1997) illustrating the changes in the concentration of the limiting nutrient, the net accumulation rate of dissolved organic carbon (DOC), heterotrophic bacteria, phototrophs and microzooplankton according to the trophic gradient.

increase in heterotrophs. Grazing pressure is usually higher on bacteria than on phototrophs due to reduced defensive skills of bacteria. In turn, the increase in phototrophic biomass generates a higher utilization of inorganic nutrients and bacteria must compete with phototrophs for their use. As nutrient content increases, the bacteria-phototrophs coupling weakens, the bacterial growth rate decreases due to nutrient limitation and a net accumulation of DOC occurs. The presence of other non-specific bacterial predators (not considered by the model in Fig. 5.4) can affect the coupling between bacteria and phototrophs either directly by grazing on bacteria, or indirectly by grazing on DOC-producing phototrophs (e.g., Duarte et al. 2005).

Once carbon is assimilated by phototrophs, several microbial processes determine their fate within the ocean. The better described pathway that involves living organisms consists of the transformation of the carbon dioxide fixed in the photic zone into POC by means of a number of processes collectively called the Biological Carbon Pump (BCP) (Volk and Hoffert 1985). The BCP forces a positive gradient of DIC towards the seafloor, acting against the diffusive and mixing forces of the ocean. The mechanism resides on the sinking of decaying organisms, either phototrophs or their planktonic grazers, cell debris and pellets by hydrodynamic processes (Fig. 5.5). Part of this POC may be either remineralized or respired in the euphotic zone, while another part is exported towards the mesopelagic zone (100–1000 m). The fraction of POC that resists remineralization escapes lateral transport and reaches the deep ocean (≥ 1000 m), does not return to the atmosphere within the next 100–1000 years, and is considered to be effectively sequestered (Legendre et al. 2015). Grazing by microzooplankton is the main conduit of nutrient regeneration in surface waters, which in turn, stimulates the growth of primary producers and bacteria. In this way, microzooplankton delay the nutrient limitation of their prey and prolongs the flow of carbon through the microbial food web.

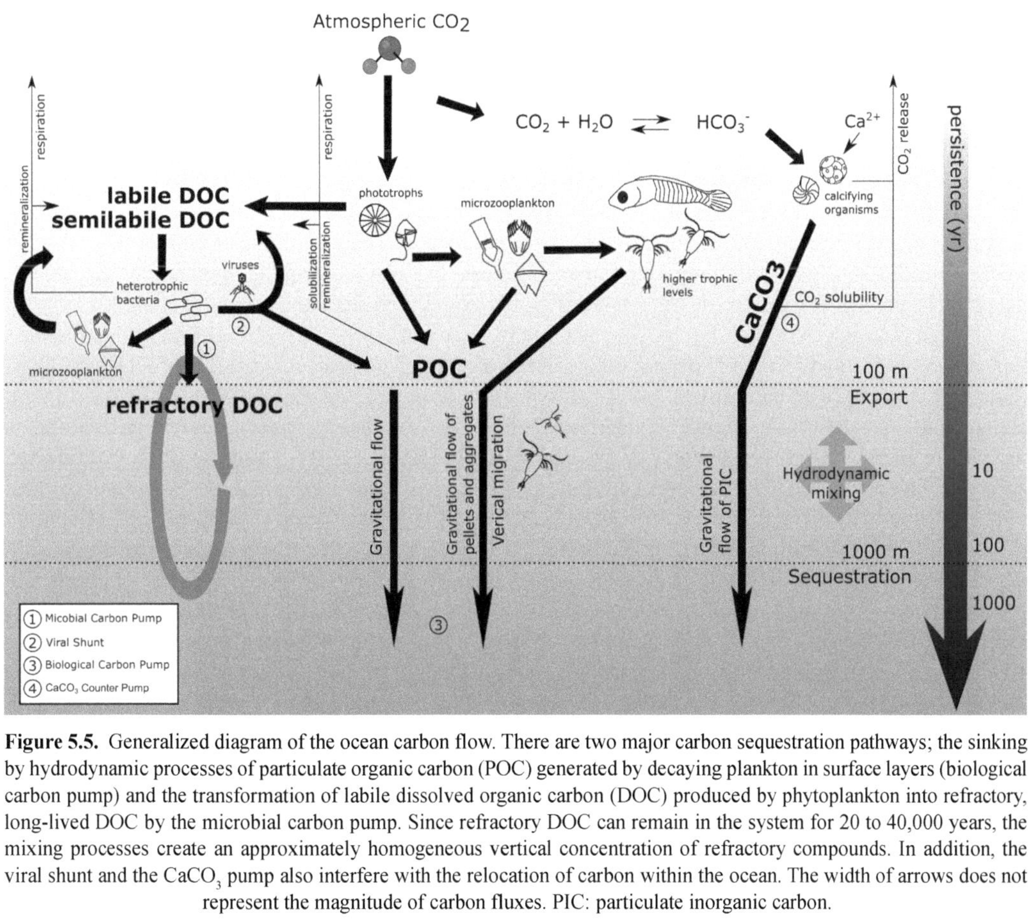

Figure 5.5. Generalized diagram of the ocean carbon flow. There are two major carbon sequestration pathways; the sinking by hydrodynamic processes of particulate organic carbon (POC) generated by decaying plankton in surface layers (biological carbon pump) and the transformation of labile dissolved organic carbon (DOC) produced by phytoplankton into refractory, long-lived DOC by the microbial carbon pump. Since refractory DOC can remain in the system for 20 to 40,000 years, the mixing processes create an approximately homogeneous vertical concentration of refractory compounds. In addition, the viral shunt and the $CaCO_3$ pump also interfere with the relocation of carbon within the ocean. The width of arrows does not represent the magnitude of carbon fluxes. PIC: particulate inorganic carbon.

Another important part of the carbon assimilated by phototrophs enters the Microbial Carbon Pump (MCP) (Jiao et al. 2010), where a low concentration of labile (lifetime of hours to days) and semilabile (average lifetime of 1.5 years) organic carbon produced by phytoplankton is transformed into a large reservoir of refractory dissolved organic carbon (rDOC) which is resistant to bacterial degradation (Ogawa et al. 2001). This rDOC is generated mainly as a result of bacterial production processes and viral lysis (Legendre et al. 2015, Weinbauer et al. 2011) and can remain in the system for thousands of years (its average lifetime is around 16,000 years). Microzooplankton also contribute to the formation of rDOC by grazing and egestion, while partially delaying stoichiometric unbalance of their prey by nutrient remineralization. However, mixotrophic protists can create a shortcut among the phototroph-bacteria-heterotroph circuit. Photosynthetic release of fresh DOC by mixotrophs stimulates bacterial growth, while simultaneous grazing on bacteria by the same organisms generates inorganic nutrient that will be otherwise unavailable. Hence, mixotrophs may potentially enhance the accumulation of rDOC by supporting photosynthesis with direct bacterial consumption which, in turn, prevents the uptake of complex DIC compounds of bacterial origin (Mitra et al. 2014).

The magnitude of the rDOC pool represents the largest carbon reservoir in the oceans and is comparable to the atmospheric CO_2 pool (Falkowski et al. 2000). The MCP is partly responsible for maintaining the enormous gradient between short (< 100 years) and long-lived (> 100 years) DOC. In fact, the long-lived DOC pool is ca. 45 times higher than that of DOC with shorter lifetimes (Hansell 2013). However, unlike the BCP and the $CaCO_3$ Counter Pump (see below) that maintain a vertical gradient of DIC, the gradient generated by the MCP is depth-independent given the prolonged lifetime of rDOC that are constantly relocated by hydrodynamic mixing. Hence, carbon sequestration by the MCP takes place not by the vertical relocation of carbon but through the formation of rDOC with a lifetime longer than 100 years (Legendre et al. 2015).

Overall, the BCP and the MCP are jointly responsible for more than 80% of the carbon pumping into the ocean, and are therefore fundamental in controlling the exchange rate between the ocean and the atmosphere (Fig. 5.5). In those environments where the microbial food web is more important than the classical phytoplankton-zooplankton-fish food web (i.e., oligotrophic environments), the carbon cycle is expected to be slower with a net accumulation of carbon. In contrast, the BCP would probably exceed the MCP in eutrophic and productive habitats (Jiao et al. 2010). At the global scale, preliminary estimates revealed that the magnitude of the BCP (POC sequestration below 1000 m) could range between 0.6 and 1.2 pg C yr^{-1}, which represents 1.2–2.4% of primary production. The magnitude of carbon sequestration in the form of rDOC with a lifetime longer than 100 years that results from the operation of the MCP is near to 0.18 pg C yr^{-1}, which represents 0.4% of global primary production (Legendre et al. 2015).

Phages also interfere with the biochemical pathways of carbon. Nearly 20% of microbial biomass including bacteria and protists is destroyed by viral lysis, which subsidises the pools of DOC and POC (Breitbart et al. 2018). The redistribution of carbon from living organisms to carbon pools due to viral action by the so called "viral shunt" (Wilhelm and Suttle 1999) implies a greater regeneration of nutrients and CO_2 respiration in surface layers (Fig. 5.5). Likewise, viral lysis can accelerate the sinking of infected cells thus favouring the BCP (Lawrence and Suttle 2004). Hence, the viral shunt can potentially control both the export (by fuelling the production of DOC) and the sequestration (by increasing the amount of decaying cells) of carbon in marine environments. Another form of carbon sequestrations relies on the bio-precipitation of $CaCO_3$ by calcifying organisms such as foraminiferans, coccolithophorids and pteropods. The formation of calcareous skeletons implies the uptake of $CaCO_3$ that results from the reaction of bicarbonate (HCO_3^-) and calcium (Ca^{2+}) ions. This reaction releases CO_2 into the atmosphere and produces precipitable bio-minerals that sink into the ocean floor. The amount of carbon sequestration by the sinking of bio-minerals, however, is counteracted by the production of gaseous CO_2 at a higher rate than that of $CaCO_3$ due to the specific buffer capacity of seawater (Zeebe 2012). Consequently, this mechanism of carbon redistribution has been termed "$CaCO_3$ counterpump" (CCP) because of its reverse effect of the BCP, which tends to reduce atmospheric CO_2.

As the above section suggests, the configuration of planktonic trophic networks largely determines carbon sequestration processes in the world ocean (Litchman et al. 2015). This highlights that under the framework of biogeochemical cycles, morphological and phylogenetic boundaries become inoperative, while the fate of carbon is better determined by properly identifying the multi-functional spectrum of microbes. The microbial spectrum, however, responds directly to nutrient balance, and nutrient balance will be profoundly affected by climate change either by increased water stratification or eutrophication. The future response of BCP and MCP to climate change is an open field of investigation, since ocean microbes have the potential to mitigate global warming by the removal of big quantities of atmospheric CO_2.

5.9 Responses to climate change

The world ocean is experiencing a sustained warming since 1970 and has taken up nearly 90% of the excess heat in the atmosphere. Since 1993, the warming trend has doubled producing a mass reduction of the cryosphere, which together with local scale freshening increased water column stability, especially in the tropics (Hoegh-Guldberg et al. 2019). Enhanced stratifications inhibit the free exchange of nutrients with deeper waters and limits of phytoplankton biomass build-up (Holt et al. 2010). In addition, ocean warming increases the prevalence of heatwaves and extreme weather events, and decreases the solubility of oxygen. The joint effects of lower oxygen solubility and enhanced organism's respiration translated to an overall decrease of 2% in the global oxygen inventory (Schmidtko et al. 2017).

The ocean has also absorbed between 20–30% of total anthropogenic CO_2 emissions since the 1980s. Warmer temperatures, along with higher atmospheric pCO_2, leads to a higher solubility of carbon dioxide in the surface ocean and the release of hydrogen ions (H^+). As a result, ocean pH has decreased by more than 0.1 units since preindustrial times (Hoegh-Guldberg et al. 2019). Under a business-as-usual carbon emission scenario, ocean trends are expected to unfold a 8.6% drop in global primary productivity (Bopp et al. 2013).

Observational data revealed that ocean protists are highly responsive to climatic trends. However, the responses of plankton to environmental changes are usually complex and vary according to the interaction between local environmental conditions and global climatic phenomena (Caron and Hutchins 2013). Overall, microzooplankton are expected to tolerate predicted climate trends by the end of the century due to specific traits that confer wide adaptability. These strategic traits are (1) their short generation time and relatively simple life history, (2) their nutritional plasticity, (3) their diverse interspecific interactions such as predator–prey and symbiotic relationships, and (4) the enormous functional diversity contained in their genetic reservoir (López-Abbate 2021). In addition, warming is expected to benefit the development of bacteria (Simon et al. 1999) and small phytoplankton (Morán et al. 2010). Consequently, model projections anticipate that a higher degree of biogenic carbon will be redirected toward microbial food webs in the future (Wohlers et al. 2009).

Some species of ciliates, dinoflagellates and foraminiferans can respond directly to warming through a poleward spreading of warm-water species (McLeod et al. 2012, Hinder et al. 2012, Langer et al. 2013). Temperature is a key factor in modulating the growth rate of protists in marine environments (Montagnes et al. 2003). In particular, microzooplankton usually have a wide range of thermal tolerance, although the mechanisms of specific ecophysiological responses have not been adequately described (Caron and Hutchins 2013). Since temperature affects the growth rate of phototrophs and heterotrophs at different rates (Rose and Caron 2007), an increase in temperature can cause a shift in the relationship between phototrophy and heterotrophy (Wohlers et al. 2009). Sea surface warming is expected to generate a greater loss of phytoplankton biomass as a result of temporal match with microzooplankton and a higher grazing pressure in eutrophic and productive environments (Aberle et al. 2007, Chen et al. 2012), although a trophic mismatch may also occur as a response to changes in phenology (Aberle et al. 2012, Chivers et al. 2017).

As documented by the match-mismatch phenomenon, the response of microzooplankton to climate change is in many cases indirect. For example, rising temperatures in the ocean can affect the dynamics of microzooplankton by changing the concentration, taxonomic composition and size of their phototrophic prey (Morán et al. 2010, Sommer and Lengfellner 2008, Daufresne et al. 2009). A differential latitudinal migration of prey and predators also indirectly affects microzooplankton communities by modifying food supply and predation risk, respectively (Chivers et al. 2017). Similarly, ocean acidification has a weak

Table 5.1. Observed and projected impacts on microzooplankton produced by main climate hazards. Since marine microbes are fundamental players in the ocean's carbon cycle, the potential impact on carbon export associated with protist's responses to climate hazards is included. For references, see López-Abbate (2021). OA: ocean acidification.

Climate hazard	Impact	Suggested effect on carbon export
Warming	Higher growth rate	Higher C remineralization
	Higher grazing rate	Higher C remineralization
	Higher trophic coupling with phytoplankton in polar and subpolar ecosystems	Increase of sinking POM in high-latitudes
	Poleward range expansion of warm-water species	Increase of sinking POM in high-latitudes
	Shifts in timing	Undetermined
	Increased predation risk on ciliates, especially in oligotrophic areas	Increase of sinking POM
	Higher relative contribution of heterotrophy to the growth of primarily heterotrophic mixotrophs	Lower CO_2 fixation
	Cell volume reduction	Higher C remineralization due to the prevalence of the MCP and slower POC sinking
	Trophic mismatch	Increase of sinking POM
OA and pCO_2 increase	Weak or no direct impact on non-calcifying organisms	Higher release of DOC at high pCO_2
	Weak or nondetectable effect on composition and diversity	Undetermined
	Weak to moderate effect on grazing rate	Undetermined
	Positive effect on biomass due to increasing prey edibility	Higher C remineralization
	Negative effect on biomass due to decreasing prey edibility and/or nutritional quality	Increase of sinking POM
	Growth inhibition of calcifying organisms	Reduction of sinking biominerals
	Growth stimulation of primarily phototrophic mixotrophs	Higher CO_2 fixation
Deoxygenation	Ciliates species replacement	Weak or under-documented effect, although a reduction in POM sinking may occur due to habitat compression of metazoans vulnerable to hypoxia
	Diversity loss	

direct impact on microzooplankton (Aberle et al. 2013), but it can alter their concentration and ingestion rate through modification of the availability and composition of prey (Rose et al. 2009, Olson et al. 2018).

Projected impacts on microzooplankton and potential consequences on the fate of the carbon cycle are summarized in Table 5.1. In the coming decades, an increase of warm-water species toward the poles and a decrease of low and mid-latitude species due to a reduction of phototrophic prey availability is anticipated. The projected increase in water column stratification in the tropics, the Arctic, the North Atlantic and the North Pacific Oceans is expected to reduce the upwelling of nutrient-rich deep waters to the euphotic zone and reduce primary productivity (Capotondi et al. 2012). The differential response of phototrophs and heterotrophs to temperature will likely produce a negative trophic amplification of climate change in low- and mid-latitudes, leading to a reduction in the ocean's capacity to remove carbon by the BCP (Chust et al. 2014). The opposite is expected to occur in high latitudes. Moreover, reduced prey availability in low- and mid-latitudes will likely favour smaller microzooplankton species, which are able to prosper under low resource concentration (Lefort et al. 2015). Under low nutrient concentration and the dominance of small prey, mixotrophic taxa are expected to dominate microbial food webs (Onda et al. 2017). The narrowing of the mixing layer will reduce the habitable niches for plankton, leading to an overall reduction in the carbon assimilation area and in the ocean's capacity to remove atmospheric CO_2 (Laufkötter et al. 2017).

Concurrent with a lower nutrient inflow into the euphotic zone at low- and mid-latitudes, a departure from near-Redfield ratios of ambient nutrients is expected with ongoing climate change. In particular, phosphorous is expected to decline at a higher rate than that of nitrogen due to biological N-fixation in the tropics (Kwiatkowski et al. 2018). Experimental approaches have revealed that the N:P ratio determines the degree of mixotrophy of nanoflagellates; P-limitation may stimulate mixotrophy over heterotrophy, while N-limitation produces the opposite effect (Fischer et al. 2017). Ciliates and dinoflagellates may allow their internal N:P ratio to vary among wider intervals compared to that of metazoans, and are thereby able to thrive on nutrient deficient prey (Marki and Pahlow 2016), while stoichiometric boundaries are generally narrower among ciliates than dinoflagellates (Golz et al. 2015). This poses a higher risk of ciliates' growth inhibition when offered nutrient-deficient prey (Carrillo et al. 2008), a situation that may arise in a future, more oligotrophic ocean.

Although most model projections assume that phototrophs are the main source of energy for microzooplankton, bacterial biomass also represents an essential food item for small ciliates and flagellates. There is some consensus on the effects of warming on bacteria, which include increasing biomass and decreasing body size (Morán et al. 2015); however, the future cascading effect on microzooplankton is largely unknown. In addition, the intensification of interactive effects between climate stressors will produce the emergence of irregular, local- and regional-scale responses of plankton that will contrast with projected trends. For instance, a multi-factorial experiment combining the effect of eutrophication, warming and acidification in a Norwegian fjord, revealed group-specific, cascading effects on microzooplankton (Calbet et al. 2014). Generally, eutrophication reduced the phototropic to heterotrophic ratio, acidification reduced ciliates' biomass due to a lower nutritional value of prey and warming promoted a fast synchronization between protist predators and their phototrophic prey. Although microzooplankton showed a high adaptability to shifting prey exposed to combined hazards, the overall community-level response denoted a lower trophic efficiency under a climate change scenario in coastal ecosystems (Calbet et al. 2014).

Another uncertainty in the prediction of microzooplankton responses to climate change resides in the ability of this group to synchronize climate trends with evolutionary timescales. That is, many trade-off considered here (e.g., cell volume reduction and undercalcification) may not be fixed if sufficient time is given to allow the emergence of evolutionary feedbacks (Ward et al. 2019). Preliminary studies suggest that microbes, with their fast generation times and diverse genetic pool, are prone to accumulate adaptations in response to climate trends. Although evolution experiments have been mostly performed with phytoplankton, particularly diatoms (Irwin et al. 2015, Schaum et al. 2018), less is known about microzooplankton species. In addition, the scarcity of sustained observations compared with other groups such as phytoplankton and mesozooplankton makes the prediction of microzooplankton responses to climate change difficult.

Summary

1. Microzooplankton are ubiquitous organisms in aquatic systems that play multifaceted trophic roles, as they may act as bacterivores, herbivores, omnivores, nutrient recyclers and prey for metazoans.
2. Given the prevalence of phototrophic plankton among primary producers, they have evolved preferentially toward an herbivore trophic mode, representing the main mortality factor of phytoplankton in most aquatic ecosystems.
3. Most species have the ability to photosynthesize either by the innate capacity to manufacture plastids (constitutive mixotrophs) or by the acquired photosynthetic capacity derived from their prey (non-constitutive mixotrophs).
4. The ubiquity of mixotrophy among microzooplankton is now recognized thanks to global recompilations, and the relative abundance of different functional types is expected to produce contrasting effects on element budgets and the ecosystem functioning.

5. Microzooplankton present specific traits, such as cell size, swimming behaviour, closer stoichiometry to that of consumers and the capacity to *de novo* synthesise complex molecules (i.e., lipids and sterols), that make them the preferred prey of mesozooplankton, particularly in oligotrophic areas.
6. At the community level, microzooplankton interfere with carbon pumps by prolonging the permanence of nutrients in the sunlit ocean through remineralization processes, thus delaying the resource-limitation of their prey.
7. Intense activity of heterotrophic and mixotrophic protists typically deviates carbon from the biological carbon pump toward the microbial carbon pump.
8. Most observations suggest that microzooplankton communities will rapidly respond and adapt to climate change due to their fast generation time, their nutritional plasticity, and the enormous functional diversity contained in their genetic reservoir.
9. Warming is expected to stimulate the growth and grazing of microzooplankton and given the differential thermal dependency of phototrophs relative to heterotrophs, the coupling between microzooplankton and phytoplankton will maximise in polar and subpolar regions.
10. Considering the projected dominance of small prey in the future ocean, microzooplankton will constitute a trophic buffer through efficiently repackaging prey inaccessible to larger consumers.

Acknowledgements

I would like to thank Dr. Suzana Leles and Dr. Nicole Aberle-Malzahn for their insightful contributions to this chapter.

References

Aberle, N., K. Lengfellner and U. Sommer. 2007. Spring bloom succession, grazing impact and herbivore selectivity of ciliate communities in response to winter warming. Oecologia. 150: 668–681.

Aberle, N., B. Bauer, A. Lewandowska, U. Gaedke and U. Sommer. 2012. Warming induces shifts in microzooplankton phenology and reduces time-lags between phytoplankton and protozoan production. Mar. Biol. 159: 2441–2453.

Aberle, N., K.G. Schulz, A. Stuhr, A.M. Malzahn, A. Ludwig and U. Riebesell. 2013. High tolerance of microzooplankton to ocean acidification in an Arctic coastal plankton community. Biogeosciences 10: 1471–1481.

Adl, S.M., D. Bass, C.E. Lane, J. Lukeš, C.L. Schoch, A. Smirnov, S. Agatha, C. Berney, M.W. Brown and F. Burki. 2019. Revisions to the classification, nomenclature, and diversity of eukaryotes. J. Eukaryotic Microbiol. 66: 4–119.

Agatha, S. 2011. Global diversity of aloricate oligotrichea (Protista, Ciliophora, Spirotricha) in marine and brackish sea water. PloS One 6: e22466.

Azam, F., T. Fenchel, J.G. Field, J.S. Gray, L.A. Meyer-Reil and F. Thingstad. 1983. The ecological role of water-column microbes in the sea. Mar. Ecol. Prog. Ser. 10: 257–263.

Be, A.W. 1977. An ecological, zoogeographic and taxonomic review of recent planktonic foraminifera. Oceanic Micropaleontology 1–100.

Berge, T., L.K. Poulsen, M. Moldrup, N. Daugbjerg and P. Juel Hansen. 2012. Marine microalgae attack and feed on metazoans. ISME J. 6: 1926–1936.

Biard, T., L. Stemmann, M. Picheral, N. Mayot, P. Vandromme, H. Hauss, G. Gorsky, L. Guidi, R. Kiko and F. Not. 2016. *In situ* imaging reveals the biomass of giant protists in the global ocean. Nature 532: 504–507.

Biard, T., E. Bigeard, S. Audic, J. Poulain, A. Gutierrez-Rodriguez, S. Pesant, L. Stemmann and F. Not. 2017. Biogeography and diversity of Collodaria (Radiolaria) in the global ocean. ISME J. 11: 1331–1344.

Bird, D.F. and J. Kalff. 1986. Bacterial grazing by planktonic lake algae. Science 231: 493–495.

Blossom, H.E., N. Daugbjerg and P.J. Hansen. 2012. Toxic mucus traps: A novel mechanism that mediates prey uptake in the mixotrophic dinoflagellate *Alexandrium pseudogonyaulax*. Harmful Algae 17: 40–53.

Boltovskoy, D., S.A. Kling, K. Takahashi and K. Bjørklund. 2010. World atlas of distribution of recent polycystina (Radiolaria). Palaeontol. Electron. 13: 1–230.

Bopp, L., L. Resplandy, J.C. Orr, S.C. Doney, J.P. Dunne, M. Gehlen, P. Halloran, C. Heinze, T. Ilyina, R. Séférian, J. Tjiputra and M. Vichi. 2013. Multiple stressors of ocean ecosystems in the 21st century: projections with CMIP5 models. Biogeosciences 10: 6225–6245.

Breitbart, M., C. Bonnain, K. Malki and N.A. Sawaya. 2018. Phage puppet masters of the marine microbial realm. Nature Microbiology 3: 754–766.

Broglio, E., M. Johansson and P.R. Jonsson. 2001. Trophic interaction between copepods and ciliates: effects of prey swimming behavior on predation risk. Mar. Ecol.: Prog. Ser. 220: 179–186.
Broglio, E., E. Saiz, A. Calbet, I. Trepat and M. Alcaraz. 2004. Trophic impact and prey selection by crustacean zooplankton on the microbial communities of an oligotrophic coastal area (NW Mediterranean Sea). Aquat. Microb. Ecol. 35: 65–78.
Burkholder, J.M., P.M. Glibert and H.M. Skelton. 2008. Mixotrophy, a major mode of nutrition for harmful algal species in eutrophic waters. Harmful Algae 8: 77–93.
Buskey, E.J. and D.K. Stoecker. 1989. Behavioral responses of the marine tintinnid Favella sp. to phytoplankton: influence of chemical, mechanical and photic stimuli. J. Exp. Mar. Biol. Ecol. 132: 1–16.
Calbet, A. 2001. Mesozooplankton grazing effect on primary production: A global comparative analysis in marine ecosystems. Limnol. Oceanogr. 46: 1824–1830.
Calbet, A. and M.R. Landry. 2004. Phytoplankton growth, microzooplankton grazing, and carbon cycling in marine systems. Limnol. Oceanogr. 49: 51–57.
Calbet, A. and E. Saiz. 2005. The ciliate-copepod link in marine ecosystems. Aquat. Microb. Ecol. 38: 157–167.
Calbet, A. 2008. The trophic roles of microzooplankton in marine systems. ICES Journal of Marine Science 65: 325–331.
Calbet, A., A.F. Sazhin, J.C. Nejstgaard, S.A. Berger, Z.S. Tait, L. Olmos, D. Sousoni, S. Isari, R.A. Martínez, J.-M. Bouquet, E.M. Thompson, U. Båmstedt and H.H. Jakobsen. 2014. Future climate scenarios for a coastal productive planktonic food web resulting in microplankton phenology changes and decreased trophic transfer efficiency. PloS One 9: e94388.
Capotondi, A., M.A. Alexander, N.A. Bond, E.N. Curchitser and J.D. Scott. 2012. Enhanced upper ocean stratification with climate change in the CMIP3 models. J. Geophys. Res.: Oceans 117: C4.
Caron, D.A. 2009. Past president's address: Protistan biogeography: Why all the fuss? J. Eukaryotic Microbiol. 56: 105–112.
Caron, D.A., A.Z. Worden, P.D. Countway, E. Demir and K.B. Heidelberg. 2009. Protists are microbes too: a perspective. ISME J. 3: 4.
Caron, D.A., P.D. Countway, A.C. Jones, D.Y. Kim and A. Schnetzer. 2012. Marine protistan diversity. Annu. Rev. Mar. 4: 467–493.
Caron, D.A. and D.A. Hutchins. 2013. The effects of changing climate on microzooplankton grazing and community structure: drivers, predictions and knowledge gaps. J. Plankton Res. 35: 235–252.
Caron, D.A., H. Alexander, A.E. Allen, J.M. Archibald, E.V. Armbrust, C. Bachy, C.J. Bell, A. Bharti, S.T. Dyhrman, S.M. Guida, K.B. Heidelberg, J.Z. Kaye, J. Metzner, S.R. Smith and A.Z. Worden. 2016. Probing the evolution, ecology and physiology of marine protists using transcriptomics. Nat. Rev. Microbiol. 15: 6.
Carrillo, P., M. Villar-Argaiz and J.M. Medina-Sanchez. 2008. Does microorganism stoichiometry predict microbial food web interactions after a phosphorus pulse? Microb. Ecol. 56: 350–363.
Chen, B., M.R. Landry, B. Huang and H. Liu. 2012. Does warming enhance the effect of microzooplankton grazing on marine phytoplankton in the ocean? Limnol. Oceanogr. 57: 519–526.
Chivers, W.J., A.W. Walne and G.C. Hays. 2017. Mismatch between marine plankton range movements and the velocity of climate change. Nature Communications 8: 14434.
Chust, G., J.I. Allen, L. Bopp, C. Schrum, J. Holt, K. Tsiaras, M. Zavatarelli, M. Chifflet, H. Cannaby, I. Dadou, U. Daewel, S.L. Wakelin, E. Machu, D. Pushpadas, M. Butenschon, Y. Artioli, G. Petihakis, C. Smith, V. Garçon, K. Goubanova, B. Le Vu, B.A. Fach, B. Salihoglu, E. Clementi and X. Irigoien. 2014. Biomass changes and trophic amplification of plankton in a warmer ocean. Glob. Change Biol. 20: 2124–2139.
Daufresne, M., K. Lengfellner and U. Sommer. 2009. Global warming benefits the small in aquatic ecosystems. Proc. Natl. Acad. Sci. USA 106: 12788–12793.
Davidson, K., A. Cunningham and K.J. Flynn. 1995. Predator-prey interactions between Isochrysis galbana and Oxyrrhis marina. III. Mathematical modelling of predation and nutrient regeneration. J. Plankton Res. 17: 465–492.
do Rosário Gomes, H., J.I. Goes, S.G.P. Matondkar, E.J. Buskey, S. Basu, S. Parab and P. Thoppil. 2014. Massive outbreaks of *Noctiluca scintillans* blooms in the Arabian Sea due to spread of hypoxia. Nature Communications 5: 4862.
Dolan, J., C. Gallegos and A. Moigis. 2000. Dilution effects on microzooplankton in dilution grazing experiments. Mar. Ecol.: Prog. Ser. 200: 127–139.
Dolan, J.R. and M.T. Pérez. 2000. Costs, benefits and characteristics of mixotrophy in marine oligotrichs. Freshwater Biol. 45: 227–238.
Dolan, J.R. 2005. An introduction to the biogeography of aquatic microbes. Aquat. Microb. Ecol. 41: 39–48.
Dolan, J.R. and R.W. Pierce. 2012. Diversity and distributions of tintinnids. pp. 214–243. *In*: Dolan, J.R., D.J. Montagnes, S. Agatha, D.W. Coats and D. Stoecker [eds.]. The Biology and Ecology of Tintinnid Ciliates: Models for Marine Plankton. Wiley-Blackwell. UK.
Dolan, J.R., E.J. Yang, S.-H. Kang and T.S. Rhee. 2016. Declines in both redundant and trace species characterize the latitudinal diversity gradient in tintinnid ciliates. ISME J. 10: 2174.
Duarte, C.M., S. Agustí, D. Vaqué, N.S. Agawin, J. Felipe, E.O. Casamayor and J.M. Gasol. 2005. Experimental test of bacteria-phytoplankton coupling in the Southern Ocean. Limnol. Oceanogr. 50: 1844–1854.
Edwards, K.F. 2019. Mixotrophy in nanoflagellates across environmental gradients in the ocean. Proc. Natl. Acad. Sci. USA 116: 6211–6220.

Falkowski, P., R. Scholes, E. Boyle, J. Canadell, D. Canfield, J. Elser, N. Gruber, K. Hibbard, P. Högberg and S. Linder. 2000. The global carbon cycle: a test of our knowledge of earth as a system. Science 290: 291–296.
Faure, E., F. Not, A.-S. Benoiston, K. Labadie, L. Bittner and S.-D. Ayata. 2019. Mixotrophic protists display contrasted biogeographies in the global ocean. ISME J. 13: 1072.
Fenchel, T. 1980a. Suspension feeding in, ciliated protozoa: structure and function of feeding organelles. Archiv für Protistenkunde. 123: 239–260.
Fenchel, T. 1980b. Suspension feeding in ciliated protozoa: functional response and particle size selection. Microb. Ecol. 6: 1–11.
Fenchel, T. 1982. Ecology of heterotrophic microflagellates. I. Some important forms and their functional morphology. Mar. Ecol.: Prog. Ser. 8: 211–223.
Fenchel, T. 1990. Adaptive significance of polymorphic life cycles in Protozoa: responses to starvation and refeeding in two species of marine ciliates. J. Exp. Mar. Biol. Ecol. 136: 159–177.
Fenchel, T. and N. Blackburn. 1999. Motile chemosensory behaviour of phagotrophic protists: mechanisms for and efficiency in congregating at food patches. Protist. 150: 325–336.
Fenchel, T. 2001. How dinoflagellates swim. Protist. 152: 329–338.
Fischer, R., H.-A. Giebel and R. Ptacnik. 2017. Identity of the limiting nutrient (N vs. P) affects the competitive success of mixotrophs. Mar. Ecol.: Prog. Ser. 563: 51–63.
Flynn, K.J., D.K. Stoecker, A. Mitra, J.A. Raven, P.M. Glibert, P.J. Hansen, E. Granéli and J.M. Burkholder. 2012. Misuse of the phytoplankton–zooplankton dichotomy: the need to assign organisms as mixotrophs within plankton functional types. J. Plankton Res. 35: 3–11.
Flynn, K.J., A. Mitra, K. Anestis, A.A. Anschütz, A. Calbet, G.D. Ferreira, N. Gypens, P.J. Hansen, U. John, J.L. Martin, J.S. Mansour, M. Maselli, N. Medić, A. Norlin, F. Not, P. Pitta, F. Romano, E. Saiz, L.K. Schneider, W. Stolte and C. Traboni. 2019. Mixotrophic protists and a new paradigm for marine ecology: where does plankton research go now? J. Plankton Res.
Foissner, W. 2006. Biogeography and dispersal of micro-organisms: a review emphasizing protists. Acta Protozool. 45: 111–136.
Gallegos, C.L. 1989. Microzooplankton grazing on phytoplankton in the Rhode River, Maryland: Nonlinear feeding kinetics. Mar. Ecol.: Prog. Ser. 57: 23–33.
Ger, K.A., L.-A. Hansson and M. Lürling. 2014. Understanding cyanobacteria-zooplankton interactions in a more eutrophic world. Freshwater Biol. 59: 1783–1798.
Gimmler, A., R. Korn, C. de Vargas, S. Audic and T. Stoeck. 2016. The Tara Oceans voyage reveals global diversity and distribution patterns of marine planktonic ciliates. Sci. Rep. 6: 33555.
Golz, A.-L., A. Burian and M. Winder. 2015. Stoichiometric regulation in micro- and mesozooplankton. J. Plankton Res. 37: 293–305.
Guasto, J.S., K.A. Johnson and J.P. Gollub. 2010. Oscillatory flows induced by microorganisms swimming in two dimensions. Phys. Rev. Lett. 105: 168102.
Guo, Z., H. Zhang and S. Lin. 2014. Light-promoted rhodopsin expression and starvation survival in the marine dinoflagellate *Oxyrrhis marina*. PloS One 9: e114941.
Hansell, D.A. 2013. Recalcitrant dissolved organic carbon fractions. Annu. Rev. Mar. 5: 421–445.
Hansen, B., P.K. Bjornsen and P.J. Hansen. 1994. The size ratio between planktonic predators and their prey. Limnol. Oceanogr. 39: 395–403.
Hansen, P.J., P.K. Bjørnsen and B.W. Hansen. 1997. Zooplankton grazing and growth: Scaling within the 2–2,000-µm body size range. Limnol. Oceanogr. 42: 687–704.
Hansen, P.J. and A.J. Calado. 1999. Phagotrophic mechanisms and prey selection in free-living dinoflagellates. J. Eukaryotic Microbiol. 46: 382–389.
Hartmann, M., C. Grob, G.A. Tarran, A.P. Martin, P.H. Burkill, D.J. Scanlan and M.V. Zubkov. 2012. Mixotrophic basis of Atlantic oligotrophic ecosystems. Proc. Natl. Acad. Sci. USA 109: 5756–5760.
Havskum, H., T.F. Thingstad, R. Scharek, F. Peters, E. Berdalet, M.M. Sala, M. Alcaraz, J.C. Bangsholt, U.L. Zweifel and Å. Hagström. 2003. Silicate and labile DOC interfere in structuring the microbial food web via algal—bacterial competition for mineral nutrients: Results of a mesocosm experiment. Limnol. Oceanogr. 48: 129–140.
Herstoff, E.M., S.B. Baines, M. Boersma and C.L. Meunier. 2019. Does prey elemental stoichiometry influence copepod movement over ontogeny? Limnol. Oceanogr. 64: 2467–2477.
Hessen, D.O., J.J. Elser, R.W. Sterner and J. Urabe. 2013. Ecological stoichiometry: An elementary approach using basic principles. Limnol. Oceanogr. 58: 2219–2236.
Hinder, S.L., J.E. Manning, M.B. Gravenor, M. Edwards, A.W. Walne, P.H. Burkill and G.C. Hays. 2012. Long-term changes in abundance and distribution of microzooplankton in the NE Atlantic and North Sea. J. Plankton Res. 34: 83–91.
Hoegh-Guldberg, O., D. Jacob, M. Taylor, T. Guillén Bolaños, M. Bindi, S. Brown, I.A. Camilloni, A. Diedhiou, R. Djalante, K. Ebi, F. Engelbrecht, J. Guiot, Y. Hijioka, S. Mehrotra, C.W. Hope, A.J. Payne, H.-O. Pörtner, S.I. Seneviratne, A. Thomas, R. Warren and G. Zhou. 2019. The human imperative of stabilizing global climate change at 1.5°C. Science 365: eaaw6974.

Holling, C.S. 1959. The Components of predation as revealed by a study of small-mammal predation of the european pine sawfly. Can. Entomol. 91: 293–320.

Holt, J., S. Wakelin, J. Lowe and J. Tinker. 2010. The potential impacts of climate change on the hydrography of the northwest European continental shelf. Prog. Oceanogr. 86: 361–379.

Irigoien, X., K. Flynn and R. Harris. 2005. Phytoplankton blooms: a 'loophole' in microzooplankton grazing impact? J. Plankton Res. 27: 313–321.

Irwin, A.J., Z.V. Finkel, F.E. Müller-Karger and L. Troccoli Ghinaglia. 2015. Phytoplankton adapt to changing ocean environments. Proc. Natl. Acad. Sci. USA 112: 5762–5766.

Isari, S., M. Antó and E. Saiz. 2013. Copepod foraging on the basis of food nutritional quality: can copepods really choose? PloS One 8: e84742.

Jakobsen, H.H. 2002. Escape of protists in predator-generated feeding currents. Aquat. Microb. Ecol. 26: 271–281.

Jakobsen, H.H., E. Halvorsen, B.W. Hansen and A.W. Visser. 2005. Effects of prey motility and concentration on feeding in *Acartia tonsa* and *Temora longicornis*: the importance of feeding modes. J. Plankton Res. 27: 775–785.

Jeong, H.J., K.A. Seong, Y.D. Yoo, T.H. Kim, N.S. Kang, S. Kim, J.Y. Park, J.S. Kim, G.H. Kim and J.Y. Song. 2008. Feeding and grazing impact by small marine heterotrophic dinoflagellates on heterotrophic bacteria. J. Eukaryotic Microbiol. 55: 271–288.

Jiao, N., G.J. Herndl, D.A. Hansell, R. Benner, G. Kattner, S.W. Wilhelm, D.L. Kirchman, M.G. Weinbauer, T. Luo and F. Chen. 2010. Microbial production of recalcitrant dissolved organic matter: long-term carbon storage in the global ocean. Nat. Rev. Microbiol. 8: 593.

Jonsson, P.R. and P. Tiselius. 1990. Feeding behaviour, prey detection and capture efficiency of the copepod *Acartia tonsa* feeding on planktonic ciliates. Mar. Ecol.: Prog. Ser. 35–44.

Kiørboe, T. and E. Saiz. 1995. Planktivorous feeding in calm and turbulent environments, with emphasis on copepods. Mar. Ecol.: Prog. Ser. 122: 135–145.

Kiørboe, T. 2008. A Mechanistic Approach to Plankton Ecology. Princeton University Press.

Kiørboe, T., A. Andersen, V.J. Langlois, H.H. Jakobsen and T. Bohr. 2009. Mechanisms and feasibility of prey capture in ambush-feeding zooplankton. Proc. Natl. Acad. Sci. USA 106: 12394–12399.

Kiørboe, T. 2011. How zooplankton feed: mechanisms, traits and trade-offs. Biol. Rev. 86: 311–339.

Kiørboe, T., H. Jiang, R.J. Gonçalves, L.T. Nielsen and N. Wadhwa. 2014. Flow disturbances generated by feeding and swimming zooplankton. Proc. Natl. Acad. Sci. USA 111: 11738–11743.

Klein Breteler, W.C.M., N. Schogt, M. Baas, S. Schouten and G.W. Kraay. 1999. Trophic upgrading of food quality by protozoans enhancing copepod growth: role of essential lipids. Mar. Biol. 135: 191–198.

Klein Breteler, W., M. Koski and S. Rampen. 2004. Role of essential lipids in copepod nutrition: no evidence for trophic upgrading of food quality by a marine ciliate. Mar. Ecol.: Prog. Ser. 274: 199–208.

Koski, M., W.K. Breteler and N. Schogt. 1998. Effect of food quality on rate of growth and development of the pelagic copepod *Pseudocalanus elongatus* (Copepoda, Calanoida). Mar. Ecol.: Prog. Ser. 170: 169–187.

Kwiatkowski, L., O. Aumont, L. Bopp and P. Ciais. 2018. The impact of variable phytoplankton stoichiometry on projections of primary production, food quality, and carbon uptake in the global ocean. Global Biogeochem. Cycles 32: 516–528.

Langer, M.R., A.E. Weinmann, S. Lötters, J.M. Bernhard and D. Rödder. 2013. Climate-driven range extension of Amphistegina (Protista, Foraminiferida): models of current and predicted future ranges. PloS One 8: e54443.

Laufkötter, C., J.G. John, C.A. Stock and J.P. Dunne. 2017. Temperature and oxygen dependence of the remineralization of organic matter. Global Biogeochem. Cycles 31: 1038–1050.

Lawrence, J.E. and C.A. Suttle. 2004. Effect of viral infection on sinking rates of Heterosigma akashiwo and its implications for bloom termination. Aquat. Microb. Ecol. 37: 1–7.

Lefort, S., O. Aumont, L. Bopp, T. Arsouze, M. Gehlen and O. Maury. 2015. Spatial and body-size dependent response of marine pelagic communities to projected global climate change. Glob. Change Biol. 21: 154–164.

Legendre, L. and F. Rassoulzadegan. 1995. Plankton and nutrient dynamics in marine waters. Ophelia. 41: 153–172.

Legendre, L., R.B. Rivkin, M.G. Weinbauer, L. Guidi and J. Uitz. 2015. The microbial carbon pump concept: Potential biogeochemical significance in the globally changing ocean. Prog. Oceanogr. 134: 432–450.

Leles, S., A. Mitra, K.J. Flynn, D.K. Stoecker, P.J. Hansen, A. Calbet, G.B. McManus, R.W. Sanders, D.A. Caron and F. Not. 2017. Oceanic protists with different forms of acquired phototrophy display contrasting biogeographies and abundance. Proc. R. Soc. London, Ser. B 284: 20170664.

Leles, S.G., L. Polimene, J. Bruggeman, J. Blackford, S. Ciavatta, A. Mitra and K.J. Flynn. 2018. Modelling mixotrophic functional diversity and implications for ecosystem function. J. Plankton Res. 40: 627–642.

Leles, S.G., A. Mitra, K.J. Flynn, U. Tillmann, D. Stoecker, H.J. Jeong, J. Burkholder, P.J. Hansen, D.A. Caron and P.M. Glibert. 2019. Sampling bias misrepresents the biogeographical significance of constitutive mixotrophs across global oceans. Glob. Ecol. Biogeogr. 28: 418–428.

Litchman, E., P. de Tezanos Pinto, K.F. Edwards, C.A. Klausmeier, C.T. Kremer, M.K. Thomas and A. Austin. 2015. Global biogeochemical impacts of phytoplankton: a trait-based perspective. J. Ecol. 103: 1384–1396.

Löder, M.G.J., C. Meunier, K.H. Wiltshire, M. Boersma and N. Aberle. 2011. The role of ciliates, heterotrophic dinoflagellates and copepods in structuring spring plankton communities at Helgoland Roads, North Sea. Mar. Biol. 158: 1551–1580.

López-Abbate, M.C. 2021. Microzooplankton communities in a changing ocean: a risk assessment. Diversity 13: 82.
Marki, A. and M. Pahlow. 2016. Microzooplankton stoichiometric plasticity inferred from modeling mesocosm experiments in the Peruvian upwelling region. Front. Mar. Sci. 3: 1–15.
Martin-Creuzburg, D., A. Wacker and E. von Elert. 2005a. Life history consequences of sterol availability in the aquatic keystone species *Daphnia*. Oecologia. 144: 362–372.
Martin-Creuzburg, D., A. Bec and E. Von Elert. 2005b. Trophic upgrading of picocyanobacterial carbon by ciliates for nutrition of *Daphnia magna*. Aquat. Microb. Ecol. 41: 271–280.
Martínez, R.A., A. Calbet and E. Saiz. 2017. Effects of small-scale turbulence on growth and grazing of marine microzooplankton. Aquat. Sci. 80: 2.
McLeod, D.J., G.M. Hallegraeff, G.W. Hosie and A.J. Richardson. 2012. Climate-driven range expansion of the red-tide dinoflagellate *Noctiluca scintillans* into the Southern Ocean. J. Plankton Res. 34: 332–337.
Menden-Deuer, S., E.J. Lessard, J. Satterberg and D. Grünbaum. 2005. Growth rates and starvation survival of three species of the pallium-feeding, thecate dinoflagellate genus Protoperidinium. Aquat. Microb. Ecol. 41: 145–152.
Meunier, C.L., M. Boersma, K.H. Wiltshire and A.M. Malzahn. 2016. Zooplankton eat what they need: copepod selective feeding and potential consequences for marine systems. Oikos. 125: 50–58.
Mitra, A., K.J. Flynn, J.M. Burkholder, T. Berge, A. Calbet, J.A. Raven, E. Granéli, P.M. Glibert, P.J. Hansen, D.K. Stoecker, F. Thingstad, U. Tillmann, S. Våge, S. Wilken and M.V. Zubkov. 2014. The role of mixotrophic protists in the biological carbon pump. Biogeosciences 11: 995–1005.
Mitra, A., K.J. Flynn, U. Tillmann, J.A. Raven, D. Caron, D.K. Stoecker, F. Not, P.J. Hansen, G. Hallegraeff, R. Sanders, S. Wilken, G. McManus, M. Johnson, P. Pitta, S. Våge, T. Berge, A. Calbet, F. Thingstad, H.J. Jeong, J. Burkholder, P.M. Glibert, E. Granéli and V. Lundgren. 2016. Defining planktonic protist functional groups on mechanisms for energy and nutrient acquisition: incorporation of diverse mixotrophic strategies. Protist. 167: 106–120.
Moigis, A.-G. 2006. The clearance rate of microzooplankton as the key element for describing estimated non-linear dilution plots demonstrated by a model. Mar. Biol. 149: 743–762.
Montagnes, D.J., S.A. Kimmance and D. Atkinson. 2003. Using Q10: Can growth rates increase linearly with temperature? Aquat. Microb. Ecol. 32: 307–313.
Montagnes, D.J., A.B. Barbosa, J. Boenigk, K. Davidson, K. Jürgens, M. Macek, J.D. Parry, E.C. Roberts and K. Simek. 2008. Selective feeding behaviour of key free-living protists: avenues for continued study. Aquat. Microb. Ecol. 53: 83–98.
Montagnes, D.J.S. 2012. Ecophysiology and behavior of tintinnids. pp. 85–121. *In*: Dolan, J.R., D.J. Montagnes, S. Agatha, D.W. Coats and D. Stoecker [eds.]. The Biology and Ecology of Tintinnid Ciliates. John Wiley & Sons, Ltd.
Moorthi, S.D., R. Ptacnik, R.W. Sanders, R. Fischer, M. Busch and H. Hillebrand. 2017. The functional role of planktonic mixotrophs in altering seston stoichiometry. Aquat. Microb. Ecol. 79: 235–245.
Morán, X.A.G., Á. López-Urrutia, A. Calvo-Díaz and W.K.W. Li. 2010. Increasing importance of small phytoplankton in a warmer ocean. Glob. Change Biol. 16: 1137–1144.
Morán, X.A.G., L. Alonso-Sáez, E. Nogueira, H.W. Ducklow, N. González, Á. López-Urrutia, L. Díaz-Pérez, A. Calvo-Díaz, N. Arandia-Gorostidi and T.M. Huete-Stauffer. 2015. More, smaller bacteria in response to ocean's warming? Proc. R. Soc. London, Ser. B. 282: 20150371.
Myung, G., H.S. Kim, J.W. Park, J.S. Park and W. Yih. 2013. Sequestered plastids in *Mesodinium rubrum* are functionally active up to 80 days of phototrophic growth without cryptomonad prey. Harmful Algae 27: 82–87.
Nagata, T., C. Tamburini, J. Arístegui, F. Baltar, A.B. Bochdansky, S. Fonda-Umani, H. Fukuda, A. Gogou, D.A. Hansell, R.L. Hansman, G.J. Herndl, C. Panagiotopoulos, T. Reinthaler, R. Sohrin, P. Verdugo, N. Yamada, Y. Yamashita, T. Yokokawa and D.H. Bartlett. 2010. Emerging concepts on microbial processes in the bathypelagic ocean–ecology, biogeochemistry, and genomics. Deep Sea Res., Part II 57: 1519–1536.
Nielsen, L.T. and T. Kiørboe. 2015. Feeding currents facilitate a mixotrophic way of life. ISME J. 9: 2117.
Ogawa, H., Y. Amagai, I. Koike, K. Kaiser and R. Benner. 2001. Production of refractory dissolved organic matter by bacteria. Science 292: 917–920.
Olson, M.B., K. Solem and B. Love. 2018. Microzooplankton grazing responds to simulated ocean acidification indirectly through changes in prey cellular characteristics. Mar. Ecol.: Prog. Ser. 604: 83–97.
Onda, D.F.L., E. Medrinal, A.M. Comeau, M. Thaler, M. Babin and C. Lovejoy. 2017. Seasonal and interannual changes in ciliate and dinoflagellate species assemblages in the Arctic Ocean (Amundsen Gulf, Beaufort Sea, Canada). Front. Mar. Sci. 4: 16.
Pécseli, H., J. Trulsen and Ø. Fiksen. 2014. Predator–prey encounter and capture rates in turbulent environments. Limnol. Oceanogr.: Fluids and Environments 4: 85–105.
Pernthaler, J. 2005. Predation on prokaryotes in the water column and its ecological implications. Nat. Rev. Microbiol. 3: 537.
Pomeroy, L.R. 1974. The ocean's food web, a changing paradigm. Bioscience 24: 499–504.
Raven, J.A., J. Beardall, K.J. Flynn and S.C. Maberly. 2009. Phagotrophy in the origins of photosynthesis in eukaryotes and as a complementary mode of nutrition in phototrophs: relation to Darwin's insectivorous plants. J. Exp. Bot. 60: 3975–3987.

Roohani, K., B.A. Haubrich, K.-L. Yue, N. D'Souza, A. Montalbano, T. Rynearson, S. Menden-Deuer and C.W. Reid. 2019. Trophic upgrading and mobilization of wax esters in microzooplankton. PeerJ. 7: e7549.
Rose, J.M. and D.A. Caron. 2007. Does low temperature constrain the growth rates of heterotrophic protists? Evidence and implications for algal blooms in cold waters. Limnol. Oceanogr. 52: 886–895.
Rose, J.M., Y. Feng, C.J. Gobler, R. Gutierrez, C.E. Hare, K. Leblanc and D.A. Hutchins. 2009. Effects of increased pCO_2 and temperature on the North Atlantic spring bloom. II. Microzooplankton abundance and grazing. Mar. Ecol.: Prog. Ser. 388: 27–40.
Saiz, E. and T. Kiørboe. 1995. Predatory and suspension feeding of the copepod *Acartia tonsa* in turbulent environments. Mar. Ecol.: Prog. Ser. 122: 147–158.
Saiz, E. and A. Calbet. 2011. Copepod feeding in the ocean: scaling patterns, composition of their diet and the bias of estimates due to microzooplankton grazing during incubations. Hydrobiologia 666: 181–196.
Schaum, C.E., A. Buckling, N. Smirnoff, D.J. Studholme and G. Yvon-Durocher. 2018. Environmental fluctuations accelerate molecular evolution of thermal tolerance in a marine diatom. Nature Communications 9: 1719.
Schmidtko, S., L. Stramma and M. Visbeck. 2017. Decline in global oceanic oxygen content during the past five decades. Nature 542: 335.
Schmoker, C., S. Hernández-León and A. Calbet. 2013. Microzooplankton grazing in the oceans: impacts, data variability, knowledge gaps and future directions. J. Plankton Res. 35: 691–706.
Schoener, D. and G. McManus. 2012. Plastid retention, use, and replacement in a kleptoplastidic ciliate. Aquat. Microb. Ecol. 67: 177–187.
Selosse, M.-A., M. Charpin and F. Not. 2017. Mixotrophy everywhere on land and in water: the grand écart hypothesis. Ecol. Lett. 20: 246–263.
Seymour, J.R., S.A. Amin, J.-B. Raina and R. Stocker. 2017. Zooming in on the phycosphere: the ecological interface for phytoplankton–bacteria relationships. Nature Microbiology 2: 17065.
Sherr, E.B. and B.F. Sherr. 2002. Significance of predation by protists in aquatic microbial food webs. Antonie van Leeuwenhoek. 81: 293–308.
Sherr, E.B. and B.F. Sherr. 2007. Heterotrophic dinoflagellates: a significant component of microzooplankton biomass and major grazers of diatoms in the sea. Mar. Ecol.: Prog. Ser. 352: 187–197.
Sherr, E.B. and B.F. Sherr. 2009. Capacity of herbivorous protists to control initiation and development of mass phytoplankton blooms. Aquat. Microb. Ecol. 57: 253–262.
Sieburth, J.M., V. Smetacek and J. Lenz. 1978. Pelagic ecosystem structure: Heterotrophic compartments of the plankton and their relationship to plankton size fractions. Limnol. Oceanogr. 23: 1256–1263.
Simon, M., F.O. Glöckner and R. Amann. 1999. Different community structure and temperature optima of heterotrophic picoplankton in various regions of the Southern Ocean. Aquat. Microb. Ecol. 18: 275–284.
Siuda, A.N. and H.G. Dam. 2010. Effects of omnivory and predator-prey elemental stoichiometry on planktonic trophic interactions. Limnol. Oceanogr. 55: 2107–2116.
Sogin, M.L., H.G. Morrison, J.A. Huber, D.M. Welch, S.M. Huse, P.R. Neal, J.M. Arrieta and G.J. Herndl. 2006. Microbial diversity in the deep sea and the underexplored "rare biosphere". Proc. Natl. Acad. Sci. USA 103: 12115–12120.
Sohrin, R., M. Imazawa, H. Fukuda and Y. Suzuki. 2010. Full-depth profiles of prokaryotes, heterotrophic nanoflagellates, and ciliates along a transect from the equatorial to the subarctic central Pacific Ocean. Deep Sea Res., Part II 57: 1537–1550.
Sommer, U. and K. Lengfellner. 2008. Climate change and the timing, magnitude, and composition of the phytoplankton spring bloom. Glob. Change Biol. 14: 1199–1208.
Steinberg, D.K. and M.R. Landry. 2017. Zooplankton and the ocean carbon cycle. Annu. Rev. Mar. 9: 413–444.
Sterner, R.W. and J.J. Elser. 2002. Ecological Stoichiometry: The Biology of Elements from Molecules to the Biosphere. Princeton University Press. Princeton, Oxford.
Stoecker, D.K., A.E. Michaels and L.H. Davis. 1987. Large proportion of marine planktonic ciliates found to contain functional chloroplasts. Nature 326: 790–792.
Stoecker, D.K. and J.M. Capuzzo. 1990. Predation on protozoa: its importance to zooplankton. J. Plankton Res. 12: 891–908.
Stoecker, D.K., M.D. Johnson, C. de Vargas and F. Not. 2009. Acquired phototrophy in aquatic protists. Aquat. Microb. Ecol. 57: 279–310.
Stoecker, D.K., P.J. Hansen, D.A. Caron and A. Mitra. 2017. Mixotrophy in the marine plankton. Ann. Rev. Mar. Sci. 9: 311–335.
Stoecker, D.K. and P.J. Lavrentyev. 2018. Mixotrophic plankton in the Polar Seas: a pan-arctic review. Front. Mar. Sci. 5: 292.
Tang, K.W. and M. Taal. 2005. Trophic modification of food quality by heterotrophic protists: species-specific effects on copepod egg production and egg hatching. J. Exp. Mar. Biol. Ecol. 318: 85–98.
Taylor, F.J.R., M. Hoppenrath and J.F. Saldarriaga. 2008. Dinoflagellate diversity and distribution. Biodivers. Conserv. 17: 407–418.
Thingstad, T.F., Å. HagstrÖm and F. Rassoulzadegan. 1997. Accumulation of degradable DOC in surface waters: Is it caused by a malfunctioning microbial loop? Limnol. Oceanogr. 42: 398–404.

Thingstad, T. 2005. Simulating the response to phosphate additions in the oligotrophic eastern Mediterranean using an idealized four-member microbial food web model. Deep Sea Res., Part II 52: 3074–3089.
Traboni, C., A. Calbet and E. Saiz. 2020. Effects of prey trophic mode on the gross-growth efficiency of marine copepods: the case of mixoplankton. Sci. Rep. 10: 12259.
Unrein, F., J.M. Gasol, F. Not, I. Forn and R. Massana. 2014. Mixotrophic haptophytes are key bacterial grazers in oligotrophic coastal waters. ISME J. 8: 164–176.
Visser, A.W. and T. Kiørboe. 2006. Plankton motility patterns and encounter rates. Oecologia. 148: 538–546.
Volk, T. and M.I. Hoffert. 1985. Ocean carbon pumps: Analysis of relative strengths and efficiencies in ocean-driven atmospheric CO_2 changes. pp. 99–110. *In*: Sundquist, E.T. and W.S. Broecker [eds.]. The Carbon Cycle and Atmospheric CO_2: Natural Variation Archean to Present. American Geophysical Union. Washington.
Ward, B.A. and M.J. Follows. 2016. Marine mixotrophy increases trophic transfer efficiency, mean organism size, and vertical carbon flux. Proc. Natl. Acad. Sci. USA 113: 2958–2963.
Ward, B.A., S. Collins, S. Dutkiewicz, S. Gibbs, P. Bown, A. Ridgwell, B. Sauterey, J.D. Wilson and A. Oschlies. 2019. Considering the role of adaptive evolution in models of the ocean and climate system. J. Adv. Model. Earth Syst. 11: 3343–3361.
Weinbauer, M.G., F. Chen and S.W. Wilhelm. 2011. Virus-mediated redistribution and partitioning of carbon in the global oceans. pp. 54–56. *In*: Jiao, N., F. Azam and S. Sanders [eds.]. Microbial Carbon Pump in the Oceans. Science/AAAS. Washington.
Wilhelm, S.W. and C.A. Suttle. 1999. Viruses and nutrient cycles in the sea: viruses play critical roles in the structure and function of aquatic food webs. Bioscience 49: 781–788.
Wirtz, K.W. 2012. Who is eating whom? Morphology and feeding type determine the size relation between planktonic predators and their ideal prey. Mar. Ecol.: Prog. Ser. 445: 1–12.
Woese, C.R., O. Kandler and M.L. Wheelis. 1990. Towards a natural system of organisms: proposal for the domains Archaea, Bacteria, and Eucarya. Proc. Natl. Acad. Sci. USA 87: 4576–4579.
Wohlers, J., A. Engel, E. Zollner, P. Breithaupt, K. Jurgens, H.G. Hoppe, U. Sommer and U. Riebesell. 2009. Changes in biogenic carbon flow in response to sea surface warming. Proc. Natl. Acad. Sci. USA 106: 7067–7072.
Worden, A.Z., M.J. Follows, S.J. Giovannoni, S. Wilken, A.E. Zimmerman and P.J. Keeling. 2015. Rethinking the marine carbon cycle: factoring in the multifarious lifestyles of microbes. Science 347: 1257594.
Zeebe, R.E. 2012. History of seawater carbonate chemistry, atmospheric CO_2, and ocean acidification. Annu. Rev. Earth Planet. Sci. 40: 141–165.

CHAPTER 6

Zooplankton: The Ocean Drifters

Danilo Luis Calliari Cuadro,[1,*] *Anabela Anahí Berasategui*[2] and *María Clara Menéndez*[2]

6.1 A general approach to zooplankton

This chapter provides an introduction to the main biological and ecological traits of marine zooplankton. It focuses on metazoan plankton (or metazooplankton; Protista are discussed in Chapter 5). The term zooplankton has etymological roots in the fusion of the Greek *zoon*, that means animal, and *planktos*, that means wandering or drifting (Hensen 1887). It identifies the free-living organisms whose swimming capacity is restricted and insufficient to oppose the movements of marine water masses. Although most zooplankton groups do possess some degree of motility that allows to control their position within the water column (for instance, to perform vertical migrations), their horizontal displacements and distribution are largely determined by marine currents. Due to their typical size, zooplankton develop their vital functions and activities (food capture, evasion of predators, vertical migrations) under a wide range of Reynolds numbers (Re)[a] that includes the hydrodynamic transition between viscous and inertial regimes (Vogel 2004, Mann and Lazier 2006).

The zooplankton is outstanding due to its functional role as a trophic link within marine communities, connecting primary producers and higher consumers like juvenile and adults of many fish species. Therefore, fluctuations in zooplankton composition, distribution and abundance influence the potential of pelagic fisheries (Lenz 2000, Bamstedt et al. 2000). Zooplankton grazing pressure may also modulate the seasonal succession and composition of phytoplankton, and consequently determine carbon and nutrient fluxes to the benthos (Miller and Wheeler 2012, Sommer et al. 2012). From an ecological standpoint, knowledge about the zooplankton community offers insights into the trophic state of the ecosystem, for instance based on its size structure, or on the feeding habits of the dominant species in the community. Some species occur over narrow temperature and/or salinity ranges. The association of those species to water masses with well-defined temperature and salinity ranges contributes to the understanding of circulation patterns in specific regions of the ocean (Boltovskoy 1981a, Bjonberg 1981, Boltovskoy et al. 1999). In addition to physical factors, zooplankton distribution is also modulated by life cycle,

[1] Facultad de Ciencias, Universidad de la República. Iguá 4225 - CP 11400, Montevideo, Uruguay.
[2] Instituto Argentino de Oceanografía. Camino La Carrindanga km 7.5, Bahía Blanca, Argentina.
Emails: aberasa@criba.edu.ar; menendez@criba.edu.ar
* Corresponding author: dcalliar@fcien.edu.uy

[a] The Reynolds (Re) number is a dimensionless quantity that describes the relative importance of inertial and viscous forces for a given process. Mathematically, Re is a ratio defined as (Zaret 1980): Re = (ρ/η) L. U; where L (m) is the typical length of the immersed object or organism moving at speed U ($m\ s^{-1}$) through a fluid of density ρ ($kg\ m^{-3}$) and viscosity η ($kg\ m^{-1}\ s^{-1}$).

reproductive behavior, daily/seasonal migrations, and feeding ecology, combined with the dynamics of their phytoplankton and microzooplankton prey.

Zooplankton can also be excellent bioindicators of water quality and climate change. Due to their short life cycles, fast rates of reproduction (Parmar et al. 2016) and sensitivity to environmental factors such as temperature, salinity and pollution, they respond rapidly to changes generated either by large scale (and global) climatic variability or to anthropic pressures at a local scale. Hence, biomass, diversity and species richness of zooplankton communities are useful indicators of marine ecosystem health.

This chapter begins with a brief description of the main systematic groups within the zooplankton, followed by a review of alternative classification schemes, a discussion of the trophic ecology of selected groups and of their role in marine food webs. It closes with brief remarks on the regulation of zooplankton secondary production.

6.1.1 Main systematic groups of zooplankton

As stated, the taxonomic diversity within the meta-zooplankton is very wide, including members of most extant metazoan phyla (from Cnidaria to Vertebrata). Since crustaceans contribute, by far, the largest component in terms of abundance, biomass and species richness, their most representative groups are presented below. Genetics has marked a new era in taxonomy with a strong impact on the established classification of marine fauna. The World Register of Marine Organisms (2020) is recommended for further reading and updates.

Phylum Arthropoda. Subphylum Crustacea:

Cladocerans (Class Branchiopoda, Order Diplostraca, Subperorder Cladocera—Latreille 1829). These are small crustaceans usually between 0.3 and 1.2 mm in body size (total length). Their lax body is partially or totally covered by a bivalve shell without a dorsal joint that leaves the cephalic portion free in most cases (Fig. 6.1a). Life cycle of cladocerans is notable among metazoans due to the prevalence of parthenogenesis: under normal environmental conditions, a given population is mostly comprised of females that reproduce asexually. Adverse conditions trigger the appearance of males and sexual reproduction and the production of resting eggs, which can remain dormant for long periods in the sediments (Ramírez and Pérez Seijas 1985).

The trophic ecology of cladocerans has been less studied than that of other planktonic groups (see Section 6.3). Some species (i.e., *Penilia avirostris*) may play an important role in microbial food webs due to their capability to reach small prey sizes. For other groups (*Podon* spp., *Pleopis* spp.), predation of large prey has been suggested through suction of prey's body fluids (Turner et al. 1988, Kim et al. 1989, Fofonoff 1994). Cladocerans are occasionally abundant and dominant in the mesozooplankton of coastal environments (Calliari et al. 2009, Onbé 1999). But contrary to the taxonomic richness of their freshwater counterparts, the number of marine cladoceran species is low, comprising only two families: Podonidae (Mordukhai-Boltovskoi 1968) and Sididae (Baird 1850).

Ostracods (Superclass Oligostraca, Class Ostracoda—Latreille 1802). Ostracods are also small organisms, between 0.5 and 30 mm. This group features a bivalve carapace that covers all soft parts of the animal and depending on the species, the shell can be soft or highly calcified, making them look like a tiny clam (Fig. 6.1b). Ostracods are a relatively little studied group. The class Ostracoda comprises four subclass, of which only two include currently living species and only one (Myodocopa—Sars 1866) contains planktonic species.

The feeding appendages of planktonic ostracods lack morphological adaptations that would indicate filter-feeding, and it has been suggested that all myodocopids are active predators on crustaceans, chaetognaths and jellyfish (Ramírez and Pérez Seijas 1981). They might also feed on salps feces, which would imply an important role in the re-cycling of material *en route* towards the deep ocean (Angel 1981). Ostracods are frequently the second most abundant group after copepods, both in the tropics and in high latitudes (below 100 m depth). Despite their abundance, their contribution to the "standing crop" may be low due to their small size.

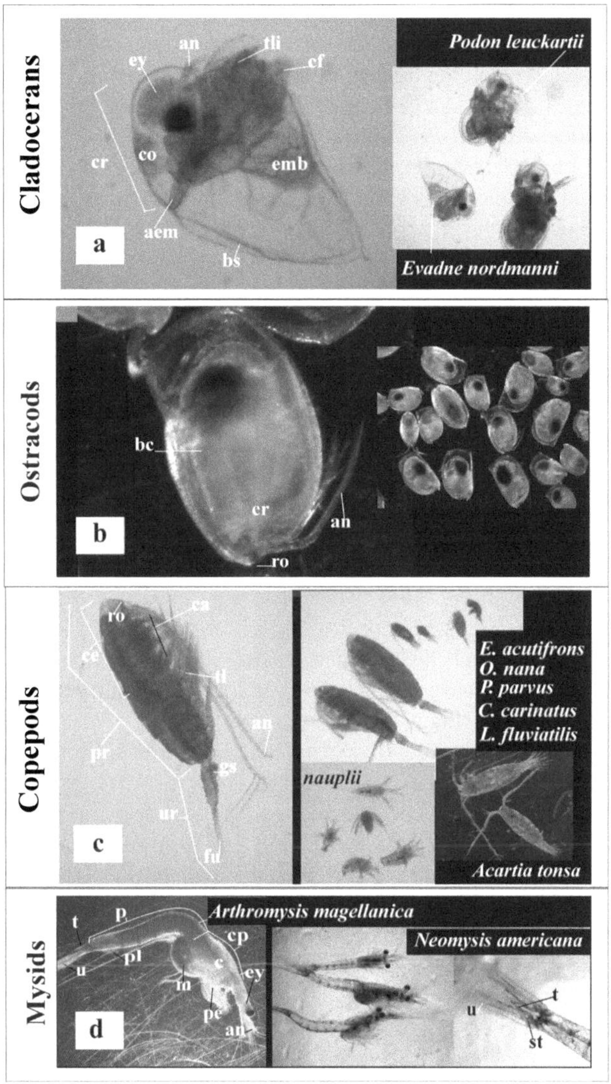

Figure 6.1. (a) *Evadne nordmanni* from Rincon front, Argentina: ey-compound eye, an-antennae, tli-thoracic limbs, cf-caudal furca, emb-embryo, bs-bivalve shell, aem-antennal elevator muscle, co-cervical organ, and cr-cephalic region. (b) Ostracod from Malvinas current: an-second antennae, ro-rostrum, cr-cephalic region, and bc-bivalved carapace. (c) *Calanoides carinatus* from Rincon front Argentina: ce-cephalosome, ro-rostrum, ca-cephalic appendages, tl-thoracic appendages, an-antennule, gs-genital somite, fu-furca, ur-urosome, and pr-prosome. Center-up, from smallest to largest size: *Euterpina acutifrons*, *Ctenocalanus vanus* (male), *Oithona nana*, *Paracalanus parvus* (female), *Calanoides carinatus* (female), *Labidocera fluviatilis* (females). Center-bottom: mixed nauplii from copepods and other crustaceans. Right-bottom: *Acartia tonsa*, from Bahía Blanca estuary, Argentina (d) *Arthromysis magellanica*, from Bahía Blanca estuary: t-telson, p-pleon, cp-cephalopereon, c-carapace, ey-eyes, an-antennule, pe-pereopods, m-marsupium, pl-pleopods, u-uropods, st-statocyst. *Neomysis americana*, from Rocha lagoon, Uruguay. Photos by authors except b (G. Cepeda, and E. Leonarduzzi).

Copepods (Class Hexanauplia, Subclass Copepoda—Milne Edwards 1840). These are generally small crustaceans, with a total length between 0.2 mm and 18 mm (Fig. 6.1c). Copepods are dioecious, with sexual dimorphism that allows relatively easy distinction between females and males. Subclass Copepoda includes three planktonic Orders: Calanoida, Cyclopoida and Harpacticoida (the latter includes many benthic or benthic-pelagic species). Fertilized eggs are released in the water column in broadcast spawners—a strategy mostly found in calanoid copepods—or they can be carried by the female in one or two ovigerous sacs (most cyclopoid and harpacticoid copepods). Egg hatching is followed by six naupliar stages and later by six copepodite stages, the last corresponding to the adult, sexual mature form.

Copepods play a crucial role in the transfer of matter and energy from primary producers to higher trophic levels (Calbet et al. 2000, Huskin et al. 2001, see also Section 6.3). Early research interest in copepods was largely associated to fisheries (Bjornberg 1981) as they are primary food items for many commercial fish species. The role of copepods in pelagic food webs and export of organic carbon is further discussed in Section 6.3. Pelagic copepods represent the main component within the mesozooplankton, comprising between 55 and 95% of zooplankton numbers in net samples (Mauchline 1998), and contributing up to 90–97% of the biomass, depending on region and season (Bradford Grieve et al. 1999).

Mysids (Class Malacostraca, Subclass Eumalacostraca, Superorder Peracarida; Order Mysida—Boas 1883) or "opossum shrimps" have body sizes in the range of 2.6 to up to 31 mm. Like all Peracarida, female mysids have extensions in the pereopods (thoracic appendages) that form a ventral marsupium or incubator chamber where offspring develop during embryogenesis and larval stages (Fig. 6.1d).

Mysids can be neritic pelagic or demersal in the deep sea, with mesopelagic and bathypelagic species. But they also inhabit very shallow coastal waters; intertidal species may burrow into the sediments and stay buried at low tide. Most pelagic mysids are omnivorous feeders on phytoplankton, detritus and zooplankton (Murano 1999). These crustaceans are usually very abundant in neritic systems like estuaries, where they are prey for many fish species in the larval stage (Ramírez 1976, Sánchez et al. 1991, Schiariti et al. 2004, Cardelli et al. 2006, Giberto et al. 2007, López Cazorla et al. 2011, Acha et al. 2018, Rodríguez-Graña et al. 2018).

Amphipods (Class Malacostraca, Subclass Eumalacostraca, Superorder Peracarida, Order Amphipoda—Latreille 1816). Pelagic amphipods have body sizes between 2 and 150 mm approximately, being laterally compressed and lacking a carapace (Fig. 6.2a).

This group of crustaceans successfully colonized the pelagic realm from benthic ancestors; in fact, most living amphipods are still benthic. This order includes four sub-orders, Hyperiidea, Gammaridea, and Caprellidea, with numerous members in epi- and mesopelagic communities (Vinogradov 1999). Hyperid species are associated with gelatinous plankton: about half of them parasite jellyfish, siphonophores or salps (Vinogradov 1999). Free-living species may even lay their eggs or larvae on gelatinous plankters (Madin and Harbison 1977, Laval 1980). Genera *Hyperia and Lestrigonus* are ectoparasites of gelatinous plankton and feed on their tissues (Vinogradov 1999). They reproduce sexually. Gammarids and some hyperiids have direct development. However, many species of hyperiids have larval stages that the female transfers from the marsupium to the host (Harbison et al. 1977, Laval 1980, Gasca and Haddock 2004).

Amphipods are important in the diet of nektonic predators including fish such as *Engraulis anchoita* (Angelescu and Anganuzzi 1981) and *Merluccius hubbsi* (Cordo 1981, Angelescu and Prenski 1987, Prenski and Angelescu 1993, Moriondo 2002), and demersal fish such as *Gadus morhua* (cod) and *Melanogrammus aeglefinus* (haddock) (Jaworski and Ragnarsson 2006). Amphipods are also a food resource for the squid *Illex argentinus* (Ivanovic and Brunetti 1994, Cabreira et al. 2011) and whales (Nemoto and Yoo 1970, Highsmith and Coyle 1992) and may play a key role in the Antarctic ecosystem supporting invertebrates, fishes, birds and mammals (Dauby et al. 2003). Indeed, it has been reported that "*the total amount of consumed amphipods was estimated to be 60 million tons per year for the whole Southern Ocean, i.e., the second animal group in importance after euphausiids.*" (Dauby et al. 2003).

Euphausiids (Class Malacostraca; Subclass Eumalacostraca; Superorder Eucarida; Order Euphausiacea—Dana 1852). Exclusively marine organisms distributed in oceanic and neritic waters globally, also known as krill. Krill can be relatively large (size 10–150 mm) and sometimes are referred to as "micro-nekton". They have an elongated and laterally compressed body vaguely resembling shrimp (Fig. 6.2b). Euphausiids are dioic and their development includes several larval stages: nauplius, a stage of metanauplius, three stages of calyptopis and a variable number (3–8) of furcilla stages, before reaching the juvenile phase (Gibbons et al. 1999, and references therein).

Food availability influences the trophic habits of krill, which can be truly omnivorous or carnivorous, preying on dinoflagellates, tintinnids, flagellates, diatoms, fish eggs and larvae, and detritus (Mauchline and Fisher 1969, Stuart and Huggett 1992). *Euphausia superba*, the Antarctic krill, is a highly abundant

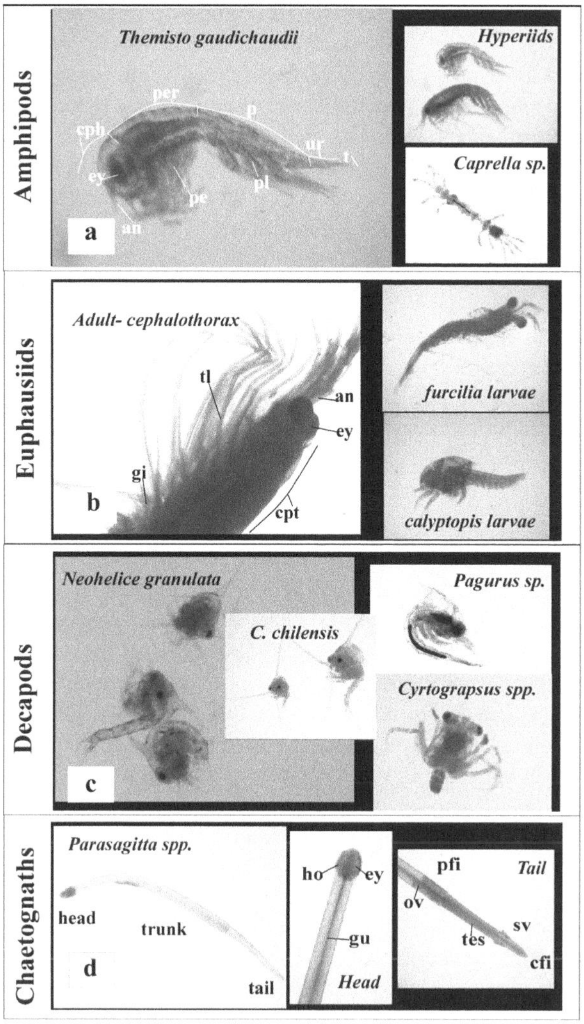

Figure 6.2. (a) *Themisto gaudichaudii* from Rincon front, Argentina: ey-eye, an-antennae, cph-cephalon, per-pereon, p-pleon, ur-urosome, t-telson, pl-pleopods, and pe-pereopods. (b) Unidentified euphausiids from Potter creek, Antarctica: an-antennae, ey-eye, cpt-cephalothorax, tl-thoracic legs, and gi-gills. (c) Decapods, from Bahía Blanca estuary. Zoea larval stage of *Neohelice granulata, Pagurus* sp., *Corystoides chilensis,* and megalopa larval stage of *Cyrtograpsus* spp. (d) *Parasagitta* spp. from Rincon front, Argentina: ho-hooks, ey-eyes, gu-gut, ov-ovary, pfi-posterior fin, tes-testes, sv-seminal vesicle, and cfi-caudal fin. Photos by authors except a and b (M. Garcia).

species that forms mono-specific "swarms" (e.g., Ramírez 2016, Cepeda et al. 2018, and references therein). It is considered a "keystone" species as it consumes large amounts of phytoplankton and channels a significant fraction of the primary production. In turn, *E. superba* serves as food for a large number of predators, e.g., penguins, pinnipeds, beard whales, fish, squid, albatrosses and most species of flying seabirds.

Decapods (Class Malacostraca; Subclass Eumalacostraca, Superorder Eucarida; Order Decapoda —Latreille 1802). Decapod larvae contribute to the meroplankton of all neritic ecosystems (Fig. 6.2c); morphology and characteristics of the developmental stages differ depending on the suborder (Dendrobranchiata and Pleocyemata). Most Dendrobranchiata adults inhabit demersal-pelagic systems, integrating the macrozooplankton. For further information on decapods larvae, we recommend the reviews by Boschi (1981) and Pohle et al. (1999).

Suborder Dendrobranchiata (Spence Bate 1888). This Suborder (shrimps) includes the Superfamily Penaeoidea (Rafinesque 1815) and Sergestoidea (Dana 1852). As the name implies, adults of this group possess dendrobranchia-type gills. Adult body is laterally compressed and the first three pairs of pereopods have chelae. Specimens of both families are usually marine (demersal-pelagic or benthic), but some species inhabit brackish waters. Eggs are released directly to the sea where they contribute to the meroplankton. The first larval stage is the nauplius, followed by protozoea, mysis and decapodid stages. Most penaeoideans are important food resource for several fish species and are target of commercial fisheries. The diversity of Dendrobrachiata is highest at low latitudes, decreasing towards temperate and cold regions.

Suborder Pleocyemata (Burkenroad 1963). This Suborder includes 11 infraorders that in turn represent most crabs and lobsters. As a general characteristic, the types of gills present are phyllochranchia or tricobranchia, and females carry their eggs with the pleopods, where embryogenesis occurs. Infraorders Stenopodidea (Spence Bate 1888), Caridea (Dana 1852), Axiidea (de Saint Laurent 1979), Gebiidea (de Saint Laurent 1979), Anomura (MacLeay 1838) and Brachyura (Latreille 1802) have several stages of zoea followed by decapodid stages. Members of Infraorder Achelata (Scholtz and Richter 1995) develop through larval types Phyllosoma followed by a Decapodid stage (puerulus).

Phylum Chaetognatha. Class Sagittoidea:

Chaetognaths are an exclusively marine phylum (Chaetognatha, see, Casanova 1999, and references therein) commonly called arrow-worms owing to their elongate shape (Fig. 6.2d). Their body is transparent to semi opaque and ranges between 2 and 120 mm. They live in benthic, pelagic neritic and open sea habitats. Some species are useful as indicators of water masses. Their vertical distribution is influenced by individual age, and also by temperature and light intensity. They are voracious carnivorous on larger microzooplankton and on mesozooplankton, e.g., nauplii and copepodid stages, but also are important consumers of fish eggs and larvae. The role of chaetognaths in pelagic food webs is further discussed in Section 6.3. Chaetognaths are protandrous hermaphrodites; the position and structure of the connection between testis and ovary are features of taxonomic interest.

6.1.2 Classification of the zooplankton

As stated, zooplankton is an ataxonomic category that groups organisms with a common mode of life: free drifting. Thus, zooplankton comprises a vast and heterogeneous assemblage of organisms, and their classification is the first step to their study. Zooplankton has been classified according to various criteria: (a) considering their life cycle; (b) based on body size (Sieburth et al. 1978); (c) according to ecological attributes of the habitat; (d) according to their trophic habit; and (e) based on functional traits and role in the ecosystem.

a) Life cycle: Species that develop their entire life cycle in the pelagic realm belong to the holoplankton, such as ctenophores, pteropods, most copepods, amphipods, euphausiids, chaetognaths, salps and appendicularians, among others. Instead, those species whose life cycle is only partially developed in the water column (i.e., in general as eggs and/or larval stages) are classified as meroplankton. Larval stages of cephalopods and fish develop in the plankton while their juvenile-adult stages live as nekton. Other typical meroplankton are the larval stages of benthic animals, like polichaets, decapods, bivalves and gastropods. Most hydrozoans and scyphozoa have alternating generations with pelagic medusae as the sexual phase and benthic polyps as the asexual phase.
b) Body size: Marine zooplankton comprise a very wide spectrum of body sizes, from tiny protozoans a few micrometers (μm) long to giant jellyfish with sizes in the order of meters. Individual body size is arguably the single most important trait that defines the attributes and role of organisms in ecological networks (Elton 1927, Woodward et al. 2005). Attributes linked to size include the rates of vital functions and the general metabolism of individuals (Brown et al. 2004), longevity (Calder 1983, Holm et al. 2016) and trophic position (Arim et al. 2010, Segura et al. 2015). The most accepted size classification for plankton was proposed by Sieburth et al. (1978). According to that, individuals

smaller than 200 μm are classified as nanozooplankton (0.2–20 μm) and microzooplankton (20–200 μm); those include dinoflagellates, ciliated and flagellated protozoans. The microzooplankton includes most larval forms from the meroplankton and embryonic and larval stages from the holoplankton (e.g., eggs and nauplii of crustaceans). The next size category is the mesozooplankton (0.2–20 mm), which includes adult holoplanktonic crustaceans like copepods, chaetognaths, appendicularians, small hydromedusae and ctenophores, doliolids, and advanced larval stages of certain meroplankton (e.g., fish eggs and larvae). Most metazooplankton species fall into this category and for this reason, plankton nets fitted with meshes of pore sizes between 200 and 500 μm have been traditionally used to collect them (Boltovskoy 1981b). Finally, the two larger size categories are the macrozooplankton (2–20 cm) and the megazooplankton (20–200 cm). Largest specimens of gelatinous zooplankton (ctenophores, hydromedusae, siphonophores, scyphomedusae), fish larvae and some crustaceans like mysids, amphipods and euphausiids, are classified as macrozooplankton, while most of the megazooplankton belongs to the gelatinous plankton (e.g., pelagic tunicates and large medusae). In this chapter, we will see how body size is a widely pervasive trait linked to most important aspects in the ecology of the zooplankton.

c) Habitat: In a first approximation, we may identify two main large systems: neritic and oceanic, with marked physicochemical differences that condition the specific composition of the zooplankton. The neritic system extends from the littoral zone up to the external limit of the continental shelf (the shelf-break, corresponding to a depth of approx. 200 m); further offshore lies the deeper oceanic system. Neritic zooplankton tend to be better adapted to stronger temperature and salinity variability, turbulence, and to a generally higher trophic status. Due to the proximity to the benthos, the presence of meroplankton is more significant in neritic systems. Oceanic zooplankton is generally adapted to more stable temperature and salinity regimes, low turbidity, stronger vertical stratification, and lower nutrients and biomass concentrations.

d) Trophic relationships: Metazoan zooplankton are heterotrophic, and they are generally thought go be opportunistic consumers (Miller and Wheeler 2012). Therefore, except for some near-exclusively herbivorous groups, most zooplankters are omnivorous, although they exhibit tendencies towards either herbivorous or carnivorous behavior (Turner 1984). That matter will be discussed further in Section 6.3 (Feeding and Food webs). As an example, hyperiid amphipods, ostracods of the genus *Conchoecia*, chaetognaths of the genus *Sagitta*, and the copepod *Labidocera fluviatilis*, all have mandibular appendages with sharp and serrated edges typical of carnivorous organisms (Ramírez and Pérez Seijas 1981, Mauchline 1998, Ramírez 2007).

e) Functional classification: This approach is based on individual functional traits, i.e., the phenotypic attributes that modulate fitness and define the contribution of an organism to ecosystem functioning (Violle et al. 2007). This approach takes into consideration key traits related to the behavior, physiology and biogeography of the organisms. Some traits, such as body size and motility, transcend several functions and are important determinants of the ecological role of zooplankton (Barnett et al. 2007, Litchman et al. 2013, Benedetti et al. 2016, 2018, Kiørboe et al. 2018).

Functional approaches have a strong tradition in phytoplankton studies (e.g., Reynolds 1980, Reynolds et al. 2002, Litchman and Klausmeier 2008, Kruk et al. 2010, see also Chapter 3 in this book), but are less explored for the zooplankton. Functional groups can be defined from a given pool of species with known functional traits using statistical methods (e.g., cluster analysis, or analogous). Identification of the relevant traits is naturally of utmost importance. Litchman et al. (2013) proposed a very complete set of functional traits for zooplankton that encompasses a combination of biological functions (feeding, growth and reproduction, and survival), and types (morphological, physiological, behavioral, and life history). They noted that most traits are not independent of one another but tend to be inter-correlated due to fundamental physiological constraints that result in trade-offs which are reflected in the emergence of ecological strategies. They identified an emerging pattern of two main life history types among zooplankton that feed mainly on pico- *vs.* nano-microplankton. It can be noted that those life-strategies match well-defined environmental, biological (taxonomic and/or functional structure) and biogeochemical scenarios. Litchtman et al. (2013) also considered that models focused on traits could be a more rewarding avenue of research than the construction of functional groups, as models would allow the emergence of novel combinations of traits under changing environmental conditions. Relevant directions for further development identified by these same authors are the exploration of traits variability and their distribution along environmental gradients (e.g., temperature, trophic status).

Interest in functional classifications is clearly sparked by their potential to contribute to ecosystem models where a highly diverse community may be properly represented by a few, functionally homogeneous groups. Currently, marine ecosystem models still largely rely on coarse representations of the zooplankton in size categories which do not capture their intrinsic biological diversity. Thus, functional trait-based approaches may constitute a step towards the incorporation of biological diversity into more realistic conceptual and quantitative ecosystem models, within reasonable degrees of complexity. In turn, traits-based modeling could teach about how patterns and traits along the ecological hierarchy emerge from trade-offs, from interactions with the environment and between traits at different levels in the biological hierarchy. Such functional trait-based models may contribute a new take on currently pressing issues, like the effects of environmental variability and human-induced alterations in marine ecosystems—including the effects of climate change—and on the ocean's capacity to continue providing services to humankind in a sustainable way.

To illustrate the use of functional classification, we refer to recent examples that explore the association between functional groups and environmental conditions. Benedetti et al. (2016) built a database of functional traits of the Mediterranean copepod assemblage (191 species) and through multiple correspondence analysis clustered species into six functional groups: I: large cruise feeding carnivores; II: smaller sac-spawning ambush feeders (carnivores); III: small omnivorous broadcasters with mixed feeding strategies; IV: filter feeding, omnivorous-herbivorous; V: small ambush feeding omnivores (mostly *Oithona*); VI: omnivores and omnivore-herbivorous and cruising detritivores, mainly small sac-spawners (*Oncaea* spp., *Microsetella* spp.). Building on those results, Benedetti et al. (2018) investigated the distribution of functional groups along environmental gradients, finding that carnivorous groups were more important under oligotrophic and tropical conditions, while large and small current-feeding herbivores were associated with colder, more productive and seasonal environments. Other (small cruising detritivores and current-feeding herbivores) showed no clear preferences. It can be expected that in the near future similar descriptions will be available for other regions of the world ocean.

6.2 Spatio-temporal distribution of marine zooplankton in relation to the environment

Functional traits mediate the fundamental life missions (Violle et al. 2007) and determine the fitness of individuals under given biotic and abiotic conditions (Westoby and Wright 2006, Litchman and Klausmeier 2008, Litchman et al. 2013). *Body size* represents the main morphological trait that affects all major life missions (Litchman et al. 2013, see also Section 6.3.2). For instance, as a very general rule, predation in the pelagic ecosystem is size dependent, i.e., predators are larger than their typical prey in a more or less constant ratio. For organisms in the size range of mesozooplankton, predator-prey ratio ranges between 5 to > 50 (18:1 for copepods, 50:1 for cladocerans and meroplankton larvae) (Hansen et al. 1994).

The variability of fitness along environmental gradients is a major driver of temporal and spatial distribution of zooplankton species at large scales, that is, of the biogeography of zooplankton (Brun et al. 2016), but also at smaller scales as we will see below. Certain individual trait combinations are favored under specific environmental conditions, and together with the environment they shape higher level traits like abundance of organisms and magnitude of fundamental ecosystem fluxes, e.g., primary and secondary production, grazing, etc. Oceanographic variability relevant for the zooplankton occurs at the full spectrum of scales, from basin-wide to the micro-scale. Next, we discuss how selected oceanographic patterns affect the distribution and activity of the zooplankton at different spatial scales.

6.2.1 Macroscale distribution: Biogeography of marine zooplankton

This section discusses aspects related to the large-scale distribution of the zooplankton, broadly considered as that corresponding to scales of thousands of km. A consequence of size-dependent predation is that zooplankton communities are strongly size-structured (Barton et al. 2013). The oligotrophic subtropical gyres are dominated by small phytoplankton that are grazed by small herbivorous zooplankton, which is in turn eaten by larger zooplankton (Calbet and Landry 2004). However, large and predominantly herbivorous zooplankton dominate the strongly seasonal subpolar and arctic regions characterized by spring blooms of large diatoms (Barton et al. 2013). Accordingly, the average zooplankton body size decreases in the oligotrophic seas and increases in more nutrient-rich regions (San Martin et al. 2006). That is also linked to the interaction between ontogeny and environmental conditions, notably temperature and food.

Development time is the time taken to develop from egg to a reproducing adult, and it plays an important role in the rate of population growth. The timing of maturity can be especially important in organisms that reproduce once in their life history, or in organisms that depend on seasonal food production cycles (e.g., blooms). Development rate depends primarily on temperature, particularly under non-food-limited conditions (Mauchline 1998). However, growth and development rates are not strongly coupled and the disproportionate influence of temperature on development produces a strong negative relation between temperature and adult body size, both within and between species (Forster et al. 2011). Copepod species have longer development times in polar and subpolar regions than in tropical and subtropical systems, and individual size of local species is larger in polar regions and decreases towards the Equator (Barton et al. 2013). That pattern conforms to Bergmann's rule, which for marine crustaceans was explained as a result of increased cell size and life span in colder environments (Timofeev 2001). On the other hand, the *relative offspring size* also changes with latitude, being largest in the tropics and subtropics (Brun et al. 2016). Large offspring size is often seen as an adaptation to harsh environments (Segers and Taborsky 2011), as it may contribute to minimize predation and to improve competition for resources.

The development of overwintering strategies in some zooplankton species also has biogeographical implications, as it is the case of high latitude copepod species which undergo dormant stages to survive periods of unfavorable conditions (Johnson et al. 2008). Diapausing organisms are generally larger and can be critical to the functioning of highly productive ecosystems in polar, subpolar, and upwelling regions due to lipid accumulations (Lee et al. 2006). A discussion of biogeochemical implications of overwintering in zooplankton is presented in Box 1.

Finally, it should be noted that the relationship between food availability and abundance of predators determines the spatio-temporal distribution of the *feeding traits* of the zooplankton along environmental gradients (Barton et al. 2013). Zooplanktonic organisms may employ different feeding behaviors in response to changing environmental conditions and prey abundance (Mariani et al. 2013).

6.2.2 Meso- to small scale distribution: Frontal ecosystems

An oceanographic front is a region of the ocean where the properties of the water column change abruptly in the horizontal plane (Mann 2000; Chapter 2). Some well-studied types are upwelling fronts, shelf-break and tidal fronts, estuarine fronts and river plumes, as well as fronts associated to eddies and abrupt topographic features. Fronts are frequently associated to increased primary and secondary production (Olson and Backus 1985) favored by complementary biogeochemistry of intervening water masses, and by vertical transport of nutrient-rich waters associated with frontal circulation. More generally, the condition leading to enhanced biological production at fronts can be thought of as a delicate balance of physical processes under which primary producers manage to optimize the basic trade-off between factors that control individual growth: nutrients and light. This idea can be traced back to nearly a century ago (Gran and Braarud 1923, see also Section 6.3.2). The physical balance is represented by the mixing of the water column—which fertilizes illuminated layers near the surface—and stratification that allows the primary producers to stay within the illuminated layers (Cushing 1989, Mann and Lazier

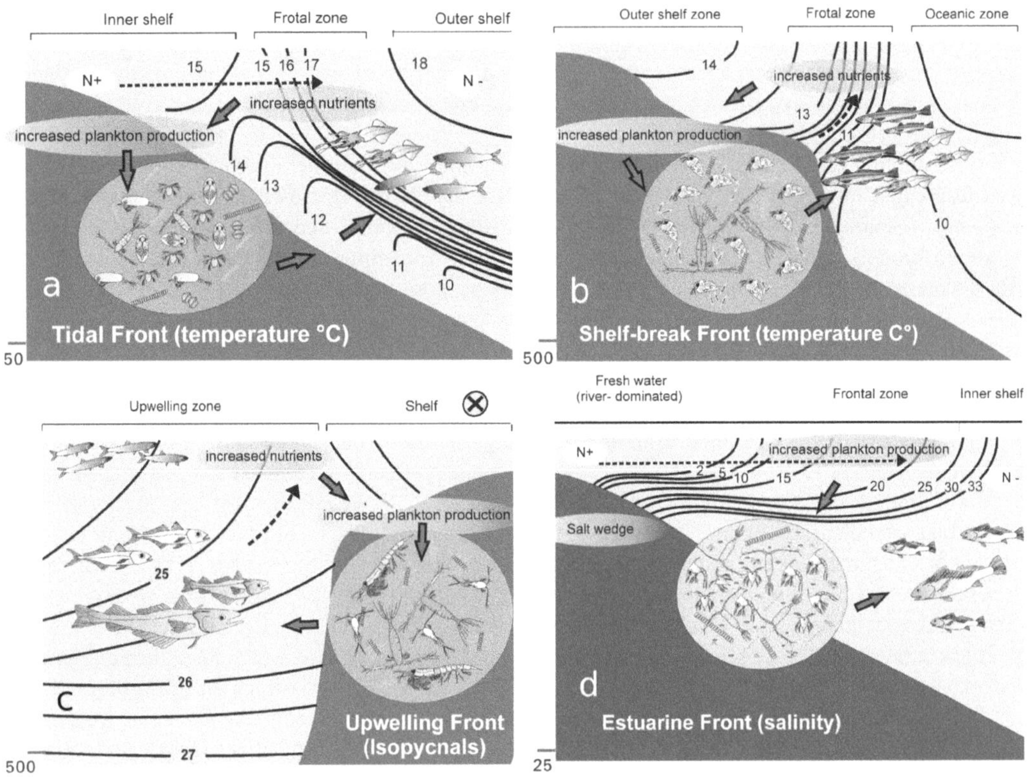

Figure 6.3. Schematic representation of the physical configuration and biological responses under different types of oceanographic frontal systems: (a) Tidal front; (b) Shelf-brak front; (c) Upwelling front, and (d) Estuarine front. Dashed arrows indicate the general direction of nutrient fluxes. Thick blue arrows depict the consequences at sequentially higher trophic levels of increased primary production. Note that isolines represent different variables at different types of fronts.

2006). The alternation of stable (stratified) and unstable (un-stratified) conditions take place at spatial (i.e., oceanographic fronts) and temporal transitions (i.e., formation/destruction of a seasonal thermocline leading to the "spring bloom"). Such transitions (i.e., the "weakly stratified water column", Cushing 1989) represent particularly favorable conditions for primary producers, and their response to the changing balance in the physical scenario reverberates along marine food webs, and first of all among the zooplankton. We review next the patterns resulting from zooplankton responses to physical modulation in the main types of oceanographic fronts.

Tidal fronts (or Shelf-Sea fronts, Fig. 6.3a) tend to be highly productive areas that sustain rich fisheries (Bakun 1996). These systems develop in transitions where the water column changes from tidally mixed to stratified. Mixing results from the action of tidal currents over shallow topography in continental shelves subjected to ample tidal ranges (Bakun 1996, Mann 2000). When a seasonal, shallow thermocline develops, turbulent mixing is able to overcome density (thermal) stratification in shallower waters, but not at deeper offshore waters, defining a frontal area in-between those regions (Simpson and Hunter 1974, Pingree et al. 1975, Le Févre 1986, Chapter 2). Primary production usually peaks in the frontal area, or in the stratified side and along the pycnocline, stimulated by nutrient fluxes from the mixed side of the front (Kiørboe 1993, Bakun 1996). Higher microzooplankton densities may occur at the frontal area (Viñas and Ramírez 1996). Zooplankton respond to that scenario in several ways. Copepod reproduction and growth follow the spatial pattern of the primary producers (Kiørboe and Johansen 1986, Kiørboe 1993, Viñas and Ramírez 1996). Zooplankton biomass may also peak at the frontal area and nearby stratified waters (Sabatini and Martos 2002, Derisio et al. 2014, Temperoni et al. 2014), but that is not always the case (Kiørboe 1993). Copepod community structure also changes across these fronts. For example, in the

Peninsula Valdes tidal front, in well-mixed waters and at the frontal interface dominant calanoid copepods are *Drepanopus forcipatus, Ctenocalanus vanus, Paracalanus parvus, Calanus australis* and *Calanoides carinatus*, while in stratified waters the cyclopoids *Oithina aff. helgolandica* and *O. atlantica* dominate. *C. carinatus* is progressively replaced offshore by *Calanus simillimus* and *C. australis* (Sabatini and Martos 2002, Temperoni et al. 2014).

The velocity structure in the frontal area is characterized by surface warm and near-bottom cold shoreward flows on the offshore side of the front which converge in the frontal zone. A compensating returning (seaward) flow of mixed water develops at intermediate depths (Bakun 1996, Mann and Lazier 2006). Plankton able to control their vertical position, i.e., copepods and particularly fish larvae, may avoid offshore advection by using this stratified flow structure (Iles and Sinclair 1982). In the Gulf of Maine – Georges Bank, haddock larvae seem to concentrate and experience faster growth rates in the pycnocline region immediately offshore of the mixed zone (Buckley and Lough 1987), probably as a consequence of better trophic conditions there.

Shelf-Break fronts (Fig. 6.3b) run along the shelf-break and upper continental slope, and from a biological perspective constitute one of the most important types of oceanographic fronts (Mann and Lazier 2006). Shelf-break fronts are particularly conspicuous in the North West and South West Atlantic, i.e., between Cape Hatteras and Davis strait (ca. 35°N–65°N), and off the Argentinean Patagonia in the Southern Hemisphere (the Patagonian Shelf brake front, ca. 38°S–55°S, Franco et al. 2008, Carranza et al. 2017). Each rise and fall of the tide forces water to move on and off the shelf; the interaction of tidal currents with the shelf-break induces internal waves at the interface between upper and lower layers of a stratified water column (Mann 2000) that propagate both towards and away from shore (Mann 2000). The turbulence associated with the passing of these waves over the shelf-brake can mix cool and nutrient-rich water from the lower layer upward. Also, along-front wind variability on timescales typical of atmospheric storms (2–10 days) modulate upwelling of sub-surface nutrient-rich water at the front (Carranza et al. 2017), that stimulates primary and secondary production. Further up in the trophic webs, high zooplankton production favors dense aggregations of both pelagic and benthic consumers (Mann and Lazier 2006).

Upwelling fronts. Upwelling systems result from wind action blowing parallel to the coastline (Fig. 6.3c). Description of the physical drivers and mechanisms can be found in Chapter 2, and previous contributions, e.g., Bakun (1996), Open University (2004), and Mann and Lazier (2006). Canonical examples are the upwelling systems along the Eastern margins of the main oceanic basins, i.e., those in the Perú-Chile, California, Benguela and Canary Islands regions. Action of upwelling—favorable wind can be variable on different timescales, notably seasonal and shorter, which leads to alternating periods of active and relaxed upwelling. Dynamical processes that force seawater upward can substantially enhance biological productivity. Upwelled water is normally colder and more nutrient-rich than surrounding waters, generating strong horizontal gradients and stimulating primary production. Major coastal upwelling systems occupy only about 1% of the world's oceans but on average account for 11% of new primary production (Lalli and Parsons 2006).

Mesozooplankton biomass in coastal upwelling systems is usually dominated by few copepods species of medium to large size-especially the genus *Calanus*-, one or two euphausiid species, and few chaetognaths (Mackas et al. 2006). In systems where primary production is intermittent, blooms typically form and decline within a period of 6 to 12 days. Copepods and euphausiids require 14 to 50 or more days to complete the generation, so their numbers do not respond to the short-term upwelling events. Euphausiids are also a significant component of the zooplankton community, for example, *Euphausia lucens* in the Benguela system (Anderson and Lucas 2008), *E. mucronata* in the Humboldt Current system (Antezana 2010), and *E. distinguenda* in the California Current system (Herrera et al. 2019). The success of euphausiids to thrive and dominate in upwelling ecosystems seems to derive from their large size and a long-life cycle compared to other zooplanton, as well as key adaptations to cope with intermittent food supply and low oxygen conditions typical of upwelling systems (Anderson and Lucas 2008, Antezana 2009). Physical transport away from upwelling centers remains a problem, and therefore, some planktonic organisms perform diel vertical migrations in order to maintain their position in a stratified flow field. The larger size makes euphausiids a key prey for large consumers, including whales that are temporary residents of upwelling systems.

6.2.3 Meso- to microscale distribution: Estuarine waters

The physical structure and functioning of estuarine ecosystems are strongly influenced by three main factors: *freshwater input*, *tides* and *wind regime*. Less dense fresh water tends to ride over more dense saltier waters favoring vertical stratification, whereas tidal movements and winds tend to homogenize the water column (Mann 2000). The presence of diluted seawater and a strong variability of salinity in time and space are probably the most prominent environmental features in estuaries, and that with the heaviest impact on the structure and distribution of the inhabiting fauna. Mixing/stratification regimes define different types of estuaries and consequently, the characteristics of zooplankton communities that inhabit them. Estuaries with high freshwater input and low tidal amplitude tend to be stratified. On the contrary, estuaries receiving little freshwater and subject to strong tidal/wind action are vertically mixed. The balance between these drivers may change seasonally. According to the circulation and saline structure, four types of estuaries are usually recognized:

- *Salt-wedge estuaries* (Fig. 6.3d): Mixing between the two layers is limited, and the vertical profile shows a sharp halocline. Seawater enters the estuary as a bottom layer of progressively shorter height, which in an axial section resembles the shape of a (salt) wedge. At the horizontal interface, water resulting from the mix of fresh and seawater is entrained in the out-flowing surface layer and the resulting compensatory flow contributes to further the advancement of the tip of the wedge into the estuary. The penetration of this wedge also changes with river flow condition.
- *Partially mixed*: Strong tides and/or winds cause partial vertical mixing of salt and fresh waters in the water column.
- *Vertically homogeneous*: The river flow is weak and the tidal flow strong. There is a longitudinal gradient from low salinity at the river mouth to high salinity near the sea, but no vertical gradients are present.

Estuarine waters are in a permanent dynamic balance and, consequently, zooplankton communities vary in time and spatially. The general pattern of faunal distribution along estuarine gradients was earlier described by Remane and Schlieper (1971) and refined by Attrill (2002), who highlighted the overriding importance of average salinity and its time variability for sessile species, or those with limited mobility. The pattern seems to apply for plankton as well. Generally, the highest number of species is found in the more saline water near the mouth, and richness tends to decrease with decreasing salinity. The number of species increases again at the freshwater zone (albeit, to a lower extent compared to marine waters). That pattern results from the interplay between salinity variability and the tolerance to such conditions by the zooplankton. Estuarine species are classified according to their tolerance to salinity in: *stenohaline*, those which tolerate only narrow ranges of salinity, and are generally found near the mouth of the estuary, and *euryhaline*, those which tolerate salinity fluctuations within a wider range, and can thus penetrate further up the estuary. *True estuarine* species are sometimes considered those confined to estuaries.

The net seaward flux characteristic of estuaries is a challenge for sustaining local zooplankton populations within the estuary. In stratified and partially mixed estuaries, surface low-salinity water typically flows seaward and a bottom compensatory flow of more saline bottom water moves up the estuary. One strategy zooplankton have evolved to avoid export off the estuarine realm is through vertical migration synchronized with tides, instead of the diel cycle (Naylor 2006). Apparently, zooplankton take advantage of the vertically structured, two-layered circulation pattern in order to remain within the estuarine range and avoid offshore advection by adjusting the timing of their vertical movements. While residing in near-surface waters, zooplankton is carried towards the estuarine mouth; by migrating downwards, zooplankton ride on the bottom flow that takes them back towards the head of the estuary. Thus, by performing vertical migrations with the proper timing, zooplankton manage to keep populations in the estuary year-round, or entering and leaving the estuary on a seasonal basis (e.g., Castel and Veiga 1990, Morgan et al. 1997). For instance, the copepod *Acartia longipatella* in South-African estuaries

(Wooldridge and Erasmus 1980) and the cladoceran *Podon polyphemoides* in Chesapeake Bay (Bosch and Taylor 1973) undergo vertical migrations with tidal periodicity. However, this mechanism is unlikely to work in well-mixed systems (Menéndez et al. 2012), where the behavioral basis in the case of the copepod *Eurytemora affinis* may reside in the avoidance of adverse currents when in its preferred salinity zone (De Pauw 1973), active anadromous migration (Heckman 1986) and tide-related vertical migrations (e.g., Devreker et al. 2008). Additionally, Morgan et al. (1997) observed higher densities of *E. affinis* upper in the water column during the flood and lower during the ebb in the Columbia River Estuary, suggesting that settling or swimming to a lower position in the water column could occur prior to the ebb in order to avoid advective losses. The alternative of a passive transport has been also found for different copepod species (e.g., Castel and Veiga 1990); for example, *E. affinis* in the Gironde Estuary seems to be maintained through the same hydrological processes that resuspend, advect, trap and concentrate suspended particles (Castel and Veiga 1990). In the Bahia Blanca estuary (38°43′S, South West Atlantic (SWA)), a well-mixed estuary, increased abundance of *Acartia tonsa* during the ebb has been associated with a lateral movement of this copepod to areas of decreased flushing such as channel margins (Menéndez et al. 2011, 2012), as also suggested for other copepod species (e.g., Cronin et al. 1962).

Coastal (or choked) lagoons are a particular type of estuary in which a coastal indentation is partially isolated from the sea by a sand barrier. Lagoon and sea keep a connection through one or more restricted inlets which remain open at least intermittently (Kjerfve et al. 1994, see also Chapter 1).

The restricted movement of water and organisms between the lagoon and the sea is a typical feature that differentiates coastal lagoons from other estuaries, and which imposes characteristic variability patterns in physico-chemical drivers (temperature, salinity), biogeochemistry, plant and animal densities. Coastal lagoons seldom exceed a few meters depth. A lagoon may or may not be subject to tidal mixing, and salinity can range between that found in coastal freshwater lakes to hypersaline, depending on the hydrologic balance (Kjerfve et al. 1994). Also, *spatial environmental gradients* and *tidally and wind-driven circulation* are central determinants in the ecology of coastal lagoons. Due to the shallow depth, restricted connection with the sea, and relatively long residence time, coastal lagoons may be highly susceptible to anthropogenic stressors such as chemical pollution, exotic species, physical deterioration, and maritime traffic (Chaalali et al. 2013). Eutrophication, changes in land use, and wastewater discharge represent further pressures in these habitats, on top of climatic variability (Kennish and Paerl 2010).

Zooplankton represent a very sensitive tool for monitoring environmental variations in coastal lagoons (Branco et al. 2007). *Salinity* is also here one of the main forces influencing zooplankton communities. The existence of a connection with the sea *vs.* permanent cut off clearly affects the community (Kozlowsky-Suzuki and Bozelli 2004, Branco et al. 2007, Santangelo et al. 2007): when the connection is restricted, the zooplankton is mostly comprised of freshwater taxa. With an open communication, the community is taxonomically richer and composed mainly of marine eurihaline zooplankton. When the communication with the sea has been cut off recently and waters are mesohaline, the community is comprised of fewer taxa, at high abundances; in some cases, brackish conditions may persist despite extended periods of isolation from the sea, and under these circumstances the zooplankton can be dominated by one or very few highly eurihaline taxa (e.g., the copepod *Acartia tonsa*; Calliari et al. 2009). In Patos Lagoon (Brazil), the world's largest "choked" lagoon (Kjerfve 1986), wind and hydrology modulate the composition of dominant zooplankton in a very predictable way (Montú et al. 1997, Muxagata et al. 2012): marine copepods *A. tonsa*, *Euterpina acutifrons*, and larvae of cirripede *Balanus improvisus* enter the estuary during seawater inflow, and during flushing freshwater copepod *Notodiaptomus incompositus* and cladocerans *Eubosmina tubicen* and *Diaphanosoma sarsi* arrive from tributaries and from the freshwater reaches of the main water body.

6.2.4 Meso- to small scale distribution: Surf zone waters of sandy beaches

Sandy shores consist of three entities: surf zones, beaches, and dunes linked by the interchange of materials and comprising a single geomorphic system, *the littoral active zone* (McLachlan and Defeo 2018). That zone is characterized by wave- and wind-driven sand transport, and its extension lies between the outer limit of wave action on bottom stability and the landward limit of aeolian sand transport (McLachlan and

Defeo 2018). The main exchange of organic materials and nutrients takes place between the littoral zone and the sea, the pathway for such exchanges being the surf zone (McLachlan and Defeo 2018). Within this geomorphic unit, two different ecological systems can be identified: the marine beach/surf zone, which is mainly influenced by *wave energy,* and the terrestrial dune influenced by the wind. In the beach/surf zone ecosystem, the resident fauna is of marine origin.

Besides its importance in shaping the beach and transporting materials, the water envelope of the surf zone supports a varied fauna of zooplankton and fishes. However, the general ecology of beach zooplankton is still poorly known. McLachlan and Defeo (2018) described the surf-zooplankton as composed mainly by crustaceans, with gelatinous forms as intermittent visitors. Resident species in surf waters may be truly planktonic or bentho-planktonic (demersal), while non-residents may be part of the holoplankton or the meroplankton. It should be noted that the term "residence" indicates regular rather than permanent occurrence. True planktonic residents in surf zones include copepods, mysid shrimps, small prawns, and larval stages of beach fauna. Bentho-planktonic forms include aquatic isopods, amphipods, and mysids associated to the bottom, but which regularly migrate into the plankton. According to these authors, non-resident holoplankton of surf zones includes microzooplankton (tintinnid ciliates and copepod nauplii) and meso- and macrozooplankton, including copepods, cladocerans, chaetognaths, and jellyfishes. Finally, the meroplankton is composed of larval stages of non-planktonic animals from other areas (e.g., larvae of crabs, bivalves, polychaetes, and fishes from nearby estuaries, rocky shores, or offshore ecosystems). Non-residents are mainly small in size, with copepods as the most characteristic forms. Copepods contribute most species in surf zooplankton samples and may also be numerically dominant (Pinheiro et al. 2013, Oliveira-Santos et al. 2016). Non-residents are highly variable in occurrence, their sporadic appearance in the surf zone is linked to *advection by wind-, wave-, and tide-induced currents*. Given the incipient development of the field of zooplankton beach ecology, it could be expected that some of the current considerations are revised and corrected in the near future.

The surf zone can be thought of as a physically dynamic and harsh environment where inhabitants are linked by strong trophic interactions. Despite the perceived severe hydrodynamic conditions, surf waters host rich mesozooplankton communities typical of neritic or estuarine waters (Costa et al. 2011, Pinheiro et al. 2013, Oliveira-Santos et al. 2016, Stull et al. 2016, Menéndez et al. 2019). In Monte Hermoso sandy beach (Argentina, SWA, 38°59′S), two calanoid copepods dominate the surf waters: the estuarine *A. tonsa* and the neritic *Paracalanus parvus* (Menéndez et al. 2019). In Brazilian beaches, common and occasionally dominant copepods include Paracalanidae, *Ctenocalanus vanus*, and the bentho-pelagic *E. acutifrons* (Avila et al. 2009, Costa et al. 2011, Pinheiro et al. 2011, 2013), probably linked to high concentrations of suspended particulate matter and strong resuspension in surf zones (Pinheiro et al. 2013). Surf-mesozooplankton may be subjected to strong seasonality associated to the temperature cycle, as seen in the SWA (Argentina; Menéndez et al. 2019) and in the Mediterranean Sea (Egypt; Aboul-Ezz et al. 2014). In turn, in surf waters of tropical and subtropical Brazilian beaches seasonal variation in precipitation and associated changes in salinity seem to be the main drivers of zooplankton community structure (Avila et al. 2009, Costa et al. 2011, Pinheiro et al. 2011, 2013).

Waves, local winds, and associated turbulence are key factors that influence zooplankton dynamics at sub-seasonal and small spatial scales. Local hydrology concentrates detritus, phytoplankton, and zooplankton in the surf zone. For instance, chlorophyll concentration in the surf-zone of Monte Hermoso beach is higher than in open shelf areas, suggesting better food supply compared to the nearby sea (Menéndez et al. 2019). High turbidity is another typical feature of surf-zones due to particulate matter in suspension. Despite particulate organic carbon concentrations being high in surf waters (Menéndez et al. 2019), most of that carbon is detrital (55–85%) and non-live material (13–43%) (Talbot and Bate 1989). It has been speculated that these could constitute alternative, albeit low quality, food sources for zooplankton (Menéndez et al. 2019).

Strong turbulence typical in surf zone implies that resident zooplankton communities must cope somehow with high-energy regimes, although it is unclear how that may work. Excessive levels of turbulence and suspended particulate matter can be deleterious for zooplankton feeding and for their general eco-physiological performance, particularly for copepods (Saiz and Kiørboe 1995, Alcaraz 1997, Saiz et al. 2003, Shadrin and Litvinchuk 2005). To date, the potential role of those environmental processes in the modulation of the biogeochemistry and trophodynamic fluxes in surf-zone ecosystems has been little explored. Clearly, the ecology of surf-zone zooplankton is a topic that deserves further attention.

6.3 Feeding and food webs

Mesozooplankton play critical functional roles in marine ecosystems. The functional significance of mesozooplankton stems from the range they occupy in the spectrum of individual body sizes (200–2000 μm), and from their role as energetic shunt between a wide variety of basal prey and predator types. In few words, the assemblage of planktonic organisms within the mesozooplankton size category channels the energy stored in the biomass of the primary producers of the ocean to larger consumers, either directly or indirectly. Larger consumers means shrimp, fish, marine reptiles, birds and mammals, among others. But, as we will see next, the importance of mesozooplankton extends beyond that, as they also influence the structure and dynamics of prey and predator populations and the global biogeochemistry of key elements in the biosphere, like carbon.

As previously mentioned, copepods comprise between 70 and 90% of the numerical abundance of marine mesozooplankton (Turner 2004). They rank among the most abundant metazoans on the planet (Humes 1994) and are exceptionally well adapted to the marine pelagial (Kiørboe 2010). Copepods are armed with a resistant exoskeleton and are thus relatively easy to collect from nature, and they preserve well in several standard fixing solutions. They are quite resistant to handling and several species have been successfully kept in laboratory cultures for many generations. Such attributes render copepods particularly relevant and well suited for field and experimental investigations that allow characterization of vital rates, ecophysiological responses to environmental variability and their role in marine ecosystems. Copepods were early adopted as model mesozooplankton organisms and for several decades constituted a privileged focus of investigations within the fields of Marine Ecology and Biological Oceanography. For such reasons, knowledge on the feeding mechanisms and functional role of copepods preceded that for most other zooplanktonic groups.

This section begins with a revision of feeding mechanisms and strategies, including behavioral aspects. A second subsection deals with the role of mesozooplankton within marine food webs, and how the conceptualization of that role has changed over time.

6.3.1 Feeding mechanisms

The fundamental problem for mesozooplankton, as it regards feeding, is to locate and pursue enough amounts of food to sustain metabolic needs, grow and reproduce within a nutritionally diluted medium like the ocean. "Nutritionally diluted" means that finding and concentrating food in the ocean is a task far from trivial: an individual zooplankter must process (i.e., clear of food particles) a volume of water in the order of 10^5–10^7 times (between one-hundred thousand and ten million times!) its own body volume in order to obtain its daily ration (Kiørboe 2011).

Mesozooplankton consume living and non-living particles in a size range from nearly 1 μm to several thousands of μm. The prey spectrum includes bacteria, algae, microheterotrophs (i.e., protozoans like ciliates), other mesozooplankton, fecal pellets, and macro-detritus like carcasses and marine snow, among many potential items. Such an ample range of potential food types results from a diversity of feeding structures, strategies and behaviors associated with the search, localization, selection (in some cases) and capture of food. Also, at the species level, diet composition tends to be rather flexible. Excellent reviews exist on the feeding in marine plankton, their constraints, trade-offs, and fundamental mechanisms involved (Kiørboe 2011, 2016, Conley et al. 2018).

Zooplankton have evolved four main strategies in order to locate and collect food (Kiørboe 2011): passive and active ambush feeding, current feeding and cruise feeding (Fig. 6.4). Passive ambushers rely on the movement of the prey in order to encounter, intercept and capture food; active ambushers rely on prey movement for the encounter, and attack the prey for capture; current feeders move the water relative to their own body (i.e., generate feeding currents) and collect prey immersed in the feeding

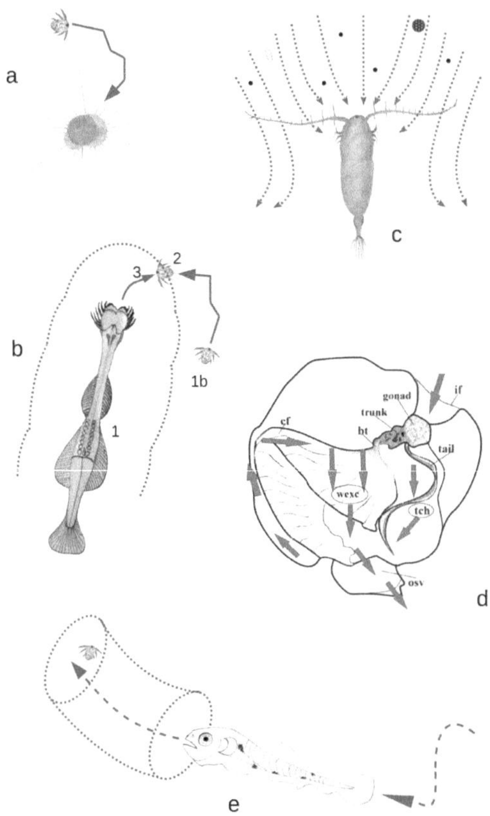

Figure 6.4. Cartoon representation of the four main feeding strategies by planktonic predators. (a) Passive ambush feeding: Chance encounter between predator and prey is mediated by movement of the prey, which is retained by sticky structures upon contact. In the example, a foraminiferan preys on a copepod nauplii. (b) Active ambush feeding: encounter is mediated by prey movement; the predator remains motionless ("silent"); "contact" occurs once the prey enters the perception volume of the predator, and is followed by a rapid attack. In the example, an arrow worm (Chaetognath) preys on a copepod nauplii. Dotted lines represent the perception volume. Numbers indicate the sequence of events leading to successful pursuit of prey (1) the predator is still while prey swims in the surrounding environment; (2) the prey enters the detection volume; (3) predator attacks. (c) Feeding by filtering (sensing) currents: food particles are entrained in a flow field generated by the copepod. The example shows the flow lines of a "sensing current" which en-trains different potential prey types; those are detected in the near-field by mechano and/or chemosensors in the mouth-parts. (d) Feeding by filtering currents: appendicularia develop a radically different type of feeding by filtering currents (see text). Thick arrows indicate the water flow (redrawn based on Fenaux 1986, and Lombard et al. 2011). (e) Cruise predation: the predator swims while scanning the environment for prey. Once a prey item is detected within the perception volume (dotted lines), a rapid attack follows (drawn loosely based on Sundby 1997).

current by one of several different mechanisms: interception, filtration, or by perceiving and (selectively) capturing individual prey. Cruise feeders swim continuously (i.e., they cruise) to encounter prey, which are detected by chemical, hydromechanical or visual cues. Below we review feeding strategies of selected ecologically important groups within the mesozooplankton: copepods, cladocerans, appendicularians, and chaetognaths in order to further detail their feeding modes.

Copepods. Feeding in copepods is a process at low Re that relies on a fine balancing of inertial and viscous hydrodynamical forces (Yen 2000). The body structures involved in feeding are cephalic appendages: antennules, antennae, mandibles, maxillules and maxillae (Mauchline 1988), but how each of these appendages acts depends to a large extent on the specific feeding strategy developed. Copepods were long considered as herbivorous, and the canonical grazers of the primary production in the ocean (see Section 6.3.2 below), but they are currently known to have a much richer feeding repertoire. Trophic habits can be schematically grouped into: I: Omnivorous-herbivorous, when diet is mostly constituted by algae, plus varying amounts of other small (single-celled) organisms like autotrophic or mixotrophic nanoflagellates and small autotrophic, mixotrophic or heterotrophic ciliates and dinoflagellates. II: Omnivorous-carnivorous, which feed on algae but also ciliates, dinoflagellates, invertebrate eggs, other mesozooplankton like nauplii (including cannibalism) and metazoans like rotifers and larvae of other marine invertebrates. Omnivorous-carnivorous species often actively select heterotrophic prey, so the contribution of these items to their diet can be higher than the proportion of such type of prey in seawater; i.e., non-algal prey tend to be over-represented in the diet of omnivorous-carnivorous copepods. III: Carnivorous copepods, whose diet is composed of relatively large organisms, typically other copepods or mesozooplankton in general.

The above classification is loosely based on the trophic level (i.e., primary producer *vs.* consumer) and size of the prey. Another classification scheme categorizes feeding by the relative size of prey to that of the predator. Lowest prey size for copepods is usually between 5 and 10 µm (as Equivalent Spherical Diameter-ESD; Bartram 1982, Hansen et al. 1994, Kiørboe 2010), and predator to prey size ratio ranges between 3:1 to about 180:1, with optimal ranges between 10:1 and 30:1 (Hansen et al. 1994). According to the predator to prey size ratio, copepods are divided into: I: Microfeeders, which feed on prey that are much smaller than themselves (high predator:prey size ratio), and II: Macrofeeders, when the predator:prey size ratio is lower. Yet another classification considers the behavior of the predator during feeding: I: Suspension feeders, II: Ambush feeders and III: Cruise feeders. All these classifications are complementary, non-mutually exclusive, and some combinations are most frequent. For instance, a typical omnivorous-herbivorous copepod is usually a microfeeder that collects particles using feeding currents. A common species with this feeding strategy is *P. parvus*. Carnivore species display ambush or cruise swimming behavior and their prey are relatively large compared to their own body size. Examples are copepods of genus *Euchaeta*, *Paraeuchaeta* and *Tortanus* (cruise predators) but also small species of genus *Oithona* (e.g., *O. davisae*, an ambush predator). At least some Omnivorous-carnivorous like *A. tonsa* are flexible in terms of feeding and may switch between suspension and ambush feeding modes according to the relative abundance of prey (Jonsson and Tiselius 1990), presumably selecting that type which provides the highest energetic reward (Kiørboe et al. 1996). It is interesting to dig a little further on the behavioural repertoire of copepods when it comes to food aquisition, as represented by their main feeding modes.

Suspension feeding involves generation of feeding currents, so that suspended particles entrained in the feeding current are directed towards the mouthparts of the copepod. This is considered the most efficient mechanism for collecting non-motile, non-evasive prey (Kiørboe 2011). Feeding currents result from high frequency beating of antennae and maxillipeds (Cowles and Strickler 1983, Price and Paffenhofer 1986). The asynchronous beating of cephalic appendages generates a laminar water flow in front of the copepod (see Fig. 6.4c) that is actually a scanning current, as copepods use sensory structures to perceive the prey (Kiørboe 2010). The flow field of the feeding current is double sheared as water velocity falls away from the longitudinal body axis and also with distance ahead of the mouth area (Strickler 1982). The core of the feeding current is dragged towards the mouth area and a fraction of the small particles carried with the flow are deposited onto the maxillae due to boundary layer effects (Vogel 2004) and eventually ingested. Larger food particles like diatoms are captured individually, and in this case active selection can take place. Food selection of individual particles by suspension feeders is achieved based on detection and fast reaction to chemical or mechanical signals before physical contact with the particles thanks to sensory antennular receptors (Koehl and Strickler 1981, Yen 2000). It is an open issue whether prey detection in the near-field (ca. 100 µm from the sensing appendages) relies on mechanical *vs.* chemical detection, with encountered arguments for the former (e.g., Goncalvez and Kiørboe 2015, Kiørboe 2016) and the latter (e.g., Paffenhôfer and Jiang 2016). The argument for chemical detection is based on the fact that phytoplankton (or another potential food particle) leak organic exudates which form a phycosphere or "odor cloud" around the cell with valuable information for the predator about the nature and composition of the particle. Once entrained in the feeding current, the shear stress stretches the phycosphere, advancing the leading edge of the cloud relative to the cell. The cloud thus "arrives earlier" than the prey itself to the sensing organs of the copepod, allowing some time for the predator to react and capture (or reject) it. High-speed video records of particles ingestion by free swimming copepods showed that reaction to the presence of a food particle occurs at a short distance (<< 100 μm) of the second antennae or maxillipeds (Tiselius et al. 2013), that is much less than proposed by the chemical cue theory; also, it is argued that the intensity of the chemical signal would need to be too strong for the proposed mechanism to be viable in most cases, and it may work only for large cells (i.e., > 35 μm ESD; Tiselius et al. 2013, Kiørboe et al. 2016).

In any case, the reaction of the copepod to the presence of a food particle is extremely fast, in the order of 35 ms (and rejection of unwanted cells take ca. 61 ms; Tiselius et al. 2013). The reaction involves a reorientation of the body that can be recognized macroscopically as a "feeding bout" (Tiselius 1992).

Ambush feeding is a mode in which the predator relies on the detection of hydrodynamic signals within its detection volume, i.e., disturbances in the flow field produced by moving prey. In this case, signal perception relies on mechanosensors. These are setae-like structures that sit along the surface of the antennules, plus a tuft of sensing setae which is usually found near their distal end (Yen 2000). Fluid disturbances induce mechanical stimuli by displacing these setae. Typically, an ambush copepod scans the surrounding environment with the antennules fully extended while quietly sinking. During sinking, the flow field is undisturbed, so the copepod is "hydrodynamically silent", maximizing the detection of signals emitted by potential prey. The speed of water flow around the copepod increases with distance from the body, and the sensory array seems to be suited to such flow structure so that the distal sensors detect the movement of predators (and thus elicit escape responses), while sensors nearer to the proximal end detect disturbances produced by moving prey (Yen 2000).

Cruise feeding (or Cruise and sink mode) involves continuous swimming along more or less convoluted paths while searching for prey, with occasional interruptions of the swimming during which the copepod sinks before resuming the food search. There are two main modes within this feeding strategy (Kiørboe 2011): cruise feeders on small prey (smaller than the predator), and those that search for large particles (marine snow). Cruise predators on small particles may rely on chemical, hydromechanical or visual cues to detect prey. For copepods, probably chemical and mechanical cues are more important (Doall et al. 2002, Utieri et al. 2008), although the efficiency of those has been questioned: the signals (chemical and mechanical) would normally be weak, with limited extension or with a low signal-to-noise ratio. Vision is the ideal sense for cruise feeders on small particles. But even if several copepod species have visual organs that detect light, forms and movement, true vision is likely not present to help prey detection in copepod cruise feeders (Kiørboe 2011). Only fish larvae, with well-developed eyes and true vision, would conform to ideal cruise feeders among the zooplankton.

Macroscopic aggregates (marine snow) can be used as a food source by *cruise feeders specialized in large particles*. Aggregates are organic conglomerates of particles of diverse origin (notably, discarded *Oikopleura* houses, see below). Bacteria colonize these aggregates and feed on the organic matrix, while heterotrophic protozoa also colonize the aggregates and feed on bacteria. Thus, aggregates represent hotspots of biological activity and food for larger predators. Since the number of aggregates per unit volume of seawater is very low, random encounter is not a viable mechanism to locate them. Instead, copepods seem to use mechanical and chemical cues for finding aggregates. While aggregates sink, they produce a hydrodynamical disturbance (wake). They also leak metabolites from microorganisms living in it and from decomposing organic matter, leaving a chemical trail that copepods can detect and follow to reach the aggregate (Kiørboe 2011).

Cladocerans. Knowledge of the feeding biology of marine cladocerans lagged behind that of copepods and until the 1990s—with few exceptions (i.e., Pavlova 1959)—information was scarce and inconclusive regarding their feeding modes, prey types and sizes (Turner 1984).

Current information suggests that Podonidae are carnivorous-omnivorous that prey upon diverse types of organisms. Early investigations considered Podonidae as raptorial predators based on the structural analysis of their feeding apparatus (Nival and Raven 1979) and the presence of a conspicuous compound eye (Russell-Hunter 1979). Qualitative observations of freshly caught *Evadne nordmanni* reported individuals holding prey within the microzooplankton size range by grasping with the endites of the second and third pair of legs or pressed between labrum and mouth (Bainbridge 1958). Incidence of *E. nordmanni* holding prey was higher during day *vs.* night, suggesting that vision was a significant mechanism for locating and capturing prey. Prey included dinoflagellates *Ceratium* spp., *Peridinium* spp., and ciliates (*Tintinnopsis* spp.), and marginally copepod eggs and nauplii. These observations were supported by White (1980) who provided clear evidence of toxin trophic transfer from the dinoflagellate *Alexandrium tamarense* (*Gonyaulax excavata* in the original article) to fish *Clupea harengus harengus* via the intermediate consumer *E. nordmanni*. Qualitative observations similar to those of Bainbridge (1958) were reported for *Podon polyphemoides* (Bosch 1970) grasping copepod nauplii and copepodids. However, scanning electronic microscopy analyses of gut contents from field collected *E. nordmanni*, *E. tergestina*, *P. polyphemoides* and *Podon leuckarti* indicated that centric diatoms and dinoflagellates were the dominant prey for all species (Kim et al. 1989). It is likely that in this case the method of analysis prevented identification of food types lacking hard structures. The same approach (microscopic analysis of gut contents) of field caught *P. intermedius* provided similar results to those of Kim et al. (1989), and under laboratory incubations that species appeared to consume the largest of several phytoplankton species offered (Jagger et al. 1988), leading authors to characterize that species as a "raptorial-feeding herbivore".

Penilia avirostris is a suspension feeder relying on feeding currents. The species preys upon smaller particle sizes compared to Podonidae. Early experiments suggested that *P. avirostris* ingests diverse types of food (and non-food) particles < 8 µm (Pavlova 1959). Ingestion is non-selective, irrespective of whether particles are provided as single or mixed suspensions (Pavlova 1959). The range of prey sizes consumed by *P. avirostris* lies between ca. 2 and 20 µm (or up to 35 µm; Kim et al. 1989), the upper threshold increasing with grazer individual age (size) (Paffenhoffer and Orcut 1986). Such size range includes small nanoflagellates (2–5 µm), phytoplanktonic algae and dinoflagellates, but not naturally occurring bacteria which seem to be too small (< 1 µm) for efficient capture by *P. avirostris* (Turner et al. 1988, Kim et al. 1989, Atienza et al. 2006). Under natural conditions, *P. avirostris* develops high grazing rates on phytoplankton (as measured by the gut fluorescence method), even higher mass specific rates than co-occurring copepods, although their impact on primary producers is moderate or low (Wong et al. 1992, Calliari and Tiselius 2009), with few exceptions (Lipej et al. 1997). Still, available evidence is consistent in pointing to a lower size threshold in particle capture for *P. avirostris* compared to copepods. Complementary prey selectivity (either active or passive) (Katechakis et al. 2004, Atienza et al. 2006) contributes to niche separation that would favor coexistence between cladocerans and copepods.

Appendicularians. This group of planktonic tunicates is ubiquitous in coastal and oceanic waters and can reach "bloom" densities under favorable environmental conditions (Capitanio et al. 2018). They stand out among the zooplankton for their remarkably high metabolic rates and feeding ecology (Sato et al. 2001, Tiselius et al. 2003). Appendicularians are sometimes referred to as mucous feeders, i.e., predators which use mucous structures to collect their prey (Fig. 6.4). They share that attribute with other planktonic groups like *Thecosome* pteropods (Mollusca) and Thaliacea (Urochordata) (Conley et al. 2018).

A remarkable characteristic regarding feeding in appendicularians is their prey size range, which reaches the sub-µm range, being able to collect particles down to colloid size (Flood et al. 1992, Conley et al. 2018). That includes free bacteria and viruses, at the lowest size threshold for particulate matter, i.e., in the operational limit between particulate and dissolved organic matter (Gowing 1993, Lawrence et al. 2017, Conley et al. 2018).

Such capability results from a striking feeding apparatus, which is mostly composed of an external cellulose-mucous structure known as house (Fig. 6.4d). Houses are secreted by the appendicularian and surround the individual. The role of the house is to assist in the formation of feeding currents, in the conduction of the water through different filters, and in the collection and carrying of food particles to the buccal tube. Collection of food relies on three different filters: the inlet filter, which is a first, coarser mesh at the entrance of the house that excludes larger particles (species of genus *Fritillaria* lack inlet filter); a food-concentrating filter that retains small particles down to ca. 2 µm; and a pharyngeal filter that retains ingested particles (Lombard et al. 2011). Sinusoidal beating of the tail pumps water into the house and drives it through the filters.

Given that particle collection is based on mechanical filtering, appendicularians were traditionally considered non-selective feeders. However, different selection mechanisms have been shown to operate. Passive selection is performed by the inlet filter which determines the upper size threshold for food particles, while food-concentrating and pharyngeal filters set the lower particle threshold size (Conley et al. 2018). It should be remembered though that these mechanisms operate at low Re, so viscosity also plays a key role in particles' retention. Appendicularians can also actively reject (and thus, select not to ingest) individual particles, especially large ones. Relatively large particles that manage to enter the house, for instance large, elongated cells of width < inlet filter mesh size and that for some reason are not edible, can be rejected out of the mouth by flow reversal. Selection is based on the size and the chemical characteristics of the particle, the latter likely related to nutritious quality (Fenaux 1986, Deibel 1986, Lombard et al. 2011). Other rejection mechanisms reported include "pipe smoking", which involves backflushing of the concentration filter, so particles are rejected into the exit chamber and spouted out of the house (Acuña and Kiefer 2000, Selander and Tiselius 2003, Lombard et al. 2011, Conley et al. 2018). That type of behavior typically occurs at very high particle densities when filters clog. However, in the long run, accumulation of particles and detritus affect the efficiency of food collection. Appendicularians regularly discard the house and a new one is rapidly produced. That can happen several times a day if the environmental concentration of large particles is high. As they sink, discarded houses concentrate and continue scavenging organic particulate matter (phytoplankton, pellets, other detritus), thus constituting a fast and efficient vehicle for downwards carbon transport (Kiørboe 1996). A single appendicularian (*Oikopleura dioica*) can produce about 50 houses in a short lifetime (ca. ten days), with a daily house production corresponding to up to ca. 300% of its own biomass, while discarded house materials may reach 490–1100% of their biomass (Sato et al. 2001). These numbers show that appendicularians can be a key player in oceanic carbon fluxes.

Chaetognaths. They are strict ambush predators, and copepods constitute their main prey type (Huliselan 2002, Tönnesson and Tiselius 2005). Appendicularians and other chaetognaths (including cannibalism) seem to support metabolic needs when copepod prey are in short supply (Øresland 1987, Tönnesson and Tiselius 2005).

Prey detection in chaetognaths relies on sensing hydrodynamical disturbances produced by moving prey, similarly to what ambush copepods do. For that, chaetognaths are armed with longitudinal and transversal arrays of setae located on the body surface according to species-specific patterns that act as mechanoreceptors (Feigenbaum and Reeve 1977, Feigenbaum 1978). Experimentally, planktonic chaetognaths (*Sagitta hispida*) were shown to locate and respond to prey at distances of less than 3 mm (Feigenbaum and Reeve 1977). Location of a prey is followed by a rapid attack, as it is usually the case for ambush predators (Kiørboe 2011). Around their head, chaetognaths feature a crown of spines which serve to grasp the prey, and one to three rows of teeth that puncture the body of the prey and allow the penetration of toxins produced in papillary pores (Foster 2006, fide Le Roux 2011).

Reported feeding rates are very variable, between less than 1 and more than 10 prey predator^{-1} d^{-1} (Feigenbaum 1982, Giescke et al. 2010, Vergara et al. 2010). Biomass-specific predation increases with temperature and decreases with the size of the predator (Feigenbaum 1982). Their impact on copepod populations can be significant (up to 48% of individuals removed daily for given species). Chaetognaths can be significant contributors to carbon fluxes within the pelagial, and to downward export of organic matter thanks to their fast-sinking pellets (Froneman et al. 2002, Giescke et al. 2010, Kruse et al. 2010).

6.3.2 The evolving concept of mesozooplankton's role in marine food webs

Mesozooplankton were once thought to be the main grazers of marine phytoplankton and, as referred above, mesozooplankton meant mostly copepods. Such a view was held until the 70s–early 80s and was multi-causal. First, knowledge on marine zooplankton derived mostly from the analysis of samples collected using plankton nets fitted with more or less fine meshes (usually between 90 and 300 μm). The first quantitative net samplers were introduced by Hensen (1887) and, with relatively minor variations, are still used today (for a nice overview of the development of zooplankton sampling, see Wiebe and Benfield 2003). Copepods are commonly the most abundant group in the catch of such nets, i.e., they constitute the bulk of "net zooplankton". Thus, it was only natural that the investigations which targeted the structure and functioning of marine food webs had a privileged focus on primary producers and on those that were perceived as their direct consumers: copepods. Second, Fisheries science was one important driver for the development of Biological Oceanography during the 20th century. Early in the development of Fisheries science, copepods and their nauplii were recognized as key prey for the survival and growth of larval stages of most marine fish (Fuiman and Werner 2002). The resulting conceptualization of marine pelagic food webs was that of a linear array where energy and matter were transferred from relatively large phytoplankton (e.g., diatoms) to their copepod grazers, and from those to fish directly or through some intermediary consumer (Kiørboe 1993). This is sometimes referred to as the classical food chain: phytoplankton – copepod – fish. The cornerstone book "The Structure of Marine Ecosystems" by John Steele (1974) was a very influential synthesis of knowledge at that stage.

The introduction of new observational approaches, e.g., epifluorescence microscopy, incorporation of labeled substrates, allowed for improved enumeration of autotrophic and heterotrophic bacteria and other picoplankton, and for quantification of activity rates of organisms in size categories corresponding to micro-, nano- and pico-plankton. From those it became evident that the smallest organisms were far more abundant and developed higher metabolic rates than so far believed. Predator–prey dynamics appeared tightly coupled: in spite of high production rates, bacterial numbers were kept at relatively constant levels (as opposed to the case of the larger phytoplankton, for instance), suggesting that efficient top-down controls were in action. The resulting picture showed that small organisms so far overlooked were responsible for a large fraction of energy and matter fluxes in the marine pelagial. That led to the formalization of the "microbial loop", an energy and matter pathway alternative to the "classical food chain" (Fenchel 1982, Azam et al. 1983, Jumars et al. 1989) constituted by pico-sized primary producers and osmotrophic heterotrophs (i.e., bacteria) living on dissolved organic matter, plus a diverse array of nano-sized mixotrophic and heterotrophic flagellates which feed on picoplankton, but many of which also have the capacity to photosynthesise and produce their own organic matter, depending on environmental conditions and specific requirements. Nano- and micro-sized heterotrophic and mixotrophic ciliates and dinoflagellates were later identified as top-predators within such microbial web.

By early 90s, conceptualization of pelagic food webs was that of coexistent classical and microbial trophic paths, their relative importance modulated by the physical environment (e.g., a physical-biological coupling; Cushing 1989, Legendre and LeFevre 1989, Kiørboe 1993, Bakun 1996). According to that, the stability of the water column (i.e., meso-scale turbulence) regulates the flux of new nitrogen to upper layers and the possibility of relatively large and heavy non-motile phytoplankton (i.e., diatoms) to stay within illuminated and favorable depth ranges. Ecosystems characterized by weak stratification represent a favorable compromise between nutrition and irradiance. In those ecosystems, high biomass and production rates of large phytoplankton cells support a short and efficient classical food chain with abundant copepods and other mesozooplankton, high export fluxes and fish production. Under strong turbulence (well mixed water column), primary production and plankton biomass are at a minimum due to light limitation, whereas under strong stratification primary production is dominated by pico-sized cells living on regenerated nutrients within the surface mixing layer. The dominant pathway under those conditions is the microbial-loop, and pico-phytoplankton is subjected to strong top-down control that limits biomass accumulation.

Weak stratification is not a stable or equilibrium condition; rather, it is associated with transitions along a time axis, for instance during the temperate latitude seasonal cycle of stratification–de-stratification of the water column, or along a spatial axis, represented by sharp horizontal changes in water column conditions at marine fronts.[b] This is analogous to the cycle's stabilization/destabilization of the water column, generalized as Gran effect (Mann and Lazier 2006). Diverse oceanographic processes may lead to weak stratification and a coupled biological response. Some well-known examples are upwelling fronts, shelf-break fronts, and estuaries, as reviewed above. Since they result from the

[b] The vernal bloom of phytoplankton at temperate latitudes is a classic example of physical-biological coupling. The conceptual evolution of the mechanism that results in these blooms is a very interesting and current topic. The traditional Critical Depth Hypothesis (Sverdrup 1953), with adjustments, was generally accepted until not long ago. Recently, it has been strongly questioned and a new Disturbance Recovery Hypothesis (DRH) has been put forward (Beherenfeld 2010, Beherenfeld and Boss 2018). Both references are strongly recommended for a review of conceptual issues and a discussion of evidence regarding Sverdrup's and other theories on the spring bloom, including the presentation of DRH by its proponents.

interplay of physical forcing and local conditions (i.e., bathymetry, topography, coastal configuration), they occur with characteristic timing and extension at different sectors of the ocean. This view is thus consistent with an ocean characterized by ample regions of low productivity regimes (e.g., central ocean gyres) with specific regions where intermittent hot spots of high biological production are more frequent and/or spatially extended, and which confer overall higher biological productivity, particularly of large organisms.

That understanding was enriched and partly modified in the 2000s by a series of meta-analyses of global data on predation pressure within classical and microbial networks (Calbet 2001, Calbet and Landry 2004, Calbet and Saiz 2005). Those indicated that herbivore consumption by copepods increases from low productivity ecosystems (e.g., central oceanic gyres) to highly productive (e.g., estuaries, coastal areas). But predation pressure—defined as the percentage of primary production consumed by copepods—was shown to be overall low (mean of 22%, median 6%). Furthermore, herbivory pressure by copepods decreases with increasing ecosystem productivity. Those findings are opposite to the expected dominance of the classical food chain in more productive ecosystems. In turn, meta-analyses focused on the microbial loop showed that consumption by ciliates (representative of the micro-grazers community) corresponds to ca. 60% of primary production. Also, in the case of the micro-grazers, herbivore consumption increases with system's productivity, and again herbivory pressure decreases with ecosystem productivity.

Such results provided strong evidence of a global supremacy of the microbial web over the classical food chain, and lower predation pressure by both micro and mesozooplankton in more productive systems is consistent with a higher scope for organic carbon export in those ecosystems. In less productive ecosystems, where pico-sized producers are strongly dominant, the main grazers of the microbial web seem to be heterotrophic nanoflagellates, rather than ciliates (or dinoflagellates) (Calbet 2008). The overall pattern thus suggests an overriding importance of the microbial over the classical trophic pathway, irrespective of the productivity of the system under consideration.

Both pathways coexist and cannot be regarded as disconnected from each other. As discussed above, mesozooplankton and particularly copepods are omnivorous. Among many heterotrophic prey, their diet includes ciliates and dinoflagellates (e.g., Jonsson and Tiselius 1990, Dam et al. 1994, see 6.3.1 above). In natural waters and under almost every condition, ciliates are less abundant than phytoplankton. The meta-analysis by Calbet and Saiz (2005) revealed that the contribution of ciliates to the diet of copepods is higher than their share in the prey assemblage, particularly in low productivity ecosystems. That is a clear indication of an active selection of ciliates over phytoplankton as preferred prey of copepods. Selectivity of ciliates over algal cells can be due to different complementary reasons, like improved prey detection due to higher motility of ciliates, optimal size and better nutritive quality. The contribution of ciliates to copepods' diet decreased from an average of nearly 50% in less productive ecosystems to 22% in most productive ones. This ciliate-copepod connection thus constitutes a significant functional link between microbial and classical food webs, where the top predators of the former enter as prey in the latter.

A final remarkable finding was the generally high weight-specific clearance rate of ciliate prey by copepods, enough to match the growth rate of ciliates' populations (Calbet and Saiz 2005), which implies a potential for copepods to keep ciliates' populations in check. That opens the possibility to the occurrence of trophic cascades, where copepod grazing may favor the increase of (small) phytoplankton by limiting population size (and thus grazing pressure) of the main micro-grazers. Indications of such trophic cascades are often found as negative copepod grazing rates in experimental set-ups where copepods are incubated with natural assemblages of microzooplankton and phytoplankton[c] (Nejstgaard et al. 2001) and have been confirmed in specific experiments where copepod additions released predation pressure on picoeukaryotes, *Synechococcus*, *Prochlorococcus* (Armengol et al. 2017). It must be noted that patterns resembling trophic cascades may also occur within the microbial web at the interaction between top predators (ciliates, heterotrophic dinoflagellates), heterotrophic nanoflagellates (intermediate predators)

[c] The phrase "negative grazing" should not be taken literally, as negative grazing rates are not possible. However, under certain conditions, the estimates derived from experimental incubations can get negative values if the role of the intermediate predator (microzooplankton) and the existence of a trophic cascade is not taken into account.

and pico-autotrophs as the basal prey in the framework of dilution experiments (Calbet and Saiz 2013, 2018). These interactions may result in autotrophic net growth patterns which are difficult to interpret (again, suggesting "negative grazing" in less diluted treatments).

Behavioral responses also contribute to shape and modulate the intensity of trophic interactions. Best known cases are those where predation risk—as perceived by a potential prey with flexible feeding behavior—alters its feeding activity due to the presence of a predator. Those are known as top-down behaviorally mediated indirect interactions (top-down BMIIs). Since feeding comes at a cost of increased predation risk (Tiselius et al. 1997, Titelman 2001, Almeda et al. 2017), presence of a predator may inhibit (or diminish, or change) feeding patterns by the potential prey (Werner and Peacor 2003). One classical example is the development of circadial vertical migrations and feeding patterns in the zooplankton (Huntley et al. 1992, Pearre 2003, Ohman and Romagnan 2016).

A different, novel case is that of bottom-up BMIIs, for which experimental evidence was first presented recently (van Someren Gréve et al. 2019). BMIIs arise when changes in one group of organisms (referred to as the 'initiator' of the BMII) evoke a behavioral response in a second species or group (the 'transmitter'), which in turn affects a third species (the 'receiver'). To test that generic mechanism, van Someren Gréve et al. (2019) used a simple experimental model food chain: phytoplankton (*Rhodomonas salina*, the 'initiator')—intermediate predator (copepod nauplii, 'transmitter')—top predator (copepod *Centropages hamatus*, 'receiver'). The experimental design included transmitters of different species with either ambush or active swimming strategies. Theory and empirical evidence indicate that activity of ambush grazers tends to be invariant under changing food concentration, but that of active grazers is modulated by food availability. Results showed that algal food availability induced large changes in the feeding effort of active grazers (higher activity under low food conditions), but the changes were less evident in ambush grazers. Predation mortality mimicked changes in feeding effort, i.e., ingestion by the top predator *C. hamatus* was negatively affected by decreased motility of the active grazers. This bottom

Box 1: Role of zooplankton in vertical carbon fluxes

Through their activity, zooplankton become important players in the biological carbon pump. This term refers to the net downward transport of organic matter produced in the surface ocean to deeper layers (Turner 2015). The biological pump is a key component of the global carbon cycle, and as such it contributes to the regulation of Earth's climate (see also Chapters 2 and 3). Excess carbon dioxide (CO_2) in the atmosphere is transferred to the ocean in a process governed by gas solubility and differential partial CO_2 pressure between compartments. Photosynthetic primary producers take up dissolved CO_2 in surface layers of the ocean to build organic molecules that are passed on to the zooplankton by trophic interactions. Zooplankton contributes to the downward flux of particulate organic carbon (POC) through two main mechanisms: I: production of fecal pellets and body structures with relatively high sinking potential (gravitational fluxes); and II: diel vertical migration coupled to circadial feeding rhythms.

Gravitational fluxes. The first mechanism implies the re-packaging of POC initially contained in small, slowly sinking particles (i.e., phytoplankton cells) into larger particles with higher sinking velocities and increased potential to reach deep ocean layers. Such "larger particles" can be of different nature and origin. These include fecal pellets, moults, carcasses, and discarded somatic structures.

Copepods are long known to be key contributors to pellets fluxes. They produce compact pellets enclosed in a peritrophic membrane secreted by the lining cells in the posterior mid-gut (Gauld 1957), which are resistant to mechanical fragmentation and constitute an efficient vehicle for vertical transport. The contribution of fecal pellets to sinking POC is strongly variable, modulated by plankton community structure and notably by the abundance and size of dominant species (Tiselius et al. 2012, Stamieszkin et al. 2015). Pellets' fluxes peak during periods of high abundance of large copepods (like *Neocalanus* sp.), euphausiids, and during population bursts of salps (Class Thaliacea—Van der Haeven 1850), which also produce compact and very large, fast-sinking pellets (Steinberg and Landry 2017).

Box 1 Contd. ...

...Box 1 Contd.

The flux of pellets is further regulated by bacterial degradation, consumption, and re-packaging by mesopelagic plankton while they sink *en route* to the sea floor (Steinberg and Landry 2017).

Discarded somatic structures, notably appendicularian houses and feeding structures produced by other mucous-mesh grazers like Thecosomata (pelagic molluscs) (Conley et al. 2018), also make a very significant contribution to vertical POC fluxes. Houses and mucous structures conform a sticky matrix that continues scavenging particles as they sink, and can attain extremely fast sinking velocities in the order of 800 m d^{-1} (Noji et al. 1997, Robison et al. 2005). Due to consumption and remineralization processes, POC fluxes decrease with increasing depth.

Non-predatory copepod mortality is now known to be a widespread phenomenon (Martinez et al. 2013, Tang et al. 2014, 2019) and thus carcasses in the water column can be abundant, in the order of 10 to 20% of total copepod numbers. Carcasses may constitute another potentially important vehicle for downward POC fluxes. However, copepod carcasses are rapidly colonized and decomposed by bacteria (Tang et al. 2006). Even for rather shallow locations in the Japanese Sea (between 9 and 32 m), an average of 50% of the carbon in the carcass was estimated to be respired (remineralized) in the water column, and only the remaining 50% reached the sea floor (Tang et al. 2019). A special case of carcass flux is the episodic massive sinking of gelatinous plankton (medusae, salps and large colonial pyrosomes). These events are known as jelly-falls (Steinberg and Landry 2017). In spite of the low specific-carbon content of gelatinous plankton, their fast sinking may imply an important contribution to vertical POC fluxes. However, the episodic character of jelly-falls makes direct calculations of the flux associated to these events difficult. In summary, carcasses are a potentially important vehicle for vertical POC flux, but their actual contribution is still unclear.

Active POC transport. Active fluxes are mediated by the active movement of organisms between depth layers, as opposed to passive gravitational sinking of non-living particles. These fluxes result from the interplay of diel vertical migration and circadial patterns in feeding activity; i.e., organisms feed near the surface at night and descend to deeper layers during the illuminated part of the day where part of the acquired carbon is lost. That loss includes respiration, dissolved organic carbon excretion, fecal pellet production and mortality.

Active POC transport is also highly variable, its magnitude depending on biomass, composition and activity rates of migrant populations (Takahashi et al. 2009). Its relative contribution to the biological pump increases with depth, in part as a consequence of the rapid decrease of gravitational fluxes. Interestingly, the C:N:P stoichiometry of dissolved inorganic and organic matter produced at depth by migrating zooplankton can deviate from Redfield ratios, contributing to observed deep nutrient-ratio anomalies (Steinberg et al. 2002, Steinberg and Landry 2017).

A special case of active POC flux is that associated with the seasonal migration of *Calanus* spp. populations in high latitudes (*C. finmarchicus, C. helgolandicus, C hiperboreus, C. glacialis*). These are long lived, large copepods with a characteristic 1-year life history coupled to ontogenic vertical migration. Synchronized, overwintering populations in C5 development stage (late juveniles) ascend from near bottom layer to near surface waters during early spring, moult into adults and reproduce. During development, offspring store nutritive reserves in the form of lipids which will sustain metabolism during overwintering. Lipid reserves—mostly wax esters (WE)—can take up a substantial portion of the whole body volume, ca. 50% of individual body mass, and are very conspicuous under low magnification. By the end of the productive season, late juvenile *Calanus* spp. migrate to depths between 600 and 1400 m, where they reside until the next spring. For the North Atlantic, Jónasdóttir et al. (2015) showed that the contribution of the seasonally migrating *C. finmarchicus* population can be highly significant in terms of carbon export, roughly equivalent to the gravitational flux of POC below the permanent thermocline, and termed this mechanism lipid pump. A very interesting feature of this lipid pump is that, given WE are nearly devoid of N, P and Si (core macro nutrients for phytoplankton production), transport by seasonally migrating copepods de-couples carbon and nutrients' export fluxes, acting as a lipid shunt in the regulation of export fluxes (Jónasdóttir et al. 2015). Further recommended readings on vertical carbon fluxes and the role of zooplankton are the reviews by Turner (2015) and Steinberg and Landry (2017).

up BMII suggests an intriguing mechanism where diminished energetic transfer from phytoplankton to mesozooplankton is linked to increased basal resources. It is a nice example of the complexities involved in ecosystem-level functional traits arising from a rich repertoire of biological behaviors. Such results also suggest behavioral traits as important factors to understand and predict the dynamics of energy and matter fluxes, besides population densities as considered so far in most modeling approaches.

6.4 Secondary production

Secondary production (SP) by the zooplankton can be conceptualized as one of the fundamental fluxes of energy and organic matter in marine ecosystems. The biomass accumulated by the zooplankton constitutes the energetic resource and the organic building blocks available to sustain the metabolic requirements of their predators, and indirectly, of all consumer higher up in the trophic web. Thus, knowledge of the typical levels of SP, their patterns of variability and environmental drivers is central for a proper understanding of marine ecosystem functioning. SP can be defined as the quantity of tissue generated in a certain area or volume of water within a given time period. Usually, SP is expressed in biomass or energy flux units, e.g., mg C m^{-3} d^{-1}, i.e., amount of organic carbon contributed by the zooplankton per unit of volume and time. The new biomass produced is interpreted as the energy assimilated and stored as somatic structures by all individuals in the community. New somatic structures may be lost or discarded after a relatively short period, e.g., appendicularian houses, or copepods' eggs, but nonetheless contribute to SP and, in fact, may represent a very significant fraction of total SP.

Temperature and food availability are primary drivers of SP as they govern the growth of organisms. Zooplankton are poikilotherms and temperature affects their metabolism and physiological rates (e.g., respiration rate) according to the general form (Brown et al. 2004, López-Urrutia and Morán 2007):

$$B = b_0 * \exp(\frac{-Eh}{k * T}),$$

where B is the respiration rate, Eh is the activation energy for heterotrophic respiration, and k is Boltzmann's constant (8.62×10^{-5} eV*K^{-1}). In turn, food availability determines the amount of energy that will be available for growth, as it regulates food intake according to some type of functional response (e.g., Besiktepe and Dam 2002). "Functional response" is used as shorthand for "variability pattern of food consumption rate along a wide range of food concentration". Functional responses describe the shape of such relationship, i.e., food consumption (e.g., ug C ind^{-1} h^{-1}) *vs.* concentration (e.g., ug C L^{-1}). Three basic types of functional response encompass nearly all patterns found in nature: type-I describes a linear increase in food consumption; type-II describes a nearly linear increase at low food levels, which progressively saturates, reaching an asymptote at high concentrations; type-III describes a sigmoid pattern. For the zooplankton, there are examples of the three types, but the overwhelming majority of cases follow a type-II, or a "modified type-II" response, where there is an offset along the abscissa (i.e., the curve does not pass through the origin), which is interpreted as a concentration threshold to elicit feeding behavior. Type-II responses are mathematically represented using Ivlev's model, Holling Disk equation, or a Monod function.

In middle and high latitudes, temperature and food vary seasonally, imposing strong temporal variability on SP. Also, recalling discussions along previous sub-sections referring to environmental modulation of zooplankton abundance and activity rates at oceanographic fronts and through the modulation of food web structure, it is clear that SP should be strongly variable at a wide range of spatial and time scales. Indicative examples of the ranges of SP that can be found at different regions of the ocean are presented in Table 6.1.

Strong variability of SP implies that a proper characterization of this fundamental ecosystem flux at the relevant scales and with sufficient replication remains a big challenge. On top of that, it should be noted that the measurement of SP can be a highly demanding task in itself. Several approaches have been devised for the estimation of SP in the zooplankton. Those in which actual growth is measured are referred to as "direct methods"; examples include Cohort, Artificial cohort and Egg production methods. Indirect methods provide some proxy for growth, for example those based on biochemical indicators like

Table 6.1. Examples of mesozooplankton secondary production estimates for different regions of the ocean. (A) Cohort incubation models; Kimmerer and McKinnon (1987). (B) Empirical equation; Hirst et al. (1999); Hirst and Bunker (2003). (C) Egg production method; Berggreen et al. (1988). (D) Chitobiase enzyme activity; Sastri and Dower (2009). (E) Empirical equation; Huntley and Lopez (1992). *Production expressed in aerial units, not directly comparable with volumetric estimates.

Reference	Sampling area	Species/Group	SP	Method
Peterson et al. 1991	Skagerrak, Denmark	*Centropages typicus* *Calanus finmarchicus* *Temora longicornis* *Paracalanus parvus* *Acartia longiremis*	mean 4.6 (mgC m^{-3} day^{-1})	A
Ara 2002	Cananéia Lagoon, Brazil	*Temora turbinata*	from 0.0002 ± 0.0006 to 1.115 ± 0.261 (mgC m^{-3} d^{-1})	B
McKinnon and Duggan 2003	Exmouth Gulf, Australia (18–75 m)	Paracalanidae Clausocalanidae Oithonidae	mean 12.6 (mgC m^{-2} d^{-1})* ~ 3.4 (gC m^{-2} y^{-1})*	A
Calliari et al. 2004	Río de la Plata, Uruguay	*Acartia tonsa*	0.023–6.633 (mgC m^{-3} d^{-1})	C
Renz et al. 2008	North Sea - German Bight (14–44 m)	*Pseudocalanus elongatus*	mean 26.2 (mgC m^{-2} d^{-1})*	A
Calliari et al. 2009	Rocha Lagoon, Uruguay	*Acartia tonsa*	mean 1.767 (mgC m^{-3} d^{-1})	C
Temperoni et al. 2010	EPEA, Argentina	*Oithona nana*	0.01–0.05 (mgC m^{-3} d^{-1})	C
Avila et al. 2012	Patos Lagoon, Brazil	*Acartia tonsa* *Notodiaptomus incompositus* Cladocera Cyclopoida	2.35–5.55 (mgC m^{-3} d^{-1})	D
Muxagata et al. 2012	Patos Lagoon, Brazil	*Acartia tonsa*	0.40–3.65 (mgC m^{-3} d^{-1})	B
Basedow et al. 2014	Barents Sea, Polar Arctic	small zooplankton medium-sized copepods *Calanus* spp.	134 (mgC m^{-3} d^{-1}) 3.6 (mgC m^{-3} d^{-1}) 0.9 (mgC m^{-3} d^{-1})	B B B
Rato et al. 2014	Berlengas Archipelago, Portugal	mesozooplankton	mean 0.053 ± 0.126 (mgC m^{-3} d^{-1})	E
Cepeda et al. 2015	Río de la Plata, Argentina-Uruguay	*Oithona helgolandica* *Oithona nana*	0.003–0.014 (mgC m^{-3} d^{-1}) 0.005–0.13 (mgC m^{-3} d^{-1})	C C
Teixeira-Amaral et al. 2017	Patos Lagoon Brazil	*Acartia tonsa* *Notodiaptomus incompositus*	mean 1.17 (mgC m^{-3} d^{-1}) mean 0.13 (mgC m^{-3} d^{-1})	B B
Medellín-Mora et al. 2019	Concepcion Chile (80–90 m)	mesozooplankton	mean 80 (gC m^{-x0002_2} y^{-1})*	B

nucleic acids (DNA:RNA ratio) or enzymatic activities (e.g., chitobiase, aminoacyl-tRNA synthetases); a third (also indirect) approach is the estimation of SP from empirical (statistical) models derived from the meta-analyses of copepods' growth rates. Several excellent reviews exist, some rather recent, on methodological approaches for the estimation of SP in zooplankton. Interested readers are thus referred to Postel et al. (2000), Yebra et al. (2016), Hirts (2017), Setubal et al. 2020.

6.5 Perspectives

Increased knowledge on marine zooplankton is required for addressing several of current most pressing issues globally, notably the evolution of the structure and functioning of marine ecosystems under a changing climate system, and the sustainability of marine populations contributing to support human well-being. How zooplankton responds to the several facets of climate change will be critical for the efficiency of the biological carbon pump and for the sustainability of marine populations at higher trophic levels, including those harvested for human consumption. These constitute core scientific problems that will likely guide research efforts in years to come.

Direct responses of the zooplankton to environmental changes associated to oceanic warming, eutrophication, deoxigenation and acidification involve alteration of vital rates, physiological functions and life-history traits (Dam and Baumann 2018). Also, changes in the vertical structure of the ocean (e.g., thermal stratification) will reduce the flux of new nutrients to the euphotic zone, the rate of primary production, and the taxonomic and size structure of phytoplankton communities (Boyd and Doney 2002, Behrenfeld et al. 2006, Finkel et al. 2009). In that way, changes in the physical-biological coupling at the base of the trophic webs are expected to have indirect effects on zooplankton (Dam and Baumann 2018). A better understanding of the changes already under way will require advancing the knowledge on multiple aspects of the zooplankton including, but not limited to:

- the trophic structure near basal levels in marine webs, with focus on the interactions between phytoplankton, microzooplankton, mixotrophic and metazoan plankton, the role of zooplankton in mesopelagic remineralization processes and its influence in the stoichiometry between carbon and other elements (Steinberg and Landry 2017).
- an improved quantification of the efficiency, dynamics and drivers of the biological pump (Turner 2015).
- the scaling of evolutionary rates to environmental change in order to predict if zooplankton will be able to adapt to the pace of ongoing changes. That will require elucidating the genetic contribution to phenotypic variability and the heritability of traits linked to fitness: survival, fecundity, growth and developmental rates, size and age at maturity (Dam 2013).
- reduce the uncertainties regarding regional impacts of climate change, and bottom-up responses to changing temperature, physics, and nutrient delivery (Steinberg and Landry 2017).
- the synergistic impact of other forms of contamination and habitat degradation, typically at local or regional scales, on top of global change.

Summary

1. Zooplankton comprises a highly diverse assemblage of organisms in terms of life history, body size, trophic habits, functional traits and role within the ecosystem.
2. Spatial and temporal patterns in the distribution of zooplankton abundance and traits tend to be directed by large-scale gradients in temperature and nutrition regimes, and by recurrent oceanographic features like fronts that modulate productivity and the structure of planktonic assemblages.
3. Zooplankton have evolved four main strategies for successful feeding on a wide spectrum of prey within a nutritionally diluted medium: passive and active ambush feeding, current feeding and cruise feeding. In most cases detection, pursuit and ingestion of prey requires a fine balancing of hydrodynamics at low and intermediate Re.
4. The classical view of zooplankton as a simple herbivorous link between primary producers and higher order consumers within a linear trophic array has given way to a more complex depiction of planktonic food webs where mixotrophy is pervasive among the smaller components of the microbial loop, and omnivory among metazoan zooplankton, and particularly among copepods. Microbial and metazoan trophic pathways coexist, are interlinked and their relative importance is modulated by bio-physical coupling; on average, microbial grazers consume the largest fraction of marine primary production in the global ocean.

References

Aboul-Ezz, S.M., A.M.H. Heneash and S.M. Gharib. 2014. Variability of spatial and temporal distribution of zooplankton communities at Matrouh beaches, southeastern Mediterranean Sea, Egypt. Egyptian J. Aquat. Res. 40: 283–290.

Acha, E.M., M.D. Ehrlich, J.H. Muelbert, M. Pájaro, D. Bruno, L. Machinandiarena and M. Cadaveira. 2018. Ichthyoplankton associated to the frontal regions of the Southwestern Atlantic. pp. 219–246. *In*: Hoffmeyer, M., M. Sabatini, F.P. Brandini, D.L. Calliari and N.H. Santinelli [eds.]. Plankton Ecology of the Southwestern Atlantic from the Subtropical to the Subantarctic Realm. Springer, Switzerland.

Acuña, J.L. and M. Kiefer. 2000. Functional response of the appendicularian *Oikopleura dioica*. Limnol. Oceanogr. 45: 608–618.

Alcaraz, M. 1997. Copepods under turbulence: grazing, behavior and metabolic rates. Sci. Mar. 61(1): 177–195.

Almeda, R.H., G. van Someren and T. Kiørboe. 2017. Behavior is a major determinant of predation risk in zooplankton. Ecosphere doi.org/10.1002/ecs2.1668.

Anderson, T.R. and M.I. Lucas. 2008. Upwelling ecosystems. pp. 3651–3661. *In*: Jorgensen, S.E. and B. Fath [eds.]. Encyclopedia of Ecology. Elsevier Science, New York, USA.

Angel, M. 1981. Ostracoda. pp. 543–585. *In*: Boltovskoy, D. [ed.]. Atlas del zooplancton del Atlántico sudoccidental y métodos de trabajo con el zooplancton marino. INIDEP Press, Mar del Plata, Argentina.

Angelescu, V. and A. Anganuzzi. 1981. Resultados sobre la alimentación de la anchoíta (*Engraulis anchoita*) en el área explorada por el B/I “Shinkai Maru” durante las campañas VI y VIII en el mar epicontinental argentino. Ser. Cont. INIDEP 383: 281–298.

Angelescu, V. and L.B. Prenski. 1987. Ecología trófica de la merluza común del Mar Argentino (Merluccidae, *Merluccious hubbsi*). Ser. Cont. INIDEP 56: 1–205.

Antezana, T. 2009. Species-specific patterns of diel migration into the oxygen minimum zone by euphausiids in the Humboldt current ecosystem. Prog. Oceanogr. 83(1-4): 228–236.

Antezana, T. 2010. *Euphausia mucronata*: A keystone herbivore and prey in the Humboldt current system. Deep Sea Res. II 57(7-8): 652–662.

Ara, K. 2002.Temporal variability and production of *Temora turbinata* (Copepoda: Calanoida) in the Cananéia Lagoon estuarine system, São Paulo, Brazil. Sci. Mar. 66(4): 399–406.

Arim, M., S.R. Abades, G. Laufer, M. Loureiro and P.A. Marquet. 2010. Food web structure and body size: trophic position and resource acquisition. Oikos. 119: 147–153.

Armengol, L., G. Franchy, A. Ojeda, A. Santana-del Pino and S. Hernández-León. 2017. Effects of copepods on natural microplankton communities: do they exert top-down control? Mar. Biol. 164: 1–136.

Atienza, D., E. Saiz and A. Calbet. 2006. Feeding ecology of the marine cladoceran *Penilia avirostris*: natural diet, prey selectivity and daily ration. Mar. Ecol. Prog. Ser. 315: 211–220.

Attrill, M.J. 2002. A testable linear model for diversity trends in estuaries. Journal of Animal Ecology 71: 262–269.

Avila, T.R., C.S. Pedrozo and J.G.F. Bersano. 2009. Variação temporal do zooplâncton da Praia de Tramandaí, Rio Grande do Sul, com ênfase em Copepoda. Iheringia 99(1): 18–26.

Avila, T.R, A.A. de Souza Machado and A. Bianchini. 2012. Estimation of zooplankton secondary production in estuarine waters: Comparison between the enzymatic (chitobiase) method and mathematical models using crustaceans. J. Exper. Mar. Biol. Ecol. 416-417: 144–152.

Azam, F., T. Fenchel, J.G. Field, J.S. Grey, L.A. Meyer-Reil and F. Thingstad. 1983. The ecological role of water column microbes in the sea. Mar. Ecol. Prog. Ser. 10: 257–263.

Bainbridge, V. 1958. Some observations on *Evadne nordmanni,* Loven. J. Mar. Biol. Ass. U.K. 37: 349–370.

Bakun, A. 1996. Patterns in the Ocean. Ocean processes and marine population dynamics. California Sea Grant College System/NOAA/Centro de Investigaciones Biológicas del Noroeste, La Paz, México.

Bamstedt, U., D.J. Gifford, A. Irigoien, A. Atkinson and M. Roman. 2000. Feeding. pp. 297–399. *In*: Harris, R.P., P.H. Wiebe, J. Lenz, H.R. Skojoldal and M. Huntley [eds.]. Zooplankton Methodology Manual. ICES Academic Press. London, UK.

Barnett, A.J., K. Finlay and B.E. Beisner. 2007. Functional diversity of crustacean zooplankton communities: towards a trait-based classification. Freshw. Biol. 52: 796–813.

Barton, A.D., A.J. Pershing, L. Litchman, N.R. Record, K.F. Edwards, Z.V. Finkel, T. Kiørboe and B.A. Ward. 2013. The biogeography of marine plankton traits. Ecol. Lett. 16(4): 522–534.

Bartram, W.C. 1982. Experimental development of a model for the feeding of neritic copepods on phytoplankton. J. Plankton Res. 3: 25–51.

Basedow, S., M. Zhou and S. Tande. 2014. Secondary production at the Polar Front, Barents Sea. J. Mar. Syst. 130: 147–159.

Behrenfeld, M.J., R.T. O’Malley, D.A. Siegel, C.R. McClain, J.L. Sarmiento, G.C. Feldman, A.J. Milligan, P.G. Falkowski, R.M. Letelier and E.S. Boss. 2006. Climate-driven trends in contemporary ocean productivity. Nature 444: 752–755.

Behrenfeld, M.J. 2010. Abandoning Sverdrup’s Critical Depth Hypothesis on phytoplankton blooms. Ecology 91: 977–989.

Behrenfeld, M.J. and E.S. Boss. 2018. Student’s tutorial on bloom hypotheses in the context of phytoplankton annual cycles. Glob Change Biol. 24: 55–77.

Benedetti, F., S. Gasparini and S.D. Ayata. 2016. Identifying copepod functional groups from species functional traits. J. Plankton Res. 38: 159–166.
Benedetti, F., M. Vogt, D. Righetti, F. Guilhaumon and S.D. Ayata. 2018. Do functional groups of planktonic copepods differ in their ecological niches? J. Biogeogr. 45(3): 604–616.
Berggreen, U., B. Hansen and T. Kiørboe. 1988. Food size spectra, ingestion and growth of the copepod *Acartia tonsa* during development: implications for determination of copepod production. Mar. Biol. 99: 341–352.
Besiktepe, S. and H.G. Dam. 2002. Coupling of ingestion and defecation as a function of diet in the calanoid copepod *Acartia tonsa*. Mar. Ecol. Prog. Ser. 229: 151–164.
Bjornberg, T.K.S. 1981. Copepoda. pp. 587–679. *In*: Boltovskoy, D. [ed.]. Atlas del zooplancton del Atlántico sudoccidental y métodos de trabajo con el zooplancton marino. INIDEP Press, Mar del Plata, Argentina.
Boltovskoy, D. 1981a. Zooplancton del Atlántico Sudoccidental. pp. 255–259. *In*: Boltovskoy, D. [ed.]. Atlas del zooplancton del Atlántico sudoccidental y métodos de trabajo con el zooplancton marino. INIDEP Press, Mar del Plata, Argentina.
Boltovskoy, D. 1981b. Métodos de trabajo con el zooplancton marino. pp. 1–107. *In*: Boltovskoy, D. [ed.]. Atlas del zooplancton del Atlántico sudoccidental y métodos de trabajo con el zooplancton marino. INIDEP Press, Mar del Plata, Argentina.
Boltovskoy, D. 1999. South Atlantic Zooplankton. Backhuys Publisher, Leiden, The Netherlands.
Boltovskoy, D., M.J. Gibbons, L. Hutchings and D. Binet. 1999. General biological features of the South Atlantic. pp. 1–42. *In*: Boltovskoy, D. [ed.]. South Atlantic Zooplankton Vol.1. Backhuys Publishers, Leiden, Netherlands.
Bosch, H.F. 1970. Ecology of *Podon polyphemoides* (Crustacea, Branchiopoda) in Chesapeake Bay. Technical Reports Chesapeake Bay Institute of John Hopkins University 66: 1–77.
Bosch, H.F. and W.R. Taylor. 1973. Diurnal vertical migration of an estuarine cladoceran *Podon polyphemoides* in the Chesapeake Bay. Mar. Biol. 19(2): 161–171.
Boschi, E. 1981. Larvas de Crustacea Decapoda. pp. 699–758. *In*: Boltovskoy, D. [ed.]. Atlas del zooplancton del Atlántico sudoccidental y métodos de trabajo con el zooplancton marino. INIDEP Press, Mar del Plata, Argentina.
Boyd, P.W. and S.C. Doney. 2002. Modelling regional responses by marine pelagic ecosystems to global climate change. Geophys. Res. Let. Doi10.1029/2001GL014130.
Bradford-Grieve, J.M., E.L. Markhaseva, C.E.F Rocha and B. Abiahy. 1999. Copepoda. pp. 869–1098. *In*: Boltovskoy, D. [ed.]. South Atlantic Zooplankton Vol.2. Backhuys Publishers, Leiden, Netherlands.
Branco, C.W.C., B. Kozlowsky-Suzuki and F.A. Esteves. 2007. Environmental changes and zooplankton temporal and spatial distribution in a disturbed Brazilian coastal lagoon. Braz. J. Biol. 67(2): 251–262.
Brown, J.H., J.F. Gillooly, A.P. Allen, V.M. Savage and G.B West. 2004. Toward a metabolic theory of Ecology. Ecology 85: 1771–1789.
Brun, P., M.R. Payne and T. Kiørboe. 2016. Trait biogeography of marine copepods—an analysis across scales. Ecol. Lett. 19(12): 1403–1413.
Buckley, L.J. and R.G. Lough. 1987. Recent growth, biochemical composition, and prey field of larval haddock (*Melanogrammus aeglefinus*) and Atlantic cod (*Gadus morhua*) on Georges Bank. Canad. J. Fish. Aquat. Sci. 44(1): 14–25.
Cabreira, A.G., A. Madirolas and N.E. Brunetti. 2011. Acoustic characterization of the Argentinean short-fin squid aggregations. Fish. Res. 108: 95–99.
Calbet, A., M.R. Landry and R.D. Scheinberg. 2000. Copepod grazing in a subtropical bay: species-specific responses to a midsummer increase in nanoplankton standing stock. Mar. Ecol. Prog. Ser. 193: 15–84.
Calbet, A. 2001. Mesozooplankton grazing effect on primary production: A global comparative analysis in marine ecosystems. Limnol. Oceanogr. 46: 1824–1830.
Calbet, A. and M.R. Landry. 2004. Phytoplankton growth, microzooplankton grazing, and carbon cycling in marine systems. Limnol. Oceanogr. 49: 51–57.
Calbet, A. and E. Saiz. 2005. The ciliate-copepod link in marine ecosystems. Aquat. Microb. Ecol. 38: 157–167.
Calbet, A. 2008. The trophic roles of microzooplankton in marine systems. ICES J. Mar. Sci. 65: 325–331.
Calbet, A. and E. Saiz 2013. Effects of trophic cascades in dilution grazing experiments: from artificial saturated feeding responses to positive slopes. J. Plankton Res. 35: 1183–1191.
Calbet, A. and E. Saiz. 2018. How much is enough for nutrients in microzooplankton dilution grazing experiments? J. Plankton Res. 40(2): 109–117.
Calder, W.A. 1983. Body size, mortality. J. Theor. Biol. 102: 135–144.
Calliari D., G. Cervetto and R. Castiglioni. 2004. Summertime herbivory and egg production by *Acartia tonsa* at Montevideo coast, Río de la Plata. Ophelia 58: 115–128.
Calliari, D. and P. Tiselius. 2009. Organic carbon fluxes through the mesozooplankton and their variability at different time-scales in the Gullmarsfjord, Sweden. Estuar. Coast. Shelf Sci. 85: 107–117.
Calliari, D., A. Britos and D. Conde. 2009. Testing the relationship between primary production and *Acartia tonsa* grazing pressure in an estuarine lagoon. J. Plankton Res. 31: 1045–1058.
Capitanio, F.L., M.L. Spinelli, M.L. Presta, G.E. Aguirre, G. Cervetto, M. Pájaro and C.M. Derisio. 2018. Ecological role of common appendicularian species from shelf waters off Argentina. pp. 201–218. *In*: Hoffmeyer, M., M. Sabatini, F.P.

Brandini, D.L. Calliari and N.H. Santinelli [eds.]. Plankton Ecology of the Southwestern Atlantic from the Subtropical to the Subantarctic Realm. Springer, Switzerland.
Cardelli, N.V., P.M. Cervellini and M.C. Piccolo. 2006. Abundancia estacional y distribución espacial de Mysidacea en el Atlántico sudoccidental, estuario de Bahía Blanca (38°42′–39°26′ S y 62°28′–61°40′ W). Rev. Biol. Mar. Oceanogr. 41: 177–185.
Carranza, M.M., S.T. Gille, A.R. Piola, M. Charo and S.I Romero. 2017. Wind modulation of upwelling at the shelf-break front off Patagonia: observational evidence. J. Geophys. Res. Oceans 122(3): 2401–2421.
Castel, J. and J. Veiga. 1990. Distribution and retention of the copepod *Eurytemora* aff. *hirundoides* in a turbid estuary. Mar. Biol. 107(1): 119–128.
Cepeda, G.D., R.P. Di Mauro, M.C. Hozbor, D. Cucchi Colleoni, D. Hernandez and M.D. Viñas. 2015. Spatial variation in life-history traits of *Oithona* spp. in a shallow temperate estuarine system (Río de la Plata, south-west Atlantic) during spring. Mar. Freshw. Res. 66(9): 795–804.
Cepeda, G.D., B. Temperoni, M.E. Sabatini, M.D. Viñas, C.M. Derisio, B.A. Santos, J.C. Antacli and L.N. Padovani. 2018. Zooplankton communities of the argentine continental shelf (SW Atlantic, ca. 34°–55°S), an overview. pp. 171–199. *In*: Hoffmeyer, M., M. Sabatini, F.P. Brandini, D.L. Calliari and N.H. Santinelli [eds.]. Plankton Ecology of the Southwestern Atlantic from the Subtropical to the Subantarctic Realm. Springer, Switzerland.
Chaalali, A., G. Beaugrand, V. Rayboud, E. Goberville, V. David, P. Boet and B. Sautour. 2013. Climatic facilitation of the colonization of an estuary by *Acartia tonsa*. PLoS One doi.org/10.1371/journal.pone.0074531.
Conley, K.F., F. Lombard and K.R. Sutherland. 2018. Mammoth grazers on the ocean's minuteness: a review of selective feeding using mucous meshes. Proceedings of the Royal Society B: Biol. Scien. 285: 1–10.
Cordo, H.D. 1981. Resultados sobre la alimentación de la merluza del Mar Argentino (*Merluccius hubbsi*). Análisis biológico y estadístico de los datos obtenidos de las campañas de los B/I "Shinkai Maru" y "Walther Herwig" (1978–1979). Contrib. Inst. Nac. Inv. Desarr. Pesq. 383: 299–312.
Costa, K.G., P.R.S. Pinheiro, C.A.R. Melo, S.M.O. Oliveira, L.C.C. Pereira and R.M. Costa. 2011. Effects of seasonality on zooplankton community dynamics in the macrotidal coastal zone of the Amazon region. J. Coast. Res. 64: 364–368.
Cowles, T.J. and J.R. Strickler. 1983. Characterization of feeding activity patterns in the planktonic copepod *Centropages typicus* Kroyer under various food conditions. Limnol. Oceanogr. 28: 106–115.
Cronin, L.E., J.C. Daiber and M. Hulbert. 1962. Quantitative seasonal aspects of zooplancton in the Delaware River Estuary. Chesap. Sci. 3(2): 63–93.
Cushing, D.H. 1989. A difference in structure between ecosystems in strongly stratified waters and in those that are only weakly stratified. J. Plankton Res. 11: 1–13.
Dam, H.G., W.T. Peterson and D.C. Bellantoni. 1994. Seasonal feeding and fecundity of the calanoid copepod *Acartia tonsa* in Long Island Sound: is omnivory important to egg production? Hydrobiologia 292(293): 191–199.
Dam, H.G. 2013. Evolutionary adaptation of marine zooplankton to global change. Ann. Rev. Mar. Sci. 5: 349–370.
Dam, H.G. and H. Baumann. 2018. Climate change, zooplankton and fisheries. pp. 851–874. *In*: Phillips, B.F. and M. Pérez-Ramírez [eds.]. Climate Change Impacts on Fisheries and Aquaculture: A Global Analysis, Volume II, First Edition. John Wiley and Sons Ltd.
Dauby, P., F. Nyssew and C. De Broyer. 2003. Amphipods as food sources for higher trophic levels in the Southern Ocean: a synthesis. pp. 29–134. *In*: Huiskes, A.H.L., W.W.C. Gieskes, R.M.L. Schorno, S.M. Van der Vies and W. Wolff [eds.]. Antarctic Biology in a Global Context. Backhuys Publishers, Leiden, The Netherlands.
De Pauw, N. 1973. On the distribution of *Eurytemora affinis* (Poppe) (Copepoda) in the Western Scheldt Estuary. Internationale Vereinigung für Theoretische und Angewandte Limnologie: Verhandlungen 18: 1462–1472.
Deibel, D. 1986. Feeding mechanism and house of the appendicularian *Oikopleura vanhoeffeni*. Mar. Biol. 93: 429–436.
Derisio, C., M. Braverman, E. Gaitán, C. Hozbor, F. Ramírez, J. Carreto, F. Botto, D.A. Gagliardini, E.M. Acha and H. Mianzan. 2014. The turbidity front as a habitat for *Acartia tonsa* (Copepoda) in the Río de la Plata, Argentina-Uruguay. J. Sea Res. 85: 197–204.
Devreker, D., S. Souissi, J.C. Molinero and F. Nkubito. 2008. Trade-offs of the copepod *Eurytemora affinis* in mega-tidal estuaries: insights from high frequency sampling in the Seine Estuary. J. Plankton Res. 30(12): 1329–1342.
Doall, M.H., J.R. Strickler, D.M. Fields and J. Yen. 2002. Mapping the free-swimming attack volume of a planktonic copepod, *Euchaeta rimana*. Mar. Biol. 140: 871–879.
Elton, C. 1927. Animal Ecology. Sedgewick and Jackson Publisher, London.
Feigenbaum, D. and M.R. Reeve. 1977. Prey detection in the Chaetognatha: Response to a vibrating probe and experimental determination of attack distance in large aquaria. Limnol. Oceanogr. 22: 1052–1058.
Feigenbaum, D. 1978. Hair-fan patterns in the Chaetognatha. Canadian J. Aqua. Sci. 56: 536–546.
Feigenbaum, D. 1982. Feeding by the chaetognath, *Sagitta elegans*, at low temperatures in Vineyard Sound, Massachusetts. Limnol. Oceanogr. 27: 699–706.
Fenaux, R. 1986. The house of *Oikopleura dioica* (Tunicata, Appendicularia): structure and functions. Zoomorphology 106: 224–231.
Fenchel, T. 1982. Ecology of heterotrophic microflagellates. Quantitative occurrence and importance as bacterial consumers. Mar. Ecol. Prog. Ser. 9: 35–42.

Finkel, Z.V., J. Beardall, K.J. Flynn, A. Quigg, T.A.V. Rees and J.A. Raven. 2009. Phytoplankton in a changing world: cell size and elemental stoichiometry. J. Plankton Res. 32: 119–137.

Flood, P.R., D. Deibel and C.C. Morris. 1992. Filtration of colloidal melanin from seawater by planktonic tunicates. Nature 355: 630–632.

Fofonoff, P.W. 1994. Marine Cladocerans in Narragansett Bay. Ph.D. Thesis, Univ. Rhode Island.

Forster, J., A.G. Hirst and D. Atkinson. 2011. How do organisms change size with changing temperature? The importance of reproductive method and ontogenetic timing. Funct. Ecol. 25: 1024–1031.

Foster, L. 2006. Chaetognaths (arrow worms). pp. 235–343. *In*: Lewbart, G.A. [ed.]. Invertebrate Medicine. Blackwell Publishing, Oxford.

Franco, B.C., A.R. Piola, A.L. Rivas, A. Baldolini and J.P. Pisoni. 2008. Multiple thermal fronts near the Patagonian shelf break. Geoph. Res. Lett. doi.org/10.1029/2007GL032066.

Froneman, P.W., E.A. Pakhomov, L.J. Gurney and B.P.V. Hunt. 2002. Predation impact of carnivorous macrozooplankton in the vicinity of the Prince Edward archipelago (Southern Ocean) in austral autumn 1998. Deep-Sea Res. II 49: 3243–3254.

Fwiman, L.A. and R.G. Werner. 2002. Fishery Science. The Unique Contributions of Early Life Stages. Blackwell Science Ltd.

Gasca, R. and S.H.D. Haddock. 2004. Associations between gelatinous zooplankton and hyperiids amphipods (Crustacea: *Peracarida*) in the Gulf of California. Hydrobiologia 530(531): 529–535.

Gauld, D.T. 1957. A peritrophic membrane in calanoid copepods. Nature 179: 325–326.

Gibbons, J.M., V.A. Spiridonov and G.A Tarling. 1999. Euphausiacea. pp. 1241–1279. *In*: Boltovskoy, D. [ed.]. South Atlantic Zooplankton Vol.2. Backhuys Publishers, Leiden, Netherlands.

Giberto, D.A., C.S. Bremec, E.M. Acha and H.W. Mianzan. 2007. Feeding of the whitemouth croaker *Micropogonias furnieri* (Sciaenidae; Pisces) in the estuary of the Río de la Plata and adjacent Uruguayan coastal waters. Atlantica 29: 75–84.

Giesecke, R., H.E. González and U. Bathmann. 2010. The role of the chaetognatha *Sagitta gazellae* in the vertical carbon flux of the Southern Ocean. Polar. Biol. 33 (3): 293–304.

Gonçalves, R.J. and T. Kiørboe. 2015. Perceiving the algae: How feeding-current feeding copepods detect their non-motile prey. Limnol. Oceanogr. 60: 1286–1297.

Gowing, M.M. 1993. Large virus-like particles from vacuoles of *Phaeodarian radiolarians* and from other marine samples. Mar. Ecol. Prog. Ser. 101: 33–44.

Gran, H.H. and T. Braarud. 1935. A quantitative study of the phytoplankton in the Bay of Fundy and the Gulf of Maine (including observations on hydrography, chemistry and turbidity). J. Biol. Board Can. 1: 279–467.

Hansen, B.W., P.K. Bjornsen and P.J. Hansen. 1994. The size ratio between planktonic predators and their prey. Limnol. Oceanogr. 39: 395–403.

Harbison, G.R., D.C. Biggs and L.P. Madin. 1977. The associations of Amphipoda Hyperiidea with gelatinous zooplankton-II. Associations with Cnidaria, Ctenophora and Radiolaria. Deep-Sea Res. 24: 465–468.

Heckman, C.W. 1986. The anadromous migration of a calanoid copepod, *Eurytemora affinis* (Poppe, 1880) in the Elbe Estuary. Crustaceana 5(2): 176–181.

Hensen, V. 1887. Uber die Bestimmung des Plankton's oder des im Meere treibenden Materials an Pflanzen und Thieren. Funfter Bericht der Kommission zur Wissenschaftliche Untersuchung der deutschen Meere in Kiel, fur die Jahre, Berlin.

Herrera, I., L. Yebra, T. Antezana, A. Giraldo, J. Färber-Lorda and S. Hernández-León Yebra. 2019. Vertical variability of *Euphausia distinguenda* metabolic rates during diel migration into the oxygen minimum zone of the Eastern Tropical Pacific off Mexico. J. Plankton Res. 41(2): 165–176.

Highsmith, R.C. and K.O. Coyle. 1992. Productivity of arctic amphipods relative to gray whale energy requirements. Mar. Ecol. Prog. Ser. 83: 141–150.

Hirst, A.G., M. Sheader and J.A. Williams. 1999. Annual pattern of calanoid copepod abundance, prosome length and minor role in pelagic carbon flux in the Solent, U.K. Mar. Ecol. Prog. Ser. 177: 133–146.

Hirst, A. and A. Bunker. 2003. Growth of marine planktonic copepods: global rates and patterns in relation to chlorophyll a, temperature, and body weight. Limnol. Oceanogr. 48: 1988–2010.

Hirst, A., J. Roff and R. Lampitt, 2003. A synthesis of growth rates in marine epipelagic invertebrate zooplankton. Adv. Mar. Bio. 44: 1–14.

Hirts, A.G. 2017. Zooplankton productivity. pp 35–41. *In*: Castellani, C. and M. Edwards [eds.]. Marine Plankton: A Practical Guide to Ecology, Methodology and Taxonomy. Oxford University Press, 101093/acprofioso9780199233261001001.

Holm, S., R.B. Davis, J. Javois, E. Ounap, A. Kaasik, F. Molleman and T. Tammaru. 2016. A comparative perspective on longevity: the effect of body size dominates over ecology in moths. J. Evol. Biol. 29: 2422–2435.

Huliselan, N.V. 2002. The role of zooplankton predator, chaetognaths (*Sagitta* spp.) in Baguala Bay waters, Ambon Islands. J. Coast. Develop. 6: 9–21.

Huntley, M.E. and D.G. López Mai. 1992. Temperature-dependent production of marine copepods: a global synthesis. The American Naturalist 140(2): 201–242.

Huskin, I., R. Anadan, G. Medina, R.N. Head and R.P. Harris. 2001. Mesozooplankton distribution and copepod grazing in the subtropical Atlantic near the Azores: influence of mesoscale structures. J. Plankton Res. 23: 671–691.

Iles, T.D. and M. Sinclair. 1982. Atlantic Herring: Stock discreteness and abundance. Science 215(4533): 6227–633.

Ivanovic, M.L. and N.E. Brunetti. 1994. Food and feeding of *Illex argentinus*. Antarct. Sci. 6: 185–193.

Jagger, R.A., W.J. Kimmerer and G.P. Jenkins. 1988. Food of the cladoceran *Podon intermedius* in a marine embayment. Mar. Ecol. Prog. Ser. 43: 245–250.

Jaworski, A. and S.A. Ragnarsson. 2006. Feeding habits of demersal fish in Icelandic waters: a multivariate approach. ICES J. Mar. Syst. 63: 1682–1694.

Johnson, C.L., A.W. Leising, J.A. Runge, E.J.H. Head, P. Pepin, S. Plourde and E.G. Durbin. 2008. Characteristics of *Calanus finmarchicus* dormancy patterns in the Northwest Atlantic. ICES J. Mar. Syst. 65: 339–350.

Jónasdóttir, S.H., A.W. Vissera, K. Richardson and M.R. Heath. 2015. Seasonal copepod lipid pump promotes carbon sequestration in the deep North Atlantic. PNAS 112(39): 12122–12126.

Jonsson, P.R. and P. Tiselius. 1990. Feeding behavior, prey detection and capture efficiency of the copepod *Acartia tonsa* feeding on planktonic ciliates. Mar. Ecol. Prog. Ser. 60: 35–44.

Jumars, P.A, D.L. Penry, J.A. Baross, M.J. Perry and B.W. Frost. 1989. Closing the microbial loop: dissolved carbon pathway to heterotrophic bacteria from incomplete ingestion, digestion and absorption in animals. Deep-Sea Res. 36: 483–495.

Katechakis, A., H. Stibor, U. Sommer and T. Hansen. 2004. Feeding selectivities and food niche separation of *Acartia clausi, Penilia avirostris* (Crustacea) and *Doliolum denticulatum* (Thaliacea) in Blanes Bay (Catalan Sea, NW Mediterranean). J. Plankton Res. 26: 589–603.

Kennish, M.J. and H.W. Paerl. 2010. Coastal lagoons critical habitats of environmental change. pp. 1–16. *In*: Kennish, M.J. and H.W. Paerl [eds.]. Coastal Lagoons: Critical Habitats of Environmental Change (Marine Science Series). CRC Press, Boca Raton.

Kim, S.W., T. Ombé and Y.H. Yoon. 1989. Feeding habits of marine cladocerans in the Inland Sea of Japan. Mar. Biol. 100: 313–318.

Kimmerer, W.J. and A.D. McKinnon. 1987. Growth, mortality, and secondary production of the copepod *Acartia tranteri* in Westernport Bay, Australia. Limnol. Oceanogr. 32: 14–28.

Kiørboe, T. and J. Johansen. 1986. Studies of a larval herring (*Clupea harengus* L.) patch in the Buchan area. IV. Zooplankton distribution and productivity in relation to hydrographic features. Dana 6: 37–51.

Kiørboe, T. 1993. Turbulence, phytoplankton cell size and the structure of pelagic food webs. Advances Mar. Biol. 29: 1–72.

Kiørboe, T. 1996. Material flux in the water column. Coast Est. Stud. 52: 67–94.

Kiørboe, T., E. Saiz and M. Viitasalollo. 1996. Prey switching behaviour in the planktonic copepod *Acartia tonsa*. Mar. Ecol. Prog. Ser. 143: 65–75

Kiørboe, T. 2010. How zooplankton feed: mechanisms, traits and trade-offs. Biol. Rev. 86: 311–339.

Kiørboe, T. 2011. What makes pelagic copepods so successful? J. Plankton Res. 33: 677–685.

Kiørboe, T. 2016. Foraging mode and prey size spectra of suspension-feeding copepods and other zooplankton. Mar. Ecol. Prog. Ser. 558: 15–20.

Kiørboe, T., A. Visser and K.H. Andersen. 2018. A trait-based approach to ocean ecology. J. Mar. Sci. 75: 1849–1863.

Kjerfve, B. 1986. Comparative oceanography of coastal lagoons. pp. 63–81. *In*: Wolfe, D.A. [ed.]. Estuarine Variability. Academic Press, New York.

Kjerfve, B. 1994. Coastal lagoons. pp. 1–8. *In*: Kjerfve, B. [ed.]. Coastal Lagoon Processes. Elsevier Science Publishers, Amsterdam.

Koehl, M.A.R. and J.R. Strickler. 1981. Copepod leading currents: food capture at low Reynolds number. Limnol. Oceanogr. 26: 1062–1073.

Kozlowsky-Suzuki, B. and R.L. Bozelli. 2004. Resilience of a zooplankton community subjected to marine intrusion in a tropical coastal lagoon. Hydrobiologia 522: 165–177.

Kruk, C., V. Huszar, E. Peeters, S. Bonilla, L. Costa, M. Lurling, C.S. Reynolds and M. Scheffer. 2010. A simple morphological classification captures much of the functional variety of phytoplankton. Freshw. Bio. 55: 614–627.

Kruse, S., T. Brey and U. Bathmann. 2010. Role of midwater chaetognaths in Southern Ocean pelagic energy flow. Mar Ecol. Prog. Ser. 416: 105–113.

Lalli, C.R. and T.R. Parsons. 2006. Biological Oceanography an Introduction (Second Edition). The Open University, Oxford-UK.

Laval, P. 1980. Hyperiid amphipods as crustacean parasitoids associated with gelatinous zooplankton. Oceanogr. Mar. Biol. Ann. Rev. 18: 11–56.

Lawrence, J., J. Töpper, E. Petelenz-Kurdziel, G. Bratbak, A. Larsen, E. Thompson, C. Troedsson and J.L. Ray. 2017. Viruses on the menu: the appendicularian *Oikopleura dioica* efficiently removes viruses from seawater. Limnol. Oceanogr. 63: 244–253.

Le Févre, J. 1986. Aspects of the biology of frontal systems. Adv. Mar. Bio. 23: 163–299.

Le Roux, M. 2011. Chaetognatha Arrow Worms. Invertebrates of the Coral, University of Queensland, Australia.

Lee, R.F., W. Hagen and G. Kattner. 2006. Lipid storage in marine zooplankton. Mar. Ecol. Prog. Ser. 307: 273–306.

Legendre, L. and J. Le Fèvre. 1989. Hydrodynamic singularities as controls of recycled *vs.* export production in oceans. pp. 44–63. *In*: Berger, W.H., V.S. Smetacek and G. Wefer [eds.]. Productivity of the Ocean: Present and Past. John Wiley and Sons press, Chichester.

Lenz, J. 2000. Growth and metabolism. pp. 11–13. *In*: Harris, R.P., P.H. Wiebe, J. Lenz, H.R. Skojoldal and M. Huntley [eds.]. Zooplankton Methodology Manual. London: ICES Academic Press. London.

Lipej, L., P. Mozetič, V. Turk and A. Malej. 1997. The trophic role of the marine cladoceran *Penilia avirostris* in the Gulf of Trieste. Hydrobiologia 360: 197–203.

Litchman, E. and C.A. Klausmeier. 2008. Trait-based community ecology of phytoplankton. Annu. Rev. Ecol. Evol. Syst. 39: 615–39.

Litchman, E., M.D. Ohman and T. Kiørboe. 2013. Trait-based approaches to zooplankton communities. J. Plankton Res. 35: 473–484.

Lombard, F., E. Selander and T. Kiørboe. 2011. Active prey rejection in the filter-feeding appendicularian *Oikopleura dioica*. Limnol. Oceanogr. 56: 1504–1512.

Lopes, R.M., M. Katsuragawa, J.F. Dias, M.A. Montú, J.S. Muelbert, C. Gorri and F.P. Brandini. 2006. Zooplankton and ichthyoplankton distribution on the southern Brazilian shelf: an overview. Scient. Mar. 70(2): 189–202.

López Cazorla, A., R.E. Pettigrosso, L. Tejera and R. Camina. 2011. Diet and food selection by *Ramnogaster arcuata* (Osteichthyes, Clupeidae). J. Fish Biol. 78(7): 2052–2066.

López-Urrutia, A., J.L. Acuña, X. Irigoien and R. Harris. 2003. Food limitation and growth in temperate epipelagic appendicularians (Tunicata). Mar. Eco. Prog. Ser. 252: 143–157.

López-Urrutia, A. and X.A.G. Morán. 2007. Resource limitation of bacterial production distorts the temperature dependence of oceanic carbon cycling. Ecology 88: 817–822.

Mackas, D., P.T. Strub, A.C. Thomas and V. Montecino. 2006. Eastern ocean boundaries pan-regional view. pp. 21–60. *In*: Robinson, A.R. and K.H. Brink [eds.]. The Sea. Harvard Press Ltd., Boston.

Madin, L.P. and G.T. Harbison. 1977. The association of Amphipoda Hyperiidea with gelatinous zooplankton. Associations with Salpidae. Deep. Sea Res. 24: 449–464.

Mann, K.H. 2000. Ecology of Coastal Waters: With Implications for Management, 2nd edition. Wiley-Blackwell, United Kingdom.

Mann, K.H. and J.R.N. Lazier. 2006. Dynamics of Marine Ecosystems: Biological-Physical Interactions in the Oceans. Blackwell Publishing, USA.

Mariani, P., K.H. Andersen, A.W. Visser, A.D. Barton and T. Kiørboe. 2013. Control of plankton seasonal succession by adaptive grazing. Limnol. Oceanogr. 58: 173–184.

Martinez, M., N. Espinosa and D.L. Calliari. 2013. Incidence of dead copepods and factors associated with non-predatory mortality in the Río de la Plata estuary. J. Plankton Res. 36(1): 265–270.

Mauchline, J. and L.R. Fisher. 1969. The biology of mysids and euphausiids. Adv. Mar. Biol. 7: 1–454.

Mauchline, J. 1998. The Biology of Calanoid Copepods. Academic Press. The Netherlands.

McKinnon, A.D. and S. Duggan. 2001. Summer egg production rates of Paracalanidae copepods in subtropical waters adjacent to Australia's North West Cape. Hydrobiologia 453/454: 121–132.

McKinnon, A.D. and S. Duggan. 2003. Summer copepod production in subtropical waters adjacent to Australia's North West Cape. Mar. Bio. 143: 897–907.

McLachlan, A. and O. Defeo. 2018. The Ecology of Sandy Shores, 3rd edition. Academic Press.

Medellín-Mora, J., A. Atkinson and R. Escribano. 2019. Community structured production of zooplankton in the eastern boundary upwelling system off central/southern Chile (2003–2012). J. Mar. Sci. doi:10.1093/icesjms/fsz193.

Menéndez, M.C., M.C. Piccolo, M.S. Hoffmeyer and M. Sassi. 2011. Estuarine mesozooplankton dynamics on a short-term time scale: Role of semidiurnal tidal cycle. Braz. J. Oceanogr. 59(3): 281–286.

Menéndez, M.C., M.C. Piccolo and M.S. Hoffmeyer. 2012. Short-term variability on mesozooplankton community in a shallow mixed estuary (Bahía Blanca, Argentina): Influence of tidal cycles and local winds. Estuar. Coast. Shelf Sci. 112: 11–22.

Menéndez, M.C., C.A. Baleani, M. Amodeo, E.M. Acha and M.C. Piccolo. 2019. Assessment of surf zone zooplankton dynamics in a Southwestern Atlantic sandy beach: seasonal cycle and tidal height influence. Estuar. Coast. Shelf Sci. 227: 106–307.

Miller, C.B. and P.A. Wheeller. 2012. Biological Oceanography. Willey Blackweell, Oxford, USA.

Montú, M.A., I.M. Gloeden, A.K. Duarte and C. Resgalla Jr. 1997. Zooplankton. pp. 110–114. *In*: Seeliger, U., C. Odebrecht and J.P. Castello [eds.]. Subtropical Convergence Environments. The Coast and Sea in the Southwestern Atlantic. Springer, Berlin Heidelberg, New York.

Morgan, C.A., J.R. Cordell and C.A. Simenstad. 1997. Sink or swim? Copepod population maintenance in the Columbia River estuarine turbidity-maxima region. Mar. Biol. 129(2): 309–317.

Moriondo, P. 2002. Alimentación de larvas y juveniles de merluza (*Merluccius hubbsi*) en su área de cría nor-patagónica. Seminario de Licenciatura, Facultad de Ciencias Exactas y Naturales, Universidad Nacional de Mar del Plata.

Murano, M. 1999 Mysidacea. pp. 1098–1140. *In*: Boltovskoy, D. [ed.]. South Atlantic Zooplankton Vol.2. Backhuys Publishers, Leiden, Netherlands.

Muxagata, E., W.J.A. Amaral and C.N. Barbosa. 2012. *Acartia tonsa* production in the Patos Lagoon estuary, Brazil. ICES J. Mar. Syst. 69(3): 475–482.

Naylor, E. 2006. Orientation and navigation in coastal and estuarine zooplankton. Mar. Freshw. Behav. Physiol. 39(1): 13–24.

Nejstgaard, J.C., L.J. Naustvoll and A. Sazhin. 2001. Correcting for underestimation of microzooplankton grazing in bottle incubation experiments with mesozooplankton. Mar. Ecol. Prog. Ser. 221: 59–75.

Nemoto, T. and K.I. Yoo. 1970. The amphipod (*Parathemisto gaudichaudii*) as a food of the Antarctic Sei whale. Scient. Rep. Whales Res. Inst. Tokyo 22: 153–158.

Nival, S. and S. Ravera. 1979. Morphological study of the appendages of the marine cladoceran *Evadne spinifera* Muller by means of the scanning electron microscope. J. Plankton Res. 1: 207–213.

Noji, T.T., U.V. Bathmann, B. von Bodungen, M. Voss, A. Antia, M. Krumbholz, B. Klein, I. Peeken, C.I. Noji and F. Rey. 1997. Clearance of picoplankton-sized particles and formation of rapidly sinking aggregates by the pteropods *Limacina retroversa*. J. Plankton Res. 19: 863–875.

Ohman, M.D. and J.B. Romagnan. 2016. Nonlinear effects of body size and optical attenuation on diel vertical migration by zooplankton. Limnol. Oceanogr. 61: 765–770.

Oliveira-Santos, N.M., T. Martins-Garcia and M. Oliveira-Soares. 2016. Micro- and mesozooplankton communities in the surf zone of a tropical sandy beach (Equatorial Southwestern Atlantic). Latin American J. Aquat. Res. 44(2): 247–255.

Olson, D.B. and R.H. Backus. 1985. The concentrating of organisms at fronts: a cold-water fish and a warm-core Gulf Stream ring. J. Mar. Res. 43: 113–137.

Onbe, T. 1999. Ctenopoda and Onychopoda (Cladocera). pp. 797–813. *In*: Boltovskoy, D. [ed.]. South Atlantic Zooplankton. Backhuys Publishers, Leiden, Netherlands.

Open University. 2004. Ocean Circulation. Butterworth-Heinemann. 286 p.

Øresland, V. 1987. Feeding of the chaetognaths *Sagitta elegans* and *S. setosa* at different seasons in Gullmars fjorden, Sweden. Mar. Ecol. Prog. Ser. 39: 69–79.

Paffenhofer, G.A. and J.D. Orcutt. 1986. Feeding, growth and food conversion of the marine cladoceran *Penilia avirostris*. J. Plankton Res. 8: 741–754.

Paffenhöfer, G.A. and H. Jiang. 2016. On phytoplankton perception by calanoid copepods. Limnol. Oceanogr. 61: 1161–1168.

Parmar, T.K., D. Rawtani and Y.K. Agrawal. 2016. Bioindicators: the natural indicator of environmental pollution. Front. Life Sci. 9(2): 110–118.

Pavlova, E.V. 1959. *O pitanii* and *Penilia avirostris* Dana. On grazing by *Penilia avirostris* Dana. pp. 63–71. *In*: Izdatelstvo Akademii Nauk SSSR [ed.]. Trudy Sevastoporskoi Biologicheskoi Stantsii. Academy of Science USSR, Moscow.

Pearre, S. 2003. Eat and run? The hunger satiation hypothesis in vertical migration: history, evidence and consequences. Biol. Rev. Camb. Philos. Soc. 78: 1–79.

Peterson, W.T., P. Tiselius and T. Kiørboe. 1991. Copepod egg production, moulting and growth rates, and secondary production, in the Skagerrak in August 1988. J. Plankton Res. 13(1): 131–154.

Pingree, R.D., P.R. Pugh, P.M. Holligan and G.R. Forster. 1975. Summer phytoplankton blooms and red tides along tidal fronts in the approaches to the English Channel. Nature 258: 672–677.

Pinheiro, S.C.C., N.R. Leite, V.B. Costa, K.G.L. Costa, C.C. Pereira and R.M. Costa. 2011. Spatial-temporal influence of hydrological variables on the diversity and abundance of copepods on an equatorial macrotidal beach in the Brazilian Amazon region. J. Coast. Res. SI64: 425–429.

Pinheiro, S.C.C., A. Magalhaes, V.B. Costa, C.C. Pereira and R.M. Costa. 2013. Temporal variation of zooplankton on a tropical Amazonian beach. J. Coast. Res. SI65: 1838–1843.

Pohle, G., F.L.M. Mantelatto, M.L. Negreiros-Fransozo and A. Fransozo. 1999. Larva Decapoda (Brachyura). pp. 1281–1351. *In*: Boltovskoy, D. [ed.]. South Atlantic Zooplankton Vol.2. Backhuys Publishers, Leiden, Netherlands.

Postel, L., H. Fock and W. Hagen. 2000. Biomass and abundance. pp. 83–192. *In*: Harris, R.P., P.H. Wiebe, J. Lenz, H.R. Skjoldal and M. Huntley [eds.]. ICES Zooplankton Methodology Manual. Academic Press, San Diego, USA.

Prenski, L.B. and V.A. Angelescu. 1993. Ecología trófica de la merluza común del Mar Argentino (*Merluccius hubbsi*). Consumo anual de alimento a nivel poblacional y su relación con la explotación de las pesquerías multiespecífica. INIDEP Doc. Cient. 3: 1–118.

Price, H.J. and G.A. Paffenhofer. 1986. Effects of concentration on the feeding of a marine copepod in algal monocultures and mixtures. J. Plankton Res. 8: 119–128.

Ramírez, F. 2007. Distribución y alimentación del zooplancton. El mar argentino y sus recursos pesqueros. INIDEP 5: 45–69.

Ramírez, F. 2016. Eufáusidos. Los crustáceos de interés pesquero y otras especies relevantes en los ecosistemas marinos. pp. 15–28. *In*: Boschi, E.E. [ed.]. El Mar Argentino y sus Recursos Pesqueros. INIDEP Press, Mar del Plata, Argentina.

Ramírez, F.C. 1976. Contribución al conocimiento del espectro alimenticio de la sardina fueguina (Clupeidae, *Sprattus fuegensis Jenyns*). Neotrópica 22(68): 137–141.

Ramírez, F.C. and M. Pérez Seijas. 1981. Observaciones sobre desarrollo post naupliar, estructura poblacional y ciclo reproductivo del ostrácodo planctónica *Conchoecia serrulata*, Claus. Physis 40: 15–32.

Ramírez, F.C. and M. Pérez Seijas. 1985. New data on the ecological distribution of cladocerans and first local observations on reproduction of *Evadne nordmanni* and *Podon intermedius* (Crustacea, Cladocera) in Argentina seawaters. Physis 43: 131–143.

Rato Lénia, D., P. Maranhão Paulo, F.P. Avelelas, C. Velez, D. Marquesel, F. Queiroga and M. Sérgio. 2014. Zooplanktonic biomass and secondary production in Berlengas Biosphere Reserve (UNESCO)-Portugal. Front. Mar. Sci. Conference Abstract: IMMR | International Meeting on Marine Research 2014. doi10.3389/conf.fmars.2014.02.00104.
Remane, A. and C. Schlieper. 1971. Biology of Brackish Water. E. Schweiserbart'sche Verlagsbuchhandlung, Stuttgart.
Renz, J., D. Mengedoht and H. Hans-Jürgen. 2008. Reproduction, growth and secondary production of *Pseudocalanus elongatus* Boeck (Copepoda, Calanoida) in the southern North Sea. J. Plankton Res. 5: 511–528.
Reynolds, C.S. 1980. Phytoplankton associations and their periodicity in stratifying lake systems. Holartic Ecol. 3: 141–159.
Reynolds, C.S., V. Huszar, C. Cruk, L. Naselli-Flores and S. Melo. 2002. Towards a functional classification of the freshwater phytoplankton. J. Plankton Res. 24: 417–28.
Rigler, F.H. and J.A. Downing. 1984. The calculation of secondary productivity. pp. 19–58. *In*: Downing, J.A. and F.H. Rigler [eds.]. A Manual on Methods for the Assessment of Secondary Productivity in Fresh Waters. Blackwell, Oxford.
Robison, B.H., K.R. Reisenbichler and R.E. Sherlock. 2005. Giant larvacean houses: rapid carbon transport to the deep seafloor. Science 308: 1609–11.
Rodríguez-Graña, L., M. Vera, G. Cervetto and D.L. Calliari. 2018. Trophic ecology of the white croaker (*Micropogonias furnieri* Desmarest, 1823) and Rough Scad (*Trachurus lathami* Nichols, 1920) Larvae in the Río de la Plata Estuary. pp. 349–371. *In*: Hoffmeyer, M., M. Sabatini, F.P. Brandini, D.L. Calliari and N.H. Santinelli [eds.]. Plankton Ecology of the Southwestern Atlantic from the Subtropical to the Subantarctic Realm. Springer, Switzerland.
Russell-Hunter, W.D. 1979. A Life of Invertebrates. Macmillan Publishing Co., New York.
Sabatini, M. and P. Martos. 2002. Mesozooplankton features in a frontal area off northern Patagonia (Argentina) during spring 1995 and 1998. Scien. Mar. 66(3): 215–232.
Saiz, E. and T. Kiørboe. 1995. Predatory and suspension feeding of the copepod *Acartia tonsa* in turbulent environments. Mar. Ecol. Prog. Ser. 122: 147–158.
Saiz, E., A. Calbet and E. Broglio. 2003. Effects of small-scale turbulence on copepods: The case of *Oithona davisae*. Limnol. Oceanogr. 48(3): 1304–1311.
San Martin, E., R.P. Harris and X. Irigoien. 2006. Latitudinal variation in plankton size spectra in the Atlantic Ocean. Deep Sea Res. II 53: 1560–1572.
Sánchez, F., N. Mari, C. Lasta and A. Giangiobbe. 1991. Alimentación de la corvina rubia (*Micropogonias furnieri*) en la Bahía Samborombón. Frente Marítimo 8: 43–50.
Santangelo, J.M., A.M. Rocha, R.L. Bozelli, L.S. Carneiro and F.A. Esteves. 2007. Zooplankton responses to sandbar opening in a tropical eutrophic coastal lagoon. Est. Coast. Shelf Sci. 71: 657–668.
Sastri, A.R. and J.F. Dower. 2009. Interannual variability in chitobiase based production rates of the crustacean zooplankton community in the Strait of Georgia. Mar. Ecol. Prog. Ser. 388: 147–157.
Sato, R., Y. Tanaka and T. Ishimaru. 2001. House production by *Oikopleura dioica* (Tunicata, Appendicularia) under laboratory conditions. J. Plankton Res. 21: 415–423.
Schiariti, A., H.W. Mianzan and F.C. Ramírez. 2004. New records of *Mysidopsis tortonesei* Băcescu, 1968 and *M. rionegrensis* Hoffmeyer, 1993 (Mysidacea) from the Río de la Plata estuary and Buenos Aires coastal waters. Crustaceana 77(7): 887–893.
Segers, F.H.I.D. and B. Taborsky. 2011. Egg size and food abundance interactively affect juvenile growth and behaviour. Funct. Ecol. 25: 166–176.
Segura, A.M., V. Franco, P. Franco and M. Arim. 2015. Gape and energy limitation determine a humped relationship between trophic position and body size. Can. J. Fish. Aquat. Sci. 72: 198–205.
Selander, E. and P. Tiselius. 2003. Effects of food concentration on the behaviour of *Oikopleura dioica*. Mar. Biol. 142: 263–270.
Setubal, R.B., R. Araújo do Nascimento and R.L. Bozelli. 2020. Zooplankton secondary production: main methods, overview and perspectives from Brazilian studies. Int. Aquat. Res. 12: 85–99.
Shadrin, N. and L. Litvinchuk. 2005. Impact of increased mineral particle concentration on the behavior, suspension-feeding and reproduction of *Acartia clausi* (Copepoda). *In*: Dame, R.F. and S. Olenin [eds.]. The Comparative Roles of Suspension-Feeders in Ecosystems. NATO Science Series IV: Earth and Environmental Series, vol. 47. Springer, Dordrecht.
Sieburth, J.M.C.N., V. Smetacek and J. Lenz. 1978. Pelagic ecosystem structure: Heterotrophic compartments of the plankton and their relationship to plankton size fractions. Limnol. Oceanogr. 23: 1256–1263.
Simpson, J.H. and J.R. Hunter. 1974. Fronts in the Irish Sea. Nature 250: 404–406.
Sommer, U., R. Adrian, L.D. Domis, J.J. Elser, U. Gaedke, B. Ibelings, E. Jeppesen, M. Lurling, J.C. Molinero, W.M. Mooij, E. Van Donk and M. Winder. 2012. Beyond the plankton ecology group (PEG) model: mechanisms driving plankton succession. Annu. Rev. Ecol. Evol. Syst. 43: 429–448.
Stamieszkin, K., A.J. Pershing, N.R. Record, C.H. Pilskaln, H.G. Dam and L.R. Feinberg. 2015. Size as the master trait in modeled copepod fecal pellet carbon flux. Limnol. Oceanogr. 60: 2090–2107.
Steinberg, D.K., S.A. Goldthwait and D.A. Hansell. 2002. Zooplankton vertical migration and the active transport of dissolved organic and inorganic nitrogen in the Sargasso Sea. Deep-Sea Res. I 49: 1445–1461.
Steinberg, D.K. and M.R. Landry. 2017. Zooplankton and the ocean carbon cycle. Annu. Rev. Mar. Sci. 9: 413–44.

Strickler, J.R. 1982. Calanoid copepods, feeding currents, and the role of gravity. Science 218: 158–160.
Stuart, V. and J.A. Huggett. 1992. Prey selection by *Euphausia lucens* (Hansen) and feeding behaviour in response to a mixed algal and animal diet. J. Exper. Mar. Biol. Ecol. 164: 117–133.
Stull, K.J., L.B. Cahoon and T.I. Lankford. 2016. Zooplankton abundance in the surf zones of nourished and unnourished beaches in Southeastern North Carolina, U.S.A. J. Coast. Res. 32(1): 70–77.
Sverdrup, H.U. 1953. On conditions for the vernal blooming of phytoplankton. J. Cons. Perm. Int. l'Explor. Mer. 18: 287–295.
Takahashi, K., A. Kuwata, H. Sugisaki, K. Uchikawa and H. Saito. 2009. Downward carbon transport by diel vertical migration of the copepods *Metridia pacifica* and *Metridia okhotensis* in the Oyashio region of the western subarctic Pacific Ocean. Deep-Sea Res. 56(10): 1777–1791.
Talbot, M.M.B. and G.C. Bate. 1989. Beach morphodynamics and surf-zone diatom populations. J. Exper. Mar. Biol. Ecol. 129: 231–241.
Tang, K.W., K.M.L. Hutalle and H.P. Grossart. 2006. Microbial abundance, composition and enzymatic activity during decomposition of copepod carcasses. Aquat. Microb. Ecol. 45: 219–27.
Tang, K.W., M.I. Gladyshev, O. Dubovskaya, G. Kirillin and H.P. Grossart. 2014. Zooplankton carcasses and non-predatory mortality in freshwater and inland sea environments. J. Plankton Res. 36: 597–612.
Tang, K.W., J.A. Ivory, S. Shimode, Y. Nishibe and K. Takahashi. 2019. Dead heat: copepod carcass occurrence along the Japanese coasts and implications for a warming ocean. ICES J. of Mar. Sci. doi:10.1093/icesjms/fsz017.
Teixeira-Amaral, P., W.J. Appolinário Amaral, D. Ortiz, V.O. Agostini and E. Muxagata. 2017. The mesozooplankton of the Patos Lagoon Estuary, Brazil: trends in community structure and secondary production. Mar. Biol. Res. 13(1): 48–61.
Temperoni, B., M.D. Viñas, N. Diovisalvi and R. Negri. 2010. Seasonal production of *Oithona nana* Giesbrecht, 1893 (Copepoda: Cyclopoida) in temperate coastal waters off Argentina. J. Plankton Res. 33(5): 729–740.
Temperoni, B., M.D. Viñas, P. Martos and M. Marrari. 2014. Spatial patterns of copepod biodiversity in relation to a tidal front system in the main spawning and nursery area of the Argentine hake *Merluccius hubbsi*. J. Mar Syst. 139: 443–445.
Timofeev, S.F. 2001. Bergmann's principle and deep-water gigantism in marine crustaceans. Biol. Bull. 28: 646–650.
Tiselius, P. 1992. Behavior of *Acartia tonsa* in patchy food environments. Limnol. Oceanogr. 37: 1640–1651.
Tiselius, P., P.R. Jonsson, S. Kaartvedt, M.E. Olsen and T. Jarstad. 1997. Effects of copepod foraging behavior on predation risk: An experimental study of the predatory copepod *Pareuchaeta norvegica* feeding on *Acartia clausi* and *A. tonsa* (Copepoda). Limnol. Oceanogr. 42: 164–170.
Tiselius, P., J.K. Petersen, T.G. Nielsen, M. Maar, E.F. Møller, S. Satapoomin, K. Tönnesson, T. Zervoudaki, E. Christou, A. Giannakourou, A. Sell and C. Vargas. 2003. Functional response of Oikopleura dioica to house clogging due to exposure to algae of different sizes. Mar. Biol. 142: 253–261.
Tiselius P., B.W. Hansen and D. Calliari. 2012. Fatty acid transformation in zooplankton: from seston to benthos. Mar. Ecol. Prog. Ser. 446: 131–144.
Tiselius, P., E. Saiz and T. Kiørboe. 2013. Sensory capabilities and food capture of two small copepods, *Paracalanus parvus* and *Pseudocalanus* sp. Limnol. Oceanogr. 58: 1657–1666.
Titelman, J. 2001. Swimming and escape behavior of copepod nauplii: implications for predator-prey interactions among copepods. Mar. Ecol. Prog. Ser. 213: 203–213.
Tönnesson, K. and P. Tiselius. 2005. Diet of the chaetognaths *Sagitta setosa* and *S. elegans* in relation to prey abundance and vertical distribution. Mar. Ecol. Prog. Ser. 289: 177–190.
Tranter, D.J. 1976. Herbivore production. pp. 186–224. *In*: Cushing, D.H. and J.J. Walsh [eds.]. Ecology of the Seas. W.B. Saunders, Philadelphia.
Turner, J.T. 1984. The feeding ecology of some zooplankters that are important prey items of larval fish. NOAA Technical Report NMFS 7.
Turner, J.T. P.A. Tester and R.L. Ferguson. 1988. The marine cladocerans *Penillia avirostris* and the microbial loop of the pelagic food webs. Limnol. Oceanogr. 33: 245–255.
Turner, J.T. 2004. The importance of small planktonic copepods and their roles in pelagic marine food webs. Zool. Stud. 43(2): 255–66.
Turner, J.T. 2015. Zooplankton fecal pellets, marine snow, phyto-detritus and the ocean's biological pump. Progr. Oceanogr. 130: 205–48.
Uttieri, M., G.A. Paffenhöfer and M.G. Mazzocchi. 2008. Prey capture in *Clausocalanus furcatus* (Copepoda: Calanoida). The role of swimming behaviour. Mar. Biol. 153: 925–935.
van Someren Gréve, H., T. Kiørboe and R. Almeda. 2019. Bottom-up behaviourally mediated trophic cascades in plankton food webs. Proc. R. Soc. B 286: 20181664. doi: 10.1098/rspb.2018.1664.
Vergara, O., D. Calliari, P. Tiselius, R. Escribano, M.L. González and S. Soto. 2010. Functional response of *Sagitta setosa* (Chaetognatha) and *Mnemiopsis leidyi* (Ctenophora) under variable food concentration in the Gullmar fjord, Sweden. Rev. Biol. Mar. Oceanogr. 45: 35–42.
Viñas, M.D. and F.C. Ramírez. 1996. Gut analysis of first-feeding anchovy larvae from Patagonian spawning area in relation to food availability. Arch. Fish. Mar. Res. 43: 231–256.
Vinogradov, G. 1999. Amphipoda. pp. 1141–1240. *In*: Boltovskoy, D. [ed.]. South Atlantic Zooplankton Vol.2. Backhuys Publishers, Leiden, Netherlands.

Violle, C., M.L. Navas, D. Vile, E. Kazakou, C. Fortunel, I. Hummel and E. Garnier. 2007. Let the concept of trait be functional. Oikos 116: 882–892.
Vogel, S. 2004. Life in Moving Fluids. Second edition Princeton University Press.
Werner, E.E. and S.D. Peacor. 2003 A review of trait mediated indirect interactions in ecological communities. Ecology 84: 1083–1100.
Westoby, M. and I.J. Wright. 2006. Land-plant ecology on the basis of functional traits. Trends in Ecology and Evolution 21: 261–268.
White, A.W. 1980. Recurrence of kills of Atlantic herring (*Clupea harengus harengus*) caused by dinoflagellate toxins transferred through herbivorous zooplankton. Can. J. Fish. Aquat. Sci. 37: 2262–2265.
Wiebe, P.H. and M.C. Benfield. 2003. From the Hensen net toward four-dimensional biological oceanography. Progr. Oceanogr. 56: 127–136.
Wong, C.K., A.L.C. Chan and K.W. Tang. 1992. Natural ingestion rates and impact of the marine cladoceran *Penilia avirrostris* Dana in Tolo Harbour, Hong Kong. J. Plankton Res. 14: 1757–1765.
Woodward, G., B. Ebenman, M. Emmerson, J.M. Montoya, J.M. Olesen, A. Valido and P.H. Warren. 2005. Body size in ecological networks. Tren. Ecol. Evol. 20: 402–409.
Wooldridge, T. and T. Erasmus. 1980. Utilization of tidal currents by estuarine zooplankton. Est. Coast. Shelf Sci. 12: 107–114.
World Register of Marine Organisms. 2020. http://www.marinespecies.org. Last accessed 11th November 2020.
Yebra, L., T. Kobari, A.R. Sastri, F. Gusmão and S. Hernández-León. 2016. Advances in biochemical indices of zooplankton production. Adv. Mar. Biol. 76: 157–240.
Yen, J. 2000. Life in transition: balancing inertial and viscous forces by planktonic copepods. Biol. Bull. 198: 213–224.
Zhou, M., F. Carlotti and Y. Zhu. 2010. A size-spectrum zooplankton closure model for ecosystem modelling. J. Plankton Res. 32: 1147–1165.

CHAPTER 7
Gelatinous Zooplankton

Miodeli Nogueira Júnior,[1,*] *Everton Giachini Tosetto,*[2] *Letícia Cazarin Baldoni,*[3] *Sofia Dutto,*[4] *Mitsuko Hidaka,*[5] *Dhugal J. Lindsay*[5] and *Renato Mitsuo Nagata*[3]

7.1 Introduction

The diversity and ecological importance of zooplankton assemblages, as well as their main functional traits have been previously introduced in this book. Along with the great taxonomic diversity of zooplankton, there are many functional classifications (e.g., according to size, life history, feeding, habitat, vertical distribution, etc.; this volume, Chapter 6). One such functional classification considers the amount of water in the tissues, classifying them as gelatinous, semi-gelatinous and non-gelatinous zooplankton (Larson 1986, Haddock 2004). Classically, in gelatinous taxa the dry weight represents < 5% of the wet weight and the carbon and nitrogen content is usually < 15% and < 5% of the dry weight, respectively; in semi-gelatinous taxa, dry weight represents between 3.5–30% of the wet weight and carbon nitrogen weight represents 15–40% and 5–10% of the dry weight, respectively, while in the non-gelatinous taxa the dry weight represents between 10–30% of the wet weight and carbon and nitrogen weight represents 30–60% and 5–10% of the dry weight, respectively (Larson 1986). As expected, different species of the same higher taxon may have different amounts of water, but in general, cnidarians, ctenophores, tunicates, and some mollusks are the typical gelatinous zooplankton (GZ; Fig. 7.1), apart from some protists, such as radiolarians, the poorly known nemerteans and larvae of some groups, such as echinoderms, sipunculids, and hemichordates. Semi-gelatinous zooplankton includes mainly chaetognaths, polychaetes, and some mollusks as well, while the many crustacean groups and fishes are the main non-gelatinous taxa (Hamner et al. 1975, Larson 1986, Haddock 2004).

The GZ taxa do not share a direct phylogenetic relationship and have independently developed aqueous tissues (Haddock 2004). Such a large number of higher groups with convergent characteristics suggests evolutionary advantages, including: (i) better buoyancy due to the large amount of water, resulting in less energy expenditure to maintain their position in the water column (Bidigare and Biggs 1980, Tsukamoto et al. 2009); (ii) transparency to make them less visible, in order to camouflage them from visual predators in the relatively homogeneous pelagic habitat, with no hiding places (Johnsen 2001); (iii) low carbon investment, thus only requiring a relatively low metabolic cost, allowing them to achieve larger sizes (Larson 1986, Acuña et al. 2011); indeed, some GZ species often are amongst the largest planktonic organisms (e.g., Esnal 1999a, Mapstone 2014, 2015); (iv) larger sizes increase

[1] Departamento de Sistemática e Ecologia, Universidade Federal da Paraíba, João Pessoa, PB, Brasil.
[2] Universidade Federal de Pernambuco, Departamento de Oceanografia, Recife, PE, Brasil.
[3] Instituto de Oceanografia, Universidade Federal do Rio Grande, RS, Brasil.
[4] Instituto Argentino de Oceanografía, CONICET, Bahía Blanca – Argentina.
[5] Japan Agency for Marine-Earth Science and Technology (JAMSTEC), Yokosuka, Kanagawa, Japan.
* Corresponding author: miodeli@gmail.com

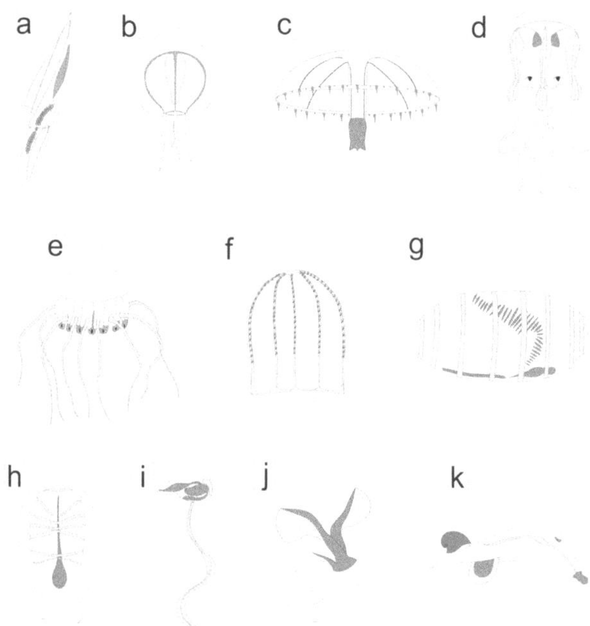

Figure 7.1. General sketch of main gelatinous zooplankton higher taxa: siphonophora (a), hydromedusae (b, c), cubomedusae (d), scyphomedusae (e), ctenophore (f), doliolid (g), salp (h), appendicularia (i), pteropod (j) heteropod and (k) mollusks. Drawings by E.G. Tosetto after several sources.

encounter chances as well as the size range of possible prey and, in addition, reduce the chances of being predated on by smaller organisms (Larson 1986, Arai 1997, Kremer 2002); (v) rapid individual growth and production, which is associated with high reproductive output (asexual and/or sexual) resulting in high population growth sustaining large blooms (e.g., Paffenhöfer and Lee 1987, Acuña et al. 2011); (vi) great body plasticity, allowing them to grow, shrink and grow again according to environmental conditions and food availability (Arai 1997, Miyake et al. 1997, Lilley et al. 2014).

Gelatinous organisms are mostly solitary, but colonial forms are also common, such as the siphonophores and pyrosomes. Their sizes are highly variable with most species measuring a few millimeters. Yet, frequently the GZ are amongst the largest zooplanktonic organisms. Many species reach several centimeters, with some cnidarians, ctenophores, and pyrosomes reaching a few meters and some siphonophores and the pyrosome *Pyrostremma spinosum* having giant colonies more than 20 m in length (Esnal 1999a, Mapstone 2014, 2015, Purushothaman et al. 2018).

Currently, there are around 2,500 gelatinous and semi-gelatinous valid metazoan species, distributed over more than 640 genera, ~ 200 families, 32 orders, and 11 classes (Fig. 7.2). The most diverse gelatinous taxa across all taxonomic levels are the cnidarians, particularly the hydromedusae of the orders Anthoathecata and Leptothecata, which are mainly meroplanktonic with a benthic polypoid and a pelagic medusoid life history stage (Bouillon et al. 2006). Most of the other phyla have between 100 and 200 gelatinous/semi-gelatinous species currently described, in less than 50 genera and 24 families (Fig. 7.2). These numbers certainly are underestimated since most of these taxa are poorly known in virtually all oceanic and deep-sea areas and many coastal areas as well, as is evident by the constant finding of new species, genera and families (Gershwin et al. 2010, Lindsay et al. 2017, Pugh et al. 2018, Tosetto et al. 2020). Moreover, these numbers disregard benthic groups, which are highly diversified, and where the larvae may be gelatinous and/or semi-gelatinous, such as echinoderms, polychaetes, mollusks, hemichordates, and others. Information on the life history, and thus the presence or suppression of a planktonic larval stage, in particular taxa within these groups, is largely unknown, hindering accurate estimates.

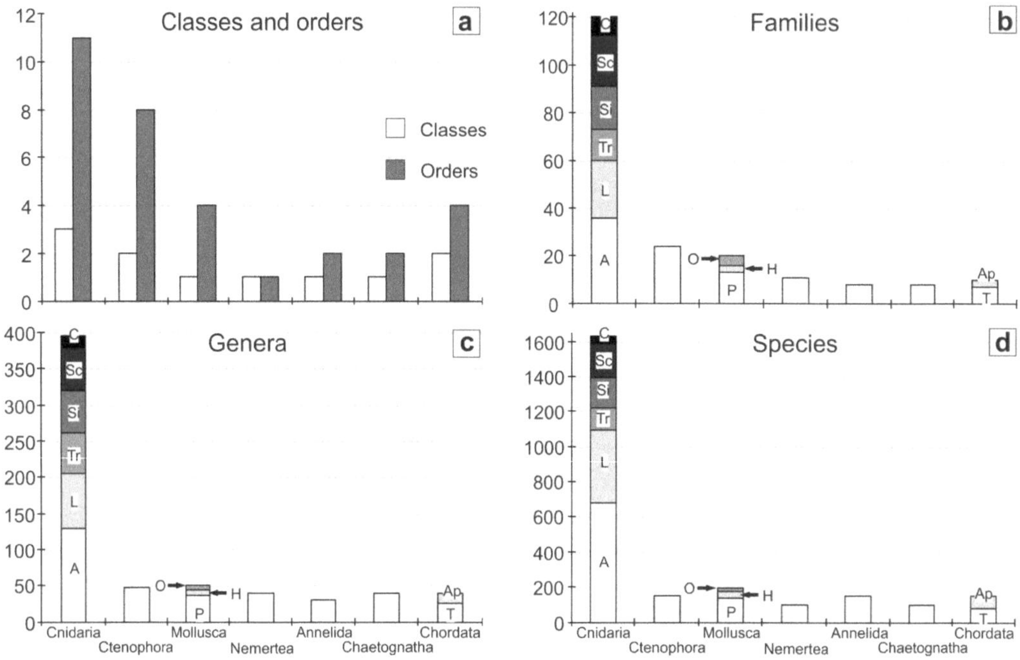

Figure 7.2. Number of current valid (a) classes and orders, (b) families, (c) genera and (d) species of main metazoan gelatinous and semi-gelatinous zooplankton phyla. Taxa with gelatinous larvae and benthic adults (e.g., echinoderms, polychaete larvae) have not been included since these stages are largely unknown. Data compiled from World Register of Marine Species website (http://www.marinespecies.org/; consulted November 2020). For hydrozoans, we used Bouillon et al. (2006) to check whether each taxon includes a free-living planktonic medusa in the life history. A = anthomedusae; L = leptomedusae; Tr = trachylina; Si = siphonophora; Sc = scyphomedusae; C = cubomedusae; P = pteropods; H = heteropods; O = other (adult) planktonic mollusks; T = thaliaceans; Ap = appendicularians.

In spite of their ecological, economic and societal importance, GZ are quite understudied compared to other pelagic groups, such as crustaceans and fishes. This is largely due to their fragility, causing difficulties with their sampling and preservation in good condition, a prerequisite for accurate taxonomic classification and sometimes even just to count them. For instance, ctenophores have often been considered amongst the most difficult marine animals to study, representing a great challenge to conventional quantitative oceanographic sampling due to their fragility (Harbison et al. 1978, Mianzan 1999), and the colonial siphonophores are often found broken into many pieces in plankton net samples, discouraging their study by those who do not like "jigsaw puzzles" for organisms. Moreover, the large size of many gelatinous taxa (e.g., pyrosomes, salps, scyphomedusae, some ctenophores, siphonophores), along with the often-patchy nature of their distribution, make their quantitative sampling with standard plankton nets, such as the widely used WP2 and/or Bongo nets, difficult (e.g., UNESCO 1968, Harris et al. 2001). These difficulties in sampling most gelatinous taxa are reflected in the lack of studies for many of these groups, particularly in the 20th century, when a quantitative view dominated oceanographic studies and thus a "crustaceo-centric" view predominated (Haddock 2004). In the last few decades, the perception of GZ blooms in many regions around the world and their associated impact to ecosystems, along with the general view of gelatinous animals as plagues for fisheries and other human activities at sea (e.g., Condon et al. 2012, 2013, Opadal et al. 2019), have stimulated increased research and the development of many sampling and study techniques for these delicate animals (Lindsay et al. 2012, Kelley et al. 2016, Hosia et al. 2017). In this chapter, we will summarize some of the most important functional aspects of these strange and beautiful organisms.

7.2 Population dynamics, reproduction, growth and blooms

One of the most remarkable features of GZ is the ability of populations to suddenly appear in large and, sometimes, conspicuous aggregates, that rapidly decline and disappear after a few weeks (Boero et al. 2008). Although plankton populations are known to have a fluctuating nature, this feature is even more pronounced among GZ. This is explained because many tunicates, ctenophores, and planktonic cnidarians have adaptive characteristics of opportunistic-colonizing species, rapidly increasing their populations in response to favorable conditions. Such features include: (i) faster individual growth rates (Hirst et al. 2003, Pitt et al. 2013); (ii) short generation times (Heron 1972, Fenaux 1976, Costello et al. 2012); and (iii) high fecundity rates by means of a set of reproduction pathways, such as alternation of generations and multiple types of modular reproduction (Mackie et al. 1987, Esnal and Daponte 1999a, b, Boero et al. 2008, Deibel and Lowen 2012). In addition, some species have a broad physiological tolerance to variations in their physical environment (e.g., salinity, temperature, dissolved oxygen; Purcell 2012) and versatile dietary niches that allow consumption of a variety of nutritional sources (Mills 1995, Arai 2005, Southerland et al. 2010). These features may have contributed to the wide geographical range of many species (Mianzan 1999, Oliveira et al. 2016), with numerous examples of invasive success (Lotan et al. 1992, Mills and Rees 2000, Costello et al. 2012) and their propagation under conditions that adversely affect other pelagic taxa (Purcell 2012).

7.3 General life history of the main groups

Planktonic cnidarians are the GZ taxa with the highest diversity (Fig. 7.2) and greatest variety in life history patterns (Boero et al. 1997). Although holoplanktonic and meroplanktonic cnidarian life forms exist, the basal medusozoan life history involves alternation between a benthic asexual reproducing stage (the polyp) and the planktonic sexual reproducing stage (the medusa; e.g., Collins 2002). Different lineages, however, evolved numerous life history variations including: (i) reductions and suppression of the planktonic stage in some Anthoathecata and Leptothecata; (ii) suppression of the benthic stage in Siphonophorae, Narcomedusae, Trachymedusae and in some Anthoathecata and Scyphomedusae; (iii) multiple asexual reproduction through fissions or buddings by both polyps and medusae; (iv) resting stages to survive (i.e., podocysts in Scyphozoa and stolons in Hydrozoa); and (v) reverse development in some hydrozoans and scyphozoans (Boero et al. 1997, Mianzan and Cornelius 1999, Collins 2002, de Vito et al. 2006, Kubota 2011, He et al. 2015).

Siphonophores are complex and highly polymorphic cnidarian colonies, composed of many polypoid and medusoid individuals (Mackie et al. 1987). Most siphonophores for which development has been studied produce their gonads on reproductive zooids known as gonophores, which are budded from the colony stem. These gonophores release gametes in the water column and can remain attached to the colony (in Physonectae and Cystonectae), or can be released from the colony as free-swimming medusoids, as gonophores of *Nanomia bijuga* reared in laboratory (Carré 1969), or as independent colonies known as eudoxids in most Calycophorae (Mackie et al. 1987, Mapstone 2009, 2015).

Ctenophores hatch the eggs as a cydippid larva (except the beroids; Mianzan 1999). Most ctenophore species are simultaneous hermaphrodites able to self-fertilize (Reeve and Walter 1978, Sasson and Ryan 2016), and some species may reproduce sexually at early larval stages (Martindale 1987, Jaspers et al. 2012, Martindale and Henry 2015). Many ctenophore species have the remarkable ability to regenerate and reconstitute a viable animal when cut in half or one-fourth (Martindale and Henry 2015). Chaetognaths are in general protandrous hermaphrodites, i.e., the testes develop earlier than the ovaries. Most species have cross-fertilization, but self-fertilization may occur (Reeve and Walter 1972, Casanova 1999). Species of *Eukrohnia* exhibit parental care, with eggs and newly hatched individuals kept inside two marsupia, protected by part of the lateral fins (Terazaki and Miller 1982).

Holopelagic mollusks typically have internal fertilization with copulation, being either dioecious with sexual dimorphism such as the heteropods, or hermaphrodites such as the pteropods, nudibranchs and epitoniids. In the hermaphrodite, taxa self-fertilization may occur, although pteropods are mostly protandric (Lalli and Gilmer 1989, Richter and Seapy 1999, Beu 2017, Calabrò et al. 2019). In many cases, it is not clear whether sex change occurs one or more times during the life history, or if the animals may act as males and females simultaneously (Lalli and Gilmer 1989, Calabrò et al. 2019). In some *Janthina* populations, males are smaller than females, suggesting a protandric hermaphroditism (Ankel 1930). However, different patterns may occur, such as similar-sized females, males and sterile individuals, and very large males, suggesting complex systems where the sex change may occur more than once that may be associated with different environmental conditions (Laursen 1953). Fertilized eggs are usually released in the water column, and in pteropods and heteropods eggs typically form floating strings ranging from 600 to 6,000 eggs that may remain attached to the female body in heteropods. They typically go through a planktonic veliger stage, but there is very limited information on the development and metamorphosis of these mollusks (Thiriot-Quiévreux 1973, 1975, Lalli and Gilmer 1989, Richter and Seapy 1999, Calabrò et al. 2019).

Among planktonic tunicates, the appendicularian life history is comparatively simple, with direct development by solitary protandric hermaphrodites (Deibel and Lowen 2012). By contrast, the thaliacean life history strategies include alternation of sexual (aggregate or blastozooid) and asexual (solitary or oozoid) generations in salps (Alldredge and Madin 1982, Esnal and Daponte 1999b). Doliolids have a sexual blastozooid and a complex polymorphism in budding production by the asexual generation (i.e., phorozooids and trophozooids; Esnal and Daponte 1999a, Deibel and Lowen 2012). In thaliaceans, the oozoid stage is the main trigger of blooms through the production of chains with numerous blastozooid individuals in salps or by the production of phorozooids in doliolids (Deibel 1985, Esnal and Daponte 1999a, b, Paffenhöfer and Gibson 1999, Paffenhöfer and Köster 2011).

7.4 Patterns of population dynamics and factors affecting population growth

The cyclical occurrence of favorable conditions is the trigger for rapid increases in GZ populations. Holoplanktonic GZ, such as ctenophores, tunicates, chaetognaths, siphonophores, and some medusae reproduce continuously, alternating periods of rapid increases under favorable conditions, followed by sharp declines in their populations. Somewhat similarly, the benthic stages of meroplanktonic GZ, represented by hydro, scypho, and cubomedusae, release their medusae in response to external stimuli, commonly seasonally changing factors in temperate and higher latitude environments (Boero et al. 2008). Our capacity to predict such population pulses in relation to the temporal environmental variability is limited, in part due to some irregular occurrences, but also because of the lack of knowledge on the life history of most species.

In mid and high-latitude ecosystems, seasonal changes in abiotic (e.g., temperature, stratification, solar radiation, photoperiod, and ice cover) and biotic (e.g., food availability, predation pressure, parasitism) components of the environment are pronounced and may drive the population dynamics of all plankters, including the GZ (Boero et al. 2008). Although inter-annual variability exists (e.g., Condon et al. 2013), GZ populations usually follow the seasonal productivity cycles. Their occurrences and succession are relatively more predictable and mostly constrained to the period from early spring to autumn depending on the species, latitude, and environmental conditions (Buecher et al. 1997, Purcell and Decker 2005, Hosia and Båmstedt 2007, Licandro et al. 2012, Lucas et al. 2012). Populations of carnivorous GZ, such as cnidarians, chaetognaths, and ctenophores, increase following a rise in temperature and availability of crustacean zooplankton (Purcell and Decker 2005, Batistic et al. 2007, Condon and Steinberg 2008, Nogueira Júnior et al. 2019). In some cases, these GZ may control crustacean populations, whereas larger tertiary GZ consumers, such as ctenophores (*Beroe* spp.) and medusae (*Chrysaora* spp. and *Cyanea* spp.) may control populations of secondary GZ consumers, such as scypho- and hydromedusae, and ctenophores (Condon and Steinberg 2008, Bandara et al. 2016).

Some holopelagic high latitude GZ are not limited to the productive season and may occur during winter, such as ctenophores (e.g., *Mertensia ovum* and *Beroe cucumis*), siphonophores (e.g., *Dimophyes arctica* and *Lensia conoidea*), and medusae (e.g., *Periphylla periphylla*; Hosia et al. 2007, 2008, Bandara et al. 2016). Although most GZ taxa have life spans ranging from weeks to several months and are restricted to one season in the year, some larger medusae (e.g., *Aurelia*, *Chrysaora* spp. and *P. periphylla*) may live more than one year, based on cohort analysis or on the developmental time of early medusae stages (Miyake et al. 1997, Jarms et al. 1999, Brodeur et al. 2008). Under this view, it has been speculated that mature medusae could overwinter close to the bottom and resume their development at the onset of the next productive season, joining new cohorts (Ceh et al. 2015, Purcell et al. 2018). This unorthodox view of life history strategies has been criticized (Morandini et al. 2016) and no conclusive evidence (e.g., true jellyfish populations in diapause) has been shown in favor of the "overwintering idea" so far.

In warmer environments, seasonal changes in light and temperature are less pronounced and event-scales (i.e., days to weeks) may play a major role driving GZ pulses. Complex interplay between local hydrographic processes and fast responses by some species with shorter lifespans (< week) become an important feature. In tropical regions, growth rates tend to be at their temperature-dependency maxima (Båmstedt et al. 1999, Hopcroft and Roff 1999) and generation times tend to be shorter (Deibel and Lowen 2012). This allows rapid responses of GZ populations to mesoscale processes (e.g., intrusions of cold nutrient-rich waters, coastal upwelling, continental runoff) that fuel food webs. Filter-feeding planktonic tunicates with their large food-size spectrum and faster growth rates are commonly the first metazoans to take advantage of these processes (Deibel 1985, Paffenhöffer et al. 1987, 1995, Nakamura et al. 1997, Esnal 1999a, b, Esnal and Daponte 1999a, b). Blooms of planktonic tunicates over hundreds of kilometers in coastal areas may form within a few days, triggered by events such as increasing concentrations of detritus, heterotrophic bacteria, and phytoplankton (Deibel 1985, Paffenhöffer and Lee 1987, Valentin et al. 1987, Deibel and Paffenhofer 2009). Although thaliacean blooms are ephemeral, usually lasting for less than one month (Deibel and Paffenhofer 2009, Henschke et al. 2014), copepod recruitment and zooplankton concentrations can be significantly decreased because of the extremely high feeding rates by these gelatinous competitors (Deibel 1985). Larger GZ species from warm waters are much less studied with regard to their population dynamics (Condon et al. 2013). Isolated studies demonstrate extended periods of recruitment, multiple size classes and population peaks throughout the year for some scyphomedusae (e.g., *Aurelia* spp., *Lychnorhiza lucerna, Mastigias* sp.) and hydromedusae (e.g., *Olindias sambaquiensis*, *Rhacostoma atlanticum*, and *Liriope tetraphylla*; Dawson and Martin 2001, Nogueira Júnior et al. 2010, Lucas et al. 2012, Nagata et al. 2014, Nogueira Júnior and Haddad 2017).

7.5 Possible biological and physical drivers of blooms

As previously stated, environmental variability drives population dynamics of GZ by affecting life history traits (i.e., growth, reproduction, settlement, and survivorship). However, GZ blooms cannot be explained solely by the intrinsic high reproductive capacities and short life spans, but by a combination between multiple biological and physical forcing, that promote the concentration or translocation of individuals (Graham et al. 2001, Costello et al. 2006, Goldstein and Riisgård 2016). Thus, linking life history and environmental forcing is a central issue for developing a predictive framework (Verity and Smetacek 1996, Schnedler-Meyer et al. 2018).

Some GZ, such as planktonic tunicates, have the highest growth rates of all metazoans, reaching growth rates of up to 2.1 day^{-1} (Heron 1972, Hirst et al. 2003), and generation times usually shorter than a month (Paffenhöfer and Gibson 1999), but also shorter than one week, and as short as two days in the salp *Thalia democratica* (Deibel and Lowen 2012). Some ctenophores, such as *Mertensia ovum*, may undergo precocious reproduction and develop functional gametes within days after development is completed (Jaspers et al. 2012). The cydippid *Pleurobrachia bachei* may start sexual reproduction as early as ca. 25 days (Hirota 1972), while the lobate *Mnemiopsis leidyi* within 13 days after hatching (Martindale 1987, Costello et al. 2012). They commonly have extremely high fecundity, with *M. leidyi* producing between 2,000 (Costello et al. 2006) and 12,000 eggs day^{-1} (Kremer 1976). The species continues to release viable eggs under low food concentrations and under starvation for up to 12 days (Jaspers et al. 2015). Because of these features, *M. leidyi* is considered a colonizing and very successful invader that has expanded its distribution range from the Atlantic coast of South and North America, to the Black, North, Mediterranean, and Baltic seas (Costello et al. 2012).

For meroplanktonic cnidarians, the production of medusae by benthic polyps is a keystone process often invoked to explain population fluctuations and bloom dynamics (Lucas 2001). This process may be triggered by increased or specific temperatures, increased food availability, changes in photoperiod, salinity, and possible combinations of these factors (Werner 1963, Carré and Carré 1990, Arai 1992, Lucas et al. 2012, Purcell et al. 2012). Processes related to polyp-to-polyp asexual reproduction, such as the production of motile or non-motile buds/planuloids and polyp fragmentation, must also be taken into account (Lucas et al. 2012, Schiariti et al. 2014). Many of these processes are induced by variations in temperature and food availability and may affect fecundity, medusae growth, morphology and investment *per* offspring (Schiariti et al. 2014, Chiaverano and Graham 2018). Benthic polyps can also produce resting stages to survive unfavorable conditions that may trigger new blooms during the next favorable season (Boero et al. 1997, Arai 2009, Kawahara et al. 2012). This encystment can occur under normal conditions (Schiariti et al. 2014), but is, in general, induced by stressing factors such as temperature extremes, low salinities, and hypoxia (Kawahara et al. 2012). Therefore, factors taking place both in the benthic and pelagic realms will influence the size of next planktonic population.

Transport by currents plays an important role in metapopulation dynamics of some GZ species, such as in the ctenophores *M. leidyi* and *Pleurobrachia pileus*. These species are very seasonal, and during non-favorable conditions they disappear locally from some areas due to advection losses, but persist in areas of low advection and exchange rates (e.g., embayments and deeper waters; Schneider 1987, Costello et al. 2012). These latter areas represent the source of new individuals to non-overwintering areas (Schneider 1987, Costello et al. 2006).

True blooms, i.e., caused by population increases, are often found in coastal enclosed or semi-enclosed areas (e.g., lagoons, estuaries, bays and fjords), where advective losses are low and productivity is high (Mills 2001, Condon and Steinberg 2008, Purcell 2012). Particular patterns of physical processes, such as currents and winds, may also promote and enhance jellyfish aggregations along coastal areas (Graham et al. 2001). This may lead to mass stranding events as are common in pleustonic gelatinous organisms, such as *Physalia physalis, Velella velella* and *Glaucus atlanticus* (Flux 2008, Pietro et al. 2015, Pinotti et al. 2019), large scyphomedusae (Houghton et al. 2007, Canepa et al. 2014, Mianzan et al. 2014), small calycophoran siphonophores (Guerrero et al. 2018), and hydroids (Genzano et al. 2008). Although the role of biological production and physical transport is unquestionable, behavioral factors, such as oriented swimming, known particularly in large medusae, may also enhance aggregations (Zavodnik 1987, Shanks and Graham 1987, Fossette et al. 2015). Large cubomedusae have robust swimming (Colin et al. 2013) and a set of very sophisticated eyes used for oriented swimming toward their preferred environment, using terrestrial visual cues (Garm et al. 2011). The neurological basis of environmental perception remains unclear in large scyphomedusae (Graham et al. 2001, Fossette et al. 2015), but patterns of oriented swimming, such as counter-current or toward surface swimming have been described, and these may facilitate swarm formation and stranding avoidance (Zavodnik 1987, Shanks and Graham 1987, Albert 2009, Fossette et al. 2015). Because oriented swimming seems phylogenetically widespread in Scyphozoa, it may represent an important adaptive trait.

7.6 Trophic role and impact on biogeochemical cycles

a. Gelatinous zooplankton as consumers: From feeding biomechanics to the impact on natural assemblages

GZ are important consumers in the pelagic realm both at lower and higher trophic levels and, when abundant, can have significant effects on the structure of the pelagic community (Alldredge 1981, Alldredge and Madin 1982, Madin and Kremer 1995, Purcell et al. 2001, Marshalonis and Pinckney 2008, McNamara et al. 2014, Henschke et al. 2016). Most gelatinous species are documented as primarily carnivorous, eating a wide variety of available zooplankton, including microzooplankton (e.g., tintinnids, copepod nauplii, dinoflagellates), meso-, and macrozooplankton components (e.g., copepods, decapods, chaetognaths, cladocerans, other jellyfish, eggs, larvae, and juveniles of fish; Seapy 1980, Stoecker et al. 1987, Arai 1988, Mills 1995, Purcell 1997, Colin et al. 2005, McNamara et al. 2010, Kamiyama 2013). However, some gelatinous species may feed on bacteria, phytoplankton, and detritus (Boero et al. 2007, Sutherland et al. 2010, Lombard et al. 2010, McNamara and Lonsdale 2014, Morais et al. 2015). Indeed, planktonic tunicates (thaliaceans and appendicularians) are among the main metazoan taxa preying on nano and picoplankton, and thus have a paramount importance in oligotrophic areas and in the microbial food web (Kremer and Madin 1992, Madin and Kremer 1995, Esnal 1999a, b, Esnal and Daponte 1999a, b). In addition, some GZ may possess photosynthetic endosymbionts from which they receive organic compounds (see later in this chapter), and a few have been shown to take up dissolved and fine particulate organic matter directly (Fergusson 1988, Skikne et al. 2009, Casanova et al. 2012).

Gelatinous organisms feed through different techniques. Many medusae obtain their food by filtering sea water though the flow created during bell pulsation, bringing prey into contact with the tentacles and/or other nematocysts-bearing surfaces, i.e., capture surfaces (Fraser 1969, Mianzan and Cornelius 1999). The type of swimming and the flow used for prey capture and the capture surface morphology vary among species and, among other factors, determine feeding selectivity (Costello and Colin 1995, Purcell 1997, Colin and Costello 2002). Some medusae are cruising foraging predators, which spend the majority of their time swimming, while other species are known to be ambush predators, which remain motionless with their tentacles extended to entangle actively swimming prey (Colin and Costello 2002, Costello and Colin 2002). Some others present sophisticated fishing techniques such as the cubozoans, which presumably employ their complex visual structures in prey capture (Buskey 2003, Courtney et al. 2015), and some siphonophores that attract their prey by luring them with structures imitating other zooplankton (Purcell 1980, Mackie et al. 1987).

Among ctenophores, the feeding mode also varies according to the animal's morphology; it may range from using tentacles like fishing lines to catch food with the colloblasts in an analogous way to medusae, to using lobes to generate feeding currents with their auricles, to feeding by swallowing the whole prey in a raptorial feeding mode (Haddock 2007). Prey selection occurs across this phylum as well, and there are several examples of ctenophores feeding extensively on one prey type (Purcell 1997, Haddock 2007).

Holopelagic gastropods are mostly carnivorous, preying on different zooplankton taxa and size spectra. They typically use their image-forming eye and an extensible radula to find and capture prey. The gymnosomatous pteropods are highly specialized, feeding mostly or exclusively on a few thecosomatous species (Lalli and Gilmer 1989, Richter and Seapy 1999, van der Spoel and Dadon 1999). Heteropods are usually opportunistic predators, but some taxa such as *Carinaria* may selectively predate on small-sized chaetognaths and thaliaceans (Seapy 1980, Richter and Seapy 1999). *Janthina* and *Recluzia* species (Epitonidae) prey mostly on cnidarians, particularly the floating colonies of *Physalia*, *Velella*, and *Porpita* in the case of the former, and floating sea anemones for *Recluzia* (Lalli and Gilmer 1989, Beu 2017). *Janthina* can also prey on other neustonic animals, as well as being cannibalistic (Beu 2017), much like other predatory pelagic gastropods (Seapy 1980). Contrastingly, the Pseudothecosomata and Euthecosomata are omnivorous, feeding mostly on microplankton. They produce a mucous web up to 100 times larger than the animal to which protists get adhered. Eventually they eat the web with all the attached organisms and produce a new web (Gilmer 1972, Gilmer and Harbison 1986, Lalli and Gilmer 1989, van der Spoel and Dadon 1999).

Conversely, pelagic tunicates are filter-feeding animals. Thaliaceans filter their food by rhythmically pumping water into their sac-like bodies through the action of circular muscle bands in salps and/or by ciliary action in doliolids (Alldredge and Madin 1982, Madin and Kremer 1995, Esnal and Daponte 1999a, b). Food particles within a wide size range, from small bacteria (< 1 μm) to large protists and occasionally small crustaceans (> 1 mm), can be filtered with different efficiencies. The particles entering the pharyngeal chamber are strained through a mucous filtering mesh-net and rolled into a food strand that moves posteriorly towards the esophagus resulting in ingestion of adhered particles (Esnal and Daponte 1999a, b, Sutherland et al. 2010, Ishak et al. 2017). Appendicularians, on the other hand, harvest small planktonic food particles by a unique trait, "the external filter house", an external mucous net secreted by a specialized gland that sieves and concentrates a wide range of particle sizes up to 30 μm, from bacteria to microplankton (Esnal 1999b, Vargas and González 2004, Berline et al. 2011, Lombard et al. 2011). The inlet filters serve as a selective mesh for too large particles and prevents clogging. Still, if the filter system of the "house" is clogged, the appendicularian discards it and secretes a new one. Depending on the species, temperature, particle concentration and probably also the nutritional condition of the animal,

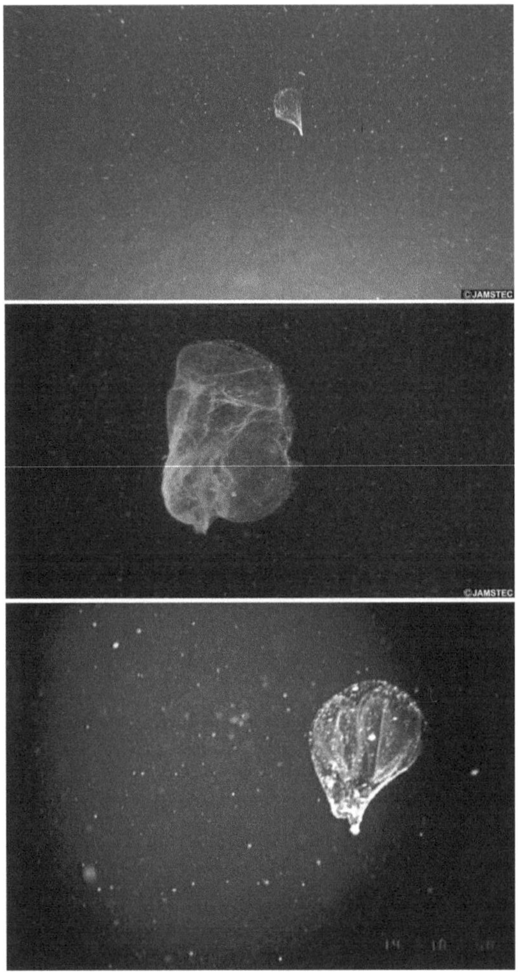

Figure 7.3. Examples of appendicularian discarded houses sinking in the water column. The images are *in situ* framegrabs and no scale bars are available. Photographic credits: D.J. Lindsay.

up to 40 houses may be secreted per day (Sato et al. 2003). The discarded houses sink through the water column (Fig. 7.3) and have great importance in biogeochemical cycles and in the microbial loop.

It is clear that GZ employ a diverse array of food acquisition techniques. The feeding impact of these animals is often substantial, although its magnitude depends on their population size (Purcell and Arai 2001). When a bloom occurs, the feeding impact of GZ on the pelagic community is enhanced and its effect can be overwhelming on the pelagic community, particularly if the ecosystem is under human exploitation (Boero 2013). Medusae and ctenophores not only consume the early developmental stages of fish (eggs, larvae, and/or juveniles; McNamara et al. 2010), but can also compete for the same potential prey with larval and/or adult fish (Boero 2013). Tunicates can also have a great impact on other pelagic populations, as when they are abundant they can quickly deplete smaller particles from the water column due to their high grazing rates and low assimilation efficiency (Alldredge and Madin 1982, Madin and Kremer 1995, Gibson and Pafenhöffer 2000, 2002, Pafenhöffer and Köster 2005), reducing available food in the lower trophic levels. Moreover, planktonic tunicates often are the main zooplankton groups to take advantage of any increase in bacterial biomass.

GZ can have daily predation values that exceed their own body weight in terms of carbon (Stibor and Tokle 2003). The impact on prey populations may be high, as in the example of the ctenophore *M. leidyi* that may daily remove more than 40% of copepod standing stocks (Purcell and Decker 2005), and more than 70% of the total ichthyoplankton (Tzikhon-Lukanina et al. 1993). A well-documented case was the introduction of *M. leidyi* to the Black and Azov Seas, which was followed by a significant decimation of the zooplankton and a consequent drop in planktivorous fish numbers, due to competition and consumption of their eggs and larvae (Shiganova et al. 2001). Other known cases include: (i) the calycophoran siphonophore *Bassia bassensis* removing up to 70% of small copepod abundance (Pagès et al. 2001); (ii) the leptothecate hydromedusa *Aequorea victoria* decreasing the herring population by predating on 73% of newly-hatched larvae per day (Purcell and Arai 2001); (iii) blooms of the giant scyphozoan *Nemopilema nomurai*, that may consume up to 24% of the total mesozooplankton biomass per day, and directly affect the recruitment of fish populations by preying on fish eggs and larvae, threatening the local fisheries, and leading to social and economic disturbances (Uye 2008).

On the other hand, the grazing impact of salps and appendicularians can be equal or even higher than that of copepods, showing one of the highest individual ingestion rates of all marine suspension feeders (Alldredge 1981, Alldredge and Madin 1982, Hopcroft and Roff 1995, 1998, Hopcroft et al. 1998, Esnal 1999b, Esnal and Daponte 1999a, b, Henschke et al. 2016). Because they feed heavily on microbial components and comprise an important energy source to pelagic predators, they have an important role in the energy transfer of the short-lived microbial carbon into longer-lived harvestable resources. Moreover, they produce fast-sinking fecal pellets, which frequently contain undigested material, at high rates, substantially contributing to the vertical flux of carbon to deep waters and enriching the benthos (Fortier et al. 1994, Madin and Kremer 1995, Yoon et al. 1996, Paffenhöfer and Köster 2005, Conley and Sutherland 2017).

b. Who eats jellies? A new perspective on gelatinous zooplankton as prey

Although GZ have historically been considered as trophic dead-ends (e.g., Verity and Smetacek 1996), many animals, including other gelatinous species, corals, mollusks, chaetognaths, crustaceans, fish, turtles, birds, and even humans, feed occasionally, regularly, or exclusively on GZ (Mianzan et al. 1996, 2001, Omori and Nakano 2001, Purcell et al. 2005, Alamaru et al. 2009, Cardona et al. 2012, Hoeksa and Waheed 2012, Ates 2017, Henschke et al. 2016, Ayala et al. 2018). In addition, micropredators/parasites using GZ as hosts are also quite common (see later in this chapter). Considering the difficulties of recognizing gelatinous remains in stomach contents (Diaz Briz et al. 2017), the recognized number of predators of GZ, and the ecological importance of jellies as prey, may increase with increasing studies using alternative approaches, such as molecular and imaging techniques. These methods can help generate a wider understanding of GZ as a food source and illuminate their roles in trophic webs.

In fact, gelatinous prey may be more nutritious than previously thought. Their usually low energy content (due to their high-water content) may be compensated for by higher rates of digestion and assimilation by their predators (Arai 2005). Moreover, certain gelatinous species may have similar energetic values to those of phytoplankton or even small copepods which, when considered with their typically slow movements that facilitate their capture, make gelatinous organisms an efficient food source for pelagic predators (Purcell et al. 2005, Henschke et al. 2016). Also, the high concentrations that gelatinous animals sometimes reach in certain regions facilitate their inclusion in the diet of predators. The herbivorous juveniles of the green turtle *Chelonia mydas*, for instance, show an adaptive foraging behavior in the temperate SW Atlantic, preying more on GZ rather than on herbivorous items (González Carman et al. 2014). Many commercially important fish species prey heavily on GZ (Arai 1988, Purcell and Arai 2001, Henschke et al. 2016). There are specialized fish species, which have developed morphological and physiological adaptations to feed on jellies (Arai 1988, Purcell and Arai 2001), and other non-specialized ones, usually not considered as gelatinous predators, which may feed occasionally on GZ, apparently when nothing better is available (Mianzan et al. 1996, 1997, 2001, Diaz Briz et al. 2018). Along the continental shelves of Argentina and Uruguay, for instance, at least 39 fish species prey, with different frequencies, on different GZ taxa, suggesting that such trophic interactions may be more common than previously recognized (Diaz Briz et al. 2017, 2018). Providing further evidence of the regular incorporation of jellyfishes in fish diets, GZ commonly host the larval stages of helminths, the adults of which are fish-parasites and rely on trophic relationships to complete their life history (Marcogliese 1995).

c. *Fecal pellets, carcasses (jelly-falls), discarded houses and implications for vertical carbon flux and marine snow*

When GZ populations collapse, their dead bodies contribute to biogeochemical fluxes. Their prolific body mass, external houses, and fecal pellets constitute relevant components of the locally-produced particulate organic matter (POM), providing an important pathway for carbon and nitrogen export to deeper waters (Morris 1988, Yoon et al. 1996, Lebrato et al. 2012, 2013) and driving secondary production and many benthic processes (Ruhl et al. 2008, Smith et al. 2008). In a jellyfish-dominated ecosystem in western Norway, the maximum flux rate of the gelatinous carcasses or "jelly-fall" to the sea floor was estimated to be equivalent to 96 and 160% of the phytodetrital C and N flux (Sweetman and Chapman 2015), pointing out the significant contribution to detrital C and N fluxes that jelly-falls may represent. Furthermore, decomposing *Chrysaora quinquecirrha* may raise organic carbon and nutrients by 1–2 orders of magnitude (Frost et al. 2012). The large gelatinous species probably have a particularly significant role in such processes.

It has long been recognized that pelagic tunicates have particular and substantial roles in biogeochemical cycles. In comparison with other zooplankton, salps and appendicularians contribute disproportionately to carbon flux, producing large, resistant, fast-sinking, and carbon-rich fecal pellets frequently with partially undigested material (Fortier et al. 1994, Yoon et al. 1996, Paffenhöfer and Köster 2005, Henschke et al. 2016, Conley and Sutherland 2017). Discarded appendicularian houses have particular ecological implications as well. They contain concentrated, non-ingested prey, thus constituting a major source of organic-rich marine snow and dominating vertical carbon fluxes, typically contributing up to 39% of total particulate organic carbon exported to the seafloor (Conley and Sutherland 2017).

Because food is a restrictive factor in the deep ocean, gelatinous carrion emerges as a crucial source of energy. Although, in comparison with other marine taxa, gelatinous material generally has low energy content (Doyle et al. 2007), this may be compensated for by the reduced energy spent searching for food and the easy digestion of the gelatinous material (Lebrato et al. 2012). Given that GZ biomass appears to be increasing in certain areas of the world's oceans (Purcell 2012), the transfer of jelly-POM to the ocean floor may be amplified accordingly and, in consequence, the magnitude and relevance of the biogeochemical and ecological processes associated with jelly-falls may be boosted (Lebrato et al. 2012, 2013).

7.7 Living on jellies

One incomparable characteristic of the pelagic environment is the lack of solid substrate and, thus, the lack of physical/topological structure and sites where it is possible to shelter and hide from predators. Consequently, several species have adapted associations in the pelagic environment and all organisms can become potential hosts. This creates an intricate net ranging from accidental short-term to obligatory long-term relationships within a wide array of different sorts of interactions, most of them largely unknown. In this sense, those organisms with larger sizes tend to be better hosts and harbor a larger number of associated individuals and/or species. Consequently, the GZ, which frequently are amongst the largest marine plankton, may play a particular role, influencing the nature of the pelagic habitat and the distribution and behavior of other organisms, contributing to the increase and maintenance of local biodiversity by providing structural complexity to pelagic habitats. Moreover, cnidarians may confer additional advantages for their symbionts, since their cnidae can provide further protection from possible predators. Indeed, several taxa, ranging from bacteria to fishes, have adapted to associate in many different ways with GZ (Ohtsuka et al. 2009).

Only for a few cases is the role of each organism involved in the association known (Laval 1980, Towanda and Thuesen 2006, Niggl et al. 2010, Chiaverano et al. 2015), while for the vast majority of cases the advantages and/or disadvantages for each partner involved have only been hypothesized.

Moreover, most of such studies rely on net-collections, which frequently complicate the interpretation of the association since the organisms may be artificially induced to form short-term associations due to stress and/or artificial accumulation within the nets. This may be particularly true for organisms that do not live attached to their jelly-host, but rather keep swimming around it, as is the case with larval/juvenile fishes (Kingsford 1993). A general-view and summary on some of the best-known cases are presented below.

a. Viruses and bacteria

There are relatively few studies dealing with such associations, even though they are likely to be common, given the widespread nature of these microbes in marine environments, including in zooplankton communities (Kinne 1990, Tang et al. 2009, 2010, Grossart et al. 2010, Martinelli-Filho et al. 2020). High persistence and prevalence rates of viruses have been recently detected in some coastal ctenophores, which suggests that these are regular infections that may have important roles controlling host population dynamics (Breitbart et al. 2015), although further studies are necessary to test this hypothesis. Unidentified bacterial infections, associated with wounds, have been reported for gelatinous organisms (Heeger et al. 1992, Mills 1993), and this is likely to be common in spite of the few reports. Apart from parasitic bacteria, GZ may also harbor commensal and mutualistic taxa, also potentially serving as vectors of human and/or animal pathogens, though in most cases the nature of such associations is not clear (Kinne 1980, 1990, Ferguson et al. 2010, Delannoy et al. 2011, Dinasquet et al. 2012, Daley et al. 2016, Martinelli-Filho et al. 2020). In any case, the microbes on GZ typically differ from those in the surrounding seawater and show differences between different host-species, populations, life-cycle stages, body parts and seasons, suggesting some degree of specificity (Arai 1997, Hao et al. 2015, Daley et al. 2016, Kos Kramar et al. 2019). Pyrosomes depend on the presence of luminous bacteria in an obligate relationship, being acquired by the host *via* transfer from generation to generation (Kinne 1990).

b. Protists

One of the most emblematic cases is the symbiosis of the dinoflagellate *Symbiodinium* (i.e., "zooxanthella") with scyphozoan polyps and/or medusae, and with some floating colonies of *V. velella* (Théodoridés 1989, Arai 1997, Djeghri et al. 2019). The primary advantage for the host is nutrition. The algae transfer to their hosts between 25 and 60% of the fixed carbon and recycle nitrogenous waste from them (Goodson et al. 2001). Zooxanthellae, in turn, receive protection and a constant supply of CO_2 and nitrogen (ammonia), both produced by the metabolism of the host (Muscatine and Marian 1982, Arai 1997, Goodson et al. 2001). The scyphozoan *Mastigias* also has a behavioral adaptation, performing vertical migration to the chemocline to uptake environmental ammonium to provide for its symbiont microalgae (Muscatine and Marian 1982). The carbon produced by the algae may satisfy a great part of the metabolic demand of the host and, in some cases, even generate an excess (Kremer et al. 1990, McCloskey et al. 1994). Indeed, heterotrophic feeding was estimated to be a minor source of carbon for both *Mastigias* sp. (McCloskey et al. 1994) and *Linuche unguiculata*, but a major source of nitrogen and phosphorus for the latter species (Kremer 2005). The zooxanthellae also have an important role in strobilation for some scyphopolyps, which seem not to be able to strobilate in the absence of the algae (Sugiura 1964, 1969). In most cases, the algae seem to be acquired from the environment rather than from parental inheritance (Arai 1997). However, *Mastigias papua* polyps asexually budded from zooxanthellate-harboring polyps inherit the algae (Sugiura 1964).

A very peculiar case is the apostome ciliate *Vampyrophrya pelagica*, which lives encysted on copepod exoskeletons. When the host's exoskeleton is breached it excysts, enters the copepod-body through a wound, and quickly feeds on its tissues (Rohde 2005). Thus, when these copepods are ingested by ctenophores and/or medusae, *V. pelagica* consumes the copepods' tissues faster than the predator (Ohtsuka et al. 2004, 2009). Consequently, the jellyfishes do not gain nutrients while feeding on these copepods and actually expend energy by capturing them (Ohtsuka et al. 2004, 2009). These ciliates may be observed on copepods from different depths and oceans (Sewell 1951, Ohtsuka et al. 2004, 2009), but their possible influence on jellies' energetic expenditure has not been evaluated in detail and it may be larger than currently recognized. Other apostomes are known to live on different gelatinous taxa, but their roles have not yet been evaluated (Théodoridès 1989, Rohde 2005).

c. Metazoans

Metazoans from many different groups, including cnidarians, ctenophores, platyhelminths, nematodes, mollusks, pycnogonids, many different crustacean taxa, echinoderms and chordates are known to associate in some way with GZ (Figs. 7.4, 7.5). The nature of such associations varies widely, including, but not limited to, phoresy/inquilinism, different types of parasitism (i.e., endo, ecto, klepto, parasitoidism), mutualism and commensalism. While attempts to categorize the associations have often been made, most associations with GZ have not been analyzed in depth, hindering the precise use of many of these ecological terms. Moreover, in several cases, considerable overlap exists between one grade and another. This includes the copepod *Rhincalanus gigas*, which may enter the body cavity of *Salpa thompsoni* and

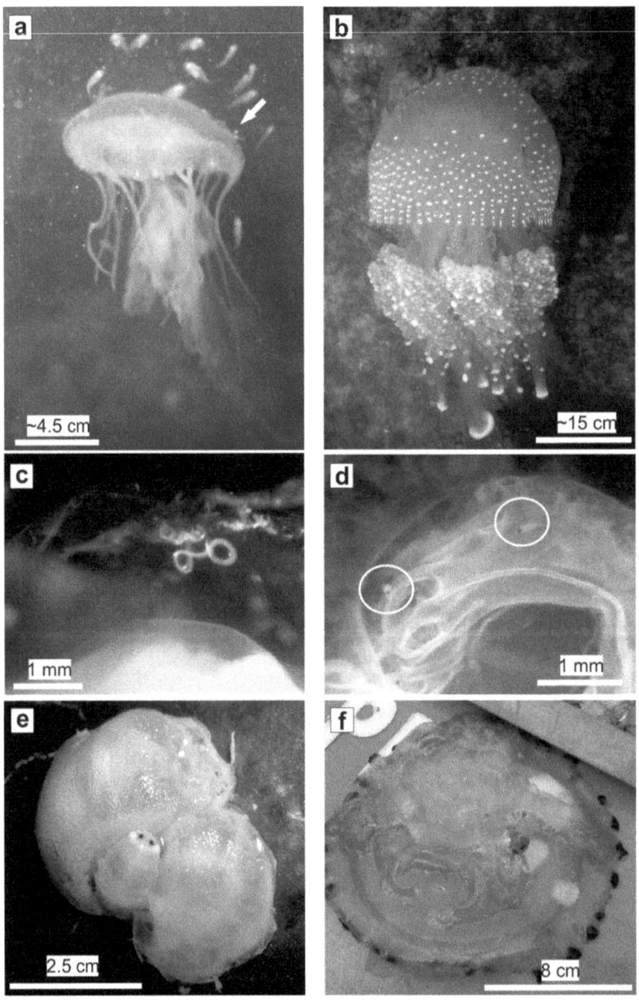

Figure 7.4. Examples of metazoans associated with gelatinous zooplankton. (a) carangid fish and the cleaner-shrimp *Periclimenes paivai* (arrow) associated with the scyphozoan *Chrysaora lactea*; (b) carangid fishes associated with the scyphozoan *Phyllorhiza punctata*; (c) anisakid-like nematode parasitizing the hydromedusae *Bougainvillia pagesi*; (d) *Opechona* sp. metacercaria (circles) on the gonadal cavity of *C. lactea*; (e) the cymothoid isopod *Cymothoa catarinesis* parasitizing *C. lactea*; (f) the spider crab *Libinia ferreirae* associated with the scyphozoan *Lychnorhiza lucerna*. All from Brazilian subtropical (~ 25–27°S) shallow waters. Photographic credits: M. Nogueira Júnior.

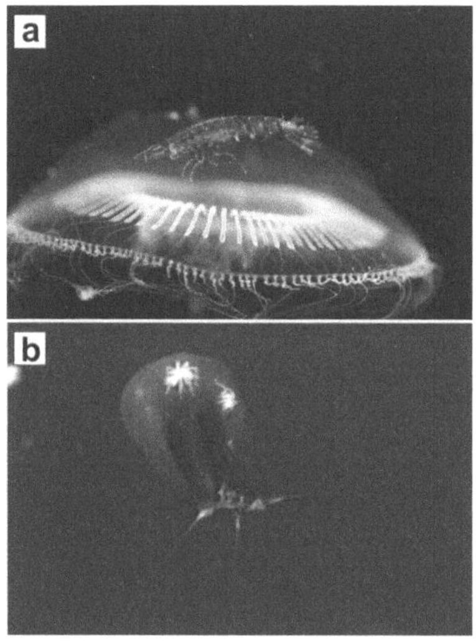

Figure 7.5. Examples of metazoans associated with gelatinous zooplankton: (a) the hyperiid *Oxycephalus clausi* on the hydromedusa *Aequorea* sp. from Suruga Bay, Japan, superficial waters; (b) one adult and one juvenile (?) pycnogonid attached to the hydromedusa *Pandea rubra* offshore Japan at 913 m depth during ROV HyperDolphin Dive 105. The upper image was taken in a shipboard photographic aquarium after scooping the specimen with a bucket. The lower image is an *in situ* framegrab and *Aequorea* sp. was roughly estimated to have around 5 cm in bell diameter and *P. rubra* around 10 cm in height. Photographic credits: D.J. Lindsay.

feed on the microplankton concentrated in the food strand of the salp. Although this could be seen as a case of opportunistic kleptoparasitism, the copepods apparently also benefit the salps by cleaning their clogged filter apparatus (Perissinotto and Pakhomov 1997). Likewise, the crab *Metacarcinus gracilis* (as *Cancer gracilis*) may be kleptoparasitic on scyphomedusae during its megalopa stage and first-crab-instars, and may gradually become a facultative cleaner, eating the jellyfish's parasites as it grows (Towanda and Thuesen 2006). Along with the type, the strength of the associations also varies widely. They may be ephemeral, with some organisms starting the association by "accident" and/or opportunity, as seems the case for some isopods, copepods, pycnogonids, mysids and fishes, to obligatory, since the association is part of the life history of some symbionts as is the case with some cnidarians, helminthes, and hyperiid amphipods. Many cases fall in-between these two extremes.

In most cases, these metazoans living in association with GZ seek shelter, protection from predators and may also benefit by saving energy otherwise invested in swimming. Additionally, they may gain nutrition by feeding directly on the tissues of the host, by eating mucus, eating parasites of the jellies, stealing food before the host ingests it, or stealing partially digested food from the gastric system of the host. Disadvantageous effects to the host may be due to (i) direct feeding on the gelatinous host tissue (i.e., micropredators/parasites); (ii) loss of captured or ingested food such as with the endobiont sea anemone *Edwardsiella lineata* on ctenophores (Bumann and Puls 1996); (iii) an excessive energy expenditure from carrying the hitchhiker, as in the example of some pandalid shrimps piggybacking on hydromedusae almost the same size as themselves (Marliave and Mills 1993). However, in some cases the gelatinous host may also benefit, mostly when their associates are eating their parasites, as is the case for the juvenile crabs described above, or with many fish-gelatinous zooplankton associations (e.g., Kingsford 1993, Castro et al. 2002, Masuda et al. 2008).

Many larval/juvenile stages of helminths (digenean, cestodes and nematodes) are known to use GZ taxa as intermediate hosts. In spite of particularities in the life-histories of these different parasites, their definitive hosts are usually vertebrates, typically fishes, and the parasites rely on trophic interactions for transmission. Thus, the frequently high infection rates observed in many GZ populations indicate that these gelatinous taxa are regularly eaten by the vertebrate definitive host, and parasitism can be an indirect way to trace trophic relationships between GZ and their vertebrate predators (Stunkard 1980, Køie 1993, Shih and Jeng 2002, Gaevskaya et al. 2012, Nogueira Júnior et al. 2015).

Probably the best-documented case of metazoan-GZ association is with hyperiid amphipods. The nature and extent of these associations is variable, ranging from ephemeral to more permanent. In any case, the associations are considered to be detrimental to the host and all hyperiids are usually assumed to be dependent on their gelatinous host, at least for part of their lives (Harbison 1976, Harbison et al. 1977, Laval 1980, Dittrich 1988, 1992, Fleming et al. 2014). Hyperiids may hatch as a larva or a juvenile, and in either case the young, or in some cases the eggs, are transferred from the marsupium to the host by their mother, and the individuals typically remain on/in the host until reaching maturity. Some juveniles may consume either the host's tissues or its food items, typically devouring the host when reaching the adult-stage and becoming predatory, and then likely searching for a new host (Laval 1980, Dittrich 1988). Alternatively, adults of the hyperiid *Vibilia* consume exclusively the food strand of the salps and not their tissues (Madin and Harbison 1977).

The Phronimidae are known to modify their gelatinous host (mostly thaliaceans) into a "barrel" into which they attach and feed their larvae on prey brought back inside the barrel by the mother, with the barrel serving like a platform, giving them both protection and buoyancy (Laval 1978, 1980, Diebel 1988, Davenport 1994, Hirose et al. 2005, Aoki et al. 2013). Another peculiar case is *Hyperiella* from the Antarctic Ocean, which captures and carries the pteropods *Clione limacina antarctica* or *Spongiobranchaea australis* as a "backpack" for anti-predatory behavior, since the pteropods produce a toxin that works efficiently against predatory fishes (McClintock and Janssen 1990, Havermans et al. 2019).

7.8 Global changes and gelatinous zooplankton: A challenge for the future

Human activities have modified marine ecosystems at all possible scales, from small estuaries to the global ocean, and the perspective is that this scenario will worsen in the next few decades (see more details in Maenza and Pan, Chapter 14 of this volume). The changes may result in significant modifications in the marine biota inhabiting these systems, which is particularly true for GZ. While some pelagic cnidarians, ctenophores and tunicates may benefit from the new conditions and resources of modified ecosystems, blooming (see above) and dominating the pelagic community (Purcell et al. 2007, Purcell 2012), other gelatinous and semi-gelatinous organisms, such as mollusks with shells, may face physiological constraints that may preclude their existence (Fabry et al. 2008, Manno et al. 2017).

In this section, we briefly discuss the known and potential impacts of present and projected anthropogenic changes in marine systems on GZ. It is important to note that although general trends are assumed, different species in each group, or even different populations of the same species, often present distinct responses to similar phenomena, calling for more experiments and long-term field observations to properly predict the functioning of future pelagic systems.

a. Climate changes and ocean acidification

Temperature has a fundamental role on molecular kinetic energy and cellular processes and even small changes may increase the metabolic rates of species, determining life history traits, population growth and ecosystem functioning (discussed at length in Salinas, Chapter 13 of this volume). Rises in atmospheric CO_2 concentrations have increased the global average temperature by ~ 0.2°C per decade over the past 30 years. This excess heat is absorbed by the oceans, where surface temperature increased, on average, by 0.6°C over the past 100 years, and climate models project similar rates for coming decades (IPCC 2014). Other processes, such as increasing water column vertical stratification, acidification, desalinization and oxygen reduction commonly occur concomitantly (Bednaršek et al. 2016, Manno et al. 2017). Facing this scenario, species outside their physiological optima will have their fitness reduced and populations will decline (Hoegh-Guldberg and Bruno 2010). Also, species re-distribution towards higher latitudes is expected, as exemplified by the recent range expansion of the tropical and subtropical *Pyrosoma atlanticum* appearing in high abundances in temperate and subpolar latitudes of the northeast Pacific (Brodeur et al. 2018, Sutherland et al. 2018), and the northwards expansion of the deep-sea jellyfish *Periphylla periphylla* invading fjords (Tiller et al. 2016).

Several studies reported GZ blooming or their occurrence at higher abundances in warmer and/or saltier water conditions. These cases led to a concern from part of the scientific community, inflated by the popular media, that the global populations of pelagic cnidarians would increase globally with higher ocean temperatures (reviewed by Purcell et al. 2007, Purcell 2012). Although representing only a small portion of the diverse gelatinous fauna, many species considered in these studies are typically dominant and/or widespread, such as the scyphomedusa *Aurelia aurita*, the hydromedusae *L. tetraphylla*, *Solmundella bitentaculata*, *Aglaura hemistoma*, *Aglantha digitale* and *Clytia hemisphaerica*, the siphonophores *Chelophyes appendiculata* and *Abylopsis tetragona* and the ctenophore *M. leidyi* (see Supplementary Table 2 in Purcell et al. 2012). This seems to indicate that at least some dominant gelatinous species may benefit from future changes in ocean temperature.

On closer look, however, these positive correlations of temperature and abundance of GZ are not ubiquitous. Many studies reported reduced abundance in warmer conditions, even for species whose abundance increased in other locations, such as *A. aurita* and *Clytia hemisphaerica*. Other common species that presented negative correlations between temperature and abundances were the scyphomedusae *Chrysaora fuscescens* and *Chrysaora hysoscella*, the hydromedusae *Leuckartiara octona* and *Obelia* spp. and the ctenophores *Pleurobrachia pileus*, *Pleurobrachia rhodopis* and *Beroe* spp. (Supplementary Table 2 in Purcell et al. 2012).

The few *in vitro* studies on the subject have indicated that some pelagic cnidarian species could benefit from warmer waters. For instance, the development of *Pelagia noctiluca* ephyrae was twice as fast at 19°C as at 13.5°C (Avian 1986). Warmer temperatures also increased asexual reproduction of the scyphomedusae *Chrysaora quinquecirrha* and *Aurelia labiata*, and the hydromedusa *Moerisia lyonsi* (Purcell et al. 1999a, Ma and Purcell 2005, Purcell 2007).

In addition to ocean warming, climate change will contribute to an increased vertical stratification of the water column that may reduce biological production in tropical latitudes (Bednaršek et al. 2016), and affect oceanographic processes such as current circulation patterns (IPCC 2014). The distribution of many GZ is closely associated with water masses, and current intrusions significantly affect the communities' structure (Pagès and Gili 1991, 1992, Pagès 1992, Nogueira Júnior et al. 2014). Blooms of *Pelagia noctiluca* in the Adriatic Sea and of the siphonophores *Muggiaea atlantica* and *Apolemia uvaria* in the North Sea were related to changes in local hydrography (Mills 2001). Advection and accumulation are in part responsible for the formation of swarms and blooms of tunicates as well (Paffenhöfer and Lee 1987, Graham et al. 2001, Chae et al. 2008). Therefore, changes in circulation patterns may result in GZ invading new areas and blooming due to incursion, retention and accumulation phenomena.

Absorption of anthropogenic CO_2 by the oceans also decreases its pH, with significant impact on the calcium carbonate saturation state, reducing or even precluding calcification in marine organisms (Doney et al. 2009). Many cnidarians and ctenophores have microscopic calcium concretions in statocysts, which could be impaired with ocean acidification; however, the effects of decreasing pH on the secretion of these structures are still unknown (Purcell et al. 2007). In one experiment, after 12 h exposure to 7.5-pH seawater (i.e., the projected value for 2300), around 35% of the small-sized (< 3 cm) cubozoan *Carybdea xaymacana* population died and most of the surviving individuals showed various degrees of alteration in behavior and/or morphology (Chuard et al. 2019).

The holoplanktonic gastropods that have aragonite shells are particularly vulnerable to ocean acidification, and likely to be good indicators of its extent (Bednaršek and Ohman 2015, Riebesell and Gattuso 2015, Wall-Palmer et al. 2016). In pteropods, there is a direct influence of the aragonite saturation level on shell-condition (Weisberg et al. 2016, Manno et al. 2017). Calcification may be reduced in these animals due to elevated pCO_2 and temperature, which may degrade/dissolve their shell and/or reduce their growth and lead to death (Comeau et al. 2010a, b, Lischka et al. 2011). The shells of these live pteropods provide species-specific substrates for the polyps of hydromedusa taxa and a run-on effect has been suggested/illustrated due to these hydromedusae also serving as substrates for other organisms in turn (Lindsay et al. 2008). Moreover, the combined effect of acidification with other concomitant processes, such as increase in temperature and deepening/strengthening of the thermocline, may alter the locomotory activity of pelagic gastropods (Manno et al. 2012), as well as affect their vertical migration and community structure (Bednaršek and Ohman 2015). The Pterotracheoidea are also vulnerable to the effects of ocean acidification, since they also may have aragonite shells; however, such effects have not been evaluated (Wall-Palmer et al. 2016, Burridge et al. 2017).

b. Overfishing

Currently, more than 33% of global marine fish stocks are captured at biologically-unsustainable levels (FAO 2018). Zooplanktivorous forage fish and fish larvae are potential competitors for gelatinous predators due to their dietary overlap (Purcell and Sturdevant 2001) and the reduction or depletion of their stocks may directly affect food availability for cnidarians and ctenophores. Increases in the abundance of the scyphomedusae *C. hysoscella* and *Chrysaora fulgida*, the hydromedusa *Aequorea forskalea*, the siphonophore *Nanomia cara* and the ctenophore *M. leidyi* were suggested to be a consequence of the reduction of zooplanktivorous fish competitors (Lynam et al. 2006, Shiganova 1998, Mills 2001, Daskalov 2002, Daskalov et al. 2007, Roux et al. 2013). On the other hand, many predatory fish and by-catch species, such as sea turtles, include GZ in their diet and the reduction in their populations may bring additional benefits for the gelatinous prey.

c. Eutrophication

Eutrophication of estuaries and coastal areas due to agricultural fertilizers and sewage discharge is a major pollution problem. Food availability is commonly the main trigger for the asexual reproduction and bloom formation of tunicates (Paffenhöfer and Lee 1987), and increases in sexual and asexual reproduction of cnidarians (Purcell et al. 1999b, Lucas 2001). Therefore, nutrient enrichment may lead to higher phytoplankton production and greater biomass, scaling up to GZ (Paffenhöfer and Lee 1987, Daskalov 2002, Henschke et al. 2014). Eutrophication not only enhances phytoplankton productivity, but also causes complex changes in the food web. The low levels of silica in anthropogenic nutrient discharge hinders the proliferation of diatoms; instead, non-siliceous phytoplankton, such as the smaller flagellates may proliferate (Harashima et al. 2006), usually creating a longer and less efficient food-web. The size reduction in lower levels of the trophic web is detrimental to fishes, turtles, marine mammals and seabirds, which usually are visual predators and have high-energy requirements. On the other hand, GZ have lower energy requirements and do not rely on visual cues to capture food, feeding on a wide range of prey with respect to size and type (Sommer et al. 2002). For instance, many pelagic tunicates can efficiently filter micrometer and sub-micrometer sized-particles and benefit from smaller-sized organisms in the lower trophic levels.

d. Anoxia and hypoxia

A side effect of eutrophication is the reduction of dissolved oxygen in the bottom layers due to the decomposition of particulate organic matter produced by phytoplankton (Diaz and Rosenberg 2008). While the fitness of most fish is reduced in waters with oxygen levels < 2–3 mg.L^{-1}, many pelagic cnidarians tolerate these hypoxic conditions. Species such as *Rhopalonema velatum*, *Lensia meteori*, *Persa incolorata* and *S. bitentaculata* have been found in oxygen minimum layers (Santhakumari and Nair 1999, Batistić et al. 2004, Apablaza and Palma 2006, Bravo et al. 2011), thereby being benefited by the absence of natural competitors and even benefiting through more efficient feeding on impaired fish larvae (Shoji et al. 2005). However, the tolerance of gelatinous species to low levels of dissolved oxygen is not ubiquitous, with *L. tetraphylla*, *M. atlantica* and *Obelia* spp. only being found in the well-oxygenated upper layer off Chile (Apablaza and Palma 2006). Space competition has a major role in the structure of hard-bottom communities (Jackson 1977), thus cnidarian polyp development is also benefited by their tolerance to hypoxic waters, since larval settlement may occur in low oxygen areas where recruitment is compromised for other groups (Ishii et al. 2008, Ishii and Katsukoshi 2010).

e. Other anthropogenic impacts

Other anthropogenic disturbances may become significant in structuring GZ communities. For example, the installation of artificial structures over the continental shelf, such as harbors, aquaculture tanks, breakwaters, oil platforms and artificial reefs that provide additional suitable surface for the settlement of cnidarian larvae could lead to polyp proliferation and production of pelagic medusae. This was demonstrated for aquaculture rafts off Taiwan, oil platforms in the Gulf of Mexico and a floating pier off Japan (Graham 2001, Lo et al. 2008, Makabe et al. 2014). Aquaculture practices not only provide hard substrate for polyps, but also demand a large amount of fishmeal, utilized as food supply, contributing to the overfishing of forage fish, and releasing effluents into seawater, enhancing eutrophication, and therefore, combining multiple impacts in a single activity (Purcell 2012). In fact, all disturbances discussed herein usually occur simultaneously in coastal areas and estuaries all over the world, where human populations and their impacts on ecosystems have historically been concentrated. Consequently, their effects on pelagic gelatinous populations are difficult to be identified separately, but rather occur in response to multiple causes.

7.9 Are gelatinous populations increasing globally?

The concern that GZ populations are increasing globally and that blooms are becoming more frequent due to human impacts on marine systems has been widespread in the scientific literature and popular media in recent decades (Purcell et al. 2007, Condon et al. 2012). The concept took shape after large blooms of jellyfish were reported in several regions of the world and a review concerning the subject was published (Mills 2001). After that, the impact in the scientific and public media generated a demand for research in the field.

However, blooms are also natural phenomena that have occurred even in the early evolution of GZ. Dense accumulations of scyphomedusae in shallow marine habitats are preserved in the fossil record as early as the late Cambrian (Hagadorn et al. 2002). Historical literature from the past two centuries has also mentioned large occurrences of hydromedusae, scyphomedusae and salps (e.g., Herdman 1888, Mayer 1910, Russell 1953).

Nevertheless, most of the published information and public perception regarding recent increases in GZ blooms, and relating it to anthropogenic causes, are based on the immediate past of observers, where episodic and seasonal events, with high interannual variability, are common (Condon et al. 2012, 2013). They are also largely based on local and regional evaluations concerning few species, mostly in enclosed seas and bays (Purcell et al. 2007, Purcell 2012, Condon et al. 2012). Therefore, the modern perception of increased gelatinous populations may not reflect the global patterns of abundance and/or the overall responses of the group to changing environmental conditions.

Even with the lack of strong evidence, the idea of a global increase in GZ populations ruled by anthropogenic drivers was bought by the popular media and a part of the scientific community. After the first papers and review concerning the subject (Mills 2001), hundreds of papers have linked jellyfish blooms to anthropogenic stressors (Pitt et al. 2018), six instances of the "International Jellyfish Blooms Symposium" happened, television documentaries were produced, and the mainstream media reported it widely as fact. Although data from some regions indeed do indicate blooms are increasing, flawed citation practices and contradictory statements with variable conclusions have lent credence to the untested hypothesis that GZ blooms are increasing globally (Sanz-Martins et al. 2016, Pitt et al. 2018). Notable events such as "jellyfish blooms" are more remarkable than "low-levels of gelatinous zooplankton", resulting in a bias of popular and public media toward reporting these events (Condon et al. 2012); thus, this hypothetical assumption has acquired the status of a global ocean calamity (Duarte et al. 2015).

Two attempts to test the hypothesis of the global increase of gelatinous populations through meta-analysis have been performed up until the present time; both of them are somewhat inconclusive due to the scarcity of information on the tropics, open ocean gyres and the Southern hemisphere as a whole. Brotz et al. (2012), integrated data from 45 large marine ecosystems and concluded that 62% of them showed evidence of increased blooms. Condon et al. (2013) were more pragmatic and only considered long-term monitoring programs of more than 10 years' duration; among the 37 datasets, 27% showed significant evidence of increasing abundance in GZ populations, while 16% of them experienced decreases in abundance. In the global analysis, a weak but significant increment was observed over the past 40 years, which could be correlated to global changes. However, a stronger recurrent pattern was decadal oscillations over the entire time-series, explaining 15% of the linear data variance. For now, environmental drivers of these oscillations and large-scale spatial and temporal ecological baselines concerning GZ abundance and distribution patterns are still poorly known. This hampers our ability to properly understand GZ responses to climate patterns, regime shifts or even annual variations, as well as know whether, if, and how GZ will take advantage of changing future oceans and proliferate.

Summary

1. Gelatinous zooplankton (GZ) is a non-taxonomic term for jelly-like organisms that have a high proportion of water in body tissues.
2. Advantages in being gelatinous include better buoyancy, transparency, larger sizes with lower carbon investment, rapid growth and body plasticity.
3. Most GZ taxa have different pathways of asexual reproduction that may quickly boost their populations.
4. Thus, many GZ are opportunistic-colonizing species that rapidly increase their populations in response to favorable conditions, often resulting in extensive agglomerations (i.e., blooms).
5. The bloom dynamics depends on multiple environmental drivers, such as increased temperature, food availability, predation, eutrophication, etc., and can be enhanced by physical transport.
6. GZ blooms have implications for vertical flux of particulate matter *via* production of fecal pellets and carcasses.
7. Feeding mechanisms and trophic niches of GZ are diverse, ranging from filter-feeding very small particles to voracious predators. In any case, GZ are known for their high feeding rates that can impact prey populations.
8. GZ have been historically considered trophic dead ends; however, a variety of animals feed occasionally, regularly, or exclusively on these organisms, including other GZ, mollusks, crustaceans, fish, turtles, birds and humans.
9. The large bodies of GZ species offer microhabitats for numerous organisms ranging from virus and bacteria to crustaceans and fishes. Possible advantages to symbionts include protection from predators, transport, and food supply.
10. Human activities have modified marine ecosystems. While some GZ species may benefit from these modifications, shelled mollusks may face physiological constraints that may preclude their existence.
11. It has been hypothesized that GZ blooms are increasing globally, in response to anthropogenic stressors such as climate change, eutrophication, hypoxia, and overfishing.
12. Although GZ blooms have been recurrently reported and some species may thrive in conditions that are detrimental to other metazoans, evidence to support a global GZ increase is relatively weak when compared to evidence supporting the existence of other changes in marine environments (e.g., overfishing).
13. Most data on GZ are spatially, temporally, and taxonomically limited, cover few populations, and cannot account for the role of multiple natural drivers (e.g., ocean-atmosphere oscillations).

References

Acuña, J.L., Á. López-Urrutia and S. Colin. 2011. Faking giants: the evolution of high prey clearance rates in jellyfishes. Science 333: 1627–1629.

Addad, S., J.-Y. Exposito, C. Faye, S. Ricard-Blum and C. Lethias. 2011. Isolation, characterization and biological evaluation of jellyfish collagen for use in biomedical applications. Mar. Drugs 9: 967–983.

Alamaru, A., O. Bronstein, Y. Loya and G. Dishon. 2009. Opportunistic feeding by the fungiid coral *Fungia scruposa* on the moon jellyfish *Aurelia aurita*. Coral Reefs 28: 865.

Albert, D.J. 2009. *Aurelia labiata* (Scyphozoa) jellyfish in Roscoe Bay: their spatial distribution varies with population size and their behaviour changes with water depth. J. Sea Res. 61: 140–143.

Alldredge, A.L. 1981. The impact of appendicularian grazing on natural food concentrations *in situ*. Limnol. Oceanogr. 26: 247–257.

Alldredge, A.L. and L.P. Madin. 1982. Pelagic tunicates: unique herbivores in the marine plankton. BioScience 32: 655–663.

Ankel, W.M. 1930. Die Atypische Spermatogenese von *Janthina* (Prosobranchia, Ptenoglossa). Cell Tissue Res. 11: 491–608.

Aoki, M.N., C. Matsumoto-Ohshima, E. Hirose and J. Nishikawa. 2013. Mother–young cohabitation in *Phronimella elongata* and *Phronima* spp. (Amphipoda, Hyperiidea, Phronimidae). J. Mar. Biol. Assoc. U.K. 93(6): 1553–1556.

Apablaza, P. and S. Palma. 2006. Efecto de la zona de mínimo oxígeno sobre la migración vertical de zooplancton gelatinoso en la bahía de Mejillones. Investig. Mar. 34: 81–95.

Arai, M.N. 1988. Interactions of fish and pelagic coelenterates. Can. J. Zool. 66: 1913–1927.

Arai, M.N. 1992. Active and passive factors affecting aggregations of hydromedusae: a review. Sci. Mar. 56: 99–108.

Arai, M.N. 1997. A Functional Biology of Scyphozoa. Chapman and Hall, London. UK.

Arai, M.N. 2005. Predation on pelagic coelenterates: a review. J. Mar. Biol. Assoc. U.K. 85: 523–536.

Arai, M.N. 2009. The potential importance of podocysts to the formation of scyphozoan blooms: a review. Hydrobiologia 616: 241–246.

Ates, R.M.L. 2017. Benthic scavengers and predators of jellyfish, material for a review. Plankton Benthos Res. 12(1): 71–77.

Avian, M. 1986. Temperature influence on *in vitro* reproduction and development of *Pelagia noctiluca* (Forskål). Boll. Zool. 53: 385–391.

Ayala, D.J., P. Munk, R.B.C. Lundgreen, S.J. Traving et al. 2018. Gelatinous plankton is important in the diet of European eel (*Anguilla anguilla*) larvae in the Sargasso Sea. Sci. Rep. 8: 61–56.

Båmstedt, U., J. Lane and M.B. Martinussen. 1999. Bioenergetics of ephyra larvae of the scyphozoan jellyfish *Aurelia aurita* in relation to temperature and salinity. Mar. Biol. 135: 89–98.

Bandara, K., Ø. Varpe, J. Søreide, J. Wallenschus, J. Berge and K. Eiane. 2016. Seasonal vertical strategies in a high-Arctic coastal zooplankton community. Mar. Ecol. Prog. Ser. 555: 49–64.

Batistić, M., F. Kršinić, N. Jasprica, M. Carić, D. Viličić and D. Lučić. 2004. Gelatinous invertebrate zooplankton of the South Adriatic Species composition and vertical distribution. J. Plankton Res. 26: 459–474.

Bastić, M., N. Jasprica, M. Carić and D. Lučić. 2007. Annual cycle of the gelatinous invertebrate zooplankton of the eastern South Adriatic coast (NE Mediterranean). J. Plankton Res. 29: 671–686.

Bednaršek, N. and M. Ohman. 2015. Changes in pteropod distributions and shell dissolution across a frontal system in the California current system. Mar. Ecol. Prog. Ser. 523: 93–103.

Bednaršek, N., C.J. Harvey, I.C. Kaplan, R.A. Feely and J. Možina. 2016. Pteropods on the edge: Cumulative effects of ocean acidification, warming, and deoxygenation. Prog. Oceanogr. 145: 1–24.

Berasategui, A., M.C. Menendez and D. Calliari. 2021. Mesozooplankton: functional groups and ecological topics. *In*: Pan, J. and P.D. Pratolongo [eds.]. Marine Biology: A Functional Approach to the Oceans and their Organisms. Science Publishers, CRC Press-Taylor and Francis Group.

Berline, L., L. Stemmann, M. Vichi, F. Lombard and G. Gorsky. 2011. Impact of appendicularians on detritus and export fluxes: a model approach at DyFAMed site. J. Plankton Res. 33: 855–872.

Beu, A.G. 2017. Evolution of *Janthina* and *Recluzia* (Mollusca: Gastropoda: Epitoniidae). Rec. Aust. Museum. 69: 119–222.

Bidigare, R.R. and D.C. Biggs. 1980. The role of sulfate exclusion in buoyancy maintenance by siphonophores and other oceanic gelatinous plankton. Comp. Biochem. Physiol. B 66A: 467–471.

Boero, F., J. Bouillon, S. Piraino and V. Schmid. 1997. Diversity of hydrozoan life cycles: ecological implications and evolutionary patterns. pp. 53–62. *In*: den Hartog, J.C. [ed.]. Proceedings of the 6th International Conference on Coelenterate Biology. Leiden, The Netherlands.

Boero, F., C. Bucci, A. Colucci, C. Gravili and L. Stabili. 2007. *Obelia* (Cnidaria, Hydrozoa, Campanulariidae): a microphagous, filter-feeding medusa. Mar. Ecol. 28: 178–183.

Boero, F., J. Bouillon, C. Gravili, M.P. Miglietta, T. Parsons and S. Piraino. 2008. Gelatinous plankton: irregularities rule the world (sometimes). Mar. Ecol. Prog. Ser. 356: 299–310.

Boero, F. 2013. Review of jellyfish blooms in the Mediterranean and Black Sea. Studies and Reviews. General Fisheries Commission for the Mediterranean, 92 FAO, Rome.

Bouillon, J., C. Gravili, F. Pagès, J.-M. Gili and F. Boero. 2006. An introduction to Hydrozoa. Memoirs du Muséum national d'Histoire naturelle, Tome 194, Paris.

Bravo, V., S. Palma and N. Silva. 2011. Seasonal and vertical distribution of medusae in Aysen region, southern Chile. Lat. Am. J. Aquat. Res. 39: 359–377.

Breitbart, M., B.E. Benner, P.E. Jernigan et al. 2015. Discovery, prevalence, and persistence of novel circular single-stranded DNA viruses in the ctenophores *Mnemiopsis leidyi* and *Beroe ovata*. Front. Microbiol. 6: 1427.

Brodeur, R.D., M.B. Decker, L. Ciannelli, J. Purcell et al. 2008. The rise and fall of jellyfish in the Bering Sea in relation to climate regime shifts. Prog. Oceanogr. 77: 103–111.

Brodeur, R., I. Perry, J. Boldt, L. Flostrand, M. Galbraith, J. King, J. Murphy, K. Sakuma and A. Thompson. 2018. An unusual gelatinous plankton event in the NE Pacific: The Great Pyrosome Bloom of 2017. PICES Press 26: 7.

Brandon, M. and C.E. Cutress. 1985. A new *Dondice* (Opisthobranchia: Favorinidae), predator of *Cassiopea* in southwest Puerto Rico. Bull. Mar. Sci. 36(1): 139–144.
Brotz, L., W.W.L. Cheung, K. Kleisner, E. Pakhomov and D. Pauly. 2012. Increasing jellyfish populations: trends in large marine ecosystems. Hydrobiologia 690: 3–20.
Buecher, E., J. Goy, B. Planque, M. Etienne and S. Dallot. 1997. Long-term fluctuations of *Liriope tetraphylla* in Villefranche Bay between 1966 and 1993 compared to *Pelagia noctiluca* populations. Oceanol. Acta 20: 145–157.
Bumann, D. and G. Puls. 1996. Infestation with larvae of the sea anemone *Edwardsia lineata* affects nutrition and growth of the ctenophore *Mnemiopsis leidyi*. Parasitology 113: 123–128.
Burridge, A.K., E. Goetze, D. Wall-Palmer, S.L. Le, J. Huisman and K.T.C.A. Peijnenburg. 2017. Diversity and abundance of pteropods and heteropods along a latitudinal gradient across the Atlantic Ocean. Prog. Oceanogr. 158: 213–223.
Buskey, E.J. 2003. Behavioral adaptations of the cubozoan medusa *Tripedalia cystophora* for feeding on copepod (*Dioithona oculata*) swarms. Mar. Biol. 142: 225–232.
Calabrò, C., A. Rindone, C. Bertuccio and S. Giacobbe. 2019. Hermaphroditism in a violet snail, *Janthina pallida* (Gastropoda, Caenogastropoda): a contribution. Biologia (Bratisl) 74: 509–513.
Cardona, L., I. Álvarez de Quevedo, A. Borrell and A. Aguilar. 2012. Massive consumption of gelatinous plankton by mediterranean apex predators. PLoS ONE 7: e31329.
Carré, D. 1969. Etude histologique du développement de *Nanomia bijuga* (Chiaje, 1841), Siphonophore Physonecte, Agalmidae. Cah. Biol. Mar. 8: 325–341.
Carré, D. and C. Carré. 1990. Complex reproductive cycle in *Eucheilota paradoxica* (Hydrozoa: Leptomedusae): medusae, polyps and frustules produced from medusa stage. Mar. Biol. 104: 303–310.
Casanova, J.P. 1999. Chaetognatha. pp. 1353–1374. *In*: Boltovskoy, D. [ed.]. South Atlantic Zooplankton. Leiden, Backhuys Publishers.
Casanova, J.P., R.M. Barthélémy, M. Duvert and E. Faure. 2012. Chaetognaths feed primarily on dissolved and fine particulate organic matter, not on prey: implications for marine food webs. Hypotheses in the Life Sciences 2: 20–29.
Castro, J.J., J.A. Santiago and A.T. Santana-Ortega. 2002. A general theory on fish aggregation to floating objects: An alternative to the meeting point hypothesis. Rev. Fish Biol. Fish. 11: 255–277.
Ceh, J., J. Gonzalez, A.S. Pacheco and J.M. Riascos. 2015. The elusive life cycle of scyphozoan jellyfish—metagenesis revisited. Sci. Rep. 5: 12037.
Chae, J., H.W. Choi, W.J. Lee, D. Kim and J.H. Lee. 2008. Distribution of a pelagic tunicate, *Salpa fusiformis* in warm surface current of the eastern Korean waters and its impingement on cooling water intakes of Uljin nuclear power plant. J. Environ. Biol. 29: 585–590.
Chiaverano, L., W.M. Graham and J. Costello. 2015. Parasites alter behavior, reproductive output, and growth patterns of *Aurelia* medusae in a marine lake. Mar. Ecol. Prog. Ser. 540: 87–98.
Chiaverano, L.M. and W.M. Graham. 2017. Morphological plasticity in *Aurelia* polyps, with subsequent effects on asexual fecundity and morphology of young medusae. Mar. Ecol. Prog. Ser. 582: 79–92.
Chuard, P.J.C., M.D. Johnson and F. Guichard. 2019. Ocean acidifcation causes mortality in the medusa stage of the cubozoan *Carybdea xaymacana*. Sci. Rep. 9: 5622.
Collins, A.G. 2002. Phylogeny of Medusozoa and the evolution of cnidarian life cycles. J. Evol. Biol. 15: 418–432.
Colin, S.P. and J.H. Costello. 2002. Morphology, swimming performance and propulsive mode of six co-occurring hydromedusae. J. Exp. Biol. 205: 427–37.
Colin, S.P., R. Williams, J.H. Costello, W. Graham and J. Higgins. 2005. Omnivory by the small cosmopolitan hydromedusa *Aglaura hemistoma*. Limnol. Oceanogr. 50: 1264–1268.
Colin, S.P., J. Costello, K. Katija, J. Seymour and K. Kiefer. 2013. Propulsion in Cubomedusae: mechanisms and utility. PLoS ONE 8(2): e56393.
Comeau, S., G. Gorsky, S. Alliouane and J.P. Gattuso. 2010a. Larvae of the pteropod *Cavolinia inflexa* exposed to aragonite undersaturation are viable but shell-less. Mar. Biol. 157: 2341–2345.
Comeau, S., R. Jeffree, J.L. Teyssie and J.P. Gattuso. 2010b. Response of the arctic pteropod *Limacina helicina* to projected future environmental conditions. PLoS One 5: 1–7.
Condon, R.H. and D.K. Steinberg. 2008. Development, biological regulation, and fate of ctenophore blooms in the York River estuary, Chesapeake Bay. Mar. Ecol. Prog. Ser. 369: 153–168.
Condon, R.H., W. Graham, C. Duarte et al. 2012. Questioning the rise of gelatinous zooplankton in the world's oceans. BioScience 62: 160–189.
Condon, R.H., C.M. Duarte, K.A. Pitt et al. 2013. Recurrent jellyfish blooms are a consequence of global oscillations. Proc. Natl. Acad. Sci. U.S.A. 110: 1000–1005.
Conley, K.R. and K.R. Sutherland. 2017. Particle shape impacts export and fate in the ocean through interactions with the globally abundant appendicularian *Oikopleura dioica*. PLoS ONE 12: e0183105.
Costello, J., K. Bayha, H.W. Mianzan, T. Shiganova and J. Purcell. 2012. Transitions of *Mnemiopsis leidyi* (Ctenophora: Lobata) from a native to an exotic species: a review. Hydrobiologia 690: 21–46.
Costello, J.H. and S.P. Colin. 1995. Flow and feeding by swimming scyphomedusae. Mar. Biol. 124: 399–406.

Costello, J.H. and S.P. Colin. 2002. Prey resource use by coexistent hydromedusae from Friday Harbor, Washington. Limnol. Oceanogr. 47: 934–942.
Costello, J.H., B.K. Sullivan, D.J. Gifford, D. Van Keuren and L.J. Sullivan. 2006. Seasonal refugia, shoreward thermal amplification and metapopulation dynamics of the ctenophore *Mnemiopsis leidyi* in Narragansett Bay, Rhode Island. Limnol. Oceanogr. 51: 1819–1831.
Courtney, R., N. Sachlikidis, R. Jones and J. Seymour. 2015. Prey capture ecology of the cubozoan *Carukia barnesi*. PLoS ONE 10: e0124256.
Daley, M.C., J. Urban-Rich and P.H. Moisander. 2016. Bacterial associations with the hydromedusa *Nemopsis bachei* and scyphomedusa *Aurelia aurita* from the North Atlantic Ocean, Mar. Biol. Res. 12: 1088–1100.
Daly, N.L., J. Seymour and D. Wilson. 2014. Exploring the therapeutic potential of jellyfsh venom. Future Med. Chem. 6: 1715–1724.
Daskalov, G. 2002. Overfishing drives a trophic cascade in the Black Sea. Mar. Ecol. Prog. Ser. 225: 53–63.
Daskalov, G.M., A.N. Grishin, S. Rodionov and V. Mihneva. 2007. Trophic cascades triggered by overfishing reveal possible mechanisms of ecosystem regime shifts. Proc. Natl. Acad. Sci. U.S.A. 104: 10518–10523.
Davenport, J. 1994. Observations on the locomotion and buoyancy of *Phronima sedentaria* (Forskål, 1775) (Crustacea: Amphipoda: Hyperiidea). J. Nat. Hist. 28: 787–793.
Dawson, M.N. and L.E. Martin. 2001. Geographic variation and ecological adaptation in *Aurelia* (Scyphozoa, Semaeostomeae): some implications from molecular phylogenetics. Hydrobiologia 451: 259–273.
De Vito, D., S. Piraino, J. Schmich, J. Bouillon and F. boero. 2006. Evidence of reverse development in Leptomedusae (Cnidaria, Hydrozoa): the case of *Laodicea undulata* (Forbes and Goodsir 1851). Mar. Biol. 149: 339.
Deibel, D. 1985. Blooms of the pelagic tunicate, *Doliolella gegenbauri*: are they associated with Gulf Stream frontal eddies? J. mar. Res. 43(1): 211–236.
Deibel, D. and B. Lowen. 2012. A review of the life cycles and life-history adaptations of pelagic tunicates to environmental conditions. ICES J. Mar. Sci. 69(3): 358–369.
Delannoy, C.M.J., J. Houghton, N. Fleming and H.W. Ferguson. 2011. Mauve stingers (*Pelagia noctiluca*) as carriers of the bacterial fish pathogen *Tenacibaculum maritimum*. Aquaculture 311: 255–257.
Diaz, R.J. and R. Rosenberg. 2008. Spreading dead zones and consequences for marine ecosystems. Science 321: 926–929.
Diaz Briz, L.M., F. Sánchez, N. Marí, H. Mianzan and G. Genzano. 2017. Gelatinous zooplankton (ctenophores, salps and medusae): an important food resource of fishes in the temperate SW Atlantic Ocean. Mar. Biol. Res. 13: 630–644.
Diaz Briz, L.M., F. Sánchez, N. Marí and G. Genzano. 2018. Seasonal variation of gelatinous plankton consumption by fish in the South-western Atlantic Ocean: a question of strategy. Mar. Biol. Res. 14: 739–751.
Diebel, C.E. 1988. Observations on the anatomy and behavior of *Phronima sedentaria* (Forskal) (Amphipoda: Hyperiidea). J. Crust. Biol. 8: 79–90.
Djeghri, N., P. Pondaven, H. Stibor and M.N. Dawson. 2019. Review of the diversity, traits, and ecology of zooxanthellate jellyfishes. Marine Biology 166: 147.
Dinasquet, J., L. Granhag and L. Riemann. 2012. Stimulated bacterioplankton growth and selection for certain bacterial taxa in the vicinity of the ctenophore *Mnemiopsis leidyi*. Front. Microbiol. 3: 302.
Dittrich, B. 1988. Studies on the life cycle and reproduction of the parasitic amphipod *Hyperia galba* in the North Sea. Helgol. Meeresunters. 42: 79–98.
Dittrich, B. 1992. Functional morphology of the mouthparts and feeding strategies of the parasitic amphipod *Hyperia galba* (Montagu, 1813). Sarsia 77: 11–18.
Doney, S.C., V.J. Fabry, R.A. Feely and J.A. Kleypas. 2009. Ocean acidification: the other CO_2 problem. Annu. Rev. Mar. Sci. 1: 212–250.
Doyle, T.K., J. Houghton, R. McDevitt, J. Davenport and G.C. Hays. 2007. The energy density of jellyfish: estimates from bomb-calorimetry and proximate-composition. J. Exp. Mar. Biol. Ecol. 343: 239–252.
Doyle, T.K., G. Hays, C. Harrod and J.D.R. Houghton. 2013. Ecological and Societal Benefits of Jellyfish. pp. 105–127. *In*: Pitt, K.A. and C. Lucas [eds.]. Jellyfish Blooms, Springer Science+Business Media Dordrecht.
Duarte, C.M., R.W. Fulweiler, C.E. Lovelock, P. Martinetto, M.I. Saunders, J.M. Pandolfi, S. Gelcich and S.W. Nixon. 2015. Reconsidering ocean calamities. BioScience 65: 130–139.
Esnal, G.B. 1999a. Pyrosomatida. pp. 1401–1408. *In*: Boltovskoy, D. [ed.]. South Atlantic Zooplankton. Leiden, Backhuys Publishers.
Esnal, G.B. and M.C. Daponte. 1999a. Doliolida. pp. 1409–1421. *In*: Boltovskoy, D. [ed.]. South Atlantic Zooplankton. Leiden, Backhuys Publishers.
Esnal, G.B. 1999b. Appendicularian. pp. 1375–1399. *In*: Boltovskoy, D. [ed.]. South Atlantic Zooplankton. Leiden, Backhuys Publishers.
Esnal, G.B. and M.C. Daponte. 1999b. Salpida. pp. 1423–1444. *In*: Boltovskoy, D. [ed.]. South Atlantic Zooplankton. Leiden, Backhuys Publishers.
Fabry, V.J., B.A. Seibel, R.A. Feely and J.C. Orr. 2008. Impacts of ocean acidification on marine fauna and ecosystem processes. ICES J. Mar. Sci. 65: 414–432.
FAO. 2018. The State of World Fisheries and Aquaculture 2018—Meeting the Sustainable Development Goals. FAO, Rome.

Fenaux, R. 1976. Cycle vital, croissance et production chez *Fritillaria pellucida* (Appendicularia), dans la baie de Villefranche-sur-Mer, France. Mar. Biol. 34: 229–238.

Ferguson, H.W., C. Delannoy, S. Hay, J. Nicolson, D. Sutherland and M. Crumlish. 2010. Jellyfish as vectors of bacterial disease for farmed salmon (*Salmo salar*). J. Vet. Diagn. Invest. 22: 376–382.

Ferguson, J.C. 1988. Autoradiographic demonstration of the use of free amino acid by Sargasso Sea zooplankton. J. Plankton Res. 10: 1225–1238.

Fleming, N.E.C., C. Harrod, D. Griffin, J. Newton and J. Houghton. 2014. Scyphozoan jellyfish provide short-term reproductive habitat for hyperiid amphipods in a temperate near-shore environment. Mar. Ecol. Prog. Ser. 510: 229–240.

Flux, J.E.C. 2008. First mass stranding of *Velella velella* in New Zealand. Mar. Biodivers. Rec. 1: 1–2.

Fortier, L., J. Le Fèvre and L. Legendre. 1994. Export of biogenic carbon to fish and to the deep ocean: the role of large planktonic microphages. J. Plankton Res. 16: 809–839.

Fossette, S., A. Gleiss, J. Chalumeau, T. Bastian, C.D. Armstrong, S. Vandenabeele, M. Karpytchev and H. Hays. 2015. Current-oriented swimming by jellyfish and its role in bloom maintenance. Curr. Biol. 25: 342–347.

Fraser, J.H. 1969. Experimental feeding of some medusae and chaetognatha. J. Fish. Res. Board Can. 26: 1743–1762.

Frost, J., C.A. Jacoby, T.K. Frazer and A.R. Zimmerman. 2012. Pulse perturbations from bacterial decomposition of *Chrysaora quinquecirrha* (Scyphozoa: Pelagiidae). Hydrobiologia 690: 247–256.

Gaevskaya, A.V., A. Temnykh, N.V. Pronkina and J. Korniychuk. 2012. The Black Sea ctenophores and chaetognaths in life cycle of nematode *Hysterothylacium aduncum* (Nematoda: Anisakidae). Mar. Ecol. J. 1: 25–28 (in Russian with English abstract).

Garm, A., M. Oskarsson and D.-E. Nilsson. 2011. Box jellyfish use terrestrial visual cues for navigation. Curr. Biol. 21: 798–803.

Genzano, G., H. Mianzan, L. Briz and C. Rodriguez. 2008. On the occurrence of *Obelia* medusa blooms and empirical evidence of unusual massive accumulations of *Obelia* and *Amphisbetia* hydroids on the Argentina shoreline. Lat. Amer. J. Aquat. Res. 36(2): 301–307.

Gershwin, L., W. Zeidler and P.J.F. Davie. 2010. Ctenophora of Australia. Memoirs of the Queensland Museum 54(3): 1–45.

Gilmer, R.W. 1972. Free floating mucus webs: a novel feeding adaptation for the open ocean. Science 176: 1239–1240.

Gilmer, R.W. and G.R. Harbison. 1986. Morphology and field behaviour of pteropod molluscs: feeding methods in the families Cavoliniidae, Limacinidae and Peraclididae (Gastropoda: Thecosomata). Mar. Biol. 91: 47–57.

Goldstein, J. and H.U. Riisgård. 2016. Population dynamics and factors controlling somatic degrowth of the common jellyfish, *Aurelia aurita*, in a temperate semi-enclosed cove (Kertinge Nor, Denmark). Mar. Biol. 163: 1–12.

González Carman, V., F. Botto, E. Gaitan, D. Albareda, C. Campagna and H. Mianzan. 2014. A jellyfish diet for the herbivorous green turtle *Chelonia mydas* in the temperate SW Atlantic. Mar. Biol. 161: 339–349.

Goodson, M.S., L.F. Whitehead and A.E. Douglas. 2001. Symbiotic dinoflagellates in marine Cnidaria: diversity and function. Hydrobiologia 461: 79–82.

Graham, W.M. 2001. Numerical increases and distributional shifts of *Chrysaora quinquecirrha* (Desor) and *Aurelia aurita* (Linné) (Cnidaria: Scyphozoa) in the northern Gulf of Mexico. Hydrobiologia 451: 97–111.

Graham, W.M., F. Pagès and W.M. Hamner. 2001. A physical context for gelatinous zooplankton aggregations: a review. Hydrobiologia 451: 199–212.

Grossart, H.P., C. Dziallas, F. Leunert and K. Tang. 2010. Bacteria dispersal by hitchhiking on zooplankton. Proc. Natl. Acad. Sci. U.S.A. 107: 11959–11964.

Guerrero, E., K. Kienberger, A. Villaescusa, J.-M. Gili, G. Navarro and L. Prieto. 2018. First record of beaching events for a calycophoran siphonophore: *Abylopsis tetragona* (Otto, 1823) at the Strait of Gibraltar. Mar. Biodiv. 49: 1587–1593.

Haddock, H.D. 2004. A golden age of gelata: past and future research on planktonic ctenophores and cnidarians. Hydrobiologia 530/531: 549–556.

Haddock, S.H. 2007. Comparative feeding behavior of planktonic ctenophores. Integr. Comp. Biol. 47: 847–53.

Hagadorn, J.W., R.H. Dott and D. Damrow. 2002. Stranded on a late Cambrian shoreline: medusae from central Wisconsin. Geology 30: 147.

Hamner, W.M., L.P. Madin, A.L. Alldredge, R.W. Gilmer and P. Hamner. 1975. Underwater observations of gelatinous zooplankton: sampling problems, feeding biology, and behavior. Limnol. Oceanogr. 20: 907–917.

Hao, W., G. Gerdts, J. Peplies and A. Wichels. 2015. Bacterial communities associated with four ctenophore genera from the German Bight (North Sea). FEMS Microbiol. Ecol. 91(1): 1–11.

Harashima, A., T. Kimoto, T. Wakabayashi and T. Toshiyasu. 2006. Verification of the silica deficiency hypothesis based on biogeochemical trends in the aquatic continuum of Lake Biwa-Yodo River-Seto Inland Sea, Japan. Ambio 35: 36–42.

Harbison, G.R. 1976. Development of *Lycaea pulex* Marion, 1874 and *Lycaea vincentii* Stebbing, 1888 (Amphipoda, Hyperiidea). Bull. Mar. Sci. 26(2): 152–164.

Harbison, G.R., D. Biggs and L.P. Madin. 1977. The associations of Amphipoda Hyperiidea with gelatinous zooplankton-II. Associations with Cnidaria, Ctenophora and Radiolaria. Deep-Sea Res. 24: 465–488.

Harris, R.P., P. Wiebe, J. Lenz, H. Skjoldal and M. Huntley. 2001. ICES Zooplankton Methodology Manual. Academic Press, London/San Diego.

Havermans, C., W. Hagen, W. Zeidler, C. Held and H. Auel. 2019. A survival pack for escaping predation in the open ocean: amphipod–pteropod associations in the Southern Ocean. Mar. Biodiv. 49: 1361–1370.

He, J., L. Zheng, W. Zhang and Y. Lin. 2015. Life cycle reversal in *Aurelia* sp.1 (Cnidaria, Scyphozoa). PLoS One 10: e0145314.

Heeger, T., U. Piatkowski and H. Möller. 1992. Predation of jellyfish by the cephalopod *Argonauta argo*. Mar. Ecol. Prog. Ser. 88: 293–296.

Henschke, N., J. Everett, M. Doblin, K. Pitt, A. Richardson and I.M. Suthers. 2014. Demography and interannual variability of salp swarms (*Thalia democratica*). Mar. Biol. 161: 149–163.

Henschke, N., J.D. Everett, A.J. Richardson and I.M. Suthers. 2016. Rethinking the role of salps in the ocean. Trends Ecol. Evol. 31: 720–733.

Herdman, W.A. 1888. Report on the tunicata collected by H.M.S. Challenger during the years 1873–1876. Dartmouth College, DE.

Heron, A.C. 1972. Population ecology of a colonizing species: the pelagic tunicate *Thalia democratica*. 2. Population growth rate. Oecologia 10: 294–312.

Hirose, E., M. Aoki and J. Nishikawa. 2005. Still alive? Fine structure of the barrels made by *Phronima* (Crustacea: Amphipoda). J. Mar. Biol. Assoc. U.K. 85: 1435–1439.

Hirst, A.G., J. Roff and R.S. Lampitt. 2003. A synthesis of growth rates in marine epipelagic invertebrate zooplankton. Adv. Mar. Biol. 44: 1–142.

Hirota, J. 1972. Quantitative natural history of *Pleurobrachia bachei* in La Jolla Bight. Fish. Bull. 72(2): 295–335.

Hoegh-Guldberg, O. and J.F. Bruno. 2010. The impact of climate change on the world's marine ecosystems. Science 328: 1523–1528.

Hoeksema, B.W. and Z. Waheed. 2012. It pays to have a big mouth: mushroom corals ingesting salps at northwest Borneo. Mar. Biodiv. 42: 297–302.

Hopcroft, R.R. and J. Roff. 1995. Zooplankton growth rates: extraordinary production by the larvacean *Oikopleura dioica* in tropical waters. J. Plankton Res. 17: 205–220.

Hopcroft, R.R. and J. Roff. 1998. Production of tropical larvaceans in Kingston Harbour, Jamaica: are we ignoring an important secondary producer? J. Plankton Res. 20: 557–569.

Hopcroft, R.R., J. Roff and H.A. Bouman. 1998. Zooplankton growth rates: the larvaceans *Appendicularia*, *Fritillaria* and *Oikopleura* in tropical waters. J. Plankton Res. 20: 539–555.

Houghton, J.D.R., T. Doyle, J. Davenport, M.K. Lilley, R. Wilson and G.C. Hays. 2007. Stranding events provide indirect insights into the seasonality and persistence of jellyfish medusae (Cnidaria: Scyphozoa). Hydrobiologia 589: 1–13.

Hosia, A. and U. Båmstedt. 2007. Seasonal changes in the gelatinous zooplankton community and hydromedusa abundances in Korsfjord and Fanafjord, western Norway. Mar. Ecol. Prog. Ser. 351: 113–127.

Hosia, A. and U. Båmstedt. 2008. Seasonal abundance and vertical distribution of siphonophores in western Norwegian fjords. J. Plankton Res. 30: 951–962.

Hosia, A., T. Falkenhaug, E.J. Baxter and F. Pagès. 2017. Abundance, distribution and diversity of gelatinous predators along the northern Mid-Atlantic Ridge: A comparison of different sampling methodologies. PloS ONE 12(11): e0187491.

Hsieh, Y.H.P., F. Leong and J. Rudloe. 2001. Jellyfish as food. Hydrobiologia 451: 11–17.

IPCC. 2014. Climate Change 2014: Synthesis Report. Contribution of Working Groups I, II and III to the Fifth Assessment Report of the Intergovernmental Panel on Climate Change. IPCC, Geneva, CH.

Ishak, N.H., L. Clementson, R. Eriksen, R. Enden, G.D. Williams and K. Swadling. 2017. Gut contents and isotopic profiles of *Salpa fusiformis* and *Thalia democratica*. Mar. Biol. 164: 144.

Ishii, H., T. Ohba and T. Kobayashi. 2008. Effects of low dissolved oxygen on planula settlement, polyp growth and asexual reproduction of *Aurelia aurita*. Plankton Benthos Res. 3: 107–113.

Ishii, H. and K. Katsukoshi. 2010. Seasonal and vertical distribution of *Aurelia aurita* polyps on a pylon in the innermost part of Tokyo Bay. J. Oceanogr. 66: 329–336.

Jackson, J.B.C. 1977. Competition on marine hard substrata: the adaptive significance of solitary and colonial strategies. Am. Nat. 111: 743–767.

Jarms, G., U. Båmstedt, H. Tiemann, M.B. Martinussen and J.H. Fosså. 1999. The holopelagic life cycle of the deep-sea medusa *Periphylla periphylla* (Scyphozoa, Coronatae). Sarsia 84: 55–65.

Jaspers, C., M. Haraldsson, B. Bolte, T.B.H. Reusch, U.H. Thygesen and T. Kiørboe. 2012. Ctenophore population recruits entirely through larval reproduction in the central Baltic Sea. Biol. Lett. 8: 809–812.

Jaspers, C., L.F. Møller and T. Kiørboe. 2015. Reproduction rates under variable food conditions and starvation in *Mnemiopsis leidyi*: significance for the invasion success of a ctenophore. J. Plankton Res. 37: 1011–1018.

Johnsen, S. 2001. Hidden in plain sight: the ecology and physiology of organismal transparency. Biol. Bull. 201: 301–318.

Kamiyama, T. 2013. Planktonic ciliates as food for the scyphozoan *Aurelia aurita* (s.l.): effects on asexual reproduction of the polyp stage. J. Exp. Mar. Biol. Ecol. 445: 21–28.

Kawahara, M., K. Ohtsu and S.-I. Uye. 2012. Bloom or non-bloom in the giant jellyfish *Nemopilema nomurai* (Scyphozoa: Rhizostomeae): roles of dormant podocysts. J. Plankton Res. 35: 213–217.

Kelley, C., T. Kerby, P.M. Sarradin, J. Sarrazin and D.J. Lindsay. 2016. Submersibles and remotely operated vehicles. pp. 285–305. *In*: Clark, M.R. [ed.]. Biological Sampling in the Deep Sea, First Edition. John Wiley and Sons.
Kingsford, M.J. 1993. Biotic and abiotic structure in the pelagic environment: importance to small fishes. Bull. Mar. Sci. 53(2): 393–415.
Kinne, O. 1980. Diseases of Marine Animals. Vol. I: General Aspects, Protozoa to Gastropoda. John Wiley and Sons.
Kinne, O. 1990. Diseases of Marine Animals. Vol. III: Introduction, Cephalopoda, Annelida, Crustacea, Chaetognatha, Echinodermata, Urochordata. John Wiley and Sons.
Køie, M. 1993. Aspects of the life cycle and morphology of *Hysterothylacium aduncum* (Rudolphi, 1802) (Nematoda, Ascaridoidea, Anisakidae). Can. J. Zool. 71: 1289–1296.
Kos Kramar, M., T. Tinta, D. Lučić, A. Malej and V. Turk. 2019. Bacteria associated with moon jellyfish during bloom and post-bloom periods in the Gulf of Trieste (northern Adriatic). PLoS ONE 14(1): e0198056.
Kremer, P. 1976. Population dynamics and ecological energetics of a pulsed zooplankton predator, the ctenophore *Mnemiopsis leidyi*. pp. 197–215. *In*: Wiley, M.L. [ed.]. Estuarine Processes. Academic Press, New York, USA.
Kremer, P., J. Costello, J. Kremer and M. Canino. 1990. Significance of photosynthetic endosymbionts to the carbon budget of the scyphomedusa *Linuche unguiculata*. Limnol. Oceanogr. 35: 609–624.
Kremer, P. and L.P. Madin. 1992. Particle retention efficiency of salps. J. Plankton Res. 14: 1009–1015.
Kremer, P. 2002. Towards understanding of salp swarm dynamics. CM 12: 1–15.
Kremer, P. 2005. Ingestion and elemental budgets for *Linuche unguiculata*, a scyphomedusa with zooxanthellae. J. Mar. Biolog. Assoc. U.K. 85: 613–625.
Kubota, S. 2011. Repeating rejuvenation in *Turritopsis*, an immortal hydrozoan (Cnidaria, Hydrozoa). Biogeography 13: 101–103.
Lalli, C.M. and R.W. Gilmer. 1989. Pelagic Snails: The Biology of Holoplanktonic Gastropod Mollusks. Stanford University Press, Stanford, USA.
Larson, R.J. 1986. Water content, organic content, and carbon and nitrogen composition of medusae from the northeast Pacific. J. Exp. Mar. Biol. Ecol. 99: 107–120.
Laursen, D. 1953. The genus *Janthina*. A monograph. Dana Reports 38: 1–40.
Laval, P. 1978. The barrel of the pelagic amphipod *Phronima sedentaria* (Forsk.) (Crustacea: Hyperiidea). J. Exp. Mar. Biol. Ecol. 33: 187–211.
Laval, P. 1980. Hyperiid amphipods as crustacean parasitoids associated with gelatinous zooplankton. Oceanogr. Mar. Biol. Annu. Rev. 18: 11–56.
Lebrato, M. and D.O.B. Jones. 2009. Mass deposition event of *Pyrosoma atlanticum* carcasses off Ivory Coast (West Africa). Limnol. Oceanogr. 54(4): 1197–1209.
Lebrato, M., K. Pitt, A. Sweetman et al. 2012. Jelly-falls historic and recent observations: a review to drive future research directions. Hydrobiologia 690: 227–245.
Lebrato, M., P. Mendes, D. Steinberg, J. Cartes et al. 2013. Jelly biomass sinking speed reveals a fast carbon export mechanism. Limnol. Oceanogr. 58: 1113–1122.
Licandro, P., S. Souissi, F. Ibanez and C. Carre. 2012. Long-term variability and environmental preferences of calycophoran siphonophores in the Bay of Villefranche (northwestern Mediterranean). Prog. Oceanogr. 97: 152–163.
Lilley, M.K., A. Elineau, M. Ferraris et al. 2014. Individual shrinking to enhance population survival: quantifying the reproductive and metabolic expenditures of a starving jellyfish, *Pelagia noctiluca*. J. Plankton Res. 36: 1585–1597.
Lindsay, D.J., F. Pagès, J. Corbera et al. 2008. The anthomedusan fauna of the Japan Trench: preliminary results from *in situ* surveys with manned and unmanned vehicles. J. Mar. Biol. Assoc. U.K. 88(8): 1519–1539.
Lindsay, D.J., H. Yoshida, K. Uemura et al. 2012. The untethered remotely-operated vehicle PICASSO-1 and its deployment from chartered dive vessels for deep sea surveys off Okinawa, Japan, and Osprey Reef, Coral Sea, Australia. Mar. Tech. Soc. J. 46(4): 20–32.
Lindsay, D.J., M. Grossmann, B. Bentlage et al. 2017. The perils of online biogeographic databases: a case study with the ‘monospecific’ genus *Aegina* (Cnidaria, Hydrozoa, Narcomedusae). Mar. Biol. Res. 13(5): 494–512.
Lischka, S., J. Büdenbender, T. Boxhammer and U. Riebesell. 2011. Impact of ocean acidification and elevated temperatures on early juveniles of the polar shelled pteropod *Limacina helicina:* Mortality, shell degradation, and shell growth. Biogeosciences 8: 919–932.
Lo, W.-T., J.E. Purcell, J.-J. Hung, H.-M. Su and P.-K. Hsu. 2008. Enhancement of Jellyfish (*Aurelia aurita*) populations by extensive aquaculture rafts in a coastal lagoon in Taiwan. ICES J. Mar. Sci. 65: 453–461.
Lotan, A., R. Ben-Hillel and Y. Loya. 1992. Life cycle of *Rhopilema nomadica*: a new immigrant scyphomedusan in the Mediterranean. Mar. Biol. 112: 237–242.
Lombard, F., F. Renaud, C. Sainsbury, A. Sciandra and G. Gorsky. 2008. Appendicularian ecophysiology. 1. Food concentration dependent clearance rate, assimilation efficiency, growth and reproduction of *Oikopleura dioica*. J. Mar. Systems 78: 606–616.
Lombard, F., D. Eloire, A. Gobet, L. Stemmann, J. Dolan and G. Gorsky. 2010. Experimental and modeling evidence of appendicularian-ciliate interactions. Limnol. Oceanogr. 55: 77–90.

Lombard, F., E. Selander and T. Kiørboe. 2011. Active prey rejection in the filter-feeding appendicularian *Oikopleura dioica*. Limnol. Oceanogr. 56: 1505–1512.

Lucas, C.H. 2001. Reproduction and life history strategies of the common jellyfish, *Aurelia aurita*, in relation to its ambient environment. Hydrobiologia 451: 229–246.

Lucas, C.H., W.M. Graham and C. Widmer. 2012. Jellyfish life histories: the role of polyps in forming and maintaining scyphomedusa populations. Adv. Mar. Biol. 63: 33–196.

Lynam, C.P., J. Gibbons, B.E. Axelsen, C.A.J. Sparks, B.G. Heywood and A.S. Brierley. 2006. Jellyfish overtake fish in a heavily fished ecosystem. Curr. Biol. 16: R492.

Ma, X. and J.E. Purcell. 2005. Temperature, salinity, and prey effects on polyp versus medusa bud production by the invasive hydrozoan *Moerisia lyonsi*. Mar. Biol. 147: 225–234.

Mackie, G.O., P.R. Pugh and J. Purcell. 1987. Siphonophore biology. Adv. Mar. Biol. 24: 97–262.

Madin, L.P. and G.R. Harbison. 1977. The association of Amphipoda Hyperiidea with gelatinous zooplankton-I. Associations with Salpidae. Deep-Sea Res. 24: 449–463.

Madin, L.P. and P. Kremer. 1995. Determination of the filter-feeding rates of salps (Tunicata, Thaliacea). ICES J. Mar. Sci. 52: 583–595.

Malej, A., T. Kogovšek and S. Uye. 2014. Medusae—pest or resource for humans? Biol. Mar. Mediterr. 21(1): 1–29.

Makabe, R., R. Furukawa, M. Takao and S. Uye. 2014. Marine artificial structures as amplifiers of Aurelia aurita s.l. blooms: a case study of a newly installed floating pier. J. Oceanogr. 70: 447–455.

Manno, C., N. Morata and R. Primicerio. 2012. *Limacina retroversa*'s response to combined effects of ocean acidification and sea water freshening. Estuar. Coast. Shelf Sci. 113: 163–171.

Manno, C., N. Bednaršek, G.A. Tarling et al. 2017. Shelled pteropods in peril: assessing vulnerability in a high CO_2 ocean. Earth-Sci. Rev. 169: 132–145.

Mapstone, G.M. 2009. Siphonophora (Cnidaria: Hydrozoa) of Canadian Pacific Waters. NRC Research Press. Ottawa, Canada.

Mapstone, G.M. 2014. Global diversity and review of Siphonophorae (Cnidaria: Hydrozoa). PLoS ONE 9(2): e87737.

Mapstone, G.M. 2015. Systematics of Siphonophores. pp. 1–37. *In*: Gopalakrishnakone, P. and A. Malhotra (eds.). Evolution of Venomous Animals and Their Toxins, Springer Science+Business Media Dordrecht.

Marcogliese, D.J. 1995. The role of zooplankton in the transmission of helminths parasites to fish. Rev. Fish Biol. Fish. 5: 336–371.

Marliave, J.B. and C.E. Mills. 1993. Piggyback riding by pandalid shrimp larvae on hydromedusae. Can. J. Zool. 71: 257–263.

Martindale, M.Q. 1987. Larval reproduction in the ctenophore *Mnemiopsis mccradyi* (order Lobata). Mar. Biol. 94: 409–414.

Martindale, M.Q. and J.Q. Henry. 2015. Ctenophora. Evol. Dev. Biol. Inv. 1: 179–201.

Marshalonis, D. and J.L. Pinckney. 2008. Grazing and assimilation rate estimates of hydromedusae from a temperate tidal creek system. Hydrobiologia 606: 203–211.

Martinelli Filho, J.E., R.R. Colwell, A.F.S. Queiroz, I.N.G. Rivera and R. Lopes. 2020. *Vibrio cholerae* O139 attached to zooplankton: reservoir diversity and distribution over an Estuarine-coastal gradient. J. of Coastal Research 95(sp1): 92–96.

Masuda, R., Y. Yamashita and M. Matsuyama. 2008. Jack mackerel *Trachurus japonicas* juveniles use jellyfish for predator avoidance and as a prey collector. Fish. Sci. 74: 276–284.

Mayer, A.G. 1910. Medusae of the World, vol. 1–3. Carnegie Institution of Washington, Washington, EUA.

McClintock, J.B. and J. Janssen. 1990. Pteropod abduction as a chemical defence in a pelagic Antarctic amphipod. Nature 346: 424–426.

McCloskey, L., L. Muscatine and F. Wilkerson. 1994. Daily photosynthesis, respiration, and carbon budgets in a tropical marine jellyfish (*Mastigias* sp.). Mar. Biol. 119(1): 13–22.

McNamara, M.E., D.J. Lonsdale and R.M. Cerrato. 2010. Shifting abundance of the ctenophore *Mnemiopsis leidyi* and the implications for larval bivalve mortality Mar. Biol. 157: 401–412.

McNamara, M.E. and D.J. Lonsdale. 2014. Interannual differences in microplankton drive changes in the feeding and fecundity of *Mnemiopsis leidyi* in a Long Island estuary Estuar. Coast. Shelf Sci. 144: 1–7.

McNamara, M.E., D.J. Lonsdale and R.M. Cerrato. 2014. Role of eutrophication in structuring planktonic communities in the presence of the ctenophore *Mnemiopsis leidyi*. Mar. Ecol. Prog. Ser. 510: 151–165.

Mianzan, H.W., N. Marí, B. Prenski and F. Sanchez. 1996. Fish predation on neritic ctenophores from the Argentine continental shelf: a neglected food resource? Fish. Res. 27: 69–79.

Mianzan, H.W., M. Pájaro, L. Machinandiarena and F. Cremonte. 1997. Salps: possible vectors of toxic dinoflagellates? Fish. Res. 29: 193–197.

Mianzan, H.W. 1999. Ctenophora. pp. 561–573. *In*: Boltovskoy, D. [ed.]. South Atlantic Zooplankton. Leiden, Backhuys Publishers.

Mianzan, H.W. and P.F.S. Cornelius. 1999. Cubomedusae and Scyphomedusae. pp. 513–559. *In*: Boltovskoy, D. [ed.]. South Atlantic Zooplankton. Leiden, Backhuys Publishers.

Mianzan, H.W., M. Pájaro, G. Álvarez Colombo and A. Madirolas. 2001. Feeding on survival-food: gelatinous plankton as a source of food for anchovies. Hydrobiologia 451: 45–53.

Mianzan, H.W., J. Quiñones, S. Palma, A. Schiariti, M. Acha, K. Robinson and W.M. Graham. 2014. *Chrysaora plocamia*: a poorly understood jellyfish from South American waters. pp. 219–236. *In*: Pitt, K.A. and C.H. Lucas [eds.]. Jellyfish Blooms: Springer, Leiden, The Netherlands.

Mills, C.E. 1993. Natural mortality in the Pacific coast hydromedusae: grazing predation, wound healing and senescence. Bull. Mar. Sci. 53(1): 194–203.

Mills, C.E. 1995. Medusae, siphonophores, and ctenophores as planktivorous predators in changing global ecosystems. ICES J. Mar. Sci. 52: 575–581.

Mills, C.E. and J.T. Rees. 2000. New observations and corrections concerning the trio of invasive hydromedusae *Maeotias marginata* (=*M. inexpectata*), *Blackfordia virginica*, and *Moerisia* sp. in the San Francisco Estuary. Sci. Mar. 64: 151–155.

Mills, C.E. 2001. Jellyfish blooms: are populations increasing globally in response to changing ocean conditions? Hydrobiologia 55–68.

Miyake, H., K. Iwao and Y. Kakinuma. 1997. Life history and environment of *Aurelia aurita*. South Pacific Study 17: 273–285.

Morandini, A.C., A. Schiariti, S.N. Stampar, M.N. Maronna, I. Straehler-Pohl and A.C. Marques. 2016. Succession of generations is still the general paradigm for scyphozoan life cycles. Bull. Mar. Sci. 92: 343–351.

Morais, P., M. Parra, R. Marques et al. 2015. What are jellyfish really eating to support high ecophysiological condition? J. Plankton Res. 37: 1036–1041.

Morris, R.J., Q. Bone, R. Head, J. Braconnot and P. Nival. 1988. Role of salps in the flux of organic matter to the bottom of the Ligurian Sea. Mar. Biol. 97: 237–241.

Muscatine, L. and R.E. Marian. 1982. Dissolved inorganic nitrogen flux in symbiotic and nonsymbiotic medusae. Limnol. Oceanogr. 27(5): 910–917.

Nagata, R.M., M. Haddad and M. Nogueira Jr. 2009. The nuisance of jellyfish to shrimp trawls in central part of Southern Brazilian Bight, from the perspective of artisanal fishermen. Pan. J. Aquat. Sci. 4: 312–325.

Nagata, R.M., M. Nogueira Júnior, F.P. Brandini and M.A. Haddad. 2014. Spatial and temporal variation of planktonic cnidarian density in subtropical waters of the Southern Brazilian Bight. J. Mar. Biolog. Assoc. U.K. 94: 1387–1400.

Nakamura, Y., K. Suzuki, S.-Y. Suzuki and J. Hiromi. 1997. Production of *Oikopleura dioica* (Appendicularia) following a picoplankton 'bloom' in a eutrophic coastal area. J. Plankton Res. 19: 113–124.

Niggl, W., M.S. Naumann, U. Struck, R. Manasrah and C. Wild. 2010. Organic matter release by the benthic upside-down jellyfish *Cassiopea* sp. fuels pelagic food webs in coral reefs. J. Exp. Mar. Biol. Ecol. 384: 99–106.

Nishikawa, J., MN. Thu, T. Ha and P.T. Thu. 2008. Jellyfish fisheries in northern Vietnam. Plankton Benthos Res. 3(4): 227–234.

Nishikawa, J., S. Ohtsuka, M.N. Mujiono, D.J. Lindsay, H. Miyamoto and S. Nishida. 2015. A new species of the commercially harvested jellyfish *Crambionella* (Scyphozoa) from central Java, Indonesia with remarks on the fisheries. J. Mar. Biol. Assoc. U.K. 95(3): 471–481.

Nogueira Júnior, M., R.M. Nagata and M.A. Haddad. 2010. Seasonal variation of macromedusae (Cnidaria) at North Bay, Florianópolis, southern Brazil. Zoologia 27: 377–386.

Nogueira Júnior, M., F.P. Brandini and J.C.U. Codina. 2014. Distribution of planktonic cnidarians in response to South Atlantic Central Water intrusion in the South Brazilian Bight. Cont. Shelf Res. 89: 93–102.

Nogueira Júnior, M., L.M. Diaz Briz and M.A. Haddad. 2015. Monthly and inter-annual variations of *Opechona* sp. (Digenea: Lepocreadiidae) parasitizing scyphomedusae off southern Brazil. Mar. Biol. 162(2): 391–400.

Nogueira Júnior, M. and M.A. Haddad. 2017. Seasonal distribution, abundance and biomass of large medusae in subtropical coast of Brazil. pp. 3–26. *In*: Mariotini, L. [ed.]. Jellyfish: Ecology, Distribution Patterns and Human Interactions. Nova Publishers, New York, USA.

Nogueira Júnior, M., E. Tosetto and E.L. Razzolini. 2019. The influence of environmental conditions on *Parasagitta friderici* (Chaetognatha) abundance in a subtropical estuary (south Brazil). J. Mar. Biol. Assoc. U.K. 99: 785–793.

Ohtsuka, S., M. Hora, T. Suzaki, M. Arikawa, G. Omura and K. Yamada. 2004. Morphology and host-specificity of the apostome ciliate *Vampyrophrya pelagica* infecting pelagic copepods in the Seto Inland Sea, Japan. Mar. Ecol. Prog. Ser. 282: 129–142.

Ohtsuka, S., K. Koike, D. Lindsay, J. Nishikawa et al. 2009. Symbionts of marine medusae and ctenophores. Plankton Benthos Res. 4(1): 1–13.

Ohtsuka, S., Y. Kondo, Y. Sakai et al. 2010. *In-situ* observations of symbionts on medusae occurring in Japan, Thailand, Indonesia and Malaysia. Bull. Hiroshima Univ. Mus. 2: 9–18.

Oliveira, O.M.P., T.P. Miranda, E.M. Araújo et al. 2016. Census of Cnidaria (Medusozoa) and Ctenophora from South American marine waters. Zootaxa 4194: 1–256.

Omori, M. and E. Nakano. 2001. Jellyfish fisheries in southeast Asia. Hydrobiologia 451: 19–26.

Opadal, A.F., R.D. Brodeur, K. Cieciel, G.M. Daskalov, V. Mihneva, J.J. Ruzicka, H.M. Verheye and D.L. Aksnes. 2019. Unclear associations between small pelagic fish and jellyfish in several major marine ecosystems. Sci. Rep. 9: 2997.

Paffenhöfer, G.A. and T.N. Lee. 1987. Development and persistence of patches of Thaliacea. Afr. J. Mar. Sci. 5: 305–318.
Paffenhöfer, G.-A., L.P. Atkinson, T.N. Lee, P.G. Verity and L.R. Bulluk. 1995. Distribution and abundance of thaliaceans and copepods off the southeastern USA during winter. Cont. Shelf. Res. 15: 255–280.
Paffenhöfer, G-A. and D.M. Gibson. 1999. Determination of generation time and asexual fecundity of doliolids (Tunicata, Thaliacea). J. Plankton Res. 21: 1183–1189.
Paffenhöfer, G.-A. and M. Köster, M. 2005. Digestion of diatoms by planktonic copepods and doliolids. Mar. Ecol. Prog. Ser. 297: 303–310.
Paffenhöfer, G.A. and M. Köster. 2011. From one to many: on the life cycle of *Dolioletta gegenbauri* Uljanin (Tunicata, Thaliacea). J. Plankton Res. 7: 1139–1145.
Pagès, F. and J.-M. Gili. 1991. Vertical distribution of epipelagic siphonophores at the confluence between Benguela waters and the Angola Current over 48 hours. Hydrobiologia 216-217: 355–362.
Pagès, F. 1992. Mesoscale coupling between planktonic cnidarian distribution and water masses during a temporal transition between active upwelling and abatement in the northern Benguela system. South Afr. J. Mar. Sci. 12: 41–52.
Pagès, F. and J.-M. Gili. 1992. Influence of Agulhas waters on the population structure of planktonic Cnidarians in the southern Benguela Region. Sci. Mar. 56: 109–123.
Pagès, F., H. González, M. Ramón, M. Sobarzo and J-M. Gili. 2001. Gelatinous zooplankton assemblages associated with water masses in the Humboldt Current System, and potential predatory impact by *Bassia bassensis* (Siphonophora: Calycophorae). Mar. Ecol. Prog. Ser. 210: 13–24.
Perissinotto, R. and E.A. Pakhomov. 1997. Feeding association of the copepod *Rhincalanus gigas* with the tunicate salp *Salpa thompsoni* in the southern ocean. Mar. Biol. 127: 479–483.
Pinotti, R.M., F.C. Bom and E. Muxagata. 2019. On the occurrence and ecology of *Glaucus atlanticus* Forster, 1777 (Mollusca: Nudibranchia) along the Southwestern Atlantic coast. An. Acad. Bras. Ciênc. 91: e20180154.
Pitt, K.A., C.M. Duarte, C.H. Lucas et al. 2013. Jellyfish body plans provide allometric advantages beyond low carbon content. PLoS ONE 8(8): e72683.
Pitt, K.A., C.H. Lucas, R.H. Condon, C.M. Duarte and B. Stewart-Koster. 2018. Claims that anthropogenic stressors facilitate jellyfish blooms have been amplified beyond the available evidence: a systematic review. Front. Mar. Sci. 5: 1–11.
Prieto, L., D. Macías, A. Peliz and J. Ruiz. 2015. Portuguese Man-of-War (*Physalia physalis*) in the Mediterranean: a permanent invasion or a casual appearance? Sci. Rep. 5: 11545.
Pugh, P.R., C. Dunn and S.H.D. Haddock. 2018. Description of *Tottonophyes enigmatica* gen. nov., sp. nov. (Hydrozoa, Siphonophora, Calycophorae), with a reappraisal of the function and homology of nectophoral canals. Zootaxa 4415: 452–472.
Purcell, J., J. White, D. Nemazie and D. Wright. 1999a. Temperature, salinity and food effects on asexual reproduction and abundance of the scyphozoan *Chrysaora quinquecirrha*. Mar. Ecol. Prog. Ser. 180: 187–196.
Purcell, J. and M. Sturdevant. 2001. Prey selection and dietary overlap among zooplanktivorous jellyfish and juvenile fishes in Prince William Sound, Alaska. Mar. Ecol. Prog. Ser. 210: 67–83.
Purcell, J. 2007. Environmental effects on asexual reproduction rates of the scyphozoan *Aurelia labiata*. Mar. Ecol. Prog. Ser. 348: 183–196.
Purcell, J., S. Uye and W. Lo. 2007. Anthropogenic causes of jellyfish blooms and their direct consequences for humans: a review. Mar. Ecol. Prog. Ser. 350: 153–174.
Purcell, J.E. 1980. Influence of Siphonophore behavior upon their natural diets: evidence for aggressive mimicry. Science 209: 1045–1047.
Purcell, J.E. 1997. Pelagic cnidarians and ctenophores as predators: selective predation, feeding rates, and effects of prey populations. Ann. Inst. Oceanogr. Paris 73: 125–137.
Purcell, J.E., U. Båmstedt and A. Båmstedt. 1999b. Prey, feeding rates, and asexual reproduction rates of the introduced oligohaline hydrozoan *Moerisia lyonsi*. Mar. Biol. 134: 317–325.
Purcell, J.E. and M.N. Arai. 2001. Interaction of pelagic cnidarians and ctenophores with fish: a review. Hydrobiologia 451: 27–44.
Purcell, J.E., W. Graham and H.J. Dumont. 2001. Jellyfish Blooms: Ecological and Societal Importance. Kluwer Academic Publisher, The Netherlands.
Purcell, J.E. and M.B. Decker. 2005. Effects of climate on relative predation by scyphomedusae and ctenophores on copepods in Chesapeake Bay during 1987–2000. Limnol. Oceanogr. 50: 376–387.
Purcell, J.E., M. Sturdevant and C.P. Galt. 2005. A review of appendicularians as prey of invertebrate and fish predators. pp. 359–435. *In*: Gorsky, G., M.J. Youngbluth and D. Deibel [eds.]. Response of Marine Ecosystems to Global Change: Ecological Impact of Appendicularians. Éditions Scientifiques, Paris.
Purcell, J.E. 2012. Jellyfish and ctenophore blooms coincide with human proliferations and environmental perturbations. Annu. Rev. Mar. Sci. 4: 209–235.
Purcell, J.E., A.R. Juhl, M.J. Mańko and C.F. Aumack. 2018. Overwintering of gelatinous zooplankton in the coastal Arctic Ocean. Mar. Ecol. Prog. Ser. 591: 281–286.
Purushothaman, J., S. Hansda, J. Dey, S. Mohan, A. Basu and K. Venkataraman. 2018. An annotated checklist of Thaliaceans (Chordata: Tunicates). Mar. Biodiv. 48: 1903–1930.

Raskoff, K.A., F. Sommer, W. Hamner and K.M. Cross. 2003. Collection and culture techniques for gelatinous zooplankton. Biol. Bull. 204: 68–80.
Reeve, M.R. and M.A. Walter. 1972. Observations and experiments on methods of fertilization in the chaetognath *Sagitta hispida*. Biol. Bull. 143: 207–214.
Reeve, M.R. and M.A. Walter. 1978. Nutritional ecology of ctenophores—a review of recent research. Adv. Mar. Biol. 15: 249–287.
Richter, G. and R.R. Seapy. 1999. Heteropoda. pp. 621–647. *In*: Boltovskoy, D. [ed.]. South Atlantic Zooplankton. Leiden, Backhuys Publishers.
Riebesell, U. and J.P. Gattuso. 2015. Lessons learned from ocean acidification research. Nat. Clim. Chang. 5: 12–14.
Rohde, K. 2005. Marine Parasitology. CSIRO Publishing. Victoria, Australia
Roux, J.-P., C.D. Van Der Lingen, M.J. Gibbons, N.E. Moroff, L.J. Shannon, A.D. Smith and P.M. Cury. 2013. Jellyfication of marine ecosystems as a likely consequence of overfishing small pelagic fishes: lessons from the Benguela. Bull. Mar. Sci. 89: 249–284.
Ruhl, H.A., J. Ellena and K.L. Smith. 2008. Connections between climate, food limitation, and carbon cycling in abyssal sediment communities. Proc. Natl. Acad. Sci. U.S.A. 105: 17006–17011.
Russell, F.S. 1953. The Medusae of the British Isles. Cambridge University Press, London, UK.
Santhakumari, V. and V.R. Nair. 1999. Distribution of hydromedusae from the exclusive economic zone of the west and east coasts of India. Indian J. Mar. Sci. 28: 150–157.
Sasson, D.A. and J.F. Ryan. 2016. The sex lives of ctenophores: the influence of light, body size, and self-fertilization on the reproductive output of the sea walnut, *Mnemiopsis leidyi*. PeerJ 4: e1846.
Sanz-Martín, M., K.A. Pitt, R.H. Condon, C.H. Lucas, C. Novaes de Santana and C.M. Duarte. 2016. Flawed citation practices facilitate the unsubstantiated perception of a global trend toward increased jellyfish blooms: Flawed citation practices in jellyfish population trends. Ecol. Biogeogr. 25: 1039–1049.
Sato, R., Y. Tanaka and T. Ishimaru. 2003. Species-specific house productivity of appendicularians. Mar. Ecol. Prog. Ser. 259: 163–72.
Schiariti, A., A.C. Morandini, G. Jarms, R.G. Paes, D. Franke and H.W. Mianzan. 2014. Asexual reproduction strategies and blooming potential in Scyphozoa. Mar. Ecol. Prog. Ser. 510: 241–253.
Schnedler-Meyer, N.A., T. Kiørboe and P. Mariani. 2018. Boom and bust: life history, environmental noise, and the (un) predictability of jellyfish blooms. Front. Mar. Sci. 5: 257.
Schneider, G. 1987. Role of advection in the distribution and abundance of *Pleurobrachia pileus* in Kiel Bight. Mar. Ecol. Prog. Ser. 41: 99–102.
Seapy, R.R. 1980. Predation by the epipelagic heteropod mollusk *Carinaria cristata* forma *japonica*. Mar. Biol. 30: 137–146.
Sewell, R.B.S. 1951. The epibionts and parasites of the planktonic Copepoda of the Arabian Sea. Scientific Reports of the John Murray Expedition 9: 255–394.
Shanks, A.L. and W.M. Graham. 1987. Orientated swimming in the jellyfish *Stomolophus meleagris* L. Agassiz (Scyphozoan: Rhizostomida). J. Exp. Mar. Biol. Ecol. 108: 159–169.
Shiganova, T.A. 1998. Invasion of the Black Sea by the ctenophore *Mnemiopsis leidyi* and recent changes in pelagic community structure. Fish. Oceanogr. 7: 305–310.
Shiganova, T.A., Z.A. Mirzoyan, E.A. Studenikina et al. 2001. Population development of the invader ctenophore *Mnemiopsis leidyi*, in the Black Sea and in other seas of the Mediterranean Basin. Mar. Biol. 139: 431–445.
Shih, H.-H. and M.-S. Jeng. 2002. *Hysterothylacium aduncum* (Nematoda: Anisakidae) infecting a herbivorous fish, *Siganus fuscescens*, off the Taiwanese Coast of the Northwest Pacific. Zool. Stud. 41: 208–215.
Shoji, J., R. Masuda, Y. Yamashita and M. Tanaka. 2005. Effect of low dissolved oxygen concentrations on behavior and predation rates on red sea bream *Pagrus major* larvae by the jellyfish *Aurelia aurita* and by juvenile Spanish mackerel *Scomberomorus niphonius*. Mar. Biol. 147: 863–868.
Skikne, S.A., R. Sherlock and B.H. Robison. 2009. Uptake of dissolved organic matter by ephyrae of two species of scyphomedusae. J. Plankton Res. 31: 1563–1570.
Smith, C.R., F.C. De Leo, A. Bernardino, A.K. Sweetman and P. Martinez-Arbizu. 2008. Abyssal food limitation, ecosystem structure and climate change. Trends Ecol. Evol. 23: 518–28.
Sommer, U., H. Stibor, A. Katechakis, F. Sommer and T. Hansen. 2002. Pelagic food web configurations at different levels of nutrient richness and their implications for the ratio fish production: Primary production. Hydrobiologia 484: 11–20.
Stibor, H. and N. Tokle. 2003. Feeding and asexual reproduction of the jellyfish *Sarsia gemmifera* in response to resource enrichment. Oecologia 135: 202–208.
Stoecker, D.K., P. Verity, A. Michaels and L.H. Davis. 1987. Feeding by larval and postlarval ctenophores on microzooplankton. J. Plankton Res. 9: 667–683.
Stunkard, H. 1980. Successive hosts and developmental stages in the life history of *Neopechona cablei* sp. n. (Trematoda: Lepocreadiidae). J. Parasitol. 66: 636–641.
Suganthi, K., S. Bragadeeswaran, N. Sri Kumaran, S. Thangaraj and T. Balasubramanian. 2011. Biological and pharmacological activities of jelly fish *Crambionella stuhlmanni* (Chun, 1896) and *Chrysaora quinquecirrha* (Desor, 1848). Int. J. Pharm. Pharm. Sci. 3: 230–236.

Sugiura, Y. 1964. On the life-history of rhizostome medusae II. Indispensability of zooxanthellae for strobilation in *Mastigias papua*. Embryologia 8: 223–233.

Sugiura, Y. 1969. On the life-history of rhizostome medusae V. On the relation between zooxanthellae and the strobilation of *Cephea cephea*. Bull. Mar. Biol. Stn. Asamushi 13: 227–233.

Sutherland, K.R., L. Madin and R. Stocker. 2010. Filtration of submicrometer particles by pelagic tunicates. Proc. Natl. Acad. Sci. U.S.A. 107: 15129–15134.

Sutherland, K.R., H.L. Sorensen, O.N. Blondheim, R.D. Brodeur and A.W.E. Galloway. 2018. Range expansion of tropical pyrosomes in the northeast Pacific Ocean. Ecology 99: 2397–2399.

Sweetman, A.K. and A. Chapman. 2015. First assessment of flux rates of jellyfish carcasses (jelly-falls) to the benthos reveals the importance of gelatinous material for biological C-cycling in jellyfish-dominated ecosystems. Front. Mar. Sci. 2: 47.

Tang, K., C. Dziallas, K. Hutalle-Schmelzer and H.P. Grossart. 2009. Effects of food on bacterial community composition associated with the copepod *Acartia tonsa* Dana. Biol. Lett. 5: 549–53.

Tang, K.W., V. Turk and H.P. Grossart. 2010. Linkage between crustacean zooplankton and aquatic bacteria. Aquat. Microb. Ecol. 61: 261–77.

Terazaki, M. and C.B. Miller. 1982. Reproduction of meso- and bathypelagic chaetognaths in the genus *Eukrohnia*. Mar. Biol. 71: 193–196.

Théodoridès, J. 1989. Parasitology of marine zooplankton. Adv. Mar. Biol. 25: 117–177.

Thiel, M.E. 1976. Wirbellose Meerestiere als Parasiten, Kommensalen oder Symbionten in oder an Scyphomedusen. Helgoland. Wiss. Meer. 28: 417–446.

Thiriot-Quiévreux, C. 1973. Heteropoda. Ocean. Mar. Biol. Ann. Rev. 11: 237–261.

Thiriot-Quiévreux, C. 1975. Observations sur les larves et les adultes de Carinariidae (Mollusca: Heteropoda) de l'Océan Atlantique Nord. Mar. Biol. 32: 379–388.

Tiller, R.G., Å.L. Borgersen, Ø. Knutsen, J. Bailey, H.V. Bjelland, J. Mork, L. Eisenhauer and Y. Liu. 2017. Coming soon to a Fjord near you: future jellyfish scenarios in a changing climate. Coast. Manage. 45: 1–23.

Tinta, T., T. Kogovšek, K. Klun, A. Malej, G.H. Herndl and V. Turk. 2019. Jellyfish-associated microbiome in the marine environment: exploring its biotechnological potential. Mar. Drugs 17: 94.

Tosetto, E.G., S. Neumann-Leitão and M. Nogueira Júnior. 2020. New species of Eirenidae (Hydrozoa: Leptothecata) from the Amazonian coast (northern Brazil). Sci. Mar. 84(4): doi.org/10.3989/scimar.05051.14A.

Towanda, T. and E.V. Thuesen. 2006. Ectosymbiotic behavior of *Cancer gracilis* and its trophic relationships with its host *Phacellophora camtschatica* and the parasitoid *Hyperia medusarum*. Mar. Ecol. Prog. Ser. 315: 221–236.

Tsukamoto, K., Y. Yamada, A. Okamura et al. 2009. Positive buoyancy in eel leptocephali: an adaptation for life in the ocean surface layer. Mar. Biol. 156: 835–846.

Tzikhon-Lukanina, E.A., O. Reznichenko and T.A. Lukasheva. 1993. Ecological variation of comb-jelly *Mnemiopsis leidyi* (Ctenophora) in the Black Sea. Zh. Obshch. Biol. 54: 713–724.

UNESCO. 1968. Monographs on oceanographic methodology. Zooplankton Sampling vol 2. UNESCO, Paris: 174 pp.

Uye, S. 2008. Blooms of the giant jellyfish *Nemopilema nomurai*: a threat to the fisheries sustainability of the East Asian Marginal Seas. Plankton Benthos Res. 3: 125–131.

Valentin, J.L., W.M. Monteiro-Ribas, M.A. Mureb and E. Pessotti. 1987. Sur quelques zooplanctontes abondants dans l'upwelling de Cabo Frio (Brésil). J. Plankton Res. 9: 1195–1216.

van der Spoel, S. and J.R. Dadon. 1999. Pteropoda. pp. 649–706. *In*: Boltovskoy, D. [ed.]. South Atlantic Zooplankton. Leiden, Backhuys Publishers.

Vargas, C.A. and H.E. González. 2004. Plankton community structure and carbon cycling in a coastal upwelling system. I. Bacteria, microprotozoans and phytoplankton in the diet of copepods and appendicularians. Aquat. Microb. Ecol. 34: 151–164.

Verity, P.G. and V. Smetacek. 1996. Organism life cycles, predation, and the structure of marine pelagic ecosystems. Mar. Ecol. Prog. Ser. 130: 277–293.

Yoon, W.D., J.-C. Marty, D. Silvayn and P. Nival. 1996. Degradation of faecal pellets in *Pegea confoederata* (Salpidae, Thaliacea) and its implication in the vertical flux of organic matter. J. Exp. Mar. Biol. Ecol. 203: 147–177.

Wall-Palmer, D., C.W. Smart, R. Kirby, M.B. Hart, K.T.C.A. Peijnenburg and A.W. Janssen. 2016. A review of the ecology, palaeontology and distribution of atlantid heteropods (Caenogastropoda: Pterotracheoidea: Atlantidae). J. Molluscan Stud. 82: 221–234.

Weisberg, S.B., N. Bednarsek, R.A. Feely, F. Chan, A.B. Boehm, M. Sutula, J.L. Ruesink, B. Hales, J.L. Largier and J.A. Newton. 2016. Water quality criteria for an acidifying ocean: challenges and opportunities for improvement. Ocean Coast. Manag. 126: 31–41.

Werner, B. 1963. Effect of some environmental factors on differentiation and determination in marine Hydrozoa, with a note on their evolutionary significance. An. New York Acad. Sci. 105: 461–488.

Zavodnik, D. 1987. Spatial aggregations of the swarming jellyfish *Pelagia noctiluca* (Scyphozoa). Mar. Biol. 94: 265–269.

CHAPTER 8

Soft-bottom Marine Benthos

Jerónimo Pan[1,*] and *Paula D. Pratolongo*[2]

8.1 Introduction to the benthos and some definitions

In the opening to this book, we described the division of the oceans and defined the two-dimensional habitat in which bottom-dwellers are found as the benthic environment, a word that stems from the Greek term *benthos*, meaning 'depth of the sea'. In turn, the word benthos is used by scientists to refer to the organisms that inhabit benthic habitats, and these, as we will see in this and other chapters, may range from bacteria, to microscopic algae, to rooted vascular plants, to a plethora of invertebrates and vertebrates (e.g., endobenthic, epibenthic and demersal fish, such as pleuronectiforms, myxinids, gunnels, blennies, among others).

The benthic realm can be broadly split into two kinds: the so-called 'hard' and 'soft' bottoms. However, geologically speaking, all bottom types are made of solid rock, what varies is the size of particles. As it was introduced in Chapter 1, some oceanic basins, or *depositional environments*, promote the accumulation of sediment (i.e., rock and mineral fragments of various sizes, origin and chemical composition); after all, a sediment particle is, by definition, matter that settles to the bottom in a fluid medium. On the other hand, several types of hard bottoms are composed of rocky substrates of lithogenic or hydrogenic (abiotic) origin or biogenic nature, that have not been eroded or weathered and provide an extensive, harder, more rigid substrate for settlement or attachment than soft bottoms. These considerably basic differences in bottom type (hard *vs.* soft) make for very different structuring of benthic communities and different prevalence of ecological and biogeochemical processes. Accordingly, they will be discussed in two different chapters, beginning with this chapter, devoted to soft-bottom benthos, and following with Chapter 10, focused on marine hard-substrate communities.

Sediments cover about 80% of the sea floor (i.e., the benthic habitat; James 2005); thus, the areal extent of soft bottoms make them dominant at a global scale. Their distribution does not show any distinct latitudinal pattern of variation, but they tend to be more frequently found in estuaries and coastal zones (see Chapter 1), and along continental platforms. In Chapter 1, we discussed the composition and extent of *biogenic oozes*, composed of the remains of once-living pelagic organisms. Furthermore, living organisms inhabit a large proportion of the soft-bottom benthic environment. To put it bluntly, "sediments are not inert". Even what might appear to be a barren patch of sand, as found on a beach, harbors a substantial diversity of bacterial groups (Watermann et al. 1999, Degenhardt et al. 2020), microeukaryotes such as diatoms (Stal and de Brouwer 2005, Pan et al. 2017), protozoans, small interstitial invertebrates known as meiobenthos, meiofauna or mesopsammon, and larger organisms like amphipods, isopods,

[1] National Scientific and Technical Research Council, Argentina (CONICET); Instituto de Geología de Costas y del Cuaternario (IGCyC); Instituto de Investigaciones Marinas y Costeras (IIMyC).

[2] Centro de Recursos Naturales Renovables de la Zona Semiárida (CERZOS-CONICET-UNS); Dto. Biología, Bioquímica y Farmacia-Universidad Nacional del Sur (BByF-UNS); National Scientific and Technical Research Council, Argentina (CONICET).

* Corresponding author: jeronimopan@gmail.com

and mollusks (mostly bivalves) that compose the macrofauna (see McLachlan and Brown 2006, for a thorough treatment). At the μm scale, free-living marine bacteria present remarkably constant densities in surface sediments, on the order of 1 cell μm^{-2} (Dale 1974) or 10^9 cells ml^{-1} of sediment (Schmidt et al. 1998). At the sub-millimeter scale, soft-bottom marine sediments also harbor significant zoological diversity; for example, many unique phyla and clades of metazoans inhabit the 10–1000 μm-scale interstices created between sand grains (interstitial meiofauna) but passed unnoticed in sediment samples, until their discovery in the 1950s, and being more thoroughly studied over the last few decades (Rundell and Leander 2010).

Marine sediments are a very dilute medium with respect to organic matter (OM). But before we continue with this topic, it is important to ponder spatial and temporal scales. Almost all OM produced on Earth will eventually end up being incorporated into marine sediments, primarily along continental margins (Hedges and Keil 1995). The processes leading up to carbon burial are not measured in an organism's lifetime, that roughly spans from a few days, in the case of microalgae, to a few years in the case of marine crustaceans, but rather, over geological timescales (hundreds to thousands to millions of years). Thus, the burial of OM in marine sediments constitutes the main link between 'active' pools of carbon in the oceans, the atmosphere and land, and those carbon pools that cycle on much longer, geological timescales (Burdige 2007). Continental shelves and slopes harbor most of the active (surficial) pools of OM in marine sediments. Sill they represent all but a quarter of the dissolved organic carbon (DOC) active pool in seawater (0.15 *vs.* 0.60 10^{18} g C, respectively; data compiled by Hedges and Keil 1995). Perhaps the most striking figure to bear in mind is that less than 0.5% of the total OM produced on Earth (on land and in the oceans) gets preserved in the marine environment. In other words, ~ 99.5% of the photosynthetic production is remineralized (or 'respired'), returning either to the atmosphere or to the ocean dissolved inorganic carbon (DIC) pool (Burdige 2007).

The previous paragraph meant to provide you with a sense of scale, concerning the fate of OM in biogeochemical cycles. What follows in a large part of this chapter will deal with the transformations that 'active' surface pools of OM in marine sediments may experience in shorter timescales (roughly an organism's lifetime).

Let's start with the sources of OM—what are they? OM can be both locally produced (autochthonous) and remotely produced (allochthonous, Pedrosa-Pàmies et al. 2015). The greatest portion of marine sedimentary OM is originally produced in the form of particulate matter through photosynthesis in surface waters and then becomes incorporated into sediments through vertical transport and deposition. Also, in shallow, well-illuminated sediments, OM might be produced *in situ,* as in the case of the exudates and new cells produced by microalgae of the microphytobenthos (MPB, see Section 8.5.2). But most of the OM in surface 'active' pools comes from allochthonous sources (Mann 2009), such as estuarine primary production or terrigenous (soil, plant debris) inputs and atmospheric deposition (Pedrosa-Pàmies et al. 2015). Consequently, sedimentary OM constitutes a heterogeneous and complex mixture of organic compounds with varied physico-chemical properties.

Bacteria not only remineralize OM from these various sources, but they also constitute a significant component of the living biomass of benthic microbes, as previously mentioned. A most significant pool of OM comes from microbial secretion and accumulation of extracellular polymeric substances (EPS). EPS is a generic term for a group of compounds of various chemical identities that are purposefully produced by microbes. These secretions may secure microbial attachment and favor the intercellular communication of microbes within biofilms (Flemming and Wuertz 2019). They may also be secreted as metabolic excess or waste products (Decho and Gutierrez 2017), or as the result of unbalanced growth, when the supply of nutrients lags behind photosynthetic CO_2 fixation (Staats et al. 2000a, b, Stal 2010). Chemically, EPS are complex molecules among which highly-hydrated polysaccharides make up the majority. Other components include simple sugars, aminoacids, proteins, lipids, and lipopolysaccharides, as well as uronic acids which make EPS acidic (Stal and de Brouwer 2005, Stal 2010, Decho and Gutierrez 2017).

In addition to microbially produced EPS, virtually all benthic animals secrete copious amounts of mucus to accompany many physiological processes that include locomotion, protection (Calow 1977), burrowing and coating of burrows (Lalonde et al. 2010, Crane and Merz 2017), laying egg masses or gelatinous spawn (a mixed benthic and planktonic development strategy, Strathmann 1985). This mucus production represents a significant energy investment (> 25%; Kideys and Hartnoll 1991). Mucus is also profusely employed in suspension-feeding by benthic metazoans (Section 8.6.1). With respect to its chemical nature, mucus consists primarily of water (> 96% by weight for mollusks), whereas its salt-free dry weight constitutes only 0.7% of the total weight (Hannides and Aller 2016). Polychaete mucus, on the other hand, is comprised of proteins, carbohydrates, and lipids in an approximate ratio of 200:50:1 (in the dry weight fraction), and the mucin-like material has abundant carboxyl, phosphoryl, hydroxyl, amino, and thiol functional groups that are geochemically important for driving metal adsorption (Lalonde et al. 2010). It has been suggested that benthic animal-secreted mucus plays a functional role in promoting sediment OM remineralization processes through a 'priming' effect (Hannides and Aller 2016), much like spreading marmalade on an old toast to make it more appealing. For example, mucopolysaccharides have been shown to promote the growth of specific microbes that drive nitrogen cycling, thus enhancing nitrification rates (Dale et al. 2019).

8.2 Factors influencing soft-bottom benthos

Research on animal-sediment relationships suggests that benthos distribution is controlled by a combination of factors such as sediment granulometry, dissolved O_2, OM content, as well as various disturbance processes and pollution events (e.g., Ellis et al. 2012). Early studies correlated benthic species distributions with sediment grain size, leading to the generalization of distinct associations between animals and specific sediment types. However, animal-sediment relationships are complex and variable and there is, in fact, little evidence that sedimentary grain size, by itself, is the primary determinant of benthic species distributions (see Snelgrove and Butman (1994) for a comprehensive review). In addition to grain size, other factors have been proposed to affect the distribution of benthic animals, such as sediment OM content, microbial biomass, food supply and trophic interactions, and no single mechanism entirely explains species composition and distribution patterns observed across many different environments.

The sedimentary and hydrodynamic environment affects soft-sediment communities through many complex interactions. Grain size covaries with a number of factors that influence benthic fauna, including porewater chemistry, OM content, and the composition of microbial communities, and all these variables, in turn, are modified by the near-bed flow regime. Thus, the hydrodynamic and sediment-transport regime in the benthic boundary layer is largely responsible for sediment grain size distribution, as well as larval and food supply, and exert a strong influence on the species composition. For instance, studying relationships among hydrodynamics, sediment transport and benthic biology showed the relevance of permeability, an often-neglected characteristic of sandy sediments. Permeable sandy bottoms allow the advective flow of porewater. Water flow above a rippled sandy bed creates pressure gradients, pumping water carrying organic-rich materials and dissolved O_2 into the sediment (Huettel et al. 2003), facilitating transport (Thibodeaux and Boyle 1987). In contrast, transport of porewater constituents in finer-grained, cohesive sediments is exclusively diffusion-controlled and driven by concentration gradients, resulting in a much slower mixing rate.

In turn, O_2 availability, along with OM depositional flux (vertical advection and diffusion), largely control OM mineralization. The porewater content of dissolved O_2 decreases rapidly beneath the water-sediment interface, creating a concentration gradient for the downward diffusion of O_2 and for the concentration of ions that have different solubilities in oxygenated and reduced (anoxic) porewater. There is a certain depth below the sediment-water interface that represents the boundary between oxic and anoxic sediments. In coarse-grained sandy bottoms, this boundary is normally deeper than in mud, but the depth of O_2 penetration may fluctuate in response to factors such as OM flux to the sediment surface, O_2 content in the overlying bottom water, and biological activity (Katsev et al. 2006). In soft bottoms of intertidal areas, intermittent or temporary flooding alternate with periods of aerial exposure. Once the sediment surface becomes flooded, O_2 diffusion diminishes, and anaerobic conditions develop as soon as porewater O_2 is depleted by microorganisms. The alternance between oxidized and reduced states, typical of intertidal mudflats and sandflats, enhances the coupling of different biogeochemical reactions associated with organic matter degradation. As we will see in Section 8.5.1, these coupled reactions increase the capacity for nitrogen removal, which is a valuable ecosystem function in coastal areas affected by excessive nitrogen loading from anthropogenic sources (Mitsch and Gosselink 2000).

Redox reactions transfer electrons among atoms and molecules. Any complete redox reaction consists of an oxidation and a reduction component called half-reactions. Oxidation is the loss of electrons from an atom, while reduction is the gain of electrons by another atom. As a result of the oxidation and reduction, elements often change their speciation causing, for instance, previously dissolved elements to precipitate, solid minerals to dissolve or dissolved ions to turn to gases. In surface sediments, O_2 availability is limited to the upper layer and extensive anoxic areas commonly develop below the surface. The transition zone from oxygenated to anoxic conditions in sediments is called the redox potential discontinuity (RPD) layer. More accurately, the RPD layer marks the boundary between predominantly oxidizing and reducing conditions (Fenchel and Riedl 1970). In anoxic conditions, the accumulation of reduced compounds produces a greyish, green or black layer below the surface that can be readily differentiated from the oxygenated reddish-brown oxygenated layer above it, which owes its color to iron oxides and hydroxides (Valdemarsen et al. 2009). This visual boundary is often used as a relative measure of O_2 penetration (Diaz and Trefry 2006). The depth of the RPD layer can vary extensively in time and space (from a few millimeters to nearly half a meter) and is influenced by physical and biological factors like particle size, hydrodynamics, OM content, bacterial activity, and the presence of burrowing animals (see Section 8.3.4 below). Finer sediments, higher OM content, and enhanced microbial activity promote anaerobic conditions. Faunal irrigation and bioturbation, on the other hand, allow oxygenated water to penetrate deeper into the sediment, creating a three-dimensional mosaic of redox conditions (Kristensen 2000, Solan and Kennedy 2002, Hunting and Kampfraath 2013).

The basic oxidation half-reactions in bottom sediments are promoted by heterotrophic microorganisms, mainly bacteria, in OM decomposition. During bacterial respiration, organic compounds are the main source of electrons (also called electron donors), and when these compounds oxidize the electrons released are used for the reduction half-reactions, where different compounds act as electron acceptors (Vepraskas and Faulkner 2001). O_2 is the primary electron acceptor in aerobic sediments whereas, under anaerobic conditions, when O_2 is unavailable for OM respiration, different groups of microorganisms use alternate electron acceptors in a sequence of thermodynamic favorability to continue decomposition. In this sequence, after O_2 disappears, NO_3^- is reduced, followed by Mn^{4+} (in the form of MnO_2), Fe^{3+} oxides, SO_4^{2-}, and finally CO_2 (which is reduced to methane). As we will discuss in Section 8.5.1, different groups of decomposers use different compound within this sequence. For example, if NO_3^-, SO_4^{2-}, and OM are present in the absence of O_2, then a group of bacteria called denitrifiers will reduce all the available NO_3^- before the sulfate reducers proceed (Reddy and DeLaune 2008), due to the greater amount of ATP that can be extracted *per* mole of substrate from the denitrification process compared to the sulfate-reduction process. In the absence of macrofauna, this leads to a vertically-stratified dominance of specific groups of decomposers, typically coinciding with the chemical profile of O_2, NO_3^-, Mn^{2-}, Fe-oxides, SO_4^{2-} and CH_4 concentrations (Froelich et al. 1979).

Nepheloid layers are zones of increased turbidity (from the Greek word *nephos*, meaning cloud), caused by larger concentrations of suspended particulate matter. The bottom water of the ocean usually has a higher turbidity than water at intermediate depths, and such a benthic nepheloid layer may extend several hundred m above the seafloor. Turbidity in benthic nepheloid layers is caused by local resuspension, but it is also maintained by fine particles transported over long distances, kept in suspension because of the enhanced turbulence and shear near the seafloor (McCave 2009). Long-range particle transport in the benthic boundary layer plays a significant role in the export of shelf-derived particles across the continental margin. This lateral, off-slope transport of suspended particles has been shown to be an important supply of OM to benthic communities (Peinert et al. 2000). Besides the long-range advection of fine terrigenous particles, there is a dynamic cycling of fresh material in the benthic boundary layer mediated by living organisms. The local resuspension loop (Graf 1997) includes settling of fresh particles, resuspension, and aggregation-disaggregation cycles. Within this resuspension loop, bacterial activity promotes the aggregation of dissolved OM into particulate OM, thus providing a supplementary food source for benthic fauna. In general, suspended particles in a benthic nepheloid layer, especially in deep-sea benthic communities, represent an important, high-quality food source, and many suspension-feeding benthic organisms (e.g., crinoids, sea feathers, and soft corals) have anatomical and behavioral adaptations to capture these nutritious particles at some height above the sea floor (Piepenburg et al. 2001).

8.3 Functional classification of soft-bottom marine benthos

The ecological study of biological communities is often simplified by defining taxonomic functional groups that share similar functional attributes or exploit a common resource. This approach has been particularly valuable in benthic ecology and is useful to understand the effect of different environmental variables (i.e., depth, OM enrichment, or grain size) on the distribution of organisms. For example, one useful application of functional group classification is its use as a basis of indices for the assessment of pollution.

The various physical and biological factors that influence life in the soft bottom benthic environment result in many different functional adaptations in benthic organisms. Although different species are adapted to unique combinations of environmental conditions, benthic organisms can be grouped according to common functional attributes regarding, for instance, habitat preference, resource partitioning and utilisation, strategies for predator avoidance, or breeding requirements. Food resources are recognized fundamental attributes and, therefore, the mode in which benthic animals acquire food has been most frequently used to define functional groups in the soft bottom benthos. One principal and obvious division regarding resource use is that which classifies organisms as primary producers or consumers. Thus, photoautotrophic microalgae of the MPB, treated in Section 8.5.2, is a functional group that can be distinguished from heterotrophic organisms of the zoobenthos in terms of its source of carbon.

Another distinction considers size that broadly mirrors different ecological functions. The microbenthos comprises microscopic benthic organisms that are less than 42 μm in size. Besides photosynthetic organisms of the MPB, other members of the microbentos include heterotrophic and chemoautotrophic bacteria, and archaea that participate in biogeochemical cycles (see Section 8.5.1). Meiobenthos or benthic meiofauna (42–500 μm) refers to a diverse group of heterotrophic organisms that inhabit the interstitial space between sediment grains. Typical examples of meiofaunal organisms are nematodes, benthic (harpacticoid) copepods, ostracods and single-celled foraminifera that play a critical trophic role as a link between bacterial food sources and larger consumers. In addition to these taxa, there is a rich diversity of other, lesser-known taxa, some of which occur exclusively in the meiobenthos, including kinorhyncha, gnathostomulida, or gastrotricha. Macrobenthos is the group of benthic organisms larger than 500 μm, which is generally larger than the average sediment particle diameter of the matrix they inhabit and includes a large number of species and functional variations. Macrofauna tend to be the most conspicuous functional community, not only visually but also in terms of biomass. Further, macrofauna typically influence both bed roughness and sediment aggregation, with significant impacts on erodibility. In the following sections, we will discuss some common divisions within macrobenthos, regarding habitat preference with respect to the seafloor, feeding modes, and eco-geomorphic function.

8.3.1 Habitat preference: Epifauna vs. infauna

Based on their seafloor habitat preference, epifaunal organisms are those that live on the surface, while infauna live beneath the surface and burrow in the sediments. At the sediment-water interface, the increasing effects of viscosity in the boundary layer and bed shear stress dominate. Soft-bottom epifauna include urchins, starfishes, sea cucumbers, shrimps, crabs, snails, and some worms. For epibenthic organisms that exploit the sediment-water interface, bed flow rate is a critical variable. Thus, feeding modes, predator-prey interactions and spawning behavior are adapted to the speed of water movement over the surface. There is a critical flow velocity at which soft bottoms are composed of coarse grain sediments (gravel to coarse sands), and only a small number of epibenthic animals, usually sessile organisms, or species with strong musculature and capable of squatting (e.g., crabs, echinoderms, or large snails), are adapted to thrive under such unstable conditions. Common sessile epifaunal taxa include bivalves and sponges, many of which grow attached to shells or cobbles, firmly cementing themselves to the substrate. In areas where flow rates are greater, soft sediments are replaced by hard substrata (pebbles to rock) that severely restrict burrowing, such that benthic fauna is confined to the benthic boundary zone. Below the sediment water interface, within the sedimentary layers, burrowing organisms or tube-builders dominate the infauna which, accordingly, tend to be slender and equipped with effective peristaltic musculature that facilitates burrowing and irrigation of burrows. For infaunal organisms, grain size, as well as sediment processing by other organisms (particle capture, subduction, ingestion, and defecation) have the greatest influence on their abundance (Pearson 2001).

Macrozoobenthos pertaining to these two living habits may present distinctive feeding strategies, even within the same taxonomic group. Bacon et al. (1998) experimentally manipulated diet quantity (concentration of seston) and quality (different POM levels) to test the hypothesis that infaunal and epifaunal suspension-feeding bivalves can effectively regulate their ingestion rates at varying particle concentrations. They found that infaunal bivalves showed relatively high clearance rates at low concentrations and significantly reduced them as the concentration of food particles increased. Infaunal macrozoobenthos also relied more on post-ingestive selection of food particles (a strategy tending to maximize food intake), whereas epifaunal species relied primarily on preingestive selection, maintaining high organic ingestion rates at higher concentrations. This indicates that among bivalves, infaunal species which may experience large and frequent fluctuations in the concentration and quality of their food supply, are more opportunistic than epifaunal ones.

Physical characteristics such as bottom topography and hydrography (Norén et al. 1999) also play significant roles, determining and structuring benthic communities. Besides, the use of the habitat that macrobenthos makes depends on the prevailing environmental conditions. For example, OM supply is a key determinant of the dominant functional traits and relative abundance of different functional communities, such as suspension feeding or deposit feeding (Lessin et al. 2019), as explained in the following section.

8.3.2 Feeding strategies: Shredders, grazers, suspension- and deposit feeders

The mechanisms of food acquisition by soft-bottom macrobenthos makes for very different strategies, which have produced different morpho-physiological adaptations, even within same taxonomic groups of metazoans. Moreover, the same feeding strategy in different species can result in the ingestion of very different food items. Macroinvertebrates grouped as shredders wander the sediment surface looking for plant detritus. They use their tearing mouthparts to rip and shred coarse particulate matter such as leaf litter and feed to a large extent on the associated microbial biomass. Mud snails like Littoraria *irrorata* and Peringia (Hydrobia) *ulvae*, which will be treated in more detail in Chapter 9, are typical examples within this group. In turn, grazers (also known as scrapers) specialize on shearing attached algae from surface particles. This group is widespread in hard substrate communities, and includes mollusks (e.g., gastropods and polyplacophorans), sea urchins, amphipods, and crabs (see Chapter 12), but it also includes soft-bottom infauna that graze on biofilms, using their rasping mouthparts to scrape detritus and MPB off the sediments.

Suspension and deposit feeders are the most common feeding guilds in soft-bottom benthic communities. In the first case, animals have devised feeding strategies to make use of the large number of particles in suspension in seawater, trapping and sorting them, before funneling them to their mouths for ingestion, thereby diverting POM in suspension to the benthos. This mode of feeding used to be called filter-feeding but given that mucus-producing glands and adhesive particle capture are involved rather than true filtration through a sieve, the term suspension feeding is currently favored. On the other hand, animals may consume the 'active' pool of OM deposited on or buried in surface sediments and extract it from this most diluted medium by means of specific digestive enzymes and *via* different ingestion mechanisms. As they make use of OM that has been *deposited* in a sedimentary lattice, this process is called deposit feeding.

Within deposit feeders, a further distinction is often made between surface- and borrowing deposit feeders. Surface deposit feeders, such as fiddler crabs and snails, feed on the sediment surface and their

Figure 8.1. Bioturbating effects of the fiddler crab, *Uca uruguayensis*, feeding on surface sediments. Fiddler crabs use their minor chelae to bring the organic matter-rich sediment to their buccal region, where specialized mouth parts sort particles for ingestion or rejection as feeding pellets. On the photo a male specimen is shown, easily identifiable by his asymmetrically large left chela (the "fiddle"), which he tends to wave in a fashion that attracts female fiddler crabs and discourages male competitors. Photographic credits by J. Pan.

effects are commonly limited to the upper sediment layers (2–3 cm depth; Fig. 8.1). Burrowing fauna, in turn, live and feed deeper within the sediment matrix, and may disturb the sedimentary structure up to 30-cm depth in a process known as bioturbation (see Section 8.3.4). Besides borrowing, deposit feeders can alter sediment properties through selective removal of particles and fecal deposition. Conveyor-belt deposit-feeding organisms (head-down deposit feeders) live in vertically-oriented burrows, removing sediments at depth and ejecting the processed material at the sediment surface. For instance, the lugworm *Arenicola marina* lives head-down in J-shaped burrows up to 50-cm depth and ingests bulk sediments below the sediment-water interface within a temporary feeding pocket. After digesting the detrital material, lugworms defecate coils of sediment-rich feces (worm casts) on the seabed surface. In benthic environments under steady and slow sedimentation, head-down deposit feeders consume relatively refractory material that has not been remineralized by the time it reaches the feeding depth. On the other hand, reverse conveyor-belt feeders are head-up, surface deposit feeders that feed on microalgae-enriched sediments and fresh OM at the surface and defecate processed material at depth within their burrows. Surface deposit-feeding animals that extend feeding appendages from a burrow are quite common at all depths in the ocean. For example, the cirratulid polychaete *Cirriformia grandis* feeds on surface sediments using its tentacles and defecates at several cm in depth (see Jumars et al. (1990) for a comprehensive review on deposit feeding strategies).

Some deposit- and suspension feeders can switch feeding mode throughout life-history stages. For example, the isotopic signatures of the preferential diets of the co-occurring clam *Macoma balthica* (a facultative deposit feeder) and the cockle *Cerastoderma edule* (an obligatory suspension feeder) were studied at different stages of their life history. It was found that the cockle diet was consistently composed of phytoplankton, whereas *M. balthica* could feed on a mixed diet of MPB and phytoplankton, switching to phytoplankton primarily after attaining larger sizes (Rossi et al. 2004). In general, suspension feeders respond rapidly (~ 1 week) to pelagic production (phytoplankton blooms), whereas deposit feeders typically exhibit a lag in their feeding response of between 1 and 4 months with respect to the peak in pelagic production (Lessin et al. 2019).

A mechanistic approach to deposit- and suspension-feeding will be further discussed in Sections 8.5.3 and 8.6.1, respectively; as well as some of the ecosystemic effects of these trophic strategies in soft-bottom benthic environments.

8.3.3 Predators and scavengers

Predators in soft-bottom habitats include snails, crabs, worms, fishes, and birds. In intertidal flats, predators can be extremely abundant and shape the distribution of their prey species (and the species dependent on these) through top-down controls and trophic cascades (see Chapter 9). Large carnivorous epibenthic predators (such as bottom-feeding fishes, crabs, and birds) often prey on benthic infauna and control their abundances (see review by Peterson 1979). Many infaunal species, however, are themselves predators (e.g., nemertean and polychaete worms). This predatory infauna can influence the abundances of non-predatory infauna by preying on or reducing larval settlement and juvenile recruitment. For instance, common predatory polychaete families, like glycerids and nereids, prey on other polychaetes, amphipods, isopods, gastropods, and bivalves (Ockelmann and Vahl 1970, Ambrose 1983), and have a negative effect on larval settlement and juvenile survivorship (Oliver et al. 1982, Ambrose 1983). Most predatory infauna do not form protective tubes, and those that have burrowing habits often temporarily expose parts of their bodies on the sediment surface (e.g., nereids, phyllodocids), making predatory infauna more readily available as prey for epibenthic predators. On the other hand, non-predatory infauna are less vulnerable to predation from epibenthic predators since they construct protective tubes and burrows and rarely venture outside of them (e.g., capitellids, chaetopterids, and maldanids).

Detritus is commonly treated as a uniform source of organic carbon to detrital food webs, while in reality, detrital quality varies significantly, from low-quality dead plant material to high-quality carrion (dead animal tissue). Scavenging (carrion consumption) represents a significant form of energy transfer between trophic levels, different from predation, since it exploits exceptional food quality without the expenditure of great amounts of energy required for chasing or subduing a prey (Wilson and Wolkovich 2011). Because of the high food quality of carrion, scavenging is widespread, and many scavenger guilds dominate the carnivore trophic level.

The death of mobile consumers leads to downward flux of carbon and substantial carrion subsidies increase benthic secondary production. For example, when a whale carcass (megacarrion) falls onto the seabed, it becomes a point-source organic input to the benthos and a food source to large motile scavenging fauna. Shortly after a cetacean carcass arrives to the seafloor, the soft tissue is first consumed in a matter of weeks to years by large scavenging fauna (typical of whale falls worldwide) including ophidiid and zoarcid fishes, and galatheid and munidopsid decapods. Large mobile scavengers are attracted to the carcass by its odor plume and potentially by mechanosensory cues (Klages et al. 2002). These scavengers are responsible for making food available to smaller organisms in the surrounding area (Smith and Baco 2003) *via* their sloppy eating.

This first stage is followed by the enrichment-opportunist stage, characterized by different faunal assemblages attracted to the organically-enriched sediments surrounding the skeleton and includes bone epifauna as well. The sediments within 1–3 m of a carcass commonly harbor extremely high-density, low-diversity assemblages of macrofauna, similar to macrofaunal community patterns around point sources of organic enrichment in shallow water. These assemblages include dorvilleid polychaetes and cumaceans (small crustaceans), while bones may be densely covered with writhing masses of the chrysopetalid annelid *Vigtorniella ardabilia* (Smith and Baco 2003). While it is recognized that carcass falls are consumed by a distinct succession of scavenging species, the larger cetacean carcasses not only supply food for rapidly-responding scavengers but also attract secondary predators that do not utilise the food fall directly, but rather prey on scavengers (Kemp et al. 2006). Eventually, large whale carcasses become zones of prolonged interaction between scavenging and non-scavenging species that in time contribute to the development of anaerobic conditions. This is called the sulfophilic stage, which occurs when anaerobic microbial degradation of the lipid-rich skeleton creates reduced compounds that fuel a chemosynthetic community that can last for decades (Amon et al. 2013).

8.3.4 Bioturbators and stabilizers

Through activities such as deposit feeding, locomotion through sediment and burrowing, some soft-bottom animals promote sediment mixing which, albeit limited in extent at the organismal level, can be considerable when extrapolated to ecosystem-wide scales. This sediment mixing is known as bioturbation, and the animals that cause it are called bioturbators. Bioturbators may be motile or sedentary, and some examples include nemerteans, sipunculids, echiurans, priapulids, clams, polychaetes, holothuroids, and crabs. The physical reworking of surface sediments by the fiddler crab, *Uca uruguayensis*, is illustrated in Fig. 8.1. Fiddler crabs use their minor chelae to bring OM-rich sediment to their buccal region, where specialized mouth parts sort particles for ingestion or rejection as feeding pellets (Miller 1961).

In estuaries, bioturbation contributes significantly to OM recycling; this may happen either directly through OM consumption (largely driven by macrofauna; Sturdivant and Shimizu 2017), or by the *indirect* effects on bacterial community and metabolism that arise from the modification of physico-chemical properties of the bioturbated sediment. One direct consequence of bioturbation is the 'irrigation' of anoxic sediments by O_2-containing seawater. This oxygenation modifies sedimentary redox potential and alters the environment for microbial communities. The rate and depth of bioturbation performed by a macroinfaunal community, and its relative contribution to OM degradation, may vary seasonally, due to seasonality in metabolic activity (Sturdivant and Shimizu 2017).

Trites et al. (2005) studied diatom communities and sediment properties in two mudflats in the Bay of Fundy; one of them had experienced a marked population decline of *Corophium volutator*, a keystone amphipod in mudflats, some 20 yr prior to the study. The mudflat lacking *Corophium* presented variable depths of the oxygenated layer and several anoxic patches, which were attributed to the lack of bioturbation. On the other hand, it had a comparatively larger density of epipelic diatoms, due to the lack of top-down controls, otherwise exerted by *Corophium*.

Furthermore, the redistribution of sediments caused through bioturbation disrupts stratification patterns and other sedimentary structures normally created by sediment deposition. In fact, the re-working and redistribution of sediments have marked effects on sediment stability. Bioturbation first evolved with the advent of complex metazoans during the Ediacaran period (635 to 542 MYA), when the oceans were becoming fully oxygenated. The traces left by these early bioturbators were shallow and parallel to the sediment surface. It has been suggested that these ancient bioturbators not only exploited microbially-colonized sediments as a food source, but also aimed at exploiting O_2-rich oases created by microbial oxygenic photosynthesis within an otherwise low-O_2 setting (Gingras et al. 2011).

Contrary to bioturbation, the physiological activity, modes of living and growth habit of certain organisms may stabilize soft-bottom sediments. This applies both to primary producers and metazoans. Section 8.5.2 discusses biostabilization by MPB and Box 1 describes the ways in which seagrasses do so in sandy bottoms. In reference to metazoan stabilizers, some reef-building species may become established on soft bottoms and significantly alter sedimentation, thereby acting as ecosystem engineers. For example, a study from a temperate coastal lagoon in Argentina (Schwindt et al. 2004) described the ecological role of an introduced serpulid polychaete in affecting sedimentation and water circulation. The reef-building activity of the serpulid (forming calcareous reefs up to 7 m in diameter and 0.5 m in height) caused a significant change in the lagoonal geomorphology, affecting bedload transport and sediment deposition. In combination with tidal currents and wind-induced circulation, the reefs act as efficient sediment traps and favor deposition of mud around them as well as the distribution and abundance of infaunal organisms (e.g., nematodes and ostracods) in the lagoon, possibly generating refuge for several soft-bottom organisms (i.e., ecological facilitation). Other important sediment-stabilizers in soft bottoms are oyster reefs and mangrove roots.

8.4 Deep-sea benthic ecology

The *HMS Challenger* Expedition (1872–1876) is regarded as the origin of modern oceanography, and the scientific knowledge of deep-sea benthic ecology has since advanced prodigiously, in parallel with technological developments. Deep-sea ecosystems support some of the largest biodiversity on the planet, but our knowledge is largely constrained by their remoteness and the technical challenges for their exploration. The discovery of high deep-sea benthos diversity has promoted intense research, intended to explain its patterns. The seminal work by Sanders (1968) proposed the stability-time hypothesis that has shaped much of contemporary research on deep-sea ecology. This hypothesis is based on the assumptions that (i) the deep-sea environment is ecologically stable, allowing coexistence through niche differentiation; and that (ii) stable conditions persisted long enough to allow extensive evolutionary diversification. However, the assumption of environmental stability in the deep sea has been challenged by recent exploration, and continuous advances in paleoceanography that contest the view of a long history of stability. Moreover, patterns of diversity have been shown to vary locally, regionally, and globally, suggesting that different environmental variables affect biodiversity at varying spatial scales (Stuart et al. 2003). Dayton and Hessler (1972) proposed that a high species diversity would rather arise as a result of continued biological disturbance than of niche diversification to avoid competition. According to their cropper-disturbance hypothesis, most deep-sea deposit feeders would consume living- as well as dead particles (a feeding mode defined as 'cropping'). Intense cropping, including the juvenile stages of other deposit feeders, would prevent competitive exclusion, allowing a high overlap in the utilization of food resources. However, there is little evidence for cropping in the deep sea, and most organisms do not present adaptations to avoid predation.

More recent hypotheses on deep-sea benthic biodiversity make a point on the fact that the deep sea has a quite high spatial heterogeneity at all scales. According to the more accepted spatial-temporal mosaic model (Grassle and Morse-Porteous 1987), low frequency disturbance or episodic introduction of OM (e.g., whale falls and pulses of phytodetritus resulting from phytoplankton blooms) would create patches that persist for long periods of time, compared to the slow recolonization rates of the deep-sea benthos. Species composition within patches depends on the length of time following disturbance and the identity of colonizing species. Thus, differences in species composition within patches and the surrounding environment remain for years, creating an environmental mosaic. Since the late 1970s, disturbance arose as an important factor shaping biodiversity patterns in all kind of environments. The intermediate disturbance hypothesis (Connell 1978) postulates a broad causal relationship between disturbance and biodiversity, and the relevance of disturbance in shaping ecosystems has been tested and widely accepted across multiple marine and terrestrial environments. In the case of deep-sea benthic communities, the spatial-temporal mosaic hypothesis can be experimentally tested in the field, while the stability-time hypothesis entails an evolutionary timescale and cannot be tested by experiments done in ecological time. Observations and results from field experiments are consistent with the spatio-temporal mosaic hypothesis for high deep-sea diversity, but it is a forthcoming challenge for benthic ecologists to test the relative importance of ecological and evolutionary timescale processes in the deep sea.

Over the past few decades, many new, highly-diverse deep-sea habitats have been described, including canyons, seamounts, vents, ridges, deep-water coral reefs, cold seeps, pockmarks, mud volcanoes, carbonate mounds, brine pools, gas hydrates, fractures, and trenches, all hosting rich, unique and diverse biological assemblages. In relation to the energy source that fuels them, deep-sea ecosystems can be divided into two major groups: heterotrophic and chemosynthetic habitats.

Most deep-sea benthic organisms are heterotrophs sustained by OM photosynthetically produced at the ocean's surface. They comprise heterotrophic communities on continental margins from the shelf break to 3,000 m depths, and the vast abyssal plains extending between 3,000 m and 6,000 m depths that cover the greatest proportion of the ocean floor. In comparison, continental margins cover around 13% of the world's seafloor (Wollast 2002) and harbor up to 90% of terrigenous sediments (McCave 2002). At large spatial scales, abiotic factors such as geology, temperature, currents, and water masses drive the composition and abundance of faunal assemblages. On the other hand, factors like downward flux of OM, O_2 concentration, sediment type, and episodic or catastrophic events modulate species distribution at spatial scales of 1 to 100 km (Gage 2002). Yet, it is at smaller scales that biological interactions become most relevant. As a general pattern, the epifaunal and suspension-feeding benthic biomass decreases with sediment and water depth, whereas burrowing animals are more abundant in deeper sediment layers and deep-sea sediments, where deposit-feeding dominates. Analogously, species composition tends to shift from deposit feeders dominating deep-sea sediment slopes, to a greater species richness of epifaunal organisms dominated by suspension feeders in shallower waters, where higher current velocities and suspended particle concentrations prevail. On the continental slope, close to the shelf, the increase in grain size correlates with a decrease in the number of organisms and species, and a peak in biodiversity at mid-slopes has been observed (Stuart et al. 2003). Finally, canyons are deep incisions on the continental margins and are often considered as biodiversity hotspots that act as major pathways for OM export from the continent to the deep sea. Mobile sediments on the canyon floor commonly support a high proportion of endemic species, but their irregular topography and the difficulty of sampling have prevented their detailed investigation until recently (e.g., De Leo et al. 2020).

Abyssal plains lie beyond the continental slope, and are covered by a flat, thick layer of fine sediments. These extensive environments are apparently stable over short- and long timescales. However, physical disturbances like tidal variations in the flow of cold dense water close to the seafloor, and high-energy, unpredictable events such as benthic storms or turbidity currents may affect biological communities in these remote areas. Abyssal plains support a very high biodiversity at the ecosystem scale (gamma diversity), composed mainly of meiofauna (dominated by nematodes and foraminifera; Gooday 1996), and macrofauna (dominated by polychaetes, with small peracarid crustaceans, mollusks, nemerteans, sipunculans, echiurans, and enteropneusts; Grassle and Maciolek 1992). Larger macrofauna (body size >> 1 cm) is composed of holothurians, asteroids, echinoids, decapod crustaceans, and fish (Gage and Tyler 1991). Even though abyssal plains have been sampled since the times of the *HMS Challenger* Expedition, only a small fraction of the vast extent of these ecosystems has been studied to date. Some latest research suggests that present-day climate change is already impacting deep-seafloor environments of the abyssal plains, as evidenced by increased deep-sea temperature (see Meinen et al. 2020), deoxygenation, lowered pH of intermediate deep-waters, and altered POC flux to the seafloor (see Sweetman et al. 2017 and references therein for recent studies on deep-sea habitats worldwide). Overall, heterotrophic deep-seafloor ecosystems depend on the OM photosynthetically produced at the ocean's surface and the ongoing climate-driven environmental changes may be disturbing their functioning with still poorly understood consequences.

Deep-water chemosynthetic ecosystems have been known and studied for only a few decades, since the discovery of hydrothermal vents in 1977. Although these ecosystems comprise an environmental mosaic of soft- and hard bottom substrates, where tectonic and geothermal activity is of paramount importance, we will discuss them only briefly here. In these ecosystems, primary producers are chemoautotrophic microorganisms that use reduced inorganic compounds, such as H_2S or CH_4, to synthesize OM. Hydrothermal vents occur at the spreading centers of mid-ocean ridges, where cold seawater permeates through cracks in the recently formed crust. As water moves through the mantle, it gets heated, depleted of O_2, and enriched in reduced minerals (many of which are metal sulfides). When the superheated fluid seeps and mixes with seawater, the metals precipitate and form the dense 'black smoke' that characterizes hydrothermal vents. Dense biological populations associate with hydrothermal vents and sustain complete autochthonous (chemolithotrophic) food webs based on the energy of reduced compounds from the Earth's interior (Karl et al. 1980, Jannasch and Mottl 1985).

Hydrothermal vent microorganisms form symbiotic relationships with many invertebrate groups, such as worms and bivalves (Felbeck et al. 1981) that, in turn, present specific adaptations. For instance, the giant tubeworm *Riftia pachyptila* does not have a mouth or digestive system but a sack filled with chemoautotrophic bacteria called trophosome. *Riftia* worms take up CO_2 and H_2S from the hydrothermal fluid, and blood vessels irrigate the trophosome, where the sulfide-oxidizing bacteria synthesize OM. Symbiotic relationships also appear in other groups, such as clams, mussels, shrimps, crabs and other polychaetes, with different degrees of dependency. Hydrothermal vents support dense populations of large invertebrates, but with a low diversity, like other ecosystems with extreme chemical characteristics. However, the biomass of these communities sustained by an abundant supply of reduced compounds, exceeds 20–30 kg m^{-2}, which is much higher than the most productive coastal ecosystems (Lalli and Parsons 1993).

Cold seep communities were discovered in 1983, in the Gulf of Mexico, at about 500 m depths (Paull et al. 1984). In cold seeps, there is a seepage of cold fluid from the sea floor, containing a high concentration of CH_4, that may be of biological or thermogenic origin (Levin et al. 2016). These high methane concentrations tend to co-occur with high concentrations of H_2S, produced by the bacterial anaerobic oxidation of methane (see Section 8.5.1). Both CH_4 and H_2S sustain highly productive cold seep communities through chemoautotrophy by free-living and symbiotic bacteria (Paull et al. 1984, Barry et al. 1997). Cold seep habitats are more stable systems than hydrothermal vents, which only exist on the order of several decades, and there is a transition of physical and chemical factors between the seep habitat and the heterotrophic surrounding system, allowing for a higher biodiversity that includes bivalves (mytilids, vesicomyids, lucinids, and thyasirids) and tubeworms (siboglinid polychaetes), with sponges, gastropods, and shrimps sometimes also abundant (Levin et al. 2016).

Finally, chemosynthesis-dominated communities can also develop in anoxic areas where mid-water oxygen minima (< 0.5 ml l^{-1} dissolved O_2) intercept the continental margin. These seafloor Oxygen Minimum Zones (OMZ) form in areas of high pelagic primary production and poor water circulation, where the biological degradation of the sinking OM results in O_2 depletion (Rogers 2000, Helly and Levin 2004). Seafloor OMZs typically occur between 200–1,000 m depths in the eastern Pacific, NW Pacific margin, Philippines area, Bay of Bengal, Arabian Sea and SW Africa beneath the Benguela current (Rogers 2000, Helly and Levin 2004). The high OM concentrations sustain large numbers of sulfide-oxidizing bacteria and many heterotrophs in OMZs consume chemoautotrophic bacteria by grazing on bacterial mats (Gallardo et al. 1995). Endosymbiotic sulfide-oxidizing bacteria are also widespread in foraminifera, flagellates, ciliates, some polychaetes, and some bivalves (Levin et al. 2016). The details of the metabolic interactions between host and symbiont remain unknown, as does the extent to which chemosynthesis provides nutrients to the OMZ benthos. Ongoing and future research will, no doubt, extend our knowledge on these types of relationships, and elucidate the evolutionary links with fauna from other deep-water chemosynthetic ecosystems (Ramírez-Llodra and Billett 2006).

8.5 The role of biota influencing soft sediment properties

8.5.1 Microorganisms and biogeochemistry

From an ecosystem perspective, a major function of soft-bottom ecosystems is that of nutrient regeneration, through the processing of OM that comes from different autochthonous and allochthonous sources (Sundback et al. 2003). In the oxic layer, OM is rapidly colonized by heterotrophic microorganisms, using the available O_2 as the primary electron acceptor. Degradation of nitrogen-containing organic compounds (Fig. 8.2a) involves two biologically mediated transformations: ammonification and nitrification. Ammonification is the process by which microorganisms mineralize dissolved organic molecules using O_2, nitrate or nitrite as electron and producing NH_4^+ in the process (either by deamination of nitrogen-containing organic molecules and/or by reducing oxidizing inorganic nitrogen species). Ammonification occurs in either aerobic or anaerobic conditions and is the intermediary step between the biological degradation of large organic molecules and nitrification (Reddy and DeLaune 2008). Nitrification is the biological oxidation of NH_4^+ to NO_2^- followed by the oxidation of the NO_2^- to NO_3^-. There are two functionally distinct groups of nitrifiers: those that oxidize NH_4^+ to NO_2^- (ammonia-oxidizing bacteria and archaea) and those that oxidize NO_2^- to NO_3^- (nitrite-oxidizing bacteria). These microbes are all aerobes, and predominantly chemolithoautotrophic that depend on CO_2 as their major carbon source and use NH_4^+ or NO_2^- as the sole source of reducing power for their autotrophic nutrition (Ward 2008).

In anaerobic conditions, different groups of bacteria use alternative electron acceptors to decompose OM and liberate energy. The most important groups are fermenters, NO_3^- reducers, SO_4^{2-} reducers, and methanogens. In fermentation, water-soluble substances are degraded to low molecular weight organic compounds like lactate, acetate, propionate, or different alcohols. Fermenting bacteria are mostly obligate anaerobes (meaning that they require anaerobic conditions and are poisoned by O_2) that can derive

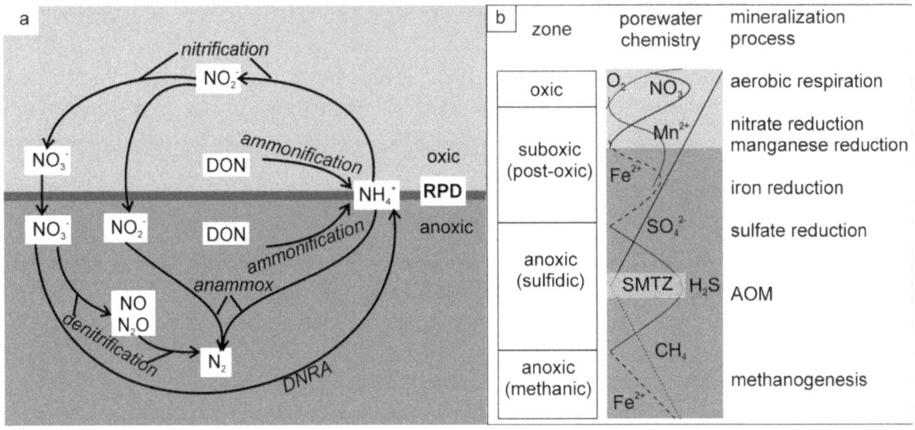

Figure 8.2. (a) The marine biogeochemical nitrogen cycle and the main redox reactions involved. RPD: redox potential discontinuity; DON: dissolved organic nitrogen; DNRA: dissimilatory nitrate reduction to ammonium; (b) Schematic representation of the biogeochemical zonation in marine sediments, according to Froelich et al. (1979) and Berner (1981; in parenthesis), and relevant dissolved species in the porewater. The depth, peak heights and concentration scales are arbitrary, but reflect the depth sequence of the dominant OM mineralization processes. SMTZ: sulfate-methane transition zone; AOM: anaerobic oxidation of methane (modified from Froelich et al. 1979).

a relatively small amount of energy through partial oxidation of organic substrates using as electron acceptors selected organic compounds present in the bacterial cell (Molongoski and Klug 1976).

Nitrate reduction may follow two different pathways: denitrification and dissimilatory nitrate reduction to ammonia (DNRA). Most denitrifying bacteria are facultative aerobic heterotrophs that switch from aerobic respiration to denitrification when O_2 is unavailable. In anaerobic conditions, denitrifying bacteria reduce NO_3^- and ultimately produce N_2 through a series of intermediate gaseous nitrogen oxide products. This reaction occurs under moderately reduced conditions, in the absence of O_2 (Ljundahl and Erickson 1985). In mudflats, coupling of aerobic and anaerobic processes is enhanced, and much of the NO_3^- formed during nitrification under aerobic conditions is rapidly consumed by denitrifiers in adjacent anaerobic environments. Furthermore, denitrification can occur in aerobic environments as well. For instance, Marchant et al. (2017) reported that a bacterial species, *Paracoccus denitrificans*, participates in denitrification under both aerobic- and anoxic conditions simultaneously, with a high activity in sandy intertidal sediments. It reinforces the role of intertidal soft bottoms in the mitigation of pollution by NO_3^- resulting from excessive agricultural or residential fertilizers.

Dissimilatory nitrate reduction to ammonia (also known as DNRA or nitrate/nitrite ammonification) is the result of anaerobic respiration by heterotrophic microbes using NO_3^- as an electron acceptor for OM oxidation. Unlike denitrification, DNRA preserves bioavailable nitrogen in the system, producing soluble NH_4^+ rather than unreactive N_2 gas. Intertidal sediments are hotspots for DNRA, and recent studies suggest that redox-oscillation due to tidal pumping may sustain this nitrate-reducing pathway (Zheng et al. 2016).

Nitrogen reduction under anoxic conditions can also occur through a process called anaerobic ammonium oxidation (*anammox*). In the presence of NO_2^-, the anammox process also produces N_2, by using NO_2^- as an electron acceptor to oxidize NH_4^+ (Keunen 2008). Anammox makes a large contribution to N_2 formation on the continental shelf (Thamdrup and Dalsgaard 2002). In intertidal environments, a coupling between nitrification and anammox was proposed, similar to nitrification-denitrification coupling, in which anammox would be enhanced by high rates of nitrification in aerobic zones and accumulation of NO_3^-, which can diffuse into anaerobic zones. If denitrification is NO_3^- saturated, some of the NO_2^- formed may support anammox (Dalsgaard et al. 2005).

The DNRA coupled with anammox is likely to be responsible for significant nitrogen losses in areas with no detectable denitrification, such as in OMZs off the coast of Chile, Peru and Namibia. While denitrification is more energetically favorable than DNRA, there is evidence that bacteria using DNRA gain more energy than denitrifiers, allowing them to grow faster. Thus, *via* DNRA-anammox coupling, bacteria using DNRA and anammox may be stronger competitors for substrates than denitrifiers (Jensen et al. 2011).

Sulfate-reducing bacteria and sulfate-reducing archaea are sulfidogenic microorganisms that can perform anaerobic respiration by reducing SO_4^{2-} to H_2S. Many organisms reduce small amounts of SO_4^{2-} to synthesize sulfur-containing cell components in a process called assimilatory sulfate reduction. By contrast, dissimilatory sulfate reduction is the reduction of large quantities of SO_4^{2-} by sulfate-reducing microorganisms to obtain energy. Most of them are anaerobes, but there are examples of sulfate-reducing microorganisms that are O_2-tolerant, and some of them can even perform aerobic respiration (Kjeldsen et al. 2004). A toxic waste product of SO_4^{2-} reduction is H_2S, responsible for the sulfurous odor of anoxic sediments. Further, H_2S can react with metal ions in water to produce metal sulfides such as ferrous sulfide (FeS), an insoluble, black compound, leading to the dark color of anoxic sludge (Muyzer and Stams 2008). SO_4^{2-} is a much less favorable electron acceptor than NO_3^-, Mn^{4+} and Fe-oxides, but sulfate occurs at high concentrations in marine environments (~ 2.7 mg l^{-1} for seawater). Thus, sulfur cycling largely regulates many microbial communities and redox reactions.

Finally, methanogenesis is the dominant process in OM degradation in anaerobic sediments where SO_4^{2-} is depleted. All known methanogens belong to the domain of archaea and produce CH_4 by using CO_2 as the terminal electron acceptor in respiration at the final step in OM decomposition. The sulfate-methane transition zone (SMTZ) located from 1 to 100s of m below the sediment surface is the region where SO_4^{2-} is almost depleted and CH_4 concentration starts to markedly increase (Komada et al. 2016). Biogeochemical cycles in this SMTZ are increasingly gaining attention because of a unique microbial community whose main form of metabolism is the anaerobic oxidation of methane (AOM). The AOM metabolic pathway uses SO_4^{2-} and CH_4 to produce certain carbon species (mainly HCO_3^-) and H_2S (Fig. 8.2b). It is thought that AOM could be a process that effectively controls CH_4 emission from many anaerobic environments into the atmosphere (Thomsen et al. 2001).

8.5.2 *The role of microphytobenthos in well-illuminated sediments*

The microscopic, unicellular eukaryotic algae and cyanobacteria, which grow in the uppermost few mm of illuminated sediments and in association with sedimentary particles, are collectively known as epipelic microphytobenthos (Vos and de Wolf 1993, MacIntyre et al. 1996). For brevity, microphytobenthos will be abbreviated with the acronym MPB. The term epipelic makes reference to their growth habit associated with surface sediments (*epi* = on top of; *pelos*, Greek word for clay); when small diatoms live attached to *individual* sand grains, their habit is known as epipsammic (Greek: *psammos* = sand; Stal 2010). Epipelic microalgae are highly motile, migrate vertically through the sediment in order to optimize light harvesting for photosynthesis and often form a conspicuous biofilm layer on the sediment surface (Fig. 8.3).

Surface processes are of paramount importance to photosynthetic microbial communities, particularly with respect to anchoring, light-harvesting, and nutrient exchange. On the one hand, (i) the high surface area/volume ratio of fine-grained sediments offers a vast and structurally complex landscape for colonization by microbes that rely on surface-adhesion processes (Pan et al. 2018). Furthermore, (ii) sediment grain size can determine the type of dominant MPB community that becomes established. On the other hand, as MPB is autotrophic, (iii) they need abundant solar radiation to carry out photosynthesis and net primary production. In sediments, the penetration of the different wavelengths of PAR (photosynthetically active radiation, 400–700 nm) is very different than in seawater; blue wavelengths become extinguished in the first 1 mm of sediment, whereas orange and red wavelengths penetrate down to 2.5–3 mm, while the pattern is the exact opposite in seawater (see Fig. 3.1 in this volume). Thus, differential wavelength penetration plays an organizing role in the distribution of photo-autotrophic microbes. Finally, (iv) the supply of nutrients is largely governed by tidal flooding, as tidal sediment-water exchange of nutrients is several orders of magnitude larger than diffusive fluxes through porewater (Ospina-Alvarez et al. 2014).

Living MPB communities modify sediment properties as part of their normal physiology, especially with regards to stabilization processes. Epipelic diatoms (Fig. 8.3b–e) exhibit a remarkable motility and excrete extracellular polymeric substances (EPS, see Section 8.1; de Brouwer et al. 2005, Decho and Gutierrez 2017), which not only facilitates the gliding motion of individual cells through the sediment matrix but, on an ecosystem scale, contribute to particle adhesion and agglutination, ultimately having

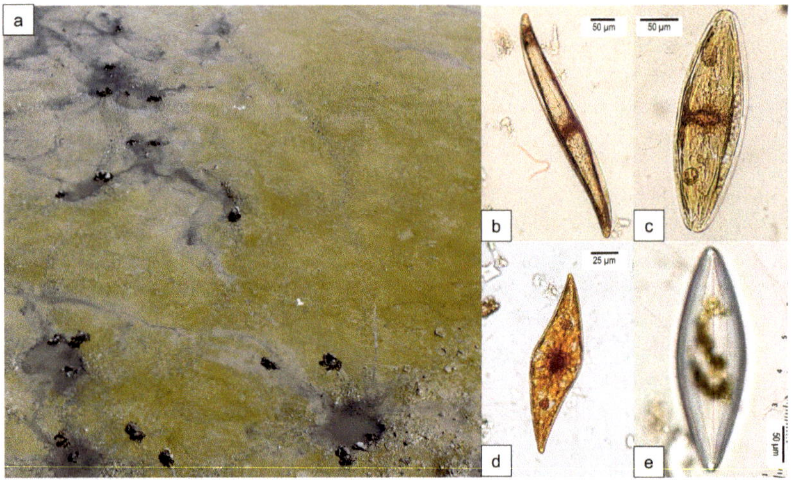

Figure 8.3. Microphytobenthos in intertidal mudflats at Mar Chiquita, a coastal lagoon in the SW Atlantic Ocean (37°45′ S). Note the golden-greenish coloration imparted to the sediment surface by the dominant diatom biofilm, and the traces left by crab (*Neohelice granulata*) foraging excursions. A collection of light micrographs of common epipelic diatoms from the same location: (b) *Gyrosigma spencerii*; (c) *Tryblionella circumsuta*; (d) *Pleurosigma* sp.; (e) *Caloneis bivittata*. Photo credits (a), (b) and (c) by J. Pan; (d) and (e) courtesy of M.F. Yorlano.

marked effects on biostabilization and geomorphology (Stal 2010). A meta-analysis, integrating data from three European mudflats, found a linear positive fit between EPS concentration as a function of chlorophyll *a*, leaving no doubt that the largest EPS fraction is excreted by photoautotrophs (Stal 2010). Conversely, the same study plotted EPS concentration as a function of the median grain size of the sediment and produced a negative hyperbolic function, which may relate to a higher biomass of microorganisms, or adsorption processes in fine-grained sediments that make them better capable of retaining the EPS.

Despite being typically overlooked as primary producers, the MPB often is the dominant autochthonous source of photosynthetically fixed C in estuaries, and when adjusted for their large extent within an estuary, unvegetated habitats comprise the majority of the estuarine carbon pool (Bulmer et al. 2020). The distribution and abundance patterns of MPB are not spatially-uniform but rather patchy across sand- or mudflats. Several factors interact and juxtapose to determine the community structure of MPB, such as seasonality of physical (e.g., temperature, light, resuspension) and chemical parameters (e.g., nutrient and O_2 concentration, pH and redox potential; Jesus et al. 2009) and species-specific characteristics of the microbial consortia. This patchiness is consequently reflected at macro- (Riethmüller et al. 2000, Fidalgo e Costa et al. 2002) and microscales (Pan et al. 2017) with a resulting environmental mosaic. Soft-bottom benthic animals make differential use of this MPB mosaic, accordingly.

8.5.3 Mechanistic and ecological aspects of deposit-feeding

What animals ingest and how they digest it has always puzzled biologists. Let us now apply these timeless biological questions to soft-bottom marine sediments. We know by now that marine sediments represent quite a dilute medium with respect to OM, and we also know that sedimentary OM has many chemical signatures and varying levels of digestibility (or refractoriness, if looked at the other way around). Despite these apparent first-hand constraints, many metazoans pertaining to diverse taxonomic groups and various body size classes (i.e., macro-, meso- and microfauna) and life habits (i.e., epi-, infaunal and meiofaunal species) make use of the 'active' pools of OM in surface marine sediments, as we described in previous sections.

Sediments ingested by benthic animals can be expected to experience a unique environment controlled by the digestive chemistry of the organism (Mayer et al. 1996). A sediment particle's passage through the gut of a deposit-feeding invertebrate initiates with ingestion and physical trituration, followed by enzymatic solubilization *via* digestive fluids in the fore- and midgut, and ends with selective absorption in the hindgut or, alternatively, egestion (defecation) of unassimilated solid material. Solubilization is a necessary step in the digestive process as many organic compounds in marine sediments occur in association with (adsorbed to) solids. Solubilizing agents include amino acids, surfactants and enzymes. Mayer et al. (1995) applied an experimental approach that used an *in vitro* biomimetic incubation of microbial proteases and fresh sediments and found that amino acids were yielded at a similar rate as from gut juices from a deposit-feeding holothuroid. Microbial extracellular proteolytic activity is regulated by a slow hydrolysis step in which small peptides and amino acid are separated from refractory substrates.

It is important to make a clear distinction between two main pathways to enzymatic degradation of OM: the microbial one (*extracorporeal*, Mayer et al. 2001), in which heterotrophic bacteria secrete enzymes to an external (sedimentary) medium and make labile breakdown products of OM readily available for uptake by microbes and animals; and the *intracorporeal* (Mayer et al. 2001), in which enzymes of animal or microbial origin (the latter obtained through symbiotic associations) are secreted into the gut lumen of benthic invertebrates, and so the digestive process is confined to an internal compartment. To test the efficiency of these two main enzymatic degradation modes, Mayer et al. (1997) compared the enzyme activities in 19 species of echinoderms and polychaetes (including deposit- and suspension feeders, and carnivores), with those of extracellular (bacterial) enzymes from ambient sediments. Enzyme activities ranged over five orders of magnitude, with averages decreasing in the order polychaetes > holothuroids > sediment. Deposit feeders had the most intense surfactancy, which was to be expected, considering that the digestible substrate is diluted by adsorption to mineral grains. Furthermore, deposit feeder digestive fluids solubilized sedimentary food substrates more readily than sedimentary extracellular enzymes, owing to either more powerful solubilizing agents, or to gut environments which better channel chemical condensation as well as solubilization reactions (Mayer et al. 2001). A marked difference resides in extracellular hydrolytic enzymes being *sorptive* for bacteria and *dissolved* for animals, enhancing both speed and net digestive gain for the latter.

In a very graphical analogy, Mayer et al. (2001) likened the guts of deposit-feeding invertebrates to laundry washing machines: the gut resembles the inside cylinder of the machine into which dirty clothes with stains are thrown. In the case of a deposit feeder, dirty clothes with stains would be like sediment grains with adsorbed OM, thus is, a mixture of digestible and indigestible components. Through the process of laundering, the detergent extracts some components of the 'mixture' (say, lipids and proteins adsorbed onto the textile of the garment) and clean water induces desorption of remaining adsorbed dirt and cleaning agents, carrying it away and leaving the washed clothes with little or nothing more to extract. In the guts of deposit-feeding invertebrates, the digestible fraction would be the OM adsorbed, diluted or encapsulated in the sediment, and surfactants and specific digestive enzymes that are secreted into the fore- and midguts may be likened to the detergent used in washing machines. Gut wall absorption occurs throughout the gut but is completed in the hindgut by rinse cycles induced by either back-flushing with clean water from the posterior end, or simple absorptive removal of dissolved materials.

Let us now illustrate the previous exposition with a few examples, first revisiting the relative importance of dead- and live pools of phytoplanktonic production to deposit feeders. In order to test the hypothesis that deposit-feeding animals preferentially consume a freshly settled supply of planktonic diatoms, Hansen and Josefson (2004) traced the fate of the spring bloom once it settled into coastal aphotic sediments. Surprisingly, they found that the diatom spring bloom was not utilized immediately by the benthic community; in fact, a major fraction passed through the guts of deposit feeders without being digested. This finding indicates, on the one hand, that the pool of live pelagic diatoms in the sediment rather is a food source that gets buried and gradually becomes available as diatoms degrade. On the other hand, there is a considerable time lag between the input of food and the secondary production of benthic deposit feeders.

Deposit feeding may also affect planktonic recruitment. Albertsson and Leonardsson (2001) studied how a burrowing deposit-feeding amphipod reduced the numbers of benthic resting eggs of copepods by negatively impacting their hatching through predation, burial or physical injury, ultimately affecting zooplankton recruitment.

Box 1: Seagrass bed communities

Seagrass beds stand out among vegetated coastal ecosystems. *Seagrass beds* or *seagrass meadows* are composed of dense populations of flowering plants pertaining to various botanical families, including Hydrocharitaceae, Cymodoceaceae, Zosteraceae, etc. They may be composed of mono-specific assemblages, or more commonly, more than one species of seagrasses. For example, in the Indian River Lagoon on the Atlantic coast of Florida, all six species of seagrasses known from the tropical Western hemisphere occur jointly (Dawes et al. 1995). These comprise the larger and most common species *Halodule wrightii*, followed in abundance by *Syringodium filiforme* and *Thalassia testudinum* (Fig. 8.4a). Three species of the genus *Halophila* (*H. decipiens*, *H. englemannii*, and *H. johnsonii*) complete the assemblage in deeper water, while the least common *Ruppia maritima* (a freshwater species that can tolerate varying salinities) is found in shallow areas. On the other hand, in temperate and subtropical subtidal sandy sediments, eelgrasses such as *Zostera* sp. and surfgrasses (e.g., *Posidonia* sp.) occupy similar ecological niches.

A common feature of seagrass beds is that they develop on well-illuminated (often oligotrophic) soft sandy bottoms, in coastal areas with restricted circulation, such as embayments and estuaries, where calm hydrodynamic regimes prevail. In turn, seagrass beds can modify the hydrodynamic regime. For example, wave heights are attenuated by 25–49% within a temperate *Zostera marina* seagrass meadow compared to an adjacent bare site (Reidenbach and Thomas 2018). It has also been shown that the density of flexible shoots affects the hydrodynamics, and that water flow is slowed down when passing above seagrass beds (Fonseca et al. 2019). Seagrass beds are distributed across tropical and temperate latitudes worldwide. As seagrasses are vascular plants, they are rooted to the sandy substrate by an intricate radicular system and grow in modular form through rhizomes (modified shoots) that run parallel to the sediment surface (Fig. 8.4b). The only parts that grow above the sediment are modified, ribbon-shaped leaves and the reproductive structures (flowers).

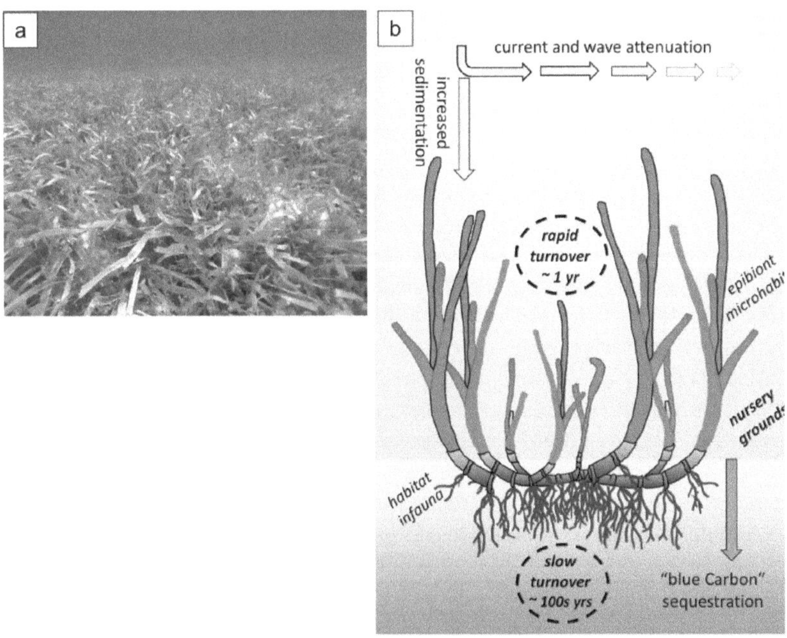

Figure 8.4. (a) *Thalassia testudinum* seagrass beds in the southwestern Caribbean (San Andrés Island, 12°S). (b) Schematic representation of the above-ground and below-ground portions of a seagrass, and its rhizomatous shoots, and the different ecosystemic processes that are influenced or determined by seagrass beds. Photo credits (a) courtesy of L. Debali.

Box 1 Contd. ...

...Box 1 Contd.

Seagrass meadows are highly productive and highly autotrophic, despite the oligotrophic environment in which they develop. A recent study found that the net daily ecosystem production (54–119 mmol m^{-2} d^{-1}) was about one-half of gross primary production, implying that they fix substantially more organic carbon than they respire (Koopmans et al. 2020). Due to their capacity to fix seawater-dissolved CO_2 through photosynthesis and incorporate it into coastal sediments where the OM is sequestered for long periods of time (100s of yr) and preserved through anaerobic conditions (forming peats), seagrass meadows are one of the so-called 'Blue Carbon' ecosystems (treated more extensively in Chapter 9). In fact, there is a global dataset on seagrass soil organic carbon storage, which is of importance for conservation policies that contemplate mitigation of the effects of anthropogenic CO_2 inputs (Bedulli et al. 2020). A study of Mediterranean seagrass (*Posidonia oceanica*) beds identified a distinctive ecological dynamic between aboveground and belowground parts (Romero et al. 1992). Seagrass leaf biomass is highly dynamic and most of the leaf material is decomposed or exported in less than one year, representing a net nutrient loss. Conversely, belowground biomass (roots and rhizomes) has turnover times in the order of centuries, with a consequent accumulation of OM in the sediment. The potential for organic carbon storage amounts to 5.1 kg C_{org} m^{-2} on average, for 0.5 m-thick seagrass soil deposits in Indian Ocean seagrass beds off Australia (Bedulli et al. 2020).

Thayer et al. (1984) reviewed the role of herbivores on seagrasses. Only a few animals graze directly on living seagrasses, a fact that is attributed to factors such as the chemical composition of the plants, featuring relatively high amounts of structural cell wall material (cellulose and lignin), and the presence of toxic or inhibitory substances in their tissues. Herbivores feeding on seagrass beds range from urchins, fishes, sirenians and green turtles in tropical systems to waterfowl in temperate systems. However few in numbers, these macroherbivores can have a profound effect on seagrass canopies, altering the nutrient content and digestibility of the plant, as well as its productivity.

Leaf litter production and decomposition in seagrass beds are very important processes from an ecosystemic point of view. Seagrasses produce significant amounts of leaf litter that undergoes microbial colonization by bacteria and fungi, as well as consumption by specific fauna (Walker et al. 2001). It has been shown that decomposing buoyant seagrass detritus is capable of supporting planktonic larvae of a commercially important fish (the blue grenadier, *Macruronus novaezelandiae*), in an unusual case of benthic-pelagic coupling (Thresher et al. 1992). Moreover, a recent review (Duarte and Krause-Jensen 2017) concludes that the contribution of seagrass meadows to carbon sequestration has been underestimated by only including carbon burial within seagrass sediments, overlooking the export of seagrass carbon beyond the meadows (either in particulate or dissolved forms), which may be 30% of the carbon sequestered within seagrass soils.

Seagrass beds offer structural complexity in an otherwise topographically simple, two-dimensional environment. Therefore, seagrasses may be considered *ecosystem engineers* (see Chapter 10 for more on this topic), providing habitat for macrozoobenthos, fishes and manatees (Dawes et al. 1995). In fact, a study by Serrano et al. (2017) notes that when eroded, the mats that form inside *Posidonia* beds result in the formation of escarpments with caves that are used as a three-dimensional habitat by reef fishes such as groupers. Moreover, several epifaunal animals find optimal microhabitats for settlement and protection from predation in seagrass canopies (Leber 1985). Many benthic and planktonic species of invertebrates and fish use seagrass meadows as nursery grounds during some part of their life history. A meta-analysis of published papers concludes that the abundance, growth, and survival of a number of species are greater in seagrasses than in unstructured habitats. Yet, there are few significant differences in abundance, growth or survival when seagrass meadows are compared to other structurally complex habitats, such as oyster reefs or macroalgal beds (Heck et al. 2003). Many fish species that use seagrass beds as nursery grounds go through ontogenetic changes in feeding habits (Carr and Adams 1973). The seagrass ecosystem offers varied trophic pathways, supplementing exclusive planktivory (copepods, and crustacean and molluskan larvae) by herbivory, detritivory and carnivory (consuming primarily benthic invertebrates). Also, detritus may be consumed by juveniles of some fish species.

Box 1 Contd. ...

...Box 1 Contd.

The overall health of seagrass beds seems to be declining worldwide as a product of anthropogenic impacts, with a reducing trend in both their areal extent and water-depth. Short-term events such as storm impacts, local water quality, clarity, salinity, and temperature are all factors that affect seagrass distribution and the relative composition of seagrass species (Fletcher and Fletcher 1995). A study on a *Zostera muelleri* intertidal bed found that elevated turbidity is associated with lower nutrient (NH_4^+) removal capacity (Bulmer et al. 2018), suggesting that the impacts of these factors are not one-directional, but connected to one another *via* feedback mechanisms. Translocation of species and biological invasion is another issue that may have profound effects on seagrass ecosystems. A recent study based on long-term (~ 10 yr) monitoring of the invasive Indian Ocean seagrass species *Halophila stipulacea* in Caribbean seagrass beds targeted some of its ecological impacts. For example, *H. stipulacea* likely displaces some native seagrasses (*Syringodium filiforme* and *Halodule wrightii*), but not others (*Thalassia testudinum*). On the other hand, due to its ability to grow well under nutrient limitation, it has lower N and protein content and higher C:N ratios than native seagrasses, and as such is of poorer nutritional quality to consumers. Thus, while maintaining a physiognomically-similar ecosystem, the replacement of native seagrasses by *H. stipulacea* may alter ecosystem functions (Muthukrishnan et al. 2020).

8.6 Benthic-pelagic coupling

The pelagic and benthic sub-systems are often conceived as separate compartments in aquatic ecosystems (freshwater and marine), with unique physical and biological characteristics that govern their functioning and structure. Despite often being treated as discrete and independent entities and traditionally studied separately, these sub-systems are intimately interconnected. Such connections may be established through physical or biologically-mediated processes, summarized under the term benthic-pelagic coupling (BPC). Take, for instance, shallow ecosystems such as coastal lagoons (a type of estuary, under the geomorphological classification presented in Chapter 1). Wind-stress induced mixing coupled with shallowness is the most important physical factor determining the functioning of coastal lagoons (Kjerfve 1994), in turn promoting a tight trophic coupling between the plankton and benthos. Vertical turbulent diffusion and horizontal transport by tidal currents make phytoplankton much more readily available to the benthos, and therefore, the macrozoobenthic community in lagoons is dominated by suspension-feeding organisms (Bazaïri et al. 2003). Likewise, benthic secondary production, respiration and biogeochemical cycles in deep-sea sediments are correlated with primary productivity in the overlying water *via* vertical carbon flow (Hansen and Josefson 2004, Lessin et al. 2019).

The study of ecological BPC has traditionally focused on trophic interactions and those that involve reproductive patterns and strategies. Through the former, benthic suspension-feeding organisms are known to exert a dominant organizing role in shallow aquatic ecosystems by diverting production from the water column to the benthos (French McCay et al. 2003). In general, planktonic systems are less mature (here used as synonym for long-lived) than benthic ones, and in the coupling of plankton and benthos there is a net transfer of energy from the plankton to the benthos (Margalef 1963, Cloern and Dufford 2005). While there are physical (e.g., plankton sedimentation) and biological mechanisms (e.g., the occurrence of benthic suspension-feeders) implied in the directionality of energy transfer from one system to the other, it is the ecological maturity that dictates the directionality of the energy flow from the less mature towards the more mature one (Margalef 1963).

We will now discuss trophic interactions and specifically focus on benthic suspension-feeding, an ecological strategy of energy acquisition that only occurs in aquatic environments. We will then close this chapter by discussing some reproductive strategies of marine organisms linking the benthic and pelagic environments.

8.6.1 Suspension-feeding in soft-bottom benthos

As you know from our introductory chapters on plankton, the oceans may be regarded as a fluid medium that harbor particles of all sizes and nature. According to particle Reynolds number, the smaller the particle the more viscous a fluid medium behaves for that particle (see Chapter 6). Representatives of a number of animal phyla have convergently evolved particle-trapping mechanisms to profit from the large amount of particles suspended in seawater, a strategy known as suspension-feeding. Suspension-feeding strategies are varied and widespread among marine organisms, spanning the gamut from unicellular ciliates to baleen whales (see Chapter 12). There are no analogous mechanisms to suspension-feeding in terrestrial ecology, with the only possible exception being spiderwebs, which are by no means analogous, but comparable. We will not discuss all kinds of morphological and physiological adaptations to suspension feeding, but rather focus on two of the most-studied mechanisms of suspension feeding, those of bivalve mollusks and serpulid polychaetes.

A key physiological feature of suspension-feeding bivalve mollusks is their high feeding rates, possibly an adaptation to living in estuarine systems that historically supported low concentrations of phytoplankton in relation to high loads of detrital and mineral particles (Jørgensen 1996, Newell 2004). Despite phytoplankton being the prime source of nutrition for bivalves, it is important to be aware that they can feed on a wide spectrum of items, ranging from bacterioplankton (Wright et al. 1982) to mesozooplankton (copepod nauplii; Zeldis et al. 2004).

Several mechanistic studies on suspension-feeding in bivalves (reviewed by Jørgensen 1996) point to feeding being physiologically regulated to maximize net energy gains in response to variations in quantity and quality of suspended matter in the environment (i.e., total particulate matter in suspension and its organic content; Bayne 1998). Filtration rates (i.e., water pumping rate) and the rates at which particles are cleared from a natural seston suspension (i.e., clearance rate) are known to vary widely among different bivalve groups, but generally, clearance increases linearly with seston concentration (Jørgensen 1996). In the same line, it has been noted that most bivalves exhibit higher clearance rates at high tide than at low, when the concentration of particles is progressively depleted. However, it has been observed that the rates of water processing are independent of particle concentration up to threshold levels that depend upon the composition of the suspended particles (~ food quality). On top of these physiological feedback mechanisms that are usually studied at the organismal level in laboratory-controlled experiments, the effects of ambient hydrodynamics should also be considered (Jørgensen 1996). In any case, there is no doubt that suspension-feeding bivalves have a notable compensation capacity to adapt to changes in their food environment (Bayne 1998).

Particle selectivity is an important species-specific process in suspension-feeding bivalves, and relatively new technologies (e.g., flow cytometry, video endoscopy, scanning electron microscopy) have contributed significantly to its understanding (reviewed by Ward and Shumway 2004). Selection of food particles can take place at three stages, namely: (i) the selective retention or uptake of particles, (ii) the preingestive processing (eventually leading to the production of pseudofeces; Fig. 8.5) and (iii) the postingestive processing. Regarding the first two stages, bivalve mollusks rely on mucus-coated

Figure 8.5. Biodeposits of the northern quahog clam, *Mercenaria mercenaria*, in the form of a string of pseudofeces. Pseudofeces are poorly-packed, preingestive rejected particles produced by the clam when feeding on natural plankton assemblages of variable nutritional quality. Photo credit: J. Pan.

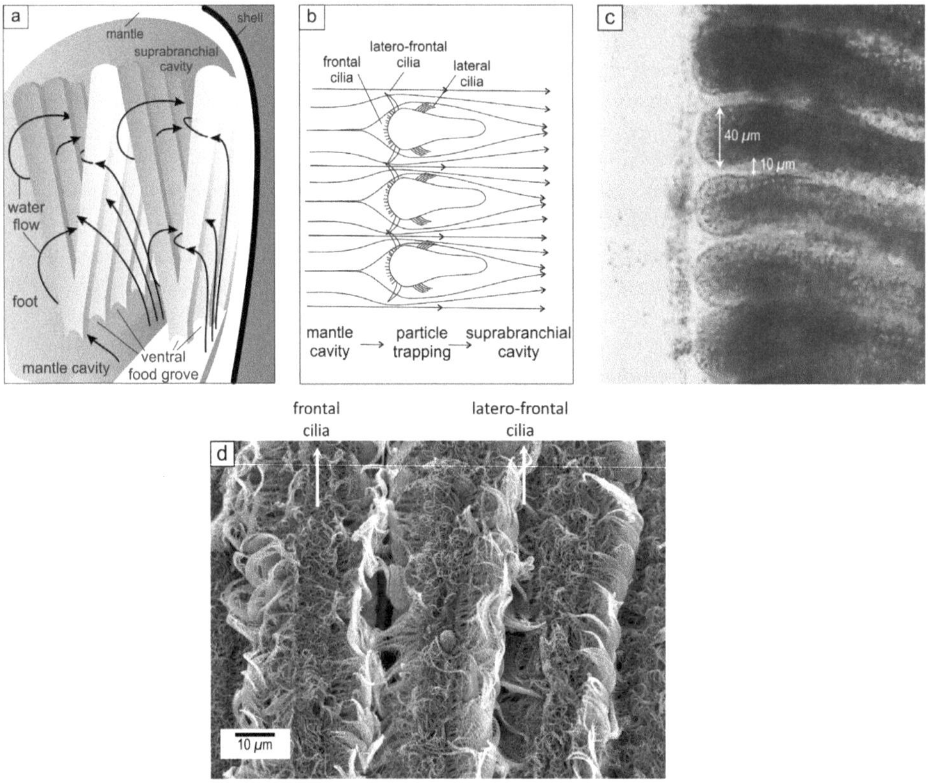

Figure 8.6. (a) Schematic representation of a lamellibranch gill (ctenidium) used in bivalve suspension feeding (modified and redrawn from Jørgensen 1990). The foot separates the mantle cavity in two-only the right half is shown for simplicity. In mytilid bivalves (commonly known as "mussels"), each gill is composed of two hemibranchial arcs of parallel-oriented filaments, forming a W pattern, and fused by interlamellar and interfilamentous junctions (not shown). Arrows show the ventral-to-dorsal water currents in mantle cavity. (b) Cross-section schematic through three gill filaments (modified and redrawn from Silvester and Sleigh 1984) and (c) light micrograph of five filaments in cross section; note the low-density mucus sludge adjacent to the frontal edge of the filaments. (d) Scanning electron (SEM) photomicrograph of gill filaments in frontal view, depicting the arrangement of frontal and latero-frontal ciliature. Photo credits (c) and (d) by J. Pan.

ciliated structures for particle collection, the ctenidia (or gills; a word derived from the Greek *ktenidion* = comb; Fig. 8.6) and the palps (Jørgensen 1996). Food particles are basically selected by their size, shape and flexibility, behavioral patterns (e.g., the swimming ability and motility of phytoplankton; Bricelj et al. 1998), their biochemical composition and chemical cues in their 'coatings' (Beninger and Decottignies 2005), and/or other qualitative characteristics such as cell stickiness (Waite et al. 1995) and cell physiological state (Brillant and MacDonald 2003). Selection mechanisms in bivalves can be active or passive (usually working in concert). Furthermore, bivalves not only process and select what they ingest, but during a third, postingestive stage, food particles are further 'sorted' in the gut (based both on physical and chemical properties), leading either to the production of intestinal (poorly digested) or glandular feces (i.e., food particles that are processed in the digestive gland; Brillant and MacDonald 2003).

Serpulid polychaete worms represent another example of suspension-feeding benthos that may drive ecosystem-scale processes (Pan and Marcoval 2014). In these animals, the filtration apparatus consists of ciliated radioles that can retain a broad range of particle sizes down to 1–2 μm (Jordana et al. 2001). Radioles are crown-shaped, bipinnate branched tissue extensions (Fig. 8.7a,b), with each grooved *pinnule* lined with cilia (Fig. 8.7c); ciliary beating creates a water current that circulates water containing suspended food particles; when trapped through a muco-ciliary process, food particles are conducted and concentrated along grooves in each pinnule, the food-mucus slurry is further transferred to the radiole arm and down to the mouth. Much like bivalve ctenidia, they play a dual role in particle trapping and sorting, and in oxygenation.

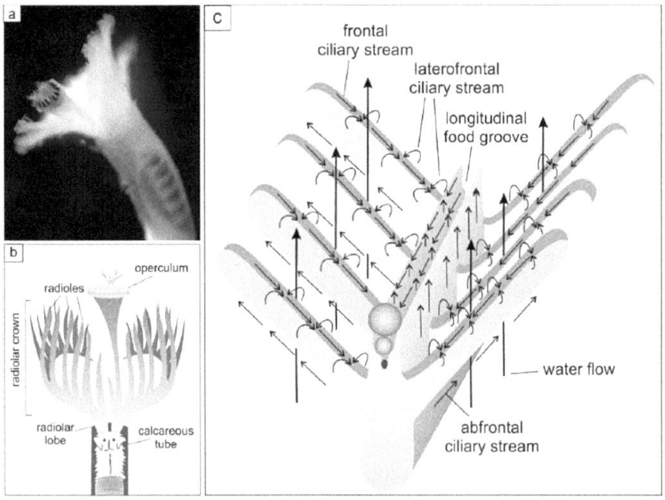

Figure 8.7. (a) The crown-shaped bipinnate radioles in a serpulid polychaete, used in suspension feeding. (b) Schematic representation of the radiole system (redrawn from Bok et al. 2017), and (c) detailed schematic of a single, grooved pinnule lined by cilia. Photo credit (a) by J. Pan.

8.6.2 Ecosystemic benefits of biologically-mediated benthic-pelagic coupling

Benthic suspension-feeding organisms, such as bivalve mollusks, are known for the positive effects they bestow on coastal ecosystems, including reduction of turbidity, prevention of harmful algal blooms (HABs), and the removal of nutrients by burial of nitrogen and phosphorus in the form of biodeposits (reviewed by Dame 1996, Prins et al. 1998, Newell et al. 2002, Newell 2004, Lonsdale et al. 2009). Despite this apparent preeminence of biological controls, some ecosystemic models indicate the significant top-down controls exerted by benthic suspension feeding function in estuaries (amounting to the consumption of > 50% of annual primary production in portions of Chesapeake Bay) are limited by water-column depth and the width of the estuary (Gerritsen et al. 1994), as the diffusion of significant amounts of pelagic food to the benthos is mostly controlled by turbulent mixing.

A number of studies comparing coastal systems that have experienced dramatic changes in abundances of suspension-feeding benthos (either through introductions or eliminations) have linked them to subsequent shifts in pelagic structure and function of the whole system (Caraco et al. 1997, Jackson et al. 2001, Pomeroy et al. 2006, Mann et al. 2009). Accordingly, studies of trophic interactions induced by BPC have examined changes in pelagic structure (e.g., standing stock, community composition; Norén et al. 1999).

A most important feature of bivalve suspension feeding is the transfer of undigested organic and inorganic material in feces and pseudofeces to the sediment surface (Newell 2004). Feces and pseudofeces are mucus- or other polysaccharide-bound (e.g., chitin) aggregates with faster sinking rates that undergo a de-watering process and gradually become incorporated into the sediment. In the process of bivalve grazing, once the nutritional needs are satisfied, excess phytoplankton cells are rejected in pseudofeces, which can account for as much as 80–90% of the total volume of filtered particulate material. Assimilation efficiencies are highly variable (ranging from ~ 20 to 90%); therefore, pseudofeces contain substantial amounts of POM. Through this mechanism, large amounts of undigested particulate organic nitrogen (PON) and phosphorous are transferred to the sediment surface. Thus, the burial of nitrogen and phosphorous and the ultimate removal of nitrogen from the ecosystem *via* denitrification (Fig. 8.2a) is enhanced by bivalve biodeposition. If the sediments contain sufficient O_2, aerobic nitrifying bacteria can oxidize nitrogenous compounds (PON and NH_4^+) to NO_2^- and NO_3^{2-}. Some of this NO_2^- and NO_3^{2-} diffuses out of the sediment and enters the water column where it becomes readily available for phytoplankton uptake. In deeper anoxic sediment layers, denitrification reactions reduce NO_2^- and NO_3^{2-} to N_2, which diffuses out of the system as a gas. MPB can limit nitrogen fluxes across the sediment-water interface through the uptake of dissolved inorganic nitrogen (DIN) that was regenerated in the upper portions of the sediment.

8.6.3 Reproductive strategies linking benthic and pelagic environments

The previous section dealt with trophic interactions involved in BPC. Here we discuss reproductive mechanisms and life-history patterns that couple the pelagic realm with soft-bottom benthos. Marine organisms have developed unique life-history strategies that differ from those of terrestrial organisms for three basic reasons. On the one hand, (i) seawater is denser and more viscous than air; this, in turn, (ii) allows a vast array of food particles to remain in suspension and be abundant, as we discussed in the previous section. Finally, (iii) desiccation is not a problem for marine organisms, while it is one of the main stressors for terrestrial ones (Strathmann 1990).

Many planktonic organisms enter a benthic resting phase and spend weeks to years in the seabed (the production of dinoflagellates cysts described in Chapter 3 is one such example); these benthic resting stages usually enable the persistence of species in the system, as they appear to be important in the recolonization and 'seeding' of the water column (Marcus and Boero 1998). Reciprocally, most macrozoobenthos in coastal systems have a planktonic larval stage that, after a variable time, settles to the bottom (a mechanism known as benthic larval recruitment). As discussed in the introduction to this section, planktonic systems are less mature than benthic ones; according to this view, animals tend to spend the adult stages of their life histories in the more mature system, but reproduce in the less mature pelagic realm. The advantage to reproduce in the less mature system resides in energy for larval growth being more readily available (Margalef 1963).

In a seminal paper, Strathmann (1985) elaborated further on the role of planktotrophic larvae in the life histories of benthic marine invertebrates. A feeding planktonic larval stage of benthic animals may be viewed as a migration into the plankton during early development for feeding and safety (the latter contrasting with Margalef's view). Strathmann considers three major consequences of this life-history strategy; on the one hand, (i) having a feeding planktonic larval stage has subsequent marked effects on the fecundity of an animal. This is not necessarily related to the allocation of energy into reproductive efforts, as species with planktotrophic larvae have more numerous, but smaller eggs than do related species with non-feeding development, such as lecitotrophic larvae-producing species. (ii) An advantageous outcome of having a planktonic larva is its associated extensive dispersal (what Strathmann refers to as an 'accidental by-product'). Larval dispersal in marine animals is often obligate with sexual reproduction (Strathmann 1990), and it increases gene flow and distribution range. However, selection for dispersal does not seem to be what maintains a feeding larval stage in life histories. Dispersal lies at the heart of the concept of supply-side ecology (aptly developed from the study of benthic communities), which states that the maintenance of populations and species requires that organisms be replaced when they die (Lewin 1986). (iii) Finally, larval dispersal often includes a precompetent period during which larvae cannot settle at good sites, yet, for a benthic invertebrate, spending a part of its life history in the plankton may be advantageous as a prolonged period in which to screen the adequate cues, and select an optimal site for sedentary adult life.

Regardless of its evolutionary advantages and associated effects, the larval stage is a critical period in the life history of marine species (this concept will be further explored, in relation to temperature, in Chapter 13). Of special interest here is the planktonic larval phase of benthic species, which spend this critical time *feeding* in this relatively less stable medium (by comparison to the benthos). The nutritional quality of dietary items on which planktonic larval stages are reared may have lasting legacies reflected in benthic juvenile and adult phases of the life history (Przeslawski et al. 2008). Likewise, the different lipidic levels and composition (concentration of and balance between different PUFAs) introduced during early ontogeny of bivalves through a phytoplankton diet greatly influences larval size, fatty acid composition and settlement (Pernet and Tremblay 2004).

Finally, in nearshore communities where physical stress is significant, this and multiple other factors influence reproductive BPC. As in the case of trophic BPC, while biological processes such as larval behavior may moderate the importance of dispersal and transport, other multiple mechanisms delivering competent larvae to shore (most notably hydrodynamics and bottom spatial variation) are superimposed to determine recruitment. For example, Morgan et al. (2017) studied larval densities and settlement of the dominant invertebrate (the Pacific mole crab, *Emerita analoga*) inside and outside the surf zone, contrasting the narrow surf zone of a reflective beach and the broader zone in a dissipative beach. They found that settlement rate in the dissipative surf zone was greater than in the reflective beach, but also concluded that surf zone hydrodynamics (as influenced by wind stress and wave height) is a key determinant of the supply of larval recruits to sandy nearshore communities.

Summary

1. The benthic environment is the two-dimensional habitat in which bottom-dwellers are found; the term benthos applies to the organisms that inhabit benthic habitats. These range in size and taxonomy from bacteria to vertebrates.
2. Organic matter (OM) can be locally (autochthonous) or remotely produced (allochthonous); the greatest portion of marine sedimentary OM is originally produced in the form of particulate matter through photosynthesis. The burial of OM in marine sediments constitutes the main link between 'active' pools of carbon, and those carbon pools that cycle on much longer, geological, timescales.
3. In addition to microbially-produced extracellular polymeric substances (EPS), virtually all benthic animals secrete copious amounts of mucus to accompany many physiological processes.
4. Benthos distribution is controlled by a combination of factors such as sediment granulometry, dissolved O_2, OM content, as well as disturbance and pollution processes. There is little evidence to a single factor being the primary determinant of benthos distributions.
5. The transition zone or boundary from predominantly oxidizing to reducing conditions in sediments is called the redox potential discontinuity (RPD) layer. The depth of the RPD layer can vary extensively in time and space.
6. The functional classification of soft-bottom marine benthos may follow different criteria, namely, size (micro-, meio- and macrobenthos); habitat preference (epifauna, infauna); feeding strategies (shredders, grazers, suspension- and deposit feeders); type of carnivory (predators and scavengers); or their contribution to sediment stability and/or erodibility (stabilizers and bioturbators).
7. Deep-sea benthic ecosystems harbor a high biological diversity; several hypothesis (either focusing on stability or disturbance as determinants) have been invoked to explain such diversity and distribution patterns. In relation to the energy fuelling them, deep-sea ecosystems can be divided into two major groups: heterotrophic and chemosynthetic habitats.
8. Hydrothermal vents, methane cold seeps, and communities developing on seafloor oxygen minimum zones (OMZs) are examples of deep-see chemolithotrophic communities, in which sulfide-oxidizing bacteria create trophic endosymbioses with the dominant invertebrates.
9. The benthos influences soft sediment properties in a number of ways and at various scales.
10. Soft-bottom ecosystems play a major role in nutrient regeneration, through the processing of OM that comes from different sources. Ammonification produces NH_4^+ and nitrification mediated by distinct groups of nitrifying bacteria yields NO_3^-. In anaerobic conditions, different groups of bacteria use alternative electron acceptors to decompose OM and liberate energy. Nitrate reduction may follow the denitrification pathway, or dissimilatory nitrate reduction to ammonia (DNRA). Nitrogen reduction under anoxic conditions can also occur through anaerobic ammonium oxidation (*anammox*). In mudflats, aerobic and anaerobic processes are tightly coupled.
11. Sulfate-reducing bacteria and archaea perform anaerobic respiration by reducing SO_4^{2-} to H_2S. Methanogenesis in anaerobic sediments is mediated by archaea and produces CH_4 by using CO_2 as the terminal electron acceptor at the final step in OM decomposition.
12. Epipelic microphytobenthos (MPB) are microscopic, unicellular eukaryotic algae and cyanobacteria that grow in the uppermost mm of illuminated sediments and in association with sedimentary

particles. MPB secretes EPS as part of their normal physiology, thus contributing to biostabilization and geomorphological processes in mudflats. Furthermore, when adjusted by the areal extent they cover, these autotrophs contribute the most to active estuarine carbon pools.

13. Sediments ingested by benthic animals experience a unique environment controlled by the digestive chemistry of the organism. Deposit-feeding implies several sequential steps initiated with ingestion and trituration, followed by solubilization (mediated by surfactants and enzymes) in the fore- and midgut, and ending with selective absorption in the hindgut and/or defecation of unassimilated solid material. The guts of deposit-feeding invertebrates could be likened to laundry washing machines.
14. Seagrass beds are composed of dense populations (mono-specific or more commonly, more than one species) of flowering plants that grow on well-illuminated (often oligotrophic) soft sandy bottoms, across tropical and temperate latitudes worldwide. They offer structural complexity in an otherwise topographically simple, two-dimensional environment and can modify the hydrodynamic regime. Seagrass meadows are highly productive and highly autotrophic; furthermore, OM is sequestered for long periods of time (100s of yr) and preserved forming anaerobic peats, thus being categorized as 'Blue Carbon' ecosystems.
15. The pelagic and benthic sub-systems are intimately interconnected. Such connections may be established through physical or biologically-mediated processes, collectively known as benthic-pelagic coupling (BPC). Ecological BPC traditionally focuses on trophic and reproductive interactions, and it generates many ecosystemic benefits.
16. Suspension-feeding is an energy-acquisition strategy convergently evolved among many phyla of marine organisms, that targets a profit from the large amount of particles suspended in seawater. There is no analogous mechanism to suspension-feeding in terrestrial ecology.
17. Marine organisms have developed unique life-history strategies that differ from those of terrestrial organisms. Many planktonic organisms enter a benthic resting phase and spend weeks to years in the seabed. Reciprocally, most macrozoobenthos in coastal systems have a planktonic larval stage that, after a variable time, settles to the bottom (benthic larval recruitment).

References

Albertsson, J. and K. Leonardsson. 2001. Deposit-feeding amphipods (*Monoporeia affinis*) reduce the recruitment of copepod nauplii from benthic resting eggs in the northern Baltic Sea. Mar. Biol. 138: 793–801.

Ambrose, W.G.J. 1983. The influence of the predatory polychaetes *Glycera dibranchiata* and *Nereis virens* on the structure of a soft-bottom community in a Maine estuary. PhD Thesis. University of North Carolina at Chapel Hill.

Amon, D.J., A.G. Glover, H. Wiklund, L. Marsh, K. Linse, A.D. Rogers and J.T. Copley. 2013. The discovery of a natural whale fall in the Antarctic deep sea. Deep Sea Res. Part II Top. Stud. Oceanogr. 92: 87–96.

Bacon, G.S., B.A. MacDonald and J.E. Ward. 1998. Physiological responses of infaunal (*Mya arenaria*) and epifaunal (*Placopecten magellanicus*) bivalves to variations in the concentration and quality of suspended particles. I. Feeding activity and selection. J. Exp. Mar. Biol. Ecol. 219: 105–125.

Bayne, B.L. 1998. The physiology of suspension feeding by bivalve molluscs: an introduction to the Plymouth "TROPHEE" workshop. J. Exp. Mar. Biol. Ecol. 219: 1–19.

Bazaïri, H., A. Bayed, M. Glémarec and C. Hily. 2003. Spatial organisation of macrozoobenthic communities in response to environmental factors in a coastal lagoon of the NW African coast (Merja Zerga, Morocco). Oceanol. Acta 26: 457–471.

Bedulli, C., P.S. Lavery, M. Harvey, C.M. Duarte and O. Serrano. 2020. Contribution of seagrass Blue Carbon toward carbon neutral policies in a touristic and environmentally-friendly island. Front. Mar. Sci. 7: 1; doi:10.3389/fmars.2020.00001.

Beninger, P.G. and P. Decottignies. 2005. What makes diatoms attractive for suspensivores? The organic casing and associated organic molecules of *Coscinodiscus perforatus* are quality cues for the bivalve *Pecten maximus*. J. Plankton Res. 27: 11–17.

Bok, M.J., M.L. Porter, H.A. Ten Hove, R. Smith and D.-E. Nilsson. 2017. Radiolar eyes of serpulid worms (Annelida, Serpulidae): structures, function, and phototransduction. Biol. Bull. 233: 39–57.

Bricelj, V.M., J.E. Ward, A.D. Cembella and B.A. MacDonald. 1998. Application of video-endoscopy to the study of bivalve feeding on toxic dinoflagellates. pp. 453–456. *In*: Reguera, B., J. Blanco, M.L. Fernández and T. Wyatt [eds.]. Harmful Algae. Xunta de Galicia and Intergovernmental Oceanographic Commission (UNESCO), Santiago de Compostela, Spain.

Brillant, M.G.S. and B.A. MacDonald. 2003. Postingestive sorting of living and heat-killed *Chlorella* within the sea scallop, *Placopecten magellanicus* (Gmelin). J. Exp. Mar. Biol. Ecol. 290: 81–91.

Bulmer, R.H., M. Townsend, T. Drylie and A.M. Lohrer. 2018. Elevated turbidity and the nutrient removal capacity of seagrass. Front. Mar. Sci. 5: 462; doi:10.3389/fmars.2018.00462.

Bulmer, R.H., F. Stephenson, H.F.E. Jones, M. Townsend, J.R. Hillman, L. Schwendenmann and C.J. Lundquist. 2020. Blue Carbon stocks and cross-habitat subsidies. Front. Mar. Sci. 7: 380; doi:10.3389/fmars.2020.00380.

Burdige, D.J. 2007. Preservation of organic matter in marine sediments: controls, mechanisms, and an imbalance in sediment organic carbon budgets? Chem. Rev. 107: 467–485.

Calow, P. 1977. Ecology, evolution and energetics: a study in metabolic adaptation. Adv. Ecol. Res. 10: 1–62.

Caraco, N.F., J.J. Cole, P.A. Raymond, D.L. Strayer, M.L. Pace, S.E.G. Findlay and D.T. Fischer. 1997. Zebra mussel invasion in a large, turbid river: phytoplankton response to increased grazing. Ecology 78: 588–602.

Carr, W.E.S. and C.A. Adams. 1973. Food habits of juvenile marine fishes occupying seagrass beds in the estuarine zone near Crystal River, Florida. Trans. Am. Fisheries Soc. 102: 511–540.

Cloern, J.E. and R. Dufford. 2005. Phytoplankton community ecology: principles applied in San Francisco Bay. Mar. Ecol. Prog. Ser. 285: 11–28.

Connell, J.H. 1978. Diversity in tropical rainforests and coral reefs. Science 199: 1302–1310.

Crane, R.L. and R.A. Merz. 2017. Mechanical properties of sediment determine burrowing success and influence distribution of two lugworm species. J. Exp. Biol. 220: 3248–3259.

Dale, H., J.D. Taylor, M. Solan, P. Lam and M. Cunliffe. 2019. Polychaete mucopolysaccharide alters sediment microbial diversity and stimulates ammonia-oxidising functional groups. FEMS Microbiol. Ecol. 95: fiy234; doi.org/10.1093/femsec/fiy234.

Dale, N.G. 1974. Bacteria in intertidal sediments: factors related to their distribution. Limnol. Oceanogr. 19: 509–518.

Dalsgaard, T., B. Thamdrup and D.E. Canfield. 2005. Anaerobic ammonium oxidation (anammox) in the marine environment. Res. Microbiol. 156: 457–464.

Dame, R.F. 1996. Ecology of Marine Bivalves, An Ecosystem Approach. CRC Press, Boca Raton. 254 pp.

Dayton, P.K. and R.R. Hessler. 1972. Role of biological disturbance in maintaining diversity in the deep sea. Deep-Sea Res. 19: 199–208.

Dawes, C.J., D. Hanisak and J.W. Kenworthy. 1995. Seagrass biodiversity in the Indian River Lagoon. Bull. Mar. Sci. 57: 59–66.

de Brouwer, J.F.C., K. Wolfstein, G.K. Ruddy, T.E.R. Jones and L.J. Stal. 2005. Biogenic stabilization of intertidal sediments: the importance of extracellular polymeric substances produced by benthic diatoms. Microb. Ecol. 49: 501–512.

Decho, A.W. and T. Gutierrez. 2017. Microbial extracellular polymeric substances (EPSs) in ocean systems. Front. Microbiol. 8: 922; doi.org/10.3389/fmicb.2017.00922.

Degenhardt, J., L. Dlugosch, J. Ahrens, M. Beck, H. Waska and B. Engelen. 2020. Seasonal dynamics of microbial diversity at a sandy high energy beach reveal a resilient core community. Front. Mar. Sci. 7: 573570; doi:10.3389/fmars.2020.573570.

De Leo, F.C., A.F. Bernardino and P.Y.G. Sumida. 2020. Continental slope and submarine canyons: benthic biodiversity and human impacts. pp. 37–72. *In*: Sumida, P.Y.G., A.F. Bernardino and F.C. De Leo [eds.]. Brazilian Deep-sea Biodiversity. Springer Nature, Cham, Switzerland.

Diaz, R.J. and J.H. Trefry. 2006. Comparison of sediment profile image data with profiles of oxygen and Eh from sediment cores. J. Mar. Syst. 62: 164–172.

Duarte, C.M. and D. Krause-Jensen. 2017. Export from seagrass meadows contributes to marine carbon sequestration. Front. Mar. Sci. 4: 13; doi:10.3389/fmars.2017.00013.

Ellis, J.I., G. Fraser and J. Russell. 2012. Discharged drilling waste from oil and gas platforms and its effects on benthic communities. Mar. Ecol. Prog. Ser. 456: 285–302.

Erickson, A.A., S.S. Bell and C.J. Dawes. 2012. Associational resistance protects mangrove leaves from crab herbivory. Acta Oecol. 41: 46–57.

Felbeck, H., J.J. Childress and G.N. Somero. 1981. Calvin-Benson cycle and sulphide oxidation enzymes in animals from sulphide-rich habitats. Nature 293: 291–293.

Fenchel, T.M. and R.J. Riedl. 1970. The sulfide system: a new biotic community underneath the oxidized layer of marine sand bottoms. Mar. Biol. 7: 255–268.

Fidalgo e Costa, P., V. Brotas and L. Cancela da Fonseca. 2002. Physical characterization and microphytobenthos biomass of estuarine and lagoon environments of the Southwest coast of Portugal. Limnetica 21: 69–79.

Flemming, H.-C. and S. Wuertz. 2019. Bacteria and archaea on Earth and their abundance in biofilms. Nat. Rev. Microbiol. 17: 247–260.

Fletcher, S.W. and W.W. Fletcher. 1995. Factors affecting changes in seagrass distribution and diversity patterns in the Indian River Lagoon complex between 1940 and 1992. Bull. Mar. Sci. 57: 49–58.

Fonseca, M.S., J.W. Fourqurean and M.A.R. Koehl. 2019. Effect of seagrass on current speed: importance of flexibility vs. shoot density. Front. Mar. Sci. 6: 376; doi:10.3389/fmars.2019.00376.

French McCay, D.P., C.H. Peterson, J.T. DeAlteris and J. Catena. 2003. Restoration that targets function as opposed to structure: replacing lost bivalve production and filtration. Mar. Ecol. Prog. Ser. 264: 197–212.
Froelich, P., G.P. Klinkhammer, M.L. Bender, N.A. Luedtke, G.R. Heath, D. Cullen, P. Dauphin et al. 1979. Early oxidation of organic matter in pelagic sediments of the eastern equatorial Atlantic: suboxic diagenesis. Geochim. Cosmochim. Acta 43: 1075–1090.
Gage, J.D. and P.A. Tyler. 1991. Deep-Sea Biology: A Natural History of Organisms at the Deep-Sea Floor. Cambridge University Press.
Gage, J.D. 2002. Benthic biodiversity across and along the continental margin: patterns, ecological and historical determinants, and anthropogenic threats. pp. 307–321. *In*: Wefer, G., D. Billett, D. Hebbeln, B.B. Jørgensen, M. Schlüter and T.C.E. van Weering [eds.]. Ocean Margin Systems. Springer, Berlin, Heidelberg, Germany.
Gerritsen, J., A.F. Holland and D.E. Irvine. 1994. Suspension-feeding bivalves and the fate of primary production: an estuarine model applied to Chesapeake Bay. Estuaries 17: 403–416.
Gingras, M., J.W. Hagadorn, A. Seilacher, S.V. Lalonde, E. Pecoits, D. Petrash and K.O. Konhauser. 2011. Possible evolution of mobile animals in association with microbial mats. Nature Geosci. 4: 372–375.
Gooday, A.J. 1996. Epifaunal and shallow infaunal foraminiferal communities at three abyssal NE Atlantic sites subject to differing phytodetritus input regimes. Deep Sea Res. Part I Oceanogr. Res. Pap. 43: 1395–1421.
Graf, G. 1997. SCOR Working Group 95: sediment suspension and sea bed properties. J. Mar. Syst. 3: 267–268.
Grassle, J.F. and N.J. Maciolek. 1992. Deep-sea species richness: regional and local diversity estimates from quantitative bottom samples. Am. Nat. 139: 313–341.
Hannides, A.K. and R.C. Aller. 2016. Priming effect of benthic gastropod mucus on sedimentary organic matter remineralization. Limnol. Oceanogr. 61: 1640–1650.
Hansen, J.L.S. and A.B. Josefson. 2004. Ingestion by deposit-feeding macro-zoobenthos in the aphotic zone does not affect the pool of live pelagic diatoms in the sediment. J. Exp. Mar. Biol. Ecol. 308: 59–84.
Heck Jr., K.L., G. Hays and R.J. Orth. 2003. Critical evaluation of the nursery role hypothesis for seagrass meadows. Mar. Ecol. Prog. Ser. 253: 123–136.
Hedges, J.I. and R.G. Keil. 1995. Sedimentary organic matter preservation: an assessment and speculative synthesis. Mar. Chem. 49: 81–115.
Helly, J.J. and L.A. Levin. 2004. Global distribution of naturally occurring marine hypoxia on continental margins. Deep Sea Res. Part I Oceanogr. Res. Pap. 51: 1159–1168.
Huettel, M., H. Roy, E. Precht and S. Ehrenhauss. 2003. Hydrodynamical impact on biogeochemical processes in aquatic sediments. Hydrobiologia 494: 231–236.
Hunting, E.R. and A.A. Kampfraath. 2013. Contribution of bacteria to redox potential (Eh) measurements in sediments. Int. J. Environ. Sci. Technol. 10: 55–62.
Jackson, J.B.C., M.X. Kirby, W.H. Berger, K.A. Bjorndal, L.W. Botsford et al. 2001. Historical overfishing and the recent collapse of coastal ecosystems. Science 293: 629–637.
James, R. 2005. Marine Biogeochemical Cycles, 2nd Ed. Elsevier/Butterworth-Heinemann, Oxford. 130 pp.
Jannasch, H.W. and M.J. Mottl. 1985. Geomicrobiology of deep-sea hydrothermal vents. Science 229: 717–725.
Jensen, M.M., P. Lam, N.P. Revsbech, B. Nagel, B. Gaye, M.S.M. Jetten and M.M.M. Kuypers. 2011. Intensive nitrogen loss over the Omani Shelf due to anammox coupled with dissimilatory nitrite reduction to ammonium. ISME J. 5: 1660–1670.
Jesus, B., V. Brotas, L. Ribeiro, C.R. Mendes, P. Cartaxana and D.M. Paterson. 2009. Adaptations of microphytobenthos assemblages to sediment type and tidal position. Cont. Shelf Res. 29: 1624–1634.
Jordana, E., A. Grémare, F. Lantoine, C. Courties, F. Charles, J.M. Amouroux and G. Vétion. 2001. Seasonal changes in the grazing of coastal picoplankton by the suspension-feeding polychaete *Ditrupa arietina* (O.F. Müller). J. Sea Res. 46: 245–259.
Jørgensen, C.B. 1990. Bivalve filter feeding: hydrodynamics, bioenergetics, physiology and ecology. Olsen & Olsen, Fredensborg. 140 pp.
Jørgensen, C.B. 1996. Review: Bivalve filter feeding revisited. Mar. Ecol. Prog. Ser. 142: 287–302.
Jumars, P.A., L.M. Mayer, J.W. Deming, J.A. Baross and R.A. Wheatcroft. 1990. Deep-sea deposit-feeding strategies suggested by environmental and feeding constraints. Phil. Trans. Roy. Soc. London 332: 85–101.
Karl, D.M., C.O. Wirsen and H.W. Jannasch. 1980. Deep-sea primary production at the Galapagos hydrothermal vents. Science 207: 1345–1347.
Katsev, S., B. Sundby and A. Mucci. 2006. Modeling vertical excursions of the redox boundary in sediments: Application to deep basins of the Arctic Ocean. Limnol. Oceanogr. 51: 1581–1593.
Kemp, K.M., A.J. Jamieson, P.M. Bagley, H. McGrath, D.M. Bailey, M.A. Collins and I.G. Priede. 2006. Consumption of large bathyal food fall, a six month study in the NE Atlantic. Mar. Ecol. Prog. Ser. 310: 65–76.
Kideys, A.E. and R.G. Hartnoll. 1991. Energetics of mucus production in the common whelk *Buccinum undatum* L. J. Exp. Mar. Biol. Ecol. 150: 91–105.
Kjeldsen, K.U., C. Joulian and K. Ingvorsen. 2004. Oxygen tolerance of sulphate-reducing bacteria in activated sludge. Environ. Sci. Technol. 38: 2038–2043.

Kjerfve, B. 1994. Coastal lagoons. pp. 1–8. *In*: Kjerfve, B. [ed.]. Coastal Lagoon Processes. Elsevier, Amsterdam, The Netherlands.

Klages, M., S. Muyakshin, T. Soltwedel and W.E. Arntz. 2002. Mechanoreception, a possible mechanism for food fall detection in deep-sea scavengers. Deep Sea Res. Part I Oceanogr. Res. Pap. 49: 143–155.

Komada, T., D.J. Burdige, H.-L. Li, C. Magen, J.P. Chanton, and A.K. Cada. 2016. Organic matter cycling across the sulfate-methane transition zone of the Santa Barbara Basin, California Borderland. Geochim. Cosmochim. Acta 176: 259-278.

Koopmans, D., M. Holtappels, A. Chennu, M. Weber and D. de Beer. 2020. High net primary production of Mediterranean seagrass (*Posidonia oceanica*) meadows determined with aquatic eddy covariance. Front. Mar. Sci. 7: 118; doi:10.3389/fmars.2020.00118.

Kristensen, E. 2000. Organic matter diagenesis at the oxic/anoxic interface in coastal marine sediments, with emphasis on the role of burrowing animals. Hydrobiologia 426: 1–24.

Kuenen, J.G. 2008. Anammox bacteria: from discovery to application. Nat. Rev. Microbiol. 6: 320–326.

Lalonde, S.V., L.T. Dafoe, S.G. Pemberton, M.K. Gingras and K.O. Konhauser. 2010. Investigating the geochemical impact of burrowing animals: proton and cadmium adsorption onto the mucus lining of Terebellid polychaete worms. Chem. Geol. 271: 44–51.

Levin, L.A., A.R. Baco, D.A. Bowden, A. Colaco, E.E. Cordes, M.R. Cunha, A.W.J. Demopoulos et al. 2016. Hydrothermal vents and methane seeps: rethinking the sphere of influence. Front. Mar. Sci. 3: 72; doi.org/10.3389/fmars.2016.00072.

Leber, K.M. 1985. The influence of predatory decapods, refuge, and microhabitat selection on seagrass communities. Ecology 66: 1951–1964.

Lessin, G., J. Bruggeman, C.L. McNeill and S. Widdicombe. 2019. Time scales of benthic macrofaunal response to pelagic production differ between major feeding groups. Front. Mar. Sci. 6: 15; doi:10.3389/fmars.2019.00015.

Lewin, R. 1986. Supply-side ecology. Science 234: 25–27.

Lonsdale, D.J., R.M. Cerrato, R. Holland, A. Mass, L. Holt, R.A. Schaffner, J. Pan and D.A. Caron. 2009. Influence of suspension-feeding bivalves on the pelagic food webs of shallow, coastal embayments. Aquat. Biol. 6: 263–279.

MacIntyre, H.L., R.J. Geider and D.C. Miller. 1996. Microphytobenthos: the ecological role of the "secret garden" of unvegetated, shallow-water marine habitats. I. Distribution, abundance and primary production. Estuaries 19: 186–201.

Mann, K.H. 2009. Ecology of Coastal Waters: With Implications for Management, 2nd Ed. Blackwell Science, Malden. 432 pp.

Mann, R., J.M. Harding and M.J. Southworth. 2009. Reconstructing pre-colonial oyster demographics in the Chesapeake Bay, USA. Estuar. Coast. Shelf Sci. 85: 217–222.

Marchant, H.K., S. Ahmerkamp, G. Lavik, H.E. Tegetmeyer, J. Graf, J.M. Klatt, M. Holtappels et al. 2017. Denitrifying community in coastal sediments performs aerobic and anaerobic respiration simultaneously. ISME J. 11: 1799–1812.

Marcus, N.H. and F. Boero. 1998. Minireview: The importance of benthic-pelagic coupling and the forgotten role of life cycles in coastal aquatic systems. Limnol. Oceanogr. 43: 763–768.

Margalef, R. 1963. On certain unifying principles in Ecology. Am. Nat. 97: 357–374.

Mayer, L.M., L.L. Schick, T. Sawyer, C.J. Plante, P.A. Jumars and R.L. Self. 1995. Bioavailable amino acids in sediments: a biomimetic, kinetics-based approach. Limnol. Oceanogr. 40: 511–520.

Mayer, L.M., Z. Chen, R.H. Findlay, J. Fang, S. Sampson, R.F.L. Self, P.A. Jumars, C. Quetél and O.F.X. Donard. 1996. Bioavailability of sedimentary contaminants subject to deposit-feeder digestion. Environ. Sci. Technol. 30: 2641–2645.

Mayer, L.M., L.L. Schick, R.F.L. Self, P.A. Jumars, R.H. Findlay, Z. Chen and S. Sampson. 1997. Digestive environments of benthic macroinvertebrate guts: enzymes, surfactants and dissolved organic matter. J. Mar. Res. 55: 785–812.

Mayer, L.M., P.A. Jumars, M.J. Bock, Y.-A. Vetter and J.L. Schmidt. 2001. Two roads to Sparagmos: extracellular digestion of sedimentary food by bacterial inoculation versus deposit-feeding. pp. 335–347. *In*: Aller, J.Y., S.A. Woodin and R.C. Aller [eds.]. Organism-sediment Interactions. University of South Carolina Press, Columbia, USA.

McCave, I.N. 2009. Nepheloid layers. pp. 484–494. *In*: Thorpe, S.A. [ed.]. Elements of Physical Oceanography: A Derivative of the Encyclopedia of Ocean Sciences, 2nd Ed. Academic Press/Elsevier, London, UK.

McLachlan, A. and A.C. Brown. 2006. The Ecology of Sandy Shores, 2nd Ed. Academic Press. 392 pp.

Meinen, C.S., R.C. Perez, S. Dong, A.R. Piola and E. Campos. 2020. Observed ocean bottom temperature variability at four sites in the Northwestern Argentine Basin: evidence of decadal deep/abyssal warming amidst hourly to interannual variability during 2009–2019. Geophys. Res. Lett. 47: e2020GL089093.

Miller, D.C. 1961. The feeding mechanism of fiddler crabs, with ecological considerations of feeding adaptations. Zoologica 46: 89–101.

Mitsch, W.J. and J.G. Gosselink. 2000. Wetlands, 3rd Ed. John Wiley and Sons, Inc., New York, USA.

Molongoski, J.J. and M.J. Klug. 1976. Characterization of anaerobic heterotrophic bacteria isolated from freshwater lake sediments. Appl. Environ. Microbiol. 31: 83–90.

Morgan, S.G., A.L. Shanks, J. MacMahan, A.J.H.M. Reniers, C.D. Griesemer, M. Jarvis and A.G. Fujimura. 2017. Surf zones regulate larval supply and zooplankton subsidies to nearshore communities. Limnol. Oceanogr. 62: 2811–2828.

Muthukrishnan, R., K.L. Chiquillo, C. Cross, P. Fong, T. Kelley, C.A. Toline, R. Zweng and D.A. Willette. 2020. Little giants: a rapidly invading seagrass alters ecosystem functioning relative to native foundation species. Mar. Biol. 167: 81; doi.org/10.1007/s00227-020-03689-8.

Muyzer, G. and A.J.M. Stams. 2008. The ecology and biotechnology of sulphate-reducing bacteria. Nat Rev Micro 6: 441–454.

Newell, R.I.E., J.C. Cornwell and M.S. Owens. 2002. Influence of simulated bivalve biodeposition and microphytobenthos on sediment nitrogen dynamics: a laboratory study. Limnol. Oceanogr. 47: 1367–1379.

Newell, R.I.E. 2004. Ecosystem influences of natural and cultivated populations of suspension-feeding bivalve molluscs: a review. J. Shellfish Res. 23: 51–61.

Norén, F., J. Haamer and O. Lindahl. 1999. Changes in the plankton community passing a *Mytilus edulis* mussel bed. Mar. Ecol. Prog. Ser. 191: 187–194.

Ockelmann, K.W. and O. Vahl. 1970. On the biology of the polychaete *Glycera alba*, especially its burrowing and feeding. Ophelia 8: 275–294.

Oliver, J.S., J.M. Oakden and P.N. Slattery. 1982. Phoxocephalid amphipod crustaceans as predators on larvae and juveniles in marine soft-bottom communities. Mar. Ecol. Prog. Ser. 7: 179–184.

Ospina-Alvarez, N., M. Caetano, C. Vale, J. Santos-Echeandía, P. Bernárdez and R. Prego. 2014. Exchange of nutrients across the sediment-water interface in intertidal ria systems (SW Europe). J. Sea Res. 85: 349–358.

Pan, J. and M.A. Marcoval. 2014. Top-down effects of an exotic serpulid polychaete on natural plankton assemblage of estuarine and brackish systems in the SW Atlantic. J. Coast. Res. 30: 1226–1235.

Pan, J., D.G. Cuadrado and C.N. Bournod. 2017. Diatom-driven recolonization of microbial-mat dominated siliciclastic tidal flat sediments. FEMS Microbiol. Ecol. 93: fix111.

Pan, J., P.D. Pratolongo and D.G. Cuadrado. 2018. Geological, physical and chemical foundations. pp. 11–42. *In*: Beninger, P.G. [ed.]. Mudflat Ecology. Springer Nature, Cham, Switzerland.

Paull, C.K., B. Hecker, R. Commeau, R.P. Freeman-Lynde, C. Neumann, W.P. Corso, S. Golubic et al. 1984. Biological communities at the Florida Escarpment resemble hydrothermal vent taxa. Science 226: 965–967.

Pearson, T.H. 2001. Functional group ecology in soft-sediment marine benthos: the role of bioturbation. Ocean. Mar. Biol. Annu. Rev. 39: 233–267.

Pedrosa-Pàmies, R., C. Parinos, A. Sanchez-Vidal, A. Gogou, A. Calafat, M. Canals, I. Bouloubassi and N. Lampadariou. 2015. Composition and sources of sedimentary organic matter in the deep eastern Mediterranean Sea. Biogeosciences 12: 7379–7402.

Peinert, R., A. Antia, E. Bauerfeind, B. v Bodungen, O. Haupt, M. Krumbholz, I. Peeken et al. 2001. Particle flux variability in the Polar and Atlantic biogeochemical provinces of the Nordic Seas. pp. 53–68. *In*: Schäfer, P., W. Ritzrau, M. Schlüter and J. Thiede [eds.]. The Northern North Atlantic. Springer, Berlin, Heidelberg, Germany.

Pernet, F. and R. Tremblay. 2004. Effect of varying levels of dietary essential fatty acid during early ontogeny of the sea scallop *Placopecten magellanicus*. J. Exp. Mar. Biol. Ecol. 310: 73–86.

Peterson, C.H. 1979. Predation, competitive exclusion, and diversity in the soft-sediment benthic communities of estuaries and lagoons. pp. 233–264. *In*: Livingston, R.J. [ed.]. Ecological Processes in Coastal and Marine Systems Springer, Boston, USA.

Piepenburg, D., A. Brandt, K. von Juterzenka, M. Mayer, K. Schnack, D. Seiler, U. Witte et al. 2001. Patterns and determinants of the distribution and structure of benthic faunal assemblages in the northern North Atlantic. pp. 179–198. *In*: Livingston, R.J. [ed.]. Ecological Processes in Coastal and Marine Systems. Springer, Boston, USA.

Pomeroy, L.R., C.F. D'Elia and L.C. Schaffner. 2006. Limits to top-down control of phytoplankton by oysters in Chesapeake Bay, USA. Mar. Ecol. Prog. Ser. 325: 301–309.

Prins, T.C., A.C. Smaal and R.F. Dame. 1998. A review of the feedbacks between bivalve grazing and ecosystem processes. Aquat. Ecol. 31: 349–359.

Przeslawski, R., P.E. Bourdeau, M.H. Doall, J. Pan, L. Perino and D.K. Padilla. 2008. The effects of a harmful alga on bivalve larval lipid stores. Harmful Algae 7: 802–807.

Ramírez-Llodra, E. and D.S.M. Billett. 2006. Deep-sea ecosystems: pristine biodiversity reservoir and technological challenges. pp. 63–92. *In*: Duarte, C.M. [ed.]. The Exploration of Marine Biodiversity: Scientific and Technological Challenges. Fundación BBVA, Bilbao, Spain.

Reddy, K.R. and R.D. DeLaune. 2008. Biogeochemistry of Wetlands: Science and Applications. CRC Press, Boca Raton, USA.

Reidenbach, M.A. and E.L. Thomas. 2018. Influence of the seagrass, *Zostera marina*, on wave attenuation and bed shear stress within a shallow coastal bay. Front. Mar. Sci. 5: 397; doi:10.3389/fmars.2018.00397.

Riethmüller, R., M. Heineke, H. Kühl and R. Keuker-Rüdiger. 2000. Chlorophyll *a* concentration as an index of sediment surface stabilisation by microphytobenthos? Cont. Shelf Res. 20: 1351–1372.

Rogers, A.D. 2000. The role of the oceanic oxygen minima in generating biodiversity in the deep sea. Deep Sea Res. Part II Top. Stud. Oceanogr. 47: 119–148.

Romero, J., G. Pergent, C. Pergent-Martini, M.-A. Mateo and C. Regnier. 1992. The detritic compartment in a *Posidonia oceanica* meadow: litter features, decomposition rates, and mineral stocks. Mar. Ecol. 13: 69–83.

Rossi, F., P.M.J. Herman and J.J. Middelburg. 2004. Interspecific and intraspecific variation of $\delta^{13}C$ and $\delta^{15}N$ in deposit- and suspension feeding bivalves (*Macoma balthica* and *Cerastoderma edule*): evidence of ontogenetic changes in feeding mode of *Macoma balthica*. Limnol. Oceanogr. 49: 408–414.

Rundell, R.J. and B.S. Leander. 2010. Masters of miniaturization: convergent evolution among interstitial eukaryotes. Bioessays 32: 430–437.

Sanders, H.L. 1968. Marine benthic diversity: a comparative study. Amer. Natur. 102: 243–282.

Schmidt, J.L., J.W. Deming, P.A. Jumars and R.G. Keil. 1998. Constancy of bacterial abundance in surficial marine sediments. Limnol. Oceanogr. 43: 976–982.

Schwindt, E., O.O. Iribarne and F.I. Isla. 2004. Physical effects of an invading reef-building polychaete on an Argentinean estuarine environment. Estuar. Coast. Shelf Sci. 59: 109–120.

Serrano, O., E. Serrano, K. Inostroza, P.S. Lavery, M.A. Mateo and E. Ballesteros. 2017. Seagrass meadows provide 3D habitat for reef fish. Front. Mar. Sci. 4: 54; doi:10.3389/fmars.2017.00054.

Silvester, N.R. and M.A. Sleigh. 1983. Hydrodynamic aspects of particle capture by *Mytilus*. J. Mar. Biol. Ass. U.K. 64: 859–879.

Smith, C.R. and A.R. Baco. 2003. Ecology of whale falls at the deep-sea floor. Oceanogr. Mar. Biol. 41: 311–354.

Snelgrove, P.V.R. and C.A. Butman. 1994. Animal-sediment relationships revisited: cause versus effect. Oceanogr. Mar. Biol. Annu. Rev. 32: 111–117.

Solan, M. and R. Kennedy. 2002. Observation and quantification of in situ animal-sediment relations using time-lapse sediment profile imagery (t-SPI). Mar. Ecol. Prog. Ser. 228: 179–191.

Staats, N., L.J. Stal, B. de Winder and L.R. Mur. 2000a. Oxygenic photosynthesis as driving process in exopolysaccharide production of benthic diatoms. Mar. Ecol. Prog. Ser. 193: 261–269.

Staats, N., L.J. Stal and L.R. Mur. 2000b. Exopolysaccharide production by the epipelic diatom *Cylindrotheca closterium*: effects of nutrient conditions. J. Exp. Mar. Biol. Ecol. 249: 13–27.

Stal, L.J. and J.F.C. de Brouwer. 2005. Diatom biofilms and the stability of intertidal mudflats. Geophys. Res. Abstr. 7: 2028.

Stal, L.J. 2010. Microphytobenthos as a biogeomorphological force in intertidal sediment stabilization. Ecol. Eng. 36: 236–245.

Strathmann, R.R. 1985. Feeding and non-feeding larval development and life-history evolution in marine invertebrates. Annu. Rev. Ecol. Evol. Syst. 16: 339–36.

Strathmann, R.R. 1990. Why life histories evolve differently in the sea. Am. Zool. 30: 197–207.

Sturdivant, S.K. and M.S. Shimizu. 2017. *In situ* organism-sediment interactions: Bioturbation and biogeochemistry in a highly depositional estuary. PLoS ONE 12: e0187800; doi.org/10.1371/journal.pone.0187800.

Stuart, C.T., M.A. Rex and R.J. Etter. 2003. Large-scale spatial and temporal patterns of deep-sea benthic species diversity. pp. 295–312. *In*: Tyler, P.A. [ed.]. Ecosystems of the Deep Oceans. Elsevier Science, Amsterdam, The Netherlands.

Sundbäck, K., A. Miles, S. Hulth, L. Pihl, P. Engström, E. Selander and A. Svenson. 2003. Importance of benthic nutrient regeneration during initiation of macroalgal blooms in shallow bays. Mar. Ecol. Prog. Ser. 246: 115–126.

Sweetman, A.K., A.R. Thurber, C.R. Smith, L.A. Levin, C. Mora, C.-L. Wei, A.J. Gooday et al. 2017. Major impacts of climate change on deep-sea benthic ecosystems. Elementa 5: 4; doi.org/10.1525/elementa.203.

Thamdrup, B. and T. Dalsgaard. 2002. Production of N_2 through anaerobic ammonium oxidation coupled to nitrate reduction in marine sediments. Appl. Environ. Microbiol. 68: 1312–1318.

Thayer, G.W., K.A, Bjorndal, J.C. Ogden, S.L. Williams and J.C. Zieman. 1984. Role of larger herbivores in seagrass communities. Estuaries 7: 351–376.

Thibodeau, P.M., L.R. Gardner and H.W. Reeves. 1998. The role of groundwater flow in controlling the spatial distribution of soil salinity and rooted macrophytes in a southeastern salt marsh, USA. Mangroves Salt Marshes 2: 1–13.

Thomsen, T.R., K.A.I. Finster and N.B. Ramsing. 2001. Biogeochemical and molecular signatures of anaerobic methane oxidation in a marine sediment Appl. Environ. Microbiol. 67: 1646–1656.

Thresher, R.E., P.D. Nichols, J.S. Gunn, B.D. Bruce and D.M. Furlani. 1992. Seagrass detritus as the basis of a coastal planktonic food chain. Limnol. Oceanogr. 37: 1754–1758.

Trites, M., I. Kaczmarska, J.M. Ehrman, P.W. Hicklin and J. Ollerhead. 2005. Diatoms from two macro-tidal mudflats in Chignecto Bay, Upper Bay of Fundy, New Brunswick, Canada. Hydrobiologia 544: 299–319.

Valdemarsen, T., E. Kristensen and M. Holmer. 2009. Metabolic threshold and sulfide-buffering in diffusion controlled marine sediments impacted by continuous organic enrichment. Biogeochemistry 95: 335–353.

Vepraskas, M.J., S.P. Faulkner and J.L. Richardson. 2001. Redox chemistry of hydric soils. pp. 85–106. *In*: Richardson, J.L. and M.J. Vepraskas [eds.]. Wetland Soils Genesis, Hydrology, Landscapes and Classification. Lewis Publishers, Boca Raton, FL, USA.

Vos, P.C. and H. de Wolf. 1993. Diatoms as a tool for reconstructing sedimentary environments in coastal wetlands; methodological aspects. Hydrobiologia 269/270: 285–296.

Waite, A.M., R.J. Olson, H.G. Dam and U. Passow. 1995. Sugar containing compounds on the cell surfaces of marine diatoms measured using concanavalin A and flow cytometry. J. Phycol. 31: 925–933.

Walker, D.I., G. Pergent and S. Fazi. 2001. Seagrass decomposition. pp. 313–323. *In*: Short, F.T. and R.G. Coles [eds.]. Global Seagrass Research Methods. Elsevier, Amsterdam, The Netherlands.

Ward, B.B. 2008. Nitrification. pp. 37–47. *In*: Capone, D.G., D.A. Bronk, M.R. Mulholland and E.J. Carpenter [eds.]. Nitrogen in the Marine Environment. Elsevier, Amsterdam, The Netherlands.

Ward, J.E. and S.E. Shumway. 2004. Separating the grain from the chaff: particle selection in suspension- and deposit-feeding bivalves. J. Exp. Mar. Biol. Ecol. 300: 83–130.
Watermann, F., H. Hillebrand, G. Gerdes, W.E. Krumbein and U. Sommer. 1999. Competition between benthic cyanobacteria and diatoms as influenced by different grain sizes and temperatures. Mar. Ecol. Prog. Ser. 187: 77–87.
Wilson, E.E. and E.M. Wolkovich. 2011. Scavenging: how carnivores and carrion structure communities. Trends Ecol. Evol. 26: 129–135.
Wollast, R. 2002. Continental margins-review of geochemical settings. pp. 15–31. *In*: Wefer, G., D. Billett, D. Hebbeln, B.B. Jørgensen, M. Schlüter and T.C.E. van Weering [eds.]. Ocean Margin Systems. Springer, Berlin, Heidelberg, Germany.
Wright, R.T., R.B. Coffin, C.P. Ersing and D. Pearson. 1982. Field and laboratory measurements of bivalve filtration of natural marine bacterioplankton. Limnol. Oceanogr. 27: 91–98.
Zeldis, J., K. Robinson, A. Ross and B. Hayden. 2004. First observations of predation by New Zealand greenshell mussels (*Perna canaliculus*) on zooplankton. J. Exp. Mar. Biol. Ecol. 311: 287–299.
Zheng, Y., L. Hou, M. Liu, Z. Liu, X. Li, X. Lin, G. Yin et al. 2016. Tidal pumping facilitates dissimilatory nitrate reduction in intertidal marshes. Sci. Rep. 6: 21338; doi.org/10.1038/srep21338.

CHAPTER 9

Salt Marshes and Mangroves: Tidal Saline Wetlands Dominated by Vascular Plants

Paula D. Pratolongo

9.1 Introduction: Coastal wetland ecosystems

Coastal wetlands encompass an ecosystem continuum between subtidal depths (to which light penetrates to support photosynthesis) and the landward edge of sea level influence. A broader definition of coastal wetlands includes, in addition to salt marshes and mangroves, a wide diversity of environments, from tidal flats, to non-tidal wetlands at the landward edge, whose hydrology is still influenced by sea level. In humid climates, freshwater seepage and high groundwater levels fed by excess rainfall provide the waterlogged conditions necessary for the establishment and growth of peat-forming wetlands. Freshwater swamps, *Pocosins*, non-tidal marshes, and bogs develop in shore zones under humid climates, beyond the limits of tidal influences (Fig. 9.1a). For these wetlands, water from terrestrial sources (groundwater seeps and often perched water tables) cause the inundation, but sea level modulates their hydrology and impedes drainage (Brinson 1991, Waller et al. 1999, Gardner et al. 2000). In the shore zone of arid climates, under a combination of high evaporation and low freshwater inputs, soils develop extremely high salinities. Extensive barren salt flats or halophytic steppes characterize arid coastal lowlands instead of freshwater swamps and bogs (Fig. 9.1b). These flat salt-crusted wetlands in arid shore zones are never flooded by tides but sea level keeps a shallow water table near the surface (Pratolongo et al. 2017, 2019).

This chapter focuses on the two ecosystems dominated by vascular plants along the intertidal zone. Tidal salt marshes dominated by herbaceous or low shrubby halophytic vegetation occur on protected shorelines from the Arctic to the subtropics. Mangrove communities dominated by halophytic trees replace salt marshes in tropical regions worldwide, although they may even extend into subtropical regions where ocean currents ameliorate temperatures. While salt marshes still occur in the tropics, they cover a smaller area than mangroves. Despite their different appearance, salt marshes and mangroves are remarkably similar in terms of their ecological functions and relevance (Bertness 2007).

Coastal wetlands deliver essential services that benefit human societies and sustain global biodiversity. Mangroves and tidal salt marshes promote sediment stabilization, and serve to protect shorelines from sea level rise, storm surges, and waves. Thus, they contribute to mitigation of erosion, and other effects associated with flooding. Tidal saline wetlands also contribute to coastal water quality improvement by removing excess nutrients, pollutants, and sediment trapping (Barbier et al. 2011). Further, salt marshes and mangroves provide essential nursery habitats and feeding grounds for numerous commercially important and endangered marine species and offer food security for many coastal communities worldwide (Kritzer and Hughes 2010).

Centro de Recursos Naturales Renovables de la Zona Semiárida (CERZOS-CONICET-UNS); Dto. Biología, Bioquímica y Farmacia-Universidad Nacional del Sur (BByF-UNS); National Scientific and Technical Research Council, Argentina (CONICET).
Email: ppratolongo@gmail.com

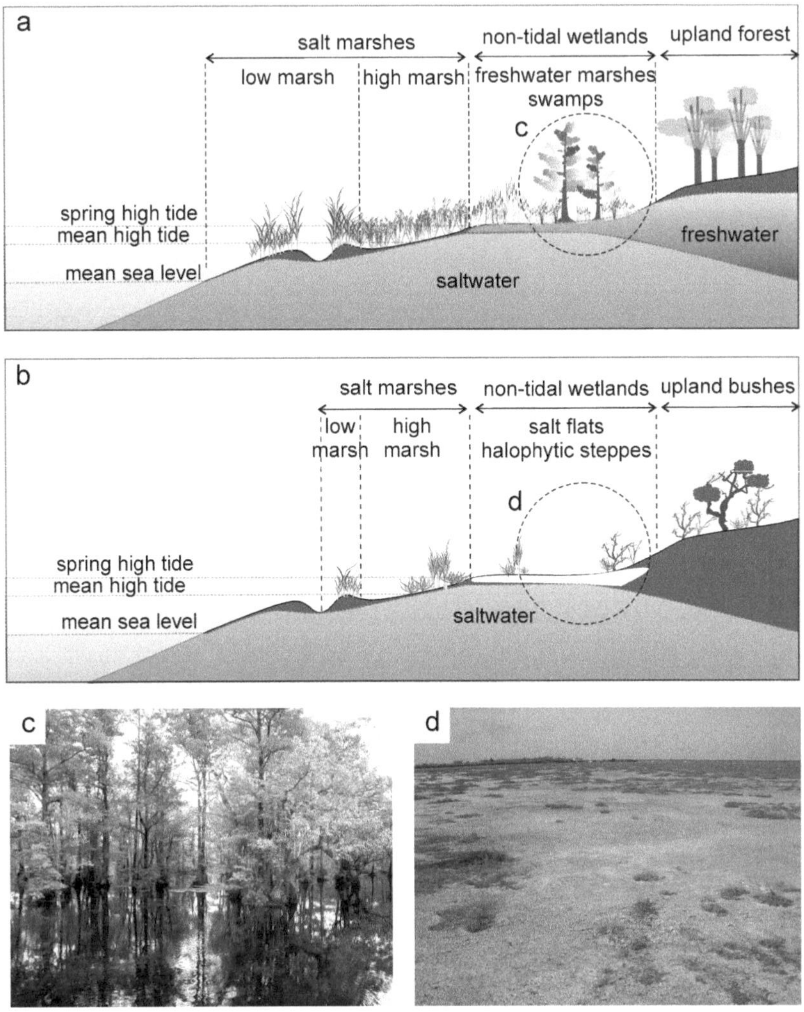

Figure 9.1. The coastal wetland continuum in humid (a) and arid climates (b), from salt marshes to non-tidal wetlands at the landward edge; (c) Raised bogs dominated by woody vegetation (*Pocosins*) in the coastal zone of North Carolina, USA; (d) Barren salt flats and halophytic steppes in the coastal zone of the Bahía Blanca Estuary, Argentina (photographs by Paula Pratolongo).

Many of these services are essential for adaptation and resilience to the impacts of climate change, especially their ability to sequester carbon (Mcleod et al. 2011). Along with seagrass meadows, salt marshes and mangroves are often referred to as Blue Carbon ecosystems (Nelleman et al. 2009). Because of their high rates of primary productivity and their ability to trap sediments, Blue Carbon ecosystems sequester large amounts of inorganic carbon from the atmosphere and oceans and store large stocks of organic carbon in their soils/sediments, known as coastal Blue Carbon (see Windham-Myers et al. (2019) for an updated review of the state of the science, economic valuation, and case studies on Blue Carbon ecosystems).

Despite the valuable services they provide and their essential role in global climate regulation, coastal wetlands are some of the most threatened ecosystems on Earth. Globally, nearly 50% of the pre-industrial extent of coastal wetlands has been lost since the 19th century (Li et al. 2018), mainly due to agriculture development, coastal settlement, and hydrological alteration. Successful implementation of national policies articulated between nations, and programs to reduce and reverse the trend of coastal wetland loss would assist climate change mitigation and will sustain the wide range of ecosystem services wetlands provide.

9.2 The physical environment

From a hydrogeomorphic (HGM) perspective, the relationship between hydrology and geomorphology defines the physical, chemical, and biological functions of wetlands. The core of the HGM approach has three components: geomorphic setting, water source, and hydrodynamics. Geomorphic setting refers to the topographic location of the wetland within the surrounding landscape. Water sources can be precipitation, surface flow, and groundwater discharge. Finally, hydrodynamics refers to flow direction and intensity of water movement within the wetland. Although all three components are at the base of the hydrogeomorphic classification system and functional assessment of wetland condition, the geomorphic setting imposes restrictions on water flows, thus landscape position is intimately linked to wetland hydrology (Brinson 1993).

According to the HGM classification, salt marshes and mangroves belong to the major class *estuarine fringe* wetlands. These wetlands typically surround estuaries, but they also fringe partially enclosed coastal water bodies, where freshwater input from rivers is negligible. Coastal settings allowing wetland development include estuaries, bays, coves, and other places of transition from land to sea, protected from the full force of ocean waves, winds and storms. In these protected shorelines, wetlands are under the influence of sea level and tidal forces; thus, bidirectional flow dominates their hydrology.

Although similar physical forces shape ecosystem functions in all estuarine fringe wetlands, there is considerable variability regarding the internal morphological characteristics, water source, and hydrodynamics. For instance, tidal wetlands frequently flood and seldom dry for significant periods, but higher elevations flood less frequently. Depending on climate, upper intertidal zones of salt marshes and mangroves can lose significant amounts of water by evaporation. Regarding water source, tidal floodwaters dominate inundation for most tidal wetlands, additional sources may be ground water discharge and precipitation. In estuaries with significant river discharge, salinity and tidal currents diminish upstream. Landward, as river flow becomes the dominant water source, salt marshes and mangroves merge with non-tidal freshwater marshes and swamps (Tiner 2018).

Overall, intertidal habitats are dynamic landscape features in space and time. This section provides an overview of the key physical forces that define the biological structure and ecological functioning of coastal wetlands.

9.2.1 Geomorphic setting

A dominant process in the long-term evolution of coastal landscapes are changes in the relative sea level. Changes in the global sea level and changes in the relative sea level due to glacio-isostatic adjustments were described in Chapter 1. A changing relative sea level produces a modification in the frequency and duration of floods and, thus in the ecological state of wetlands. In response to the different hydrologic conditions, the different plant zones within the coastal wetland continuum are expected to migrate. In coastal settings along the mid-Atlantic and Pacific coasts of North America, for instance, the relative sea level was continually rising from the last glacial maximum to the present. In these regions, Holocene estuaries drowned, and new wetlands formed landward. The process of wetlands migration upslope, in response to a rising relative sea level during the Holocene, has been widely studied along the eastern coast of North America. Coastal barrier ecosystems along the seaward margin of the Delmarva Peninsula, on the Atlantic coast of North America, underwent a sustained sea level rise during the Holocene (Oertel et al. 1989). Changes in sea level have been driving a succession of state changes in wetlands along the mainland edge of the Peninsula in which high marshes replace former upland forests (Fig. 9.2a). Following this sequence, the low marsh replaces a high marsh, and bare mudflats replace the former low marshes (Brinson et al. 1995, Christian et al. 2000).

An opposite pattern was described in high latitude coastal regions (e.g., Fennoscandia, Finland, Labrador), where the relative sea level was continually falling from the last glacial maximum to the present. In these regions, the different plant zones within the coastal wetland continuum have been typically migrating seaward. Along with the falling sea level, new land emerged, and pioneer low marsh plants colonized mudflats (Fig. 9.2b). The coastal area of western Finland rose during the Holocene because

Figure 9.2. (a) Stand of silvery trees (ghost forest) that rises out of high marshes encroaching upland forests along the mainland edge of the Delmarva Peninsula, USA, in response to a sustained sea level rise during the Holocene (photograph by Paula Pratolongo); (b) Wide tidal flats formed by coastal retreat in Iceland and seaward expansion of pioneer salt marsh plant communities in response to a falling relative sea level (photograph by Jerónimo Pan); (c) Accelerated erosion of elevated wetlands formed during the Holocene highstand and (d) soft sediment deposition at expanding marshes in the Bahía Blanca Estuary, Argentina (photographs by Paula Pratolongo).

of rapid land uplift that followed the last glacial maximum (see Chapter 1 for a detailed explanation). In the northern Gulf of Bothnia, the relative sea level fell and the shoreline displaced seaward. Shoreline retreat still occurs on this relatively flat coast, at a rate exceeding 8 mm a year, along with downward migration of plant zones (Vartiainen 1988, Ecke and Rydin 2000), and seaward expansion of pioneer plant communities (Zobel and Kont 1992). A more complex dynamic characterizes wetland environments where the relative sea level reached a maximum during the Holocene, above the elevation of the present shorelines (see Chapter 1 for a more detailed explanation). In these systems, the relative sea level fell during the late Holocene, which resulted in wide low-lying coastal landforms inherited from the former estuarine dynamics. These environments are commonly occupied by non-tidal coastal wetlands, which undergo increasing inundation under the current rising trends in relative sea level. In the Bahía Blanca Estuary, Atlantic South America, a rising sea level is a major cause of wetlands' loss in elevated Holocene surfaces (Fig. 9.2c), but the accelerated erosion of soft sediments is also the main source of suspended solids to the tidal sediment budget, allowing deposition (Fig. 9.2d) and seaward expansion of low salt marshes (Pratolongo et al. 2010, 2013).

In the case of mangroves, they occupy tropical coastlines that experienced a pattern of sea level change characterized by rapid sea level rise after the glacial maximum, decelerating, and then stabilizing about 7000 years ago. From paleoenvironmental information, it can be inferred that mangroves developed in shorelines that were up to several kilometers offshore during the last glacial maximum, when the sea was at its lowest. These mangroves have not been able to keep pace with the rapid rates of sea level rise during the early Holocene, and postglacial mangrove sediments drowned. About 7000 years ago, when relative sea level stabilized close to its present level, mangroves reestablished landward, at higher elevation. Many tropical locations have experienced a slight fall of the relative sea level during the past 7000 years (Khan et al. 2017). In the case of a falling relative sea level, progradation occurred and mangrove sediments emerged (Woodroffe 2019).

9.2.2 Water sources and hydrodynamics of tidal wetlands

Tidal waters generally represent the dominant water source to salt marshes and mangroves. The chemical composition of tidal water may be variable, depending on the connectivity to the open ocean. Salinities are commonly higher than those of freshwater, and on occasion may even be higher than salinities in the open ocean. Seasonally hypersaline estuaries often develop under arid and semiarid climates, in small and semi-isolated basins with long residence times (Largier et al. 1997). Tidal wetlands along the upper reaches of estuaries, on the other hand, may have salinities like those of freshwater systems. By comparison, rainfall water contains extremely low levels of solutes, and the relative importance of precipitation as a water source depends on the landscape position and frequency of tidal flooding. Pore-water chemistry in wetland soils at lower elevations, frequently flooded by semidiurnal tides, is like seawater. However, at higher elevations within the intertidal fringe, soils may undergo extended periods without tidal flooding. Thus, depending on factors like climate, microtopography and soil properties, both rainfall and evaporation may produce significant short-term changes in the pore-water chemistry (Wilson et al. 2015, Zhou et al. 2020).

Soluble elements are in excess in tidal water and include many required plant nutrients, such as potassium, calcium, magnesium, and sulfur. Soluble nitrogen and phosphorus concentrations, however, tend to be low, and much of the nitrogen and phosphorus required for plant growth must be imported from upland sources, either in solution, or bound to particulate material. Nitrogen-fixing microorganisms are common in salt marsh soils and contribute to plant-available nitrogen (see Jones 1974 for a pioneer work on nitrogen-fixing microorganisms). Nitrogen-fixing *Bacteria* and *Archaea*, for instance, are very active in the rhizospheres of salt marsh plants, and more recent work based on specific DNA sequences suggest that these organisms would maintain detectable populations in the rhizospheres under a wide range of marsh conditions (Davis et al. 2018). Nitrogen is also a major limiting nutrient in mangroves, and nutrient conservation strategies are quite common, such as a high nitrogen resorption efficiency in senescent leaves (Lin et al. 2010). Also, species from the Rhizophoraceae contain large amounts of tannins that bind proteins during leaf decomposition. These tannin-protein complexes are recalcitrant against microbial degradation and slow down the loss of inorganic nitrogen from the system (Laanbroek et al. 2018).

The hydrologic regime of a wetland is often quantified as a combination of tidal inundation frequency, depth, and duration, known as 'hydroperiod' (French 1993) or 'hydropattern' (Acosta and Perry 2001). The hydrologic regime of tidal wetlands is largely determined by site position within the intertidal range. Lower elevation results in more frequent and deeper flooding. Tides represent frequent and predictable flooding events in these wetlands, especially for the low-elevation and regularly inundated portions. At higher elevations, there is an irregularly flooded zone were spring tides and storm surges are responsible for inundation patterns (see Chapter 2 for details on tide dynamics). Because of the reduced frequency of flood tides reaching the irregularly flooded zones, precipitation becomes of greater significance in contributing to hydroperiod (Brinson 1991).

The hydrodynamics in tidal wetlands is also affected by microtopography, especially by tidal channels. On a ramp model of a tidal salt marsh, tidal water basically has a uniform ebb and flow across the area. However, the evolution of an intertidal area includes the incision and maturation of a tidal channel network, in a tightly coupled morphological-ecological process (Fagherazzi et al. 2012). Tidal channel initiation can be triggered by the concentration of tidal fluxes, often favored by vegetation patches and flux concentration produces local scour and channel incision which, in turn, enhances a further flux concentration (Fig. 9.3). Once tidal channels appear, most of the water movement becomes restricted to channelized flow within them. Salt marsh plants out of the channel promote accretion in those areas, while at the same time, tidal scour maintains or deepens tidal creeks (Long and Mason 1983). Similarly, mangroves are often dissected by tidal channel networks that form by the same positive feedback between erosion and channel formation, and the biophysical coupling with mangrove growth (van Maanen et al. 2015).

As tidal wetlands mature, dense networks of tidal channels develop, often following dendritic patterns, and most of the water enters and leaves the system via these creeks. In mature salt marshes, water overflows the banks only during spring tides or storm surges, allowing the marsh to become

Figure 9.3. Tidal channels in a young (a) and mature (b) salt marsh of *Spartina alterniflora* (photograph by Paula Pratolongo).

fully submerged. Thus, the soil pore-water chemistry near creek banks is like that of the tidal water, but its composition changes at short distances from the channel. The bidirectional flow characterizes tidal channels, with water moving against the bed gradient during the flood tide and along the bed gradient during the ebb. Although the water flow in tidal marshes is mainly bidirectional, vertical and unidirectional flow may also occur. At higher elevations in the marsh, rainfall infiltration may produce a significant vertical water movement and downward leaching of soluble elements. In estuarine systems associated to large rivers, heavy rainfall in the catchment area may lead to flood conditions of the marsh platform dominated by unidirectional flow (Brinson et al. 1995).

Flows within extensive mangrove forests are also driven through channels, and channel connectivity has a large impact on the water and sediment transport through the mangrove platform (McLachlan et al. 2020). Mangrove trees have a large effect on the flow structure, inducing extra flow resistance, drag-force, and blockage responsible for the enhanced tidal asymmetry and ebb-dominance. A net down-stream flow through the mangrove forest contributes to the maintenance of a deep self-sustaining drainage channel (Wolanski et al. 1980). See D'Alpaos et al. (2019) for a more comprehensive review on the formation and long-term morphological evolution of tidal channel networks.

9.3 The main primary producers: Salt marsh plants and mangrove trees

9.3.1 Adaptations to growth conditions

Conditions in intertidal habitats are stressful, and wetland plants are often adapted to a wide range of water quality conditions, such as elevated salinity, varying temperature, tidal currents, flooding, and low or no oxygen in soils. Adaptations to the harsh wetland environment include functional modifications that allow salt marsh plants and mangrove trees to survive and tolerate stress. These organisms are called tolerators or resisters. Regulators (or avoiders), on the other hand, actively avoid stress. Salt marsh plants and mangroves that successfully occupy intertidal fringes have evolved different structural and physiological adaptations that allow tolerating or regulating environmental stress (Cronk and Fennesy 2001).

One of the most severe stress for plants is the absence of oxygen in soils. Once soils get flooded, the available oxygen in pore-water is rapidly depleted by aerobic organisms that, under normal circumstances, use oxygen as the terminal electron acceptor in respiration. The rate of oxygen diffusion in water is slow and cannot supply the oxygen demand of organisms in submerged soils. Hydrophytes are flood tolerant plant species that evolved different structural and/or physiological adaptations that allow them either to tolerate or to avoid root hypoxia or anoxia (Messina and Conner 2019).

A widespread morphological (i.e., structural) adaptation in hydrophytes is the development of aerenchyma, an air space-enriched tissue, in roots and stems. Most research on the formation of aerenchyma has been performed on herbaceous wetland plants, such as Spartina grasses. Although less studied, aerenchyma is also important for root aeration and survival of mangrove species like *Avicennia marina* (Allaway et al. 2001). Aerenchyma facilitates oxygen diffusion from the aerial parts of plants into the flooded roots. In plants with well-developed aerenchyma, a main source of oxygen to root cells is no longer diffusion from the surrounding soil but downward transport through the stems (Maricle and Lee 2002). Development of adventitious roots is another common morphological adaptation of hydrophytes that allow plants to overcome soil anoxia. These roots form on the stem, above the water line, and function under aerobic conditions (Visser et al. 2000). Among mangroves, *Rhizophora* spp. grows on characteristic arched prop roots (Fig. 9.4a). These specialized roots have numerous pores (lenticels) above the water level and extend in elongated air-filled submerged roots that maintain a high oxygen concentration within the anoxic substrate. Similarly, mangrove trees of the genus *Avicennia* produce pneumatophores that protrude about 30 cm from the main root and get exposed during low tides (Cronk and Fenessy 2001).

Besides morphological adaptations, wetland plants may have several physiological adaptations to cope with changes that anoxia impose on the chemical environment of roots, especially the accumulation of reduced forms of iron, manganese, and sulfur. The magnitude of oxygen diffusion into the roots is quite large in many wetland plants, and oxygen diffuses out of the roots oxidizing the adjacent soil. A widely distributed wetland plant, *Spartina alterniflora* (see Box 1), commonly produces brown deposits of oxidized iron and manganese around the roots. These deposits form when the reduced ions get in contact with the oxidized rhizosphere (Howes and Teal 1994). Similar oxidizing capacity was described for mangrove species, denoted by diminished pore-water sulfide concentrations in the presence of prop roots or pneumatophores (Mckee and Patrick 1988). Sulfide is toxic to plant tissues and there are several mechanisms for detoxification. These include the oxidation by root oxygen diffusion, and the conversion

Figure 9.4. (a) In the foreground, pneumatophores of *Avicennia germinans* at a higher elevation within the intertidal fringe, and in the background prop roots of *Rhizophora mangle* (photograph by Jerónimo Pan); (b) Propagules of *R. mangle* growing before dropping from the parent plant; (c) Succulent stems of *Sarcocornia ambigua* (photograph by Paula Pratolongo); (d) Salt secretion in leaves of *Limonium brasiliense* (photograph by Paula Pratolongo).

to gaseous hydrogen sulfide, carbon disulfide and dimethylsulfide, that diffuse to the atmosphere (Devai and DeLaune 1995). Sulfur-oxidizing microorganisms colonizing the rhizosphere of sulfide-tolerant plants like *S. alterniflora* can couple sulfur-oxidation to carbon fixation, therefore releasing the energy trapped in sulfide or other reduced sulfur compounds. The lower concentrations of sulfide often reported from salt marshes suggest high rates of sulfide turnover, related to a tight interaction between microorganisms and plants. Anaerobic respiration is the quickest metabolic response of plant cells to anoxia. Anaerobic metabolism allows energy production under anoxic conditions, and plants can withstand brief periods in anaerobiosis. However, anaerobic respiration leads to the accumulation of ethanol and organic acids, and pH reduction in plant cells. If anoxia is prolonged, plants must develop metabolic tolerance to these changes (Colmer and Flowers 2008).

Apart from anoxia, salt marsh plants and mangrove trees withstand high levels of soil salinity. Vivipary is a common adaptation in seeds of various and phylogenetically distant mangrove tree genera, suggesting it to be an adaptive trait to the mangrove habitat. Viviparous propagules become fully mature plants before dropping off the parent tree and being dispersed by water (Kathiresan and Bingham 2001). This allows seeds to germinate before they are exposed to salt water, which might kill them. As a viviparous plant, *Rhizophora mangle* creates a propagule that is a living tree. Though resembling an elongated seed pod, the fully grown propagule on the mangrove is capable of rooting and producing a new tree (Hogarth 2015). Thus, the young tree undergoes no dormant stage as a seed, but rather progresses to a live plant before leaving its parent tree (Fig. 9.4b).

Plants adapted to successfully grow in saline environments are known as halophytes and need to be able to acquire water and avoid accumulating salts in excess. Some plants can concentrate salts in their cell solutions, that results in a high osmotic pressure. Through accumulation of sodium chloride or other solutes, these plants maintain an osmotic gradient that allows water uptake from saline sediments. This increased internal solute concentration can damage the plant. Thus, to avoid toxic concentrations within the cytoplasm, there is a compartmentalization of the cell, and solutes get sequestered in vacuoles. Succulence is an increase in cell size induced by salts, due to the large vacuole volume (Fig. 9.4c). Well-developed water storing tissues in the cortex and pith characterize succulent stems (Grigore and Toma 2017).

Halophytes can also avoid or tolerate salts through exclusion, secretion, and shedding (Cronk and Fennessy 2001). Salt exclusion is the most important mechanism to cope with high salt concentrations. In many mangrove species, nearly 99% of water salinity is excluded at the root level. However, for most species, salt exclusion at the roots is not enough, and other mechanisms arise. Many halophytes, like several Spartina and Limonium species, have salt glands on their leaves for salt secretion (Fig. 9.4d). These secreted salts are released back to the environment by leaf shedding or by wind and rain.

Box 1: The emblematic species *Spartina alterniflora*

The genus *Spartina* encompasses several salt marsh species distributed worldwide (Ainouche and Gray 2016). In 2014, a DNA-based phylogenetic study confirmed the paraphyly of the grass subtribe Sporobolinae and the creation of a large monophyletic genus *Sporobolus* was proposed, which includes species previously included in the genus Spartina (Peterson et al. 2014). The name Spartina designates a morphologically well-circumscribed group of grasses, representing a monophyletic clade. Spartina species have a tremendous impact in many scientific and non-scientific fields, provided their global ecological importance (Bortolus et al. 2019). Therefore, for the sake of simplicity, the newly designated subgenus *Spartina* will be treated here as the genus *Spartina*.

Spartina grasses frequently play a critical ecological role in coastal environments, and several *Spartina* species have a long history of human-mediated introductions outside their native range, followed by a rapid expansion, and dramatic environmental changes. In Willapa Bay, in the Pacific coast of North America, *S. alterniflora* was introduced during the late 1800s. During the first 50 years, salt marshes expanded throughout the bay, resulting in radical changes to the ecosystem functions (Simenstad and Thom 1995). After the implementation of an aggressive eradication campaign that included widespread application of herbicides, *S. alterniflora* has virtually disappeared from Willapa

Box 1 Contd. ...

...Box 1 Contd.

Bay (Strong and Ayres 2016). More recently, *S. alterniflora* was introduced in coastal to stabilize tidal flats. Since the mid 1990s, this species began to spread in the Yangtze Estuary gradually invading native marshes (Gao et al. 2014).

In Atlantic South America, *S. alterniflora* has been historically described as a native species. However, recent work proposed that this species would be either an exotic invader in this region, or an infrequent native plant, whose modern abundance and geographic expansion is due to habitat changes (Schwindt et al. 2018). Early records and indirect observations indicate that *S. alterniflora* may have expanded from Rio de Janeiro, Brazil, where it was first collected in 1817. In Argentina, *S. alterniflora* was traditionally regarded as a native species and it does not appear in the national database of exotic species, but this species was first collected in the Bahía Blanca Estuary in 1902 (Bortolus et al. 2015). At present, salt marshes of *S. alterniflora* are aggressively expanding along mudflats of the Bahía Blanca Estuary (Pratolongo et al. 2013, 2021), and marsh expansions were also documented in Peninsula Valdes, at the southernmost limit of its distribution (Bortolus et al. 2015).

Based on several historical, environmental, biological, ecological, geographic, vector, physiological, and morphological criteria, Bortolus et al. (2015) proposed that *S. alterniflora* would be a non-native species to Atlantic South America, that was accidentally introduced from either North America or Europe, sometime prior to 1817. In case that additional evidence corroborates that *S. alterniflora* is an introduced species in South America, this would be one of the largest biological invasions involving this species, with these alien marshes entirely reshaping coastal systems (Bortolus et al. 2015).

9.3.2 Plant zonation

Salt marsh and mangrove plants commonly display distinct patterns of zonation through the intertidal zone, driven by individual species' tolerance to physical stress and biological interactions acting across abiotic gradients (Pennings and Bertness 2001). The elevation gradient within the marsh establishes flood frequency, depth, and duration (the hydroperiod). Thus, abiotic factors linked to hydroperiod, like anaerobic conditions and salt concentration, commonly correlate with elevation (Moffett et al. 2012). Significant biotic interactions, like competition and facilitation, also perform in close interaction with abiotic stress (Bertness 1991, Pennings et al. 2005). For instance, the low oxygen availability at lower elevations clearly excludes non-hydrophytes. However, the physical stress does not explain why hydrophytes are usually restricted to flooded soils. In general, the lower limit of occurrence of a hydrophyte species is determined by physical stress, while its upper limit is set by competition with less tolerant species.

While physical stress and competition are the dominant factors, positive interactions also play a role in structuring plant communities. Positive interactions are common under more stressful physical conditions. For instance, at low marsh elevations, the oxidized rhizosphere produced by neighbors may benefit plant associations (Howes and Teal 1994). Another example of a positive interaction is that of ribbed mussels (*Geukensia demissa*) and cordgrass (*S. alterniflora*). The mussels partially embedded in the sediment or in superficial aggregations of individuals derive water-column nitrogen through excretion and biodeposition to the cordgrass roots they are attached to by byssal threads. This has led the ribbed mussel to be deemed a keystone species in the *Spartina* marshes fringing coastal embayments (Bertness 1984).

The interaction between stress gradients and biological response results in a shore-parallel zonation of species. At the seaward margin of a salt marsh, saltwater regularly inundates vegetation, tidal flushing prevents the accumulation of salts, and soil salinity is similar to tidal water. A few resistant genera such as *Sarcocornia, Suaeda, Aster*, and *Spartina* can tolerate the frequent tidal inundation (Doody 1992). These low marsh species must be adapted to substrate instability, anoxia, and sulfide toxicity (Mendelssohn and Morris 2000, Silliman and Bortolus 2003).

Soil salinity is more variable landward, depending on climate. In warm, humid regions, freshwater input lowers soil salinity in the high marsh (Fig. 9.5a). Thus, salts concentrate by evaporation at intermediate marsh elevations, where tidal flushing is inefficient. Highly saline mid-marshes support

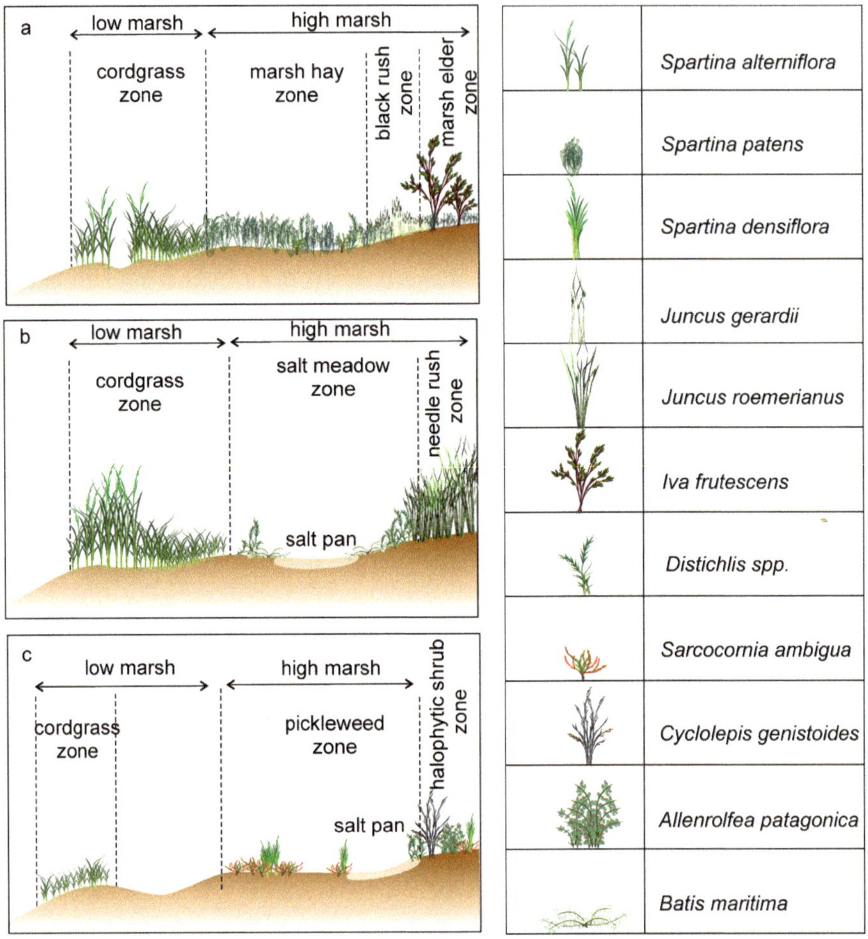

Figure 9.5. (a) Plant zonation in salt marshes of Georgia, a warm, humid region of Atlantic North America; (b) Plant zonation in New England marshes, in northern North America, under cold humid climate; (c) Salt marsh zonation in the Bahía Blanca Estuary, central Argentina.

characteristic salt-tolerant species. Permanent bare areas with low plant cover, known as salt pans, often develop in response to extremely high soil salinities (Davy and Costa 1992).

In cold humid climates, salinity decreases with increasing elevation (Pennings and Bertness 1999). A mid-marsh zone of unvegetated patches may also appear, but these pans result from poor drainage at intermediate elevations and are characterized by waterlogged anoxic soils. At higher elevations, close to the terrestrial border, common species are less tolerant to flooding and saline soils (Fig. 9.5b). *Elymus athericus* is a typical high marsh species in northern Europe, and the rush *Juncus gerardii* grows in New England marshes, in northern North America, in high marsh elevations less affected by flooding and soil salinities. A narrow strip of *Iva frutescens* often delineates the terrestrial border (Bertness et al. 2002), and brackish marshes of *Zizania aquatica, Scirpus americanus, Typha* spp., and the invasive *Phragmites australis* often appear where there is enough freshwater input (Nixon 1982).

In arid climates, a high evaporation produces an opposite saline gradient that increases with elevation. Thus, anoxia and salt concentration are uncoupled stress factors, and there is less likelihood of finding an elevation gradient that has one benign end. In these kind of environments (e.g., along southern California, northern Chile, and central Argentina), a few species occupy a narrow fringe at intermediate elevations, where both flooding frequency and salinity diminish (Callaway et al. 1990, Pratolongo et al. 2016, Fariña et al. 2018). In the Tagus estuary, on the west coast of Europe, almost 40% of the estuarine area is covered by mudflats. Typical zonation includes homogeneous stands of *Spartina maritima* in the lower marsh, with pure stands of *Sarcocornia perennis* at higher elevations. *Sarcocornia fruticosa* and *Halimione portulacoides* are also common in the upper salt marsh (Duarte et al. 2010). Most of the intertidal fringe in the Bahía Blanca Estuary (on the Atlantic coast of South America, at the northern limit of Patagonia) is covered by extensive barren mudflats (Fig. 9.5c). Through the shallow inner zone of the estuary, seasonally hypersaline conditions commonly develop because of the higher evaporation rates, and marshes of *Sarcocornia ambigua* are restricted to elevations close to the mean high tide level (Pratolongo et al. 2016, 2017). Lower marshes of *S. alterniflora* appear in the middle reach of the estuary, in the area affected by frequent tidal flushing (Pratolongo et al. 2010). Above the elevation of the mean high tide, in the zone irregularly inundated by sea water, vegetation is composed of a mosaic of salt flats, halophytic steppes and shrubs (Pratolongo et al. 2016).

Like salt marshes, mangroves also show general zonation patterns with elevation. In Florida and the Caribbean, the red mangrove *Rhizophora mangle* typically dominates seaward, followed by the black mangrove *Avicennia germinans* landward, and then the white mangrove *Laguncularia racemosa*, often accompanied by different tree species (Lugo and Snedaker 1974). Mangroves occurring along the east coast of Africa through the central Pacific Ocean are more diverse, with more than 30 species. A zonation can be recognized in open shorelines, with a seaward zone of *Sonneratia* spp. A zone of *Rhizophora* spp., *Ceriops* spp., and *Bruguiera* spp. often appears landward, and then, a zone of *Lumnitzera* spp. and *Excoecaria* spp. among others. Several species of *Avicennia* can be found throughout the upper intertidal zone.

Mangroves are associated with tropical climates worldwide, but there is an ongoing climate-driven range expansion of mangroves into salt marshes. On multiple continents, the poleward range limit of mangrove forests is controlled by winter temperatures, and climate change is expected to promote the poleward migration of mangrove forests at the expense of salt marshes (Gabler et al. 2017). In Australia, mangroves occur in temperate coastlines, where the southernmost mangroves in the world (*Avicennia marina*) occur in Corner Inlet, Victoria (38°54′25′′ S) (Saintilan et al. 2014). Mangrove expansion is a consistent pattern within the humid coasts of Australia and New Zealand (Stokes et al. 2010), and about 30% of salt marshes have been replaced by mangroves (Saintilan et al. 2014). Along the Atlantic coast of Florida, mangrove area is increasing at the expense of salt marshes (Cavanaugh et al. 2014). Overall, mangrove expansion would be associated with increased accretion, carbon burial, and higher residence time of buried soil carbon (Vaughn et al. 2020).

9.3.3 Primary production

Salt marshes and mangroves are recognized by their high primary production and their contribution to estuarine and coastal waters. Early work by Teal (1962) in Sapelo Island, Georgia, indicated that salt marsh primary production was larger than the system respiration, and the excess production would be exported to coastal waters, fueling offshore productivity. This led to the concept of *outwelling*, first used by Odum (1968) to describe the exportation of nutrients and organic detritus from coastal systems like salt marshes and mangroves, and their role in sustaining fisheries. Later work on different coastal systems either supported the outwelling hypothesis or not (e.g., Haines 1977, Nixon 1980), leading to the idea that salt marshes and mangroves may have different abilities for exportation, depending on their productivity and extent, tidal amplitude, and geomorphology. Also, exports may occur in pulses, with large outputs during high spring tides, storm surges, or heavy rainstorms (Odum 2000). The export of detritus and faunal biomass from mangroves has long been considered as an important support for offshore biological production. Most studies suggest that export is common from tidal mangroves, but many of them focused on particulate matter while it seems that dissolved organic compounds may play a more important role in matter exchange (Koch et al. 2010).

Nitrogen limitation is a major control for the primary productivity of wetland plants, and the experimental relief of nitrogen limitation may result in an increased aboveground plant height and biomass. However, one major human modification introduced to coastal environments is the increased supply of nutrients (e.g., Nixon 1995), and there are many complex interactions driving coastal wetland response. This human-driven fertilization of coastal wetlands has the potential to alter performance and competitive ability of wetland plant species, thus changing patterns of plant zonation. In New England, the effects of nutrient-rich runoff from human activities has been pointed as a major cause in the observed landward expansion of *S. alterniflora* and the seaward invasion of *Phragmites australis* (Silliman and Bertness 2004). In northern Europe, the native clonal grass *Elymus athericus* has been aggressively invading mid and low marshes since the mid-1980s. This species often forms dense monospecific stands replacing the natural *Atriplex portulacoides* marshes, and water enrichment with nitrogenous compounds would be facilitating *Elymus* invasions (Valéry et al. 2016). Nutrient enrichment can also be a driver of salt marsh loss, through increased above-ground leaf biomass, decreased biomass of bank-stabilizing roots, and increased microbial decomposition of organic matter (Deegan et al. 2012).

9.4 Consumers: The relative importance of grazing and detrital food webs

9.4.1 Decomposers of macrophyte detritus: Saprobes and detritivores

Detritus derived from vascular plants, like marsh grasses and mangroves, are rich in lignocellulose, a complex of crystalline cellulose, branching hemicellulose, and amorphous lignin. This complex is resistant to enzymatic degradation and only a limited range of organisms have evolved the ability to digest vascular plant detritus (Cragg et al. 2015). This resistance to degradation, known as recalcitrance, differentiates it from phytoplankton detritus, since lignocellulose-rich plant material must undergo fragmentation and colonization by bacteria and fungi before being consumed by most animals (Mann 1986).

Although the terms detritivore and decomposer are often used interchangeably, not all decomposers play the same role in decomposition. Detritivores are mostly invertebrate and vertebrate heterotrophs that obtain nutrients by ingesting detritus (see Chapter 8). By doing so, detritivores contribute to decomposition, but they should be distinguished from saprobes, such as many species of bacteria, fungi, and protists, which are unable to consume discrete lumps of matter. Instead, saprobes obtain nutrients by extracellular digestion, a process known as saprotrophic or lysotrophic nutrition.

Decomposition of salt marsh detritus has been well studied in *S. alterniflora*. Decay begins on still standing dead leaves and stems, and fungi are the main decomposers at this early stage. Filamentous saprobic fungi are well suited to pervade and lyse macrophyte tissues due to their rigid cell walls, apical growth, and production of lignocellulolytic enzymes (Newell 1993, 2001). Once dead tissues fall onto the marsh surface, bacteria start to play a more significant role in decomposition (Newell and Porter 2000). In the case of mangroves, the transfer of carbon and energy to animal consumers would also occur via bacterial and fungal transformation of the lignocellulosic detritus. Therefore, factors influencing rates of microbial degradation of the leachable and lignocellulosic components of mangrove detritus directly control the rates of heterotrophic microbial production, nutrient regeneration, and rates of secondary production by animals (Nordhaus et al. 2017).

To increase their biomass, microbes take up inorganic nutrients from seawater. Based on a series of experiments using ^{14}C labeled plant material, Fenchel (1977) observed that inorganic nutrient additions accelerated the growth rate and reproduction of microbes. Decomposition also resulted in a nutritive enrichment of substrates, with an increasing protein content of the detritus as particle size decreased. Thus, during macrophyte decomposition, saprobic fungi and bacteria may use plant material as a main source of carbon but obtain nitrogen and phosphorus from seawater. By doing this, decomposers retain and convert inorganic nitrogen into microbial biomass, immobilizing nutrients from the surrounding environment (Van Ryckegem et al. 2006).

Macrophyte detritus is also colonized by a diverse arrange of protozoa, rotifers, ciliates, turbellarians, nematodes, and small crustaceans that graze on microbes, especially bacteria. These animals, collectively known as microbivores, can be distinguished from detritivores as they only consume living microorganisms and their presence, like nutrient additions, accelerates decomposition. Fenchel (1977) estimated that microbivores biomass may be in the same order of magnitude as bacterial biomass and may constitute a major source of proteins to larger detritivores.

Detritivores are animals larger than protozoa that feed on detrital particles colonized by microbes and microbivores. They ingest large particles, extract some food resources, mostly associated to microbial biomass, and pass large volumes of undigested material to feces, which further contribute to the detritus pool. Once in the water column, feces get exposed to microbial colonization, allowing fungi and bacteria to use detrital carbon and nitrogen from seawater to build up microbial biomass.

Repeated passage through detritivore guts, followed by microbial recolonization, is the dominant mechanism by which marine organisms make use of macrophyte detritus as a nutritious source (Mann 2009). Thus, detritivores act as trophic links to higher order consumers. However, microbial respiration may release a large proportion of the detrital carbon to the atmosphere as CO_2. The proportion of macrophyte carbon that gets efficiently incorporated into food webs was studied in microcosms experiments using the polychaete worm *Capitella capitata* (Tenore 1977, Tenore and Hanson 1980). According to these experiments, the carbon balance between trophic transfer and respiration losses mainly depends on the time allowed for the microbial colonization. The lower the environmental nutrient level, and the higher the recalcitrance of plant material, the lower the transfer from primary production to higher order consumers. Although a rather small portion of macrophyte detritus may effectively enter coastal food webs, microbial biomass associated with detritus provides an important nutritious source for planktonic and benthic animals.

Like detritus of salt marsh macrophytes, mangrove litter is also an important trophic link between mangrove ecosystems and coastal waters. Benthic fauna, and especially infaunal and epifaunal invertebrates, use detritus as the locally available food source and significantly alter their quality through consumption or mineralization. Intertidal leaf-eating crabs largely control benthic food webs in mangroves. Leaf litter is a major food source for these crabs and leaf litter processing through ingestion and gut passage facilitates colonization by bacteria and thus facilitate decomposition and recycling of nutrients. Freshly fallen mangrove leaves are relatively nutrient-poor; thus, many crabs prefer brown decaying leaves, with improved nutritive quality due to leaching and microbial colonization (Kristensen et al. 2017). However, different crab species differentiate in their level of cellulase activity, and the assimilation efficiency of cellulose-rich detritus may control trophic and spatial segregation among crab species in mangroves (Kawaida et al. 2019).

9.4.2 Herbivores and predators

It has long been considered that herbivory was relatively unimportant in salt marshes and mangrove ecosystems. In his seminal paper, *Energy Flow in the Salt Marsh Ecosystem of Georgia,* John Teal (1962) observed that primary production in salt marshes at Sapelo Island was greater than community respiration. Teal hypothesized that a large proportion of marsh macrophyte production is exported as detritus, leading to the paradigm that salt marshes have detritus-based food webs (Odum 1980). This early view was partly supported because tissues from vascular plants are mainly composed of structural lignocellulose and this refractory material would be poorly assimilated by herbivores. Thus, most consumers would follow an indirect trophic pathway by grazing the detritus-associated microbial community supported by post-senescent plant material.

However, some salt marsh herbivores may have the potential to control plant productivity and composition. The borrowing crab *Neohelice granulata* occurs at high densities in temperate salt marshes of Atlantic South America. Field experiments carried out in some of these marshes suggest that *N. granulata* consumes large amounts of *Spartina* spp. green biomass (Bortolus and Iribarne 1999). Reduction of aboveground green biomass by direct consumption has been estimated in 20% for *Spartina densiflora,* the high marsh dominant cordgrass, while the intensity of herbivory can be far larger in low marshes of *S. alterniflora* (Martinetto et al. 2016). Thus, herbivory on new shoots directly affects cordgrass production, and crabs may also play a significant role in salt marsh zonation at a regional scale (Costa et al. 2003, Silliman and Bortolus 2003).

Other marsh herbivores include chewing insects like grasshoppers and beetles, and sap-feeding insects like aphids and leafhoppers. Although their diversity may be high, insect biomass is generally low, and their grazing on living vascular plants would consume less than 10% of the vascular plant production (Kreeger and Newell 2000).

Within vertebrates, grazing by snow geese *Anser caerulescens caerulescens* significantly affects salt marsh vegetation by consumption of belowground rhizomes and roots. Snow goose population in North America has increased at an average annual rate of 9% since the 1970s (Alisauskas et al. 2011). The increasing grazing pressure has led to severe degradation of salt marshes along the western Hudson Bay lowlands and impacted the regional biodiversity of plant, insect, and avian communities (Koons et al. 2019).

Besides vascular plants, there are surface-associated microalgae (the marine phytobenthos presented in Chapter 8) that serve as food resources to herbivores. Areal rates of primary production by these algae are generally lower than that for vascular plants, but algal biomass is much more nutritious and digestible, so they may also play an important role in consumer-based food webs. Grazing snails like *Littoraria irrorata*, *Hydrobia ulvae* and *Heleobia australis* are common salt marsh gastropods that feed on epiphytic and epibenthic algae (Steneck and Watling 1982). In the case of *Littoraria irrorata*, this species also feeds on fungi that infect grazer-induced wounds on green leaves of *S. alterniflora*. Higher-order predators like *Callinectes sapidus* and *Malaclemys* top-down regulate the abundance of this herbivorous snail. Thus, plant biomass and production would be largely controlled by the balance between grazers and their predators (Silliman and Bertness 2002).

Top-down controls, such as grazing and predation, have long been thought to be unimportant in controlling plant growth in salt marshes. Instead, bottom-up controls, such as porewater ammonium and oxygen availability, were thought to be the primary regulating factors. However, recent work showed that trophic cascades can be an important factor limiting the productivity of salt marshes. Besides herbivory by *L. irrorata*, another well-known example of trophic cascade involves the herbivorous crab *Sesarma reticulatum* in New England salt marshes. *S. reticulatum* densities would be naturally controlled by marine predators like the *Morone saxatilis*, *C. sapidus*, and *Mustelus canis*. However, decades of recreational fishing may have depleted predator populations locally and released *S. reticulatum* from top-down control. This has led to large increases in purple marsh crab populations, overgrazing, and widespread marsh die-offs. Trophic cascades are less well-documented in mangrove systems, but the relatively few studies available show herbivore effects on mangrove propagules and seedlings rather than trees (He and Silliman 2016).

9.5 Non-trophic biotic interactions

Non-trophic interactions are any non-consumptive connections between two species. Competition for limiting resources, for instance, is an interaction in which both species are harmed, and it has long been recognized as a strong determinant of ecosystem structure. For vascular plants in salt marshes and mangroves, where space is a limiting resource, dominant competitors displace competitive subordinates to refuge habitats. There are many examples in the literature, based on transplants, showing that competition limits the distribution of many salt marsh species, and that poor competitors would be excluded from the more stressful environments (e.g., Bertness and Ellison 1987, Bertness 1991).

Positive or facilitative interactions, on the other hand, are those that benefit at least one of the species, and none is damaged. Positive interactions are critical for the recovery of salt marshes and mangroves after disturbance through stress reduction. Such interactions are considered commensalisms when they are beneficial for one species but neutral for the other. For instance, burrows of the mud crab *Scylla serrata* create pool-like habitats in mangroves that would act as potential refuges for small gobiid fishes during low tide (Wasserman and Mostert 2014).

Mutualisms, on the other hand, are positive interactions that benefit both species. Clumped salt marsh plants benefit from oxygen seeping from the roots of neighboring plants (Silliman et al. 2015). Wetland species can also benefit from associational defenses. For example, mixed stands of mangroves defend trees from crab herbivory (Erickson et al. 2012), and *S. alterniflora* plants lose less biomass to snail grazing when growing in associations with *Juncus roemerianus* (Hughes 2012). There are also positive plant-animal interactions. For instance, red mangroves provide protection from sedimentation and facilitate oyster establishment (Aquino-Thomas and Proffitt 2014). Positive interactions often create positive feedback loops. For instance, mussels provide nutrients to salt marsh grasses (Bertness 1984), reduce sulfide stress, enhance biodiversity, and increase marsh resistance to drought. Plants, in turn, facilitate mussel establishment and reduce heat stress (Renzi et al. 2019).

The strength of interactions between species may vary in space and time depending on the environment. The stress-gradient hypothesis, proposed by Bertness and Callaway (1994), provides a theoretical framework for studying the interplay between facilitation and competition across physical and biological gradients of stress. According to this hypothesis, positive interactions dominate in harsh environments, whereas negative interactions prevail in less stressful environments. For instance, neighbor plants can facilitate each other in intertidal environments by trapping sediments, sharing oxygen, and reducing evaporation through shading. In the high marsh, on the other hand, inundation stress is lower and competition for resources may offset the benefits of having close neighbors. In coastal restoration projects, conventional techniques commonly involve planting with dispersed designs to avoid competition. However, small-scale restoration trials in intertidal habitats demonstrated that planting salt marsh plants and mangrove seedlings in dense clumps fosters positive interactions between neighbors and improves their chance of survival (Silliman et al. 2015). Planting with regular designs should then be considered only under less stressful abiotic conditions, where competitive interactions are more likely to be important (Renzi et al. 2019). Thus, consolidation of this ecological theory into practice would lead to changes in coastal wetland restoration paradigms.

9.6 *Blue Carbon* ecosystems and climate change

Coastal ecosystems dominated by vascular plants are highly efficient sinks of organic carbon (Chmura et al. 2003). The term *Blue Carbon* was first used to broadly define the large amount of carbon captured by marine organisms, which accounts for over half (55%) the total biological carbon captured in the planet (Nelleman et al. 2009). Underlying this concept is the idea that Blue Carbon ecosystems have the potential to compensate greenhouse gas emissions, contribute to climate change mitigation through long-term carbon preservation, and promote coastal adaptation to climate change. All these features are key criteria for the inclusion of a marine ecosystem within the Blue Carbon framework (Howard et al. 2017, Windham-Myers et al. 2019).

Salt marshes and mangroves are recognized Blue Carbon ecosystems as they meet the criteria of having high carbon stocks, supporting long-term carbon storage, and offering the potential to support policies for adaptation to climate change (Lovelock and Duarte 2019). Blue Carbon is stored within the soil, the living biomass (leaves, stems, roots), and the non-living biomass (litter, woody debris) (Mcleod et al. 2011). The organic carbon contained in plant biomass is sequestered for a relatively short period (years to decades), subject to consumption by herbivores, detritivores, and decomposers. Soils, on the other hand, can store large amounts of organic carbon over millenary timescales (Lovelock et al. 2017).

The low oxygen availability in saturated soils diminishes aerobic microbial respiration, and less carbon is released back to the atmosphere, compared with terrestrial soils. Moreover, the commonly high salinity limits methane production, a potent greenhouse gas (Chmura et al. 2003). Soil carbon accumulation is further enhanced in coastal ecosystems because of their commonly high rates of vertical accretion, resulting in continuous build-up of carbon over time (Kirwan and Megonigal 2013).

The carbon sequestered in Blue Carbon ecosystems can be classified as autochthonous or allochthonous, according to its origin. The autochthonous carbon is produced and deposited within the same location. For instance, decaying belowground biomass of salt marsh plants and mangroves decomposes very slowly in anaerobic conditions, storing the carbon within the organic soils held by their roots. The allochthonous carbon, on the other hand, is produced in one location and deposited in another. Intertidal environments receive large amounts of suspended loads transported by waves, tides, and currents. Plant canopies and prop roots are efficient at trapping suspended sediments as they move through the system, and the associated organic material adds to the local carbon stock. The ratio of autochthonous carbon to that trapped from external sources is variable, but surface accumulation would be driven primarily by tidal transport of allocthonous organic material, while *in situ* primary production would be the dominant source of recalcitrant carbon sequestered in the deep soil layers (Saintilan et al. 2013).

Unfortunately, natural coastal wetlands have declined in global area over the past decades, with considerable losses in many regions (Ramsar Convention on Wetlands 2018). Threats to salt marshes and mangroves include direct human transformations that convert coastal wetlands on a large scale; fragmentation, deforestation, perimeter erosion, extension and widening of tidal creeks, wetland 'reclamation' and urbanization are among the most common causes of loss of coastal wetlands (Valiela et al. 2004). Drainage of tidal flats and salt marshes can change their ecological character, and land reclamation may destroy or severely degrade ecosystem functions, including their sequestration capacity. Conversion for agriculture or aquaculture is the primary driver of mangrove loss, particularly in Southeast Asia (Richards and Friess 2016). Increasing human disturbance, along with sea level rise and erosion, trophic cascades, and invasive species enhance ecosystem degradation. Several ecosystem functions may be lost in degraded salt marshes and mangroves, leading to diminished carbon sequestration rates, or even changing from carbon sinks to sources, by releasing carbon from wetland soils to the atmosphere (Macreadie et al. 2017).

Box 2: Blue Carbon science and potential Blue Carbon ecosystems

Coastal wetlands are recognized Blue Carbon ecosystems as they meet the criteria of having high carbon stocks, supporting long-term carbon storage, and offering the potential to support policies for adaptation to climate change (Lovelock and Duarte 2019). Critical for the development of actionable projects, these ecosystems are included within greenhouse gas accounting guidance of the International Panel on Climate Change (IPCC) that provides emission factors for activities that result in loss and conversion of coastal wetlands, as well as those leading to restoration (Hiraishi et al. 2014). Coastal wetlands are also important ecosystems for climate change adaptation, as they can be used, for instance, to develop eco-engineering solutions for coastal protection (Duarte et al. 2013). As a consequence, after the Paris Agreement in 2016, many countries have included coastal wetlands in their mitigation activities within their National Determined Contributions.

In addition to coastal wetlands, there are several marine ecosystems that meet some of the Blue Carbon criteria, and there is a debate as to whether they can be considered Blue Carbon ecosystems. For instance, habitats dominated by calcifying organisms like coral reefs contribute to climate change adaptation through energy dissipation and coastal protection. However, the process of calcification releases CO_2 and thus they might function as sources rather than net carbon sinks.

Pelagic ecosystems, on the other hand, have a high primary productivity and CO_2 sequestration rates (as presented in Chapters 2 and 3). However, phytoplankton play an uncertain role in the long-term carbon storage and they still cannot be managed for climate change adaptation and mitigation. Phytoplankton were first proposed to be used for climate change mitigation through ocean fertilization in the late 1980s (Martin 1990). Fertilizing the ocean with iron stimulates pelagic primary production, but phytoplankton biomass must sink beyond the thermocline for the long-term storage of carbon. Results from fertilization experiments carried on so far have shown that only a small proportion of the fixed carbon may reach the deep sea (0.4% of the primary production according to some estimates, Libes 2009) and many adverse consequences have been identified (Yoon et al. 2018).

Similarly, kelp forests have a high primary production and much of the organic carbon may be transported to the deep sea and sequestered for thousands of years. However, additional scientific evidence is needed to quantify their carbon storage capacity and greenhouse gas fluxes. Moreover, there is limited potential for management of the fate of kelp biomass and the locations in which carbon accumulates (Duarte et al. 2017).

In sum, the Blue Carbon science is an expanding field of knowledge that involves multiple disciplines within the natural sciences but is also linked to policy and economy, allowing countries and markets to develop Blue Carbon monitoring and management tools. A range of criteria have to be met before some of the proposed Blue Carbon ecosystems can be included in climate change mitigation and adaptation strategies. To include these marine ecosystems in the Blue Carbon framework and greenhouse gas accounting protocols, researchers have to provide evidence of their carbon storage capacity, and guidance on how they can be managed for climate change mitigation and adaptation.

In response to the significant negative effects of Blue Carbon ecosystem degradation, there is an opportunity to develop coastal wetland conservation and restauration projects for mitigating climate change. Several countries are presently considering how these ecosystems can be more effectively included within existing policy frameworks, including carbon financing mechanisms like carbon markets. Hopefully, more Blue Carbon projects will be implemented worldwide, aiding and promoting countries with rich Blue Carbon resources to conserve and restore coastal wetlands while reducing carbon emissions.

Summary

1. This chapter focuses on the two ecosystems along the intertidal zone: tidal salt marshes, dominated by herbaceous or low shrubby halophytic vegetation and mangrove communities dominated by halophytic trees.
2. The interaction between stress gradients and biological interactions results in a shore-parallel zonation of species. Physical stress determines the lower limit of a species and the upper limit is set by competition with less tolerant species. Positive interactions are common under more stressful conditions.
3. The concept of outwelling is used to describe the exportation of nutrients and organic detritus from salt marshes and mangroves. Herbivory has been thought to be relatively unimportant in salt marshes and mangroves and most consumers would follow an indirect trophic pathway by grazing the detritus-associated microbial community. However, herbivores may have the potential to control plant productivity and predators top-down regulate the abundance of herbivores. Thus, trophic cascades can control productivity.
4. Coastal wetlands deliver essential services that benefit human societies and sustain global biodiversity. They sequester large amounts of inorganic carbon from the atmosphere and oceans and store large stocks of organic carbon in their soils/sediments, known as coastal Blue Carbon.

References

Acosta, C.A. and S.A. Perry. 2001. Impact of hydropattern disturbance on crayfish population dynamics in the seasonal wetlands of Everglades National Park, USA. Aquat. Conserv. Mar. Freshw. Ecosyst. 11: 45–57.

Ainouche, M. and A. Gray. 2016. Invasive Spartina: lessons and challenges. Biol. Invasions 18: 2119–2122.

Alisauskas, R.T., R.F. Rockwell, K.W. Dufour, E.G. Cooch, G. Zimmerman, K.L. Drake, J.O. Leafloor et al. 2011. Harvest, survival, and abundance of midcontinent lesser snow geese relative to population reduction efforts. Wildl. Monogr. 179: 1–42.

Allaway, W.G., M. Curran, L.M. Hollington, M.C. Ricketts and N.J. Skelton. 2001. Gas space and oxygen exchange in roots of *Avicennia marina* (Forssk.) Vierh. var. australasica (Walp.) Moldenke ex NC Duke, the grey mangrove. Wetl. Ecol. Manag. 9: 221–228.

Aquino-Thomas, J. and C.E. Proffitt. 2014. Oysters Crassostrea virginica on red mangrove *Rhizophora mangle* prop roots: facilitation of one foundation species by another. Mar. Ecol. Prog. Ser. 503: 177–194.

Barbier, E.B., S.D. Hacker, C. Kennedy, E.W. Koch, A.C. Stier and B.R. Silliman. 2011. The value of estuarine and coastal ecosystem services. Ecol. Monogr. 81: 169–193.

Bertness, M.D. and A.M. Ellison. 1987. Determinants of pattern in a New England salt marsh plant community. Ecological Monogr. 57: 129–147.

Bertness, M.D. 1991. Zonation of *Spartina* patens and *Spartina alterniflora* in a New England salt marsh. Ecology 72: 138–148.

Bertness, M.D. and R. Callaway. 1994. Positive interactions in communities. Trends Ecol. Evol. 9: 191–193.

Bertness, M.D., P.J. Ewanchuk and B.R. Silliman. 2002. Anthropogenic modification of New England salt marsh landscapes. Proc. Natl. Acad. Sci. 99: 1395–1398.

Bertness, M.D. 2007. Atlantic Shorelines: Natural History and Ecology. Princeton University Press.

Bortolus, A. 2006. The austral cordgrass *Spartina densiflora* Brong.: Its taxonomy, biogeography and natural history. J. Biogeogr. 33: 158–168.

Bortolus, A., J.T. Carlton and E. Schwindt. 2015. Reimagining South American coasts: unveiling the hidden invasion history of an iconic ecological engineer. Divers Distrib. 21: 1267–1283.

Bortolus, A., P. Adam, J.B. Adams, M.L. Ainouche, D. Ayres, M.D. Bertness, T.J. Bouma et al. 2019. Supporting *Spartina*: Interdisciplinary perspective shows *Spartina* as a distinct solid genus. Ecology 100: e02863.

Brinson, M.M. 1991. Landscape properties of pocosins and associated wetlands. Wetlands 11: 441–465.
Brinson, M.M. 1993. Changes in the functioning of wetlands along environmental gradients. Wetlands 13: 65–75.
Brinson, M.M., R.R. Christian and L.K. Blum. 1995. Multiple states in the sea-level induced transition from terrestrial forest to estuary. Estuaries 18: 648–659.
Callaway, R.M., S. Jones, W.R. Ferren Jr. and A. Parikh. 1990. Ecology of a mediterranean-climate estuarine wetland at Carpinteria, California: plant distributions and soil salinity in the upper marsh. Can. J. Bot. 68: 1139–1146.
Cavanaugh, K.C., J.R. Kellner, A.J. Forde, D.S. Gruner, J.D. Parker, W. Rodriguez and I.C. Feller. 2014. Poleward expansion of mangroves is a threshold response to decreased frequency of extreme cold events. Proc. Natl. Acad. Sci. 111: 723–727.
Chmura, G.L., S.C. Anisfeld, D.R. Cahoon and J.C. Lynch. 2003. Global carbon sequestration in tidal, saline wetland soils. Global Biogeochem. Cycles 17: 1111–1133.
Colmer, T.D. and T.J. Flowers. 2008. Flooding tolerance in halophytes. New Phytol. 179: 964–974.
Costa, C.S.B., J.C. Marangoni and A.M.G. Azevedo. 2003. Plant zonation in irregularly flooded salt marshes: relative importance of stress tolerance and biological interactions. J. Ecol. 91: 951–965.
Cragg, S.M., G.T. Beckham, N.C. Bruce, T.D.H. Bugg, D.L. Distel, P. Dupree, A.G. Etxabe et al. 2015. Lignocellulose degradation mechanisms across the Tree of Life. Curr. Opin. Chem. Biol. 29: 108–119.
Cronk, J.K. and M.S. Fenessy. 2001. Wetland Plants Biology and Ecology. Lewis Publishers, Florida.
Crotty, S.M. and M.D. Bertness. 2015b. Positive interactions expand habitat use and the realized niches of sympatric species. Ecology 96: 2575–2582.
D'Alpaos, A., S. Lanzoni, A. Rinaldo and M. Marani. 2019. Chapter 5—Salt-marsh ecogeomorphological dynamics and hydrodynamic circulation. pp. 189–220. *In*: Perillo, G.M.E., E. Wolanski, D.R. Cahoon and C.W. Hopkinson [eds.]. Coastal Wetlands 2nd Edition. Elsevier.
Davis, D.A., S.L. Malone and C.R. Lovell. 2018. Responses of salt marsh plant rhizosphere diazotroph assemblages to drought. Microorganisms 6: 27.
Davy, A.J. and C.S.B. Costa. 1992. Development and organization of saltmarsh communities. pp. 157–178. *In*: Seeliger, U. [ed.]. Coast Plant Communities Lat. Am. Academic Press, San Diego, USA.
Deegan, L.A., D.S. Johnson, R.S. Warren, B.J. Peterson, J.W. Fleeger, S. Fagherazzi and W.M. Wollheim. 2012. Coastal eutrophication as a driver of salt marsh loss. Nature 490: 388–392.
Devai, I. and R.D. DeLaune. 1995. Formation of volatile sulfur compounds in salt marsh sediment as influenced by soil redox condition. Org. Geochem. 23: 283–287.
Doody, J.P. 1992. The conservation of British saltmarshes. pp. 80–114. *In*: Allen, J.R.L. and K. Pye [eds.]. Saltmarshes Morphodynamics, Conserv. Eng. Significance. Cambridge University Press, Cambridge.
Duarte, B., M. Caetano, P.R. Almeida, C. Vale and I. Caçador. 2010. Accumulation and biological cycling of heavy metal in four salt marsh species, from Tagus estuary (Portugal). Environ. Pollut. 158: 1661–1668.
Duarte, B., I. Caçador, J.C. Marques and I.W. Croudace. 2013. Tagus estuary salt marshes feedback to sea level rise over a 40-year period: Insights from the application of geochemical indices. Ecol. Indic. 34: 268–276.
Duarte, C.M., J. Wu, X. Xiao, A. Bruhn and D. Krause-Jensen. 2017. Can seaweed farming play a role in climate change mitigation and adaptation? Front. Mar. Sci. 4: 100.
Ecke, F. and H. Rydin. 2000. Succession on a land uplift coast in relation to plant strategy theory. Ann. Bot. Fenn. 37: 163–171.
Erickson, A.A., S.S. Bell and C.J. Dawes. 2012. Associational resistance protects mangrove leaves from crab herbivory. Acta Oecologica 41: 46–57.
Fagherazzi, S., M.L. Kirwan, S.M. Mudd, G.R. Guntenspergen, S. Temmerman, A. D'Alpaos, J. van de Koppel et al. 2012. Numerical models of salt marsh evolution: Ecological, geomorphic, and climatic factors. Rev. Geophys. 50: 1–28.
Fariña, J.M., Q. He, B.R. Silliman and M.D. Bertness. 2018. Biogeography of salt marsh plant zonation on the Pacific coast of South America. J. Biogeogr. 45: 238–247.
Fenchel, T. 1977. Aspects of decomposition of seagrasses. Seagrass Ecosyst. Sci. Perspect. 195–234.
French, J.R. 1993. Numerical simulation of vertical marsh growth and adjustment to accelerated sea level rise, North Norfolk, UK. Earth Surf. Process Landforms 81: 63–81.
Gabler, C.A., M.J. Osland, J.B. Grace, C.L. Stagg, R.H. Day, S.B. Hartley, N.M. Enwright et al. 2017. Macroclimatic change expected to transform coastal wetland ecosystems this century. Nat. Clim. Chang. 7: 142–147.
Gao, S., Y. Du, W. Xie, W. Gao, D. Wang and X. Wu. 2014. Environment-ecosystem dynamic processes of *Spartina alterniflora* salt-marshes along the eastern China coastlines. Sci. China Earth Sci. 57: 2567–2586.
Gardner, L.R., H.W. Reeves and P.M. Thibodeau. 2000. Groundwater dynamics along forest-marsh transects in a southeastern salt marsh, USA: description, interpretation, and challenges for numerical modeling. Wetl. Ecol. Manag. 10: 145–159.
Grigore, M.N. and C. Toma. 2017. Anatomical Adaptations of Halophytes. Springer.
Haines, E.B. 1977. The origins of detritus in Georgia salt marsh estuaries. Oikos 254–260.
He, Q. and B.R. Silliman. 2016. Consumer control as a common driver of coastal vegetation worldwide. Ecol. Monogr. 86: 278–294.
Hiraishi, T., T. Krug, K. Tanabe, N. Srivastava, J. Baasansuren, M. Fukuda and T.G. Troxler. 2014. 2013 supplement to the 2006 IPCC guidelines for national greenhouse gas inventories: Wetlands. IPCC, Switzerland.

Howard, J., E. McLeod, S. Thomas, E. Eastwood, M. Fox, L. Wenzel and E. Pidgeon. 2017. The potential to integrate Blue Carbon into MPA design and management. Aquat. Conserv. Mar. Freshw. Ecosyst. 27: 100–115.
Hogarth, P.J. 2015. The Biology of Mangroves and Seagrasses. Oxford University Press.
Howes, B.L. and J.M. Teal. 1994. Oxygen loss from *Spartina alterniflora* and its relationship to salt marsh oxygen balance. Oecologia 97: 431–438.
Hughes, A.R. 2012. A neighboring plant species creates associational refuge for consumer and host. Ecology 93: 1411–1420.
Jones, K. 1974. Nitrogen fixation in a salt marsh. J. Ecol. 62: 553–565.
Kathiresan, K. and B.L. Bingham. 2001. Biology of mangroves and mangrove ecosystems. Adv. Mar. Biol. 40: 84–254.
Kawaida, S., K. Nanjo, N. Ohtsuchi, H. Kohno and M. Sano. 2019. Cellulose digestion abilities determine the food utilization of mangrove estuarine crabs. Estuar. Coast Shelf Sci. 222: 43–52.
Khan, N.S., E. Ashe, B.P. Horton, A. Dutton, R.E. Kopp, G. Brocard, S.E. Engelhart et al. 2017. Drivers of Holocene sea-level change in the Caribbean. Quat. Sci. Rev. 155: 13–36.
Kirwan, M.L. and J.P. Megonigal. 2013. Tidal wetland stability in the face of human impacts and sea-level rise. Nature 504: 53–60.
Koch, B.P., T. Dittmar and R.J. Lara. 2010. The biogeochemistry of the Caeté mangrove-shelf system. pp. 45–67. *In*: Saint-Paul, U. and H. Schneider [eds.]. Mangrove Dyn. Manag. North Brazil. Springer.
Koons, D.N., L.M. Aubry and R.F. Rockwell. 2019. Liberalized harvest regulations have not affected overabundant Snow Geese in Northern Manitoba. Condor 121.
Kreeger, D.A. and R.I.E. Newell. 2000. Trophic complexity between producers and invertebrate consumers in salt marshes. pp. 187–220. *In*: Weinstein, M.P. and D.A. Kreeger [eds.]. Concepts Controv. Tidal Marsh Ecol. Springer.
Kristensen, E., S.Y. Lee, P. Mangion, C.O. Quintana and T. Valdemarsen. 2017. Trophic discrimination of stable isotopes and potential food source partitioning by leaf-eating crabs in mangrove environments. Limnol. Oceanogr. 62: 2097–2112.
Kritzer, J. and A. Hughes. 2010. The role of salt marshes in sustaining Long Island fisheries. Mem. Torrey Bot. Soc. 26: 34–41.
Laanbroek, H.J., Q.F. Zhang, M. Leite, J.T.A. Verhoeven and D.F. Whigham. 2018. Effects of *Rhizophora mangle* leaf litter and seedlings on carbon and nitrogen cycling in salt marshes—Potential consequences of climate-induced mangrove migration. Plant Soil 426: 383–400.
Largier, J.L., J.T. Hollibaugh and S.V Smith. 1997. Seasonally hypersaline estuaries in Mediterranean-climate regions. Estuar. Coast Shelf Sci. 45: 789–797.
Li, X., R. Bellerby, C. Craft and S.E. Widney. 2018. Coastal wetland loss, consequences, and challenges for restoration. Anthr. Coasts 1: 1–15.
Libes, S.M. 2009. Introduction to Marine Biogeochemistry, Second edition. Elsevier/Academic Press.
Lin, Y.M., X.W. Liu, H. Zhang, H.Q. Fan and G.H. Lin. 2010. Nutrient conservation strategies of a mangrove species *Rhizophora stylosa* under nutrient limitation. Plant Soil 326: 469–479.
Long, S.P. and C.F. Mason. 1983. Saltmarsh Ecology. New York: Blackie & Sons Ltd.
Lovelock, C.E., T. Atwood, J. Baldock, C.M. Duarte, S. Hickey, P.S. Lavery, P. Masque et al. 2017. Assessing the risk of carbon dioxide emissions from Blue Carbon ecosystems. Front Ecol. Environ. 15: 257–265.
Lovelock, C.E. and C.M. Duarte. 2019. Dimensions of Blue Carbon and emerging perspectives. Biol. Lett. 15: 20180781.
Lugo, A.E. and S.C. Snedaker. 1974. The ecology of mangroves. Annu. Rev. Ecol. Syst. 5: 39–64.
Macreadie, P.I., D.A. Nielsen, J.J. Kelleway, T.B. Atwood, J.R. Seymour, K. Petrou, R.M. Connolly et al. 2017. Can we manage coastal ecosystems to sequester more Blue Carbon? Front Ecol. Environ. 15: 206–213.
Mann, K.H. 1986. The role of detritus at the land-sea boundary. pp. 123–140. *In*: Lasserre, P. and J.-M.B.T.-E.O.S. Martin [eds.]. Biogeochem. Processes at the Land-sea Boundaries Elsevier.
Mann, K.H. 2009. Ecology of Coastal Waters: With Implications for Management. Second Edition. Blackwell Science Ltd, Malden, USA.
Maricle, B.R. and R.W. Lee. 2002. Aerenchyma development and oxygen transport in the estuarine cordgrasses *Spartina alterniflora* and *S. anglica*. Aquat. Bot. 74: 109–120.
Martin, J.H. 1990. Glacial-interglacial CO_2 change: the iron hypothesis. Paleoceanogr. Paleoclimatol. 5: 1–13.
Martinetto, P., D.I. Montemayor, J. Alberti, C.S.B. Costa and O. Iribarne. 2016. Crab bioturbation and herbivory may account for variability in carbon sequestration and stocks in south west Atlantic salt marshes. Front Mar. Sci. 3: 122.
Mckee, K.L. and W.H. Patrick. 1988. The relationship of smooth cordgrass (*Spartina alterniflora*) to tidal datums: a review. Estuaries 11: 143–151.
McLachlan, R.L., A.S. Ogston, N.E. Asp, A.T. Fricke, C.A. Nittrouer and V.J.C. Gomes. 2020. Impacts of tidal-channel connectivity on transport asymmetry and sediment exchange with mangrove forests. Estuar. Coast Shelf Sci. 233: 106524.
Mcleod, E., G.L. Chmura, S. Bouillon, R. Salm, M. Björk, C.M. Duarte, C.E. Lovelock et al. 2011. A blueprint for Blue Carbon: toward an improved understanding of the role of vegetated coastal habitats in sequestering CO_2. Front Ecol. Environ. 9: 552–560.
Mendelssohn, I.A. and J.T. Morris. 2000. Ecophysiological controls on the productivity of *Spartina alterniflora*. pp. 59–80. *In*: Weinstein, M. and D.A. Kreeger [eds.]. Concepts Controv Tidal Marsh Ecol. Kluwer Academic Publishers, Boston, USA.

Messina, M.G. and W.H. Conner. 2019. Adaptations of plants to flooding and soil waterlogging. pp. 173–203. *In*: Messina, M.G. and W.H. Conner [eds.]. South For Wetl. Routledge.
Moffett, K.B., S.M. Gorelick, R.G. McLaren and E.A. Sudicky. 2012. Salt marsh ecohydrological zonation due to heterogeneous vegetation–groundwater–surface water interactions. Water Resour. Res. 48: W02516.
Nelleman, C., E. Corcoran, C.M. Duarte, L. Valdés, C. DeYoung, L. Fonseca and G. Grimsditch. 2009. Blue Carbon: The role of healthy oceans in binding carbon. UNEP/FAO/UNESCO/IUCN/CSIC.
Newell, S.Y. 1993. Decomposition of shoots of a salt-marsh grass. pp. 301–326. *In*: Jones, J.G. [ed.]. Adv. Microb. Ecol. Springer.
Newell, S.Y. 2001. Multiyear patterns of fungal biomass dynamics and productivity within naturally decaying smooth cordgrass shoots. Limnol. Oceanogr. 46: 573–583.
Newell, S.Y. and D. Porter. 2000. Microbial secondary production from salt marsh-grass shoots, and its known and potential fates. pp. 159–185. *In*: Weinstein, M.P. and D.A. Kreeger [eds.]. Concepts Controv Tidal Marsh Ecol. Springer.
Nielsen, O.I., B. Gribsholt, E. Kristensen and N.P. Revsbech. 2004. Microscale distribution of oxygen and nitrate in sediment inhabited by *Nereis diversicolor*: spatial patterns and estimated reaction rates. Aquat. Microb. Ecol. 34: 23–32.
Nixon, S.W. 1980. Between coastal marshes and coastal waters—a review of twenty years of speculation and research on the role of salt marshes in estuarine productivity and water chemistry. pp. 437–525. *In*: Hamilton, P. and K.B. Macdonald [eds.]. Estuar. Wetl. Process. Springer.
Nixon, S.W. 1982. The Ecology of New England High Salt Marshes: A Community Profile. Washington, D.C.
Nixon, S.W. 1995. Coastal marine eutrophication: a definition, social causes, and future concerns. Ophelia 41: 199–219.
Nordhaus, I., T. Salewski and T.C. Jennerjahn. 2017. Interspecific variations in mangrove leaf litter decomposition are related to labile nitrogenous compounds. Estuar. Coast Shelf Sci. 192: 137–148.
Odum, E.P. 1968. A research challenge: evaluating the productivity of coastal and estuarine water. pp. 63–64. *In*: Proc. Second Sea Grant Conf. University of Rhode Island.
Odum, E.P. 2000. Tidal marshes as outwelling/pulsing systems. pp. 3–7. *In*: Weinstein, M.P. and D.A. Kreeger [eds.]. Concepts Controv Tidal Marsh Ecol. Springer.
Pennings, S.C. and M.D. Bertness. 1999. Using latitudinal variation to examine effects of climate on coastal salt marsh pattern and process. Curr. Top Wetl. Biogeochem. 3: 100–111.
Pennings, S.C. 2001. Salt marsh communities. Mar. Community Ecol. 289–316.
Pennings, S.C., M.B. Grant and M.D. Bertness. 2005. Plant zonation in low-latitude salt marshes: disentangling the roles of flooding, salinity and competition. J. Ecol. 93: 159–167.
Peterson, P.M., K. Romaschenko, Y.H. Arrieta and J.M. Saarela. 2014. A molecular phylogeny and new subgeneric classification of *Sporobolus* (Poaceae: Chloridoideae: Sporobolinae). Taxon 63: 1212–1243.
Pratolongo, P., G.M.E. Perillo and M.C. Píccolo. 2010. Combined effects of waves and plants on a mud deposition event at a mudflat-saltmarsh edge in the Bahía Blanca estuary. Estuar. Coast Shelf Sci. 87: 207–212.
Pratolongo, P., C. Mazzon, G. Zapperi, M.J. Piovan and M.M. Brinson. 2013. Land cover changes in tidal salt marshes of the Bahía Blanca estuary (Argentina) during the past 40 years. Estuar. Coast Shelf Sci. 133: 23–31.
Pratolongo, P.D., M.J. Piovan and D.G. Cuadrado. 2016. Coastal environments in the Bahía Blanca Estuary, Argentina. pp. 205–224. *In*: Khan, A.M., B. Boër, M. Őzturk, M. Clüsener-Godt, B. Gul and S.-W. Breckle [eds.]. Sabkha Ecosyst. Vol. V Am. Springer International Publishing.
Pratolongo, P., M.J. Piovan, D.G. Cuadrado and E.A. Gómez. 2017. Coastal landscape evolution on the western margin of the Bahía Blanca Estuary (Argentina) mirrors a non-uniform sea-level fall after the mid-Holocene highstand. Geo-Marine Lett. 37: 373–384.
Pratolongo, P., N. Leonardi, J.R. Kirby and A. Plater. 2019. Temperate coastal wetlands: morphology, sediment processes, and plant communities. pp. 105–152. *In*: Perillo, G.M.E., E. Wolanski, D.R. Cahoon and C.W. Hopkinson [eds.]. Coastal Wetlands. Elsevier.
Pratolongo, P., F. Funk, M.J. Piovan, C. Celleri and V. Negrin. 2021. Coastal wetlands of the Bahía Blanca Estuary. Landscape structure and plant associations. pp. 275–289. *In*: Fiori, S. and P. Pratolongo [eds.]. Bahía Blanca Estuary Ecol. Biodivers. Springer Nature.
Renzi, J.J., Q. He and B.R. Silliman. 2019. Harnessing positive species interactions to enhance coastal wetland restoration. Front Ecol. Evol. 7: 131. doi: 10.3389/fevo.2019.00131.
Richards, D.R. and D.A. Friess. 2016. Rates and drivers of mangrove deforestation in Southeast Asia, 2000–2012. Proc. Natl. Acad. Sci. 113: 344–349.
Saintilan, N., K. Rogers, D. Mazumder and C. Woodroffe. 2013. Allochthonous and autochthonous contributions to carbon accumulation and carbon store in southeastern Australian coastal wetlands. Estuar. Coast Shelf Sci. 128: 84–92.
Saintilan, N., N.C. Wilson, K. Rogers, A. Rajkaran and K.W. Krauss. 2014. Mangrove expansion and salt marsh decline at mangrove poleward limits. Glob. Chang. Biol. 20: 147–157.
Schwindt, E., N. Battini, C.B. Giachetti, K.L. Castro and A. Bortolus. 2018. Especies exóticas marino-costeras. Vazquez Mazzini. Argentina.
Simenstad, C.A. and R.M. Thom. 1995. Spartina alterniflora (smooth cordgrass) as an invasive halophyte in Pacific Northwest estuaries. Hortus Northwest 6: 9–12.

Silliman, B.R. and M.D. Bertness. 2002. A trophic cascade regulates salt marsh primary production. Proc. Natl. Acad. Sci. 99: 500–505.

Silliman, B.R. and A. Bortolus. 2003. Underestimation of *Spartina* productivity in western Atlantic marshes: Marsh invertebrates eat more than just detritus. Oikos 101: 549–554.

Silliman, B.R. and S.Y. Newell. 2003. Fungal farming in a snail. Proc. Natl. Acad. Sci. 100: 15643–15648.

Silliman, B.R. 2004. Shoreline development drives invasion of *Phragmites australis* and the loss of plant diversity on New England salt marshes. Conserv. Biol. 18: 1234–1424.

Silliman, B.R., E. Schrack, Q. He, R. Cope, A. Santoni, T. van der Heide, R. Jacobi et al. 2015. Facilitation shifts paradigms and can amplify coastal restoration efforts. Proc. Natl. Acad. Sci. 112: 14295–14300.

Steneck, R.S. and L. Watling. 1982. Feeding capabilities and limitation of herbivorous molluscs: A functional group approach. Mar. Biol. 68: 299–319.

Stevenson, J.C., L.G. Ward and M.S. Kearney. 1986. Vertical accretion in marshes with varying rates of sea level rise. pp. 241–259. *In*: Wolfe, D. [ed.]. Estuar. Var Proc. Eighth Bienn Int. Estuar. Res. Conf. Elsevier.

Stokes, D.J., T.R. Healy and P.J. Cooke. 2010. Expansion dynamics of monospecific, temperate mangroves and sedimentation in two embayments of a barrier-enclosed lagoon, Tauranga Harbour, New Zealand. J. Coast Res. 26: 113–122.

Strong, D.R. and D.A. Ayres. 2016. Control and consequences of *Spartina* spp. invasions with focus upon San Francisco Bay. Biol. Invasions 18: 2237–2246.

Teal, J.M. 1962. Energy Flow in the salt marsh ecosystem of Georgia. Ecology 43: 614–624.

Tenore, K.R. 1977. Growth of *Capitella capitata* cultured on various levels of detritus derived from different sources. Limnol. Oceanogr. 22: 936–941.

Tenore, K.R. and R.B. Hanson. 1980. Availability of detritus of different types and ages to a polychaete macroconsumer, *Capitella capitata*. Limnol. Oceanogr. 25: 553–558.

Tiner, R.W. 2018. Hydrology of coastal wetlands. pp. 217–231. *In*: Finlayson, C.M., M. Everard, K. Irvine, R.J. McInnes, B.A. Middleton, A.A. van Dam and N.C. Davidson [eds.]. The Wetland Book: I: Structure and Function, Management, and Methods. Springer

Valéry, L., A. Radureau and J.C. Lefeuvre. 2016. Spread of the native grass Elymus athericus in salt marshes of Mont-Saint-Michel bay as an unusual case of coastal eutrophication. J. Coast Conserv. 1–13.

Valiela, I., D. Rutecki and S. Fox. 2004. Salt marshes: biological controls of food webs in a diminishing environment. J. Exp. Mar. Biol. Ecol. 300: 131–159.

van Maanen, B., G. Coco and K.R. Bryan. 2015. On the ecogeomorphological feedbacks that control tidal channel network evolution in a sandy mangrove setting. Proc. R Soc. A Math Phys. Eng. Sci. 471: 20150115.

Van Ryckegem, G., G. Van Driessche, J.J. Van Beeumen and A. Verbeken. 2006. The estimated impact of fungi on nutrient dynamics during decomposition of *Phragmites australis* leaf sheaths and stems. Microb Ecol. 52: 564–574.

Vaughn, D.R., T.S. Bianchi, M.R. Shields, W.F. Kenney and T.Z. Osborne. 2020. Increased organic carbon burial in northern Florida mangrove—salt marsh transition zones. Global Biogeochem. Cycles 34: e2019GB006334.

Visser, E.J.W., T.D. Colmer, C. Blom and L. Voesenek. 2000. Changes in growth, porosity, and radial oxygen loss from adventitious roots of selected mono- and dicotyledonous wetland species with contrasting types of aerenchyma. Plant Cell Environ. 23: 1237–1245.

Waller, M.P., A.J. Long, D. Long and J.B. Innes. 1999. Patterns and processes in the development of coastal mire vegetation: Multi-site investigations from Walland Marsh, Southeast England. Quat. Sci. Rev. 18: 1419–1444.

Wasserman, R.J. and B.P. Mostert. 2014. Intertidal crab burrows as a low-tide refuge habitat for a specific gobiid: preliminary evidence for commensalism. Mar. Freshw. Res. 65: 333–336.

Wilson, A.M., T. Evans, W. Moore, C.A. Schutte, S.B. Joye, A.H. Hughes and J.L. Anderson. 2015. Groundwater controls ecological zonation of salt marsh macrophytes. Ecology 96: 840–849.

Windham-Myers, L., S. Crooks and T. Troxler, T. (eds.). 2019. A Blue Carbon Primer. CRC Press.

Wolanski, E., M. Jones and J.S. Bunt. 1980. Hydrodynamics of a tidal creek-mangrove swamp system. Mar. Freshw. Res. 31: 431–450.

Woodroffe, C.D. 2019. Chapter 2—The morphology and development of coastal wetlands in the tropics. pp. 79–103. *In*: Perillo, G.M.E., E. Wolanski, D.R. Cahoon and C.W. Hopkinson [eds.]. Coastal Wetlands 2nd Edition. Elsevier.

Yoon, J.E., K.C. Yoo, A.M. Macdonald, Y. Ho-Il, K.T. Park, E.J. Yang, H.C. Kim et al. 2018. Reviews and syntheses: Ocean iron fertilization experiments—past, present, and future looking to a future Korean Iron Fertilization Experiment in the Southern Ocean (KIFES) project. Biogeosciences 15: 5847.

Zhou, T., P. Xin, L. Li, D.A. Barry and J. Šimůnek. 2020. Effects of large macropores on soil evaporation in salt marshes. J. Hydrol. 124754.

Zobel, M. and A. Kont. 1992. Formation and succession of alvar communities in the Baltic land uplift area. Nord. J. Bot. 12: 249–256.

CHAPTER 10

Marine Hard Substrate Communities

Jorge L. Gutiérrez,[1,2,*] *Sally D. Hacker,*[3] *Martin A. Coombes,*[4] *Christian Wild,*[5] *Guilherme H. Pereira-Filho*[6] and *M. Gabriela Palomo*[1,7]

10.1 Introduction

Hard-substrate habitats include rock, the hard parts of organisms, and anthropogenic materials that vary in composition, physical properties, and spatial extent (Taylor and Wilson 2003, Wahl 2009). Hard substrates provide stable surfaces on which benthic organisms (i.e., the organisms that live on, in, or near the seabed) can attach or move across. This is in contrast to the extensive areas of unconsolidated sediments that characterize the seafloor (presented in Chapter 8 of this book); these "soft-bottom" environments are less stable and are typically colonized by species that burrow or attach within the sediment surface itself (Little and Little 2000).

Globally, hard substrates cover a relatively minor proportion of the marine benthic environment when compared to soft substrates. Nevertheless, they make up a significant fraction of the world's coastal areas (see Section 10.3 below), support high biodiversity (e.g., Connell 1978, Knowlton 2001), and provide critical ecosystem services to humans (see Section 10.8 below). Furthermore, the early scientific study of coastal hard substrate communities facilitated the development of many key methods, concepts, and theories in experimental community ecology (Connell 1961a, Paine 1966; see Section 10.7 below). Subsequently, knowledge of the structure and functioning of hard substrate communities has expanded across geographic regions, hard substrate types, and ocean depths (e.g., Clarke and Johnston 2003, Smith and Baco 2003, Roberts et al. 2006, Young 2009, Clark et al. 2010). Today, research in this field extends beyond its initial focus on the trophic organization of communities, to encompass trophic and non-trophic effects and their functional linkages within and across habitat boundaries (see Sections 10.4 and 10.5 below).

[1] Instituto de Geología de Costas y del Cuaternario (IGCyC, UNMdP-CIC) & Instituto de Investigaciones Marinas y Costeras (IIMyC, CONICET-UNMdP), Universidad Nacional de Mar del Plata, Funes 3350, Mar del Plata B7602AYL, Argentina.

[2] Grupo de Investigación y Educación en Temas Ambientales (GrIETA), Estación Biológica Las Brusquitas, San Eduardo del Mar B7783ADE, Argentina.

[3] Department of Integrative Biology, Oregon State University, 3029 Cordley Hall, Corvallis, OR 97331, USA.

[4] School of Geography and the Environment, University of Oxford, South Parks Road, Oxford, OX1 3QY, UK.

[5] Department of Marine Ecology, Faculty of Biology and Chemistry, Universität Bremen, Leobener Str. 6, 28359 Bremen, Germany.

[6] Instituto do Mar, Universidade Federal de São Paulo, Rua Carvalho de Mendonça 144, 11070-100 Encruzilhada, Santos, SP, Brazil.

[7] Museo Argentino de Ciencias Naturales "Bernardino Rivadavia" (MACN-CONICET), Av. Ángel Gallardo 470, C1405DJR Buenos Aires, Argentina.

Emails: hackers@oregonstate.edu; martin.coombes@ouce.ox.ac.uk; christian.wild@uni-bremen.de; pereira.filho@unifesp.br; gpalomo@macn.gov.ar

* Corresponding author: jgutierrez@grieta.org.ar

This chapter reviews the current state of knowledge on the organization and functioning of marine hard substrate communities. First, we provide an overview of the major kinds of coastal hard-substrate habitats and their global and regional prevalence. Second, we focus on the diversity of organisms that inhabit hard substrates giving particular attention to "foundation species" (i.e., species that by virtue of their structural or functional attributes, create and define entire ecological communities or ecosystems and their accompanying biota; Dayton 1972, Ellison 2019). Third, we illustrate the major roles of hard substrate species in organic matter flux and trophic interactions across food webs, and as physical ecosystem engineers (i.e., their roles in the physical modification of environments; *sensu* Jones et al. 1994). Fourth, we summarize the central theories on the organization of hard substrate communities with a focus on rocky shores and subtidal rocky reefs. Finally, we consider the services that humans derive from marine hard substrate communities.

To limit the scope of this chapter, we have set two major restrictions. First, we focus on hard substrate communities occurring in coastal seas, from the shoreline to ca. 100 m water depth. Second, whilst the variability in hard-substrate habitats is broadly considered, we place the major emphasis on rocky shores, coral reefs, and rhodolith beds (i.e., aggregations of calcified nodules of free-living coralline algae) (see Section 10.2 below). Organisms and their functions at deeper ocean depths and/or other particular kinds of hard substrates are discussed in detail in more specialized reviews (e.g., Gutiérrez et al. 2003, Smith and Baco 2003, Roberts et al. 2006, Young 2009, Clark et al. 2010).

10.2 Types of hard-substrate habitat in the coastal ocean

Taylor and Wilson (2003) classify marine hard substrates as abiogenic (various types of rocks), biogenic (organism hard parts), and anthropogenic (human-made structures and materials), the structure and function of which are considered in the following sections.

10.2.1 Abiogenic hard-substrate habitat

The continuous and episodic erosion of coastlines gives rise to a range of rocky landscape types that are the foundation of a variety of hard substrate communities. Broadly speaking, hard coastal habitats can be separated into vertical plunging cliffs and rocky slopes, although the combination of cliffs fronted by gently sloping or near-horizontal rock platforms is common. Platforms develop at, or close to, current sea level as the result of both wave erosion and subaerial weathering processes (Trenhaile 1987, Kennedy et al. 2014a). Subtidal formations such as volcanic seamounts can also function as important rocky habitats, typically in association with plate boundaries and island clusters (Clark et al. 2010).

The particular characteristics of a coastal rocky habitat result from a complex combination of boundary conditions including geology, oceanography, climate, and organisms (including humans) that vary in relative significance across space and over time (Naylor et al. 2010). Arguably of greatest significance are the geological characteristics of the bedrock, which directly and indirectly control the erodibility and weatherability of the exposed surfaces (Sunamura 1992, Kennedy et al. 2014b). Broadly speaking, coastlines formed of resistant rock types, such as granite and basalt, are more conducive to cliff formation, whilst soft and more easily erodible lithologies including mudstone and some types of limestone and sandstone tend to give rise to shore platforms (Kennedy et al. 2014b). Spatial variation (at a range of scales) in rock type dictate the physical complexity of the shoreline through differential rates of weathering and erosion. The British Isles provide a good example of how a diversity of 'hard' and 'soft' rocky types—encompassing cohesive clays, sedimentary limestones, and igneous granites—can give rise to an astonishing array of rocky coast landscapes ranging from horizontal shore platforms pitted by nano-scale (nm) and micro-scale (μm) morphologies (Moses 2014) to high vertical cliffs hundreds of meters high (Lim 2014). Similarly, geological structure (i.e., joints, bedding planes, and other discontinuities) further influence the modes and rates of coastal retreat (Naylor and Stephenson 2010). For example, the formation of caves, arches, stacks, and other characteristic rocky coast landforms is strongly influenced by the presence and orientation of planes of weakness (i.e., parting surfaces in stratified sedimentary rocks and weaknesses resulting from recrystallization in metamorphic rocks). Furthermore, structural characteristics affect the size and shape of material produced by erosion and the rate at which it is produced. Wave climate and tidal regimes further influence the characteristics and dynamics of this eroded material, controlling the transport and accumulation of boulder and cobble to beaches and other depositional landforms (McKenna 2005). These unconsolidated coastal features provide important habitats for a range of organisms (Herkül et al. 2018), although colonization is typically limited on highly dynamic shorelines composed of mobile sediment.

10.2.2 Biogenic hard-substrate habitat

Calcareous structures formed by biomineralization are the most diverse and widespread kind of biogenic hard substrate in the oceans. These structures include the skeletons, shells, and tubes of living and dead invertebrates (e.g., corals, mollusks, polychaetes, and barnacles; Fig. 10.1a, b, e), living and dead calcified algae (e.g., branching, encrusting, and rodolith-forming coralline algae; Fig. 10.1c), and sunken vertebrate skeletons (e.g., whales). Calcifying invertebrates and macroalgae often form "reefs" (e.g., corals, oysters, vermetid gastropods, and serpulid polychaetes) or "beds" (e.g., mussels and rhodoliths) that can extend over large areas (see Section 10.4.1.2 below; Fig. 10.1a, b, c). Reefs are structures that rise vertically from the substrate to the water surface and can be asymmetrical, mounded, or ridged (see Cummings (1932) and Wilson (1950) for full definitions). In contrast to reefs, beds consist of one or a few layers of individuals (typically less than a few tens of centimeters thick) and cover the seafloor to varying extents (up to tens of square meters or more). Calcified biogenic structures can also persist over centuries after organismal death, and form beds as a result of localized mass mortality or accumulation by hydrodynamic forces (e.g., shell beds; see Kidwell 1986, Gutiérrez and Iribarne 1999). Although these large aggregations are important biogenic habitats, individual structures scattered at low densities across soft substrates can also be important. For example, single shells can enhance the overall species richness of the seafloor, sometimes enabling colonization by sessile species that create additional hard substrate (e.g., Schwindt and Iribarne 2000, Yakovis and Artemieva 2017).

Other biogenic hard substrates in the coastal oceans include the tubes of sabellariid polychaetes, which are made of sand and shell fragments cemented together by proteinaceous secretions (Stewart et al. 2004), and those of chaetopterid polychaetes (Fig. 10.1d), which are non-rigid proteinaceous structures encrusted by sands (Bailey-Brock 1979). These tubeworms are often gregarious, forming hummocks or even large reefs (e.g., Bailey-Brock 1979, Aviz et al. 2019). Although relatively less abundant in the oceans, wood in the form of logs, branches, and the prop roots of mangrove trees can host a variety of sessile invertebrates and macroalgae in the coastal tropics (e.g., Rodriguez and Stoner 1990, Ellison et al. 1996). Coarse woody debris also provides substrate for biofilm formation and gastropod grazing in intertidal rocky shores (Storry et al. 2006). Even though large woody debris can be locally abundant in coastal areas (Gonor et al. 1988, Everett and Ruiz 1993), it is generally considered a marginal source of hard substrate when compared to sunken wood in the deep sea, which has been postulated to function as an evolutionary stepping stone in deep-sea mussel diversification (see Distel et al. 2000, Samadi et al. 2010, Judge and Barry 2016).

Finally, many other marine organisms provide surfaces for colonization including ascidians (e.g., Rimondino et al. 2015), crustaceans (e.g., decorator crabs; Sallam and Wicksten 2011), macrophytes (e.g., Hacker and Steneck 1990, Arkema and Samhouri 2019, Chava et al. 2019, Hayduk et al. 2019), and even sea turtles and whales (e.g., Scarff 1986, Pfaller et al. 2008). Whilst these habitats do not strictly conform to the conventional definition of hard substrates, they provide similar habitat with, in some cases, the added possibility of mobility (see Fig. 10.1e, f, g).

Figure 10.1. Types of marine hard substrate. (a) Basaltic rocky intertidal shore at Yachats, Oregon, USA. Mussels (*Mytilus californianus*) provide secondary hard substrate for gooseneck (*Pollicipes polymerus*) and acorn barnacles (*Balanus glandula* and *Semibalanus cariosus*) with a seastar predator (*Pisaster ochraceus*) in the foreground. (b) Coral reef at Queimada Grande Island, São Paulo, Brazil (the southernmost reef in the Atlantic Ocean; 14 m depth). The reef framework is built by a single coral species, *Madracis decactis*, and provides secondary substrate for crustose and geniculate coralline algae (e.g., *Amphiroa* spp. and *Jania* spp.), ascidians, and sponges. (c) Rhodolith bed at Fernando de Noronha Marine National Park, Pernambuco, Brazil (45 m depth). The rhodoliths (i.e., free-living nodules built by coralline red algae) are mostly covered by brown fleshy algae (e.g., Dictyotales), coralline algae (e.g., *Amphiroa* spp. and *Jania* spp.), tunicates, and sponges. (d) Proteinaceous tubes of the polychaete *Chaetopterus variopedatus* (ca. 50 cm height) in the Beagle Channel, Tierra del Fuego, Argentina (6–8 m depth). Their surfaces are colonized by ascidians (possibly *Didemnium studeri*, visible as white colonies), hydrozoans, and red algae. (e, f, g) A hairy snail (*Fusitriton magellanicum*), giant kelp, (*Macrocystis pyrifera*), and ascidians (*Paramolgula gregaria*), in the Beagle Channel, Tierra del Fuego, Argentina (1–10 m depth). Although seaweeds, ascidians, and mobile animals do not strictly conform to conventional definitions of hard substrate, their surfaces can provide comparable habitats. Here, hairy snail shells (e) are colonized by barnacles (possibly *Balanus laevis*) and bryozoans (*Alcyonidium australe*, see black colonies at the posterior end of the shell); kelp blade surfaces (f) are colonized by the bryozoan *Membranipora isabelleana* (white colonies); and *P. gregaria* (g) is colonized by red algae (*Rhodymenia* sp.), hydrozoans, bryozoans, and other ascidians (*Sycozoa gaimardi*; bottom left). (h) Hard rock (orthoquartzite) revetment built to slow down the retreat of an erodible consolidated cliff near Miramar, Buenos Aires, Argentina. (i) Urban rocky seawall in Lavender Bay, New South Wales, Australia. Photo credits: (a) Zechariah Meunier; (b) Rodrigo Melo; (c) Zaira Matheus/All Angle; (d, f, g) María Bagur; (e) Mariano Rodríguez/ Argentina Submarina; (h) Jorge Gutiérrez; (i) Gabriela Palomo. All photos reproduced with permission.

10.2.3 Anthropogenic hard-substrate habitat

Development, industry, trade, and tourism over the last 50 years have driven the widespread conversion of natural shorelines to artificially armored coastlines (Firth et al. 2016). This includes the construction of jetties, breakwaters, rock revetments and groynes, piers, pilings, seawalls, and other port and harbor infrastructure (see Fig. 10.1h, i). Land reclamation and the creation of artificial islands, the development of offshore and onshore energy infrastructure (i.e., wind farms and oil and gas platforms), and the existence of shipwrecks and aquaculture infrastructure are all transforming the distribution and types of hard substrates available for colonization by intertidal and subtidal species (Perkol-Finkel et al. 2012, Firth et al. 2016). More recently, artificial reefs made of oyster shells are being built to actively support marine biodiversity conservation.

Natural rock is widely used for coastal armoring (i.e., rubble-mound and low-crested breakwaters) with resistant rock types such as granite and basalt often favored for engineering purposes. These materials are readily colonized but may have lower habitat value than relatively softer materials such as limestone, which naturally erode over time to create more physically complex surfaces (Coombes et al. 2011). However, the overwhelming global trend is that concrete has become increasingly common as a novel substrate for marine hard substrate communities (Thompson et al. 2002). This includes its use in both pre-cast and site-cast forms, and the emplacement of interlocking but permeable layers of concrete armor units (e.g., cubes, tetrapods, and accropodes) to dissipate high-energy waves (van der Meer 1999).

The ecological function of onshore and offshore artificial structures as habitat is the subject of on-going ecological interest. Artificial structures provide less physical complexity than natural shorelines and, as a result, typically support a lower abundance and diversity of species (Chapman and Bulleri 2003). The role of artificial structures in facilitating the introduction and spread of non-native species has also been highlighted (Airoldi et al. 2015). There is, therefore, increasing effort to better design coastal infrastructure to maximize ecological potential—as a form of 'ecological engineering' (Borsje et al. 2011)—with the goal to develop, promote, and implement more ecologically sensitive coastal engineering methods that maximize the biodiversity value of common hard 'gray' infrastructure when 'greener' alternatives are not possible (Naylor et al. 2018).

Considerable progress is being made in ecologically-informed design, with increasing ecological evidence of benefits to local and regional biodiversity. For example, the abundance and diversity of marine invertebrates colonizing artificial hard substrates (and the fish and bird communities that they support) can be manipulated in numerous ways. This includes informed material choice (e.g., Naylor et al. 2017) and the inclusion of physical complexity (Strain et al. 2017). Surface texture at a range of spatial scales (Coombes et al. 2015, Evans et al. 2019) and water-retaining features (Firth et al. 2013) are particularly important for sustaining biodiversity on artificial shorelines (Firth et al. 2014, Hall et al. 2018, Perkol-Finkel et al. 2018).

10.3 Prevalence and distribution of hard-substrate habitats in the coastal ocean

10.3.1 Abiogenic hard-substrate habitat

Natural rocky coast habitats are typically cited as composing approximately 80% of the world's coastline (Emery and Kuhn 1982, Naylor et al. 2010); however, continued improvements in remote sensing technology have enabled more detailed assessments. For example, recent surveys estimate that approximately 47% of the European coastline is composed of rocky shores (Salman et al. 2004) with an equivalent value of 41% in the United States (Hammar-Klose and Thieler 2001). Most recently, Young and Carilli (2019) calculated that approximately 52% of the global shoreline consists of cliffs and associated rocky substrates, which are more common in the mid-latitudes compared to tropical low latitudes and occur in approximately 93% of the world's geographical regions overall. It has also been estimated that more than 25 million subtidal seamounts (with a height of at least 100 m) exist worldwide (Wessel et al. 2010).

10.3.2 Biogenic hard-substrate habitat

Coral reefs are one of the most important types of biogenic hard-substrate habitats, hosting 25% of marine biodiversity worldwide (e.g., Reaka-Kudla 1997, Moberg and Folke 1997) but covering just 0.1%–0.2% of the ocean area (Crossland et al. 1991, Spalding and Grenfell 1997). Coral reefs are restricted to warm, shallow (up to 30 m) tropical and subtropical waters where they cover approximately 15% of the area (Crossland et al. 1991). There is a range of different reef types including atoll and platform reefs with enclosed lagoons, and coast-parallel fringing and barrier reefs with open lagoons. Barrier reefs are the most extensive type, with examples including the Australian Great Barrier Reef, which extends over 2,300 km in length, and the Mesoamerican Barrier Reef, which is almost 1,100 km in length.

The vast majority (91%) of known shallow-water coral reefs are located in the Indo-Pacific. The remaining coral reefs are within the much younger Atlantic Ocean, with Bermuda (32°N) and Queimada Grande (Brazil, 24°S) at the northernmost and southernmost limits, respectively (Spalding et al. 2001, Pereira-Filho et al. 2019). Several reef formations are known in the Southwestern Atlantic Ocean (e.g., the Abrolhos Bank, approximately 9,000 km^2; Moura et al. 2013), and some have only recently been discovered (see Pereira-Filho et al. 2019). No true coral reefs are known in the Southeastern Atlantic Ocean.

Cold-water reefs have also gained increasing attention, which occur at higher latitudes and greater water depths (typically 100–2,000 m) than warm-water reefs. Cold-water reefs are similarly complex as warm-water reefs but are composed of comparatively few coral species (less than 10 species compared to several hundred in warm-water reefs; Roberts et al. 2006). Because of light attenuation at greater water depths, hard corals that build cold-water reefs do not usually harbor the symbiotic algae of their warm-water counterparts. In contrast to warm-water corals that acquire a high proportion of carbon from photosynthesis by symbiotic algae (autotrophy), cold-water corals acquire carbon exclusively heterotrophically and their growth rates are considerably lower as a result. Nevertheless, given the lower rates of wave erosion at greater water depths, cold-water corals can build fragile, but long-lasting calcareous reef structures (Freiwald and Roberts 2005).

In addition to coral reefs, rhodolith (or maërl) beds (i.e., aggregations of calcified nodules of free-living coralline algae) are also a significant source of hard substrate in shallow seas at a global scale (see Section 10.4.1 below). These features occur from polar to tropical latitudes, across all oceans, and at depths ranging from the intertidal zone to 270 m, but are more frequent at depths shallower than 150 m (Foster 2001, Foster et al. 2013, Amado-Filho et al. 2012). Although global estimates of areal coverage are not available for rhodolith beds, they can be extensive and regionally important. For example, the world's largest rhodolith bed is located on the Abrolhos Shelf of Brazil and covers approximately 20,900 km^2, which is comparable to the area covered by coral reefs in the Caribbean Sea (ca. 21,600 km^2) and the Australian Great Barrier Reef (see Amado-Filho et al. 2012). Large concentrations of rhodoliths also occur in the Mediterranean Sea, northeastern Atlantic Ocean, eastern Caribbean Sea, eastern Indian Ocean, northwestern Pacific Ocean, and the Gulf of California (Foster 2001). Rhodoliths are also major structural components of an extensive (9,500 km^2) deep reef system (90–120 m depth) that was recently discovered at the Amazon River mouth (Moura et al. 2016).

10.3.3 Anthropogenic hard-substrate habitat

Given that almost half of the world's population is located within 100 km of the coast (Tibbetts 2002), the prevalence of artificial habitats has increased with coastal development. While an accurate global estimate of artificial habitat remains a challenge, there is good evidence of the extent of shoreline conversion in many regions of the world. For example, in Europe, a significant proportion of the hard substrate coastal habitat is artificial (Airoldi and Beck 2007), including nearly 1,500 km of the Adriatic coast (Romano and Zullo 2014). In South Korea, human activity and land reclamation have converted more than 50% of the natural shoreline (Choi and Seong 2014). Similarly, in other parts of Asia and The Americas, more than 50% of shorelines are protected by artificial structures (Bishop et al. 2017). Furthermore, using Google Earth imagery, Waltham and Connolly (2011) estimate that over 4,000 km of artificial waterways exist worldwide, of which a proportion is tidally influenced. Beyond the coastal zone, the rapid growth of trade, commercial shipping, and leisure travel continues to drive the expansion of the global network of marinas, wharves, and docks (Dafforn et al. 2015). Most recently, Bugnot et al. (2020) synthesized a range of information to provide the first global-scale assessment of the extent of marine artificial structures, projecting an increase from a footprint of at least 32,000 km^2 in 2018 to 39,400 km^2 by 2028.

10.4 The biota of marine hard-substrate habitats

Three major groups of hard substrate organisms can be distinguished based on how they relate to their physical habitat: (a) foundation species, or those that by virtue of their size and/or abundance form the physical structure of the habitat by adding a three-dimensional component to the basal substrate, (b) accompanying benthic species, or bottom-dwelling species that inhabit the physical habitat matrix formed by foundation species and the basal substrate, and (c) temporary inhabitants, that is, mobile species that use hard-substrate habitats for restricted periods of time (e.g., on a daily or seasonal basis).

10.4.1 Foundation species

Marine hard substrate communities are usually dominated by foundation species (*sensu* Dayton 1972), which have a large effect on a community as a consequence of their abundance or size (e.g., coral and oyster reefs, kelp forests, coralline algal turfs, and mussel and rodolith beds). By contributing to the three-dimensional structure of their local habitat, foundation species determine the diversity of associated taxa and modulate fluxes of materials and energy at multiple control points in the ecosystems they define (Ellison 2019). Foundation species are typically sessile or have low mobility, and can influence communities across a variety of spatial scales (e.g., the Great Barrier Reef, a kelp forest, mussel beds, or macroalgae in a tidepool). Indeed, organisms associated with foundation species (and accompanying species; see Section 10.4.2 below) can be secondary foundation species by creating habitats and determining the local conditions at relatively smaller scales (e.g., understory and epibiotic algae as secondary habitats within kelp beds; Thomsen and South 2019). Foundation species can colonize pre-existing hard substrates (e.g., kelp, corals, mussels, and tunicates), be secondary hard substrates (e.g., mussels and corals developing on rock bottoms), and form hard substrates in their own right (e.g., rodolith-forming coralline algae). The following sections describe the main groups of foundation species in the coastal ocean.

10.4.1.1 Macrophyte foundation species

Macrophytes vary in size from a few millimeters to tens of meters (e.g., up to 45 m in the case of the thalli of giant kelp, *Macrocystis pyrifera*, Frye et al. 1915) and their morphology takes a wide range of forms. For example, macroalgae vary from simple crustose, foliose, and filamentous shapes to more complex forms with highly specialized structures for light capture, support, flotation, reproduction, and attachment to the seafloor (Steneck and Dethier 1994). Here we highlight four functional groups of macrophytes and their importance as foundation species.

10.4.1.1.1 Canopy-forming species

This term is generally used for large macroalgae (usually longer than 1 m) that form the upper layer of many hard substrate communities. Notable examples in this group are the brown seaweeds (Laminariales and Fucales) that are archetypical examples of foundation species (e.g., *Macrocystis*, *Nereocystis*, *Laminaria, Fucus, Ascophyllum, Cystoseira*; see also Ólaffson 2016). By virtue of their size, these species add significant three-dimensional complexity to hard-substrate habitats and create an understory environment in which light, heat, and water flow are modulated (see Section 10.5.2.2 below).

10.4.1.1.2 Turf-forming species

In contrast to canopy-forming species, turf-forming macroalgae are generally short (usually 1–10 cm long), filamentous, and densely packed (Connell et al. 2014). The group includes several types of macroalgae that share an extensive low-lying morphology (e.g., branched coralline algae and filamentous red algae; Connell et al. 2014). Sediment trapping is a common feature of many macroalgal turfs (Hay 1981a, see Section 10.5.2.3 below), though this is not necessarily a defining feature of this group (see Connell et al. 2014). Macroalgal turfs can also be important secondary foundation species on coral and under canopy-forming macroalgal foundation species on rocky shores and reefs (e.g., Benedetti-Cecchi et al. 2001, Milne and Griffiths 2014, Barner et al. 2016).

10.4.1.1.3 Rodolith-forming coralline algae

Rhodoliths have received increasing attention as foundation species in recent years (e.g., Foster et al. 2007, Brasileiro et al. 2016, McConnico et al. 2018). Living rhodolith beds occur as a layer of pigmented, rounded coralline thalli that usually overlay carbonatic deposits formed *via* the long-term accumulation of dead thalli (Foster et al. 2013). Rhodoliths are hard substrates themselves, with surfaces that can be colonized by sessile epibionts (e.g., corals, sponges, other algae; Fig. 10.1c) and areas between branches that host interstitial invertebrates (Steller et al. 2003, Foster et al. 2013). Epibionts on rhodoliths include sponges that agglutinate individual rhodoliths into a framework of algal reefs (Pereira-Filho et al. 2015a). Dense rhodolith beds also trap and retain sediment and detritus in the interstices between individual thalli (Foster et al. 2013, Gabara 2020).

10.4.1.1.4 Seagrasses and other vascular plants

Seagrasses are important foundation species in marine communities. Surfgrasses (*Phyllospadix* spp.) are the most common vascular plant species that grow on hard substrates, having evolved adaptations to attach to marine hard substrates including the production of adhesive root-hairs (Gibbs 1902, Cooper and McRoy 1988). Surfgrasses occur in the low intertidal and shallow subtidal zones (10–15 m water depth) on the wave-exposed rocky shores of the North Pacific and are dominant space occupiers that compete with macroalgae and sessile invertebrates (Turner 1985, Menge et al. 2005, 2015, Moulton and Hacker 2011). These species usually form dense patches that moderate turbulence and accrete sediments from the water column (Gibbs 1902, Moulton and Hacker 2011).

Despite their prominent role as foundation species in wave-sheltered soft bottom areas (*cf.*, surfgrasses; previously presented and discussed at length in Chapter 8 of this book), some seagrasses can also colonize hard substrates. For example, *Posidonia oceanica* forms meadows both on rocky and sedimentary bottoms of the Mediterranean Sea up to a water depth of 40 m (see Montefalcone et al. 2016). As with surfgrasses, adhesive root-hairs have also been described in this species (Badalamenti et al. 2015). Other seagrass species (e.g., *Thalassodendron* spp.) have hard, strong roots that allow them to penetrate friable consolidated substrates and the gaps in hard calcareous reef floors (Kuo and den Hartog 2006). In other cases, seagrasses colonize sediments retained in hard substrates, though their roles as foundation species would be considered secondary in these cases as they typically depend on the presence of primary foundation species that trap and retain sediments (e.g., macroalgae in intertidal coral platforms; Coppejans et al. 1992). In the same vein, cordgrasses (*Spartina* spp.) typically form extensive marshes in intertidal muddy areas (discussed at length in Chapter 9) but can also be foundation species on wave-protected rocky shores and cobble beaches if sediments are available for rooting (e.g., into crevices, tidepools, or the interstitial spaces among cobbles; see Altieri et al. 2007, Bortolus et al. 2009).

10.4.1.2 Invertebrate foundation species

Foundation invertebrates include species from various taxa including sponges, hard and soft corals, tube-forming polychaete worms, mussels, oysters, sea squirts, acorn and gooseneck barnacles, and an array of rock-boring taxa. Most can be grouped into the following four broad categories based on the morphology of their aggregations.

10.4.1.2.1 Reef-forming species

A variety of sessile invertebrates create reefs including hard corals, oysters, vermetid gastropods, and tube-forming polychaetes. Reef-forming species typically produce calcified exoskeletons, shells, or tubes (but see the sand reefs formed by sabelariid tubeworms; Stewart et al. 2004). Reefs can project several meters from the subtidal seafloor (e.g., barrier coral reefs and atolls) or several meters seaward from the shore (e.g., fringing coral reefs, Crossland et al. 1991). Reefs can form once the larvae of reef-forming species settle on hard substrates and expand *via* clonal colony growth and the subsequent larval colonization of the hard surfaces of the reef invertebrates themselves. In this way, reefs often dramatically multiply the availability of benthic hard substrates (e.g., polychaete reefs developing from clam or gastropod shells may attain 12 m^2; see Schwindt and Iribarne 2000).

10.4.1.2.2 Bed-forming species

In contrast to reefs, sessile invertebrate beds consist of one or a few layers of individuals (typically less than a few tens of centimeters thick) that cover the seafloor to varying extents (up to tens of square meters or more). Mussels, barnacles, sponges, and ascidians are typical bed-forming organisms on hard substrates (e.g., Suchanek 1985, Navarrete and Wieters 2000, Maldonado et al. 2016, Rius et al. 2017). Ascidian, sponge, and barnacle beds are essentially made of a single layer of individuals attaching to primary hard substrates (Navarrete and Wieters 2000, Maldonado et al. 2016, Rius et al. 2017). Mussel beds, in contrast, can either be single or multilayered, with individuals in the outer layers attaching to their underlying conspecifics (up to six layers in *Mytilus californianus*; Suchanek 1985). However, the potential for vertical accretion is generally limited due to, for example, dislodgement by waves and heat-mediated mortality in the intertidal zone (Gutiérrez et al. 2015, Mislan and Wethey 2015).

10.4.1.2.3 Garden- or thicket-forming species

Diverse sessile invertebrates may form "gardens" (e.g., gorgonians and sponges; Marliave et al. 2009, Boavida et al. 2016) or thickets (e.g., erect bryozoans and corals; Batson and Probert 2000, Vargas-Ángel et al. 2003) on marine hard substrates. Semantics aside, both terms usually refer to solitary or colonial organisms that form discrete three-dimensional structures that are sparsely or patchily distributed on the seafloor at scales of meters to tens of meters. Despite covering a relatively low proportion of the bottom surface compared to the continuous cover of bed-forming species, garden- or thicket-forming species function as structurally complex habitat islands that increase the overall species richness of the benthic landscape (Marliave et al. 2009).

10.4.2 Accompanying benthos

Four sub-habitats can be defined within the matrix of foundation species and their basal substrates, namely, surfaces (including those of the basal hard substrate and foundation species), three-dimensional solid spaces (i.e., the interior of foundation species and the basal hard substrate), interstitial spaces (of both biotic and abiotic origin, e.g., rock crevices or the gaps between neighboring individuals in mussel beds), and sediments (e.g., those retained within the interstitial spaces of the foundation species). Clearly, these four sub-habitats are not equivalently suitable for species that differ in traits such as size, mobility, adhesion, dependence on sediments, and ability to bore into rock or other solids. Here we consider broad categories of species that differ in their habitat use including surface dwellers, interstitial species, infauna, and boring organisms. Given their functional traits (e.g., high mobility and size variation) and environmental conditions (e.g., periods of tidal flooding and exposure on rocky shores), these categories are not mutually exclusive as species can exploit multiple habitats.

10.4.2.1 Surface dwellers

Surface dwellers are sessile or mobile species that occur on the surfaces of a basal substrate, foundation species, and/or other organisms. These mobile species are generally adhesive (e.g., limpets and sea urchins) and/or large and negatively buoyant so as to withstand exposure to ambient flows (e.g., benthic fishes and large mobile crustaceans).

Inhabitants of basal substrates other than foundation species include algae and invertebrates in the understory of canopy-forming macroalgae (e.g., Clark et al. 2004, Watt and Scrosati 2013, Barner et al. 2016), the species within bed-forming species (e.g., algae, barnacles, and grazers colonizing exposed rock surfaces within mussel beds; Paine and Levin 1981, Tokeshi and Romero 1995), and those in the spaces beneath individuals or colonies of garden- and thicket-forming species (e.g., Valisano et al. 2016). The surfaces of foundation species can be colonized by epibionts (i.e., sessile species showing permanent adhesion mechanisms; e.g., macroalgae, bryozoans, and barnacles adhering to mussel shells and kelp blades; Gutiérrez and Palomo 2016, Thomsen et al. 2019), mobile adhesive invertebrates (e.g., limpets that graze on kelp epiphytes, Black 1976), and mobile non-adhesive animals (e.g., amphipods that can temporarily cling to algal thalli with specialized appendages; Hacker and Steneck 1990). Surface dwellers can sometimes be considered as secondary foundation species, such as understory and epibiotic algae that act as secondary habitats within kelp beds (Thomsen and South 2019).

10.4.2.2 Interstitial species

Interstitial species are typically small mobile invertebrates and fish that depend on small-scale (millimeters to a few tens of centimeters) three-dimensionally complex habitats to avoid predation, sources of physiological stress (e.g., extreme temperatures and desiccation in the intertidal zone), and being detached by waves and currents. Examples include the spaces within kelp holdfasts, the gaps between adjoining individuals in mussel beds, invertebrate boreholes, the secluded microhabitats within reefs, and the crevices in basal rocky substrates (Garrity 1984, Ojeda and Santelices 1984, Suchanek 1985, Almany 2004, Bagur et al. 2016). Interstitial habitat users often include the recruits and juveniles of relatively large species that occur elsewhere as adults (e.g., king crabs, Pirtle et al. 2012). Some sessile species also depend on interstitial space to grow and survive as is the case of nestling bivalves, which are obligate inhabitants of cavities (e.g., crevices and boreholes made by other organisms), but which lack the ability to excavate cavities themselves (see Bagur et al. 2014; *cf.* boring species in Section 10.4.2.4).

10.4.2.3 Infauna

The term infauna is used for organisms that live within sediments in aquatic environments, including annelids, crustaceans, mollusks, ribbon worms, and flat worms. Hard substrate infauna are typically confined to the sediments retained in the interstitial spaces of three-dimensionally complex habitats (e.g., the sediment within mussel beds, Suchanek 1985; rhodolith beds, Grall et al. 2006; and surfgrass rhizomes, Moulton and Hacker 2011) and, thus, can be considered a particular example of interstitial species. As the volume of sediment trapped within these structures is generally limited, the size of hard substrate infauna is typically constrained relative to soft bottom habitats (e.g., large infaunal bivalves are often excluded from hard-substrate habitats, but see Grall et al. 2006).

10.4.2.4 Boring organisms

Some microorganisms (algae, cyanobacteria, and fungi) and a variety of invertebrates (e.g., bivalves, sponges, polychaetes, sipunculids, and crustaceans) can bore into relatively soft rock (e.g., limestone and sandstone) and biogenic hard substrates (e.g., coral skeletons, and oyster and mussel shells) *via* mechanical or chemical means or a combination of both (e.g., Cobb 1969, Scott et al. 1988, Pinn et al. 2008, Bagur et al. 2013, 2014, Ndhlovu et al. 2019; see Section 10.5.2.4 below). Chemical borers dissolve carbonatic substrates by means of acidic secretions (e.g., cyanobacteria, sponges, and bivalves including *Lithophaga* spp.; Cobb 1969, Bagur et al. 2014). Whilst they are typically constrained by substrate hardness (e.g., Bagur et al. 2014), mechanical borers inhabit a wider range of substrate mineralogies and even colonize non-mineral substrates, such as teredinid bivalves that bore into living and dead wood (Nair and Saraswathy 1971) and some crustaceans (isopods and amphipods) that bore into kelp holdfasts (Conlan and Chess 1992, Thiel 2003). Vacant invertebrate boreholes often host a significant portion of the biodiversity of hard substrate habitats (e.g., Schoppe and Werding 1996, Pinn et al. 2008, Bagur et al. 2016, 2019).

10.4.3 Temporary inhabitants of hard-substrate habitats

10.4.3.1 Nekton

Nekton, as originally defined by Haeckel (1891), includes the actively swimming organisms in the pelagic zone of aquatic environments. Examples of marine nekton temporarily using hard substrates include fishes (excluding strictly benthic species that live on the seafloor, such as gobies and flounders), shrimp, squid, sea turtles, penguins, and pinnipeds (see Aleyev 1977). Hard-substrate habitats are used as foraging grounds by pelagic and benthopelagic fishes as well as sea turtles, diving birds, and pinnipeds (e.g., Branch 2008, Karczmarski et al. 2000, Schmid et al. 2003). Some fish species also use hard-substrate habitats as resting sites and for shelter during inactive periods (e.g., ornate sharks, *Orectolobus ornatus*, and blacksmiths, *Chromis punctipinnis*, a pelagic planktivorous fish that shelters in rocky reefs at night; Bray et al. 1981, Carraro and Gladstone 2006). Furthermore, pinnipeds use nearshore rocky habitats as breeding, pupping, and resting sites (e.g., Dendrinos et al. 2007). As temporary users of hard substrates, nekton often plays an important role in the transport of organic matter and nutrients across habitat boundaries (e.g., Bray et al. 1981).

10.4.3.2 Low-tide users of intertidal hard substrates

Periodic aerial exposure makes intertidal hard substrates accessible to seabirds, shorebirds, and terrestrial vertebrates and arthropods. A variety of seabirds and shorebirds use rocky shores as resting (e.g., terns) and/or feeding sites (e.g., gulls, plovers, knots, and oystercatchers; e.g., Skead 1966, Bahamondes and Castilla 1986). Invertebrate consumption by some of these birds (e.g., gulls and oystercatchers) often significantly affects the overall composition of rocky shore communities (Wootton 1995, Kurle et al. 2008; see Section 5.1.4 below). In addition, a variety of opportunistic terrestrial consumers forage on intertidal hard substrates at low tide including some birds (e.g., crows; Wootton 1995), mammals (e.g., rats, foxes, coyotes, and racoons; Navarrete and Castilla 1993, Carlton and Hodder 2003), reptiles (e.g., lizards; Fariña et al. 2008), and invertebrates (e.g., wolf spiders; Morse 1997). These terrestrial consumers are potentially important vectors of marine resources to inland habitats (see Paetzold et al. 2008, Pizarro et al. 2012).

10.5 Functional roles of hard substrate organisms

The following sections describe the major functional roles of organisms on hard substrates involved in organic matter fluxes and trophic interactions across food webs (e.g., primary and secondary production, herbivory, predation, suspension feeding, and production and consumption of non-living organic matter; see Section 10.5.1) and physical ecosystem engineering (i.e., the structural modification of environments by organisms, *sensu* Jones et al. 1994; e.g., the formation of secondary habitat, modulation of abiotic factors, sediment trapping and binding, and bioerosion and bioprotection: see Section 10.5.2 below).

10.5.1 Organic matter fluxes and trophic interactions

10.5.1.1 Hard substrate primary production

Macrophytes (i.e., macroalgae and seagrasses) and microalgae are the main primary producers in hard substrate communities (but see Section 10.4.1.1 for vascular plant examples). Cebrian (1999) conducted an extensive review of primary production by macroalgae and microalgae in many habitat types and found that this varied by four orders of magnitude (10^{-4} to 10 g C cm^{-2} d^{-1}). Estimates of macrophyte and microalgal production in hard-substrate habitats are typically at the upper end of this range (10^{-1} to 10 g C m^{-2} d^{-1}; Table 10.1).

Gross macroalgal production in rocky shore reefs generally exceeds community respiration (i.e., net autotrophy; McQuaid and Branch 1985, Barrón et al. 2003, Branch 2008, Kraufvelin et al. 2010), where only a fraction of production is recycled *in situ via* herbivory, detritivory, and decomposition. Because of the action of waves and currents, most organic matter produced by macroalgae in rocky shore and reef habitats is exported out of the system (Branch 2008, Barrón et al. 2003; see Section 10.5.1.6 below). The fate of microalgal production on rocky shores is much less well known. Using experimental rockpools, Golléty and Crowe (2013) illustrated that despite their relatively low biomass compared to macroalgae, microbial biofilms significantly contribute to gross primary production (accounting for 5–70% of total productivity depending on the presence and type of co-occurring macroalgae). Low standing biomass relative to production suggests that the ratio between production and consumption is more balanced in microalgae compared to macroalgae (Cebrián 1999).

In the case of coral reefs, gross primary productivity is generally similar to community respiration rates (Hatcher 1988, Crossland et al. 1991) such that most primary production is recycled within the

Table 10.1. Net primary production (NPP; g C m^{-2} d^{-1}) in marine hard substrate communities.

Producer/Habitat	NPP	Comments	References
Microphytobentos on a rocky intertidal shore in southern Hokkaido, Japan	0.05	Average of monthly values during a 14-month period.	Fuji et al. (1991)
Macroalgae on a rocky intertidal shore in San Clemente Island, California, USA	0.4-3.1	Range of values for the 18 dominant species.	Littler and Murray (1974)
Kelps, in the rocky subtidal zone of St. Margaret's Bay, Nova Scotia, Canada	4.8	Summed contribution of three species (*Laminaria digitata*, *Saccharina longicruris*, and *Agarum clathratum*).	Mann (1972)
Macroalgal assemblages at subtidal rocky reefs in the eastern Canary Islands, Spain	1.6–10.6	Range of values for 5 communities varying in species composition and depth (0–15 m).	Johnston (1969)
Turf and foliose macroalgal assemblages in a subtidal rocky reef off Santa Barbara, California, USA	0.4, 0.9	Values for turf and foliose macroalgal assemblages, respectively.	Miller et al. (2009)
Macroalgal turf assemblages on subtidal rocky reefs in West Island, South Australia	0.1–2.9	Range of values from across seasons and depths (4 and 10 m).	Copertino et al. (2005)
Macroalgal turf assemblages in a coral reef at St. Croix, US Virgin Islands	2.1–3.1	Minimum-maximum seasonal values.	Carpenter (1985)
Crustose coralline algae in a fringing coral reef in Waikiki, Hawaii, USA	5.7	Summed contribution of the five dominant coralline species.	Littler (1973)
Coral (*Stylophora pistillata*) zooxanthellae in the Gulf of Eilat, Red Sea	0.6, 2.7	Values from 35 and 3 m depth, respectively.	McCloskey and Muscatine (1984)
Maerl (*Lithothamnion corallioides*) beds in the shallow subtidal zone of the Bay of Brest, France	0.1–1.6	Rates vary depending on depth (1–10 m) and algal densities.	Martin et al. (2006)

system and excess (net) production is typically low (ca. 3% on a global basis; see Crossland et al. 1991). Although corals and algae (i.e., the main reef primary producers) release up to 50% of their fixed carbon as mucus (Tanaka et al. 2008, Haas et al. 2010), most of that carbon (80%) immediately dissolves in seawater and is assimilated by sponges and microbes, ultimately being transferred across the reef food web (Wild et al. 2004a, b, de Goeij et al. 2013).

10.5.1.2 Hard substrate secondary production

As in any other habitat, secondary production in hard substrate communities (i.e., the formation of living mass by consumers over a period of time) is lower than primary production (e.g., Newell et al. 1982, Norderhaug and Christie 2011, Paar et al. 2014; see Table 10.2). One reason for this is that the feeding efficiency of algal grazers is relatively low, with a portion of the carbon and energy they assimilate being used in respiration and released as feces. The few available estimates of secondary production by grazing assemblages or dominant grazers on hard substrates are less than 2% of the total macroalgal production (Table 10.2). Nevertheless, secondary production by grazers can still represent an important proportion of overall secondary production on rocky shores (or rocky shore zones) that lack of other trophic groups (e.g., Cabral-Oliveira et al. 2014; see Table 10.2).

High secondary production is usually associated with habitats in which relatively fast-growing suspension feeders are common (e.g., barnacles and mussels; Wu and Levings 1979, Newell et al. 1982, Gardner and Thomas 1987, Golléty et al. 2008). Suspension feeder production can be equivalent to 15% of primary production in kelp beds (Newell et al. 1982) and accounts for 20–80% of the secondary production in rocky shore and reef habitats (Table 10.2). This is in contrast to habitats dominated by suspension feeders, such as gorgonians that, despite their high biomass, have a low metabolism and relatively long lifespans (see Mitchell et al. 1992, Weinbauer and Velimirov 1995).

Because of low consumption and assimilation efficiency, predators have lower secondary production than herbivores. For example, Newell et al. (1982) estimated that the secondary production of predators

Table 10.2. Secondary production of marine hard substrate biota, including whole-community estimates (a) and estimates for whole trophic groups or particular taxa within these groups (b–d). Values are expressed in energy (Kcal m^{-2} yr^{-1}) or mass units (g DW m^{-2} yr^{-1}) according to the original sources. Secondary production estimates are also shown as percentage of primary production (see percentages between parentheses; calculated only for studies where local estimates of primary production are reported). The proportional contribution of trophic groups or species to overall secondary production is shown between brackets (only for studies where secondary production was estimated at the community-level and the trophic group and/or species level). The list is illustrative, not exhaustive. DW: Dry weight, AFDW: Ash-free dry weight.

Species/Study system	Production	Comments	References
(a) Community-level			
Kelp forest at Cape peninsula, South Africa	1717 Kcal m^{-2} yr^{-1} (18.8%)	Suspension feeder production adds to 76.7% of total production.	Newell et al. (1982)
Kelp forests of Møre coast, Norway	68–308 g DW m^{-2} yr^{-1} (3–8%)	Range of values from three sites. Based on 30 dominant macroinvertebrate species (96% of the collected specimens).	Norderhaug and Christie (2011)
(b) Grazers			
Grazing assemblage in a kelp forest at Cape peninsula, South Africa	74.6 Kcal m^{-2} yr^{-1} (0.8%) [4.1%]	Sea urchins (*Parechinus angulosus*) comprise 86% of grazer biomass.	Newell et al. (1982)
Sea urchin (*Strongylocentrotus droebachiensis*) in kelp beds at St. Margaret's Bay, Canada	49.8 Kcal m^{-2} yr^{-1} (0.3%) [No data]	Dominant grazer in the study area	Miller and Mann (1973)
Sea urchin (*Diadema antillarum*) in fringing coral reefs at the west coast of Barbados	76.9 Kcal m^{-2} yr^{-1} (1.8%) [No data]	Densities > 20 ind m^{-2}	Hawkins and Lewis (1982)
Grazing gastropod assemblage at subtidal rocky reef in Cape Rodney, New Zealand	5.3 g AFDW m^{-2} yr^{-1} (No data) [10.6%]	Sampled habitats include macroalgal forests (*Ecklonia*, *Carpophyllum*), coralline algal turf, and urchin barrens.	Taylor (1998)
Scarid fishes at three coral reefs in Philippines (Sumilon, Apo, and Pamilacan Islands)	0.1–0.3 g C m^{-2} yr^{-1} (0.05–0.3%) [No data]	Range of estimates obtained by two methods.	Russ and St. John (1988)
Intertidal snail (*Melarhaphe neritoides*) in the upper shore at Peniche peninsula, Portugal	< 0.2 g AFDW m^{-2} yr^{-1} (No data) [100%]	Data from three sites differing in sewage inputs. Only secondary producer at the upper shore.	Cabral-Oliveira et al. (2014)
Intertidal limpets (*Patella* spp.) in the middle shore at Peniche peninsula, Portugal	0.2–0.3 g AFDW m^{-2} yr^{-1} (No data) [20–80%]	Data from three sites differing in sewage inputs. Higher contribution at unpolluted sites.	Cabral-Oliveira et al. (2014)
Grazing assemblage in the low intertidal zone at Peniche peninsula, Portugal	0.2–0.4 g AFDW m^{-2} yr^{-1} (No data) [40–60%]	Data from three sites differing in sewage inputs. Higher contribution at unpolluted sites.	Cabral-Oliveira et al. (2014)
(c) Suspension-feeders			
Suspension-feeding assemblage in a kelp bed at Cape peninsula, South Africa	1392.2 Kcal m^{-2} yr^{-1} (15.2%) [76.7%]	Ribbed mussels (*Aulacomya atra*) comprise 46% of suspension feeder biomass.	Newell et al. (1982)
Bivalves (*Hiatella artica*, *Mytilus edulis*, and *Modiolus modiolus*) in kelp forests of Møre coast, Norway	30–98 g DW m^{-2} yr^{-1} (1.3–3.2%) [25.0–44.6%]	Data from three sites across a gradient of wave exposure. Higher production at intermediate exposure.	Norderhaug and Christie (2011)

Table 10.2 Contd. ...

...Table 10.2 Contd.

Species/Study system	Production	Comments	References
Suspension feeders (mainly barnacles, *Chthamalus montagui*) in the mid intertidal zone at Peniche peninsula, Portugal	0.1–0.8 g AFDW m^{-2} yr^{-1} (No data) [20–80%]	Data from three sites differing in sewage inputs. Higher contribution at polluted sites.	Cabral-Oliveira et al. (2014)
Suspension feeders (mainly mussels, *Mytilus galloprovincialis* and *Modiolus modiolus*) in the low intertidal zone at Peniche peninsula, Portugal	0.1–0.7 g AFDW m^{-2} yr^{-1} (No data) [30–60%]	Data from three sites differing in sewage inputs. Higher contribution at polluted sites.	Cabral-Oliveira et al. (2014)
(d) Predators			
Rock lobster (*Jasus lalaandi*), sparid fish (*Pachymetopon blochii*) and other large predators in a kelp bed at Cape peninsula, South Africa	151.1 Kcal m^{-2} yr^{-1} (1.6%) [8.3%]	Rock lobsters and *P. blochii* comprise 47 and 32% of large predator biomass, respectively.	Newell et al. (1982)
Isopod *Cirolana imposita* in a kelp bed at Cape peninsula, South Africa	197.4 Kcal m^{-2} yr^{-1} (2.2%) [10.9%]	Small species. Low biomass per area unit but high turnover rates relative to large predators (*cf.* previous example).	Newell et al. (1982)
Predatory fish assemblage at subtidal rocky reef in Cape Rodney, New Zealand	2.8 g AFDW m^{-2} yr^{-1} (No data) [5.6%]	Sampled habitats include macroalgal forests (*Ecklonia*, *Carpophyllum*), coralline algal turf, and urchin barrens.	Taylor (1998)

was equivalent to 4% of the primary production in a kelp forest, and that their overall contribution to secondary production was approximately 19% (See Table 10.2 for details). It is worth noting, however, that more than half of this production was attributed to a single small species (the isopod *Cirolana imposita*), which indicates the low biomass but fast turnover of this species relative to larger predators.

10.5.1.3 Hard substrate herbivory

On average, approximately 30% of global net primary productivity in coral reefs and other macroalgae-dominated ecosystems is transferred to herbivores, although local estimates are highly variable and range from negligible to more than 80% (see Duarte and Cebrian 1996). Major groups of herbivores and grazers in hard substrate communities include mollusks (e.g., gastropods and polyplacophorans), sea urchins, mobile crustaceans (e.g., amphipods and crabs), and fishes (e.g., surgeon and rabbit fishes in coral reefs). Vertebrate grazers occur on a few rocky shores including marine iguanas, which exclusively forage on the intertidal rocky shores and shallow subtidal rocky reefs of the Galapagos Islands, and terrestrial lizards in northern Chile, which facultatively graze in rocky shores during low tide (Shepherd and Hawkes 2005, Fariña et al. 2008). Grazing can have major effects on hard substrate communities. For example, a global meta-analysis conducted by Poore et al. (2012) indicated that grazers profoundly reduce the abundance of primary producers in hard-substrate habitats, with the strongest effects on rocky intertidal shores (an average reduction of 74%), subtidal rocky reefs (a reduction of 54%), and coral reefs (a 39% reduction).

Grazing impacts on rocky intertidal shores and subtidal rocky reefs are mostly caused by mollusks and sea urchins, whereas coral reefs are more intensely grazed by fishes (see Poore et al. 2012). For example, fish grazing in coral reefs can restrict colonization by macroalgal species or, in the case of browsing fish, remove established macroalgae (e.g., Puk et al. 2015). This herbivory can prevent and even reverse shifts from coral-dominated to macroalgal-dominated states that suppress coral fecundity, recruitment, and survival (see Bellwood et al. 2006, Hughes et al. 2007). Likewise, sea urchins, when released from predation by sea otters, can decimate nearshore kelp populations leading to their replacement by other species (e.g., mussels and barnacles in Alaska and the Aleutian Islands; see Estes and Duggins 1995). Seasonal and interannual variations in the abundance and/or activity of grazers also contribute to temporal variations in macroalgal and microalgal biomass (Cubit 1984, Estes and Duggins 1985). Grazing also affects the distribution of algal species leading to patterns of zonation with water depth on both rocky shores and in coral reefs (e.g., Hay 1981b, Underwood and Jernakoff 1984, Lewis 1985, Witman 1987). Besides their effects on macroalgal abundance and distribution, grazers can determine the taxonomic and functional composition of primary producers on hard substrates, usually favoring the occurrence of grazing-resistant (e.g., encrusting coralline algae) and less profitable species (e.g., Paine and Vadas 1969, Sammarco 1983, Lewis 1985, Filbee-Dexter and Scheibling 2014).

10.5.1.4 Hard substrate predation

Predators in hard substrate communities include both invertebrates and vertebrates that can be residents of such habitats (e.g., sea stars, crabs, lobsters, fishes, and gastropods) or nektonic species that temporarily use them as foraging grounds (e.g., fishes, cormorants, and seals, see Section 10.4.3.1 above). In the case of intertidal communities, predators also include terrestrial animals (e.g., rats, crows, and spiders; see Section 10.4.3.2 above).

Predation can affect several attributes of prey populations in marine hard substrate communities including abundance, size distribution, and recruitment (e.g., Moreno et al. 1986, Robles et al. 2001, Kurle et al. 2008). In the case of coral reefs, some predators consume the coral itself (e.g., crown-of-thorns starfish and butterflyfishes) and, thus, concomitantly reduce overall colony growth and reproduction rates (Rotjan and Lewis 2008). At the community level, the predation of dominant sessile species, such as mussels, usually reduces their coverage and generates vacant space for colonization by algae and other sessile species (e.g., Paine 1966). Similarly, the lesions on coral tissues caused by predators are

Table 10.3. Consequences of predation in hard-substrate habitats.

Study system	Predator effect	References
Rocky intertidal shore in Mehuin, Chile	Muricid gastropods, *Concholepas concholepas*, reduce mussel and barnacle cover.	Moreno et al. (1986)
Rocky intertidal shore in Santa Catalina Island, California, USA	Size-selective predation by spiny lobsters, *Panulirus interruptus*, alters age/size distribution of mussels but not cover.	Robles et al. (2001)
Rocky intertidal shore in Mukkaw Bay, Washington, USA	Predation by seastars, *Pisaster ochraceus*, prevents space monopolization by mussels allowing other species to colonize the basal rock substrate.	Paine (1966)
Rocky intertidal shores in the Aleutian archipelago, USA	Glaucous-winged gulls, *Larus glaucescens*, and black oystercatchers, *Haematopus bachmani*, reduce densities of herbivorous snails and limpets with cascading, positive effects on fleshy algal cover. Such a trophic cascade is modulated by introduced Norway rats, *Rattus norvegicus*, which prey on seabird eggs and fledglings.	Kurle et al. (2008)
Rocky shallow subtidal zone in Pemaquid Point, Maine, USA	Large predators (fishes, crustaceans) reduce densities of juvenile green sea urchins and other small benthic organisms.	Ojeda and Dearborn (1991)
Shallow rocky subtidal zones in the Aleutian archipelago, USA	Sea otters, *Enhydra lutris*, reduce the density and biomass of grazing sea urchins, which favors the formation of dense, nearshore kelp beds.	Estes and Duggins (1995)
Great Barrier Reef, Australia	Corallivorous butterflyfishes, *Chaetodon plebeius*, cause 27–53% reductions in the tissue biomass of their host colonies.	Cole and Pratchett (2011)
Great Barrier Reef, Australia	Piscivorous fishes control fish community composition *via* their effects on small and herbivorous species.	Boaden and Kingsford (2015)
Coral reefs in Kenya (multiple sites)	Red-lined triggerfish, *Balistapus undulatus*, reduce sea urchin densities, with concomitant increases in the abundance of fleshy algae.	McClanahan (2000)
Coral reefs in Moorea, French Polynesia	Outbreaks of predatory crown-of-thorns starfish, *Acanthaster planci*, dramatically reduce coral cover and diversity as well as the density and diversity of coral-feeding fishes. Turf forming algae proliferates on dead coral colonies.	Kayal et al. (2012)

susceptible to colonization by algae, sponges, and other sessile invertebrates (e.g., Bak and Steward-van Es 1980; see Rotjan and Lewis 2008 for a review). On the other hand, the predation of grazers can have cascading positive effects on algal biomass *via* reduced grazing pressure (Estes and Duggins 1995, Wootton 1995, McClanahan 2000). Detailed examples of the consequences of predation in hard substrate communities are shown in Table 10.3.

10.5.1.5 Hard substrate suspension feeding

Suspension feeding (i.e., the capture and ingestion of food particles that are suspended in the water column, previously discussed in Chapter 8) is widespread among hard substrate invertebrates. All major invertebrate taxa include representative suspension feeders (Hentschel and Shimeta 2019) and many function as foundation species including bivalves, sea squirts, sponges, corals, polychaetes, and bryozoans (see Section 10.4.1.2 above). The particles consumed by suspension feeders include phytoplankton, zooplankton, and bacteria as well as particulate and dissolved organic matter (Gili and Coma 1998, Hentschel and Shimeta 2008). There are varying degrees of particle selectivity and food specialization among suspension feeders, although dietary composition can vary as a function of flow velocity and the availability of distinct food items (e.g., Ribes et al. 1999, Bracken et al. 2012).

The suspension feeders that form aggregations on hard substrates (e.g., reefs and beds; see Section 10.4.1.2 above) consume vast amounts of suspended particles on a per volume basis, thereby affecting their composition (e.g., effects on planktonic community structure; van Duyl et al. 2002) and overall abundance (e.g., suspended particle depletion in the water layers immediately above these aggregations; Glynn 1973, Fabricius and Dommisse 2000). Moreover, the aggregations formed by suspension feeders are critical for benthic-pelagic coupling, driving important fluxes of carbon and nutrients (e.g., Fabricius and Dommisse 2000) and regulating the carbon balance of hard-substrate habitats (see Table 10.4, which is complementary to the discussion of this topic for soft-bottom benthos in Chapter 8). Sponges are especially important in these fluxes on coral reefs, feeding on dissolved organic

Table 10.4. Carbon fluxes from the pelagic zone to suspension feeding assemblages in marine hard substrates. POC: Particulate organic carbon, DOC: Dissolved organic carbon.

Study system	C input ($g\ m^{-2}\ d^{-1}$)	Observations	References
Suspension-feeding assemblages in Puerto Rican coral reefs	0.18	Total input considering only plankton larger than 76 µm. Equivalent to 4–13% of net ecosystem metabolism.	Glynn (1973)
Suspension-feeding assemblages in coral reef flat at Eilat, Red Sea, Israel	0.22	Total inputs from phytoplankton are equivalent to daily phytoplankton production in the oligotrophic waters off the reef.	Genin et al. (2009)
Cryptic fauna in the cavities of coral reefs at Curaçao	4.10	Total DOC flux. DOC removal rates are comparable to the average gross primary production rates of the entire coral reef ecosystem.	de Goeij and van Duyl (2007), de Goeij et al. (2013)
Nearshore reef communities dominated by soft corals in the Great Barrier Reef, Australia	2.5	Total input from suspended particles < 10 µm. Estimated soft coral contribution to C flux is > 80%.	Fabricius and Dommisse (2000)
Sponge *Mycale lingua* at Amrnen Rock Pinnacle, Gulf of Maine, USA	0.03	Total input from heterotrophic and autotrophic plankton < 10 µm.	Pile et al. (1996)
Caribbean giant barrel sponge *Xestospongia muta* on Conch Reef, Florida Keys, USA	1.58	Total input considering POC and DOC consumption (the latter comprises 55% of the sponge diet). Largest value registered for a single sponge species.	McMurray et al. (2017)
Sponge assemblages in rocky reefs at the south coast of Wellington, New Zealand	0.07–3.50	Total POC inputs from picoplankton.	Perea-Blazquez et al. (2012)

matter and transforming it into particulate organic matter that is accessible to higher trophic levels and, thereby, increasing its retention and cycling within the reef community (de Goeij et al. 2013).

10.5.1.6 Production of non-living organic matter

The fraction of organic matter produced by hard substrate organisms that is not ingested and assimilated by consumers is released to the environment as detritus and other waste products (e.g., feces and urine), exudates (e.g., mucus), and dead tissues. Such releases can represent an important flux of organic matter, especially from those species that are dominant. For example, rocky shore mussels (*Mytilus edulis*) in Japan were estimated to release ca. 25 g C m^{-2} d^{-1} in the form of feces (*cf.* rates of primary production in Table 10.1) and ca. 7 g C m^{-2} d^{-1} in the form of pseudofeces (i.e., aggregates of ingested particles that are rejected before digestion; see Tsuchiya 1980). Similarly, dense sea urchin aggregations produce feces at a rate of 9.17 g m^{-2} d^{-1} (Sauchyn et al. 2011), while corals secrete ca. 0.12–0.25 g C m^{-2} d^{-1} as mucus (Wild et al. 2004a). As for the production of dead tissues, globally, kelp produces detritus at an average rate of 706 g C m^{-2} yr^{-1} *via* incremental blade erosion and fragmentation, and dislodgement of entire fronds and thalli, which accounts for 82% of annual kelp productivity (Krumhansl and Scheibling 2012).

As hard-substrate habitats are typically affected by currents, the organic matter released by their inhabiting organisms is often exported elsewhere. Coral mucus, for example, is partly transported to adjoining sandy bottom systems where it is rapidly degraded and remineralized by microbes (Wild et al. 2005a, b). Similarly, kelp detritus travels long distances and frequently forms mats on sandy beaches, rocky shores, tidal flats, and subtidal habitats at various depths, providing a significant resource for communities tens of meters to hundreds of kilometers from the source of production (Krumhansl and Scheibling 2012). Nevertheless, local retention can be significant when non-living organic matter is negatively buoyant, currents are weak, and physical structures intercept and trap particles (e.g., see Sauchyn and Scheibling 2009, Krumhansl and Scheibling 2012, and Section 10.5.2.3 below).

10.5.1.7 Consumption of non-living organic matter

Scavengers (i.e., animals that feed on carrion) and detritivores (i.e., animals that feed on detritus) play critical roles in the recycling of dead organic matter. In hard-substrate habitats, scavengers include fishes, gastropods, polychaetes, sea anemones, and crustaceans such as crabs, amphipods, and isopods (Keable 1995, McClanahan 1995, Davenport et al. 2011, Morton 2011). Scavengers are usually facultative as they can shift from scavenging to predation or other feeding modes depending on carrion availability, although obligate scavenging seems likely among nassariid gastropods and lysianassoid amphipods (see Kaiser and Moore 1999). Many hard substrate scavengers rely on material left by predators (e.g., green crabs that forage on leftovers of mussels consumed by whelks, Morton 2011). Some scavengers thrive in environments that are stressful to their carrion sources and where mortality rates are consequently high (Nielsen and Gosselin 2011). The hydrodynamic environment (i.e., currents, turbulence, and wave action) and transport properties of the carrion (e.g., size, attachment, and buoyancy) ultimately determine whether material is scavenged *in situ* or exported and recycled elsewhere (such as beach carrion).

The relatively smaller and/or more buoyant organic matter consumed by detritivores can accumulate in quiescent, interstitial, and/or secluded environments. An array of small invertebrate detritivores is associated with these microhabitats including amphipods in the epilithic algal layers of coral reefs, and deposit-feeding polychaetes and other infauna in the interstitial spaces of mussel beds (Klumpp et al. 1988, Tokeshi and Romero 1995). Other important detritivores in hard substrate communities include ‘cryptogenic’ fishes on coral and rocky reefs (i.e., cryptic benthic species of typically < 5 cm in length; Depczynski and Bellwood 2003) and herbivorous fishes that feed on the epilithic algal matrix and which incorporate significant amounts of detritus to their diets (Mendes et al. 2009). As much of the detritus produced in hard-substrate habitats is exported elsewhere, the consumption of detritus is unlikely an important energy-flow pathway in these systems. Possible exceptions are rhodolith beds that retain detritus within their interstitial spaces (Gabara 2020), act as net importers of particulate organic matter (Martin et al. 2007), and support abundant and diverse detritivore assemblages (Grall 2006, Sciberras et al. 2009, Gabara 2020).

The mucus released by corals (see Section 10.5.1.6 above) is an important source of energy and nutrients for benthic microbes. Because of its sticky character, this mucus traps small organic and inorganic particles from the water column (Wild et al. 2004a), increasing its organic C and N content by a factor of 26 and 79, respectively, within just three hours of being released (Huettel et al. 2006). These enriched mucus-particle aggregates rapidly sink to the seafloor and are degraded by the abundant bacteria associated with carbonate reefs sands (Wild et al. 2004b, 2006).

10.5.2 Physical ecosystem engineering

10.5.2.1 Formation of secondary (or higher-order) habitat

The macroalgae and invertebrates colonizing hard substrates often provide secondary substrate for the attachment of other organisms, i.e., *basibionts* and *epibionts*, respectively. In the same vein, organisms attaching to secondary substrates can provide attachment sites for other organisms and so forth, giving rise to tertiary and higher-order biogenic substrates (e.g., Gutiérrez and Palomo 2016, Yakovis and Artemieva 2017, Thomsen and South 2019). All these are examples of autogenic engineering (i.e., the engineers changes the environment via their own physical structures; *sensu* Jones et al. 1994).

In general, the basal, secondary and higher-order biogenic substrates differ in their suitability for the attachment of other species (e.g., Rimondino et al. 2015, Gutiérrez and Palomo 2016, Thomsen and South 2019). Moreover, differences in the function of biogenic surfaces as attachment sites can even occur within a single species (e.g., individuals of different ages; Chava et al. 2019) and within individuals (e.g., kelp holdfasts, stipes, and blades; Teagle et al. 2017). Differences in substrate suitability occur because organismal adhesion and attachment strength depend on the physical (e.g., microtopography and friability) and chemical (e.g., polarity) properties of the substrate, which may vary between substrate types (Fletcher and Callow 1992, Waite 2002). In addition, the external surfaces of organisms often have mechanical, physical, and chemical defenses that reduce potential colonization (i.e., antifouling defenses such as spicules, mucus, and other surface-associated compounds; see Wahl 1989, Krug 2006).

Despite their role as attachment surfaces, calcified sessile invertebrates (e.g., corals and bivalves) and some kelp species also serve as substrates for boring invertebrates and microalgae (see Section 10.4.2.4 above). These biogenic three-dimensional solid substrates differ in their suitability for boring organisms owing to differences in hardness, colonizable volume, acid solubility, and defense mechanisms (see Geffen and Rosenberg 2005, Bagur et al. 2014).

Hard substrate organisms also create secondary three-dimensional living space that favors the development of interstitial assemblages. Indeed, sessile invertebrates and macroalgae add to the three-dimensional relief of marine hard substrates to varying degrees (e.g., barnacle tests *vs.* coral reefs). Other hard substrate organisms also create, maintain, and modify three-dimensional structures *via* their activities (i.e., allogenic engineering; *sensu* Jones et al. 1994) including the construction of rodolith mounds by sand tilefish, the maintenance of sediment-free karst dilution holes by excavating red grouper, and simplification of coral architecture by coral-eating fishes and invertebrates (Bozec et al. 2015, Pereira-Filho et al. 2015b, Ellis 2019). Structural legacies resulting from organismal mortality (e.g., dead coral) and past activities (e.g., vacant invertebrate boreholes) also contribute to the three-dimensionality of hard-substrate habitats (Bozec et al. 2015, Bagur et al. 2016, 2019). Moreover, the fragments resulting from structural decay (e.g., coarse carbonate sands produced by the erosion of dead coral skeletons) contribute to three-dimensionality at relatively smaller spatial scales including those relevant to microbes (see Wild et al. 2006).

A variety of three-dimensional structures serve as refuges for a diversity of organisms by limiting enemy access (Table 10.5). The association of organisms with biogenic three-dimensional refuges can either be a consequence of increased colonization (i.e., active habitat selection by larvae or mobile adults, e.g., Hurlbut 1991), reduced mortality (e.g., due to limited predator access; Pawlik 1998), or a combination of both (e.g., Shepherd and Turner 1985).

Overall, the secondary (or higher-order) habitats created and/or maintained by organisms are usually critical for the biodiversity of hard substrate communities. The combination of basal substrates with different kinds of biogenic substrates and three-dimensional structures produces patchworks of colonizable

Table 10.5. Three-dimensional structures made by physical ecosystem engineers and their functions as enemy-free space for other organisms.

Engineered structure/Site	Function	References
Algae, *Coralllina officinalis*, thalli in rocky shores of Wellington Bay, New Zealand	Protect copepods and other meiofauna from blenny (*Helcogramma medium*) predation	Coull and Wells (1983)
Filamentous algae, *Cladophora columbiana*, clusters of mussels, *Mytilus* spp., and dense assemblages of the barnacle *Semibalanus cariosus* in rocky shores at Barkley Sound, British Columbia, Canada	Protect newly hatched snails, *Nucella emarginata*, from hermit crab, *Pagurus* spp., and shore crab, *Hemigrapsus nudus*, predation.	Gosselin and Chia (1995)
Empty shells from dead oysters, *Saccostrea kegaki*, in a rocky shore at Izu Peninsula, Shizuoka Prefecture, Japan	Protect limpets, *Cellana nigrolineata*, from predation by the muricid snail, *Reishia clavigera*.	Tomatsuri and Kon (2017)
Mussel, *Modiolus modiolus*, beds in the rocky subtidal zone off the Isles of Shoals, New Hampshire, USA	Protect bivalves, *Hiatella arctica*, sea urchins, *Strongylocentrotus droebachiensis*, and ophiuroids, *Ophiopholis aculeata*, from predation by crabs, lobsters, and fishes. Reduce the effect of disturbance by grazing *S. droebachinesis* on overall benthic community composition.	Witman (1987)
Crevices in coral reefs on the east coast of Palau, Micronesia	Protect coral, *Porites lobata*, recruits from predation by butterflyfishes (*Chaetodon lunulatus, C. punctatofasciatus*), triggerfishes (*Balistapus undulatus, Balistoides viridescens*), and wrasse (*Cheilinus fasciatus*).	Gallagher and Doropoulos (2017)
Cryptic habitats in coral reefs off Grand Bahama and Acklins Islands, Bahamas, and off Key Largo, Florida, USA	Allow the occurrence of 15 sponge species that would otherwise be extirpated from reefs by fish predation.	Pawlik (1998)
Coral reefs of contrasting complexity in Lee Stocking Island, Bahamas	Complex corals, *Agaricia tenuifolia*, reduce the detrimental effects of resident reef fishes (i.e., predators, adult damselfish) on post-settlement damselfish, *Stegastes leucostictus*, survival.	Almany (2004)

habitat on the seafloor (Yakovis and Artemieva 2017, Thomsen and South 2019). The distinctive habitats created by organisms often favor the establishment of species that differ from those colonizing basal rock, thereby enhancing overall species richness (e.g., Miyamoto and Noda 2004, Castilla et al. 2004, Bagur et al. 2016, 2019, Thomsen and South 2019).

10.5.2.2 Modulation of abiotic conditions

The positive responses of organisms to biogenic three-dimensional structures are not necessarily or exclusively driven by limited enemy access and predation. Indeed, the attraction of some mobile species to three-dimensional structure is not conditional on the presence of enemies or other threats (Crooks et al. 2016). Commonly, species respond to the favorable abiotic conditions and resource levels resulting from particular structural features, including the modulation of heat, light, and water flow, and the trapping of particulate materials (see Section 10.5.2.3).

Algal and sessile invertebrate cover modulates heat transfer and light irradiance in intertidal settings and, in this way, reduces the impacts of extreme temperatures and desiccation in interstitial and understory habitats (e.g., Bertness et al. 1999, Jurgens and Gaylord 2017, Gutiérrez et al. 2019). Animal-created habitats, such as bivalve boreholes, are also potential sites of reduced temperature and desiccation stress in the intertidal zone (Bagur et al. 2016). Similarly, the complex three-dimensional structures of coral reefs (e.g., vertical surfaces, caves, tunnels, and overhangs) intercept light and produce shaded environments that markedly differ in species composition from open reef areas (Dinesen 1983).

Seaweeds also cast shade, in part due to photon uptake (i.e., assimilation), but also because their physical structure absorbs and reflects photons (i.e., a physical ecosystem engineering effect; see Jones and Gutiérrez 2007). Shading by large seaweeds (e.g., kelp) also affects understory algal composition and biomass (Clark et al. 2004).

With regards to water flow, the effects of three-dimensional structure range from the modification of boundary-layer flows by isolated structural elements to the collective and often large-scale effects of biogenic structures forming reefs and beds (e.g., corals, oysters, mussels, kelp, and coralline algae). Both protruding biogenic structures, such as mussels and algal thalli, and depressions, such as bivalve boreholes and sea urchin burrows, can significantly affect boundary layer flows. For example, individual protruding structures can create localized zones of increased flow velocity immediately upstream and adjacent to the structure whereas flow velocity can be reduced immediately downstream (Nowell and Jumars 1984). Similarly, flows are generally slowed as they pass over and into topographic depressions (Yager et al. 1993). There is, however, a threshold density of structural elements above which their effects on flow are no longer isolated but become interactive (Nowell and Jumars 1984). Examples include the slowing-down of flows in the interstices of species aggregations of mussel beds (e.g., Carrington et al. 2008) and around the fringes of reefs and other large species aggregations including kelp forests (e.g., Gaylord et al. 2007).

The variable flow conditions that develop around biogenic structures and their aggregations influence benthic recruitment patterns by affecting propagule and larval delivery, substrate choice, and their attachment and persistence (Walters and Wethey 1996, Wright and Boxshall 1999, Reidenbach et al. 2009). The interstitial space between neighboring structural elements, the leeward side of protruding structures, and various kinds of biogenic cavities (e.g., vacant boreholes and empty shells and tubes) are often used as "flow refuges" by adult invertebrates to avoid being dislodged or washed away by currents (e.g., Bagur et al. 2019). On the other hand, the zones of higher flow velocity around individual structural elements and aggregations are usually favorable for suspension feeders as they are characterized by increased fluxes of suspended food particles (e.g., Gili and Coma 1999, Monismith et al. 2010). If structures are flexible enough to be bent over by flows (e.g., foliose algal thalli and surfgrass), their effects on flows may be accompanied by mechanical abrasive effects that can dislodge neighboring organisms and epibionts (i.e., "whiplash effect", Kiirikki 1996, Leonard 1999).

10.5.2.3 Particle trapping and binding

Particle trapping by biogenic structural elements contributes a major fraction of the sediment and particulate organic matter stocks in hard substrate communities (Bailey-Brock 1979, Stamski and Field 2006, Bellwood and Fulton 2008) and can be an important component of sediment budgets in these systems (see Stamski and Field 2006). The trapping and retention of sediments by biogenic three-dimensional structures creates conditions for the establishment of infaunal invertebrates, such as polychaetes, nematodes, nemerteans, and amphipods (see Section 10.4.2.3 above). These organisms often make up an important proportion of the invertebrate fauna inhabiting rocky shores (Tokeshi and Romero 1995, Bagur et al. 2016, 2019) and would not otherwise exist (see Alongi and Christoffersen 1992 for a coral reef example).

Biogenic structures can enhance the deposition of particulate materials in hard substrates in two ways: first, by the direct interception of flow (e.g., interception by macroalgal blades; Schaffelke 1999) and second, by baffling currents and creating zones of reduced flow velocity (e.g., within interstitial space; see Section 10.5.2.2 above) where sedimentation exceeds resuspension (Nowell and Jumars 1984). Biogenic structures are expected to trap particles *via* a combination of these two mechanisms, although their relative importance has seldom been quantified (but see Chen et al. 2018 for a soft-bottom example).

In addition to the trapping and retention of sediments *via* their own physical structures, some hard substrate organisms bind particles together by means of adhesive, mucus secretions, thus forming aggregates that are larger and less prone to hydrodynamic transport (Ward and Kach 2009). Particle binding is inherent to the biodeposition of suspended particles by suspension-feeding invertebrates (see Section 10.5.1.6 above). Mucus is produced by these organisms at different stages of feeding and substantially contributes to the cohesion of feces and pseudofeces and enhances their settlement rates (Haven and Morales-Alamo 1972, Cognie and Barille 1999). Kautsky and Evans (1987) estimated that biodeposition by mussels (*Mytilus edulis*) represents 24% of the sediments deposited annually in a shallow (5-m depth) rocky bottom habitat at the Baltic Sea. Particle trapping mechanisms are also expected to facilitate the retention of biodeposits within dense beds of mussels and other suspension feeders (see Kent et al. 2017).

The large amounts of mucus produced by corals (Wild et al. 2004a, b, Naumann et al. 2010, Bythell and Wild 2011) also trap inorganic and organic particles from the water column (e.g., Wild et al. 2004a, Naumann et al. 2009), thereby forming aggregates that eventually detach from the coral (Wild et al. 2004a, Huettel et al. 2006, Mayer and Wild 2010). Coral mucus aggregates typically fall out of suspension quickly (Mayer and Wild 2010), but this is dependent on water depth (i.e., in shallow lagoons, aggregates reach the seafloor faster than on steep reef slopes), current regimes (i.e., strong tidal currents transport aggregates further away and waves may keep them in suspension longer), and the weight of the trapped particles (i.e., trapping of inorganic particles leads to faster deposition than trapping of organic particles).

10.5.2.4 Bioerosion and bioprotection

In a geomorphological context, bioerosion encompasses a suite of processes involving the removal of hard substrate by organisms. Bioerosion occurs at small scales but may also facilitate larger-scale wave erosion by weakening rock joints (Naylor et al. 2012). Bioerosion occurs in two main ways. First, a range of organisms are involved in the direct mechanical removal of rock (a process called bioabrasion) and has been estimated to lower calcareous rock surfaces at rates ranging from less than 0.2 mm yr^{-1} to more than 2 mm yr^{-1} (Andrews and Williams 2000, Fornós et al. 2006). Examples include shoreline grazers (i.e., mollusks, snails, and chitons) actively scraping away layers of rock as they feed, some reef-dwelling fishes (e.g., scarids) that actively remove rock when grazing, and some species of mollusks (e.g., *Pholas dactylus*), worms (e.g., *Boccardia* spp.), and urchins (e.g., *Strongylocentrotus* spp.) that mechanically bore into rock to seek protection from predators and waves (e.g., Trudgill 1987, Russell et al. 2018).

The second means of bioerosion involves the chemical removal of rock (called 'biocorrosion' and sometimes 'bioweathering'), which can occur actively *via* the secretion of organic acids and/or passively by the moderating effects of organisms on pH, humidity, and temperature at the rock surface. Biocorrosion is especially important at a microbial scale, with many species of 'microborers' (i.e., algae, fungi, and bacteria) roughening rock surfaces *via* the creation of micrometer-scale holes at considerable densities; however, several macroinvertebrates are also chemical borers including bivalves (e.g., *Lithophaga* spp.; Kleemann 1996), sponges (e.g., *Cliona* spp.; Cobb 1969), and sipunculans (Williams and Margolis 1974).

The processes through which organisms retard the removal of material from their basal substrate are termed bioprotection (Carter and Viles 2005). Bioprotection can involve the active cementation and consolidation of substrates (including loose sediment as well as rock). Perhaps the clearest examples involve encrusting species including encrusting algae (e.g., *Lithothamnion* spp.), barnacles, and mussels, which likely function as a 'protective layer' on the surfaces they colonize. The efficacy of bioprotection with respect to wave impact and abrasion has not yet been conclusively demonstrated, although there is evidence that this may be important on some coasts (e.g., Moura et al. 2012). As well as active encrustation, microbial activity can facilitate the formation of hardened layers *via* biochemical transformations, which has been quantified on both rock and concrete (Coombes et al. 2013a). Abundant biofilms may also reduce the likelihood of salt crystallization—an important erosion process in the littoral zone (Mottershead et al. 2003).

Arguably, the passive protective effects of hard-bottom communities may be more important than their active impacts. For example, some epilithic organisms alter rock-surface microclimates in such a way as to reduce the frequency and efficiency of weathering processes. This has been clearly demonstrated for seaweed canopies, the shading and moisture-retaining effects of which not only facilitate the survival of many other species (see Section 10.5.2.2 above) but also limit the wetting-drying and heating-cooling cycles that act to gradually weaken intertidal rock (Stephenson and Kirk 2000, Coombes et al. 2013b). For example, laboratory simulations have shown that for relatively soft rock like mudstone, a cover of seaweed may reduce physical disintegration by salt crystallization by almost 80% relative to exposed rock after only 100 thermal cycles (Gowell et al. 2015). Encrusting species can also have similar effects, by moderating temperature and moisture cycles below the rock surface (Coombes et al. 2017).

10.6 Zonation patterns in hard substrate communities

Whether considering rocky shores, subtidal rocky reefs, coral reefs, or even anthropogenic structures, all hard substrate communities show some zonation pattern. Zonation is simply the variation in the distribution and abundance of organisms as a consequence of topographical gradients and their effects on physical and biological conditions, the general patterns of which are described in the following sections for rocky shores and coral reef systems.

10.6.1 Rocky shore zonation

Rocky shores around the world usually show comparable patterns of zonation despite differences in species richness and composition from one geographic location to another. A universal pattern of zonation was first proposed by Stephenson and Stephenson (1949) based on observations first made on the coasts of the United Kingdom and then expanded to include other regions of the world comprising the Mediterranean, North America (both the Atlantic and Pacific coasts), South Africa, the Indian Ocean and the Red Sea, and Australia. The pattern shows a regular ordering of zones consisting of similar dominant taxonomic groups of species in these regions of the world.

In Stephenson and Stephenson's (1949) scheme, the highest elevation on the shoreline is the upper intertidal zone (also known as the splash zone), which is rarely submerged even by the highest tides (see Figure 3 in Stephenson and Stephenson 1949). This zone is mostly wetted by wave splash and fine mists, and its extent is determined by wave exposure. The primary producers that inhabit this zone are mostly lichens, cyanobacteria, and encrusting or ephemeral algae. Gastropod herbivores such as limpets and periwinkles (e.g., *Littorina*) can graze in this zone and are eaten by predatory shore crabs, birds, and mammals such as raccoons, river otters, and martens.

The middle intertidal zone occurs between the high and low tide marks and can be divided into two subzones depending on the amount of daily tidal exposure (see Figure 3 in Stephenson and Stephenson 1949). The upper middle intertidal zone is dominated by more stress-tolerant species such as acorn barnacles (e.g., *Chthamalus*, *Balanus*, and *Semibalanus*), rockweeds (e.g., *Fucus* and *Pelvetia*), and predatory gastropod whelks, which can withstand low tide exposure. In contrast, the lower middle intertidal zone is dominated by species with a somewhat lower water-stress tolerance including mussels, acorn and gooseneck barnacles, and sea anemones. The lower end of this zone also supports macrophytes (i.e., macroalgae and surfgrass) that can live attached to mussels (e.g., *Postelsia*) or the bare patches created by wave disturbance. Predatory sea stars (e.g., *Pisaster, Asterias,* and *Stichaster*) and gastropod whelks feed on mussels and barnacles in the lower end of this zone.

The low intertidal zone is immersed most of the time except during spring low tides (see Figure 3 in Stephenson and Stephenson 1949). Macrophytes thrive in the zone, with kelps and surfgrasses extending down into the subtidal zone. There are hardy macroalgae canopy species such as kelps (e.g., *Saccharina, Ascophyllum,* and *Laminaria*) and more turf-like, filamentous, foliose, and branched and unbranched coralline understory algal species. Predators in the low intertidal include sea stars, fishes, and crabs, which exclude mussels from this zone. Herbivores are also common including sea urchins, limpets, chitons, and snails.

The types of organisms that live in the low intertidal zone continue to be present in the high to middle subtidal zones. However, depending on the steepness of the rocky shore, restrictions in light can soon limit green and brown macroalgae (to 5–15 m water depth; Hanelt et al. 2003). Because it uses green light found at depth, red macroalgae can photosynthesize in deeper waters (15–30 m). Large kelp forests are often a dominant feature of nearshore temperate hard-bottom subtidal environments, where photosynthetic activity occurs in the plant parts that float near the surface (Dayton 1985). Typically, the understory consists of crustose coralline algae and other deep-water red macroalgae (e.g., Littler et al.

1986). Sea urchins, abalone, crabs, lobster, sea cucumbers, and rockfishes and finfishes are common species associated with kelp forests, many of which are of commercial importance.

Despite similarity in the order of rocky shore zones across different regions, the species present and the width of the zones can vary depending on local factors such as the slope and aspect of the shore, tidal amplitude, and wave exposure. Moreover, deviations from the above pattern also occur as variations in climate, oceanography, and historical processes can affect the composition, diversity, and abundance of intertidal and subtidal biota. For example, there are neither fucoids nor canopy-forming macroalgae on rocky shores along much of the temperate southwestern Atlantic, and there were no balanoid barnacles until *Balanus glandula* was introduced in the late 1960s (Miloslavich et al. 2016).

In tropical and subtropical regions, high temperatures and desiccation stress result in intertidal zones with sparse and smaller macroalgae and invertebrate species restricted to cracks and crevices. Here the middle intertidal is colonized by cyanobacteria and filamentous algae that support a large guild of herbivores including limpets, chitons, gastropods, fishes, crabs, and isopods and amphipods (Underwood and Denley 1984). Likewise, in polar regions, intertidal organisms experience freezing temperatures and ice scouring in the winter months, resulting in more ephemeral species that flourish in the intertidal during the summer months (Underwood and Denley 1984). In the subtidal zone, however, polar region macrophytes and invertebrates thrive and most of the major taxonomic groups are present.

10.6.2 Coral reef zonation

Tropical coral reefs have zones that include the reef flat, reef edge, reef front, reef slope, and in some cases, the back reef. As with rocky shores, coral reef zones are characterized by distinct differences in key abiotic factors such as water depth, wave energy, light availability, water quality and turbidity, and temperature.

The reef flat is usually protected from wave action with a cross-shore extent ranging from meters to kilometers and water depths ranging from centimeters to a few meters. Corals and other organisms in this zone have adapted to tolerate a wide range of temperatures, oxygen availabilities, light intensities, and salinities, and many are found exclusively in this zone. Among these, vermetid gastropods that feed using mucus nets can be the dominant organisms. Most of the turf algae productivity and grazing activity takes place here (Bellwood et al. 2018). During extreme low tides, the corals and other reef flat organisms can be exposed to air. Due to these stressful living conditions, the diversity of life in this zone is much lower compared to the other zones (Spalding et al. 2001).

Reef crests are the highest point of reefs and receive the highest impact of wave energy. During low tide, the organisms living on reef crests are exposed to air and high solar radiation, and have developed mechanisms to protect them against desiccation, such as mucus release (e.g., cnidarians and mollusks) and lids (e.g., opercula of tubeworms). Corals and other invertebrates are typically encrusting or very large so that they can withstand intense wave action, high solar radiation, and aerial exposure to thrive in this zone (e.g., Goreau 1959, Dahlgren 1989). In the Indo-Pacific, the crest zone can be dominated by calcareous coralline red algae, creating the so-called algal ridge (e.g., Littler and Doty 1975).

The fore-reef or reef slope is farthest from the shore and slopes downward to water depths of more than 100 m depending on light availability. In some cases, reef fronts drop off steeply to form near-vertical walls. Coral reef diversity is highest at intermediate water depths of 5 to 20 m and declines at both shallower and deeper depths in this zone. Some corals have adapted their growth forms to survive at specific water depths, wave exposures, and light availabilities. For example, plate corals have more surface area exposed to the light and thus often occur in deeper waters where light becomes limiting (e.g., Colin et al. 1986, Guzmán and Cortés 1989). Likewise, the delicate growth forms of corals, other invertebrates, and calcifying red algae typically occur at greater depths (> 20 m) in this zone, where wave action is significantly reduced compared to the shallower parts of the reef slope (< 5 m water depth; e.g., Guzmán and Cortés 1989).

The back-reef area that sits shoreward of the crest and has a shallow slope leading to a lagoon is typical for atoll and platform reef types. This zone is often very shallow (with a common maximum water depth of only a few meters) and more protected from wave action than the fore reef. Shallow back reefs can be exposed to air during extreme low tides, and isolated patch reefs (and so-called micro atolls) along with coral rubble fields can often be found here (e.g., Goreau 1959, Smith et al. 2007).

10.7 Regulation of hard substrate communities: the rocky shore paradigm

Zonation on rocky shores and reefs is the result of three general, interacting factors: dispersal and recruitment, the abiotic environment, and species interactions. These three factors act as "filters" (Lawton 2000) to include or exclude species from local communities and the zones contained within them. Joseph Connell (1961a, b) was one of the first marine ecologists to recognize that the community structure of hard bottom systems, measured as rocky intertidal zonation in his study, was the result of the varying importance of physical and biological factors.

Connell hypothesized that environmental factors such as drying and warming (or freezing) temperatures control the abundance and distribution of organisms in the upper intertidal zone. Alternatively, biological interactions such as predation and competition were predicted to control the distribution and abundance of organisms at the lower limits of the intertidal zone, where physical stress is less intense. Over the last 60 years, Connell's simple model has been experimentally tested in different regions of the world (e.g., Paine 1969, 1974, Dayton 1971, Menge 1976, Scrosati et al. 2011), applied to other systems (e.g., Bertness and Hacker 1994, Hacker and Gaines 1997), and refined to include additional processes such as disturbance (e.g., Dayton 1971, Menge and Sutherland 1976, Connell 1978), dispersal and recruitment (e.g., Menge and Sutherland 1987), and positive interactions (e.g., Bertness and Leonard 1997). Some of these factors are considered in the following sections with specific reference to the structure and regulation of rocky shore communities.

10.7.1 Dispersal and recruitment

Since the 1980s, rocky shore systems have been thought of as "open systems" whose community structure and zonation depend on the delivery of propagules *via* ocean currents (e.g., Underwood and Denley 1984, Gaines and Roughgarden 1985, Roughgarden et al. 1988, Underwood and Fairweather 1989). At large spatial scales, the delivery of invertebrate larvae and algal spores depends on the reproduction output from source populations, growth and survival of individuals within the plankton, and oceanographic currents and/or local hydrodynamics responsible for carrying them to suitable benthic habitats. Once propagules are within settling distance of the benthos, their successful recruitment depends on small-scale hydrodynamics and, in the case of some larvae, sophisticated settlement behaviors. For example, many free-swimming larvae are photopositive, swimming near the water's surface, until they reach a settlement habitat and become photonegative, promoting settlement to lower intertidal and subtidal zones (Thorson 1964). Some species remain photopositive as they settle, thus resulting in recruitment to high intertidal zones. Once settled, the survival, growth, and reproduction of individuals, and ultimately the distribution and abundance of species, will depend on the environmental factors and species interactions encountered by these individuals.

10.7.2 Environmental factors

Environmental factors such as water depth, water and air temperature, wave action and currents, light intensity, and nutrients all play a role in the zonation patterns seen in rocky intertidal and subtidal communities (e.g., Doty 1946, Denny and Wethey 2001, Helmuth and Hofmann 2001, Menge et al. 2015). Sessile organisms, which cannot move from their settlement habitat, thrive in particular zones depending on their physiological and functional morphological responses to the environmental conditions they are exposed to. For example, sessile or slow-moving mussels, barnacles, chitons, snails, and limpets must be able to withstand desiccation and high or low temperatures during low-tide periods (e.g., Evans 1948, Helmuth 1998, Tomanek and Somero 1999). Most species of macroalgae and surfgrasses are restricted to low intertidal or subtidal zones, or tide pools higher in the intertidal zone. Mobile animals with the ability to move with the tides, shelter in cracks and crevices, or live in tide pools can actively avoid or reduce their exposure to desiccation and/or extreme temperatures (e.g., Stillman and Somero 1996). Finally, the distribution of primary producers in the subtidal zone, including kelps and other macroalgal species, is restricted to water depths that allow them access to light (Hanelt et al. 2003).

Besides environmental factors that may cause physiological stress, organisms in rocky shore habitats regularly experience disturbance, i.e., abiotic events that kill or damage individuals (Sousa 2001). This includes disturbance by waves, currents, ice scouring, and other events, which remove biomass and create open patches that can be colonized by other individuals (e.g., Dayton 1971, Dethier 1984, Witman 1987). By opening up space, disturbance not only plays a role in the distribution and abundance of species but also mitigates species interactions, such as facilitation and competition for space, which is an important limiting resource in rocky shore communities (Paine 1979, Sousa 1979a, b, 2001, Bertness and Leonard 1997).

10.7.3 Species interactions

Species interactions, such as predation, competition, facilitation, parasitism, and disease, are important determinants of the distribution and abundance of species in hard substrate communities. All species require food, nutrients, and habitat and thus interact with one another to gain these resources. Two influential studies in the 1960s, that of Connell (1961a, b) mentioned earlier, and Robert Paine's (1969), demonstrated the power that species interactions can have on rocky shore community structure. Connell studied the role of competition for space in shaping the distributions of two barnacle species living in separate intertidal zones (*Chthamalus* in the high intertidal zone and *Semibalanus* in the middle intertidal zone) in Scotland. He conducted species removal experiments and caging experiments to examine the effects of competition and predation, respectively. He found that *Chthamalus* was restricted to the high intertidal as a result of competitive exclusion by *Semibalanus*, while *Semibalanus* was restricted to the middle intertidal as a result of physiological intolerance to the high intertidal and predation by whelks in the low intertidal.

Connell's experiments motivated Robert Paine to study the role of predation in structuring rocky intertidal communities in the US Pacific Northwest. In this system, the dominant space occupier is the mussel *Mytilus californianus* in the upper middle intertidal zone. Paine experimentally tested the hypothesis that the predatory sea star, *Pisaster ochraceus* (Fig. 10.1a), although not dominant in number, could mediate the competitive dominance of mussels through predation, thereby indirectly promoting the competitively subordinate species of mussels. Paine's removal experiments showed that the lack of persistent predation by *Pisaster* not only allowed mussels to extend their distribution into the lower intertidal zone (Paine 1974) but also caused a decline in the number of species that compete for space with mussels (Paine 1969).

Paine's study established the concept of 'keystone species', defined as a species that has a disproportionally large effect on the community relative to its abundance (Paine 1969, Power et al. 1996). Keystone species can create trophic cascades, a food web dynamic in which a change in the rate of consumption at one trophic level results in a series of changes in species abundance or composition at lower trophic levels (e.g., Menge et al. 1994, Wootton 1995). Five decades of research have shown that keystone species and trophic cascades are common in both terrestrial and aquatic systems (see Ripple et al. 2014 for a review). This includes subtidal kelp forests in which sea otters function as keystone species, creating a trophic cascade by feeding on sea urchins and indirectly (positively) affecting kelp growth as a result (Estes and Duggins 1995).

Finally, facilitation is an important but often an underappreciated type of species interaction that can have profound effects on community structure (Bertness and Callaway 1993, Bertness and Leonard 1997). The most obvious type of facilitation involves ecosystem engineers, i.e., those species that create, modify, and/or maintain physical habitat for themselves and other species (Jones et al. 1994; see Section 10.5.2 above). For example, Dayton (1975) showed that in a low intertidal zone macroalgal community, the removal of the kelp canopy resulted in understory algal species mortality, presumably because of the amelioration of desiccation and temperature stress by kelp during low tide events. Bertness et al. (1999) showed similar ameliorating effects in the rocky intertidal of the Gulf of Maine, USA, for the intertidal canopy seaweed *Ascophyllum nodosum*. Other types of facilitation include associational defenses in which one species protects another from being consumed by predators or herbivores. For example, Pfister and Hay (1988) found that the palatable red alga *Gracilaria tikvahiae* was protected from grazing by sea urchins when it grew in association with *Sargassum filipendula*, which contains unpalatable secondary compounds.

10.7.4 Putting the processes together: regulation of rocky shore communities

Whilst the roles of dispersal, environmental factors, and species interactions have been considered separately, in reality, hard substrate communities are a product of multiple processes that interact to regulate their structure. Indeed, models and empirical studies show that it is the interaction and relative importance of these factors that best explain patterns of zonation and community structure. Not surprisingly, multiple models have been proposed to explain how communities are regulated. Some of these models, such as the intermediate disturbance hypothesis (IDH; Connell 1978) and the lottery model (LM; Sale 1977), were initially developed based on research conducted on rocky intertidal, subtidal kelp, and coral reef communities.

The IDH arose out of a recognition that gradients in disturbance could be important in regulating the ability of species to coexist under strong species interactions such as competition. Connell hypothesized that at low levels of disturbance (measured as a frequency or intensity), species diversity would be low because competition would decrease species coexistence. Under this condition, the best competitor should dominate, utilizing the resources of the competitively subordinate species and excluding them (e.g., similar to the effect of mussels without *Pisaster* described in the previous section). In contrast, at high levels of disturbance, species diversity should decline because many individuals would die and become locally extinct as a result. At intermediate levels of disturbance, species diversity should be maximized as a result of the balance between increased competitive disruption and lower mortality rates. The IDH has since been adapted to include predation, stress, and dispersal (Menge and Sutherland 1976, 1987, Dial and Roughgarden 1998), productivity (Huston 1979), and positive interactions (Hacker and Gaines 1997, Bruno et al. 2003). Empirical examples of the IDH include those from rocky reefs (e.g., Seapy and Littler 1982, Addessi 1994, Zacharias and Roff 2001, Pfaff et al. 2010, Scrosati et al. 2011), rocky boulders (e.g., Sousa 1979b, McGuinness 1987, McClintock et al. 2007), and tropical reefs (e.g., Connell 1978, Menge et al. 1985, England et al. 2008).

The LM emphasizes the role of chance in the maintenance of species diversity by assuming that resources in a community made available by the effects of disturbance, stress, and predation are captured at random by recruits from a larger pool of potential colonists. For this theory to work, species must have fairly similar effects on one another, population growth rates, and the ability to quickly use available resources. In the LM, it is the equal chance of all species to obtain resources that allows species coexistence. The LM was originally proposed to explain species diversity in coral reef fish communities (Sale 1977, Chesson and Warner 1981) but has since been adapted for other diverse communities including tropical rainforests (Hubbell 2001).

Finally, it is important to recognize that most models focus on local-scale processes, despite the inclusion of dispersal in some cases. More recently, rocky intertidal ecologists have increasingly considered the role of regional-scale processes in community structuring. This “meta-ecosystem” approach suggests that marine subsidies in the form of propagule supply, phytoplankton food, and nutrients, as determined by oceanographic processes such as currents and ocean upwelling, are critical factors in determining rocky intertidal and subtidal community structure (e.g., Menge et al. 1997, 2003, 2015, Witman et al. 2004, Navarrete et al. 2005, Okuda et al. 2010, Russell et al. 2006, Hacker et al. 2019).

10.8 Functions and services of hard substrate communities

People have obtained multiple important benefits, or ‘services’, from hard substrate communities and their associated functions including coastal protection, maintenance of fisheries, nutrient cycling, and tourism and recreation (Barbier et al. 2011). For example, rocky reefs, coral reefs, and rhodolith beds are all critical to commercial fisheries (Siegfried et al. 1994, Coleman et al. 2000, Kamenos et al. 2004,

Stål et al. 2008, Blamey and Bolton 2018) with coral reef fisheries representing approximately 9–12% of the world's total fisheries (Smith 1978). Reef organisms are also sources of raw materials such as agar, manure, lime, mother-of-pearl, building materials (e.g., coral blocks, mortar, and cement), and pharmaceuticals (e.g., antimicrobials, anticoagulants, and anti-inflammatories; see Moberg and Folke 1999, Moura et al. 2013, Blamey and Bolton 2018). Coral and rocky reef resources also afford food security and a means of subsistence for many coastal communities (Kyle et al. 1997, Craig et al. 2008, Blamey and Bolton 2018). For example, estimates indicate that nearly six million people depend on coral reef fisheries for their livelihoods, most of which reside in developing and low-income countries (Teh et al. 2013). Moreover, rocky reefs and coral reefs, in particular, are tourist destinations for millions of people each year. Indeed, recreational activities including SCUBA diving, snorkeling, tide-pooling, and sport fishing can be highly lucrative. For example, the earnings of ~ 100 diver operators in Hawaii were estimated at US$50–60 million for the year 2002 (van Beukering and Cesar 2004).

Rocky and coral reefs also buffer shorelines from extreme waves and sea-level rise, protecting coastal livelihoods and assets (Scyphers et al. 2011, Arkema et al. 2013, Beck et al. 2018). The annual coastal protection services provided by coral reefs have been valued at between US$8 and 40 million, with variations explained by local geomorphology, storm characteristics, coastal exposure, and economic development (Pascal et al. 2016).

Also, corals, seaweeds, rhodoliths, and other reef inhabitants store significant amounts of carbon (Kinsey and Hopley 1991, Amado-Filho et al. 2012, Krause-Jensen and Duarte 2013); however, corals and other calcifying organisms incorporate most of their carbon *via* $CaCO_3$ precipitation (i.e., skeletal or shell production), which involves CO_2 release, thus offering limited potential to mitigate anthropogenic CO_2 emissions (see Kinsey and Hopley 1991, Ware et al. 1992, Munari et al. 2013).

Seaweeds, on the other hand, show net carbon uptake (*via* photosynthesis) and export a significant fraction of their carbon production (ca. 11%) to the deep sea where it is effectively sequestered from the atmosphere (Krause-Jensen and Duarte 2016). Global estimates of carbon sequestration rates by rocky coast seaweed communities (173 Tg C yr^{-1}) exceed those of soft bottom coastal systems dominated by vascular plants (e.g., mangrove forests and seagrass meadows; Krause-Jensen and Duarte 2016) and is equivalent to approximately 6% of the carbon sequestered by terrestrial forest ecosystems (Filbee-Dexter 2020). Finally, with the decline of ocean pH due to increasing carbon dioxide in seawater, seagrasses, kelps, and other macroalgal species in rocky and coral reef habitats are likely to play an important role in buffering ocean acidification (Manzello et al. 2012, Mongin et al. 2016).

Despite the substantial services provided by rocky and coral reef communities and rhodolith beds, all are experiencing unprecedented change in the form of overfishing, pollution, diseases, hypoxia, bottom trawling and mining, the exploitation of oil and gas reserves, sea level rise, and ocean acidification and warming (see Moberg and Folke 1999, Knowlton 2001, Kamenos et al. 2003, Santos et al. 2016, Filbee-Dexter 2020). The loss of the integrity of these systems not only reduces the services they provide but also the economic value they generate, e.g., loss of at least 100,000 fishery jobs due to the overfishing and pollution of coral reefs in the Philippines (McAllister 1988). In the context of burgeoning coastal populations and increasing coastal resource demands and impacts, the future of hard bottom communities depends on sustainable management strategies that account for multiple threats while also reconciling ecological, environmental, and social priorities.

Summary

1. Marine hard-substrate habitats include rock, the hard parts of organisms, and anthropogenic materials that vary in composition, physical properties, and spatial extent.
2. Globally, hard substrates cover a relatively minor proportion of the marine benthic environment but make up a significant fraction of the world's coastal areas and support a relatively high biodiversity.
3. Marine hard substrate communities are usually dominated by foundation species (e.g., kelp, mussels, corals, and coralline algae) which, by virtue of their size and/or abundance, form the physical

structure of the habitat, determine the diversity of associated taxa, and modulate fluxes of materials and energy at multiple control points in their ecosystems.

4. Hard substrate biota also include accompanying benthic species (i.e., bottom-dwelling species that inhabit the physical habitat matrix formed by foundation species and the basal substrate) and temporary inhabitants (i.e., mobile species that use hard-substrate habitats for restricted periods of time).
5. The major functional roles of organisms on hard substrates can be broadly viewed as related to organic matter fluxes and trophic interactions across food webs (e.g., primary and secondary production, herbivory, predation, suspension feeding, and the production and consumption of non-living organic matter) and physical ecosystem engineering (e.g., the formation of secondary habitat, modulation of abiotic factors, sediment trapping and binding, and bioerosion and bioprotection).
6. All hard substrate communities show some zonation pattern (i.e., variation in the distribution and abundance of organisms as a consequence of topographical gradients and their effects on physical and biological conditions).
7. Zonation in hard substrate communities is the result of three general, interacting factors: dispersal and recruitment, the abiotic environment, and species interactions.
8. Multiple models have been proposed to explain how hard substrate communities are regulated including the intermediate disturbance hypothesis and the lottery model.
9. People obtain multiple important benefits, or 'services', from hard substrate communities and their associated functions including coastal protection, maintenance of fisheries, nutrient cycling, and tourism and recreation.
10. Hard substrate communities are experiencing unprecedented change in the form of overfishing, urbanization and coastal development, pollution, disease, hypoxia, bottom trawling and mining, the exploitation of oil and gas reserves, sea level rise, and ocean acidification and warming.
11. The future of hard bottom communities depends on sustainable management strategies that account for multiple threats while also reconciling ecological, environmental, and social priorities.

Acknowledgements

We thank Jerónimo Pan for the invitation to write this chapter and an anonymous reviewer for her/his constructive comments. We are also indebted to María Bagur, Zaira Matheus, Rodrigo Melo, Zech Meunier, and Mariano Rodríguez, who kindly provided us with many of the pictures that illustrate this chapter. JLG, MAC, and MGP were supported by National Geographic (CP-120R-17) during the preparation of this manuscript. This is a contribution to the program of GrIETA.

References

Addessi, L. 1994. Human disturbance and long-term changes on a rocky intertidal community. Ecol. Appl. 4: 786–797.

Airoldi, L. and M.W. Beck. 2007. Loss, status and trends for coastal marine habitats of Europe. Oceanogr. Mar. Biol. Annu. Rev. 45: 345–405.

Airoldi, L., X. Turon, S. Perkol-Finkel and M. Rius. 2015. Corridors for aliens but not for natives: effects of marine urban sprawl at a regional scale. Divers. Distrib. 21: 755–768.

Aleyev, Y.G. 1977. Nekton. Dr. W. Junk, The Hague. The Netherlands.

Almany, G.R. 2004. Does increased habitat complexity reduce predation and competition in coral reef fish assemblages? Oikos 106: 275–284.

Alongi, D.M. and P. Christoffersen. 1992. Benthic infauna and organism-sediment relations in a shallow, tropical coastal area: influence of outwelled mangrove detritus and physical disturbance. Mar. Ecol. Prog. Ser. 81: 229–45.

Altieri, A.H., B.R. Silliman and M.D. Bertness. 2007. Hierarchical organization via a facilitation cascade in intertidal cordgrass bed communities. Am. Nat. 169: 195–206.

Amado-Filho, G.M., R.L. Moura, A.C. Bastos, L.T. Salgado, P.Y. Sumida, A.Z. Guth, R.B. Francini-Filho, G.H. Pereira-Filho, D.P. Abrantes, P.S. Brasileiro, R.G. Bahia, R.N. Leal, L. Kaufman, J.A. Kleypas, M. Farina and F.L. Thompson. 2012. Rhodolith beds are major $CaCO_3$ bio-factories in the Tropical South West Atlantic. PLoS One 7: e35171.
Andrews, C. and R.B.G. Williams. 2000. Limpet erosion of chalk shore platforms in southeast England. Earth Surf. Process. Landf. 25: 1371–1381.
Arkema, K.K. and J.F. Samhouri. 2019. Living on the edge: variation in the abundance and demography of a kelp forest epibiont. Diversity 11: 120.
Arkema, K.K., G. Guannel, G. Verutes, S.A. Wood, A. Guerry, M. Ruckelshaus, P. Kareiva, M. Lacayo and J.M. Silver. 2013. Coastal habitats shield people and property from sea-level rise and storms. Nat. Clim. Chang. 3: 913.
Aviz, D., R.F. Da Silva and J.S. Rosa Filho. 2019. *Sabellaria wilsoni* (Polychaeta: Sabellariidae): an ecosystem engineer and promoter of zoobenthos diversity in the Brazilian Amazon coast. J. Mar. Biol. Assoc. UK 99: 1099–1109.
Badalamenti, F., A. Alagna and S. Fici. 2015. Evidences of adaptive traits to rocky substrates undermine paradigm of habitat preference of the Mediterranean seagrass *Posidonia oceanica*. Sci. Rep. 5: 8804.
Bagur, M., C.A. Richardson, J.L. Gutiérrez, L.P. Arribas, M.S. Doldan and M.G. Palomo. 2013. Age, growth and mortality in four populations of the boring bivalve *Lithophaga patagonica* from Argentina. J. Sea Res. 81: 49–56.
Bagur, M., J.L. Gutiérrez, L.P. Arribas and M.G. Palomo. 2014. Endolithic invertebrate communities and bioerosion rates in Southwestern Atlantic intertidal consolidated sediments. Mar. Biol. 161: 2279–2292.
Bagur, M., J.L. Gutiérrez, L.P. Arribas and M.G. Palomo. 2016. Complementary influences of co-occurring physical ecosystem engineers on species richness: insights from a Patagonian rocky shore. Biodiv. Conserv. 25: 2787–2802.
Bagur, M., J.L. Gutiérrez, L.P. Arribas and M.G. Palomo. 2019. Vacant bivalve boreholes maintain invertebrate species richness in a physically harsh, low intertidal platform. Diversity 11: 39.
Bahamondes, I. and J.C. Castilla. 1986. Predation of marine invertebrates by the kelp gull *Larus dominicanus* in an undisturbed intertidal rocky shore of central Chile. Rev. Chil. Hist. Nat. 59: 65–72.
Bailey-Brock, J.H. 1979. Sediment trapping by chaetopterid polychaetes on a Hawaiian fringing reef. J. Mar. Res. 37: 643–656.
Barbier, E., S.D. Hacker, C. Kennedy, E. Koch, B. Silliman and A.D. Stier. 2011. The value of estuarine and coastal ecosystem services. Ecol. Monogr. 81: 169–193.
Barner, A.K., S.D. Hacker, B.A. Menge and K.J. Nielsen. 2016. The complex net effect of reciprocal interactions and recruitment facilitation maintains an intertidal kelp community. J. Ecol. 104: 33–43.
Barrón, C., N. Marba, C.M. Duarte, M.F. Pedersen, C. Lindblad, K. Kersting, F. Moy and T. Bokn. 2003. High organic carbon export precludes eutrophication responses in experimental rocky shore communities. Ecosystems 6: 144–153.
Batson, P.B. and P.K. Probert. 2000. Bryozoan Thickets off Otago Peninsula. New Zealand. Fisheries Assessment Report 2000/46. Ministry of Fisheries, Wellington, New Zealand.
Beck, M.W., I.J. Losada, P. Menéndez, B.G. Reguero, P. Díaz-Simal and F. Fernández. 2018. The global flood protection savings provided by coral reefs. Nat. Commun. 9: 2186.
Bellwood. D.R., T.P. Hughes and A.S. Hoey. 2006. Sleeping functional group drives coral-reef recovery. Current Biol. 16: 2434–2439.
Bellwood, D.R. and C.J. Fulton. 2008. Sediment-mediated suppression of herbivory on coral reefs: Decreasing resilience to rising sea levels and climate change? Limnol. Oceanogr. 53: 2695–2701.
Bellwood, D.R., S.B. Tebbett, O. Bellwood, M. Mihalitsis, R.A. Morais, R.P. Streit and C.J. Fulton. 2018. The role of the reef flat in coral reef trophodynamics: past, present, and future. Ecol. Evol. 8: 4108–4119.
Benedetti-Cecchi, L., F. Pannacciulli, F. Bulleri, P.S. Moschella, L. Airoldi, G. Relini and F. Cinelli. 2001. Predicting the consequences of anthropogenic disturbance: large-scale effects of loss of canopy algae on rocky shores. Mar. Ecol. Prog. Ser. 214: 137–150.
Bertness, M.D. and R. Callaway. 1993. Positive interactions in communities. Trends Ecol. Evol. 9: 191–193.
Bertness, M.D. and S.D. Hacker. 1994. Physical stress and positive associations among marsh plants. Am. Nat. 144: 363–372.
Bertness, M.D. and G.H. Leonard. 1997. The role of positive interactions in communities: lessons from intertidal habitats. Ecology 78: 1976–1989.
Bertness, M.D., G.H. Leonard, J.M. Levine, P.R. Schmidt and A.O. Ingraham. 1999. Testing the relative contribution of positive and negative interactions in rocky intertidal communities. Ecology 80: 2711–2726.
Birkeland, C. 2015. Coral Reefs in the Anthropocene. Springer, Dordretch, The Netherlands.
Bishop, M.J., M. Mayer-Pinto, L. Airoldi, L.B. Firth, R.L. Morris, L.H. Loke, S.J. Hawkins, L.A. Naylor, R.A. Coleman, S.Y. Chee and K.A. Dafforn. 2017. Effects of ocean sprawl on ecological connectivity: impacts and solutions. J. Exp. Mar. Biol. Ecol. 492: 7–30.
Black, R. 1976. The effects of grazing by the limpet, *Acmaea insessa*, on the kelp, *Egregia laevigata*, in the intertidal zone. Ecology 57: 265–277.
Blamey, L.K. and J.J. Bolton. 2018. The economic value of South African kelp forests and temperate reefs: past, present and future. J. Mar. Syst. 188: 172–181.
Boaden, A.E. and M.J. Kingsford. 2015. Predators drive community structure in coral reef fish assemblages. Ecosphere 6: 1–33.

Boavida, J., J. Assis, I. Silva and E.A. Serrão. 2016. Overlooked habitat of a vulnerable gorgonian revealed in the Mediterranean and Eastern Atlantic by ecological niche modelling. Sci. Rep. 6: 36460.
Borsje, B.W., B.K. van Wesenbeeck, F. Dekker, P. Paalvast, T.J. Bouma, M.M. van Katwijk and M.B. de Vries. 2011. How ecological engineering can serve in coastal protection. Ecol. Eng. 37: 113–122.
Bortolus, A., E. Schwindt, P.J. Bouza and Y.L. Idaszkin. 2009. A characterization of Patagonian salt marshes. Wetlands 29: 772–780.
Bozec, Y.M., L. Alvarez-Filip and P.J. Mumby. 2015. The dynamics of architectural complexity on coral reefs under climate change. Global Change Biol. 21: 223–235.
Bracken, M.E.S., B.A. Menge, M.M. Foley, C.J.B. Sorte, J. Lubchenco and D.R. Schiel. 2012. Mussel selectivity for high-quality food drives carbon inputs into open-coast intertidal ecosystems. Mar. Ecol. Prog. Ser. 459: 53–62.
Branch, G.M. 2008. Trophic interactions in subtidal rocky reefs on the west coast of South Africa. pp. 50–78. *In*: McClanahan, T.R. and G.M. Branch [eds.]. Food Webs and the Dynamics of Marine Reefs. Oxford University Press, New York, NY, USA.
Brasileiro, P.S., G.H. Pereira-Filho, R.G. Bahia, D.P. Abrantes, S.M. Guimarães, R.L. Moura, R.B. Francini-Filho, A.C. Bastos and G.M. Amado-Filho. 2016. Macroalgal composition and community structure of the largest rhodolith beds in the world. Mar. Biodiv. 46: 407–420.
Bray, R.N, A.C. Miller and G.G. Geesey. 1981. The fish connection: a trophic link between planktonic and rocky reef communities? Science 214: 204–205.
Bruno, J.F., J.J. Stachowicz and M.D. Bertness. 2003. Inclusion of facilitation into ecological theory. Trends Ecol. Evol. 18: 119–125.
Bugnot, A.B., M. Mayer-Pinto, L. Airoldi, E.C. Heery, E.L. Johnston, L.P. Critchley, E.M.A. Strain, R.L. Morris, L.H.L. Loke, M.J. Bishop, E.V. Sheehan, R.A. Coleman and K.A. Dafforn. 2020. Current and projected global extent of marine built structures. Nat. Sustain. https://doi.org/10.1038/s41893-020-00595-1.
Bythell, J.C. and C. Wild. 2011. Biology and ecology of coral mucus release. J. Exp. Mar. Biol. Ecol. 408: 88–93.
Cabral-Oliveira, J., M. Dolbeth and M.A. Pardal. 2014. Impact of sewage pollution on the structure and functioning of a rocky shore benthic community. Mar. Freshwat. Res. 65: 750–758.
Carlton, J.T. and J. Hodder. 2003. Maritime mammals: terrestrial mammals as consumers in marine intertidal communities. Mar. Ecol. Prog. Ser. 256: 271–286.
Carpenter, R.C. 1985. Relationships between primary production and ii radiance in coral reef algal communities. Limnol. Oceanogr. 30: 784–793.
Carraro, R. and W. Gladstone. 2006. Habitat preferences and site fidelity of the ornate wobbegong shark (*Orectolobus ornatus*) on rocky reefs of New South Wales. Pac. Sci. 60: 207–223.
Carrington, E., G.M. Moeser, S.B. Thompson, L.C. Coutts and C.A. Craig. 2008. Mussel attachment on rocky shores: the effect of flow on byssus production. Integr. Comp. Biol. 48: 801–807.
Carter, N.E.A. and H.A. Viles. 2005. Bioprotection explored: the story of a little known earth surface process. Geomorphology 67: 273–281.
Castilla, J.C., N.A. Lagos and M. Cerda. 2004. Marine ecosystem engineering by the alien ascidian *Pyura praeputialis* on a mid-intertidal rocky shore. Mar. Ecol. Prog. Ser. 268: 119–130.
Cebrian, J. 1999. Patterns in the fate of production in plant communities. Am. Nat. 154: 449–468.
Chapman, M.G. and F. Bulleri. 2003. Intertidal seawalls—new features of landscape in intertidal environments. Landsc. Urban Plann. 62: 159–172.
Chava, A., A. Artemieva and E. Yakovis. 2019. Plant part age and size affect sessile macrobenthic assemblages associated with a foliose red algae *Phycodrys rubens* in the white sea. Diversity 11: 80.
Chen, Y., Y. Li, C. Thompson, X. Wang, T. Cai and Y. Chang. 2018. Differential sediment trapping abilities of mangrove and saltmarsh vegetation in a subtropical estuary. Geomorphology 318: 270–282.
Chesson, P.L. and R.R. Warner. 1981. Environmental variability promotes coexistence in lottery competitive systems. Am. Nat. 117: 923–943.
Choi, K.H. and Y.B. Seong. 2014. The rock coast of Korea. pp. 193–202. *In*: Kennedy, D.M., W.J. Stephenson and L.A. Naylor [eds.]. Rock Coast Geomorphology: A Global Synthesis. Geological Society Memoires, 40, London, UK.
Clark, M.R., A.A. Rowden, T. Schlacher, A. Williams, M. Consalvey, K.I. Stocks, A.D. Rogers, T.D. O'Hara, M. White, T.M. Shank and J.M. Hall-Spencer. 2010. The ecology of seamounts: structure, function, and human impacts. Annu. Rev. Mar. Sci. 2: 253–278.
Clark, R.P., M.S. Edwards and M.S. Foster. 2004. Effects of shade from multiple kelp canopies on an understory algal assemblage. Mar. Ecol. Prog. Ser. 267: 107–119.
Clarke, A. and N.M. Johnston. 2003. Antarctic marine benthic diversity. Oceanogr. Mar. Biol. Annu. Rev. 41: 55–57.
Cobb, W.R. 1969. Penetration of calcium carbonate substrates by the boring sponge, *Cliona*. Am. Zool. 9: 783–790.
Cognie, B. and L. Barillé. 1999. Does bivalve mucus favor the growth of their main food source, microalgae? Oceanol. Acta 22: 441–450.
Cole, A.J. and M.S. Pratchett. 2011. Effects of juvenile coral-feeding butterflyfishes on host corals. Coral Reefs 30: 623–630.

Coleman, F.C., C.C. Koenig, G.R. Huntsman, J.A. Musick, A.M. Eklund, J.C. McGovern, G.R. Sedberry, R.W. Chapman and C.B. Grimes. 2000. Long-lived reef fishes: the grouper-snapper complex. Fisheries 25: 14–21.

Colin, P.L., D.M. Devaney, L. Hillis-Colinvaux, T.H. Suchanek and J.T. Harrison. 1986. Geology and biological zonation of the reef slope, 50–360 m depth at Enewetak Atoll, Marshall Islands. Bull. Mar. Sci. 38: 111–128.

Conlan, K.E. and J.R. Chess. 1992. Phylogeny and ecology of a kelp-boring amphipod, *Peramphithoe stypotrupetes*, new species (Corophioidea: Amphithoidae). J. Crust. Biol. 12: 410–422.

Connell, J.H. 1961a. The influence of interspecific competition and other factors on the distribution of the barnacle *Chthamalus stellatus*. Ecology 42: 710–723.

Connell, J.H. 1961b. Effects of competition, predation by *Thais lapillus*, and other factors on natural populations of the barnacle *Balanus balanoides*. Ecol. Monogr. 31: 61–104.

Connell, J.H. 1978. Diversity in tropical rain forests and coral reefs. Science 199: 1302–1310.

Connell, S.D., M.S. Foster and L. Airoldi. 2014. What are algal turfs? Towards a better description of turfs. Mar. Ecol. Prog. Ser. 495: 299–307.

Coombes, M.A., L.A. Naylor, R.C. Thompson, S.D. Roast, L. Gómez-Pujol and R.J. Fairhurst. 2011. Colonization and weathering of engineering materials by marine microorganisms: an SEM study. Earth Surf. Proc. Land. 36: 582–593.

Coombes, M.A., A. Feal-Perez, L.A. Naylor and K. Wilhelm. 2013a. A non-destructive tool for detecting changes in the hardness of engineering materials: application of the Equotip durometer in the coastal zone. Eng. Geol. 167: 14–19.

Coombes, M.A., L.A. Naylor, H.A. Viles and R.C. Thompson. 2013b. Bioprotection and disturbance: seaweed, microclimatic stability and conditions for mechanical weathering in the intertidal zone. Geomorphology 202: 4–14.

Coombes, M.A., E.C. La Marca, L.A. Naylor and R.C. Thompson. 2015. Getting into the groove: opportunities to enhance the ecological value of hard coastal infrastructure using fine-scale surface textures. Ecol. Eng. 77: 314–323.

Coombes, M.A., H.A. Viles, L.A. Naylor and E.C. La Marca. 2017. Cool barnacles: Do common biogenic structures enhance or retard rates of deterioration of intertidal rocks and concrete? Sci. Total Environ. 580: 1034–1045.

Cooper, L.W. and C.P. McRoy. 1988. Anatomical adaptations to rocky substrates and surf exposure by the seagrass genus *Phyllospadix*. Aquat. Bot. 32: 365–381.

Copertino, M., S.D. Connell and A. Cheshire. 2005. The prevalence and production of turf-forming algae on a temperate subtidal coast. Phycologia 44: 241–248.

Coppejans, E., H. Beeckman and M. De Wit. 1992. The seagrass and associated macroalgal vegetation of Gazi Bay (Kenya). Hydrobiologia 247: 59–75.

Coull, B.C. and J.B.J. Wells. 1983. Refuges from fish predation: experiments with phytal meiofauna from the New Zealand rocky intertidal. Ecology 64: 1599–1609.

Craig, P., A. Green and F. Tuilagi. 2008. Subsistence harvest of coral reef resources in the outer islands of American Samoa: modern, historic and prehistoric catches. Fish. Res. 89: 230–240.

Crooks, J.A., A.L. Chang and G.M. Ruiz. 2016. Decoupling the response of an estuarine shrimp to architectural components of habitat structure. PeerJ 4: e2244.

Crossland, C.J., B.G. Hatcher and S.V. Smith. 1991. Role of coral reefs in global ocean production. Coral Reefs 10: 55–64

Cubit, J.D. 1984. Herbivory and the seasonal abundance of algae on a high intertidal rocky shore. Ecology 65: 1904–1917.

Dafforn, K.A., T.M. Glasby, L. Airoldi, N.K. Rivero, M. Mayer-Pinto and E.L. Johnston. 2015. Marine urbanization: an ecological framework for designing multifunctional artificial structures. Front. Ecol. Environ. 13: 82–90.

Dahlgren, E.J. 1989. Gorgonian community structure and reef zonation patterns on Yucatan coral reefs. Bull. Mar. Sci. 45: 678–696.

Davenport, J., T.V. Moloney and J. Kelly. 2011. Common sea anemones *Actinia equina* are predominantly sessile intertidal scavengers. Mar. Ecol. Prog. Ser. 430: 147–155.

Dayton, P.K. 1971. Competition, disturbance, and community organization: the provision and subsequent utilization of space in a rocky intertidal community. Ecol. Monogr. 1: 351–389.

Dayton, P.K. 1972. Toward an understanding of community resilience and the potential effects of enrichments to the benthos at McMurdo Sound, Antarctica. pp. 81–96. *In*: Parker, B.C. [ed.]. Proceedings of the Colloquium on Conservation Problems in Antarctica. Allen Press, Lawrence, KS, USA.

Dayton, P.K. 1985. Ecology of kelp communities. Annu. Rev. Ecol. Syst. 16: 215–245.

de Goeij, J.M. and F.C. Van Duyl. 2007. Coral cavities are sinks of dissolved organic carbon (DOC). Limnol. Oceanogr. 52: 2608–2617.

de Goeij, J.M., D. van Oevelen, M.J.A. Vermeij, R. Osinga, J.J. Middelburg, A.F.P.M. de Goeij and W. Admiraal. 2013. Surviving in a marine desert: the sponge loop retains resources within coral reefs. Science 342: 108–110.

Dendrinos, P., A. Karamanlidis, S. Kotomatas, A. Legakis, E. Tounta and J. Matthiopoulos, 2007. Pupping habitat use in the Mediterranean monk seal: a long-term study. Mar. Mammal Sci. 23: 615–628.

Denny, M. and D. Wethey. 2001. Physical processes that generate patterns in marine communities. pp. 3–37. *In*: Bertness, M.D., S.D. Gaines and M.E. Hay [eds.]. Marine Community Ecology, Sinauer Associates, Sunderland, MA, USA.

Depczynski, M. and D.R. Bellwood. 2003. The role of cryptobenthic reef fishes in coral reef trophodynamics. Mar. Ecol. Prog. Ser. 256: 183–191.

Dethier, M.N. 1984. Disturbance and recovery in intertidal pools: maintenance of mosaic patterns. Ecol. Monogr. 54: 99–118.

Dial, R. and J. Roughgarden. 1998. Theory of marine communities: the intermediate disturbance hypothesis. Ecology 79: 1412–1424.
Dinesen, Z.D. 1983. Shade-dwelling corals of the Great Barrier Reef. Mar. Ecol. Prog. Ser. 10: 173–185.
Distel, D.L., A.R. Baco, E. Chuang, W. Morrill, C. Cavanaugh and C.R. Smith. 2000. Do mussels take wooden steps to deep-sea vents? Nature 403: 725–726.
Doty, M.S. 1946. Critical tide factors that are correlated with the vertical distribution of marine algae and other organisms along the Pacific coast. Ecology 27: 315–328.
Duarte, C.M. and J. Cebrián. 1996. The fate of marine autotrophic production. Limnol. Oceanogr. 41: 1758–1766.
Ellis, R.D. 2019. Red grouper (*Epinephelus morio*) shape faunal communities via multiple ecological pathways. Diversity 11: 89.
Ellison, A.M., E.J. Farnsworth and R.R. Twilley. 1996. Facultative mutualism between red mangroves and root-fouling sponges in Belizean mangal. Ecology 77: 2431–2444.
Ellison, A.M. 2019. Foundation species, non-trophic interactions, and the value of being common. iScience 13: 254.
Emery, K.O. and G.G. Kuhn. 1982. Sea cliffs: their processes, profiles, and classification. Geol. Soc. Am. Bull. 93: 644–654.
England, P.R., J. Phillips, J.R. Waring, G. Symonds and R. Babcock. 2008. Modelling wave-induced disturbance in highly biodiverse marine macroalgal communities: support for the intermediate disturbance hypothesis. Mar. Freshwat. Res. 59: 515–520.
Estes, J.A. and D.O. Duggins. 1995. Sea otters and kelp forests in Alaska: generality and variation in a community ecological paradigm. Ecol. Monogr. 65: 75–100.
Evans, A.J., L.B. Firth, S.J. Hawkins, A.E. Hall, J.E. Ironside, R.C. Thompson and P.J. Moore. 2019. From ocean sprawl to blue-green infrastructure—A UK perspective on an issue of global significance. Environ. Sci. Pol. 91: 60–69.
Evans, R.G. 1948. The lethal temperatures of some common British littoral molluscs. J. Anim. Ecol. 17: 165–173.
Everett, R.A. and G.M. Ruiz. 1993. Coarse woody debris as a refuge from predation in aquatic communities. Oecologia 93: 475–486.
Fabricius, K.E. and M. Dommisse. 2000. Depletion of suspended particulate matter over coastal reef communities dominated by zooxanthellate soft corals. Mar. Ecol. Prog. Ser. 196: 157–167.
Fariña, J.M., M. Sepulveda, M.V. Reyna, P.K. Wallem and P.G. Ossa-Zazzali. 2008. Geographical variation in the use of intertidal rocky shores by the lizard *Microlophus atacamensis* in relation to changes in terrestrial productivity along the Atacama Desert coast. J. Anim. Ecol. 77: 458–468.
Filbee-Dexter, K. and R.E. Scheibling. 2014. Sea urchin barrens as alternative stable states of collapsed kelp ecosystems. Mar. Ecol. Prog. Ser. 495: 1–25.
Filbee-Dexter, K. 2020. Ocean forests hold unique solutions to our current environmental crisis. One Earth 2: 398–401.
Firth, L.B., R.C. Thompson, F.J. White, M. Schofield, M.W. Skov, S.P.G. Hoggart, J. Jackson, A.M. Knights and S.J. Hawkins. 2013. The importance of water-retaining features for biodiversity on artificial intertidal coastal defence structures. Divers. Distrib. 19: 1275–1283.
Firth, L.B., R.C. Thompson, K. Bohn, M. Abbiati, L. Airoldi, T.J. Bouma, F. Bozzeda, V.U. Ceccherelli, M.A. Colangelo, A. Evans, F. Ferrario, M.E. Hanley, H. Hinz, S.P.G. Hoggart, J.E. Jackson, P. Moore, E.H. Morgan, S. Perkol-Finkel, M.W. Skov, E.M. Strain, J. van Belzen and S.J. Hawkins. 2014. Between a rock and a hard place: environmental and engineering considerations when designing coastal defence structures. Coast. Eng. 87: 122–135.
Firth, L.B., A.M. Knights, D. Bridger, A.J. Evans, N. Mieszkowska, P.J. Moore, N.E. O'Connor, E.V. Sheehan, R.C. Thompson and S.J. Hawkins. 2016. Ocean sprawl: challenges and opportunities for biodiversity management in a changing world. Oceanogr. Mar. Biol. Annu. Rev. 54: 193–269.
Fletcher, R.L. and M.E. Callow. 1992. The settlement, attachment and establishment of marine algal spores. Brit. Phycol. J. 27: 303–329.
Fornós, J.J., G.X. Pons, L. Gómez-Pujol and P. Balaguer. 2006. The role of biological processes and rates of downwearing due to grazing organisms on Mallorca carbonate coasts (western Mediterranean). Z. Geomorphol. 144: 161–181.
Foster, M.S. 2001. Rhodoliths: between rocks and soft places. J. Phycol. 7: 659–657.
Foster, M.S., L.M. McConnico, L. Lundsten, T. Wadsworth, T. Kimball, L.B. Brooks, M. Medina-López, R. Riosmena-Rodríguez, G. Hernández-Carmona, R.M. Vásquez-Elizondo, S. Johnson and D.L. Steller. 2007. Diversity and natural history of a *Lithothamnion muelleri-Sargassum horridum* community in the Gulf of California. Cienc. Mar. 33: 367–384.
Foster, M.S., G.M. Amado-Filho, N.A. Kamenos, R. Riosmena-Rodríguez and D.L. Steller. 2013. Rhodoliths and rhodolith beds. pp. 143–155. *In*: Lange, M. [ed.]. Contribution of SCUBA Diving to Research and Discovery in Marine Environments, Smithsonian Contribution to the Marine Sciences No. 39. Smithsonian Institution Scholarly Press, Washington, DC, USA.
Freiwald, A. and J.M. Roberts. 2005. Cold-water Corals and Ecosystems. Springer. Heidelberg, Germany.
Frye, T.C., G.B. Rigg and W.C. Crandall. 1915. The size of kelps on the Pacific coast of North America. Bot. Gaz. 60: 473–482.
Fuji, A., H. Watanabe, K. Ogura, T. Noda and S. Goshima. 1991. Abundance and productivity of microphytobenthos on a rocky shore in southern Hokkaido. Bull. Fac. Fish. Hokkaido Univ. 42: 136–146.

Gabara, S.S. 2020. Trophic structure and potential carbon and nitrogen flow of a rhodolith bed at Santa Catalina Island inferred from stable isotopes. Mar. Biol. 167: 30.
Gaines, S. and J. Roughgarden. 1985. Larval settlement rate: a leading determinant of structure in an ecological community of the marine intertidal zone. Proc. Natl. Acad. Sci. USA 82: 3707–3711.
Gallagher, C. and C. Doropoulos. 2017. Spatial refugia mediate juvenile coral survival during coral–predator interactions. Coral Reefs 36: 51–61.
Gardner, J.P.A. and M.L.H. Thomas. 1987. Growth, mortality and production of organic matter by a rocky intertidal population of *Mytilus edulis* in the Quoddy Region of the Bay of Fundy. Mar. Ecol. Prog. Ser. 39: 31–36.
Garrity, S.D. 1984. Some adaptations of gastropods to physical stress on a tropical rocky shore. Ecology 65: 559–574.
Gaylord, B., J.H. Rosman, D.C. Reed, J.R. Koseff, J. Fram, S. MacIntyre, K. Arkema, C. McDonald, M.A. Brzezinski, J.L. Largier, S.G. Monismith, P.T. Raimondi and B. Mardian. 2007. Spatial patterns of flow and their modification within and around a giant kelp forest. Limnol. Oceanogr. 52: 1838–1852.
Geffen, Y. and E. Rosenberg. 2005. Stress-induced rapid release of antibacterials by scleractinian corals. Mar. Biol. 146: 931–935.
Genin, A., S.G. Monismith, M.A, Reidenbach, G. Yahel and J.R. Koseff. 2009. Intense benthic grazing of phytoplankton in a coral reef. Limnol. Oceanogr. 54: 938–951.
Gibbs, R.E. 1902. *Phyllospadix* as a beach-builder. Am. Nat. 36: 101–109.
Gili, J.M. and R. Coma. 1998. Benthic suspension feeders: their paramount role in littoral marine food webs. Trends Ecol. Evol. 13: 316–321.
Glynn, P.W. 1973. Ecology of a Caribbean coral reef. The *Porites* reef-flat biotope: Part II. Plankton community with evidence for depletion. Mar. Biol. 22: 1–21.
Golléty, C., F. Gentil and D. Davoult. 2008. Secondary production, calcification and CO_2 fluxes in the cirripedes *Chthamalus montagui* and *Elminius modestus*. Oecologia 155: 133–142.
Golléty, C. and T.P. Crowe. 2013. Contribution of biofilm to ecosystem functioning in rock pools with different macroalgal assemblages. Mar. Ecol. Prog. Ser. 482: 69–79.
Gonor, J.J., J.R. Sedell and P.A. Benner. 1988. What we know about large trees in estuaries, in the sea, and on coastal beaches. pp. 83–112. *In*: Maser, C., R.F. Tarrant, J.M. Trappe and J.F. Franklin [eds.]. From the Forest to the Sea, a Story of Fallen Trees. USDA Forest Service General Technical Report, GTR-PNW-229. Pacific Northwest Research Station, Portland, OR, USA.
Goreau, T.F. 1959. The ecology of Jamaican coral reefs I. Species composition and zonation. Ecology 40: 67–90.
Gosselin, L.A. and F.S. Chia. 1995. Distribution and dispersal of early juvenile snails: effectiveness of intertidal microhabitats as refuges and food sources. Mar. Ecol. Prog. Ser. 128: 213–223.
Gowell, M., M.A. Coombes and H.A. Viles. 2015. Rock-protecting seaweed? Experimental evidence of bioprotection in the intertidal zone. Earth Surf. Process. Landf. 40: 1364–1370.
Grall, J., F. Le Loc'h, B. Guyonnet and P. Riera, 2006. Community structure and food web based on stable isotopes (δ15N and δ13C) analysis of a North Eastern Atlantic maerl bed. J. Exp. Mar. Biol. Ecol. 338: 1–15.
Gutiérrez, J.L. and O.O. Iribarne. 1999. Role of Holocene beds of the stout razor clam *Tagelus plebeius* in structuring present benthic communities. Mar. Ecol. Prog. Ser. 185: 213–228.
Gutiérrez, J.L., C.G. Jones, D.L. Strayer and O.O. Iribarne. 2003. Mollusks as ecosystem engineers: the role of shell production in aquatic habitats. Oikos 101: 79–90.
Gutiérrez, J.L., M.G. Palomo, M. Bagur, L.P. Arribas and S.A. Soria. 2015. Wave action limits crowding in an intertidal mussel. Mar. Ecol. Prog. Ser. 518: 153–163.
Gutiérrez, J.L. and M.G. Palomo. 2016. Increased algal fouling on mussels with barnacle epibionts: a fouling cascade. J. Sea Res. 112: 49–54.
Gutiérrez, J.L., M. Bagur and M.G. Palomo. 2019. Algal epibionts as co-engineers in mussel beds: effects on abiotic conditions and mobile interstitial invertebrates. Diversity 11: 17.
Guzmán, H.M. and J. Cortés. 1989. Coral reef community structure at Caño Island, Pacific Costa Rica. Mar. Ecol. 10: 23–41.
Haas, A.F., M.S. Naumann, U. Struck, C. Mayr, M. el-Zibdah and C. Wild. 2010. Organic matter release by coral reef associated benthic algae in the Northern Red Sea. J. Exp. Mar. Biol. Ecol. 389: 53–60.
Hacker, S.D. and R.S. Steneck. 1990. Habitat architecture and the abundance and body-size-dependent habitat selection of a phytal amphipod. Ecology 71: 2269–2285.
Hacker, S.D. and S.D. Gaines. 1997. Some implications of direct positive interactions for community species diversity. Ecology 78: 1990–2003.
Hacker, S.D., B.A. Menge, K.J. Nielsen, F. Chan and T.C. Gouhier. 2019. Regional processes are stronger determinants of rocky intertidal community dynamics than local biotic interactions. Ecology 100: e02763.
Haeckel, E. 1891. Plankton-Studien. Jena. Zschr. Naturwiss. 25: 232–336.
Hall, A.E., R.J.H. Herbert, J.R. Britton and S.L. Hull. 2018. Ecological enhancement techniques to improve habitat heterogeneity on coastal defence structures. Est. Coast. Shelf Sci. 210: 68–78.

Hammar-Klose, E.S. and E.R. Thieler. 2001. Coastal vulnerability to sea-level rise: A preliminary database for the U.S. Atlantic, Pacific and Gulf of Mexico coasts, Vol. 68. US Geological Survey Digital Data Service, Woods Hole, MA, USA.

Hanelt, D., C. Wiencke and K. Bischof. 2003. Photosynthesis in marine macroalgae. pp. 413–435. *In*: Larkum, A.W.D., S.E. Douglas and J.A. Raven [eds.]. Photosynthesis in Algae. Advances in Photosynthesis and Respiration, vol. 14. Kluwer Academic Publisher, Dordretch, The Netherlands.

Hatcher, B.G. 1988. Coral reef primary productivity: a beggar's banquet. Trends Ecol. Evol. 3: 106–111.

Haven, D.S. and R. Morales-Alamo. 1972. Biodeposition as a factor in sedimentation of fine suspended solids in estuaries. Geol. Soc. Am. Memoir 133: 121–130.

Hawkins, C.M. and J.B. Lewis. 1982. Ecological energetics of the tropical sea urchin *Diadema antillarum* Philippi in Barbados, West Indies. Estuar. Coast. Shelf Sci. 15: 645–669.

Hay, M.E. 1981a. The functional morphology of turf-forming seaweeds: persistence in stressful marine habitats. Ecology 62: 739–750

Hay. M.E. 1981b. Spatial patterns of grazing intensity on a Caribbean barrier reef: herbivory and algal distribution. Aquat. Bot. 11: 97–109.

Hayduk, J., S.D. Hacker, J.S. Henderson and F. Tomas Nash. 2019. Evidence for regional scale controls on eelgrass (*Zostera marina*) and mesograzer community structure in upwelling-influenced estuaries. Limnol. Oceanogr. 64: 1120–1134.

Helmuth, B. 1998. Intertidal mussel microclimates: predicting the body temperature of a sessile invertebrate. Ecol. Monogr. 68: 51–74.

Helmuth, B. and G.E. Hofmann. 2001. Thermal heterogeneity and physiological gradients of stress in the rocky intertidal zone. Biol. Bull. 201: 374–384.

Hentschel, B.T. and J. Shimeta. 2008. Suspension feeders. pp. 3437–3442. *In*: Jørgensen, S.E. and B. Fath [eds.]. Encyclopedia of Ecology. Elsevier, London, UK.

Herkül, K., R. Aps, K. Lokko, A. Peterson and H. Tõnisson. 2018. Relating coastal geomorphology to marine benthic biodiversity. J. Coast. Res. 85: 366–370.

Hubbell, S.P. 2001. The Unified Neutral Theory of Biodiversity and Biogeography. Princeton University Press, Princeton, NJ, USA.

Huettel, M., C. Wild and S. Gonelli. 2006. The mucus trap in coral reefs: formation and temporal evolution of particle aggregates caused by coral mucus. Mar. Ecol. Prog. Ser. 307: 60–84.

Hughes, T.P., D.R. Bellwood and S.R. Connolly. 2002. Biodiversity hotspots, centres of endemicity, and the conservation of coral reefs. Ecol. Lett. 5: 775–784.

Hughes, T.P., M.J. Rodrigues, D.R. Bellwood, D. Ceccarelli, O. Hoegh-Guldberg, L. McCook, N. Moltschaniwskyj, M.S. Pratchett, R.S. Steneck and B. Willis. 2007. Phase shifts, herbivory, and the resilience of coral reefs to climate change. Current Biol. 17: 360–365.

Hurlbut, C.J. 1991. Larval substratum selection and postsettlement mortality as determinants of the distribution of two bryozoans. J. Exp. Mar. Biol. Ecol. 147: 103–119.

Huston, M. 1979. A general hypothesis of species diversity. Am. Nat. 113: 81–101.

Johnston, C.S. 1969. The ecological distribution and primary production of macrophytic marine algae in the eastern Canaries. Int. Revue Ges. Hydrobiol. 54: 473–490.

Jones, C.G., J.H. Lawton and M. Shachak. 1994. Organisms as ecosystem engineers. Oikos 69: 373–386.

Jones, C.G. and J.L. Gutiérrez. 2007. On the purpose, meaning, and usage of the ecosystem engineering concept. pp. 3–24. *In*: Cuddington, K., J.E. Byers, W.G. Wilson and A. Hastings [eds.]. Ecosystem Engineers: Plants to Protists. Academic Press, New York, NY, USA.

Judge, J. and J.P. Barry. 2016. Macroinvertebrate community assembly on deep-sea wood falls in Monterey Bay is strongly influenced by wood type. Ecology 97: 3031–3043.

Jurgens, L.J. and B. Gaylord. 2017. Physical effects of habitat-forming species override latitudinal trends in temperature. Ecol. Lett. 21: 190–196.

Kaiser, M.J. and P.G. Moore. 1999. Obligate marine scavengers: do they exist? J. Nat. Hist. 33: 475–481.

Kamenos, N.A., P.G. Moore and J.M. Hall-Spencer. 2004. Nursery-area function of maerl grounds for juvenile queen scallops *Aequipecten opercularis* and other invertebrates. Mar. Ecol. Prog. Ser. 274: 183–189.

Karczmarski, L., V.G. Cockcroft and A. McLachlan. 2000. Habitat use and preferences of Indo-Pacific humpback dolphins *Sousa chinensis* in Algoa Bay, South Africa. Mar. Mammal Sci. 16: 65–79.

Kautsky, N. and S. Evans. 1987. Role of biodeposition by *Mytilus edulis* in the circulation of matter and nutrients in a Baltic coastal ecosystem. Mar. Ecol. Prog. Ser. 38: 201–212.

Kayal, M., J. Vercelloni., T.L. De Loma, P. Bosserelle, Y. Chancerelle, S. Geoffroy, C. Stievenart, F. Michonneau, L. Penin, S. Planes and M. Adjeroud. 2012. Predator crown-of-thorns starfish (*Acanthaster planci*) outbreak, mass mortality of corals, and cascading effects on reef fish and benthic communities. PloS One 7: e47363.

Keable, S.J. 1995. Structure of the marine invertebrate scavenging guild of a tropical reef ecosystem: field studies at Lizard Island, Queensland, Australia. J. Nat. Hist. 29: 27–45.

Kennedy, D.M., W.J. Stephenson and L.A. Naylor. 2014a. Rock Coast Geomorphology: a Global Synthesis. Geological Society Memoires, 40, London, UK.
Kennedy, D.M., W.J. Stephenson and L.A. Naylor. 2014b. Introduction to the rock coasts of the world. pp. 1–5. *In*: Kennedy, D.M., W.J. Stephenson and L.A. Naylor [eds.]. Rock Coast Geomorphology: A Global Synthesis. Geological Society Memoires, 40, London, UK.
Kent, F.E., K.S. Last, D.B. Harries and W.G. Sanderson. 2017. *In situ* biodeposition measurements on a *Modiolus modiolus* (horse mussel) reef provide insights into ecosystem services. Estuar. Coast. Shelf Sci. 184: 151–157.
Kidwell, S.M. 1986. Taphonomic feedback in Miocene assemblages: testing the role of dead hardparts in benthic communities. Palaios 1: 239–255.
Kiirikki, M. 1996. Experimental evidence that *Fucus vesiculosus* (Phaeophyta) controls filamentous algae by means of the whiplash effect. Eur. J. Phycol. 31: 61–66.
Kinsey, D.W. and D. Hopley. 1991. The significance of coral reefs as global carbon sink—response to greenhouse. Palaeogeogr, Palaeoclim. Palaeoecol. 3: 363–377.
Kleemann, K. 1996. Biocorrosion by bivalves. Mar. Ecol. 17: 145–158.
Klumpp, D.W., A.D. McKinnon and C.N. Mundy. 1988. Motile cryptofauna of a coral reef: abundance, distribution and trophic potential. Mar. Ecol. Prog. Ser. 45: 95–108.
Knowlton, N. 2001. The future of coral reefs. Proc. Natl. Acad. Sci. USA 98: 5419–5425.
Kraufvelin, P., A. Lindholm, M.F. Pedersen, L.A. Kirkerud and E. Bonsdorff. 2010. Biomass, diversity and production of rocky shore macroalgae at two nutrient enrichment and wave action levels. Mar. Biol. 157: 29–47.
Krause-Jensen, D. and C.M. Duarte. 2016. Substantial role of macroalgae in marine carbon sequestration. Nat. Geosci. 9: 737–742.
Krug, P.J. 2006. Defense of benthic invertebrates against surface colonization by larvae, a chemical arms race. pp. 1–53. *In*: Fusetani, N. and A.S. Clare [eds.]. Antifouling Compounds. Springer-Verlag, Berlin, Germany.
Krumhansl, K.A. and R.E. Scheibling. 2012. Production and fate of kelp detritus. Mar. Ecol. Prog. Ser. 467: 281–302.
Kuo, J. and C. den Hartog. 2006. Seagrass morphology, anatomy, and ultrastructure. pp. 51–87. *In*: Larkum, A.W., R.J. Orth and C.M. Duarte [eds.]. Seagrasses: Biology, Ecology and Conservation. Springer, Dordretch, The Netherlands.
Kurle, C.M., D.A. Croll and B.R. Tershy. 2008. Introduced rats indirectly change marine rocky intertidal communities from algae- to invertebrate-dominated. Proc. Natl. Acad. Sci. 105: 3800–3804.
Kyle, R., B. Pearson, P.J. Fielding, W.D. Robertson and S.L. Birnie. 1997. Subsistence shellfish harvesting in the Maputaland Marine Reserve in northern KwaZulu-Natal, South Africa: rocky shore organisms. Biol. Conserv. 82: 183–192.
Lawton, J.H. 2000. Community Ecology in a Changing World. Ecology Institute, Oldendorf/Luhe, Germany.
Leonard, G.H. 1999. Positive and negative effects of intertidal algal canopies on recruitment and survival of barnacles. Mar. Ecol. Prog. Ser. 178: 241–249.
Lewis, S.M. 1985. Herbivory on coral reefs: algal susceptibility to herbivorous fishes. Oecologia 65: 370–375.
Lim, M. 2014. The rock coast of the British Isles: cliffs. pp. 19–38. *In*: Kennedy, D.M., W.J. Stephenson and L.A. Naylor [eds.]. Rock Coast Geomorphology: A Global Synthesis. Geological Society Memoires, 40, London, UK.
Little, C. and E.C.S. Little. 2000. The Biology of Soft Shores and Estuaries. Oxford University Press, New York, NY, USA.
Littler, M.M. 1973. The productivity of Hawaiian fringing-reef crustose corallinaceae and an experimental evaluation of production methodology. Limnol. Oceanogr. 18: 946–952.
Littler, M.M. and S.N. Murray. 1974. The primary productivity of marine macrophytes from a rocky intertidal community. Mar. Biol. 27: 131–135.
Littler, M.M. and M.S. Doty. 1975. Ecological components structuring the seaward edges of tropical Pacific reefs: the distribution, communities and productivity of *Porolithon*. J. Ecol. 63: 117–129.
Littler, M.M., D.S. Littler, S.M. Blair and J.M. Norris. 1986. Deep water plant communities from an uncharted seamount off San Salvador island, Bahamas: distribution, abundance and primary productivity. Deep Sea Res. 33: 881–892.
Maldonado, M., R. Aguilar, R.J. Bannister, J.J. Bell, K.W. Conway, P.K. Dayton, C. Díaz, J. Gutt, M. Kelly, E.L.R. Kenchington, S.P. Leys, S.A. Pomponi, H.T. Rapp, K. Rutzler, O.S. Tendal, J. Vacelet and C.M. Young. 2016. Sponge grounds as key marine habitats: a synthetic review of types, structure, functional roles, and conservation concerns. pp. 145–183. *In*: Rossi, S., L. Bramanti, A. Gori and C. Orejas [eds.]. Marine Animal Forests: The Ecology of Benthic Biodiversity Hotspots. Springer, Cham, Switzerland.
Mann, K.H. 1972. Ecological energetics of the sea-weed zone in a marine bay on the Atlantic coast of Canada. II. Productivity of the seaweeds. Mar. Biol. 14: 199–209.
Manzello, D.P., I.C. Enochs, N. Melo, D.K. Gledhill and E.M. Johns. 2012. Ocean acidification refugia of the Florida reef tract. PLoS ONE 7: e41715.
Marliave, J.B., K.W. Conway, D.M. Gibbs, A. Lamb and C. Gibbs. 2009. Biodiversity and rockfish recruitment in sponge gardens and bioherms of southern British Columbia, Canada. Mar. Biol. 156: 2247–2254.
Martin, S., M.D. Castets and J. Clavier. 2006. Primary production, respiration and calcification of the temperate free-living coralline alga *Lithothamnion corallioides*. Aquat. Bot. 85: 121–128.
Martin, S., J. Clavier, L. Chauvaud and G. Thouzeau. 2007. Community metabolism in temperate maerl beds. I. Carbon and carbonate fluxes. Mar. Ecol. Prog. Ser. 335: 19–29.

Mayer, F.W. and C. Wild. 2010. Coral mucus release and following particle trapping contribute to rapid nutrient recycling in a Northern Red Sea fringing reef. Mar. Freshwat. Res. 61: 1006–1014.
McAllister, D.E. 1988. Environmental, economic and social costs of coral reef destruction in the Philippines. Galaxea 7: 161–178.
McClanahan, T.R. 1995. Fish predators and scavengers of the sea urchin *Echinometra mathaei* in Kenyan coral-reef marine parks. Environ. Biol. Fish. 43: 187–193.
McClanahan, T.R. 2000. Recovery of a coral reef keystone predator, *Balistapus undulatus*, in East African marine parks. Biol. Conserv. 94: 191–198.
McClintock, J.B., R.A. Angus and F.E. McClintock. 2007. Abundance, diversity and fidelity of macroinvertebrates sheltering beneath rocks during tidal emersion in an intertidal cobble field: does the intermediate disturbance hypothesis hold for less exposed shores with smaller rocks? J. Exp. Mar. Biol. Ecol. 352: 351–360.
McCloskey, L.R. and L. Muscatine. 1984. Production and respiration in the Red Sea coral *Stylophora pistillata* as a function of depth. Proc. R. Soc. B 222: 215–230.
McConnico, L.A., G. Hernández-Carmona and R. Riosmena-Rodríguez. 2018. Nutrient production in rhodolith beds: impact of a foundation species and its associates. Mar. Ecol. Prog. Ser. 590: 53–66.
McGuinness, K.A. 1987. Disturbance and organisms on boulders. II. Causes of patterns in diversity and abundance. Oecologia 71: 420–430.
McKenna, J. 2005. Boulder Beaches. pp. 206–208. *In*: Schwartz, M.L. [ed.]. Encyclopedia of Coastal Science. Encyclopedia of Earth Science Series. Springer, Dordrecht, The Netherlands.
McMurray, S.E., J.R. Pawlik and C.M. Finelli. 2017. Demography alters carbon flux for a dominant benthic suspension feeder, the giant barrel sponge, on Conch Reef, Florida Keys. Funct. Ecol. 31: 2188–2198.
McQuaid, C.D. and G.M. Branch. 1985. Trophic structure of rocky intertidal communities: response to wave action and implications for energy flow. Mar. Ecol. Prog. Ser. 22: 153–161.
Mendes, T.C., R.C. Villaça and C.E.L. Ferreira. 2009. Diet and trophic plasticity of an herbivorous blenny *Scartella cristata* of subtropical rocky shores. J. Fish Biol. 75: 1816–1830.
Menge, B.A. 1976. Organization of the New England rocky intertidal community: role of predation, competition and environmental heterogeneity. Ecol. Monogr. 46: 335–393.
Menge, B.A. and J.P. Sutherland. 1976. Species diversity gradients: synthesis of the roles of predation, competition, and temporal heterogeneity. Am. Nat. 110: 351–369.
Menge, B.A., J. Lubchenco and L.R. Ashkenas. 1985. Diversity, heterogeneity and consumer pressure in a tropical rocky intertidal community. Oecologia 65: 394–405.
Menge, B.A. and J.P. Sutherland. 1987. Community regulation: variation in disturbance, competition, and predation in relation to environmental stress and recruitment. Am. Nat. 130: 730–757.
Menge, B.A., E. Berlow, C. Blanchette, S. Navarrete and S. Yamada. 1994. The keystone species concept: variation in interaction strength in a rocky intertidal habitat. Ecol. Monogr. 64: 250–286.
Menge, B.A., B.A. Daley, P.A. Wheeler and P.T. Strub. 1997. Rocky intertidal oceanography: an association between community structure and nearshore phytoplankton concentration. Limnol. Oceanogr. 42: 57–66.
Menge, B.A., J. Lubchenco, M.E.S. Bracken, F. Chan, M.M. Foley, T.L. Freidenburg, SD. Gaines, G. Hudson, C. Krenz, H. Leslie, D.N.L. Menge, R. Russell and M.S. Webster. 2003. Coastal oceanography sets the pace of rocky intertidal community dynamics. Proc. Natl. Acad. Sci. USA 100: 12229–12234.
Menge, B.A., G.W. Allison, C.A. Blanchette, T.M. Farrell, A.M. Olson, T.A. Turner and P. van Tamelen. 2005. Stasis or kinesis? Hidden dynamics of a rocky intertidal macrophyte mosaic revealed by a spatially explicit approach. J. Exp. Mar. Biol. Ecol. 314: 3–39.
Menge, B.A., T.C. Gouhier, S.D. Hacker, F. Chan and K. Nielsen. 2015. Are metaecosystems organized hierarchically? A model and test in rocky intertidal habitats. Ecol. Monogr. 85: 213–233.
Miller, R.J. and K.H. Mann. 1973. Ecological energetics of the seaweed zone in a marine bay on the Atlantic coast of Canada. III. Energy transformations by sea urchins. Mar. Biol. 18: 99–114.
Milne, R. and C. Griffiths. 2014. Invertebrate biodiversity associated with algal turfs on a coral-dominated reef. Mar. Biodiv. 44: 181–188.
Miloslavich, P., J.J. Cruz-Motta, A. Hernández, C. Herrera, E. Klein, F. Barros, G. Bigatti, M. Cárdenas, A. Carranza, A. Flores, P. Gil, J. Gobin, J.L. Gutiérrez, M. Krull, J.F. Lazarus, E. Londoño, T. Lotufo, E. Macaya, E. Mora, S. Navarrete, M.G. Palomo, M. Parragué, F. Pellizzari, R. Rocha, L. Romero, R. Retamales, R. Sepúlveda, M.C. Silva and S. Soria. 2016. Benthic assemblages in South American intertidal rocky shores: biodiversity, services, and threats. pp. 83–137. *In*: Riosmena-Rodríguez, R. [ed.]. Marine Benthos: Biology, Ecosystem Functions and Environmental Impact. Nova Science Publishers, Hauppauge, NY, USA.
Mislan, K.A.S. and D.S. Wethey. 2015. A biophysical basis for patchy mortality during heat waves. Ecology 96: 902–907.
Mitchell, N.D., M.R. Dardeau, W.W. Schroeder and A.C. Benke. 1992. Secondary production of gorgonian corals in the northern Gulf of Mexico. Mar. Ecol. Prog. Ser. 87: 275–281.
Miyamoto, Y. and T. Noda. 2004. Effects of mussels on competitively inferior species: competitive exclusion to facilitation. Mar. Ecol. Prog. Ser. 276: 293–298.

Moberg, F. and C. Folke. 1999. Ecological goods and services of coral reef ecosystems. Ecol. Econ. 29: 215–233.

Mongin, M., M.E. Baird, S. Hadley and A. Lenton. 2016. Optimising reef-scale CO_2 removal by seaweed to buffer ocean acidification. Environ. Res. Lett. 11: 034023.

Monismith, S.G., K.A. Davis, G.G. Shellenbarger, J.L. Hench, N.J. Nidzieko, A.E. Santoro, M.A. Reidenbach, J.H. Rosman, R. Holtzman, C.S. Martens, N.L. Lindquist, M.W. Southwell and A. Genin. 2010. Flow effects on benthic grazing on phytoplankton by a Caribbean reef. Limnol. Oceanogr. 55: 1881–1892.

Montefalcone, M., M. Vacchi, C. Carbone, R. Cabella, C.F. Schiaffino, F.M. Elter, C. Morri, C.N. Bianchi and M. Ferrari. 2016. Seagrass on the rocks: *Posidonia oceanica* settled on shallow-water hard substrata withstands wave stress beyond predictions. Estuar. Coast. Shelf Sci. 180: 114–122.

Moreno, C.A., K.M. Lunecke and M.I. Lépez. 1986. The response of an intertidal *Concholepas concholepas* (Gastropoda) population to protection from man in southern Chile and the effects on benthic sessile assemblages. Oikos 46: 359–364.

Morse, D.H. 1997. Distribution, movement, and activity patterns of an intertidal wolf spider *Pardosa lapidicina* population (Araneae, Lycosidae). J. Arachnol. 25: 1–10.

Morton, B. 2011. Predator-prey-scavenging interactions between *Nucella lapillus*, *Carcinus maenas* and *Eulalia viridis* all exploiting *Mytilus galloprovincialis* on a rocky shore recovering from tributyl-tin (TBT) pollution, J. Nat. Hist. 45: 39–40.

Moses, C.A. 2014. The rock coast of the British Isles: shore platforms. pp. 39–56. *In*: Kennedy, D.M., W.J. Stephenson and L.A. Naylor [eds.]. Rock Coast Geomorphology: A Global Synthesis. Geological Society Memoires, 40, London, UK.

Mottershead, D.N., A. Gorbushina, G. Lucas and J. Weight. 2003. The influence of marine salts, aspect and microbes in the weathering of sandstone in two historic structures. Build. Environ. 38: 1193–1204.

Moulton, O.M. and S.D. Hacker. 2011. Congeneric variation in surfgrasses and ocean conditions influence macroinvertebrate community structure. Mar. Ecol. Prog. Ser. 433: 53–63.

Moura, D., S. Gabriel, S. Gamito, R. Santos, E. Zugasti, L. Naylor, A. Gomes, A.M. Tavares and A.L. Martins. 2012. Integrated assessment of bioerosion, biocover and downwearing rates of carbonate rock shore platforms in southern Portugal. Cont. Shelf Res. 38: 79–88.

Moura, R.L., N.A. Secchin, G.M. Amado-Filho, R.B. Francini-Filho, M.O. Freitas, C.V. Minte-Vera, J.B. Teixeira, F. Lopes-Thompson, G. Fraga-Dutra, P.Y. Gomes-Sumida, A. Zigliatti-Guth, R. Mendes-Lopes and A. Cardoso-Bastos. 2013. Spatial patterns of benthic megahabitats and conservation planning in the Abrolhos Bank. Cont. Shelf Res. 70: 109–117.

Moura, R.L., G.M. Amado-Filho, F.C. Moraes, P.S. Brasileiro, P.S. Salomon, M.M. Mahiques, A.C. Bastos, M.G. Almeida, J.M. Silva Jr., B.F. Araujo, F.P. Brito, T.P. Rangel, B.C.V. Oliveira, R.G Bahia, R.P. Paranhos, R.J.S. Dias, E. Siegle, A.G. Figueiredo Jr., R.C. Pereira, C.V. Leal, E. Hajdu, N.E. Asp, G.B. Gregoracci, S. Neumann-Leitão, P.L. Yager, R.B. Francini-Filho, A. Fróes, M. Campeão, B.S. Silva, A.P.B. Moreira, L. Oliveira, A.C. Soares, L. Araujo, N.L. Oliveira, J.B. Teixeira, R.A.B. Valle, C.C. Thompson, C.E. Rezende and F.L. Thompson. 2016. An extensive reef system at the Amazon River mouth. Sci. Adv. 2: e1501252.

Munari, C., E. Rossetti and M. Mistri. 2013. Shell formation in cultivated bivalves cannot be part of carbon trading systems: a study case with *Mytilus galloprovincialis*. Mar. Environ. Res. 92: 264–267.

Nair, N.B. and M. Saraswathy. 1971. The biology of wood-boring teredinid molluscs. Adv. Mar. Biol. 9: 335–509.

Naumann, M.S., C. Richter, M. el-Zibdah and C. Wild. 2009. Coral mucus as an efficient trap for picoplanktonic cyanobacteria—implications for pelagic-benthic coupling in the reef ecosystem. Mar. Ecol. Prog. Ser. 385: 65–76.

Naumann, M.S., A.F. Haas, U. Struck, C. Mayr, M. el-Zibdah and C. Wild. 2010. Organic matter release by the dominant hermatypic corals of the Northern Red Sea. Coral Reefs 29: 649–659.

Navarrete, S.A. and J.C. Castilla. 1993. Predation by Norway rats in the intertidal zone of central Chile. Mar. Ecol. Prog. Ser. 92: 187–199.

Navarrete, S.A. and E.A. Wieters. 2000. Variation in barnacle recruitment over small scales: larval predation by adults and maintenance of community pattern. J. Exp. Mar. Biol. Ecol. 253: 131–148.

Navarrete, S.A., B.R. Broitman, E.A. Wieters and J.C. Castilla. 2005. Scales of benthic-pelagic coupling and the intensity of species interactions: from recruitment limitation to top-down control. Proc. Natl. Acad. Sci. USA 102: 18046–18051.

Naylor, L.A. and W.J. Stephenson. 2010. On the role of discontinuities in mediating shore platform erosion. Geomorphology 114: 89–100.

Naylor, L.A., W.J. Stephenson and A.S. Trenhaile. 2010. Rock coast geomorphology: recent advances and future research directions. Geomorphology 114: 3–11.

Naylor, L.A., M.A. Coombes and H.A. Viles. 2012. Reconceptualising the role of organisms in the erosion of rock coasts: a new model. Geomorphology 157: 17–30.

Naylor, L.A., M. MacArthur, S. Hampshire, K. Bostock, M.A. Coombes, J.D. Hansom, R. Byrne and T. Folland. 2017. Rock armour for birds and their prey: ecological enhancement of coastal engineering. Marit. Eng. 170: 67–82.

Naylor, L.A., M.A. Coombes, H. Kippen, B. Horton, T. Gardinier, M. Roca Cordell, J. Simm and G.J.C. Underwood. 2018. Developing a business case for greening hard coastal and estuarine infrastructure: preliminary results. pp. 801–814. *In*: Burgess, K. [ed.]. Coasts, Marine Structures and Breakwaters 2017: Realising the Potential, ICE Publishing, London, UK.

Ndhlovu, A., C.D. McQuaid, K. Nicastro, N. Marquet, M. Gektidis, C.J. Monaco and G. Zardi. 2019. Biogeographical patterns of endolithic infestation in an invasive and an indigenous intertidal marine ecosystem engineer. Diversity 11: 75.

Newell, R.C., J.G. Field and C.L. Griffiths. 1982. Energy balance and significance of microorganisms in a kelp bed community. Mar. Ecol. Prog. Ser. 8: 103–113.

Nielsen, T.V. and L.A. Gosselin. 2011. Can a scavenger benefit from environmental stress? Role of salinity stress and abundance of preferred food items in controlling population abundance of the snail *Lirabuccinum dirum*. J. Exp. Mar. Biol. Ecol. 410: 80–86.

Norderhaug, K.L. and H. Christie. 2011. Secondary production in a *Laminaria hyperborea* kelp forest and variation according to wave exposure. Estuar. Coast. Shelf Sci. 95: 135–144.

Nowell, A.R.M. and P.A. Jumars. 1984. Flow environments of aquatic benthos. Annu. Rev. Ecol. Syst. 15: 303–328.

Okuda, T., T. Noda, T. Yamamoto, M. Hori and M. Nakaoka. 2010. Contribution of environmental and spatial processes to rocky intertidal metacommunity structure. Acta Oecol. 36: 413–422.

Ólafsson, E. 2016. Marine Macrophytes as Foundation Species. CRC Press, Boca Raton, FL, USA.

Ojeda, F.P. and B. Santelices. 1984. Invertebrate communities in holdfasts of the kelp *Macrocystic pyrifera* from southern Chile. Mar. Ecol. Prog. Ser. 16: 65–73.

Ojeda, F.P. and J.H. Dearborn. 1991. Feeding ecology of benthic mobile predators: experimental analyses of their influence in rocky subtidal communities of the Gulf of Maine. J. Exp. Mar. Biol. Ecol. 149: 13–44.

Paar, M., A. Voronkov, H. Hop, T. Brey, I. Bartsch, M. Schwanitz, C. Wiencke, B. Lebreton, R. Asmus and H. Asmus. 2016. Temporal shift in biomass and production of macrozoobenthos in the macroalgal belt at Hansneset, Kongsfjorden, after 15 years. Polar Biol. 39: 2065–2076.

Paetzold, A., M. Lee and D.M. Post. 2008. Marine resource flows to terrestrial arthropod predators on a temperate island: the role of subsidies between systems of similar productivity. Oecologia 157: 653–659.

Paine, R.T. 1966. Food web complexity and species diversity. Am. Nat. 100: 65–75.

Paine, R.T. 1969. A note on trophic complexity and community stability. Am. Nat. 103: 65–75.

Paine, R.T. and R.L. Vadas. 1969. The effects of grazing by sea urchins, *Strongylocentrotus* spp., on benthic algal populations. Limnol. Oceanogr. 14: 710–719.

Paine, R.T. 1974. Intertidal community structure: experimental studies on the relationship between a dominant competitor and its principal predator. Oecologia 15: 93–120.

Paine, R.T. 1979. Disaster, catastrophe, and local persistence of the sea palm *Postelsia palmaeformis*. Science 205: 685–687.

Paine, R.T. and S.A. Levin. 1981. Intertidal landscapes: disturbance and the dynamics of pattern. Ecol. Monogr. 51: 145–178.

Pascal, N., M. Allenbach, A. Brathwaite, L. Burke, G. Le Port and E. Clua. 2016. Economic valuation of coral reef ecosystem service of coastal protection: a pragmatic approach. Ecosyst. Serv. 21: 72–80.

Pawlik, J.R. 1998. Coral reef sponges: Do predatory fishes affect their distribution? Limnol. Oceanogr. 43: 1396–1399.

Perea-Blazquez, A., S.K. Davy and J.J. Bell. 2012. Estimates of particulate organic carbon flowing from the pelagic environment to the benthos through sponge assemblages. PloS One 7: e29569.

Pereira-Filho, G.H., G.S. Shintate, M.V. Kitahara, R.L. Moura., G.M. Amado-Filho, R.G. Bahia, F. Moraes, L.M. Neves, C.L.B. Francini, F.Z. Gibran and F.S. Motta. 2019. The southernmost Atlantic coral reef is off the subtropical island of Queimada Grande (24°S), Brazil. Bull. Mar. Sci. 95: 277–287.

Pereira-Filho, G.H., R.B. Francini-Filho, I. Pierozzi Jr, H.T. Pinheiro, A.C. Bastos, R.L. Moura, F.C. Moraes, Z. Matheus, R. Bahia and G.M. Amado-Filho. 2015a. Sponges and fish facilitate succession from rhodolith beds to reefs. Bull. Mar. Sci. 91: 45–46.

Pereira-Filho, G.H., P. de Cerqueira Veras, R.B. Francini-Filho, R.L. Moura, H.T. Pinheiro, F.Z. Gibran, Z. Matheu, L.M. Neves and G.M. Amado-Filho. 2015b. Effects of the sand tilefish *Malacanthus plumieri* on the structure and dynamics of a rhodolith bed in the Fernando de Noronha Archipelago, tropical West Atlantic. Mar. Ecol. Prog. Ser. 541: 65–73.

Perkol-Finkel, S., F. Ferrario, V. Nicotera and L. Airoldi. 2012. Conservation challenges in urban seascapes: promoting the growth of threatened species on coastal infrastructures. J. Appl. Ecol. 49: 1457–1466.

Perkol-Finkel, S., T. Hadray, A. Rella, R. Shirazi and I. Sella. 2018. Seascape architecture—incorporating ecological considerations in design of coastal and marine infrastructure. Ecol. Eng. 120: 645–654.

Pfaff, M.C., C. Hiebenthal, M. Molis, G.M. Branch and M. Wahl. 2010. Patterns of diversity along experimental gradients of disturbance and nutrient supply—the confounding assumptions of the Intermediate Disturbance Hypothesis. Afr. J. Mar. Sci. 32: 127–135.

Pfaller, J.B., M.G. Frick, K.J. Reich, K.L. Williams and K.A. Bjorndal. 2008. Carapace epibionts of loggerhead turtles (*Caretta caretta*) nesting at Canaveral National Seashore, Florida. J. Nat. Hist. 42: 1095–1102.

Pfister, C.A. and M.E. Hay. 1988. Associational plant refuges: convergent patterns in marine and terrestrial communities result from differing mechanisms. Oecologia 77: 118–129.

Pile, A.J., M.R. Patterson and J.D. Witman. 1996. *In situ* grazing on plankton < 10 μm by the boreal sponge *Mycale lingua*. Mar. Ecol. Progr. Ser. 141: 95–102.

Pinn, E.H., R.C. Thompson and S.J. Hawkins. 2008. Piddocks (Mollusca: Bivalvia: Pholadidae) increase topographical complexity and species diversity in the intertidal. Mar. Ecol. Prog. Ser. 355: 173–182.

Pirtle, J.L., G.L. Eckert and A.W. Stoner. 2012. Habitat structure influences the survival and predator prey interactions of early juvenile red king crab *Paralithodes camtschaticus*. Mar. Ecol. Prog. Ser. 465: 169–184.
Pizarro, J.C., C.B. Anderson and R. Rozzi. 2012. Birds as marine–terrestrial linkages in sub-polar archipelagic systems: avian community composition, function and seasonal dynamics in the Cape Horn Biosphere Reserve (54–55°S), Chile. Polar Biol. 35: 39–51.
Poore, A.G.B., A.H. Campbell, R.A. Coleman, G.J. Edgar, V. Jormalainen, P.L. Reynolds, E.E. Sotka, J.J. Stachowicz, R.B. Taylor, M.A. Vanderklift and J.E. Duffy. 2012. Global patterns in the impact of marine herbivores on benthic primary producers. Ecol. Lett. 15: 912–922.
Power, M.E., D. Tilman, J.A. Estes, B.A. Menge, W. Bond, L. Mills, G. Daily, J. Castilla, J. Lubchenco and R. Paine. 1996. Challenges in the quest for keystones: identifying keystone species is difficult but essential to understanding how loss of species will affect ecosystems. BioScience 46: 609–620.
Puk, L.D., S.C. Ferse and C. Wild. 2016. Patterns and trends in coral reef macroalgae browsing: a review of browsing herbivorous fishes of the Indo-Pacific. Rev. Fish Biol. Fish. 26: 53–70.
Reaka-Kudla, M.L. 1997. The global biodiversity of coral reefs: a comparison with rain forests. pp. 83–108. *In*: Reaka-Kudla, M.L., D.E. Wilson and E.O. Wilson [eds.]. Biodiversity II: Understanding and Protecting our Biological Resources. Joseph Henry Press, Washington, DC, USA.
Reidenbach, M.A., J.R. Koseff and M.A.R. Koehl. 2009. Hydrodynamic forces on larvae affect their settlement on coral reefs in turbulent, wave-driven flow. Limnol. Oceanogr. 54: 318–330.
Ribes, M., R. Coma and J.M. Gili. 1999. Heterogeneous feeding in benthic suspension feeders: the natural diet and grazing rate of the temperate gorgonian *Paramuricea clavata* (Cnidaria: Octocorallia) over a year cycle. Mar. Ecol. Prog. Ser. 183: 125–137.
Rimondino, C., L. Torre, R. Sahade and M. Tatian. 2015. Sessile macro-epibiotic community of solitary ascidians, ecosystem engineers in soft substrates of Potter Cove, Antarctica. Polar Res. 34: 24338.
Ripple, W.J., J.A. Estes, R.L. Beschta, C.C. Wilmers, E.G. Ritchie, M. Hebblewhite, J. Berger, B. Elmhagen, M. Letnic, M.P. Nelson, O.J. Schmitz, DW. Smith, A.D. Wallach and A.J. Wirsing. 2014. Status and ecological effects of the world's largest carnivores. Science 343: 1241484.
Rius, M., P.R. Teske, P.H. Manriquez, R. Suarez-Jimenez, C.D. McQuaid and J.C. Castilla. 2017. Ecological dominance along rocky shores, with a focus on intertidal ascidians. Oceanogr. Mar. Biol. Annu. Rev. 55: 2–33.
Roberts, J.M., A.J. Wheeler and A. Freiwald. 2006. Reefs of the deep: the biology and geology of cold-water coral ecosystems. Science 312: 543–547.
Robles, C.D., M.A. Alvarado and R.A. Desharnais. 2001. The shifting balance of littoral predator-prey interaction in regimes of hydrodynamic stress. Oecologia 128: 142–152.
Rodriguez, C. and A.W. Stoner. 1990. The epiphyte community of mangrove roots in a tropical estuary: distribution and biomass. Aquat. Bot. 36: 117–126.
Romano, B. and F. Zullo. 2014. The urban transformation of Italy's Adriatic coastal strip: fifty years of unsustainability. Land Use Policy 38: 26–36.
Rotjan, R.D. and S.M. Lewis. 2008. Impact of coral predators on tropical reefs. Mar. Ecol. Prog. Ser. 367: 73–91.
Roughgarden, J., S.D. Gaines and H. Possingham. 1988. Recruitment dynamics in complex life cycles. Science 241: 1460–1466.
Russ, G.R. and J. St John. 1988. Diets, growth rates and secondary production of herbivorous coral reef fishes. Proc. 6th Int. Coral Reef Symp. 2: 37–43.
Russell, M.P., V.K. Gibbs and E. Duwan. 2018. Bioerosion by pit-forming, temperate-reef sea urchins: history, rates and broader implications. PLoS One 13: e0191278.
Russell, R., S.A. Wood, G. Allison and B.A. Menge. 2006. Scale, environment, and trophic status: the context dependency of community saturation in rocky intertidal communities. Am. Nat. 167: 158–170.
Sale, P.F. 1977. Maintenance of high diversity in coral reef fish communities. Am. Nat. 111: 337–359.
Sallam, W.S. and M.K. Wicksten. 2011. *Hyastenus hilgendorfi* (De Man, 1887) (Brachyura, Pisidae): a heavily encrusted decorator crab inhabiting the Suez Canal. Crustaceana 84: 203–219.
Salman, A., S. Lombardo and P. Doody. 2004. Living with Coastal Erosion in Europe: Sediment and Space for Sustainability. Eurosion. Office for Official Publications of the European Communities. Luxembourg, Luxembourg. http://resolver.tudelft.nl/uuid:483327a3-dcf7-4bd0-a986-21d9c8ec274e.
Samadi, S., L. Corbari, J. Lorion, S. Hourdez, T. Haga, J. Dupont, M.C. Boisselier and B. Richer De Forges. 2010. Biodiversity of deep-sea organisms associated with sunken-wood or other organic remains sampled in the tropical Indo-Pacific. Cah. Biol. Mar. 51: 459–466.
Sammarco, P.W. 1983. Effects of fish grazing and damselfish territoriality on coral reef algae. I. Algal community structure. Mar. Ecol. Prog. Ser. 13: 1–14.
Santos, C.S.G., J.B. Lino, P. de Cerqueira Veras, G.M. Amado-Filho, R.B. Francini-Filho, F.S. Motta, R.L. Moura and G.H. Pereira-Filho. 2016. Environmental licensing on rhodolith beds: insights from a worm. Nat. Conserv. 14: 137–141.
Sauchyn, L.K. and R.E. Scheibling. 2009. Fecal production by sea urchins in native and invaded algal beds. Mar. Ecol. Prog. Ser. 396: 35–48.

Sauchyn, L.K., J.S. Lauzon-Guay and R.E. Scheibling. 2011. Sea urchin fecal production and accumulation in a rocky subtidal ecosystem. Aquat. Biol. 13: 215–223.
Scarff, J.E. 1986. Occurrence of the barnacles *Coronula diadema*, *C. reginae* and *Cetopirus complanatus* (Cirripedia) on right whales. Sci. Rep. Whales Res. Inst. 37: 129–153.
Schaffelke, B. 1999. Particulate organic matter as an alternative nutrient source for tropical *Sargassum* species (Fucales, Phaeophyceae). J. Phycol. 35: 1150–1157.
Schmid, J.R., A.B. Bolten, K.A. Bjorndal, W.J. Lindberg, H.F. Percival and P.D. Zwick. 2003. Home range and habitat use by Kemp's ridley turtles in west-central Florida. J. Wildlife Manag. 67: 196–206.
Schoppe, S. and B. Werding. 1996. The boreholes of the sea urchin genus *Echinometra* (Echinodermata: Echinoidea: Echinometridae) as a microhabitat in tropical South America. PSZN Mar. Ecol. 17: 181–186.
Schwindt, E. and O.O. Iribarne. 2000. Settlement sites, survival and effects on benthos of an introduced reef-building polychaete in a SW Atlantic coastal lagoon. Bull. Mar. Sci. 67: 73–82.
Sciberras, M., M. Rizzo, J.R. Mifsud, K. Camilleri, J.A. Borg, E. Lanfranco and P.J. Schembri. 2009. Habitat structure and biological characteristics of a maerl bed off the northeastern coast of the Maltese Islands (central Mediterranean). Mar. Biodiv. 39: 251–264.
Scott, P.J.B., H.M. Reiswig and B.M. Marcotte. 1988. Ecology, functional morphology, behaviour, and feeding in coral-and sponge-boring species of *Upogebia* (Crustacea: Decapoda: Thalassinidea). Can. J. Zool. 66: 483–495.
Scrosati, R.A., A.S. Knox, N. Valdivia and M. Molis. 2011. Species richness and diversity across rocky intertidal elevation gradients in Helgoland: testing predictions from an environmental stress model. Helgol. Mar. Res. 65: 91–102.
Scyphers, S.B., S.P. Powers, K.L. Heck Jr. and D. Byron. 2011. Oyster reefs as natural breakwaters mitigate shoreline loss and facilitate fisheries. PloS One 6: e22396.
Seapy, R.R. and M.M. Littler. 1982. Population and species diversity fluctuations in a rocky intertidal community relative to severe aerial exposure and sediment burial. Mar. Biol. 71: 87–96.
Shepherd, S.A. and J.A. Turner. 1985. Studies on southern Australian abalone (genus *Haliotis*). VI. Habitat preference, abundance and predators of juveniles. J. Exp. Mar. Biol. Ecol. 93: 285–298.
Shepherd, S.A. and M.W. Hawkes, 2005. Algal food preferences and seasonal foraging strategy of the marine iguana, *Amblyrhynchus cristatus*, on Santa Cruz, Galapagos. Bull. Mar. Sci. 77: 51–72.
Shinen, J.L. and S.A. Navarrete. 2014. Lottery coexistence on rocky shores: weak niche differentiation or equal competitors engaged in neutral dynamics? Am. Nat. 183: 342–362.
Siegfried, W.R., P.A.R. Hockey and G.M. Branch. 1994. The exploitation of intertidal and subtidal biotic resources of rocky shores in Chile and South Africa—an overview. pp. 1–15. *In*: Siegfried, W.R. [ed.]. Rocky Shores: Exploitation in Chile and South Africa. Springer, Berlin, Germany.
Skead, D.M. 1966. Birds frequenting the intertidal zone of the Cape Peninsula. Ostrich 37: 10–16.
Smith, C.R. and A.R. Baco. 2003. Ecology of whale falls at the deep-sea floor. Oceanogr. Mar. Biol. Ann. Rev. 41: 311–354.
Smith, L.W., D. Barshis and C. Birkeland. 2007. Phenotypic plasticity for skeletal growth, density and calcification of *Porites lobata* in response to habitat type. Coral Reefs 26: 559–567.
Smith, S.V. 1978. Coral-reef area and the contribution of reefs to processes and resources of the world's oceans. Nature 273: 225–226.
Sousa, W.P. 1979a. Experimental investigations of disturbance and ecological succession in a rocky intertidal algal community. Ecol. Monogr. 49: 227–254.
Sousa, W.P. 1979b. Disturbance in marine intertidal boulder fields: the nonequilibrium maintenance of species diversity. Ecology 60: 1225–1239.
Sousa, W.P. 2001. Natural disturbance and the dynamics of marine benthic communities. pp. 85–130. *In*: Bertness, M.D., S.D. Gaines and M.E. Hays [eds.]. Marine Community Ecology. Sinauer Associates, Sunderland, MA, USA.
Spalding, M.D. and A.M. Grenfell. 1997. New estimates of global and regional coral reef areas. Coral Reefs 16: 225–230.
Spalding, M.D., C. Ravilious and E.P. Green. 2001. World Atlas of Coral Reefs. Prepared at the UNEP World Conservation Monitoring Centre. University of California Press, Berkeley. CA, USA.
Stål, J., S. Paulsen, L. Pihl, P. Rönnbäck, T. Söderqvist and T. Wennhage. 2008. Coastal habitat support to fish and fisheries in Sweden: Integrating ecosystem functions into fisheries management. Ocean Coast. Manag. 51: 594–600.
Stamski, R.E. and M.E. Field. 2006. Characterization of sediment trapped by macroalgae on a Hawaiian reef flat. Estuar. Coast. Shelf Sci. 66: 211–216.
Steller, D.L., R. Riosmena-Rodríguez, M.S. Foster and C.A. Roberts. 2003. Rhodolith bed diversity in the Gulf of California: the importance of rhodolith structure and consequences of disturbance. Aquat. Conserv. Mar. Freshwat. Ecosyst. 13: S5–S20.
Steneck, R.S. and M.N. Dethier. 1994. A functional group approach to the structure of algal-dominated communities. Oikos 69: 476–498.
Stephenson, T.A. and A. Stephenson. 1949. The universal features of zonation between tide-marks on rocky coasts. J. Ecol. 37: 289–305.
Stephenson, W.J. and R.M. Kirk. 2000. Development of shore platforms on Kaikoura Peninsula, South Island, New Zealand: part II: the role of subaerial weathering. Geomorphology 32: 43–56.

Stewart, R.J., J.C. Weaver, D.E. Morse and J.H. Waite. 2004. The tube cement of *Phragmatopoma californica*: a solid foam. J. Exp. Biol. 207: 4727–4734.
Stillman, J.H. and G.N. Somero. 1996. Adaptation to temperature stress and aerial exposure in congeneric species of intertidal porcelain crabs (genus *Petrolisthes*): correlation of physiology, biochemistry and morphology with vertical distribution. J. Exp. Biol. 199: 1845–1855.
Storry, K.A., C.K. Weldrick, M. Mews, M. Zimmer and D.E. Jelinski. 2006. Intertidal coarse woody debris: a spatial subsidy as shelter or feeding habitat for gastropods? Est. Coast. Shelf Sci. 66: 197–203.
Strain, E.M.A., C. Olabarria, M. Mayer-Pinto, V. Cumbo, R.L. Morris, A.B. Bugnot, K.A. Dafforn, E. Heery, L.B. Firth, P.R. Brooks and M.J. Bishop. 2017. Eco-engineering urban infrastructure for marine and coastal biodiversity: which interventions have the greatest ecological benefit? J. Appl. Ecol. 55: 426–441.
Suchanek, T.H. 1985. Mussels and their role in structuring rocky shore communities. pp. 70–96. *In*: Moore, P.G. and R. Seed [eds.]. The Ecology of Rocky Coasts. Hodder & Stoughton Press, London, UK.
Sunamura, T. 1992. The Geomorphology of Rocky Coasts. John Wiley and Sons, Chichester, UK.
Tanaka, Y., T. Miyajima, I. Koike, T. Hayashibara and H. Ogawa. 2008. Production of dissolved and particulate organic matter by the reef -building corals *Porites cylindrica* and *Acropora pulchra*. Bull. Mar. Sci. 82: 237–245.
Taylor, P.D. and M.A. Wilson. 2003. Palaeoecology and evolution of marine hard substrate communities: Earth Sci. Rev. 62: 1–103.
Taylor, R.B. 1998. Density, biomass and productivity of animals in four subtidal rocky reef habitats: the importance of small mobile invertebrates. Mar. Ecol. Prog. Ser. 172: 37–51.
Teagle, H., S.J. Hawkins, P.J. Moore and D.A. Smale. 2017. The role of kelp species as biogenic habitat formers in coastal marine ecosystems. J. Exp. Mar. Biol. Ecol. 492: 81–98.
Teh, L.S., L.C. Teh and U.R. Sumaila. 2013. A global estimate of the number of coral reef fishers. PLoS One 8: e65397.
Thiel, M. 2003. Reproductive biology of *Limnoria chilensis*: another boring peracarid species with extended parental care. J. Nat. Hist. 37: 1713–1726.
Thompson, R.C., T.P. Crowe and S.J. Hawkins. 2002. Rocky intertidal communities: past environmental changes, present status and predictions for the next 25 years. Environ. Conserv. 29: 168–191.
Thomsen, M.S. and P.M. South. 2019. Communities and attachment networks associated with primary, secondary and alternative foundation species: a case study of stressed and disturbed stands of southern bull kelp. Diversity 11: 56.
Thorson, G. 1964. Light as an ecological factor in the dispersal and settlement of larvae of marine bottom invertebrates. Ophelia 1: 167–208.
Tibbetts, J. 2002. Coastal cities: living on the edge. Environ. Health Perspect. 110: 674–681.
Tokeshi, M. and L. Romero. 1995. Filling a gap: dynamics of space occupancy on a mussel-dominated subtropical rocky shore. Mar. Ecol. Prog. Ser. 119: 167–176.
Tomanek, L. and G.N. Somero. 1999. Evolutionary and acclimation-induced variation in the heat-shock responses of congeneric marine snails (genus *Tegula*) from different thermal habitats: implications for limits of thermotolerance and biogeography. J. Exp. Biol. 202: 2925–2936.
Tomatsuri, M. and K. Kon. 2017. Effects of dead oyster shells as a habitat for the benthic faunal community along rocky shore regions. Hydrobiologia 790: 225–232.
Trenhaile, A.S. 1987. The Geomorphology of Rocky Coasts. Oxford University Press, Oxford, UK.
Trudgill, S. 1987. Bioerosion of intertidal limestone, Co. Clare, Eire – 3: Zonation, process and form. Mar. Geol. 74: 111–121.
Tsuchiya, M. 1980. Biodeposit production by the mussel *Mytilus edulis* L. on rocky shores. J. Exp. Mar. Biol. Ecol. 47: 203–222.
Turner, T. 1985. Stability of rocky intertidal surfgrass beds: persistence, preemption, and recovery. Ecology 66: 83–92.
Underwood, A.J. and E.J. Denley. 1984. Paradigms, explanations and generalizations in models for the structure of intertidal communities on rocky shores. pp. 151–180. *In*: Strong, D.R., D. Simberloff, L.G. Abele and A. Thistle [eds.]. Ecological Communities: Conceptual Issues and the Evidence. Princeton University Press, Princeton, NJ, USA.
Underwood, A.J. and P. Jernakoff. 1984. The effects of tidal height, wave-exposure, seasonality and rock-pools on grazing and the distribution of intertidal macroalgae in New South Wales. J. Exp. Mar. Biol. Ecol. 75: 71–96.
Underwood, A.J. and P.G. Fairweather. 1989. Supply-side ecology and benthic marine assemblages. Trends Ecol. Evol. 4: 16–20.
Valisano, L., F. Notari, M. Mori and C. Cerrano. 2016. Temporal variability of sedimentation rates and mobile fauna inside and outside a gorgonian garden. Mar. Ecol. 37: 1303–1314.
van Beukering, P.J.H. and H.S.J. Cesar. 2004. Ecological economic modeling of coral reefs: evaluating tourist overuse at Hanauma Bay and algae blooms at the Kihei coast, Hawai'i. Pac. Sci. 58: 243–260.
van der Meer, J.W. 1999. Design of concrete armour layers. pp. 213–221. *In*: Balkema, A.A. [ed.]. Proceedings of the International Conference on Coastal Structures '99, Santander, Spain.
van Duyl, F.C., G.J. Gast, W. Steinhoff, S. Kloff, M.J.W. Veldhuis and R.P.M. Bak. 2002. Factors influencing the short-term variation in phytoplankton composition and biomass in coral reef waters. Coral Reefs 21: 293–306.
Vargas-Ángel, B., J.D. Thomas and S.M. Hoke. 2003. High-latitude *Acropora cervicornis* thickets off Fort Lauderdale, Florida, USA. Coral Reefs 22: 465–473.

Wahl, M. 1989. Marine epibiosis. I. Fouling and antifouling, some basic aspects. Mar. Ecol. Prog. Ser. 58: 175–189.
Wahl, M. 2009. Marine Hard Bottom Communities. Springer, Berlin, Germany.
Waite, J.H. 2002. Adhesion à la moule. Integr. Comp. Biol. 42: 1172–1180.
Walters, L.J. and D.S. Wethey. 1996. Settlement and early post-settlement survival of sessile marine invertebrates on topographically complex surfaces: the importance of refuge dimensions and adult morphology. Mar. Ecol. Prog. Ser. 137: 161–171.
Waltham, N.J. and R.M. Connolly. 2011. Global extent and distribution of artificial, residential waterways in estuaries. Estuar. Coast. Shelf Sci. 94: 192–197.
Ward, J.E. and D.J. Kach. 2009. Marine aggregates facilitate ingestion of nanoparticles by suspension-feeding bivalves. Mar. Environ. Res. 68: 137–142.
Ware, J.R., S.V. Smith and M.L. Reaka-Kudla. 1991. Coral reefs: sources or sinks of atmospheric CO_2? Coral Reefs 11: 127–130.
Watt, C.A. and R.A. Scrosati. 2013. Bioengineer effects on understory species richness, diversity, and composition change along an environmental stress gradient: experimental and mensurative evidence. Estuar. Coast. Shelf Sci. 123: 10–18.
Weinbauer, M.G. and B. Velimirov. 1995. Biomass and secondary production of the temperate gorgonian coral *Eunicella cavolini* (Coelenterata: Octocorallia). Mar. Ecol. Prog. Ser. 121: 211–216.
Wessel, P., D.T. Sandwell and S.-S. Kim. 2010. The global seamount census. Oceanography 23: 24–33.
Wild, C., M. Huettel, A. Klueter, S.G. Kremb, M.Y. Rasheed and B.B. Jørgensen. 2004a. Coral mucus functions as an energy carrier and particle trap in the reef ecosystem. Nature 428: 66–70.
Wild, C., M.Y. Rasheed, U. Werner, U. Franke, R. Johnstone and M. Huettel. 2004b. Degradation and mineralization of coral mucus in reef environments. Mar. Ecol. Prog. Ser. 267: 159–171.
Wild, C., M.Y. Rasheed, C. Jantzen, P. Cook, U. Struck, M. Huettel and A. Boetius. 2005a. Benthic metabolism and degradation of natural particulate organic matter in silicate and carbonate sands of the Northern Red Sea. Mar. Ecol. Prog. Ser. 298: 69–78.
Wild, C., H. Woyt and M. Huettel. 2005b. Influence of coral mucus on nutrient fluxes in carbonate sediments. Mar. Ecol. Prog. Ser. 287: 87–98.
Wild, C., C. Laforsch and M. Huettel. 2006. Detection and enumeration of microbial cells within highly porous calcareous reef sands. Mar. Freshwat. Res. 57: 415–420.
Wilkinson, C. 2008. Status of Coral Reefs of the World. Global Coral Reef Monitoring Network and Reef and Rainforest Research Center, Townsville, Australia.
Wilson, W.B. 1950. Reef definition. Am. Assoc. Pet. Geol. Bull. 34: 181–181.
Witman, J.D. 1987. Subtidal coexistence: storms, grazing, mutualism, and the zonation of kelps and mussels. Ecol. Monogr. 57: 167–187.
Witman, J.D., R.J. Etter and F. Smith. 2004. The relationship between regional and local species diversity in marine benthic communities: a global perspective. Proc. Natl. Acad. Sci. USA 101: 15664–15669.
Wootton, J.T. 1995. Effects of birds on sea urchins and algae: a lower intertidal trophic cascade. Ecoscience 2: 321–328.
Wright, J.R. and A.J. Boxshall. 1999. The influence of small-scale flow and chemical cues on the settlement of two congeneric barnacle species. Mar. Ecol. Prog. Ser. 183: 179–187.
Wu, R.S.S. and C.D. Levings. 1979. Energy flow and population dynamics of the barnacle *Balanus glandula*. Mar. Biol. 54: 83–89.
Yager, P.L., A.R.M. Nowell and P.A. Jumars. 1993. Enhanced deposition to pits: a local food source for benthos. J. Mar. Res. 51: 209–236.
Yakovis, E. and A. Artemieva. 2017. Cockles, barnacles and ascidians compose a subtidal facilitation cascade with multiple hierarchical levels of foundation species. Sci. Rep. 7: 237.
Young, A.P. and J.E. Carilli. 2019. Global distribution of coastal cliffs. Earth Surf. Proc. Land. 44: 1309–1316.
Young, C.M. 2009. Communities on deep-sea hard bottoms. pp. 39–60. *In*: Wahl, M. [ed.]. Marine Hard Bottom Communities. Springer, Berlin, Germany.
Zacharias, M.A. and J.C. Roff. 2001. Explanations of patterns of intertidal diversity at regional scales. J. Biogeogr. 28: 471–483.

CHAPTER 11
Fish Ecology

Hannes Baumann

11.1 Introduction

With about 28,000 known species, fishes make up more than half of all known vertebrates (Helfman et al. 2009). Over the course of their long evolutionary history, they radiated in every conceivable aquatic habitat, from the open ocean and deep-sea trenches to shelf seas, estuaries and lakes, to rivers and the smallest streams and ponds. They are found in subzero Antarctic waters, altitudes of over 4,000 m and even acidic desert springs of > 40°C (Moyle and Cech 2004). The fascinating adaptations to these habitats have produced a mind-bending diversity of form and function, a difference in size that spans more than three magnitudes (0.01–18 m), and a profusion of reproductive strategies. Apart from their diversity and unique evolutionary history, fishes are of intense scientific interest for economic reasons because they comprise the nutritional foundation for a large part of humanity (Costanza et al. 1997) and their exploitation over time has led to thriving—and warring—civilizations. Today, the impetus of sustainable fish management at a time of rapid ecological re-organization due to man-made climate change has made the study of fish ecology and fish stock productivity as urgent and important as ever.

The aim of this chapter is to provide a highly abridged introduction to select major topics in fish evolution, distribution, ecology, and fisheries science. To gather a more in-depth understanding, students of ichthyology are encouraged to seek out a number of excellent textbooks on the different subjects, for example *The Diversity of Fishes* (Helfman et al. 2009) or *An Introduction to Ichthyology* (Moyle and Cech 2004). To delve further into ecological aspects, *Ecology of Teleost Fishes* (Wootton 1998) is still the go-to textbook, while students interested in a more detailed introduction to fisheries topics are recommended *Fishery Science—The Unique Contributions of Early Life Stages* (Fuiman and Werner 2009) and *Marine Fisheries Ecology* (Jennings et al. 2009).

11.2 How fishes evolved and are distributed on Earth today

As the first vertebrates on Earth, fishes have a wonderfully meandrous evolutionary history that goes all the way back to the early Paleozoic (> 500 Mya). The lack of fossils shrouds their origin in mystery, but a plausible theory is that they evolved from sessile, filter-feeding chordates similar to today's ascidians (sea squirts). A developmental abnormality known as *paedogenesis*[a] (Pierce and Smith 1979) might have allowed the swimming larvae of those hypothetical ancestors to reproduce before settling, thus skipping the sessile stage and eventually evolving into different, fully mobile organisms (Romer and Parsons 1977, Fig. 11.1A). Segmented, V-shaped tail muscles propelled them through the water, thereby defining 'forward' as an axis of locomotion, establishing the familiar bilateral symmetry and concentrating

University of Connecticut, Department of Marine Sciences, 1080 Shennecossett Rd, 06340 Groton CT, USA.
Email: hannes.baumann@uconn.edu

[a] The phenomenon is also referred to as 'neoteny in a broad sense', which is confusing (Pierce and Smith 1979).

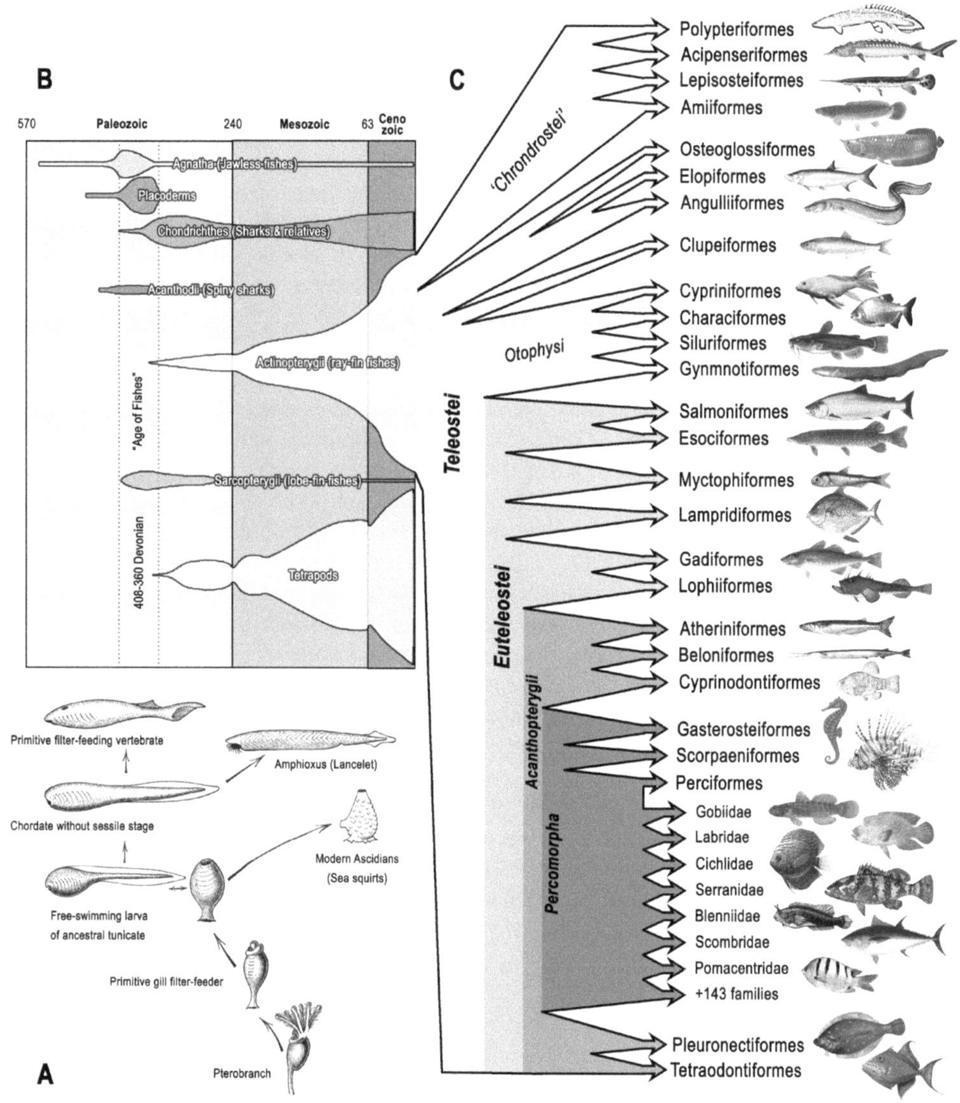

Figure 11.1. Origin, evolution, and systematics of fishes. (A) *Origin hypothesis***.** Early during chordate evolution, sessile arm feeders (pterobranchs) gave rise to gill feeders. In one line, free-swimming filter-feeding larvae lost their sessile stage and evolved into the first, gill-feeding vertebrates (redrawn after Romer and Parsons 1977). (B) *Evolution and relative abundance of major fish lines through time*. Most of today's fish groups originated in the Devonian; ray-finned fishes became the dominant fish group during the Meso- and Cenozoic (numbers refer to million years ago, Mya). (C) *Abridged overview of Actinopterygii systematics* showing select major orders (*-formes*) and Perciform families (*-idae*) sorted top to bottom from ancestral to most derived groups. Most fishes are Teleosts, and within those, most belong to the Euteleosts. Acanthopterygii evolved fin spines; the most species-rich vertebrate order are the Perciformes (after Moyle and Cech 2004).

sensory and processing neurons in the frontal, the 'head' part of their body. These precursors to a central nervous system with a brain grew in size and complexity with each evolutionary step. And because sensitive brains need protection, a cartilaginous and later ossified casing evolved, forming the cranium and vertebral column that define all *Vertebrata,* the subphylum containing fish, amphibians, reptiles, birds and mammals.

A 'fish', then, is simply a water-breathing vertebrate with gills; a broad definition that is needed to encompass a truly astounding anatomical diversity across different fish lineages. Their taxonomic relationships are elusive still, as fossils of all major groups were already present or began to appear early during the Devonian (408–360 Mya), the 'Age of Fishes' (Fig. 11.1B). Today's dominant groups were still obscure back then, but several lines of unrelated, jawless fishes already flourished. Eel-like *conodonts* left almost no record except tiny fossilized teeth (Clark 1987), and there were many different lines of small, bottom-dwelling fish covered in oddly shaped bony plates and were thus called *Ostracoderms* ('shell-skinned'). But except for hagfishes[b] and lampreys (~ 60 species), these early vertebrates vanished during the Devonian again, likely outcompeted by advanced types with a powerful new invention: the jaw (*Gnathostomata*). Evolution often 'invents' things by repurposing something old, and the jaw is a great example. Of the many cartilaginous arches that once comprised the ancestral filter-feeding pharynx, the first and second arches became flexibly attached to the skull and thus, as a jaw, allowed for wholly new and superior ways of attacking, capturing, and biting prey. Other ancestral parts were equally retooled, transforming into gill arches or various skeletal elements that these more actively swimming and chasing predators required (Moyle and Cech 2004, Romer and Parsons 1977).

The *Placoderms* ('plate-skinned') were the first jawed fishes to dominate late Devonian waters, a group of dizzyingly diverse forms with bony internal skeletons and external armor, ranging in size from small as a hand to the fearsomely large *Dunkleosteus* (9 m!). But at the end of the Devonian, they too went suddenly and mysteriously extinct (Young 1986). Another, even older group of jawed vertebrates present throughout most of the Paleozoic were the *Acanthodii* ('spiny sharks'), a scurrilous if ultimately also extinct lineage that still befuddles taxonomists with their mixture of shark and bony fish features unlike any creature found today (Jarvik 1977). But two very different designs of jawed fishes endured through time, and today they offer the most basic dichotomy in the world of fishes. As their name implies, the defining difference between these two groups is the presence (*Osteichthyes* = bony fishes) *vs.* absence (*Chondrichthyes* = cartilaginous fishes) of a skeleton made of true bone (Fig. 11.1B).

Chondrichtyes are sharks and rays (*Elasmobranchii*) plus chimaeras (*Holocephali*) that have been important top predators and scavengers ever since the late Devonian (Compagno 2001). As a group, they 'invented' inner fertilization via penis-like claspers, evolved replaceable ectodermal teeth covering not only the jaw but their entire body (*placoids*), possess 'spiral valve' intestines and regulate buoyancy via their liver's oil content (among many other characteristics; Castro 2016, Moyle and Cech 2004). They proliferated once during the Paleozoic and then a second time as so-called modern sharks and rays (*Neoselachii*) since the Jurassic (150 Mya). Among them, lamnid sharks are the most derived and exhibit fascinating convergent adaptations in caudal fin anatomy and temperature regulation with top bony fish predators like tuna (Block and Carey 1985). Today, sharks comprise ~ 1,000 almost exclusively marine, large-bodied, long-lived, and late maturing species in shallow and deep seas, whose slow rate of reproduction make them especially vulnerable to overexploitation. Both the world's largest (whale shark *Rhincodon typus*, 18 m) and longest-living fish (Greenland shark Somniosus microcephalus, 400 years, Nielsen et al. 2016) are *Chondrichthyans* (Fig. 11.1B).

The *Osteichthyes,* on the other hand, clearly won the evolutionary race by number, now comprising more than 96% of all extant fish species. Their identifying characteristics (=apomorphies) are (1) a primarily bony skeleton, (2) bony scales, (3) segmented ectodermal fin rays (*Lepidotricha*) and perhaps surprisingly (4) lungs. But contemporary appearances are deceiving because many fish groups have secondarily lost (e.g., scales), reduced, or transformed (e.g., lung → swimbladder)[c] these primary characteristics. Within the *Osteichthyes,* another early and consequential split between ray-finned (*Actinopterygii*) and lobe-finned fishes (*Sarcopterygii*) occurred during the Paleozoic, reflecting a fundamental anatomical difference between the paired pectoral and pelvic fins (Fig. 11.1B).

Lobe-finned fishes evolved pectoral and pelvic girdles connected to bony elements inside these fins, thus giving them a plumb, fleshy look as if they were limbs. There's truth to this notion because some ancestors of the lobe-finned lungfishes (*Dipnoi*)—while adapting to life in shallow, drought-stricken Devonian lakes—used these limbs to 'walk' over land and so gave rise to the *Tetrapods*, all land-living vertebrates. Those who stayed behind largely vanished before the Mesozoic ended, and today no more

[b] The taxonomy of hagfishes is still controversial; some experts consider them a sister group to the vertebrates.

[c] The homology of lung and swimbladder is controversial. Both develop from outpockets of the esophagus, but at slightly different places.

than six species of lungfishes are left on Earth. A separate line of lobe-finned fishes, the *Actinista*, thrived in fresh- and saltwater until their presumed extinction about 100 million years ago (Forey 1998). But in 1938, when Marjorie Latimer famously recognized an odd looking fish on a South African market as an *Actinist* species, the coelacanth *Latimeria chalumnae* (Smith 1939), this group's 'extinct' status suddenly needed revision. A second species, *L. menadoensis*, was described in the year 2000.

Ray-finned fishes, on the other hand, evolved only small internal bones at their fin bases, achieving stabilization instead via ectodermal *lepidotricha* (the visible lines in a fish fin). Because this allowed for thinner, more flexible fins and greater maneuverability, it ultimately proved advantageous and thus largely defines how we picture a 'fish' today (Fig. 11.1B), and although the *Actinopterygii* originated in the Devonian and rose slowly to prominence throughout the Paleo- and Mesozoic (when iconic relic fishes like sturgeon, bichirs, or gars thrived, Fig. 11.1C), the most explosive radiation happened only during the most recent 60 million years because of one group in particular: the *Teleosts* ('end-bone'). In *Teleosts*, unlike in their ancient relatives, the chorda does not extend into the caudal fin; instead it ends in a special vertebra called *urostyl*, resulting in symmetrical (homocercal) tails. *Teleosts* also always have, at least primarily, a swimbladder, elasmoid scales, fully ossified vertebrae, and a flexible jaw. Evidently, their general building plan proved so versatile that it allowed the more than 27,000 *teleost* species to attain stunningly diverse forms, exploit all food sources and conquer all aquatic habitats on this planet.

With at least 42 orders (suffix *–formes*) and 431 families (suffix *–idae*), *Teleost* systematics is intimidating even to experts and is also still subject to debate and revision (Fig. 11.1C). In broad strokes, one can distinguish ancestral *Teleosts* like bonytongues (*Osteoglossiformes*), eels (*Anguilliformes*) and herrings (*Clupeiformes*) from the slightly more derived carps (*Cypriniformes*), characins (*Characiformes*), and catfishes (*Siluriformes*), which contrast with all other groups known as 'modern' or *Euteleosts*. Those include at their base orders such as the economically important salmons (*Salmoniformes*) and cods (*Gadiformes*), followed by more advanced silversides (*Atheriniformes*), carptooths (*Cyprinodontiformes*), rockfishes (*Sebastiformes*), as well as sticklebacks and seahorses (*Gasterosteiformes*). All of those, however, pale in diversity next to the largest vertebrate order on Earth, the perch-like fishes (*Perciformes*, > 9,200 species in > 150 families), containing such well-known families as gobies (*Gobiidae*), wrasses (*Labridae*), jacks (*Carangidae*), tunas and billfishes (*Scombridae*), groupers (*Serranidae*), damselfishes (*Pomacentridae*), surgeonfishes (*Acanthuridae*), cichlids (*Cichlidae*), parrotfishes (*Scaridae*), icefishes (*Notothenoidae*) and many, many more (Fig. 11.1C). Finally, at the pinnacle of *Teleost* evolution we find flatfishes (*Pleuronectiformes*, 570 species), puffers and triggerfishes (*Tetraodontiformes*, 340 species). Along this coarse arch, evolution frequently reduced scales and the number of vertebrae and fin rays, developed fin spines (*Acanthopterygii*), led to a loss or repositioning of the pelvic fins and the fragmentation of skull and opercular bones for greater flexibility and more specialized modes of feeding. Various groups, in particular *Clupeids* and the *Otophysi* (carps, caracins, catfishes), independently developed connections between their swimbladder and the inner ear, thereby achieving a surprisingly astute sense of hearing. One of the most comprehensive works on fish systematics are Nelson's *Fishes of the World* (Nelson 2006) and Eschmeyers' Catalog of Fishes (Fricke et al. 2018).

11.2.1 Fish zoogeography

Asking how fishes distribute across Earth's aquatic habitats today is the domain of fish zoogeography, a fascinating cross-section between ecology, physiology, paleontology and taxonomy. Beyond broad strokes, however, the reasons for many distributional patterns remain elusive or hard to scientifically evaluate. In absolute terms, more fish species live in marine (58%) than freshwater environments (41%), but freshwater only constitutes ~ 0.009% of water on Earth, hence the relative diversity of freshwater fish is about 7,500 times higher (Cohen 1970). Greater habitat fragmentation likely allows for greater genetic isolation and thus speciation in freshwater, compared to saltwater, habitats. For similar reasons, freshwater fish distributions are often much better understood today than patterns in the marine realm (Moyle and Cech 2004).

Most fishes occurring in streams, rivers, lakes or estuaries are *obligatory freshwater* species because they must spent at least some part of their life history in freshwater (a common definition is for salinity to be < 25). Of those, most spend their entire life there (and would die in saltwater), which is why they are called *freshwater dispersants* (e.g., cyprinids, characins, catfishes, pikes, cichlids or guppies, to name just a few), as opposed to *saltwater dispersants* that have either larval or adult stages dispersing through marine habitats. Some fishes visiting freshwater environments are actually true marine fishes with just a high tolerance for low salinities (= *euryhaline*), as is the case for seasonally occurring whiting (*Merlangius merlangus*) in the Baltic Sea (Ojaveer et al. 2010) or bull sharks (*Carcharhinus leucas*) swimming up large freshwater rivers (Thorson et al. 1973). Rather than being euryhaline, 200–300 species or about 1% of fishes are *amphihaline*, with a physiological adaptation to switch between fresh- and saltwater as a regular part of the life history (McDowall 1997). Fish that migrate between freshwater and marine habitats in relation to maturation and spawning are called *diadromous*, which has independently evolved in many groups such as sturgeons, salmons, river herrings (which spawn in freshwater = *anadromy*), or eels (which spawn in saltwater = *catadromy*).

Dispersal, although intuitive, is not the best explanation for most large-scale zoogeographic patterns. The better alternative is called *vicariance*, a notion for geological processes like plate tectonics, Pleistocene glaciations, resultant global sea level changes, or the rise of mountain ranges promoting divergent speciation on either side of a barrier. *Vicariance*, for example, is why an otherwise dominating fish family, the cyprinids, does not occur at all in Australia and South-America (when they evolved, both continents had already separated from the African plate) or why cichlid fishes came to radiate in Asia (they 'hitched a ride' on the Indian plate that collided with the Asian plate, forming the Himalayas). It may also be invoked to explain why sturgeons or salmonids are only found across Earth's northern hemisphere or why sunfishes (*Centrarchids*) occur exclusively in North America (Moyle and Cech 2004). Vicariance is why today six broad zoogeographic regions of freshwater fishes can be distinguished, namely: (1) African (~ 3,000 species), (2) Neotropical (~ 3,600 species), (3) Oriental (~ 3,000 species), (4) Palaearctic (~ 350 species), (5) Nearctic (~ 1,100 fish species) and (6) Australian region (~ 300 species). Within these, sub-regions and ichthyological provinces are often apparent and likely maintained by dispersal processes (Moyle and Cech 2004).

Within the much larger, more dynamic, and three-dimensional marine realm, fish zoogeography is less understood and determined mostly by average thermal conditions instead of topographic features. Diversity is highest along the shallow, shelf areas of the world's oceans that frame continental margins, while less diverse, but distinct fish assemblages characterize shallow, mid (epi- and mesopelagic fishes) and the deepest parts of the open ocean (i.e., bathy-pelagic or abysso-benthic fishes). The two axes of diversity gradients in the ocean are latitude and depth (species richness decreases with increasing latitude/depth). To the extent that large-scale distributional patterns can be explained, vicariance processes dominate over dispersal processes (Moyle and Cech 2004), as recent work on the circumtropical distribution of trumpetfishes (genus Aulostomus, Bowen et al. 2001) neatly illustrates.

Because fish assemblages change predictably with latitude (temperature, really) along continental shelves, five general regions are recognized: an Arctic, a North temperate, a Tropical, a South temperate, and an Antarctic shelf region. Boundaries are still subject to debate and revision; for example, recent research proposed uniting warm-temperate and tropical regions into a single region within each ocean basin (Briggs and Bowen 2012). The species-poor Arctic contains ~ 110 species, mostly within the families *Cottidae, Zoarcidae, Pleuronectidae*, and *Cyclopteridae*, while icefishes (*Notothenoidae*) comprise the dominant fish family in the Antarctic. Here, icefishes make up a so-called 'species flock' because the family has diverged into very differently-looking species occupying all ecological niches in this extreme habitat (other famous species flocks occur in freshwater; i.e., cichlids in African Rift Lakes, cottids in Lake Baikal, or cyprinodonts in Lake Titicaca). Some icefishes have evolved anti-freezing proteins in their blood as an adaptation to life at subzero temperatures.

The four North temperate continental shelf regions are the North-American Atlantic, North-American Pacific, Asian Pacific and Mediterranean Atlantic regions (Moyle and Cech 2004). Along the North-American Atlantic region (including the northern Gulf of Mexico but not the tip of Florida), approximately 1,200 species occur, with a gradual change from families dominating in the north (e.g., *Gadidae, Pleuronectidae, Cottidae, Cyclopteridae, Salmonidae, Clupeidae*) to those dominating in the south (*Sciaenidae, Sparidae, Serranidae, Haemulidae, Labridae*). Distributions are often complicated by pronounced intra-annual patterns due to seasonal migrations (e.g., bluefish, winter skate) or the northward seasonal advection of tropical offspring with the Gulf Stream. Cape Hatteras (35°N) is one of the few clear zoogeographic breakpoints. The North American Pacific shelf region stretches from the Aleutian Islands to Baja California and is very rich in species because of its geological age. Alone within the single genus of rockfishes (*Sebastes, Scorpaenidae*) 65 species are known. Flatfishes (25 species of *Pleuronectidae*) and Cottidae (85 species) are also important, along with Pacific salmons (8 species) that have sustained civilizations for millennia. The viviparous (life-bearing) surfperches (*Embioticidae*) are endemic to this region. On the other side of the Pacific, the Asian Pacific shelf region extends from Hong Kong past the Kamchatka peninsula. Similar to the Gulf Stream in the North Atlantic, the Kuroshio Current blurs zoogeographic boundaries by facilitating a northward expansion of tropical fish taxa (10% of 900 species in southern Japan also occur in northern Japan). Finally, the Mediterranean Atlantic region encompasses not just the Mediterranean Sea, but also the entire European Atlantic shelf including its semi-enclosed shelf seas (Baltic and Black Seas). While typical north-temperate families like *Gadidae*, *Pleuronectidae*, or *Cottidae* are abundant throughout, warm water families such as *Mugilidae*, *Sparidae* or *Labridae* extend far to the north (British Isles, North Sea) because of the warming influence of the North Atlantic Gyre. In the Mediterranean Sea, approximately 600 fish species occur with a gradient toward the east of more tropical species. Interestingly, since the construction of the Suez Channel in 1869, about 54 tropical fishes from the Red Sea have managed to establish Mediterranean populations (called *Lessepsian* species), whereas only two species managed the opposite (Golani 1998).

The shallow waters of the Tropical seas (within the ~ 20°C isotherms) contain about 40% of all marine fish species (Moyle and Cech 2004), most of which are associated with coral reefs. There are four main zoogeographic regions: the Indo-Pacific, Eastern Pacific, Western Atlantic and Eastern Atlantic. Of those, the Indo-Pacific Region is the most species-rich (4,500 species), particularly within the 'Coral Sea triangle' between the islands of Malay, New Guinea and the Philippines (~ 2,800 species). The reasons for this astounding diversity are still debated, as some experts focus on its vast habitat diversity, others on the relatively old age of this area (what is known as the 'center of origin' theory), while there's also contrary evidence showing it to be a collection basin for species (Briggs 2000). In contrast, the tropical regions of the Eastern Pacific (also known as Panamanian region) are much less diverse (750 species, only 62 shared with the Indo-Pacific region), likely due to fewer shallow ocean habitats, few islands, and very few coral reefs. The regions' fish fauna was probably similar to one of the Western Atlantic until about 5 millions years ago, when the Isthmus of Panama closed during the Pliocene. The Galapagos Islands, however, stand out for their fish diversity (> 300 species), many of which are endemic species. The Western Atlantic region encompasses the southern Gulf of Mexico, the Caribbean reefs, Bahamas, down to Cape Frio in South America (23°S). Its diversity is stunning (1,500 species) but notably below the Indo-Pacific, which is generally attributed to the younger age of this area. The Eastern tropical Atlantic, finally, is the most species-poor tropical area (434 species), owing largely to the lack of coral reefs or other shallow water habitats, as well as the existence of upwelling areas along large stretches of its continental margin.

Of the three recognized South Temperate Continental Regions, the *South American region* extends from the Peruvian coast around the South American tip to approximately the latitude of Rio de Janeiro on the Atlantic side. On each side, the transitions from the tropical zone are very gradual and shifted considerably northward on the Pacific side because of extent of the cold Humboldt Current. Mirroring northern hemisphere patterns, families dominating the lower temperate latitude include *Engraulids*, *Clupeids*, *Serranids*, *Carangids*, and *Scorpaenids*, which are replaced with increasing latitude by *Gadids*, *Zoarcids* and *Pleuronectids*. Due to South America's proximity to the Antarctic, *Notothenoids* dominate too, unlike in any other temperate shelf region. The *South African temperate shelf region* is zoogeographically interesting for its pronounced lack of cold-water fishes, likely a consequence of its relatively low latitudes and the influence of the warm Agulhas Current. Icefishes (*Notothenoidae*) are missing, while eelpouts (*Zoarcidae*) and lumpsuckers (*Cyclopterids*) are constrained to the deepest, coldest waters. The high rates of endemism within the goby (*Gobiidae*), blenny (*Blenniidae*), and klippfish (*Clinidae*) families may reflect the region's long isolation. Finally, the *Australian temperate shelf region* with ~ 1,500 fish species is highly influenced by warm southward currents mixing in tropical fauna, and an ichthyologically distinct subregion around Tasmania, the southern tip of Australia and New Zealand. Of ~ 600 species inhabiting the southern Australian coast, more than 80% are endemic.

11.3 How fish consumption, metabolism and growth relate to each other, and depend on the environment

Equipped with this foray into fish evolution, systematics and distribution, let's turn to some of the fundamental processes that govern a fish's life and ask—like a true ecologist—how environmental variability influences them. For example, how fish obtain and allocate energy under changing environmental conditions is an old but still eagerly pursued question because it informs other fields such as population dynamics, aquaculture, physiology, and with increasing urgency, consequences of man-made climate change (Wootton 1998).

Fishes, as of course all metazoans, must ingest and then break down energy-rich macromolecules to provide both the fuel for their body to function and the building blocks for new tissue to grow. Ingested energy is lost via the fish's metabolism (*R*, lost via heat) and via excretion of solid feces (*F*) and liquid nitrogenous waste (= urea, *U*, mostly as ammonia). The remaining energy gets incorporated through processes called 'growth' (*P*), which includes growth in tissue (somatic growth, P_S) and the production of gametes (P_G). We can therefore posit that a fish's food consumption (*C*) needs to account for (i.e., to equal) all these energetic expenses, so if all but one is known the missing component could be estimated.

$$C = R + F + U + P_S + P_G$$

While true, actually applying this energetic equation has some practical challenges. For one, it is still impossible to quantify all components simultaneously in a single living fish. It is also difficult because processes like metabolism and somatic growth are usually measured not in energy units (i.e., Joule) but as the rate of consuming oxygen (e.g., mg O_2 kg^{-1} h^{-1}) or the gain in length/weight over time (e.g., mm d^{-1}), respectively. Nonetheless, the concept is a useful start, because it emphasizes the connection and importantly the reason for ubiquitous trade-offs between these processes. It makes clear, for example, why a fish with increased metabolic demands (e.g., while swimming in warmer water or being chased by predators) would need to grow slower if it cannot increase its food consumption. Let us now explore each one of these terms in certain depth.

11.3.1 Consumption

$$C = R + F + U + P_S + P_G$$

Fishes not only conquered all aquatic habitats on Earth, they also evolved to utilize seemingly all aquatic food sources. In contrast to terrestrial organisms, fish generally eat protein- and lipid-rich diets with few carbohydrates. As adults, most fishes are omnivorous generalists that can switch between a great variety of food items depending on their availability in the water column or the sea floor (e.g., cod *Gadus morhua*). A large number of fishes are piscivorous (e.g., tuna *Thunnus thynnus*), meaning that they consume other fish, while herbivorous (plant-eating) fishes are comparatively rare (e.g., menhaden *Brevoortia tyrannus*, rabbitfishes *Siganus* spp.). Among the many specialists, we find scavengers (e.g., hagfishes, some sharks), parasites (pearlfish *Carapidae*, lamprey *Petromyzon marinus*), detritivores (*Tilapia* spp.), coral eaters (Butterfly fishes, *Chaetodon* spp.), mammal hunters (great white shark *Carcharodon carcharias*), or in the case of the Archerfish (*Toxotes* spp.) a fish specialized on shooting at insects with a precisely aimed water jet! Fish looking for prey on or in the sediment are called benthivores, whereas those eating microscopic organisms in the water column are planktivores.

Planktivorous fishes are often small bodied, schooling fishes eaten by a diversity of bigger marine organisms and are therefore referred to as *forage fishes*. By number and mass, forage fishes are the most abundant vertebrates on Earth, but their diversity is curiously low (Bakun 2006). Their important ecological role is to transfer massive amounts of energy from zooplankton to higher (piscivorous) trophic levels in the food web (Pikitch et al. 2012). Fish larvae, on the other hand, are often a temporary part of the plankton themselves (as buoyant eggs and/or between hatch and metamorphosis) and are therefore called *ichthyoplankton*. No matter what fishes eat as adults, as larvae they rely on plankton, particularly on all stages of copepods, the world's most abundant metazoans.

While it is common that prey types change during fish ontogeny, e.g., from planktivorous to piscivorous (marlin, *Makaira nigricans*) or to benthivorous (sole, *Solea solea*), more often it is simply prey size that increases. Pursuing bigger prey is energetically more profitable for fishes, but is accompanied with higher costs of attack and handling of bigger prey. In addition, a fish's limiting gape width (most fishes only eat what they can swallow whole) place size-dependent upper constraints. These limitations were demonstrated in, for example, 15-spine sticklebacks (*Spinachia spinachia*) offered to feed on various sizes of mysid shrimp (Kislalioglu and Gibson 1976). As predicted by *optimal foraging theory* (Werner and Hall 1974), the fish chose on average the energetically most profitable prey size, which increased from ~ 1 g in 7 cm fish to 5 g for 11 cm fish (Fig. 11.2A). Optimal foraging is particularly critical for

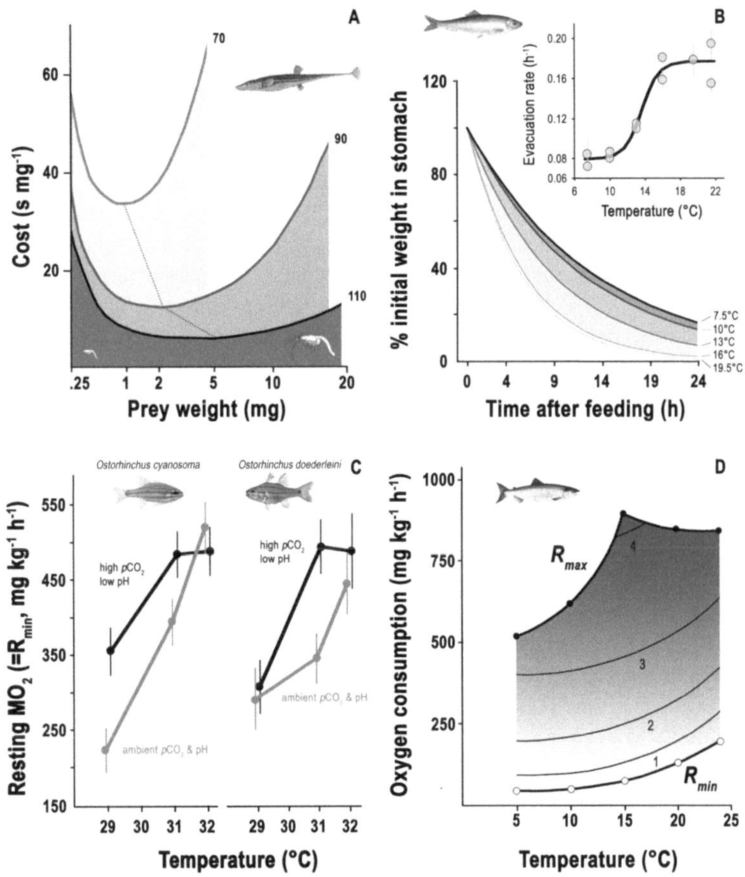

Figure 11.2. (A) Optimal foraging in juvenile *Spinachia spinachia* feeding on mysid shrimp. Cost curves estimated for 70, 90, and 110 mm fish; the dotted line indicates the upward shift in optimal prey size in larger fish (modified after Kislalioglu and Gibson 1976). (B) Temperature dependency of gastric evacuation in *Sprattus sprattus* (adapted from Bernreuther et al. 2009). Pulse-feeding and sequential sampling of 15–30 individuals per time point allowed measuring % initial stomach weight over 24 h post feeding at different temperatures. The rate of exponential evacuation (fitted from each curve) shows an S-shaped relationship with temperature (insert). (C) Interactive effects of CO_2 and temperature on resting metabolism (R_{min}) in two species of tropical reef fish. R_{min} increased with temperature and under acidified conditions (high pCO_2, low pH), except at 32°C, where fish from acidified conditions were unable to increase R_{min} further (modified after Munday et al. 2009). (D) Temperature-dependency of active (swimming) metabolism (R_A) in juvenile *Oncorhynchus nerka*. Lines show increasing R_A with temperature for 0 (=R_{min}), 1, 2, 3, and 4 body lengths per second. Maximum active metabolism (R_{max}) first increases then decreases with temperature, hence the metabolic scope (R_{max}–R_{min}) peaks at 15°C (Redrawn from Brett 1964).

fish larvae, for which searching and striking at food is energetically so costly (China and Holzman 2014) that only those few managing to quickly include larger prey in their diet ultimately succeed to survive (Dickmann et al. 2007, Voss et al. 2006).

What exactly a particular fish species or population eats and how the composition of their diet compares to prey abundance in the wild is an important but tricky question, important because fisheries are increasingly being managed within an ecosystem-based context, which requires an account of not just fishery catches but also the abundance of predator and competitor species. Those quantitative assessments need data on 'who eats whom', which are traditionally gathered from stomach content analyses of sacrificed specimens (Eriksen et al. 2020).

In its simplest form, the number of each prey item in each stomach is recorded and then expressed as the fraction of the total items. More often, diet compositions are calculated as volumetric or mass-specific fractions of whole stomachs, thereby accounting for the fact that prey items vary vastly in size (consider eating 1 copepod *vs.* 1 fish). In both cases, advanced statistical analyses are often needed to overcome 'the problem of zeroes' (fish can go without food for a surprisingly long time and thus often have empty stomachs) and the different digestion rates—and therefore detection rates—of different prey types (Wootton 1998). For example, eaten fish larvae are digested beyond recognition within hours, but crustaceans stay detectable for days. To partially overcome this, scientists have begun using genomic instead of morphological methods, analyzing genetic sequences found in stomachs and comparing them to databases of taxon-specific sequences for common enzymes or structural proteins. Such *metabarcoding* approaches are rapidly advancing, but their current ability to reveal not just what but how much was consumed is still limited (Berry et al. 2015, but see Pitchford et al. 2020, Ruppert et al. 2019).

Now, if we were able to representatively sample prey organisms in stomachs and the environment, this would allow the quantification of another ecologically important parameter: *prey selectivity*, i.e., the degree to which fishes actively seek (or avoid) the organisms they encounter in the wild. There are many related methods, but a common one is to calculate the index of prey selectivity, Chesson's α (Chesson 1983) for each food item i

$$\hat{a}_l = \frac{r_i/n_i}{\sum_{j=1}^{m}(r_j/n_j)}, \qquad i = 1,\ldots,m$$

with r_i standing for the number of prey items of type i in the diet and n_i standing for the number of prey items of type i in the environment. The index varies between perfect avoidance ($\alpha = 0$) and perfect selection ($\alpha = 1$), with values not significantly different from 0.5 indicating no preference at all. More often than not, even if fish appear to be snapping indiscriminately at anything in front of them, upon careful analysis they are surprisingly selective for their favorite prey, if it is available (Bernreuther et al. 2013).

Fish feeding rates vary, naturally, with temperature because two temperature-dependent processes govern consumption: the speed of metabolic energy turnover and the speed of stomach emptying, also known as gastric evacuation rate (Wootton 1998). Both determine a fish's hunger (= the propensity to feed) and its appetite (= how much to eat until full). In Baltic sprat (*Sprattus sprattus*), for example, this has been experimentally determined by recording the stomach fullness of individuals over 24 h after feeding and at different temperatures. The fitted data showed that the increase in gastric evacuation rate with temperature could be described with an S-shaped relationship, which suggested that sprat feed most effectively at temperatures above 16°C (Fig. 11.2B). The modeled parameter was then used to estimate feeding requirements of adult sprat in the Baltic Sea (Bernreuther et al. 2009).

11.3.2 Metabolism

$$R = C - (F + U + P_S + P_G)$$

The difference between ingested (C) and excreted energy (F+U) is called *assimilated energy*, and as highly active vertebrates fish have to spend most of it on their metabolism. Metabolic activity encompasses everything a living fish does, from muscle movement and digestion to pumping blood, repairing tissue or maintaining chemical homeostasis of intra- and extracellular fluids. Within their normal range of environmental conditions, fishes do all routine activities by *aerobic metabolism*, which is completely sustained by oxygen (i.e., energy comes from the exergonic reaction of $CH_2O + O_2 \rightarrow CO_2 + H_2O$). However, stress or fast swimming frequently demands extraordinary energy expenditures that only *anaerobic metabolism* can deliver in the short-term, while incurring an oxygen debt that must later be recovered. Interestingly, in many fishes, aerobic and anaerobic muscles are physically separated and can be distinguished by their red *vs.* white coloration, respectively.

There are three components to a fish's metabolism, known as (1) the basal or resting metabolism R_{min} (2) the swimming or active metabolism R_A and (3) the metabolic costs of digestion or *Specific Dynamic Action* (*SDA*).

Basal metabolism covers the minimum of all needed body functions, and consumption has to cover at least this much for the fish to survive. R_{min} usually scales allometrically with body size; a power relationship with weight (W) $R_{min} = a\ W^b$ and an exponent b < 1 describes it well. Simply, larger fish have an absolutely higher, but relatively lower basal metabolism than smaller fish. As expected for ectotherms, R_{min} is also highly temperature-sensitive and can be influenced by other environmental parameters including salinity and pCO_2 conditions (Fig. 11.2C, Munday et al. 2009). One difficulty with measuring R_{min} is that fish tend to swim more than rest, often spontaneously. In most cases, this can be approached by statistical extrapolation from a series of R_A measurements. In so-called swim flume respirometers that produce laminar, quantifiable current through a sealed respirometry chamber, fish are being challenged to swim against stepwise increasing currents. The recorded increase in oxygen consumption can then be extrapolated to zero swimming to give R_{min} (Fig. 11.2D). The right size of the swimming flume (they come small like a syringe for fish larvae all the way to several bathtubs large for sharks) and best experimental practices are critical for measuring R_{min} correctly.

Swim flumes are also the method of choice for quantifying the *active metabolism* R_A and its upper limit known as R_{max}. Fishes have a truly admirable capacity to increase their energetic output if needed, enabling sustained swim speeds of up to 5–6 body lengths per second (BL s^{-1}) but also short unsustainable bursts of > 20 BL s^{-1} to evade predator strikes (Wootton 1998). In the dense aquatic realm, faster swimming demands exponentially more energy. The energetically most profitable swim speed depends on fish size; it increases in absolute terms (e.g., m s^{-1}) but decreases in relative terms from ~ 6 BL s^{-1} in larvae to < 1 BL s^{-1} in most adult fish. Again, as with R_{min}, both R_A and R_{max} are highly temperature-specific, with increasing output until reaching a thermal optimum, beyond which R_{max}—and therefore metabolic scope—collapse swiftly (Fig. 11.2D)

The difference between R_{max} and R_{min}, i.e., between the maximal possible metabolic output and the minimum required for long-term survival, is called *Scope for Activity* or *Metabolic Scope*. A large metabolic scope should be adaptive (meaning: evolutionary advantageous) in fishes, as it indicates a lot of expendable energy if circumstances demand it. As the well-known case of local growth adaptation in Atlantic silversides (*Menidia menidia*, a small forage fish) shows, selection may indeed favor lower active metabolism (R_A, and therefore presumably higher metabolic scope) over fast growth in some situations (Fig. 11.3).

The final component of a fish's overall metabolism is the cost of digestion (*SDA*). Many aquarists know *SDA* from observing increased gill ventilation or agitated swimming in recently feed ornamental fish. Higher oxygen demands from the digestive tract itself are compounding SDA. Such costs can take away as much as 9–19% of all assimilated energy (Jobling 1981). In experimental settings, *SDA* can increase rapidly after a meal, and require hours and even days to return to its pre-fed baseline. *SDA* is the chief reason why metabolic experiments are done by default on unfed fish. Also, *SDA* increases with meal size and with more proteins *vs.* lipids in the diet (Wootton 1998).

When all metabolic components are estimated and their maxima summed up for a given fish size and temperature ($R_{min} + R_A + SDA$), they can be compared to independent estimates of R_{max}. This usually results in an interesting discrepancy, showing that the $R_{max} < R_{Total}$, which means that fishes cannot afford high metabolic expenditures in each compartment and have therefore to navigate trade-offs. A recently fed fish, for example, has less energy left for high-speed bursts to escape predators because SDA reduces its metabolic scope. Trade-offs like this are commonplace wherever ecologists and evolutionary biologists look, and they provide both the intrigue and the challenge to understanding complex patterns in nature.

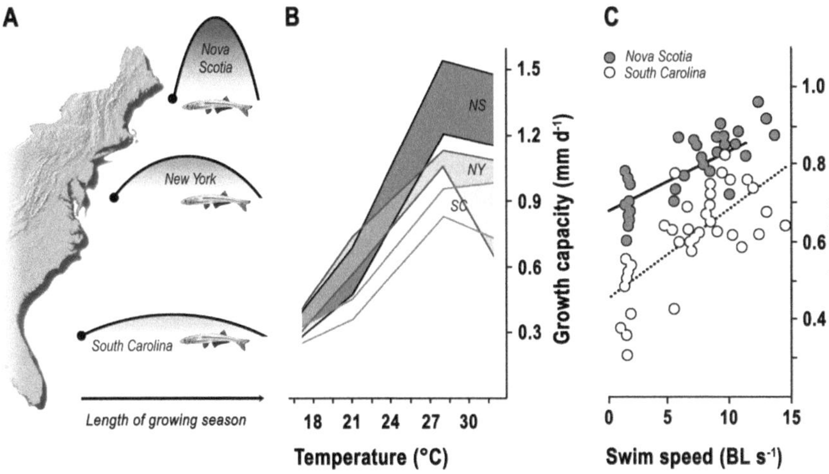

Figure 11.3. Evolution of countergradient growth variation in *Menidia menidia*. (A) Silversides are nearshore forage fishes distributed along the North American east coast. To make up for the decrease in growing season length (illustrated by decreasing widths of the arcs) with increasing latitude, northern populations grow genetically faster than their southern conspecifics (illustrated by increasing heights of the arcs), and fish are equal in size at the end of the year. (B) Lower (10th) and upper (90th) percentiles of thermal growth reaction norms for three *M. menidia* populations. Growth capacity increases faster with temperature in northern than southern populations. (C) Active metabolism (R_A) of juvenile *M. menidia* from Nova Scotia *vs.* South Carolina populations swimming at different speeds (body length per second). The slower growing southern populations have a consistently lower R_A than northern conspecifics, thus increasing the predator escape performance (modified after Arnott et al. 2006, Baumann and Conover 2011, Conover and Present 1990).

A fish's metabolism is very sensitive to variability in both abiotic (light, temperature, oxygen, salinity) and biotic environmental factors (food availability, competition, predatory risk). However, these factors exert influence in different ways, as (1) lethal, (2) controlling, (3) limiting, (4) masking or (5) directive factors that are often difficult to disentangle (Wootton 1998). A lethal factor establishes firm boundaries within which organisms can live; for example, 20°C as the upper lethal temperature for trout (*Salmo trutta*), but almost 40°C after acclimation in goldfish *Carassius aurata*. Beyond these limits, the oxygen-supply systems fail and metabolism can no longer deliver the energy needed to meet R_{min}. But temperature mostly acts as a controlling factor, i.e., an influence directly increasing or decreasing fish metabolism, with apparent adaptations in fish from different thermal regimes. As illustrated in Fig. 11.4A, R_{min} increases with temperature for temperate, tropical, or polar fishes, but the slopes and intercepts of these increases vary greatly and reflect the thermal environment the group is adapted to. For example, at a given temperature, temperate fishes have a much lower R_{min} than Arctic or even more so Antarctic fish species, and below 22°C tropical fish have a lower metabolism than temperate ones. These patterns reflect long-term adaptations of fish to their specific thermal environments. Oxygen is a good example of a limiting factor, which is defined as one that restricts supply or removal of elements in the metabolic chain. Salinity often acts as a masking factor because it modifies the effects of other factors. Finally, directive factors are those that elicit behavioral responses in fish and induce them to seek colder or more oxygen-rich environments. A good example to illustrate the interplay between many of these factors comes from Baltic sprat (*Sprattus sprattus*), a small, abundant clupeid forage fish. In winter, the Baltic Sea is not a comfortable place for a thermophilic species like sprat; the Baltic's upper, low salinity layer cools to less than 5°C. Thankfully for sprat, a warmer, more saline bottom layer of seawater always remains at depth in the Baltic Sea's basins, and this is where the fish overwinter. But the deeper basins experience prolonged periods of low or depleted oxygen (hypoxia) from microbial respiration and poor ventilation. So during winter months, sprat accumulate in dense schools in a very narrow band that is limited upward by unfavorably cold temperatures and downward by too low oxygen conditions (Stepputtis 2006, Fig. 11.4B).

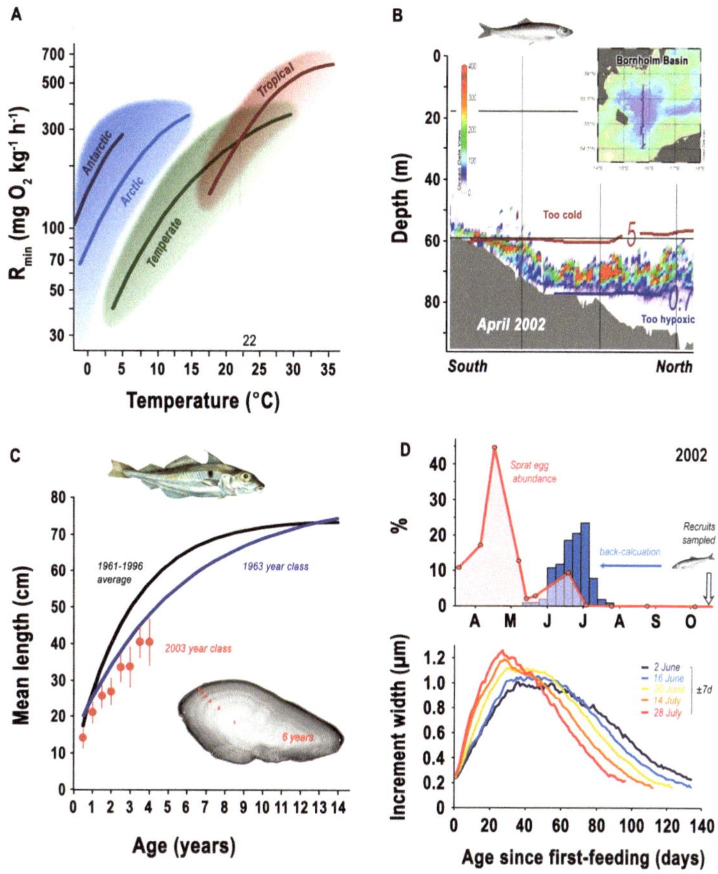

Figure 11.4. (A) Regional adaptations in resting metabolism (R_{min}) across polar, temperate, and tropical fish species. R_{min} increases with temperature in each group, but at cold temperatures polar species have absolutely higher R_{min} than temperate fishes; at temperatures > 22°C, tropical fishes have higher R_{min} than temperate fishes (after Brett and Groves 1979). (B) Temperature and oxygen as directing factors on the vertical distribution of *Sprattus sprattus* overwintering in the Bornholm Basin of the Baltic Sea. In winter, cold, low salinity water overlays warmer, more saline bottom water, but the latter is frequently affected by prolonged hypoxia. The echogram shows dense sprat aggregations in a small band sharply constrained by low temperatures upward and too hypoxic conditions downward (Stepputtis 2006). (C) Population age-length relationships for *Melanogrammus aeglefinus*, determined from otolith-based ageing of thousands of specimens and then fitted with von Bertalanffy growth curves. Strong year-classes cause intra-specific competition for food and result in slower population growth (blue: 1963 YC, red symbols: even bigger 2003 YC) compared to the long-term average (redrawn from Brodziak et al. 2008). (D) Insights from otolith microstructure analysis on newly recruited sprat. D1: back-calculated day-of-first-feeding distribution for successful recruits overlaid by the spawning effort, showing selective survival of summer-over spring-born individuals. D2: Average reconstructed otolith growth (proxy for somatic growth) over the first months of life for 5 groups born at successive 2-week intervals. Earlier born fish show slower larval but faster post-metamorphosis growth than later born fish (redrawn from Baumann et al. 2006a, 2008).

11.3.3 Somatic growth

$$P_S = C - (R + F + U + P_G)$$

After accounting for excretion and metabolic losses, fishes can use the remaining energy to grow new tissue and thus gain length and weight—this is somatic growth. Interestingly, unlike other vertebrates, fishes continue growing throughout their entire life, albeit ever more slowly. Fish ecologists and fisheries scientists study growth patterns with keen interest because they intersect with and inform other fields including fish survival, reproduction, resource use or sustainable fishing. Growth (*GR*) is a rate parameter, of course, meaning it denotes a weight (*W*) or length differential per unit time (*t*, most often: year or day), which under laboratory conditions is straightforward to determine. For example, specimens could be measured repeatedly during an experiment and the size difference divided by the number of rearing days (years) between time points to yield absolute growth.

$$GR = \frac{dW}{dt} = \frac{W_2 - W_1}{t_2 - t_1}$$

While intuitive, absolute growth is of little use when comparing fishes of different sizes. Consider a weight gain of 1 g for a 1 g juvenile *vs.* a 10 kg adult fish to understand the problem. To compare them, we need a measure of growth relative to body size, which is called *Specific Growth Rate (SGR)*. It assumes exponential weight growth over time and is expressed in% body size per unit time.

$$SGR = 100\frac{lnW_2 - lnW_1}{t_2 - t_1}$$

Typically during a fish's life, SGR increases rapidly during the larval stage, after which it begins to decline continuously until death.

Most fish are caught in the wild, however, and in this case only a single size-at-time measurement exists. Somatic growth can still be estimated, but to do that one has to first estimate a fish's age. The methods for fish ageing have been developed over more than a century and today comprise one of the most routine tools in fisheries ecology. Determining the age of fishes is generally possible thanks to the existence of hard parts such as bones, vertebrae, fin spines, scales, or otoliths (ear bones). These structures grow—just as the fish itself—in annual stanzas (Panfili et al. 2002), which is why they form discernible yearly bands or annuli that, when counted, approximate a fish's age (in years). Then, after age-size data are collected from many specimens and plotted together, the resultant population age-size data are fitted statistically, most commonly with the *von Bertalanffy growth model* that is based on bioenergetics (Fig. 11.4C).

$$L_a = L_0 + L_\infty(1 - e^{-K*a}),$$

Age (a) and length-at-age (L_a) are input, and parameters L_0 (length at time zero), L_∞ (maximum fish length) and K (the growth factor) are fitted to the data. Fitting population data to a growth curve assumes not only random growth among individuals, but also random mortality of individuals, and thus equal probabilities of representation for all fish across the age spectrum. In most cases, this assumption is likely wrong because the oldest fish in a population are often those that grew slowest (known as *Lee's Phenomenon*), while faster growing individuals tend to perish sooner, partly because they become vulnerable to fishing earlier in life. Nonetheless, more than 1 million fishes are sampled and aged annually in this way by governmental research labs all over the world because age-size data are the central ingredient for most modern fish stock assessments (Campana 2005).

Fortunately, the annuli apparent in scales and otoliths offer a second, equally important piece of information beyond age. Given that otoliths grow proportionally with the fish, otolith size reflects fish size, and the width between annuli therefore comprises a proxy for somatic growth of an individual fish at that time. In short, narrow bands mean slow growth and wide bands indicate fast growth. Crucially, this proportionality allows back-calculating previous lengths-at-age (L_a) for a single fish at hindsight. One common method to do so is called the *Biological Intercept Method* (Campana 1990)

$$L_a = L_c + \frac{(O_a - O_c)(L_c - L_o)}{(O_c - O_0)}$$

It requires corresponding measurements of length-at-catch (L_c), length-at-hatch (L_0), otolith size-at-age (O_a), otolith size-at-catch (O_c) and otolith size-at-hatch (O_0). From the back-calculated L_a values, one can then derive so-called *longitudinal growth histories* for each fish, which offer a much finer ecological insight than simple population growth curves. Longitudinal growth analyses are used to address questions of selective mortality, build growth chronologies, or investigate environmental growth determinants (Baumann et al. 2003, Fig. 11.4D).

While the existence of annual growth increments is known since the early 1900s, the existence of daily growth increments in fish otoliths was discovered later, in 1974, by the Italian researcher Giorgio

Panella. In a short article in the prestigious magazine *Science* (Pannella 1971), he reported that between the visible annual bands fish otoliths also contain between 355–370 microscopically small increments. His conclusion that these increments must have been formed on a daily basis was largely ignored first, but since the 1980s the discovery has revolutionized larval fish ecology (Stevenson and Campana 1992). The ability to back-calculate larval fish hatch dates and somatic growth histories allowed characterizing spawning times of fish stocks (Fig. 11.4D), showed that larval fish mortality is high but biased against slow growing individuals and gave insightful answers to the age-old question—why fish populations vary so greatly from one year to the next (Baumann et al. 2008, Otterlei et al. 2002, Pringle and Baumann 2019). In most fish, otolith increments begin forming already before hatch, but the reliable diel periodicity of formation appears to coincide with the time of first feeding (Heath 1992). All fish larvae, regardless of their later adult diet, are zooplanktivorous visual hunters that only feed during light hours. This triggers metabolic fluctuations and results in the cyclic deposition of calcium carbonate crystals into the otolith matrix. Once formed, increments persist as long as the fish lives, a natural miniature diary waiting to be read by a curious fish ecologist.

Growth is a highly plastic trait, meaning that it can vary greatly within and between individuals and in response to environmental parameters. A small but important source of growth variation in fishes is *endogenous*, meaning it is determined by genetic variability or by maternal or paternal contributions in the form of provisioning or epigenetic modifications (this will be further developed in Chapter 13). The majority of growth variability, however, is due to *exogenous* factors including temperature, oxygen, pH, salinity, light (= abiotic factors) or the availability of food, predation risk, or the presence of competitors (= biotic factors). Of all these, *temperature and food availability are the most important growth determinants in fishes*. Provided sufficient food, growth typically increases steadily with temperature until a maximum, beyond which it precipitously declines even in the presence of unlimited food. These species-specific *thermal growth reaction norms* are typical not just for fishes but for ectotherms in general (e.g., Fig. 11.3B).

Faster growth due to faster tissue synthesis at higher temperatures must be fueled by increased food consumption. In addition, basal and active metabolism components also increase with temperature, which sharply increases the food consumption required to fuel maximal growth with increasing temperature. This is important in fish aquaculture, where farmers want to raise their fish in the most efficient, most profitable way. To choose optimal temperature or food conditions, a farmer would thus like to know the fish's temperature-dependent *growth efficiency* (*GE*), which is the fraction of consumed food (*C*, dry weight) that gets turned into somatic growth (DW)

$$GE = 100\frac{\Delta W}{C}$$

Because the formula includes all weight (or energy) consumed as food, the result is called *gross growth efficiency*. It varies typically between 5–20% (Elliott 1975). To calculate the more precise *net growth efficiency*, one would need to use assimilated energy minus basal metabolism instead as the denominator.

11.3.4 Reproduction

$$P_G = C - (R + F + U + P_S)$$

If a fish is to be represented genetically in the next generation, at some point in life it must begin allocating time and resources to reproduction, which will determine its Darwinian fitness. This encompasses not just gamete production, which is energetically more expensive for females than males, but all potential costs that may include migration to the spawning ground, development of secondary sexual characteristics, courtship, spawning site preparation, and parental care. This additional energy must come from somewhere. Fishes could increase their food consumption, but more typically they reduce somatic growth, which therefore slows markedly after maturation. The age of maturation varies greatly in fishes. While some cyprinodonts (carp-tooth fishes) reach sexual maturity in as little as three weeks, many sharks and long-lived teleosts (e.g., Atlantic halibut *Hippoglossus hippoglossus*) only mature well into their second decade of life! Considerable differences exist within species too, for example in American plaice (*Hippoglossoides platessoides*), where age-at-maturity varies between 8–15 years. As with many plastic traits in nature, this variability in maturation arises from both external factors (e.g., temperature, food) and internal genetic differences, which reflect evolutionary constraints of different life history strategies among fishes (Roff 1992). For example, species that experience low mortality and relatively stable environmental conditions (e.g., sharks) mature late in life, whereas species experiencing large, unpredictable mortality and fluctuating environments (anchovy) tend to mature early (Southwood 1988).

Like no other vertebrate group, fishes evolved such a diversity of reproductive modes and strategies that describing, let alone quantitatively comparing, them remains a formidable challenge. Fishes could be distinguished by their *semelparous* (once a life) *vs. iteroparous* (multiple times) reproductive strategies, or by their reproductive modes ranging from laying externally fertilized eggs (*ovipary*) to depositing internally fertilized, developing embryos (*ovo-vivipary*) to giving birth to live young (*vivipary*). Different modes of energy investment characterize *capital breeders vs. income breeders* (McBride et al. 2015), with the first using energy previously stored as lipids and muscle tissue to produce gametes (cod, salmon, eel, Fig. 11.5), and the second turning food consumption directly into gamete production (anchovy). Many fish's energy investments employ aspects of both capital – income breeder modes.

Pelagic spawners release floating, buoyant eggs, whereas *benthic spawners* deposit them on sand, rocks, vegetation, or in caves. Some but not most fishes guard their eggs or carry them on their body, and some even extend energetically costly parental care to hatched offspring. Most fishes are *promiscuous* (random mating), but some have *polyandrous* (male harems), *polygynous* (female harems) or *monogamous* mating systems. For the majority of species, sex is determined at fertilization and remains unchanged throughout life (*gonochoristic*), but in some species it is determined later in ontogeny, is sensitive to temperature and photoperiod and thus called *environmental sex determination*. Several fish taxa independently evolved the ability to change sex throughout life, which can take the form of *sequential protandrous* (male to female), *sequential protogynous* (female to male), or *simultaneous hermaphroditism* (mature testes and ovaries present in one individual). At its most wondrous and extreme, in some rare cases fishes have even been found to exhibit *parthenogenesis*, which means that eggs develop either without sperm at all or with sperm triggering development but not contributing genetic material.

When exactly fishes reproduce in a year is strongly influenced by seasonality and therefore differs between high and low latitudes (Wootton 1998). In *high latitudes*, large and predictable seasonal changes in day length, temperature, primary and secondary productivity have led to the evolution of distinguishable spawning patterns, each maximizing the probability for offspring survival in their own way. Spawning from late spring to summer is the most common and timed to coincide with warming temperatures and the largest zooplankton peak. Autumn spawners often produce offspring to match a smaller fall peak in primary and secondary productivity, whereas winter spawners place offspring into cooling temperatures that delay development until feeding larvae can take advantage of the earliest zooplankton peak with little competition in the following year. A fascinating example is the Atlantic

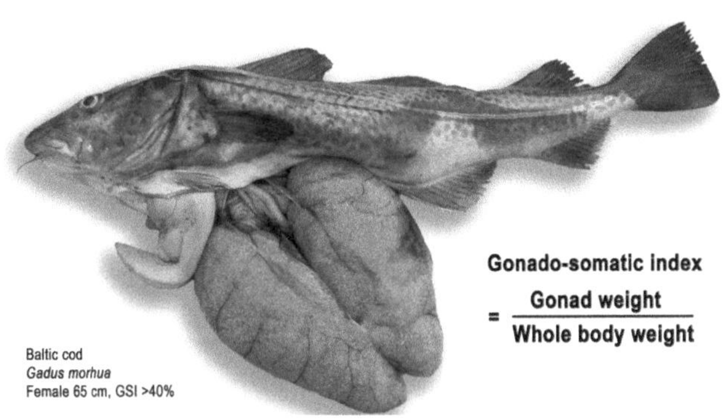

Figure 11.5. Spawning-ripe Atlantic cod *Gadus morhua* female caught in summer 2002 in the Baltic Sea. Cod are capital spawners with a typically high gonado-somatic index (GSI) (Photo: H. Baumann).

herring (*Clupea harengus*), where different localized populations evolved different spawning times and are thus referred to as spring-spawning or fall-spawning herring stocks. In *low latitudes*, on the other hand, the comparative lack of seasonality enables diverse, year round spawning patterns; tropical dolphinfish (*Coryphaena hippurus*) or ballyhoo (*Hemiramphus* spp.), for example, have been shown to spawn daily or every other day, respectively (McBride et al. 2012, McBride and Thurman 2003). Other low latitude fishes still synchronize their reproduction to lunar cycles (e.g., tarpon *Megalops atlanticus*, goliath grouper *Epinephelus itajara*) or specific seasons most conducive to offspring survival (Petersen and Warner 2002, Wootton 1998).

Fishes clearly prefer to spawn at night to minimize predation risk. For example, of 12 fish species spawning in Long Island Sound (US East coast), 8 were found to be night-spawners, 3 spawned during early morning or late afternoon, and only 1 (mackerel, *Scomber scombrus*) spawned during the day (Ferraro 1980). Coral reef fishes are overwhelmingly night spawners too (Helfman 1993). Throughout their spawning season, many fishes follow lunar or semi-lunar periodicities (rabbit fishes, damselfishes), particularly in intertidal habitats, where depositing eggs at spring high tides minimizes predation risk from aquatic predators (e.g., Atlantic silverside *Menidia menidia*, Middaugh et al. 1987).

The number of offspring produced per female varies greatly among fishes—from just a handful of young in viviparous species (guppy, shark) to almost a billion eggs in oviparous broadcast spawners such as the sunfish *Mola mola* (Forsgren et al. 2020). How many eggs a female produces is given by its *fecundity* (F) that increases allometrically with body size (W) and therefore makes larger females produce disproportionally more eggs.

$$F = aW^b \quad \text{with } b = 3.25\text{–}3.75 \text{ (range 1–5)}$$

A common proxy for fecundity is the *gonado-somatic index* (*GSI*) that is routinely recorded in fisheries research by extracting and then weighing whole gonads (W_{gon}) in order to express their weight as a percentage of body weight (W_{tot}).

$$GSI(\%) = 100\frac{W_{gon}}{W_{tot}}$$

At peak spawning, GSI values can be as low as 3%, but in broadcast spawners such as cod or salmon, female GSIs of 30% are not uncommon (Fig. 11.5), underscoring the massive cytoplasmic investment of ripening eggs, each furnished with lipid-rich yolk produced during *vitellogenesis* (eggs contain ~ 55–70% protein [illegible] ~ 10–35% lipids). Male GSIs are typically one order of magnitude smaller than female GSIs, but differences still exist between pair-mating species (close contact, little sperm needed, low GSI) and broadcast spawners (less contact, sperm competition, higher GSI).

Differences in spawning mode also influence the GSI. So-called *total spawners* have high GSIs because they prepare their eggs to be spawned in a single spawning event (salmon, sand lance). Other fishes, however, are *batch spawners* with lower GSIs because eggs are ripened up and released in batches over the course of a spawning season. In this case, ovaries contain distinct size cohorts of oocytes in different stages of development, from which researchers can estimate the *batch fecundity* (#eggs/batch) and the *breeding fecundity* (sum of all batches). Finally, researchers also use the expected reproductive life span of fishes to derive *lifetime fecundity*, a measure of the overall output per female. In all cases, fecundity appears most directly related to food consumption, although direct effects are difficult to separate from any positive effect of food on body size and therefore indirectly on fecundity.

While fecundity is a measure of how many eggs are being produced, an equally interesting aspect is how the quality of these produced eggs may differ between females of different age, size, or feeding history. *Egg size* is a trait of particular interest because larger eggs (more yolk) are known to result in larger hatchlings, which in turn have higher probabilities of early life survival. For example, egg size increases with female size in salmon and other fishes; however, the clear trade-off with fecundity (fewer, bigger *vs.* more, smaller eggs) likely prevents egg size to increase in most cases beyond ~ 1 mm, the mode of marine egg size distribution across species (Houde 1994). A second aspect is *egg quality* that increases with female size and age and appears to depend specifically on the amount of carotenoids and highly unsaturated fatty acids. Maternal food consumption plays an important role in egg composition and may also have long-lasting effects on offspring metabolism, which is known as *metabolic programming* (Mejri et al. 2020). This was shown in red drum *Sciaenops ocellatus*, where females that were fed diets containing large amounts of docosahexaenoic acid (DHA) produced offspring with faster escape responses, faster growth and better routine swimming skills and survival (Fuiman and Perez 2015).

11.4 Why do fish populations vary?

Charles Darwin once pointedly observed that all organisms produce vastly more offspring than necessary to replenish their populations. From this, he famously deduced the need for high offspring mortality, natural selection, and survival of the fittest. Fishes are prime examples to illustrate these concepts. Unlike any other vertebrate group, most fishes begin *independent* life as tiny eggs and later small, still poorly developed larvae (Houde 1997) that are eaten by almost anything bigger, including jellyfish, crustaceans, chaetognaths, other fish, birds, and even their own kind (Bailey and Houde 1989). Predation is the main source of fish mortality, responsible for just ~ 1% of all hatched fish reaching metamorphosis and even less surviving to maturity and therefore to the chance to pass on their genes (Oiestad 1985, Temming et al. 2007). The difference between 99% and 99.5% mortality, however, can precipitate large fluctuations in adult fish biomass. For example, consider a starting number of 10^{10} eggs, then a 0.5% increase in mortality would mean 50 million fewer fish. Such fluctuations are common, and ecologists and fisheries scientists would like to understand and predict them (Sissenwine 1984). It is not just scientific curiosity; it is also an economic necessity.

The process of a new cohort of fish joining the ranks of the adult stock is called *recruitment*—an oddly militaristic term traceable to one particular fish species, herring (*Clupea harengus*). Its name is derived from the German word for army ('Heer'), which was probably an apt description for the once huge herring schools that frequented European coasts and thus helped the first European trading union—the Hanseatic League—to economic and political prowess in the 12–15th century. The number of recruits determines the strength of a new year-class, which is the key piece of information for our ability to sustainably manage commercial fish stocks. If year class strength (also: recruitment strength, recruitment success) can be predicted, fishing quotas could be set in advance, and fishermen would be better able to plan ahead.

The idea that recruitment success is determined early during the larval stage was formally proposed more than a hundred years ago by the Norwegian scientist Johan Hjort, who was also the first president of the newly founded International Council for the Exploration of the Sea (ICES). He, too, was keenly interested in Atlantic herring and suggested that the notoriously large fluctuations of this stock were the result of the availability—or scarcity—of appropriate food during a specific critical period when herring larvae needed to start feeding. His *Critical Period Hypothesis* (Hjort 1914) was largely based on deduction, not empirical observation, but it provided the focus and inspiration for the work that followed later in the 20th century (Sinclair 1997), when environmental time-series and observational data had grown sufficiently large, at least for some of the most commercially-important fish stocks (herring, cod, plaice).

The British scientist David Cushing first noted that the spawning seasons of most fish populations were surprisingly fixed in time (the standard deviation of peak egg abundance in the water varied by as little as a week between years), whereas peak zooplankton abundance showed much greater interannual variability. He therefore proposed that food availability could either match or mismatch the larval abundance peak and thus result in recruitment success or failure, respectively. Over the years, the *Match/mismatch Hypothesis* (Cushing 1969, 1990) has found increasing support, but in some cases the alternative *Member/vagrant Hypothesis* (Sinclair and Iles 1989), which focused instead on the stability of the physical environment, may also be plausible. It argued that strong recruitment results from larval cohorts being retained in favorable areas by oceanographic features such as eddies or currents, while advection of cohorts into the open ocean would produce recruitment failures. Over time, researchers looking at environmental variability affecting fish early life stages have found a large number of variables that—at hindsight—statistically explain significant parts of the variability in recruitment time series. These include temperature, stage- and species-specific copepod abundance, the presence of predators or competitor species, wind-driven turbulence, stability of fronts, upwelling intensity, multi-decadal climate oscillations and many more. Curiously, the predictive power of many of these environment-recruitment relationships is still limited, and in some cases they are known to go through phases of strong and weak, or even reverse correlations (McClatchie 2012). Correlation is not causation, but also other, more complex reasons could be at play. Ecologists have long warned that searching for one factor, or even a set of factors that linearly determine fish recruitment might be a futile quest because these processes show all the characteristics of non-linear systems that have alternate phases that are seemingly unrelated (Sugihara et al. 2012). In Baltic sprat, for example, hydrodynamic simulations suggested that the degree of modeled advection *vs.* retention of sprat larvae in the central basins of the Baltic Sea strongly predicted their recruitment (Baumann et al. 2006b). However, the relationship changed suddenly in the early 1990s when the sprat spawning stock shifted from a low to a high biomass phase (Fig. 11.6).

If one ponders the odds of a small fish larva, its survival to adulthood might appear as just a matter of luck, as winning in a series of unrelated events like coin tosses (Pepin et al. 1999). But that is generally not true; *fish early life survival is mostly not random* because some combination of traits confers a higher survival probability (for a specific time and place) and therefore accumulates in the next generation, just as Darwin proposed. *Growth* is one of these important, fitness-relevant traits. From the individual perspective, being bigger means being able to pursue and swallow bigger food items, while standing a slightly higher chance of evading predation, either because of gape-limiting some predators or faster burst swimming speeds and therefore faster escapes (Anderson 1988). If the *bigger-is-better hypothesis* holds, fast larval growth should be advantageous. For Japanese anchovy, this was confirmed when researchers studied anchovy *Engraulis japonicus* larvae caught in the local Shirasu fishery. The narrow-meshed Shirasu trawls catch both larvae and their predators, which allowed sampling otoliths from larvae outside and inside predators' stomachs. Eaten larvae clearly had smaller daily increments—even at the same size—than those outside, thus suggesting the selective removal of slower growing individuals (growth-selective predation, Takasuka et al. 2003). A third advantage of faster growth is highlighted by the *stage-duration hypothesis* because faster growing fish larvae simply spend cumulatively less time within the window of their greatest vulnerability, therefore suffering cumulatively less deaths (Anderson 1988). One of nature's ubiquitous rules is therefore profoundly true for larval fish: growth is inversely related

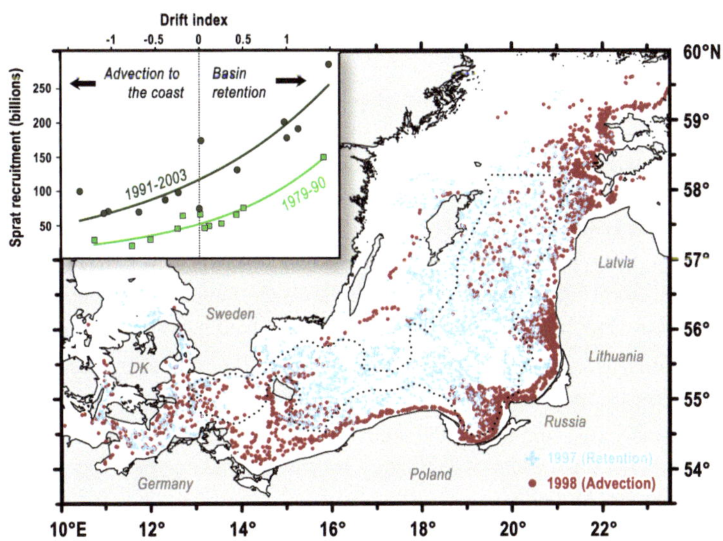

Figure 11.6. Recruitment success in Baltic sprat *Sprattus sprattus* is related to the degree of retentive *vs.* dispersive hydrodynamic circulation patterns (Baumann et al. 2006b). Map shows Lagrangian drifter positions in 1997 (blue crosses, retention year) and 1998 (red circles, advection year) modeled to drift for 50 days from their mid-basin seeding positions (circled by black dots). The resulting drift index explains 82% of the sprat recruitment variability 1979–2003, but only after accounting for periods of low and high sprat spawning stock biomass (light and dark green, respectively, insert; spawning stock biomass not shown).

to mortality. This is why most studies have found that at the time of recruitment, the new recruits consist overwhelmingly of individuals that were the fastest growing as larvae. Fast larval growth is therefore one of the most predictive *characteristics of survivors* in fishes.

While generally true, selection does not always favor fast larval growth and bigger size. For example, laboratory studies of predation (a hard thing to do) frequently report contrasting outcomes (Litvak and Leggett 1992, Pepin et al. 1992). A study on Ambon's damselfish from Lizard Island, Australia, showed that selection for fast or slow growing juveniles changed direction several times over a three-month period before, during and after settlement (Gagliano et al. 2007).

One of the best-documented examples for variable growth selection in local populations comes from the Atlantic silverside (*Menidia menidia*), an abundant, ecologically important forage fish that is widely distributed in nearshore habitats along the North American Atlantic coast (Fig. 11.3, Middaugh et al. 1987). When offspring from different populations were reared in common garden experiments (i.e., under identical conditions), intra-specific genetic population differences in growth and many other traits became apparent (Conover and Present 1990, Yamahira et al. 2006). Silverside larvae from northern populations grow much faster at all temperatures of their thermal range than their conspecifics from more southern populations—an evolutionary pattern called *countergradient variation* (Conover et al. 2009). Briefly, the shorter growing season in the north continuously selects for the fastest growing individuals in order to achieve a size that is sufficient to survive the winter. This selection for fast growth is consistent with the general growth-mortality coupling introduced above. Conversely, silverside larvae from southern populations have longer growing seasons and therefore more time to reach sufficient overwintering size, but if fast growth would be universally advantageous, it should also be selected for in the south, but it is not. The explanation lays in trade-offs with faster growth, in this case higher active swimming metabolism (Fig. 11.3C), which reduces the metabolic scope and thus results in higher vulnerabilities to predation in fast *vs.* slower growing larvae (Arnott et al. 2006, Munch and Conover 2003). We find similar trade-offs everywhere in nature, and they are important to understand the resulting evolutionary patterns on both macroscopic and molecular levels (Therkildsen et al. 2019).

Summary

This chapter gave you a peek into the far-reaching and intriguing fields of fish evolution, distribution, ecology, and fisheries science. Important points to remember from this chapter include:

1. Fishes are the oldest vertebrates with many highly divergent evolutionary lines, some of which are only known as fossils.
2. *Chondrichthes and Osteichthyes comprise the two major extant fish lines*, but only Osteichthyes have primary bone cells, and make bony skeletons. Chondrichtyes have cartilaginous endoskeletons and many other divergent features. Most fish species are Osteichthyes.
3. Within the Osteichthyes, most fish belong to the *Actinopterygii*, which have only small bones on their pelvic and pectoral fin bases. *Sarcopterygii*, however, have bony skeletal elements extending into their paired fins. One line within the *Sarcopterygii* evolved into all land vertebrates (*Tetrapoda*).
4. Most fishes are Teleosts, and within those, most belong to the Euteleosts. Acanthopterygii evolved fin spines; the most species-rich vertebrate order are the Perciformes.
5. Continental shelf margins—and within them the tropical coral reefs—house the greatest diversity of marine fish species, but the most abundant fishes by number and mass are forage fishes like herrings and anchovies in higher latitudes.
6. Large-scale distribution patterns of extant fishes are mostly the result of vicariance, i.e., of plate tectonics, glaciations, or sea level rise and fall, whereas smaller local patterns may arise from dispersal processes.
7. Fish consumption must energetically match all metabolic and excretion losses, before resources can be allocated to the production of tissue or gametes. Fish are poikilothermic organisms; hence, nearly every mentioned process is temperature-dependent, resulting in highly plastic traits such as somatic

growth, basal or maximum metabolism, or the costs of digestion. Thermal reaction norms describe the species-specific temperature-dependency of each trait.

8. Somatic growth and survival in the ocean are coupled. To study both in wild-caught fishes, daily and annual increments in otoliths and other hard-parts can be (a) counted to estimate age and (b) measured to characterize longitudinal growth histories. Fishes grow somatically their entire life, but in progressively smaller increments, after maturation.
9. Fishes evolved the most diverse reproductive patterns of all vertebrates with respect to external *vs.* internal fertilization, gonochorism *vs.* hermaphrodism, spawning times, breeding modes, parental care and many more. Fecundity, egg size, and egg quality are maternal contributions that vary with age, size, food quantity and quality.
10. For a fish larva—a member of the ichthyoplankton—the odds of survival are exceedingly small, but not random. Faster larval growth resulting from inherent, genetic advantages or external factors like more favorable temperature and feeding conditions generally confer a higher probability of survival—but exceptions exist.
11. Survival of fish early life stages is a major determinant of recruitment success, which drives the productivity of fish stocks and is therefore of greatest scientific and economic interest. To sustainably exploit fish populations, we must understand why they vary.

References

Anderson, J.T. 1988. A review of size dependent survival during pre-recruit stages of fishes in relation to recruitment. J. Northwest Atl. Fish. Sci. 8: 55–66.

Arnott, S.A., S. Chiba and D.O. Conover. 2006. Evolution of intrinsic growth rate: metabolic costs drive tradeoffs between growth and swimming performance in *Menidia menidia*. Evolution 60: 1269–1278.

Bailey, K.M. and E.D. Houde. 1989. Predation on eggs and larvae of marine fishes and the recruitment problem. Adv. Mar. Biol. 25: 1–83.

Bakun, A. 2006. Wasp-waist populations and marine ecosystem dynamics: navigating the "predator pit" topographies. Prog. Oceanogr. 68: 271–288.

Baumann, H., P. Pepin, F.J.M. Davidson, F. Mowbray, D. Schnack and J.F. Dower. 2003. Reconstruction of environmental histories to investigate patterns of larval radiated shanny (*Ulvaria subbifurcata*) growth and selective survival in a large bay of Newfoundland. ICES J. Mar. Sci. 60: 243–258.

Baumann, H., T. Grohsler, G. Kornilovs, A. Makarchouk, V. Feldmann and A. Temming. 2006a. Temperature-induced regional and temporal growth differences in Baltic young-of-the-year sprat *Sprattus sprattus*. Mar. Ecol. Prog. Ser. 319: 311–311.

Baumann, H., H.-H. Hinrichsen, C. Mollmann, F.W. Koster, A.M. Malzahn and A. Temming. 2006b. Recruitment variability in Baltic Sea sprat (*Sprattus sprattus*) is tightly coupled to temperature and transport patterns affecting the larval and early juvenile stages. Can. J. Fish. Aquat. Sci. 63: 2191–2201.

Baumann, H., R. Voss, H.-H. Hinrichsen, V. Mohrholz, J.O. Schmidt and A. Temming. 2008. Investigating the selective survival of summer-over spring-born sprat, *Sprattus sprattus*, in the Baltic Sea. Fish. Res. 91: 1–14.

Baumann, H. and D.O. Conover. 2011. Adaptation to climate change: contrasting patterns of thermal-reaction-norm evolution in Pacific versus Atlantic silversides. Proc. R. Soc. Lond., Ser. B: Biol. Sci. 278: 2265–2273.

Bernreuther, M., A. Temming and J.P. Herrmann. 2009. Effect of temperature on the gastric evacuation in sprat *Sprattus sprattus*. J. Fish Biol. 75: 1525–1541.

Bernreuther, M., J. Schmidt, D. Stepputtis and A. Temming. 2013. Vertically resolved prey selectivity and competition of Baltic herring *Clupea harengus* and sprat *Sprattus sprattus*. Mar. Ecol. Prog. Ser. 489: 177–195.

Berry, O., C. Bulman, M. Bunce, M. Coghlan, D.C. Murray and R.D. Ward. 2015. Comparison of morphological and DNA metabarcoding analyses of diets in exploited marine fishes. Mar. Ecol. Prog. Ser. 540: 167–181.

Block, B.A. and F.G. Carey. 1985. Warm brain and eye temperatures in sharks. J. Comp. Physiol. B 156: 229–236.

Bowen, B.W., A. Bass, L. Rocha, W. Grant and D.R. Robertson. 2001. Phylogeography of the trumpetfishes (Aulostomus): ring species complex on a global scale. Evolution 55: 1029–1039.

Brett, J. 1964. The respiratory metabolism and swimming performance of young sockeye salmon. J. Fish. Res. Board Can. 21: 1183–1226.

Brett, J. and T. Groves. 1979. Phyiological energetics. pp. 279–352. *In*: Hoar, D.J., D.J. Randall and J. Brett [eds.]. Fish Physiology Vol VIII, London: Academic Press.

Briggs, J.C. 2000. Centrifugal speciation and centres of origin. J. Biogeogr. 27: 1183–1188.

Briggs, J.C. and B.W. Bowen. 2012. A realignment of marine biogeographic provinces with particular reference to fish distributions. J. Biogeogr. 39: 12–30.

Brodziak, J., M.L. Traver and L.A. Col. 2008. The nascent recovery of the Georges Bank haddock stock. Fish. Res. 94: 123–132.

Campana, S.E. 1990. How reliable are growth back-calculations based on otoliths? Can. J. Fish. Aquat. Sci. 47: 2219–2227.

Campana, S.E. 2005. Otolith science entering the 21st century. Mar. Freshwater Res. 56: 485–495.

Castro, J.I. 2016. The origins and rise of shark biology in the 20th century. Mar. Fish. Rev. 78: 14–34.

Chesson, J. 1983. The estimation and analysis of preference and its relationship to foraging models. Ecology 64: 1297–1304.

China, V. and R. Holzman. 2014. Hydrodynamic starvation in first-feeding larval fishes. Proc. Natl. Acad. Sci. USA 111: 8083–8088.

Clark, D. 1987. Phylum Conodonta. pp. 636–662. *In*: Boardman, R.S., A.H. Cheetham and A.J. Rowell [eds.]. Fossil Invertebrates, Palo Alto, CA: Blackwell.

Cohen, D.M. 1970. How many recent fishes are there? Proc. Calif. Acad. Sci. 38: 341–346.

Compagno, L.J. 2001. Sharks of the World: An Annotated and Illustrated Catalogue of Shark Species Known to Date: Food & Agriculture Org.

Conover, D.O. and T.M.C. Present. 1990. Countergradient variation in growth rate: compensation for length of the growing season among Atlantic silversides from different latitudes. Oecologia 83: 316–324.

Conover, D.O., T.A. Duffy and L.A. Hice. 2009. The covariance between genetic and environmental influences across ecological gradients—reassessing the evolutionary significance of countergradient and cogradient variation. Ann. N. Y. Acad. Sci. 1168: 100–129.

Costanza, R., R. d'Arge, R. de Groot, S. Farber, M. Grasso, B. Hannon, K. Limburg, S. Naeem, R.V. O'Neill, J. Paruelo, R.G. Raskin, P. Sutton and M. van den Belt. 1997. The value of the world's ecosystem services and natural capital. Nature 387: 253–260.

Cushing, D.H. 1969. The regularity of the spawning season in some fishes. ICES J. Mar. Sci. 33: 81–92.

Cushing, D.H. 1990. Plankton production and year-class strength in fish populations: an update of the match/mismatch hypothesis. Adv. Mar. Biol. 26: 249–293.

Dickmann, M., C. Möllmann and R. Voss. 2007. Feeding ecology of Central Baltic sprat *Sprattus sprattus* larvae in relation to zooplankton dynamics: implications for survival. Mar. Ecol. Prog. Ser. 342: 277–289.

Elliott, J. 1975. The growth rate of brown trout (*Salmo trutta* L.) fed on reduced rations. J. Anim. Ecol.: 823–842.

Eriksen, E., A.N. Benzik, A.V. Dolgov, H.R. Skjoldal, M. Vihtakari, E. Johannesen, T.A. Prokhorova, F. Keulder-Stenevik, I. Prokopchuk and E. Strand. 2020. Diet and trophic structure of fishes in the Barents Sea: The Norwegian-Russian program "Year of stomachs" 2015–Establishing a baseline. Prog. Oceanogr. 183: 102262.

Ferraro, S.P. 1980. Daily time of spawning of 12 fishes in the Peconic Bays, New York. Fish. Bull. 78: 455–464.

Forey, P. 1998. History of the Coelacanth Fishes. London: Chapman & Hall.

Forsgren, K., R.S. McBride, T.M. Nakatsubo, T.M. Thys, C.D. Carson, E. Tholke, L. Kubicek and I. Potter. 2020. Reproductive biology of the ocean sunfishes. pp. 87–104. *In*: Thys, T.M., G.C. Hays and J.D.R. Houghton [eds.]. The Ocean Sunfishes: Evolution, Biology and Conservation, Boca Raton, FL, USA: CRC Press.

Fricke, R., W. Eschmeyer and R. Van der Laan. 2018. Catalog of fishes: genera, species, references. California Academy of Sciences, San Francisco, CA, USA Online version.

Fuiman, L.A. and R.G. Werner. 2009. Fishery Science: The Unique Contributions of Early Life Stages. John Wiley & Sons.

Fuiman, L.A. and K.O. Perez. 2015. Metabolic programming mediated by an essential fatty acid alters body composition and survival skills of a marine fish. Proc. R. Soc. Lond., Ser. B: Biol. Sci. 282: 20151414.

Gagliano, M., M.I. McCormick and M.G. Meekan. 2007. Survival against the odds: ontogenetic changes in selective pressure mediate growth-mortality trade-offs in a marine fish. Proceedings of the Royal Society B-Biological Sciences 274: 1575–1582.

Golani, D. 1998. Distribution of Lessepsian migrant fish in the Mediterranean. Ital. J. Zool. 65: 95–99.

Heath, M.R. 1992. Field investigations of the early life stages of marine fish. Adv. Mar. Biol. 28: 1–174.

Helfman, G. 1993. Fish behaviour by day, night, and twilight. *In*: Pitcher, T.J. [ed.]. Behaviour of Teleost Fishes. London: Chapman and Hall.

Helfman, G., B.B. Collette, D.E. Facey and B.W. Bowen. 2009. The Diversity of Fishes: Biology, Evolution, and Ecology. John Wiley & Sons.

Hjort, J. 1914. Fluctuations in the great fisheries of northern Europe viewed in the light of biological research. Rapports et procès-verbaux des réunions/Conseil permanent international pour l'exploration de la mer 20.

Houde, E.D. 1994. Differences between marine and freshwater fish larvae: implications for recruitment. ICES J. Mar. Sci. 51: 91–97.

Houde, E.D. 1997. Patterns and consequences of selective processes in teleost early life histories. pp. 173–196. *In*: Early life History and Recruitment in Fish Populations. Springer, Dordrecht.

Jarvik, E. 1977. The systematic position of acanthodian fishes. pp. 199–225. *In*: Andrews, S.M., R.S. Miles and A.D. Walker [eds.]. Problems in Vertebrate Evolution. Linn. Soc. Lond. Symp. 4.

Jennings, S., M. Kaiser and J.D. Reynolds. 2009. Marine Fisheries Ecology. Malden, MA: John Wiley & Sons.

Jobling, M. 1981. The influences of feeding on the metabolic rate of fishes: a short review. J. Fish Biol. 18: 385–400.
Kislalioglu, M. and R. Gibson. 1976. Some factors governing prey selection by the 15-spined stickleback, *Spinachia spinachia* (L.). J. Exp. Mar. Biol. Ecol. 25: 159–169.
Litvak, M.K. and W.C. Leggett. 1992. Age and size-selective predation on larval fishes: the bigger-is-better hypothesis revisited. Mar. Ecol. Prog. Ser. 81: 13–24.
McBride, R.S. and P.E. Thurman. 2003. Reproductive biology of *Hemiramphus brasiliensis* and *H. balao* (Hemiramphidae): maturation, spawning frequency, and fecundity. Biol. Bull. 204: 57–67.
McBride, R.S., D.J. Snodgrass, D.H. Adams, S.J. Rider and J.A. Colvocoresses. 2012. An indeterminate model to estimate egg production of the highly iteroparous and fecund fish, dolphinfish (*Coryphaena hippurus*). Bull. Mar. Sci. 88: 283–303.
McBride, R.S., S. Somarakis, G.R. Fitzhugh, A. Albert, N.A. Yaragina, M.J. Wuenschel, A. Alonso-Fernández and G. Basilone. 2015. Energy acquisition and allocation to egg production in relation to fish reproductive strategies. Fish Fish. 16: 23–57.
McClatchie, S. 2012. Sardine biomass is poorly correlated with the Pacific Decadal Oscillation off California. Geophys. Res. Lett. 39: L13703.
McDowall, R.M. 1997. The evolution of diadromy in fishes (revisited) and its place in phylogenetic analysis. Rev. Fish Biol. Fish. 7: 443–462.
Mejri, S., C.M. Burns, Z. Hou and L.A. Fuiman. 2020. Maternal allocation of lipids to eggs by Southern Flounder (*Paralichthys lethostigma*) after diet changes. N. Am. J. Aquacult. in press.
Middaugh, D.P., M.J. Hemmer and L.R. Goodman. 1987. Methods for spawning, culturing and conducting toxicity-tests with early life stages of four atherinid fishes: the inland silverside, *Menidia beryllina*, Atlantic silverside, *M. menidia*, tidewater silverside, *M. peninsulae* and California grunion, *Leuresthes tenuis*, ed. O.o.R.a. Development. Washington, DC: U.S. Environmental Protection Agency.
Moyle, P.B. and J.J. Cech. 2004. Fishes: An Introduction to Ichthyology. Upper Saddle River, NJ: Prentice-Hall.
Munch, S.B. and D.O. Conover. 2003. Rapid growth results in increased susceptibility to predation in *Menidia menidia*. Evolution 57: 2119–2127.
Munday, P.L., N.E. Crawley and G.E. Nilsson. 2009. Interacting effects of elevated temperature and ocean acidification on the aerobic performance of coral reef fishes. Mar. Ecol. Prog. Ser. 388: 235–242.
Nelson, J.S. 2006. Fishes of the World. Hoboken, NJ: Wiley & sons.
Nielsen, J., R.B. Hedeholm, J. Heinemeier, P.G. Bushnell, J.S. Christiansen, J. Olsen, C.B. Ramsey, R.W. Brill, M. Simon and K.F. Steffensen. 2016. Eye lens radiocarbon reveals centuries of longevity in the Greenland shark (*Somniosus microcephalus*). Science 353: 702–704.
Oiestad, V. 1985. Predation on fish larvae as a regulatory force, illustrated in mesocasm studies with large groups of larvae. NAFO Sci. Counc. Studies 8: 25–32.
Ojaveer, H., A. Jaanus, B.R. MacKenzie, G. Martin, S. Olenin, T. Radziejewska, I. Telesh, M.L. Zettler and A. Zaiko. 2010. Status of biodiversity in the Baltic Sea. PLoS One 5: e12467.
Otterlei, E., A. Folkvord and G. Nyhammer. 2002. Temperature-dependent otolith growth of larval and early juvenile Atlantic cod (*Gadus morhua*). ICES J. Mar. Sci. 59: 851–860.
Panfili, J., H. De Pontual, H. Troadec and P.J. Wright. 2002. Manual of Fish Sclerochronology, 463 pp and DVD. Published by IFREMER-IRD, Brest, France, 2002. ISBN 2-7099-1490-5.
Pannella, G. 1971. Fish otoliths: daily growth layers and periodical patterns. Science 173: 1124–1127.
Pepin, P., T. Shears and Y. De Lafontaine. 1992. Significance of body size to the interaction between a larval fish (*Mallotus villosus*) and a vertebrate predator (*Gasterosteus aculeatus*). Mar. Ecol. Prog. Ser. 81: 1–12.
Petersen, C.W. and R.R. Warner. 2002. The ecological context of reproductive behavior. Coral reef fishes: dynamics and diversity in a complex ecosystem: 103–118.
Pierce, B.A. and H.M. Smith. 1979. Neoteny or paedogenesis? J. Herpetol. 13: 119–121.
Pikitch, E.K., K.J. Rountos, T.E. Essington, C. Santora, D. Pauly, R. Watson, U.R. Sumaila, P.D. Boersma, I.L. Boyd, D.O. Conover, P. Cury, S.S. Heppell, E.D. Houde, M. Mangel, É. Plagányi, K. Sainsbury, R.S. Steneck, T.M. Geers, N. Gownaris and S.B. Munch. 2012. The global contribution of forage fish to marine fisheries and ecosystems. Fish Fish. 1: 43–64.
Pitchford, S.C., B.E. Smith and R.S. McBride. 2020. A real-time PCR assay to detect predation by spiny dogfish on Atlantic cod in the western North Atlantic Ocean. Ecol. Evol. 10: 11022–11030.
Pringle, J.W. and H. Baumann. 2019. Otolith-based growth reconstructions in young-of-year Atlantic silversides *Menidia menidia* and their implications for sex-selective survival. Mar. Ecol. Prog. Ser. 632: 193–204.
Roff, D.A. 1992. The Evolution of Life Histories: Theory and Analysis. New York: Chapman and Hall, New York, USA.
Romer, A.S. and T.S. Parsons. 1977. The Vertebrate Body. Saunders, 624pp.
Ruppert, K.M., R.J. Kline and M.S. Rahman. 2019. Past, present, and future perspectives of environmental DNA (eDNA) metabarcoding: A systematic review in methods, monitoring, and applications of global eDNA. Global Ecology and Conservation 17: e00547.
Sinclair, M. and T. Iles. 1989. Population regulation and speciation in the oceans. ICES J. Mar. Sci. 45: 165–175.

Sinclair, M. 1997. Prologue. Recruitment in fish populations: the paradigm shift generated by ICES Committee A. pp. 1–27. *In*: Chambers, R.C. and E.A. Trippel [eds.]. Early Life History and Recruitment in Fish Populations. Dordrecht: Springer Netherlands.
Sissenwine, M.P. 1984. Why do fish populations vary? pp. 59–94. *In*: May, R. [ed.]. Exploitation of Marine Communities. Springer-Verlag, Berlin.
Smith, J.L. 1939. A living fish of the Mesozoic type. Nature 143: 455–456.
Southwood, T. 1988. Tactics, strategies and templets. Oikos: 3–18.
Stepputtis, D. 2006. Distribution patterns of Baltic sprat (*Sprattus sprattus* L.)—causes and consequences, Christian-Albrechts-Universität Kiel, Germany.
Stevenson, D.K. and S.E. Campana. 1992. Otolith Microstructure Examination and Analysis: Canadian Special Publication in Fisheries and Aquatic Sciences 117.
Sugihara, G., R. May, H. Ye, C.-h. Hsieh, E. Deyle, M. Fogarty and S. Munch. 2012. Detecting causality in complex ecosystems. Science 338: 496–500.
Takasuka, A., I. Aoki and I. Mitani. 2003. Evidence of growth-selective predation on larval Japanese anchovy *Engraulis japonicus* in Sagami Bay. Mar. Ecol. Prog. Ser. 252: 223–238.
Temming, A., J. Floeter and S. Ehrich. 2007. Predation hot spots: large scale impact of local aggregations. Ecosystems 10: 865–876.
Therkildsen, N.O., A.P. Wilder, D.O. Conover, S.B. Munch, H. Baumann and S.R. Palumbi. 2019. Contrasting genomic shifts underlie parallel phenotypic evolution in response to fishing. Science 365: 487–490.
Thorson, T.B., C.M. Cowan and D.E. Watson. 1973. Body fluid solutes of juveniles and adults of the euryhaline bull shark *Carcharhinus leucas* from freshwater and saline environments. Physiol. Zool. 46: 29–42.
Voss, R., C. Clemmesen, H. Baumann and H.-H. Hinrichsen. 2006. Baltic sprat larvae: coupling food availability, larval condition and survival. Mar. Ecol. Prog. Ser. 308: 243–254.
Werner, E.E. and D.J. Hall. 1974. Optimal foraging and the size selection of prey by the bluegill sunfish (*Lepomis macrochirus*). Ecology 55: 1042–1052.
Wootton, R.J. 1998. Ecology of Teleost Fishes. Dordrecht, Netherlands: Kluwer Academic Publishers.
Yamahira, K., T.E. Lankford and D.O. Conover. 2006. Intra- and interspecific latitudinal variation in vertebral number of *Menidia* spp. (Teleostei: Atherinopsidae). Copeia 2006: 431–436.
Young, G. 1986. The relationships of placoderm fishes. Zool. J. Linn. Soc. 88: 1–57.

CHAPTER 12

Marine Megafauna
Sea Turtles, Seabirds and Marine Mammals

§*Rocío Mariano Jelicich,*[1,a,*] *Paula Berón,*[1,a] *Sofía Copello,*[1,a] *Natalia A. Dellabianca,*[2,b] *Germán García,*[1,a] *Vanessa Labrada-Martagón,*[3,c] *M. Natalia Paso Viola,*[2,b] *Jesica A. Paz,*[1,a] *Luciana Riccialdelli,*[2,b] *Analía San Martin,*[2,b] *Juan Pablo Seco Pon,*[1,a] *Mónica A. Torres*[2,b] and *Marco Favero*[1,a]

12.1 Introduction

Seabirds, marine mammals and marine turtles are assembled within the Amniota clade. The term marine megafauna is not a taxonomically defined group (Table 12.1) (Montero and Autino 2018), and mostly comprise large vertebrates strongly associated to the marine environment during the whole or a significant portion of their life-cycles. The biological features of each group are very different although many similarities can be found in their ecological roles and life history traits. Accordingly, seabirds, marine mammals and sea turtles are exposed to nearly the same threats. While many organisms still need to be on land for reproduction (e.g., seabirds, sea turtles, pinnipeds), others live in the aquatic environment (e.g., cetaceans and sirenians). Marine megafauna are distributed all over the seas from the poles to the Equator and from shallow to deep waters. Oceans are heterogeneous environments characterized by patchy prey distribution over space and time; hence, marine vertebrates are concentrated in areas of high prey abundance. These hot spots are the result of physical processes, such as upwellings, eddies, gyres or sea-ice edges, and biophysical interactions favoring productivity and prey concentration (Bost et al. 2009). Not only abundance can be affected by these physical drivers but also diversity (and species richness) which is particularly high in the Southern Ocean and the central Pacific Ocean (Jenkins and Van Houtan 2016). This chapter aims to provide a broad overview of the life history patterns of marine

[1] Instituto de Investigaciones Marinas y Costeras (IIMyC, UNMdP, CONICET), Rodríguez Peña 4046 Nivel1, Mar del Plata, Argentina (CP: B7602AYJ).

[2] Laboratorio de Ecología y Conservación de Vida Silvestre (CADIC-CONICET) Houssay 200 (V9410BFD) Ushuaia, Tierra del Fuego, Argentina.

[3] Laboratorio Ecología de la Salud, Facultad de Ciencias, UASLP, Av. Chapultepec #1570, Col. Privadas del Pedregal. CP 78295, San Luis Potosí, San Luis Potosí, México.

Emails: mpberon@mdp.edu.ar; scopello@mdp.edu.ar; ndellabianc@gmail.com; gogerman@mdp.edu.ar; vanessa.labrada@uaslp.mx; npasoviola@gmail.com; jesipaz@mdp.edu.ar; lriccialdelli@gmail.com; analiasmn@gmail.com; secopon@mdp.edu.ar; monicanatorres@yahoo.com.ar; mafavero@mdp.edu.ar

§ The authors are listed alphabetically, excluding the first and last author, and contributed equally to the writing of the chapter. Copello S., Dellabianca N.A. and Labrada-Martagón V. led the writing of seabirds, marine mammals and sea turtles respectively. a. contributed to seabirds part; b. contributed to marine mammals part; and c. contributed to sea turtles part.

* Corresponding author: rmjelic@mdp.edu.ar

Table 12.1. Taxonomic classification and phylogenetic systematics of marine megafauna (turtles, birds and mammals); representatives of this chapter are denoted in black.

Subphylum	**Class**	**Subclass**	**Clade**	**Order**	**Suborder**	**Family**
Vertebrata						
	Reptilia					
			Archelosauria			
				Testudines		
						Cheloniidae
						Dermochelydae
			Archosauria			
				Crocodilia		
	Aves			**Sphenisciformes**		Spheniscidae
				Procellariiformes		Diomedeidae Procellaridae Hydrobatidae Oceanitidae
				Suliformes		Phalacrocoracidae Fregatidae Sulidae
				Pelecaniformes		Pelecanidae
				Charadriiformes		Chioniade
						Laridae
						Alcidae
	Mammalia	Theria	Eutheria	**Carnivora**	Caniformia	Otaridae Phocidae Odobenidae Mustelidae Ursidae
				Cetartiodactyla	Mysticeti	Balaenidae Neobalaenidae Balaenopteridae Eschrichtiidae
					Odontoceti	Physeteridae Kogiidae Monodontidae Ziphiidae Delphinidae Phocoenidae Platanistidae Iniidae Lipodidae Pontoporidae
				Sirenia		Trichechidae Dugongidae

megafauna (sea turtles, seabirds and marine mammals), their functional roles in their ecosystems and conservation issues that different populations are nowadays facing.

12.2 Life history and general traits of marine megafauna

Marine turtles

Seven living marine turtle species are grouped in two families, Cheloniidae and Dermochelydae, belonging to Testudines order of class Reptilia (Table 12.1). Six species belongs to Cheloniidae family: the loggerhead turtle *Caretta caretta*, the green turtle *Chelonia mydas*, the hawksbill turtle *Eretmochelys imbricatae*, Kemp's ridley turtle *Lepidochelys kempii*, the olive ridley *Lepidochelys olivacea*, and the flatback *Natator depressus*. One extant species belong to the Dermochelydae family, the leatherback turtle *Dermochelys coriacea* (Marquez 1990, Pritchard 1997). There have been historical controversies in the definition of species and subspecies of sea turtles, mostly given by the occurrence of isolated populations as well as geographical and habitat differences across a wide distributional range (Marquez 1990, Bowen and Karl 2007), leading to differences in morphological and life history traits (Wallace 2006). However, genetic and molecular systematic studies have contributed to identifying significant taxonomic units (Karl and Bowen 1999). Regional management units have been recognized for conservational purposes, those defined as population segments of same species at multiple biological and spatial scales (Wallace et al. 2010).

Sea turtles are distributed in temperate, tropical and subtropical waters around the world (Marquez 1990), with the exception of the leatherback turtle, which is also found in temperate waters at higher latitudes due to the ability to maintain elevated body temperature even in cold waters (Paladino et al. 1990). Regarding their feeding habitats, sea turtles can be oceanic feeders (e.g., adult green turtles, leatherback turtle), hawksbills are specifically related to coral reefs, and some are often found in coastal lagoons and bays, characterized by abundance of algae, invertebrates or large seagrass meadows (Bjorndal 1999, Lopez-Mendilaharsu et al. 2002, Seminoff et al. 2014, Labrada-Martagón et al. 2017). The green turtle, olive ridley and hawksbill turtle are widely distributed in the Atlantic, Indian and Pacific Oceans (Wallace et al. 2010). While green turtles prefer temperate and subtropical habitats rich in algae or seagrass (e.g., Musick and Limpus 1997, Gonzalez Carman et al. 2011), hawksbills are more tropical and can be found in the former habitats but also prefer reef formations (Marquez 1990). The olive ridley turtle is a pantropical species, found in neritic regions of all continents (Marquez 1990). The loggerhead turtle (*C. caretta*) is a pelagic and neritic species (Peckham et al. 2011) found in the Atlantic, Pacific and Indian Ocean with defined reproductive and feeding areas (Seminoff et al. 2004, 2014, Wallace et al. 2006); ten regional management units have been proposed including the Mediterranean unit (Wallace et al. 2010). The distribution of mature Kemp's ridley turtles is restricted to the Gulf of Mexico; immature individuals may be found in temperate waters of the NW Atlantic Ocean and in Northern Europe. This species prefers bays, coastal lagoons and river mouths with sandy and muddy bottoms (Marquez 1990, Wallace et al. 2010). The flatback turtle (*N. depressus*) is endemic species from Northern Australia and south Indian Ocean (Wallace et al. 2010). This is a neritic species found in shallow coastal areas, along coral reefs (Marquez 1990). The leatherback turtle is the only pelagic sea turtle and the most widely distributed of all sea turtles; it is found in tropical and subtropical waters, and also at higher latitudes in North Atlantic where the water temperature can reach 10–20°C. Leatherbacks reach coastlines only for reproduction. Seven regional management units are recognized for this species in the Pacific, Atlantic and Indian Oceans (Wallace et al. 2010).

The life cycle of sea turtles encompasses terrestrial and marine habitats, such as neritic (coastal or inshore marine) and oceanic (open ocean) habitats, with benthic (on the sea floor) or pelagic (occupying the water column) habits (Bolten and Lutz 2003). Thus, these organisms present a complex life history experiencing ontogenetic changes in diet and habitat use throughout their life (Turner Tomaszewicz et al. 2017). Post-hatchling sea turtles leave the beach, depart from the continental shelf, traveling within currents and gyres by using magnetic cues to orient themselves at macro (to remain inside gyres) and micro scale (changes in swimming direction and paths) (Lohman et al. 2001, Mansfield et al. 2014).

During the oceanic-stage, neonates remain in sea surface, sometimes associated to pelagic rafts of Sargassum seaweed, which provide thermal benefit, food and refuge during the developmental phase (Bjorndal 1997, Mansfield et al. 2014). Hatchlings remain in the pelagic and oceanic phase until they reach a specific recruitment size to change their habitat to neritic coastal foraging and developmental grounds (Musick and Limpus 1997). Immature sea turtles can spend decades in coastal lagoons and bays where they feed and grow until mature (Chaloupka and Limpus 1997, Bjorndal et al. 2000, Seminoff et al. 2002, Balazs and Chaloupka 2004, Chaloupka et al. 2004, Labrada-Martagón et al. 2017). When mature, sea turtles perform extensive migrations of hundreds and thousands of kilometers (1050–2850 km cheloniid species, 12,000 km *D. coriacea*) from foraging to reproductive grounds and backwards (Plotkin and Lutz 2003, Hays and Scott 2013).

These organisms spend most of their lives at sea, although still depend upon land for reproduction; female sea turtles come out of the water to nest. Data about reproductive output are summarized in Table 12.2. The incubation period of eggs varies between species and populations. In the Family Cheloniidae, the incubation period lasts from 45 to 75 days and in *D. coriacea*, the duration is from 50 to 78 days (Marquez 1990). Hatchlings emerge at night and move immediately to the sea. The duration of the oceanic phase is unknown, but when hatchlings reach a minimum recruitment size they have an ontogenetic shift in habitat and diet (Bolten and Lutz 2003). Early juveniles migrate from the oceanic to the coastal region. Immature sea turtles (> 25 cm) feed and grow in neritic or benthic areas, in shallow coastal lagoons and bays with great availability of food (Bjorndal 1997). A size-dependent habitat selection has been reported for the East Pacific green turtle (*C. mydas*), where bigger sea turtles choose wide, open bays, while smaller, immature specimens remain in closed and protected areas within the same coastal lagoon (Lopez-Mendilaharsu et al. 2005).

Sea turtles are long-live reptiles with slow growing rates as for example green turtles, hawksbills and loggerheads (e.g., green turtles 0.38–14.03 cm yr^{-1} Mexican Caribbean; loggerheads 2.3–20.3 cm yr^{-1}, British Virgin Islands) (Hawkes et al. 2014, Labrada-Martagón et al. 2017). As such, immature specimens spend up to decades (13–30 years) in feeding grounds until becoming mature and reaching a minimum nesting size (Bjorndal et al. 2000, Seminoff et al. 2002, Chaloupka et al. 2004, Bjorndal et al. 2017, Labrada-Martagón et al. 2017). Mature females and males congregate near the shoreline during

Table 12.2. General life history and reproductive traits of sea turtle species (Marquez 1990).

	Age/Size at first maturity[a] (years/SCL)	**Nesting season**	**Breeding interval (years)**	**Inter-nesting interval[a] (days)**	**Clutch number (nest/season)**	**Clutch-size (eggs)**	**Mean nesting female size (SCL cm)**
Caretta caretta	28–33 > 76 cm	spring-summer	2–3	12–17	2–5	23–190	81.5–105.3
Chelonia mydas	13[b]–31[c] > 86 cm[d]	spring-summer	2	12–14	2–5	38–195	81.2–124[d]
Eretmochelys imbricata	20–25 > 79 cm	spring-summer	2–3	13–16	2.3 (average)	51–250	53–114
Lepidochelys kempii	10–16 > 65 cm	spring-beginning summer	1	20–28	1.4–1.55	102	61.4–65.7
Lepidochelys olivacea	11–16 > 66 cm	summer-autumn	1	17–45 *arribada* 14 solitary	1.5 (average)	24–155	51–75
Natator depressus	? > 91 cm	summer throughout year	1–5	16	1–4	7–76	75.1–95.1
Dermochelys coriacea	9–15 > 145 cm	autumn-winter	2–3	9–10	4–5	46–160	41–153

Abbreviations: SCL: straight carapace length; a: Spotila (2004); b: Mexican Caribbean green turtle, Labrada-Martagón et al. (2017); c: Caribbean Spotila (2004); d: measured as curved carapace length (CCL) Mexican Caribbean; Zurita et al. (1993).

the reproductive period, and courtship activity occurs in the vicinity of the nesting beach (Hamann et al. 2003). Females may mate with several males that compete for females (Alvarado and Delgado 2005). After the phases of insemination, ovulation and oviposition, adult females (the only sex and age class that ever leaves the sea) come out to the beach to nest (Hamann et al. 2003). Female sea turtles show strong philopatric behavior. Sea turtles have the capacity to return to the same geographical area where they were born, and even to their natal beaches to nest. This behavior is known as natal homing, and some hypotheses of how this occurs have been addressed (e.g., recognition of chemical cues, population genetic structure, detection of magnetic signature of coastal area), the magnetic orientation being the long distance guidance during migration (Lohmann et al. 2013, Lohmann and Lohmann 2019). Philopatry occurs also in successive nests within the same season and between years (Marquez 1990).

Sex ratio is poorly understood as a result of the methodologic complexity for sex identification (Rees et al. 2016). Generalization of life history traits and demographic parameters such as fecundity rate, remigration interval, reproductive investment and sexual maturity, are very complex to summarize given the intraspecific biological variability, given by local habitat conditions (Wallace et al. 2010); thus, the most outstanding life traits are presented in Table 12.2. Growth and reproductive rates are strongly affected by environmental factors such as local habitat conditions, seasonality, type of diet, availability and nutritional quality of food, stochastic and oceanographic events, among others (e.g., Bjorndal et al. 2000, Balazs and Chaloupka 2004, Labrada-Martagón et al. 2017). This, in turn, affects body condition and energetic reserves of reproductive females; thus, growth and reproduction output varies annually, between species and also between populations of the same species (Hamann et al. 2005) (Table 12.2).

Differences in diet between sea turtle species are evident; some species show obligate herbivory, while others are omnivores or facultative carnivores (Bjorndal 1997). Sea turtles present species-specific anatomic differences in the keratinous beaks that can be used to identify species. Sea turtle beaks have evolved and differentiated according to diet. *C. mydas*, an herbivorous turtle, presents serrations in its beak in order to forage seagrass, although carnivorous sea turtles (loggerheads and hawksbills) present more robust and keratinized narrow beaks, forming a v-shaped ridge along the buccal margin. Beaks of these animals are characterized by specialized sharp and cutting edges for eating hard shells and corals (Wyneken and Lutz 2003). The leatherback turtle buccal anatomy is specialized to grab and eat gelatinous zooplankton, presenting a w-shaped ridge in the upper jaw with a single pointed central hook in the lower jaw. Its mouth cavity and throat are covered with horny, spine-like, papillae that prevent the leak of preys (Marquez 1990).

Critical foraging habitats have been identified for some species and populations (e.g., Lopez-Mendilaharsu et al. 2005). Differences in prey preference have been reported between sea turtle populations of the same species and also, size-related differences in feeding habitat of the same population, suggesting an opportunistic selection of available resources (Hatase et al. 2002, Lopez-Mendilaharsu et al. 2003, 2005). There still remain questions and uncertainties about the feeding behavior of pelagic post-hatchling sea turtles.

Seabirds

Worldwide, seabirds comprise five Orders: Sphenisciformes, Procellariiformes, Pelecaniformes, Suliformes and certain Families among Charadriiformes, totalling some 300 species (Brooke 2002, Fig. 12.1). Other bird taxa that can be associated to marine or estuarine environments during part of their annual cycles (e.g., Anseriformes, Podicipediformes) will not be addressed in this chapter.

Sphenisciformes include 17 species of penguins, while Procellariiformes is a more diverse taxon comprising 79 species including albatrosses, petrels, shearwaters, fulmars and relatives. Some Charadriiformes and Suliformes are largely associated with freshwater and/or estuarine environments. Charadriiformes include gulls, terns, skuas, skimmers and auks with some 127 species, while Suliformes comprise tropicbirds, frigatebirds, gannets, cormorants and anhingas totalling 62 species. From the morphological and ecological perspectives, seabirds constitute a very heterogeneous group, ranging in size from very small (30 g weight and 30 cm wingspan) to very large (13 kg weight and over 3 m wingspan) species, from planktivorous to apex predators and scavengers, from restricted foraging

Figure 12.1. Representatives of four seabird orders: (a) Sphenisciformes: Gentoo penguin (*Pygoscelis papua*), (b) Suliformes: Imperial Shag (*Leucocarbo atriceps*) breeding pair in their nest, (c) Procellariformes: Southern Giant Petrel (*Macronectes giganteus*) in its nest and (d) Charadriiformes: Olrog's Gull (*Larus atlanticus*). Photo credits: Sofía Copello.

ranges to others feeding throughout the whole circumpolar region or having pantropical distribution, and from resident species to trans-equatorial migrating seabirds travelling large distances across several oceanographic systems (Brooke 2002, Oro and Martínez-Abraín 2004).

Seabird families are not restricted to the Atlantic, Indian or the Pacific Oceans and all but three seabird families can be found in both the Northern and Southern Hemispheres. The exceptions are the Spheniscidae (except few Galápagos penguin nests placed just North to the Equator) and Pelecanoididae (diving petrels) restricted to high latitudes in the South, and Alcidae (auks) limited to the Northern Hemisphere (Carboneras 1992, Martínez 1992, Nettleship 1996). Many species undertake vast seasonal migrations changing the distribution throughout the year (Newton 2008). The marine environment, the landscape and human activities are important drivers of seabird distribution. In the latter case, commercial fisheries may have a significant effect on the at-sea seabird distribution given that fishery by-products (i.e., discards and offal) may constitute a predictable source of food throughout the annual cycle (Oro et al. 2013) attracting several species in large numbers and often producing negative effects (see section Conservation of Marine Megafauna).

There are 326 seabird species, representing roughly 3% of the > 10,000 bird species currently recognized worldwide (IUCN 2019), reflecting quite a low seabird diversification, which nevertheless have adapted to almost all environments on earth spanning polar to tropical latitudes (Brooke 2002, Schreiber and Burger 2002, Oro and Martínez-Abraín 2004). The life history traits in some seabird taxa are often referred to as extreme; for example, albatrosses and large petrels show extreme lifespan and high adult survival rates, deferred maturity, small clutch-size (usually one egg, not always laid yearly), and extended chick-rearing period. Seabirds inhabit a highly heterogeneous environment and this is reflected in a large diversification in feeding tactics and strategies.

Evolution has led to a range of morphological and physiological adaptations enabling seabirds to exploit resources in coastal areas and the high seas in several ways. Seabirds are found at most trophic levels of the marine food web (Fig. 12.2); they are adapted not only to catch and handle a range of prey sizes and types, but also to reach prey in the water column, some of them with extreme diving capabilities (Fig. 12.3). Bills, feet and body shapes also show innumerable adaptations to the various lifestyles, allowing them to swim and dive using webbed feet or the actual wings to propel themselves in the water. In addition, extreme physiological adaptations should be highlighted: (a) the presence of

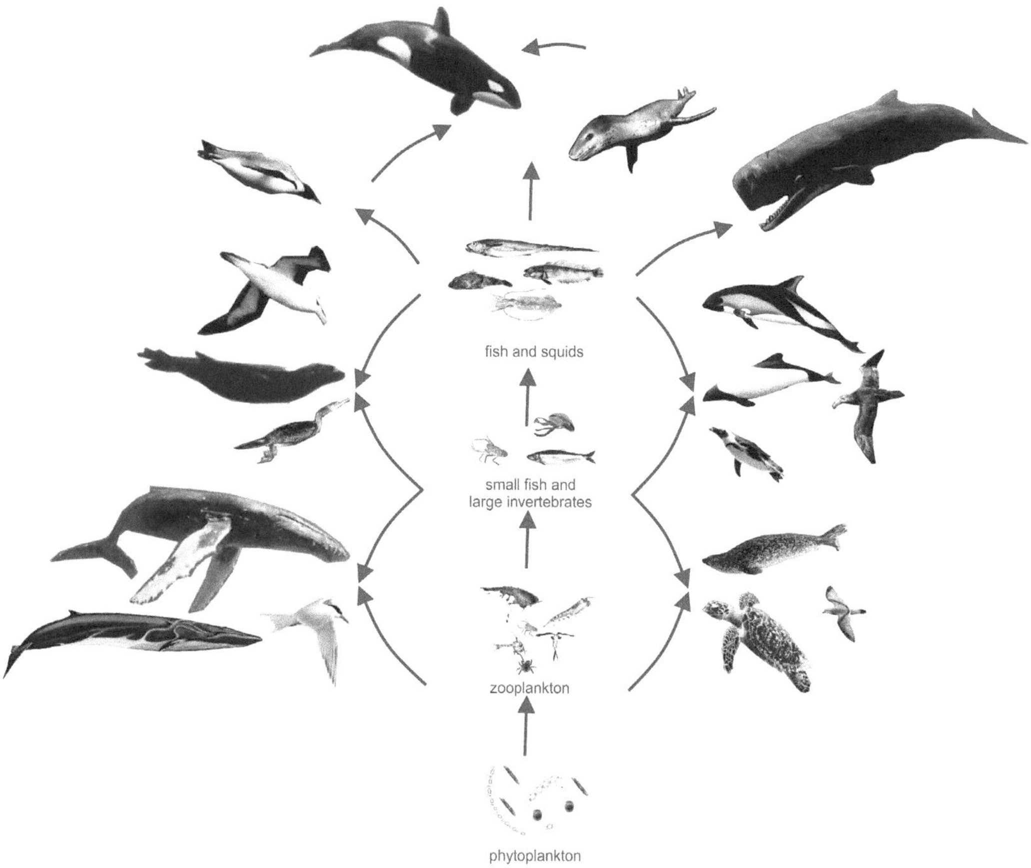

Figure 12.2. Trophic interactions and trophic levels of the marine food web occupied by marine megafauna.

salt or nasal glands, lying in shallow depressions around the orbit, above the eye, helping in coping with a salt overloaded diet; (b) the outstanding diving capabilities in many species, which are based on a combination of aerobic and anaerobic metabolism; and (c) the capacity to cope with long fasting periods in particular moments of the annual cycle, among others.

In general, seabirds breed in colonies and parents deliver food to their nestlings at/or around the nest, feeding them at intervals ranging from few hours to several days. Because they need to periodically return to their nests with food during the breeding season, seabirds are good examples of *central place foragers* (i.e., commuting long distances between colonies/nests and feeding areas) (Weimerskirch et al. 1997, Elliott et al. 2009). Their foraging activity can occur during daylight hours or at night, and these diet habits have presumably evolved in response to the behavior of their preys. The diet of seabird consists mainly of fish, crustaceans, mollusks and/or plankton. However, some species have incorporated anthropogenic items in their trophic spectrum. The food is obtained through different foraging tactics, in general linked to morphological and/or physiological features. Storm-petrels, skuas, gulls, terns, and large petrels pick food from the surface (i.e., dipping), while penguins, alcids, cormorants, and diving-petrels exploit shallow and/or deep layers whilst diving (i.e., pursuit-diving, swimming underwater and chasing prey to considerable depths). Plunge diving (i.e., plunge from the air and reach a depth a little further than its body length) is a method used by gannets, boobies, tropicbirds, terns, and pelicans. Other less common foraging tactics include kleptoparasitism, the stealing of prey from other species or conspecifics (from land through aerial pursuit), and scavenging (i.e., feeding on carrion) (Fig. 12.3). Some plankton-feeding petrels forage both by day and night, sometimes making up large flocks. Many seabirds, especially shearwaters and albatrosses, feed whilst sitting on the surface (i.e., surface seizing)

Figure 12.3. Foraging tactics and strategies described for seabirds. Adapted from Schreiber and Burger (2002).

or by forcing themselves slightly below it. In order to catch pelagic fish and large cephalopods, some shearwaters combine plunge-diving with surface-seizing (Schreiber and Burger 2002). Other seabirds specialized in plunge-diving are the sulids- their dazzling white plumage is an extraordinary effective social signal, attracting other conspecifics to the fishing area. Social fishing is widespread among seabirds, usually involving several species (Thiebault et al. 2014). Penguins are the most specialized diving birds showing the most extreme modifications (e.g., dense coat of waterproof feathers, short wings modified as paddles, modification of wings' bone microstructure). They feed largely on fish, crustacean and squid (García-Borboroglu and Boersma 2015). Emperor and King penguins (*Aptenodytes* spp.) are the largest of all diving birds and thus have exceptional diving abilities. Looking for myctophids (i.e., lanternfish) during the breeding season, King penguins may travel as far as 500 km in a seven-day round trip (Jouventin et al. 1994) reaching diving depths greater than 300 m (Pütz 1994). Emperor penguins can get even deeper, diving down to over 500 m for longer than 20 minutes. Kleptoparasitism is another particular feeding technique (Rothschild and Clay 1952). Skuas and frigatebirds can be specialized kleptoparasites during part of their annual cycle, for energy acquisition, showing adaptations that are apparently lacking in opportunist species (e.g., ability to detect and attack hosts carrying food in the proventriculous, to sustain prolonged aerial chases, and to adapt their breeding cycle to match that of their host). In contrast, gulls and terns may perform kleptoparasitism opportunistically, using a range of foraging tactics, with kleptoparasitism being context-dependent and often performed during periods of low availability of primary food sources (García et al. 2010).

Most of seabird are social and nest in colonies. The dominant mating system is monogamous with co-operation of both parents for a successful incubation and chick rearing. Some species show high mate fidelity as well. Early in the season, courtship feeding may occur in some seabird groups as gulls and terns, as a way for females to assess how good food providers their mates might be. Courtship feeding is also important to allow adequate egg formation during the pre-incubation stage, relieving the females of the time- and energy-consuming process of food acquisition. A nest is generally built in order to protect the eggs, although its structure is quite variable among seabirds. Some nests are placed on cliffs and islands, away from predators, while others can be placed in cavities to prevent attacks from aerial predators. For seabirds laying just one egg (albatrosses and petrels), a large proportion of the breeding failure occurs during brooding. For seabirds laying more than one egg, mechanisms related to egg-size asymmetry and laying/hatching asynchrony may lead to lower survival rates for the last chicks to hatch, if environmental variables (e.g., weather conditions, food availability) are unfavorable. The chick-rearing period in seabirds largely varies among groups (i.e., from four weeks in some Charadiiformes to over a year in some Sphenisciformes). Part of this variation can be explained as a result of morphological, ecological and/or life-history traits (adult body mass, foraging range and/or phylogeny, among other aspects that might affect the length of the chick-rearing period) (Dobson and Jouventin 2007). Some seabird species attempt reproduction and complete this cycle annually (e.g., the common tern, *Sterna hirundo*), while others like large albatrosses or some penguins with longer seasons breed biennially or twice every three years. After breeding, most seabirds migrate to wintering areas and migration patterns may include a range of strategies, duration and distances covered.

Marine mammals

The term marine mammals included species from different taxonomic groups unified by their dependence on marine (or freshwater) environments for feeding, and are therefore not a monophyletic group itself. There are 132 species of marine mammals grouped in the Orders Cetartiodactyla, Sirenia and Carnivora (Jefferson and Wang 2011, Dalebout et al. 2014, Berta et al. 2015, Jefferson et al. 2015) (Fig. 12.4). The marine mammals of the first Order are composed by Mysticetes (i.e., baleen whales) and Odontocetes (i.e., sea and river dolphins, porpoises, beaked whales and sperm whales, among others) with 14 and 74 species, respectively. Sirenians are represented by three species of manatees, the dugong and the now extinct Steller's sea cow (*Hydrodamalis gigas*). Finally, the Order Carnivora includes five families of marine mammals: Otariidae, Phocidae, Odobaenidae, Mustelidae and Ursidae (Table 12.1). The first three families were formerly classified as pinnipeds, including 36 species of sea lions, fur seals, true seals and walruses. Between mustelids, the marine otter (*Lontra felina*) and the sea otter are the only species considered marine, while the polar bear (*Ursus maritimus*) is the only marine ursid (Jefferson et al. 2008, Folkens and Reeves 2014, Berta et al. 2015, Würsig et al. 2018). Like seabirds, marine mammals vary substantially in weight and size (ranging from 5 kg to 180 tons and from 1 to 30 m). The body shape also differs between these groups in relation to the degree of dependence on the aquatic environment. Those which spend all their lives in the water such as cetaceans and sirenians have anterior limbs modified into flippers and have lost their posterior limbs; others such as pinnipeds are semi-aquatic and have the limbs modified for propulsion in water; while sea otters and the polar bear (less adapted to marine life) retain the four limbs characteristic of terrestrial mammals (Berta et al. 2015) (Fig. 12.4).

Marine mammals are extensively distributed around the planet, using a wide range of marine and freshwater environments. Cetaceans can be found in all oceans of the world, with representatives such as rorquals (Family *Balaenopteridae*), sperm whales (*Physeter macrocephalus*) and killer whales (*Orcinus orca*) distributed in tropical, temperate and polar waters of both hemispheres, and others confined exclusively to certain regions or ocean basins such as the Franciscana dolphin (*Pontoporia blainvillei*),

Figure 12.4. Representatives of the different groups of marine mammals: (a) Cetartiodactyla, Mysticetes: Southern right whale (*Eubalaena australis*), (b) Cetartiodactyla, odontocetes: Burmeister's porpoise (*Phocoena spinipinnis*), (c) Sirenians: West Indian manatee (*Trichechus manatus*), (d) Carnivora, pinnipeds: South American fur seal (*Arctocephalus australis*), (e) Carnivora, mustelids: Sea otter (*Enhydra lutris*), (f) Carnivora, ursids: Polar bear (*Ursus maritimus*). Photo credits: (a, b and d) Mónica A. Torres; (c) Natalia A. Dellabianca; (e) Alejandro E.J. Valenzuela; (f) Amalia Padin Goodall.

some *Cephalorhynchus* and *Phocoenidae* species and the gray whale (*Eschrichtius robustus*). Extant sirenians are restricted to coastal zones of tropical and subtropical waters of different oceans and river systems. As cetaceans, the latitudinal distribution of pinnipeds ranges from pole to pole. The polar bear has a circumpolar distribution along coastal areas of the Artic, mainly associated with sea ice. The sea otter occurs in nearshore waters of the North Pacific Ocean and the southern portion of the Bering Sea, while the marine otter is found principally along the Pacific coast of South America, although a small number of individual may still live in the southern tip of Argentina (Bastida and Rodriguez 2009, Folkens and Reeves 2014, Forcada 2018).

Marine mammal distribution is often correlated with abiotic variables, either indirectly through the oceanographic conditions which affect the distribution of their prey on a local scale (Forcada 2018) or directly given the physiological abilities and requirements of each species (Learmonth et al. 2006, Macleod 2009). In addition, anthropogenic factors such as habitat degradation, overfishing, aquaculture, farming and pollution can significantly affect the distribution of marine mammals (Forcada 2018). Changes in the distribution of marine mammals could be relatively frequent; some of them such as seasonal movements are associated to changes in the prey distribution or shifting in the environmental conditions. Others like dispersion are related to the colonization or recolonization of breeding areas, without return to the original area. Seasonal movements and dispersion have been reported for most marine mammal species; however, few species perform true migration. Migration implies regular, repeated and large-scale movements between feeding and breeding grounds (or haul-out sites for pinnipeds) (Stern and Friedlaender 2018). True migrations are well documented in sperm whales, some species of baleen whales such as humpback- (*Megaptera novaeangliae*) and gray whales, and elephant seals (*Mirounga* spp.) (Jefferson et al. 2015).

Life-history strategies are diverse among marine mammals and differ between the groups. However, most of them share some traits such as long lifespan (15–80 years), relatively long gestation period (with the exception of the marine otter), a small number (1–4) of large and precocial offspring, maternal parental care and a period of lactation that ranges from 6 to 18 months (Jefferson et al. 2015, Folkens and Reeves 2014). The age of the attainment of sexual maturity varies considerably between taxa. In cetaceans, it is delayed and seems to be correlated with longevity and body size (Chivers 2018). Pinnipeds, in comparison, become sexually mature relatively early, like the polar bear and some species of sirenians (Bonde 2018, Folkens and Reeves 2014). In polygynous species, males attain social maturity when they are large enough to compete for females and territories with other males (Folkens and Reeves 2014).

Marine mammals have evolved to feed upon a wide range of prey species at every trophic level in the marine ecosystem (Fig. 12.2). The baleen whales have baleen plates instead of teeth, to feed upon the most abundant food in the oceans, zooplankton and small schooling fishes. One of the most curious paradoxes of nature is that "the largest animals subsist on some of the smallest". Baleen whales are considered "batch feeders", filtering huge numbers of prey items at a time with their baleen plates (Heithaus et al. 2018). Most of Balaenopterids are "engulfers", which means they lunge through a prey patch, taking in great amounts of seawater by expanding their throats. Then, they close the mouth and force the water out between the baleen plates, leaving the prey trapped on the inside, and then swallow it. Right- and bowhead whales (Family Balaenidae) are defined as "skimmers", as they swim slowly through patches of prey and filter out as the water flows through the baleen plates. On the other hand, gray whales are suckers, using suction to take amphipods, mysid or other invertebrates from the bottom to the shallow waters (Berta et al. 2015). The most interesting cooperation feeding method, rarely seen in baleen whales, are bubble nets that humpback whales use to help concentrate and trap fish (Jefferson et al. 2015, Heithaus et al. 2018) (Fig. 12.5).

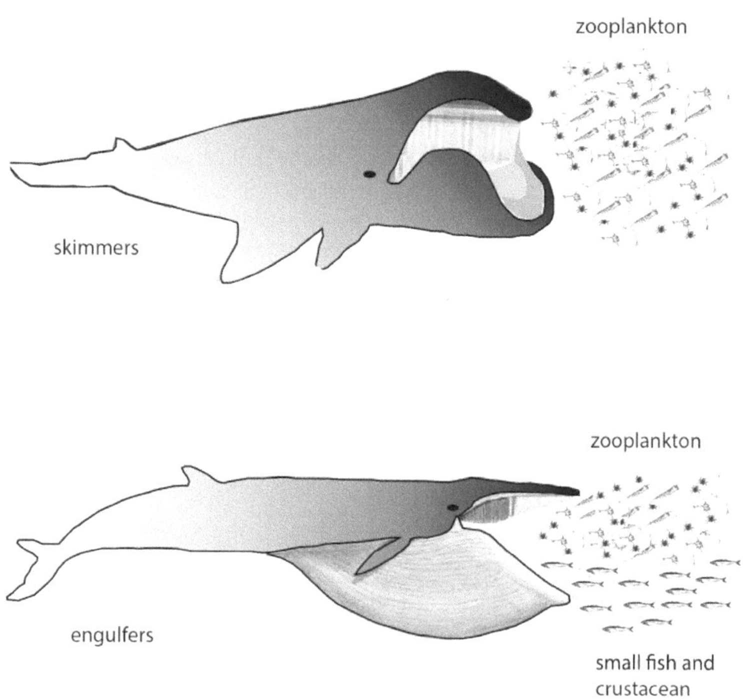

Figure 12.5. Distinct type of feeding strategies in baleen whales: (a) skimmers (Right and Bowhead whales) and (b) gulpers (Balaenopterids).

Unlike baleen whales, toothed whales (odontocete) are "raptorial feeders", as they attack and consume prey items singly, generally once at a time. The diet of dolphins and porpoises consists of larger prey than that of the misticete. Using echolocation as a method to detect prey, many species such as the beaked-, sperm- and pilot whales (*Globicephala* spp.) prey upon fish of different sizes and cephalopods, mainly squid (Folkens and Reeves 2014). They can detect relatively small objects at a considerable distance, although echolocation may not always be an effective way to detect prey given that some fish or other marine mammals can hear echolocation calls. Toothed whales may use cooperative feeding techniques, sometimes involving thousands of individuals to corral prey. The killer whale is the only cetacean species that regularly feeds on other marine mammals of all groups, from sea otters to blue whales (*Balaenoptera musculus*), while working in cooperative groups. Killer whales have a great diversity of feeding strategies, including intentional beaching to attack seals lying by the surf (Hoelzel 1991, Jefferson et al. 2008). Dolphins and porpoises are relatively shallower feeders, while beaked and sperm whales are extremely deep-divers, sometimes foraging over 1000 m from surface (Heithaus et al. 2018).

Pinnipeds are excellent hunters, feeding on a wide variety of fish (including larger species such as salmon), squid and invertebrates. The leopard seal (*Hydrurga leptonyx*) is perhaps the most famous and fearsome pinniped predator—it would attack almost anything available in the environment, including other seals, yet its most common prey types include fishes, cephalopods, seabirds (mainly penguins) and krill (Casaux et al. 2009). The crabeater seal (*Lobodon carcinophaga*) is the most specialized batch feeder pinniped and zooplankton may comprise up to 94% of its diet. Much like baleen whales, this seal has modified sieve-like teeth for feeding on krill, which are sucked into the mouth (Fig. 12.6). Seals and sea lions usually feed solitarily. There are few special adaptations for feeding in pinnipeds, such as those shown by the elephant and Weddell seals (*Leptonychotes weddellii*) that dive deeply to find food (Folkens and Reeves 2014).

Sea otters feed mostly on invertebrates, such as crabs, mussels and urchins, diving to shallows depths and often using rocks as tool to break hard shells. Polar bears (*Ursus maritimus*) prefer mostly to catch ringed seals (*Phoca hispida*), their primary prey and, to a lesser degree, bearded seals (*Erignathus barbatus*), waiting around a breathing hole through pack ice for them to come up to breathe. Polar bears are able to locate the seals by smell, even under a meter or more of compacted wind-blown snow. They also prey occasionally on walruses, belugas (*Delphinapterus leucas*), narwhals (*Monodon monceros*), waterfowl, and seabirds (Stirling 2018). Sirenians are simple grazers (and sometimes browsers) on vegetation. Manatees and dugongs are the only marine mammals that feed on plants. Dugongs feed almost exclusively on seagrasses and benthic invertebrates, while manatee feeding appears to be more diverse as they also incorporate floating or rooted vegetation (Heithaus et al. 2018).

Scavenging is a common foraging behavior on birds but it does not appear to be widespread among marine mammals except for polar bears, among others. In addition, some species of odontocetes or pinnipeds that feed on trawler discards could be considered potential scavengers. Although inter-specific behavioral interactions such as social parasitism (association of species where at least one species benefit from another species' food-finding abilities) and kleptoparasitism are common in seabirds, in marine mammals these behaviors have been observed only in polar bears and otters but have been rarely described for cetaceans. This is a consequence related to the difficulty to substantiate empirically this behavior

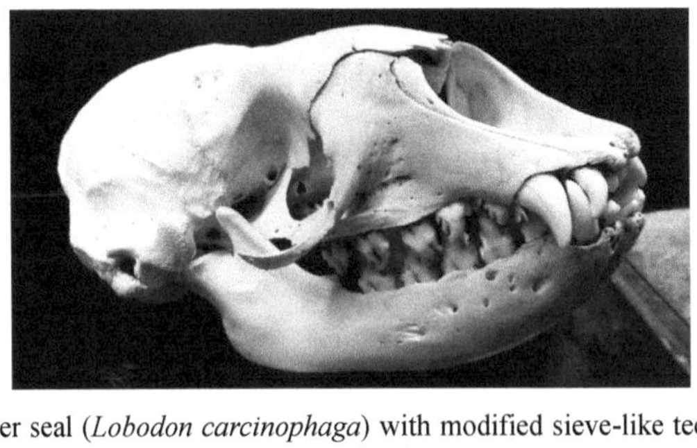

Figure 12.6. Skull of crab-eater seal (*Lobodon carcinophaga*) with modified sieve-like teeth for feeding on zooplankton, which are sucked into the mouth. Photo credit: Jerónimo Pan.

studying these animals at sea (Heithaus et al. 2018). However, some cases of kleptoparasitism have been suggested for some odontocetes involving harassment of sperm whales (*Physeter macrocephalus*) by false killer whale (*Pseurdorca crassidens*), short-finned pilot whale (*Globicephala macrorhynchus*) and Risso's dolphins (*Grampus griseus*) (Palacios and Mate 1996, Weller et al. 1996, Smultea et al. 2014). The social parasitism has been seen among bottlenose dolphins (*Tursiops truncatus*), Pantropical spotted dolphins (*Stenella attenuata*), common dolphins (*Dephinus* spp.), Northern Right Whale Dolphins (*Lissodelphis borealis*) and California sea lions (*Zalophus californianus*) (Norris and Dohl 1980, Shane 1994, Clua and Grosvalet 2001, Bearzi 2006, Cords and Würsig 2014, Smultea et al. 2014).

There is an interesting record of kleptoparasitism and social parasitism seen at the same time on three species of odontocetes in Southern California: sperm whales, Risso's dolphins and northern right whale dolphins. It was postulated that the harassment of the sperm whales by Risso's dolphins represents kleptoparasitism due to Risso's dolphins instigated aggression to induce regurgitation and potential consumption of sperm whale prey remains (squid flesh floating), while the sperm whales behaved defensively based on several events indicative of avoidance. Instead, northern right whale dolphins' association with the Risso's dolphins represents social parasitism as they generally remained on the periphery of the sperm whales and were not aggressive. It is possible that the smaller northern right whale dolphins associate with the larger Risso's dolphins to locate prey, to prey on scattered remains of Risso's dolphin prey, or for defense from predators such as sharks and killer whales. Such kleptoparasitism and social parasitism are considered a viable foraging strategy among many animal species and may be either a primary or an opportunistic-alternative foraging strategy such as the example above (Smultea et al. 2014).

The reproductive pattern of marine mammals is similar to their terrestrial relatives but there are some obvious adaptations in cetaceans and sirenians to reproduce in water. Marine mammals manifest variations in reproductive biology and mating strategies, which are related to the diversity of their habitats. Cetaceans and sirenians undertake all of their reproductive activities in the water while pinnipeds, polar bears and sea otters mate and give birth on land. The parental care is given only by females; therefore, males invest their energy trying to maximize their mating opportunities instead of guaranteeing the successful rearing of their offspring. For this reason, the most common mating system of marine mammals is polygyny, in which one male copulates with more than one female during the breeding season (Jefferson et al. 2015, Würsig et al. 2018). The gestation period is around one year for most of species. In pinnipeds and sea otters, there is a period of late implantation of the embryo, which allows them to reproduce annually. The duration of lactation is very variable between groups, ranging from few days in some seal species to several years in sperm- and killer whales. The duration of lactation is related to the composition of milk (more fat content, shorter periods). In toothed whales with long familial associations, the extended periods is also related to social bonds (Folkens and Reeves 2014).

12.3 Functional roles

Marine megafauna as top predators

The term "top predator" is synonymous with "upper-trophic level predators" and in this chapter, it will be used to refer to seabirds and marine mammals. Marine top predators play an important role in the maintenance of the ecosystem; in fact, the elimination of top predators from an ecosystem may lead to top-down processes where the next trophic level of the food web may experience overpopulation, exploiting lower trophic levels, and contributing to habitat degradation (Pauly et al. 1998). Top predators convey information about the marine environment, as they can cover very large areas in relatively short time, locate prey very efficiently, and prey on multiple trophic levels often including commercially-important resources such as fish, crustaceans and mollusks (Fig. 12.2). In this sense, top predators have been frequently used as sentinels of changes in the marine ecosystems, integrating and reflecting ecological variation across large spatial and temporal scales (Boyd et al. 2006, Moore and Zydelis 2008, Einoder 2009, Bossart 2011, Moore 2018). Top predators are widely used as bio-indicators to monitor the occurrence and ecological impacts of pollutants such as oil and mercury in the marine environment (Furness and Greenwood 1993), to detect changes in prey availability (Agnew 1997), and to reflect ecological impacts of climate and oceanographic variability such as El Niño Southern Oscillation events (e.g., Ross 2000, Diamond and Devlin 2003, Simmonds and Isaac 2007, Moore 2008, Bjorndal et al. 2017). In addition, top predators regularly exploit marine resources at similar spatial and temporal scales to those used by humans, thus increasing potential competition, and providing information about risk of human health in the consumption of marine resources (Bossart 2011). As long-lived animals, tissues of top predators are great storehouses of diverse natural and man-made pollutants in marine ecosystems (e.g., Jones and de Voogt 1999, Tanabe et al. 2004). Profiles of stable carbon isotopes and fatty acids in tissues can be used to infer the diet and broad-scale movements of marine animals (Hobson and Wassenaar 2008, Michener and Kaufman 2007, Graham et al. 2010), therefore providing evidence of changes in food webs within marine ecosystems.

Biomarkers and health

Life history (e.g., ontogenetic habitat changes, migratory behavior) and seasonal movements undertaken by marine megafauna require a multitude of physiological and biochemical responses to cope with changing environmental and local habitat conditions (e.g., Labrada-Martagón et al. 2011, Camacho et al. 2013). In addition, demanding physical conditions of the ocean, intraspecific aggression and certain activities such as prolonged fasts or extended deep dives are significant challenges to homeostasis and may elicit stress responses (e.g., Dierauf and Gulland 2001, Zenteno-Savín et al. 2010, Vázquez-Medina et al. 2006, 2011, Labrada-Martagón et al. 2011, Camacho et al. 2013).

Among natural stressors that could alter animals' health and body condition are macro- (e.g., helminths, arthropods) and micro-parasites (e.g., microbial pathogens including viruses and bacteria) (Pedersen and Fenton 2007). In sea turtles for example, most of their parasites are digenetic trematodes, annelids, nematodes, cestodes and protozoans, inhabiting the digestive tract (Work et al. 2005, Santoro and Mattiucci 2009). Ectoparasites are common as well; weak or injured individuals can show more abundant epibiotic parasites that could hinder locomotion and increase the energy expenditure (Herbst and Jacobson 1995). A variety of green, brown, and red algae has been reported growing on *C. mydas* and *C. caretta* (Senties et al. 1999), and barnacles are usually found on the carapace, plastron, mouth, hind flippers, and even the eyelids of sea turtles (Labrada-Martagón et al. 2010a). Moderate to heavy barnacle loads cannot be considered innocuous to sea turtles, as they can injure the carapace, cause visceral wounds, or obstruct the oropharynx (Herbst and Jacobson 1995). It has been proposed that the bacterial flora and fungal species isolated from sea turtles may act as opportunistic pathogens when no pathology is reported, as colonization, rather than infection, is more likely (Innis et al. 2010).

Although various diseases have been reported in sea turtle populations, fibropapillomatosis is by far the most important disease and the only one that can be considered as a disease of panzootic proportions that, together with other environmental factors, represents an additional risk for the conservation of sea turtle populations. Fibropapillomatosis is characterized by the appearance of external fibroepithelial tumors in soft tissues (e.g., skin and conjunctiva) or in hard tissues such as plastron and carapace; internal fibromas are observed occasionally. The chronic growth of tumors and their anatomical position in the individual (e.g., fins, eyelids and eyes) have behavioral consequences and a debilitating effect in sea turtles. Increasing urban development, pollution, habitat degradation and destruction have been pointed out among the main causes of emerging diseases such as fibropapillomatosis (Cunningham et al. 2017, Whilde et al. 2017) and many studies have reported that this disease can be considered as an indicator of ecosystem degradation (Jones et al. 2016, Patrício et al. 2016) (Fig. 12.7). The most affected species are the green turtle (*C. mydas*; principally the populations of Florida and Hawaii), and to a lesser extent, the loggerhead turtle (*C. caretta*). Sporadic cases have been reported in the rest of the sea turtle species. The most affected age classes are immature juveniles of medium size and, to a lesser extent, adults.

Hematology (red and white blood cell counts) and blood biochemistry tests (e.g., glucose levels, lipids, electrolytes, liver and kidney specific function indicators) are clinical tools commonly used to assess the overall health status and for diagnosing diseases of wildlife, such as birds, sea turtles and marine mammals (Campbell 1994, Dierauf and Gulland 2001, Manire et al. 2017). The analysis of blood parameters (hematology and plasma/serum biochemistry) provides information regarding the nutritional

Figure 12.7. Some examples of threats to marine megafauna: Interaction with fisheries (a); habitat degradation (b), (c), (d); fibropapillomatosis in turtles (e) and human disturbance at breeding and non-breeding sites (f). Photo credits: (a) Juan Pablo Seco Pon; (b) Vanessa Labrada-Martagón; (c and d) Fernando A. Muñoz Teneria; (e) Gisela Maldonado; (f) Sofía Copello.

status, immunity, and levels of stress at the individual and population level, and also concerns seasonal processes linked to the breeding season, such as moult (in seabird and pinnipeds) and migration (Newman et al. 1997, Dierauf and Gulland 2001, Hamann et al. 2002, 2003, 2005). Some biochemical parameters such as glucose, lipids and protein levels correlate with body condition index of green turtles, thus providing valuable information about their health and reproductive status in a simple manner (Hamann et al. 2002, 2005, Labrada-Martagón et al. 2010a, b).

On the other hand, man-made compounds such as persistent organic pollutants consisting of organochlorine pesticides (e.g., chlorpyrifos, chlordane, endosulfan, DDT, lindane and its metabolites), and industrial chemicals (e.g., polychlorinated biphenyls-PCBs, polycyclic aromatic hydrocarbons-PAHs, polybrominated diphenyl ethers-PBDEs and brominated flame retardants-BFRs), currently used in disease control, agriculture, manufacturing or industrial processes, have been found accumulated in high concentrations in different tissues of marine megafauna (e.g., liver, muscle, blubber) (e.g., Cardona-Marek et al. 2009, Labrada-Martagón et al. 2011, Castellini et al. 2012, Camacho et al. 2014, da Silva et al. 2016), causing endocrine disruption and inducing carcinogenic effects (Sparling 2016). Some of these compounds as well as heavy metals such as Hg, Pb, or Cd and organometals can be biomagnified along the food web reaching levels of toxicological importance in top predators, thus leading to various notorious health hazards and environmental effects (Burger and Gochfeld 2002, Sparling 2016). Most of these chemicals have xenoestrogenic properties that can disrupt hormone function or concentration, and can cause serious neurological, metabolic, reproductive, teratogenic, and immunological damage (O'Hara and O'Shea 2001, Keller 2013).

Nutrient cycling

Nutrient cycling is the flux of nutrients within and between the various biotic or abiotic matrices of an ecosystem (Brady and Weil 2002). Such flux of matter and energy across frontiers affects the dynamics of local populations, as well as the structure and evolution of recipient communities and food webs. In this context, the role of seabirds, turtles and marine mammals as important bio-vectors transporting nutrients across ecosystem boundaries has been largely recognized. These nutrient subsidies can fertilize recipient food webs, shaping the ecosystem structure and function of many coastal and inland sites (Caut et al. 2012).

Sea turtles have many effects on marine ecosystems as consumers (e.g., seagrass beds and invertebrate populations), as prey, as competitors, and as nutrient transporters and modifiers of landscape (Bjorndal et al. 2003). As previously mentioned, sea turtles are substrate and transport for epibionts, and they also transfer substantial quantities of nutrients and energy from nutrient-rich foraging grounds to nutrient-poor nesting beaches (Bjorndal et al. 2003). The grazing behavior of large herbivores such as green sea turtles and sirenians contributes to the maintenance of the structure, dynamics and specific composition of tropical seagrass communities (Bjorndal 1980), as the nutritional quality of plants increases when they are cropped regularly (Molina and Van Tussenbroek 2014). Sea turtles maintain grazing plots by selectively cropping residual tissues of seagrasses, favoring foliar development and nutritional quality, thus increasing nitrogen content and reducing lignin content in leaves (Bjorndal 1980, Molina and Van Tussenbroek 2014).

Seabirds are keystone species along coastlines worldwide and form dense nesting colonies that can number in the millions of individuals (Paleczny et al. 2015). They import energy and biomass into terrestrial food web through fish scraps, carcasses of chicks and egg mortality, feathers, guano, and high densities of associated ectoparasites (Polis and Hurd 1996). Through their feces, seabirds can increase the amounts of organic matter both in the soil of breeding colonies or in relatively closed marine areas such as coves, estuaries or mangroves, thus potentially affecting local animal and plant community composition and density (Powell et al. 1991). For example, dense rich growths of seaweed in the vicinities of seabird colonies indicate the enriching effects of seabird excretions on marine plants. Recent studies have highlighted the regional and global effects of seabird colonies in the cycling of elements such as nitrogen and phosphorous (Riddick et al. 2012, Otero 2018).

Beyond being considered charismatic species, marine mammals play a significant role supporting services such as ecosystem nutrient cycling (Schmitz et al. 2010). They are long-lived and actively-swimming animals that store huge amounts of nutrients in their bodies and once they are dead, their carcasses sink to the ocean floor and eventually release them. The carcasses of the great whales consist of an extraordinary amount of detritus that fall to the deep ocean (Berta et al. 2015). These whale falls can be considered as an oasis of the deep-sea since they provide for thousands of years a unique habitat and food resources for many endemic and deep-sea species (Roman et al. 2014). However, a major role frequently under-evaluated resides in how they affect the recycling of critical elements (Ratnarajah et al. 2014, Roman et al. 2014). Through a process called the "whale pump", marine mammals can deliver recycled nutrients to surface waters enhancing primary productivity (Roman and McCarthy 2010, Roman et al. 2014). The whale pump can be considered as a positive feedback loop to plankton nutrition. Marine mammals feed at different depths but in general excrete urea and release their feces at surface waters during the intervals between foraging diving. The fecal matter tends to disperse rather than sink when released, since it is defecated in a quasi-liquid state (Lavery et al. 2014). In this way, they bring to the euphotic zone the nutrients stored in their prey from deep waters (Lavery et al. 2010) in an analogous manner as the physical upwelling process. Even if marine mammals feed at superficial waters, they contribute to the retention of autochthonous nutrients in the photic zone (Lavery et al. 2014). The release of nutrient-rich excreta and feces by marine mammals in nutrient-limited waters during stratified summer conditions would benefit the production of phytoplankton by making available N, P, Fe or any other limiting nutrient (Lavery et al. 2010, 2014, Ratnarajah et al. 2014). Copepods, euphausids and fish, all prey that nourish marine mammals, are benefited from a major stock of phytoplankton but also by the direct consumption of the fine particulate fraction in the fecal plume (Roman and McCarthy 2010). The encounter of dense patches of food reinforces the aggregative foraging behavior of marine mammals and by generating turbulence, during their daily movements, they help in the mixing of water and nutrients along the water column, a process that can be especially important in stratified conditions (Roman and McCarthy 2010, Ratnarajah et al. 2014). In some regions, this pump will be most active in spring and summer when their waters are full of marine mammals feeding and nutrient levels are low. After the historical whaling operations and the exploitation of sea lion populations around the world, the recovery of the great whales and pinniped species has the potential to affect marine biogeochemistry, by enhancing primary productivity and carbon sequestration through the recycling of essential nutrients in the surface layers (Roman et al. 2014).

12.4 Conservation of marine megafauna

In the past four decades, populations of many marine turtle, seabird and marine mammal species have undergone decline, and marine habitats have suffered widespread perturbation and losses of original cover area (Convention on Biological Diversity 2010, www.cbd.int/GBO3/) through anthropogenic activity (Hoffmann et al. 2010). These losses are manifested in extinction rates that exceed baseline background rates by two to three orders of magnitude, with substantial detrimental social and economic consequences (see Hoffmann et al. 2010). In more recent years, there has been strong evidence of steady—and for some species dramatic-changes in seabird populations (Butchart et al. 2004). The International Union for Conservation of Nature's (IUCN) Red List of Threatened Species has been an internationally recognised indicator of the health of the world's biodiversity, constituting one powerful tool to monitor the conservation status of threatened species over time. Other international instruments like the Convention on Biological Diversity (CBD), the Convention on the Conservation of Migratory Species of Wild Animals (CMS), or the Convention on International Trade in Endangered Species of Wild Fauna and Flora (CITES), among others, have played a very important role in improving conservation policies worldwide. Population declines in marine megafauna have been caused by a range of threats, including bycatch (in longline, gillnet, trawl and other fisheries), pollution (oil spills, chemical pollutants, plastic and marine debris), noise (busy shipping lanes, seismic surveys, constructions and sonars), prey depletion caused by overfishing, vessel strikes, dredging, energy production and mining (McCauley et al. 2015, Sequeira et al. 2019; Fig. 12.7). On the other hand, land-based threats include the presence of invasive alien species, problematic native species (e.g., those that have become super-abundant), human disturbance, changes in land use, habitat deterioration or complete lost, commercial and residential development, and hunting, egg collection and trapping (Lutcavage et al. 1997; Fig. 12.7). Climate change and severe weather also affect species on land and at sea (Lascelles et al. 2016), mainly those which live in restricted habitats such as estuaries, rivers and shallow waters or at extremely high latitudes where climate change has already significantly altered the marine and terrestrial ecosystem.

Sea turtles, with exception of the flatback turtle (*N. depressus*), are listed under a risk category according to the IUCN Red List. The global status of hawksbill turtle and Kemp's ridley are critical endangered; the green turtle is endangered; while the loggerhead, olive ridley, and leatherback turtles are listed under the vulnerable category. The flatback turtle is listed as data deficient due to the lack of available information about its abundance and distribution, parameters required to perform a direct or indirect assessment of risk of extinction based on population status (IUCN 2012). Nesting beach numbers (adult females, eggs, hatchling numbers) is the primary source of information of population trends in sea turtles, given the philopatry of reproductive females. The problem with the former approach is that female numbers do not necessarily reflect the entire population, and alternatives to assess population size and trends are in-water surveys with tagging, capture per unit effort recovery, aerial and boat transects and strandings (Heppel et al. 2003). According to the Red List, information about population trends exists for 18 subpopulations of three species of sea turtles (IUCN 2019). A subpopulation is defined as a geographically restricted population with little demographic or genetic exchange between distinct groups in the population (IUCN 2012). The 56% of the listed subpopulations correspond to loggerhead turtle species, 39% to leatherback turtle and 6% to the green turtle species, thus suggesting that the former are the sea turtle species and populations more broadly studied. The Hawaiian green turtle subpopulation is demographically increasing and cataloged as of least concern, while female nesting numbers are declining in 32 index sites (Seminoff 2004). Five subpopulations of loggerhead turtle (in the Mediterranean, South West Indian Ocean, North Pacific, North and South West Atlantic) are increasing and of least concern, two are critically endangered (in the North West Indian Ocean and South Pacific), and trends of three subpopulations of *C. caretta* are unknown (IUCN 2019). In relation to the leatherback turtle, the Northwest and Southwest Atlantic Ocean subpopulations are increasing, and West and East Pacific Ocean subpopulations and Southwest Indian Ocean subpopulations are decreasing and critically endangered; there is no data for two subpopulations of *D. coriacea* in the Southeast and Northeast Indian Ocean (IUCN 2019).

Compared to other bird groups, the biodiversity and conservation status of seabirds is rapidly declining, currently being the most threatened bird group (Croxall et al. 2012). More than one third of the totality of currently-recognised seabird species is listed as threatened by the IUCN (17% vulnerable, 11% endangered, and 5% critically endangered; Table 12.3). The four species considered extinct in recent times (1% of the total) are the Great Auk (*Pinguinus impennis*), the Large St. Helena Petrel (*Pterodroma rupinarum*), the Small St. Helena Petrel (*Bulweria bifax*), and the Spectacled Cormorant (*Phalacrocorax perspicilliatus*). Precise quantified rates of population decline or increase are available for very few seabird species. More than half of the seabird species for which the population trend is known (275 species) are declining; only 16% of those are increasing and 27% are stable (Table 12.4). Gulls are among the species showing increased population numbers in recent decades, likely due to their capacity to exploit alternative food sources of anthropic origin (e.g., fishery discards, refuse dumps) (Fig. 12.7). Pelagic species are more threatened and their populations have declined faster than coastal species, and this difference is particularly pronounced in albatrosses and large petrels (Croxall et al. 2012). Albatrosses and petrels are susceptible to threats operating throughout their wide migratory ranges that extend across national boundaries into international waters and it is unlikely that unilateral actions will be effective in improving their global conservation status. In particular, declines among slow-breeding albatrosses and petrels have been attributed to incidental mortality associated with fisheries resulting from the growth of commercial fisheries. Although the literature primarily focuses on commercial long-line and trawl fisheries (Hoffman et al. 2010), recent investigations have demonstrated that artisanal and small-scale fisheries may have an important effect in some areas (González-Zevallos and Yorio 2006, Suazo et al. 2013). New studies have demonstrated that the conservation status of these seabirds may have worsened by the additive effects of environmental factors (i.e., climate; Pardo et al. 2016). This deterioration triggered in 1999 the negotiations for the establishment of the Agreement on the Conservation of Albatrosses and Petrels (ACAP, a CMS binding instrument), entering into force in 2004 a multilateral agreement seeking to improve the conservation status of albatrosses and petrels by coordinating international efforts.

Marine mammals have long been highly-prized resource for humans (food, bait, fur, oil) and many populations have declined seriously due to their overexploitation around the world (Merrick et al. 2018). Of the 132 species of marine mammals, 45% are cataloged with the least concern status, 5% near threatened, 11% vulnerable, 17% endangered, 2% critically endangered, 20% with data deficient and 2% are extinct (Table 12.3). Two species of pinnipeds (the Japanese sea lion, *Zalophus japonicus*, and the Caribbean monk seal, *Neomonachus tropicalis*) and one species of sirenian (Steller's sea cow) are considered extinct. In addition, three species of cetaceans are listed as critically endangered (the Atlantic humpback dolphin, *Sousa teuszii*; the vaquita, *Phocoena sinus*; and the baiji, *Lipotes vexillifer*, this latter is now possibly extinct). At date, fourteen species of marine mammals are classified as vulnerable. Between them, the small cetacean living in restricted habitats such as estuaries, rivers and shallow waters (e.g., Franciscana dolphin *Pontoporia blainvillei* and Indo-Pacific Finless Porpoise *Neophocaena phocaenoides*) are the most affected by incidental catches whereas the arctic species such as the polar bear and the walrus *Odobenus rosmarus* face the ice sea loss owing the global warming. All members of the order Sirenia are also included in this conservation status category.

The population trends of marine mammals are available only for 46 species worldwide; of them 12% are increasing, 4% are stable and 19% are declining (Table 12.4). In 1982, the International Whaling Commission (IWC) agreed to the commercial whaling moratorium (i.e., to stop commercial whaling), which remains in place today. Despite the moratorium, commercial whaling continues on small scale in some countries. Norway, Japan and Iceland still hunt whales commercially in its exclusive economic zone

Table 12.3. Total number of species and IUCN Red List categories for the more representative families of marine megafauna considered in this chapter.

Order	Families	Total number of species	DD	EX	CR	EN	VU	NT	LC
Testudines									
	Cheloniidae	6	1		2	1	2		
	Dermochelydae	1					1		
	Total	**7**	**1**		**2**	**1**	**3**		
Sphenisciformes									
	Spheniscidae	18				5	5	3	5
Procellariiformes									
	Diomedeidae	22			2	7	6	6	1
	Procellariidae	97		2	11	13	19	11	41
	Hydrobatidae	18	2		1	2	4	2	7
	Oceanitidae	9	2		1	1			6
Suliformes									
	Phalacrocoracidae	35		1	1	3	7	3	20
Charadriiformes									
	Alcidae	25		1		2	4	4	14
	Laridae	100			1	4	9	10	76
	Chionidae	2							2
	Total	**326**	**4**	**4**	**17**	**37**	**54**	**39**	**172**
Carnivora									
	Otaridae	16		1		4	1	1	9
	Phocidae	19		1		3	1		14
	Odobenidae	1					1		
	Mustelidae	2				2			
	Ursidae	1					1		
	Total	**39**		**2**		**9**	**4**	**1**	**23**
Cetartiodactyla									
	Balaenidae	4				2			2
	Neobalaenidae	1							1
	Balaenopteridae	8	1			2	1	1	3
	Eschrichtiidae	1							1
	Physeteridae	1					1		
	Kogiidae	2	2						
	Monodontidae	2							2
	Ziphiidae	22	20						2
	Delphinidae	36	4		1	2	2	4	23
	Phocoenidae	7			1	1	1	1	3
	Platanistidae	1				1			
	Iniidae	1				1			
	Lipodidae	1			1				
	Pontoporidae	1					1		
	Total	**88**	**27**		**3**	**9**	**6**	**6**	**37**
Sirenia									
	Trichechidae	3					3		
	Dugongidae	2		1			1		
	Total	**5**		**1**			**4**		

Abbreviations: DD: Data Deficient; EX: Extinct; CR: Critically Endangered; EN: Endangered; VU: Vulnerable; NT: Near Threatened; LC: Least Concern.

Table 12.4. Population trends, in number of species, listed by IUCN Red List for the more representative families of marine megafauna considered in this chapter (extinct species are excluded).

Order	Families	Total number of species	UN	ST	IN	DE
Testudines						
	Cheloniidae	6	2			4
	Dermochelydae	1				1
	Total	7	**2**			**5**
Sphenisciformes						
	Spheniscidae	18	1	4	2	11
Procellariiformes						
	Diomedeidae	22	2	4	5	11
	Procellariidae	95	8	23	11	53
	Hydrobatidae	18	5	5	1	7
	Oceanitidae	9	3	1	-	5
Suliformes						
	Phalacrocoracidae	34	11	6	5	12
Charadriiformes						
	Alcidae	24	1	1	2	20
	Laridae	100	16	31	18	35
	Chionidae	2	-	1	-	1
	Total	**322**	**47**	**75**	**44**	**156**
Carnivora						
	Otaridae	15		2	7	6
	Phocidae	18	11	2	4	1
	Odobenidae	1	1			
	Mustelidae	2				2
	Ursidae	1	1			
	Total	**37**	**13**	**4**	**11**	**9**
Cetartiodactyla						
	Balaenidae	4	2		1	1
	Neobalaenidae	1	1			
	Balaenopteridae	8	4		4	
	Eschrichtiidae	1		1		
	Physeteridae	1	1			
	Kogiidae	2	2			
	Monodontidae	2	2			
	Ziphiidae	22	22			
	Delphinidae	36	30			6
	Phocoenidae	7	4			3
	Platanistidae	1	1			
	Iniidae	1				1
	Lipodidae	1				1
	Pontoporidae	1				1
	Total	**88**	**69**	**1**	**5**	**13**
Sirenia						
	Trichechidae	3	1			2
	Dugongidae	1				1
	Total	**4**	**1**			**3**

Abbreviations: UN: Unknown; ST: Stable; IN: Increasing; DE: Decreasing.

(IWC). Consequently, some species of whales (e.g., rorquals) are now recovering from past exploitation; however, other such as the North Atlantic Right Whales (especially the Pacific subpopulations), which were brought to the brink of extinction by commercial whaling, are declining. This is a direct consequence of uncontrolled and illegal whaling, and highlights the past failure of international management to prevent such abuses (Wade 2010).

In spite of the commitment demonstrated by many nations to the conservation of biodiversity, expressed in the agenda of several multilateral efforts (e.g., CMS, CBD, IUCN, CITES, IWC, ACAP, UNCLOS), the current scenario of marine megafauna conservation is somewhat discouraging, in particular due to the poor implementation of regulations in place. Huge efforts by the scientific community in developing technologies may allow the analysis of complex systems and/or applying emerging management paradigms integrating biological, ecological, environmental, and socioeconomic data over multiple spatiotemporal scales to provide information to managers and resource users in near real time (i.e., Dynamic Ocean Management; Lewison et al. 2018). An imperative call for data-sharing within a global repository as a requirement to move analyses into the big data's realm has been launched as an opportunity to enhance sustainable management, marine megafauna conservation and to encourage collaboration to better inform policy makers (Sequeira et al. 2019). All these efforts should be highlighted and boosted, as they appear as the for-coming years' guideline to enhance marine biodiversity and, particularly, the conservation of marine megafauna, which is a globally-shared responsibility.

Summary

1. Marine megafauna, restricted in this chapter to seabirds, marine mammals and sea turtles, conform a group that mainly includes large vertebrates associated with the marine environment during the whole or a significant part of their life cycle.
2. Marine megafauna is globally distributed, concentrating in hot spot areas. Even though biological features vary significantly among groups, many similarities can be found in their ecological roles and life history traits.
3. Seabirds, marine mammals and marine turtles are mostly characterized by extreme life span, deferred maturity and slow growing rates. Most of the species within these groups perform seasonal movements or undertake large migrations.
4. Evolution has led to a range of morphological and physiological adaptations within seabirds and marine mammals that enable them to exploit resources in coastal and oceanic areas in several ways.
5. The high heterogeneity of the environment that marine megafauna inhabit is reflected in a large diversification of diet and feeding tactics.
6. Within marine megafauna, seabirds and marine mammals are considered top predators. All marine megafauna species play an important role in the maintenance of the ecosystem and are considered good environmental monitors integrating ecological variations occurring in marine ecosystems at different spatial and temporal scales.
7. Marine megafauna has an important role as bio-vectors transporting nutrients across ecosystem boundaries, shaping the structure and function of coastal and inland sites, and favoring nutrient cycling.
8. Several seabird, marine turtle and marine mammal populations have undergone and are currently facing declining trends. Marine megafauna is exposed to similar threats as bycatch, pollution, prey depletion, energy production and climate change, among others, and land-based threats as invasive alien species, human disturbance, habitat deterioration, and hunting, among others.
9. Several international instruments have played quite an important role in improving conservation policies around the world. However, there is an urgent need to develop new technologies, apply emerging management paradigms, move the analysis to the field of big data, and promote collaborations as opportunities to improve marine megafauna conservation and sustainable management.

Acknowledgments

This study was financially supported by the Universidad Nacional de Mar del Plata (National University of Mar del Plata, Argentina), Universidad Nacional de Tierra del Fuego (National University of Tierra del Fuego, Argentina), Consejo Nacional de Investigaciones Científicas y Técnicas (National Council for Scientific and Technic Research, CONICET, Argentina), the Agencia Nacional de Promoción Científica y Tecnológica (National Agency for Scientific and Technological Promotion, Argentina), Universidad Autónoma San Luis Potosí (UASLP, México); Philip & Marsha Dowd, and the Agreement on the Conservation of Albatrosses and Petrels (ACAP). Special acknowledgement to AEJ Valenzuela, A Padin Goodall, FA Muñoz Teneria and G Maldonado for the photos provided.

References

Agnew, D.J. 1997. The CCAMLR Ecosystem Monitoring Programme. Antarctic Science 9: 235–242.

Alvarado, J. and C. Delgado. 2005. Tortugas marinas de Michoacán, Historia Natural y Conservación, Primera. Universidad Michoacana de San Nicolás de Hidalgo, Morelia.

Balazs, G.H. and M. Chaloupka. 2004. Spatial and temporal variability in somatic growth of green sea turtles (*Chelonia mydas*) resident in the Hawaiian Archipelago. Mar. Biol. 145: 1043–1059.

Bastida, R. and D.H. Rodríguez. 2009. Mamíferos marinos de Patagonia y Antártida. Vazquez Mazzini, Buenos Aires.

Bearzi, M. 2006. California Sea Lions use dolphins to locate food. Journal of Mammalogy 87: 606–617.

Berta, A., J.L. Sumich and K.M. Kovacs. 2015. Introduction. pp. 1–14. *In*: Berta, A., J.L. Sumich and K.M. Kovacs [eds.]. Marine Mammals: Evolutionary Biology (third edition). Academic Press, San Diego.

Bjorndal, K.A. 1980. Nutrition and grazing behavior of the green turtle *Chelonia mydas*. Mar. Biol. 56: 147–154.

Bjorndal, K.A. 1997. Foraging ecology and nutrition of sea turtles. pp. 199–231. *In*: Lutz, P.L. and J.A. Musick [eds.]. The Biology of Sea Turtles. CRC Press, Boca Raton.

Bjorndal, K.A. 1999. Priorities for research in foraging habitats. pp. 12–15. *In*: Eckert, K.L., K.A. Bjorndal and F.A. Abreu-Grobois [eds.]. Research and Management Techniques for the Conservation of Sea Turtles. SSC/IUCN Marine Turtle Specialist Group, Pennsylvania.

Bjorndal, K.A., A.B. Bolten and M. Chaloupka. 2000. Green turtle somatic growth model: evidence for density dependence. Ecol. Appl. 10: 269–282.

Bjorndal, K.A., J.B.C. Jackson and P.L. Lutz. 2003. Roles of sea turtles in marine ecosystems: Reconstructing the past. pp. 259–270. *In*: Lutz, P.L., J.A. Musick and J. Wyneken [eds.]. The Biology of Sea Turtles, Vol II. CRC Press, Florida.

Bjorndal, K.A., A.B. Bolten, M. Chaloupka, V.S. Saba, C. Bellini, M.A.G. Marcovaldi, A.J.B. Santos, L.F.W. Bortolon, A.B. Meylan, P.A. Meylan, J. Gray, R. Hardy, B. Brost, M. Bresette, J.C. Gorham, S. Connett, B.V.S. Crouchley, M. Dawson, D. Hayes, C.E. Diez, R.P. van Dam, S. Willis, M. Nava, K.M. Hart, M.S. Cherkiss, A.G. Crowder, C. Pollock, Z. Hillis-Starr, F.A. Muñoz Tenería, R. Herrera-Pavón, V. Labrada-Martagón, A. Lorences, A. Negrete-Philippe, M.M. Lamont, A.M. Foley, R. Bailey, R.R. Carthy, R. Scarpino, E. McMichael, J.A. Provancha, A. Brooks, A. Jardim, M. López-Mendilaharsu, D. González-Paredes, A. Estrades, A. Fallabrino, G. Martínez-Souza, G.M. Vélez-Rubio, R.H. Boulon, J.A. Collazo, R. Wershoven, V. Guzmán Hernández, T.B. Stringell, A. Sanghera, P.B. Richardson, A.C. Broderick, Q. Phillips, M. Calosso, J.A.B. Claydon, T.L. Metz, A.L. Gordon, A.M. Landry, D.J. Shaver, J. Blumenthal, L. Collyer, B.J. Godley, A. McGowan, M.J. Witt, C.L. Campbell, C.J. Lagueux, T.L. Bethel and L. Kenyon. 2017. Ecological regime shift drives declining growth rates of sea turtles throughout the West Atlantic. Glob. Chang. Biol. 1–13.

Bolten, A.B. and P.L. Lutz. 2003. Variation in sea turtle life history patterns: Neritic vs. Oceanic developmental stages. pp. 243–257. *In*: Lutz, P.L., J.A. Musick and J. Wyneken [eds.]. The Biology of Sea Turtles Vol II. CRC Press, Florida.

Bonde, K.B. 2018. Sirenian life history. pp. 859–861. *In*: Würsig, B., J.G.M. Thewissen and K.M. Kovacs [eds.]. Encyclopedia of Marine Mammals (third edition). Academic Press.

Bossart, G.D. 2011. Marine mammals as sentinel species for oceans and human health. Wildl Mar. Anim. 48: 676–690.

Bost, C.A., C. Cotte, F. Bailleul, Y. Cherel, J.B. Charrassin, C. Guinet, D.G. Ainley and H. Weimerskirch. 2009. The importance of oceanographic fronts to marine birds and mammals of the southern oceans. Journal of Marine Systems 78: 363–376.

Bowen, B.W. and S.A. Karl. 2007. Population genetics and phylogeography of sea turtles. Mol. Ecol. 16: 4886–4907.

Boyd, I.L., S. Wanless and K.C.J. Camphuysen. 2006. Top Predators in Marine Ecosystems: Their Roles in Monitoring and Management. Cambridge University Press, UK.

Brady, N. and R. Weil. 2002. The Nature and Properties of Soils. 13th edition. Prentice Hall, USA.

Brooke, M.D.L. 2002. Seabird systematics and distribution: a review of current knowledge. pp. 57–85. *In*: Schreiber, A.E. and J. Burger [eds.]. Biology of Marine Birds. CRC Press, United States of America.

Burger, J. and M. Gochfeld. 2002. Effects of chemical and pollution on seabirds. pp. 486–514. *In*: Schreiber, E.A. and J. Burger [eds.]. Biology of Marine Birds. CRC Press, Boca Ratón, Florida.

Butchart, S.H.M., A.J. Stattersfield, L.A. Bennun, S.M. Shutes, H.R. Akcakaya, J.E.M. Baillie, S.N. Stuart, C. Hilton-Taylor and G.M. Mace. 2004. Measuring global trends in the status of biodiversity: Red List Indices for birds. PLoS Biol. 2: e383.

Camacho, M., O.P. Luzardo, L.D. Boada, L.F. López, M. Medina, M. Zumbado and J. Orós. 2013. Potential adverse health effects of persistent organic pollutants on sea turtles: Evidences from a cross-sectional study on Cape Verde loggerhead sea turtles. Sci. Total. Environ. 458–460: 283–289.

Camacho, M., L.D. Boada, J. Orós, P. López, M. Zumbado, M. Almeida-González and O.P. Luzardo. 2014. Monitoring organic and inorganic pollutants in juvenile live sea turtles: Results from a study of *Chelonia mydas* and *Eretmochelys imbricata* in Cape Verde. Sci. Total. Environ. 481: 303–310.

Campbell, T.W. 1994. Hematology. pp. 176–198. *In*: Ritchie, L.R., B.W. Harrison and G.J. Harrison [eds.]. Lake Worth: Wingers Publishing Inc.

Carboneras, C. 1992. Family Procellariidae (Petrels and Shearwaters). pp. 216–271. *In*: Del Hoyo, J., A. Elliot and J. Sargatal [eds.]. Handbook of the Birds of the World. Lynx Edicions, Barcelona.

Cardona-Marek, T., K.K. Knott, B.E. Meyer and T.M. O'Hara. 2009. Mercury concentrations in southern Beaufort sea polar bears: variation based on stable isotopes of carbon and nitrogen. Environ. Toxicol. Chem. 28: 1416–1424.

Castellini, J.M., L.D. Rea, C.L. Lieske, K.B. Beckmen, B.S. Fadely, J.M. Maniscalco and T.M. O'Hara. 2012. Mercury Concentrations in Hair from Neonatal and Juvenile Steller Sea Lions (*Eumetopias jubatus*): Implications Based on Age and Region in this Northern Pacific Marine Sentinel Piscivore.

Casaux, R., A. Baroni, A. Ramón, A. Carlini, M. Bertolin and C.Y. Diprincio. 2009. Diet of the leopard seal Hydrurga leptonyx at the Danco Coast, Antarctic Peninsula. Polar Biol. 32(2): 307–310, doi: 10.1007/s00300-008-0567-0.

Caut, S., E. Angulo, B. Pisanu, L. Ruffino, L. Faulquier, O. Lorvelec, J.-L. Chapus, M. Pascal, E. Vidal and F. Courchamp. 2012. Seabird modulations of isotopic nitrogen on islands. PLoS One 7: 1–7.

Chaloupka, M. and C.J. Limpus. 1997. Robust statistical modelling of hawksbill sea turtle growth rates (southern Great Barrier Reef). Mar. Ecol. Prog. Ser. 146: 1–8.

Chaloupka, M., C.J. Limpus and J. Miller. 2004. Green turtle somatic growth dynamics in a spatially disjunct Great Barrier Reef metapopulation. Coral Reefs 23: 325–335.

Chivers, S.J. 2018. Cetacean life history. pp. 186–189. *In*: Würsig, B., J.G.M. Thewissen and K.M. Kovacs [eds.]. Encyclopedia of Marine Mammals (third edition). Academic Press.

Clua, E. and F. Grosvalet. 2001. Mixed-species feeding aggregation of dolphins, large tunas and seabirds in the Azores. Aquatic Living Resources 14: 11–18.

Convention on Biological Diversity. 2010. Global Biodiversity Outlook 3, Montréal, 94 pp. www.cbd.int/doc/publications/gbo/gbo3-final-en.pdf.

Cords, M. and B. Würsig. 2014. A mix of species: Associations of heterospecifics among primates and dolphins. pp. 409–431. *In*: Yamagiwa, J. and L. Karczmarski [eds.]. Primates and Cetaceans: Field Studies and Conservation of Complex Mammalian Societies. Springer Press, Tokyo, Japan.

Croxall, J.P., S.H.M. Butchart, B. Lascelles, A.J. Stattersfield, B. Sullivan, A. Symes and P. Taylor. 2012. Seabird conservation status, threats and priority actions: a global assessment. Bird Conserv. Internat. 22: 1–34.

Cunningham, A.A., P. Daszak and J.L.N. Wood. 2017. One Health, emerging infectious diseases and wildlife: two decades of progress? Philos. Trans. R. Soc. B. Biol. Sci. 372: 20160–167.

da Silva, C.C., R.D. Klein, I.F. Barcarolli and A. Bianchini. 2016. Metal contamination as a possible etiology of fibropapillomatosis in juvenile female green sea turtles *Chelonia mydas* from the southern Atlantic Ocean. Aquat. Toxicol. 170: 42–51.

Dalebout, M.L., C.S. Baker, D. Steel, K. Thompson, K.M. Robertson, S.J. Chivers, W.F. Perrin, M. Goonatilake, R.C. Anderson, J.G. Mead, C.W. Potter, L. Thompson, D. Jupiter and T.K. Yamada. 2014. Resurrection of Mesoplodon hotaula Deraniyagala 1963: A new species of beaked whale in the tropical Indo-Pacific. Marine Mammal Sci. 30(3): 1081–1108.

Diamond, A.W. and C.M. Devlin. 2003. Seabirds as indicators of changes in marine ecosystems: Ecological monitoring on machias seal island. Environmental Monitoring and Assessment 88: 153–175.

Dierauf, L.A. and F.M. Gulland. 2001. CRC Handbook of Marine Mammal Medicine, Second Edition. CRC Press, Boca Raton, FL. USA.

Dobson, F.S. and P. Jouventin. 2007. The trade-off of reproduction and survival in slow-breeding seabirds. Canadian J. Zoology 88(9): 889–899.

Einoder, L.D. 2009. A review of the use of seabirds as indicators in fisheries and ecosystem management. Fisheries Res. 95: 6–13.

Elliott, K.H., K.J. Woo, A.J. Gaston, S. Benvenuti, L. Dall'Antonia and G.K. Davoren. 2009. Central-place foraging in an arctic seabird provides evidence for Storer-Ashmole's Halo. The Auk 126: 613–625.

Folkens, P.A. and R.R. Reeves. 2014. Guide to marine mammals of the world (No. Sirsi i9780375411410). National Audubon Society.

Forcada, J. 2018. Distribution. pp. 259–262. *In*: Würsig, B., J.G.M. Thewissen and K.M. Kovacs [eds.]. Encyclopedia of Marine Mammals (third edition). Academic Press.

Furness, R.W. and J.J.D. Greenwood. 1993. Birds as Monitors of Environmental Change. Chapman & Hall, UK.
García-Borboroglu, J.P. and P.D. Boersma. 2015. Pingüinos: historia natural y conservación. Vasquez Manzzini, Buenos Aires, Argentina.
García, G., M. Favero and A.I. Vassallo. 2010. Factors affecting kleptoparasitism by gulls in a multi-species seabird colony. Condor 112: 521–529.
González Carman, V., K.C. Álvarez, L. Prosdocimi, M.C. Inchaurraga, R.F. Dellacasa, A. Faiella, C. Echenique, R. González, J. Andrejuk, H.W. Mianzan, C. Campagna and D.A. Albareda. 2011. Argentinian coastal waters: A temperate habitat for three species of threatened sea turtles. Mar. Biol. Res. 7: 500–508.
González-Zevallos, D. and P. Yorio. 2006. Seabird use of discards and incidental captures at the Argentine hake trawl fishery in the Golfo San Jorge, Argentina. Marine Ecology and Progress Series 316: 175–183.
Graham, B.S., P.L. Koch, S.D. Newsome, K.W. McMahon and D. Aurioles. 2010. Using isoscapes to trace the movements and foraging behaviour of top preadators in oceanic ecosystems. pp. 299–318. *In*: West, J. [ed.]. Isoscapes: Understanding Movement, Pattern, and Process on Earth Through Isotope Mapping. Springer, Berlin.
Hamann, M., C.J. Limpus and J.M. Whittier. 2002. Patterns of lipid storage and movilisation in the female green sea turtle (*Chelonia mydas*). J. Comp. Physiol. B. 172: 485–493.
Hamann, M., C.J. Limpus and D.W. Owens. 2003. Reproductive cycles of males and females. pp. 135–161. *In*: Lutz, P.L., J.A. Musick and J. Wyneken [eds.]. The Biology of Sea Turtles. Vol II. CRC Press, Boca Raton.
Hamann, M., T.S. Jessop, C.J. Limpus and J.M. Whittier. 2005. Regional and annual variation in plasma steroids and metabolic indicators in female green turtle, *Chelonia mydas*. Mar. Biol. 148: 427–433.
Hatase, H., N. Takai, Y. Matsuzawa, W. Sakamoto, K. Omuta, K. Goto, N. Arai and T. Fujiwara. 2002. Size-related differences in feeding habitat use of adult female loggerhead turtles Caretta caretta around Japan determined by stable isotope analyses and satellite telemetry. Mar. Ecol. Prog. Ser. 233: 273–281.
Hawkes, L.A., A. McGowan, A.C. Broderick, S. Gore, D. Wheatley, J. White, M.J. Witt and B.J. Godley. 2014. High rates of growth recorded for hawksbill sea turtles in Anegada, British Virgin Islands. Ecol. Evol. 4: 1255–1266.
Hays, G.C. and R. Scott. 2013. Global patterns for upper ceilings on migration distance in sea turtles and comparisons with fish, birds and mammals. Funct. Ecol. 27: 748–756.
Heithaus, M.R., L.M. Dill and J.J. Kiszka. 2018. Feeding strategies and tactics. pp. 354–363. *In*: Würsig, B., J.G.M. Thewissen and K.M. Kovacs [eds.]. Encyclopedia of Marine Mammals (third edition). Academic Press.
Heppell, S.S., M.L. Snover, L. Crowder and P.L. Lutz. 2003. Sea turtle population ecology. pp. 275–306. *In*: Lutz, P.L., J.A. Musick and J. Wyneken [eds.]. The Biology of Sea Turtles. Vol II. CRC Press, Florida.
Herbst, L.H. and E.R. Jacobson. 1995. Diseases of marine turtles. pp. 593–596. *In*: Bjorndal, K.A. [ed.]. Biology and Conservation of Sea Turtles. Smithsonian Institution Press, Washington, D.C.
Hobson, K.A. and L.I. Wassenaar. 2008. Tracking Animal Migration with Stable Isotopes. Elsevier Inc., USA.
Hoelzel, A.R. 1991. Killer whale predation on marine mammals at Punta Norte, Argentina; food sharing, provisioning and foraging strategy. Behav. Ecol. Sociobiol. 29: 197–204.
Hoffmann, M., C. Hilton-Taylor, A. Angulo, M. Böhm, T.M. Brooks, S.H.M. Butchart, K.E. Carpenter, J. Chanson, B. Collen, N.A. Cox, W.R.T. Darwall, N.K. Dulvy, L.R. Harrison, V. Katariya, C.M. Pollock, S. Quader, N.I. Richman, A.S.L. Rodrigues, M.F. Tognelli, J.-C. Vié, J.M. Aguiar, D.J. Allen, G.R. Allen, G. Amori, N.B. Ananjeva, F. Andreone, P. Andrew, A.L.A. Ortiz, J.E.M. Baillie, R. Baldi, B.D. Bell, S.D. Biju, J.P. Bird, P. Black-Decima, J.J. Blanc, F. Bolaños, W. Bolivar-G, I.J. Burfield, J.A. Burton, D.R. Capper, F. Castro, G. Catullo, R.D. Cavanagh, A. Channing, N.L. Chao, A.M. Chenery, F. Chiozza, V. Clausnitzer, N.J. Collar, L.C. Collett, B.B. Collette, C.F.C. Fernandez, M.T. Craig, M.J. Crosby, N. Cumberlidge, A. Cuttelod, A.E. Derocher, A.C. Diesmos, J.S. Donaldson, J.W. Duckworth, G. Dutson, S.K. Dutta, R.H. Emslie, A. Farjon, S. Fowler, J. Freyhof, D.L. Garshelis, J. Gerlach, D.J. Gower, T.D. Grant, G.A. Hammerson, R.B. Harris, L.R. Heaney, S.B. Hedges, J.-M. Hero, B. Hughes, S.A. Hussain, J. Icochea M.R.F. Inger, N. Ishii, D.T. Iskandar, R.K.B. Jenkins, Y. Kaneko, M. Kottelat, K.M. Kovacs, S.L. Kuzmin, E. La Marca, J.F. Lamoreux, M.W.N. Lau, E.O. Lavilla, K. Leus, R.L. Lewison, G. Lichtenstein, S.R. Livingstone, V. Lukoschek, D.P. Mallon, P.J.K. McGowan, A. McIvor, P.D. Moehlman, S. Molur, A.M. Alonso, J.A. Musick, K. Nowell, R.A. Nussbaum, W. Olech, N.L. Orlov, T.J. Papenfuss, G. Parra-Olea, W.F. Perrin, B.A. Polidoro, M. Pourkazemi, P.A. Racey, J.S. Ragle, M. Ram, G. Rathbun, R.P. Reynolds, A.G.J. Rhodin, S.J. Richards, L.O. Rodríguez, S.R. Ron, C. Rondinini, A.B. Rylands, Y. Sadovy de Mitcheson, J.C. Sanciangco, K.L. Sanders, G. Santos-Barrera, J. Schipper, C. Self-Sullivan, Y. Shi, A. Shoemaker, F.T. Short, C. Sillero-Zubiri, D.L. Silvano, K.G. Smith, A.T. Smith, J. Snoeks, A.J. Stattersfield, A.J. Symes, A.B. Taber, B.K. Talukdar, H.J. Temple, R. Timmins, J.A. Tobias, K. Tsytsulina, D. Tweddle, C. Ubeda, S.V. Valenti, P. Paul van Dijk, L.M. Veiga, A. Veloso, D.C. Wege, M. Wilkinson, E.A. Williamson, F. Xie, B.E. Young, H.R. Akçakaya, L. Bennun, T.M. Blackburn, L. Boitani, H.T. Dublin, G.A.B. da Fonseca, C. Gascon, T.E. Lacher, G.M. Mace, S.A. Mainka, J.A. McNeely, R.A. Mittermeier, G.M. Reid, J.P. Rodriguez, A.A. Rosenberg, M.J. Samways, J. Smart, B.A. Stein and S.N. Stuart. 2010. The impact of conservation on the status of the World's vertebrates. Science 330: 1503.
Innis, C., C. Merigo, K. Dodge, M. Tlusty, M. Dodge, B. Sharp, A. Myers, A. McIntosh, D. Wunn, C. Perkins and others. 2010. Health evaluation of leatherback turtles (*Dermochelys coriacea*) in the northwestern Atlantic during direct capture and fisheries gear disentanglement. Chelonian Conserv. Biol. 9: 205–222.

IUCN. 2012. IUCN Red List Categories and Criteria: Version 3.1. Second edition. Gland, Switzerland and Cambridge, UK: IUCN. iv + 32pp.
IUCN. 2019. The IUCN Red List of Threatened Species.
International Whaling Commission. http://www.iwcoffice.org/Conservation & Management/ Whaling/Commercial Whaling/ htm. Accessed 3 June 2020.
Jefferson, T.A., M. Webber and R.L. Pitman. 2008. Marine Mammals of the World: A Comprehensive Guide to their Identification. Elsevier, London, 573 pp.
Jefferson, T.A. and J.Y. Wang. 2011. Revision of the taxonomy of finless porpoises (genus Neophocaena): the existence of two species. J. Mar. Animals Ecol. 4: 3–16.
Jenkins, C.N. and K. Van Houtan. 2016. Global and regional priorities for marine biodiversity protection. Biol. Conserv. 204: 333–339.
Jones, K.C. and P. de Voogt. 1999. Persistent organic pollutants (POPs): State of the science. Environmental Pollution 100: 209–211.
Jones, K., E. Ariel, G. Burgess and M. Read. 2016. A review of fibropapillomatosis in Green turtles (*Chelonia mydas*). Vet. J. 212: 48–57.
Jouventin, P., D. Capdeville, F. Cuenot-Chaillet and C. Boiteau. 1994. Exploitation of pelagic resources by a non-flying seabird: satellite tracking of the king penguin throughout the breeding cycle. Mar. Ecol. Prog. Ser. 106: 11–19.
Keller, J.M. 2013. Exposure to and effects of persistent organic pollutants. pp. 285–328. *In*: Wyneken, J., K.J. Lohmann and J.A. Musick [eds.]. The Biology of Sea Turtles. Vol III. CRC Press, Boca Raton, Florida.
Labrada-Martagón, V., L.C. Méndez-Rodríguez, S.C. Gardner, M. López-Castro and T. Zenteno-Savín. 2010a. Health indices of the Green turtle (*Chelonia mydas*) along the Pacific Coast of Baja California Sur, Mexico. I. Blood biochemistry values. Chelonian Conserv. Biol. 9: 162–172.
Labrada-Martagón, V., L.C. Méndez-Rodríguez, S.C. Gardner, V.H. Cruz-Escalona and T. Zenteno-Savín. 2010b. Health indices of the green turtle (*Chelonia mydas*) along the Pacific Coast of baja California Sur, Mexico. II. Body condition index. Chelonian Conserv. Biol. 9: 162–172.
Labrada-Martagón, V., P.A. Tenorio Rodríguez, L.C. Méndez-Rodríguez and T. Zenteno-Savín. 2011. Oxidative stress indicators and chemical contaminants in East Pacific green turtles (*Chelonia mydas*) inhabiting two foraging coastal lagoons in the Baja California peninsula. Comp. Biochem. Physiol. - C Toxicol. Pharmacol. 154: 65–75.
Labrada-Martagón, V., F.A. Muñoz Tenería, R. Herrera-Pavón and A. Negrete-Philippe. 2017. Somatic growth rates of immature green turtles Chelonia mydas inhabiting the foraging ground Akumal Bay in the Mexican Caribbean Sea. J. Exp. Mar. Bio. Ecol. 487: 68–78.
Lascelles, B.G., P.R. Taylor, M.G.R. Miller, M.P. Dias, S. Oppel, L. Torres, A. Hedd, M. Le Corre, R.A. Phillips, S.A. Shaffer, H. Weimerskirch and C. Small. 2016. Applying global criteria to tracking data to define important areas for marine conservation. Divers. and Distrib. 22: 422–431.
Lavery, T.J., B. Roudnew, P. Gill, J. Seymour, L. Seuront, G. Johnson, J.G. Mitchell and V. Smetacek. 2010. Iron defecation by sperm whales stimulates carbon export in the Southern Ocean. Proc. R. Soc. London B 277: 3527–3531.
Lavery, T.J., B. Roudnew, J. Seymour and others. 2014. Whales sustains fisheries: blue whales stimulate primary production in the Southern Ocean. Mar. Mamm. Sci. 30: 888–904. doi: 10.1111/mms.12108.
Learmonth, J.A., C.D. Macleod, M.B. Santos, G.J. Pierce, H.Q.P. Crick and R.A. Robinson. 2006. Potential effects of climate change on marine mammals. Oceanogr. Mar. Biol. 44: 431–464.
Lewison, R.L., A.F. Johnson and G.M Verutes. 2018. Embracing complexity and complexity-awareness in marine megafauna conservation and research. Front. Mar. Sci. 5: 207. doi: 10.3389/fmars.2018.00207.
Lohmann, K.J., S.D. Cain, S.A. Dodge and C.M. Lohmann. 2001. Regional magnetic fields as navigational markers for sea turtles. Science (80-) 294: 364–366.
Lohmann, K.J., C.M.F. Lohmann, J.R. Brothers and N.F. Putman. 2013. Natal homing and imprinting in sea turtles. pp. 59–77. *In*: Wyneken, J., K.J. Lohmann and J.A. Musick [eds.]. The Biology of Sea Turtles, 3rd edn. CRC Press, Boca Raton.
Lohmann, K.J. and C.M.F. Lohmann. 2019. There and back again: natal homing by magnetic navigation in sea turtles and salmon. J. Exp. Biol. 222: jeb184077.
Lopez-Mendilaharsu, M., S.C. Gardner and J.A. Seminoff. 2002. Feeding ecology of the East Pacific green turtle (*Chelonia mydas agassizii*) in Bahía Magdalena, B.C.S., Mexico. pp. 218–219. *In*: Seminoff, J.A. [ed.]. NOAA.
Lopez-Mendilaharsu, M., S.C. Gardner and J.A. Seminoff. 2003. Chelonia mydas agassizii (East Pacific Green Sea turtle) Diet. Herpetol. Rev. 34: 139–140.
Lopez-Mendilaharsu, M., S.C. Gardner, J.A. Seminoff and R. Riosmena. 2005. Identifying critical foraging habitats of the green turtle (*Chelonia mydas*) along the Pacific coast of the Baja California peninsula, Mexico. Aquat. Conserv. Mar. Freshw. Ecosyst. 15: 259–269.
Lutcavage, M.E., P. Plotkin, B. Witherington and P.L. Lutz. 1997. Human impacts on sea turtle survival. pp. 387–409. *In*: Lutz, P.L. and J.A. Musick [eds.]. The Biology of Sea Turtles. CRC Press, Boca Raton.
Macleod, C.D. 2009. Global climate change, range changes and potential implications for the conservation of marine cetaceans: a review and synthesis. Endangered Species Research 7: 125–136.

Marquez, R. 1990. Sea Turtles of the World. Species Catalogue. Food and Agriculture Organization, Rome.
Martínez, I. 1992. Orden Sphenisciformes. pp. 140–161. *In*: del Hoyo, J., A. Elliott and J. Sargatal [eds.]. Handbook of the Birds of the World - Volume 1. Ostrich to Ducks. Lynx Edicions.
Merrick, R.L., G.K. Silber and D.P. DeMaster. 2018. Endangered species and populations. pp. 313–318. *In*: Würsig, B., J.G.M. Thewissen and K.M. Kovacs [eds.]. Encyclopedia of Marine Mammals (third edition). Academic Press.
McCauley, D.J., M.L. Pinsky, S.R. Palumbi, J.A. Estes, F.H. Joyce and R.R. Warner. 2015. Marine defaunation: animal loss in the global ocean. Science 347: 1255641. doi: 10.1126/science.1255641.
Michener, R.H. and L. Kaufam. 2007. Stable isotopes ratios as tracers in marine food webs: an update. pp. 238–282. *In*: Michener, R. and Lajtha [eds.]. Stable Isotopes in Ecology and Environmental Science. 2nd ed. Balckwell Publishing Ltd, Malden, MA, USA.
Molina, A.L. and B.I. Van Tussenbroek. 2014. Patch dynamics and species shifts in seagrass communities under moderate and high grazing pressure by green sea turtles. Mar. Ecol. Prog. Ser. 517: 143–157.
Montero, R. and A.G. Autino. 2018. Sistemática y filogenia de los vertebrados, con énfasis en la fauna argentina. Tercera edition. Editorial independiente, San Miguel de Tucumán, Argentina.
Moore, J.E. and R. Zydelis. 2008. Quantifying seabird bycatch: where do we go from here? Animal Conservation 11: 257–259.
Moore, S.E. 2008. Marine mammals as ecosystem sentinels. J. Mammal. 89: 534–540.
Moore, S.E. 2018. Climate change. pp. 194–197. *In*: Würsig, B., J.G.M. Thewissen and K.M. Kovacs [eds.]. Encyclopedia of Marine Mammals, 3rd edn, Academic Press, New York.
Musick, J.A. and C.J. Limpus. 1997. Habitat utilization and migration in juveniles sea turtles. pp. 137–163. *In*: Lutz, P.L. and J.A. Musick [eds.]. The Biology of Sea Turtles. CRC, Boca Raton.
Nettleship, D.N. 1996. Familia Alcidae (Auks). pp. 678. *In*: del Hoyo, J., A. Elliott and J. Sargatal [eds.]. Handbook of the Birds of the World - Volume 3. Hoatzin to Auks. Lynx Editions.
Newman, S.H., J.F. Piatt and J.G. White. 1997. Hematological and plasma biochemical reference ranges of Alaskan seabirds: their ecological significance and clinical importance. Colonial Waterbirds 20: 492–504.
Newton, I. 2008. The Migration Ecology of Birds. Academic Press.
Norris, K.S. and T.P. Dohl. 1980. The structure and function of cetacean schools. pp. 211–261. *In*: Herman, L.M. [ed.]. Cetacean Behavior: Mechanism and Functions. Krieger Publishing Company, New York, NY. USA.
O'Hara, T.M. and T.J. O'Shea. 2001. Toxicology. pp. 471–520. *In*: Dierauf, L.A. and F.M.D. Gulland [eds.]. Handbook of Marine Mammal Medicine, Second. CRC Press, Boca Raton, Florida.
Oro, D. and A. Martínez-Abraín. 2004. Ecology and behavior of seabirds. pp. 1–11. *In*: UNESCO [ed.]. Enciclopedia of Life Support Systems. EOLSS.
Oro, D., M. Genovart, G. Tavecchia, M.S. Fowler and A. Martínez-Abraín. 2013. Ecological and evolutionary implications of food subsidies from humans. Ecology Letters 16: 1501–1514.
Otero, X.L., S. De La Peña-Lastra, A. Pérez-Alberti, T.O. Ferreira and M.A. Huerta-Diaz. 2018. Seabird colonies as important global drivers in the nitrogen and phosphorus cycles. Nature Communications 9.
Palacios, D. and B. Mate. 1996. Attack by false killer whales (*Pseudorca crassidens*) on sperm whales (*Physeter macrocephalus*) in the Galapagos Islands. Marine Mammal Science 12: 582–587.
Paladino, F.V., M.P. O'Connor and J.R. Spotila. 1990. Metabolism of leatherback turtles, gigantothermy, and thermoregulation of dinosaurs. Nature 344(6269): 858–860.
Paleczny, M., E. Hammill, V. Karpouzi and D. Pauly. 2015. Population trend of the world's monitored seabirds, 1950–2010. PLoS One 10: 1–11.
Pardo, D., J. Forcada, A.G. Wood, G.N. Tuck, L. Ireland, R. Pradel, J.P. Croxall and R.A. Phillips. 2016. Additive effects of climate and fisheries drive ongoing declines in multiple albatross species. PNAS 10829–10837, www.pnas.org/cgi/doi/10.1073/pnas.1618819114.
Patrício, A.R., C.E. Diez, R.P. van Dam and B.J. Godley. 2016. Novel insights into the dynamics of green turtle fibropapillomatosis. Mar. Ecol. Prog. Ser. 547: 247–255.
Pauly, D., V. Christensen, J. Dalsgaard, R. Froese and J.F.C. Torres. 1998. Fishing down marine food webs. Science 278: 860–863.
Peckham, S., D. Maldonado Diaz, Y. Tremblay, R. Ochoa, J. Polovina, G. Balazs, P. Dutton and W. Nichols. 2011. Demographic implications of alternative foraging strategies in juvenile loggerhead turtles Caretta caretta of the North Pacific Ocean. Mar. Ecol. Prog. Ser. 425: 269–280.
Pedersen, A.B. and A. Fenton. 2007. Emphasizing the ecology in parasite community ecology. Trends Ecol. Evol. 22: 133–139.
Plotkin, P. and P.L. Lutz. 2003. Adult migrations and habitat use. pp. 225–241. *In*: Lutz, P.L., J.A. Musick and J. Wyneken [eds.]. The Biology of Sea Turtles. Vol II. CRC Press, Florida.
Polis, G.A. and S.D. Hurd. 1996. Linking marine and terrestrial food webs: Allochthonous input from the ocean supports high secondary productivity on small islands and coastal land communities. Am. Nat. 147: 396–423.
Powell, G.V.N., J.W. Fourqurean, W.J. Kenworthy and J.C. Zieman. 1991. Bird colonies cause seagrass enrichment in a subtropical estuary. Estuarine, Coastal and Shelf Science 32: 567–579.
Pritchard, P.C.H. 1997. Evolution, phylogeny and current status. pp. 1–28. *In*: Lutz, P.L. and J.A. Music [eds.]. The Biology of Sea Turtles. CRC Press, Boca Raton.

Pütz, K. 1994. Aspects of the feeding ecology of emperor penguins (*Aptenodytes forsteri*) and king penguins (*Aptenodytes patagonicus*). Reports in Polar Research 136: 139–139.

Ratnarajah, L., A.R. Bowie, D. Lannuzel, K.M. Meiners and S. Nicol. 2014. The biogeochemical role of baleen whales and krill in Southern Ocean nutrient cycling. PLoS One 9(12): e114067. doi:10.1371/journal.pone.0114067.

Rees, A.F., J. Alfaro-Shigueto, P.C.R. Barata, K.A. Bjorndal, A.B. Bolten, J. Bourjea, A.C. Broderick, L.M. Campbell, L. Cardona, C. Carreras, P. Casale, S.A. Ceriani, P.H. Dutton, T. Eguchi, A. Formia, M.M.P.B. Fuentes, W.J. Fuller, M. Girondot, M.H. Godfrey, M. Hamann, K.M. Hart, G.C. Hays, S. Hochscheid, Y. Kaska, M.P. Jensen, J.C. Mangel, J.A. Mortimer, E. Naro-Maciel, C.K.Y. Ng, W.J. Nichols, A.D. Phillott, R.D. Reina, O. Revuelta, G. Schofield, J.A. Seminoff, K. Shanker, J. Tomás, J.P. van de Merwe, K.S. Van Houtan, H.B. Vander Zanden, B.P. Wallace, K.R. Wedemeyer-Strombel, T.M. Work and B.J. Godley. 2016. Are we working towards global research priorities for management and conservation of sea turtles? Endanger. Species Res. 31: 337–382.

Riddick, S.N., U. Dragosits, T.D. Blackall, F. Daunt, S. Wanless and M.A. Sutton. 2012. The global distribution of ammonia emissions from seabird colonies. Atmospheric Environment 55: 319–327.

Roman, J. and J.J. McCarthy. 2010. The whale pump: marine mammals enhance primary productivity in a coastal basin. PLoS One 5(10): e13255.

Roman, J., J.A. Estes, L. Morissette, C. Smith, D. Costa, J. McCarthy, J.B. Nation, S. Nicol, A. Pershing and V. Smetacek. 2014. Whales as marine ecosystem engineers. Frontiers in Ecology and the Environment 12: 377–385.

Ross, P.S. 2000. Marine mammals as sentinels in ecological risk assessment. Hum. Ecol. Risk Assess 6: 29–46.

Rothschild, M. and T. Clay. 1952. Fleas, Flukes and Cuckoos: A Study of Bird Ectoparasites. Collins Son and Co. Ltd, London, UK.

Santoro, M. and S. Mattiucci. 2009. Sea turtle parasites. pp. 507–519. *In*: Wehrtmann, I.S. and J. Cortés [eds.]. Marine Biodiversity of Costa Rica, Central America. Springer Science & Business Media.

Schmidt-Nielsen, K. 1960. The salt-secreting gland of marine birds. Circulation 21: 955–967.

Schreiber, E.A. and J. Burger. 2002. Seabirds in the marine environment. pp. 1–15. *In*: Schreiber, E.A. and J. Burger [eds.]. Biology of Marine Birds. CRC Press.

Seminoff, J.A., A. Resendiz, W.J. Nichols and T.T. Jones. 2002. Growth rates of wild green turtles *Chelonia mydas* at a temperate foraging area in the Gulf of California, Mexico. Copeia 3: 610–617.

Seminoff, J.A. 2004. Sea Turtles, Red Listing, and the Need for Regional Assessments, Marine Turtle Newsletter No. 106.

Seminoff, J.A., A. Resendiz, B. Resendiz and W.J. Nichols. 2004. Occurrence of loggerhead sea turtles (*Caretta caretta*) in the Gulf of California, Mexico: evidence of life-history variation in the Pacific Ocean. Herpetol. Rev. 35: 24–26.

Seminoff, J.A., T. Eguchi, J. Carretta, C. Allen, D. Prosperi, R. Rangel, J. Gilpatrick, K. Forney and S. Peckham. 2014. Loggerhead sea turtle abundance at a foraging hotspot in the eastern Pacific Ocean: implications for at-sea conservation. Endanger Species Res. 24: 207–220.

Senties, A.G., J. Espinoza-Avalos and J.C. Zurita. 1999. Epizoic algae of nesting sea turtles *Caretta caretta* (L.) and *Chelonia midas* (L.) from the Mexican Caribbean. Bull Mar. Sci. 64: 185–188.

Sequeira, A.M.M., G.C. Hays, D.W. Sims, V.M. Eguíluz, J.P. Rodríguez, M.R. Heupel, R. Harcourt, H. Calich, N. Queiroz, D.P. Costa, J. Fernández-Gracia, L.C. Ferreira, S.D. Goldsworthy, M.A. Hindell, M.-A. Lea, M.G. Meekan, A.M. Pagano, S.A. Shaffer, J. Reisser, M. Thums, M. Weise and C.M. Duarte. 2019. Overhauling ocean spatial planning to improve marine megafauna conservation. Front. Mar. Sci. 6: 639. doi: 10.3389/fmars.2019.00639.

Shane, S.H. 1994. Occurrence and habitat use of marine mammals at Santa Catalina Island, California from 1983–1991. Bulletin of the Southern California Academy of Sciences 93: 13–29.

Simmonds, M.P. and S.J. Isaac. 2007. The impacts of climate change on marine mammals: early signs of significant problems. Oryx 41: 19–26.

Smultea, M.A., C.E. Bacon, K. Lomac-MacNair, F. Visser and J. Bredvik. 2014. Rare mixed-species associations between sperm whales and Risso's and Northern Right Whale Dolphins off the Southern California Bight: kleptoparasitism and social parasitism? Northwest. Nat. 95(1): 43–50.

Sparling, D.W. 2016. Ecotoxicology Essentials: Environmental Contaminants and their Biological Effects on Animals and Plants. Academic Press, USA.

Spotila, J.R. 2004. Sea Turtles, A Complete Guide to their Biology, Behavior and Conservation. Jons Hopkins University Press, Baltimore, 227 p.

Stern, S.J. and A.S. Friedlaender. 2018. Migration and movement. pp. 602–608. *In*: Würsig, B., J.G.M. Thewissen and K.M. Kovacs [eds.]. Encyclopedia of Marine Mammals (third edition). Academic Press.

Suazo, C.G., R.P. Schlatter, A.M. arriagada, L. Cabezas and J. Ojeda. 2013. Fishermen's perceptions of interactions between seabirds and artisanal fisheries in the Chonos archipelago. Chilean Patagonia Fauna & Flora International, Oryx 47(2): 184–189.

Tanabe, S., M. Watanabe, T.B. Minh, T. Kunisue, S. Nakanishi, H. Ono and H. Tanaka. 2004. PCDDs, PCDFs, and Coplanar PCBs in albatross from the North Pacific and Southern Oceans: Levels, patterns, and toxicological implications. Environmental Science and Technology 38: 403–413.

Thiebault, A., R.H.E. Mullers, P. Pistorius and Y. Tremblay. 2014. Local enhancement in a seabird: reaction distances and foraging consequence of predator aggregations. Behavioral Ecology 25: 1302–1310.

Turner Tomaszewicz, C.N., J.A. Seminoff, S.H. Peckham, L. Avens and C.M. Kurle. 2017. Intrapopulation variability in the timing of ontogenetic habitat shifts in sea turtles revealed using δ15N values from bone growth rings. J. Anim. Ecol. 86: 694–704.

Vázquez-Medina, J.P., T. Zenteno-Savín and R. Elsner. 2006. Antioxidant enzymes in ringed seal tissues: potential protection against dive-associated ischemia/reperfusion. Comp. Biochem. Physiol. Part C: Toxicol. Pharmacol. 142: 198–204.

Vázquez-Medina, J.P., J.G. Soñanez-Organis, J.M. Burns, T. Zenteno-Savín and R.M. Ortiz. 2011. Antioxidant capacity develops with maturation in the deep-diving hooded seal. J. Exp. Biol. 214: 2903–2910.

Wallace, B.P., J.A. Seminoff, S.S. Kilham, J.R. Spotila and P. Dutton. 2006. Leatherback turtles as oceanographic indicators: stable isotope analyses reveal a trophic dichotomy between ocean basins. Mar. Biol. 149: 953–960.

Wallace, B.P., A.D. DiMatteo, B.J. Hurley, E.M. Finkbeiner, A.B. Bolten, M.Y. Chaloupka, B.J. Hutchinson, F.A. Abreu-Grobois, D. Amorocho, K.A. Bjorndal, J. Bourjea, B.W. Bowen, R.B. Dueñas, P. Casale, B.C. Choudhury, A. Costa, P.H. Dutton, A. Fallabrino, A. Girard, M. Girondot, M.H. Godfrey, M. Hamann, M. López-Mendilaharsu, M.A. Marcovaldi, J.A. Mortimer, J.A. Musick, R. Nel, N.J. Pilcher, J.A. Seminoff, S. Troëng, B. Witherington and R.B. Mast. 2010. Regional management units for marine turtles: a novel framework for prioritizing conservation and research across multiple scales. PLoS One 5: e15465.

Weimerskirch, H., Y. Cherel, F. Cuenot Chaillet and V. Ridoux. 1997. Alternative foraging strategies and resource allocation by male and female wandering albatrosses. Ecology 78: 2051–2063.

Weller, DW., B. Würsig, H. Whitehead, J.C. Norris, S.K. Lynn, R.W. Davis, N. Clauss and P. Brown. 1996. Observations of an interaction between Sperm Whales and Short-finned Pilot Whales in the Gulf of Mexico. Marine Mammal Science 12: 588–594.

Whilde, J., M.Q. Martindale and D.J. Duffy. 2017. Precision wildlife medicine: applications of the human-centred precision medicine revolution to species conservation. Glob. Chang. Biol. 23: 1792–1805. doi: 10.1111/gcb.13548.

Work, T.M., G.H. Balazs, J.L. Schumacher and M. Amarisa. 2005. Epizootiology of spirorchiid infection in green turtles (*Chelonia mydas*) in Hawaii. J. Parasitol. 91: 871–876.

Würsig, B., J.G.M. Thewissen and K.M. Kovacs. 2018. Encyclopedia of Marine Mammals. Academic Press.

Wyneken, J. and P.L. Lutz. 2003. The external morphology, musculoskeletal system and neuro-anatomy of sea turtles. pp. 39–76. *In*: Lutz, P.L., J.A. Musick and J. Wyneken [eds.]. The Biology of Sea Turtles. Vol. II. CRC Press, Florida.

Zenteno-Savín, T., J. St Leger and P.J. Ponganis. 2010. Hypoxemic and ischemic tolerance in emperor penguins. Comp. Biochem. Physiol. Part C: Toxicol. Pharmacol. 152: 18–23.

Zurita, J.C., R. Herrera and B. Prezas. 1993. Tortugas Marinas del Caribe. pp. 735–751. *In*: Salazar Vallejo, S.I. and N.E. González [eds.]. Biodiversidad Marina y Costera de México. Chetumal: CONABIO/CIQRO.

Section III

Topics of Current Interest

CHAPTER 13

Thermal Plasticity in Marine Organisms

Santiago Salinas[1,*] and *Nishad Jayasundara*[2]

13.1 Introduction

Temperature is one of the most important environmental variables in the life of organisms. Its influence on physiology (Angilletta 2009), life history (Atkinson 1996), macroecology (Gillooly et al. 2001), and community structure (Flanagan et al. 2019) is undeniable. Biologists have long recognized this, and have been studying temperature since Aristotle's time (Leroi 2015). Researching the effects of temperature has also recently taken on considerable importance because of anthropogenic climate change (IPCC 2018). Marine organisms may be even more vulnerable to warming than terrestrial ones (Pinsky et al. 2019) and it is critical to understand potential responses, which include migration, evolutionary change, and phenotypic plasticity.

Phenotypic plasticity, defined as the ability of a single genotype to produce different phenotypes when exposed to different environments (Pigliucci 2001), is ubiquitous in nature. Richard Woltereck, a German zoologist, first coined the term *Reaktionsnorm* in 1909 to describe the patterns in which his *Daphnia* changed when exposed to different nutrient levels—and we have been using the anglicized version ("reaction norm") ever since (Peirson 2012). The reaction norm, then, is the curve obtained when plotting an environmental variable (e.g., temperature) against a trait (e.g., growth rate) for a given genotype (in species where clones are hard to obtain, siblings are typically used in experiments). Despite this early beginning, formal study of plasticity in ecology and evolution did not blossom until the late 1980s (Forsman 2014), in part because of the emphasis on genes and gene frequencies embraced by the modern synthesis (any effect of the environment on organisms was simply dismissed as noise obscuring the fundamental process of gene selection; Sultan 1992).

Why learn about plasticity in a book about marine biology, you may ask? As Gould (1985) aptly put it, "variation is the hard reality, not a set of imperfect measures for a central tendency"—if we are to understand biology, we must understand variation. By giving plasticity our attention, we give the environment the dual role it deserves: the role of creator/inducer of phenotypic variation (plasticity) and the role of selector among the variation (selection; Whitman and Agrawal 2009). In addition to phenotypic plasticity, which occurs within the generation of an organism, there has also been a realization recently that plasticity can occur *across* generations. Termed transgenerational plasticity (TGP), it can be quite important in response to changes over a few generations (Salinas et al. 2013, Donelson et al. 2018, Donelan et al. 2020), as parents can anticipate likely environments offspring will face and help them refine their phenotypes accordingly. Before we go further, we should note that plasticity can be adaptive, neutral, or maladaptive, depending on whether the environmental change leads to a response that is closer to or away from the phenotypic optimum (Ghalambor et al. 2007, Morris and Rogers 2013).

[1] Biology Department, Kalamazoo College, 1200 Academy St. Kalamazoo, MI 49006, USA.
[2] Nicholas School of the Environment, Duke University, Durham, NC 27710, USA.
* Corresponding author: santiago.salinas@kzoo.edu

Here, we will focus on thermal plasticity in oceanic ectotherms (i.e., organisms whose internal sources of heat are negligible and rely mostly on environmental heat sources). We will eschew discussing it from a quantitative genetics perspective, choosing instead to focus on practical issues. First, we briefly describe water temperature variation in time and space as it applies to plasticity, and then explore phenotypic plasticity and transgenerational plasticity in depth. We close with a brief discussion on the role plasticity may play in our changing climate.

13.2 Variation in water temperature

That water temperature varies both temporally and spatially in the marine environment should not be surprising. Compared to air, however, water temperature is more stable because of its high specific heat capacity; in other words, air and water respond differently to the same amount of incoming solar radiation, with water taking longer to heat up. The result is that air temperature in Mar del Plata (a coastal city at 38°S in Argentina) for a midsummer day can vary by 14.3°C, while seawater temperatures only do so by 0.3°C.

To understand how the environment shapes phenotypic and transgenerational plasticity, it is important to consider two aspects of variation in a temperature time series: its amplitude, or how large the waves are (e.g., the difference from min to max, or variation around a mean) and its predictability (how well temperature can be forecasted based on previous time steps) (Kingsolver and Woods 2016).

Amplitude of the variation. Ocean temperatures fluctuate with periods of seconds to thousands of years. Of course, what is relevant to plasticity is going to depend on the lifespan of a species: Milankovitch cycles (the forcing of Earth's orbital movements on temperature—further developed in Chapter 14) will be of little relevance to a copepod's reaction norm. Throughout a day, buoy data coupled with temperature models suggest that, in the open ocean, sea surface temperature (at a depth of ~ 25 cm) fluctuates by 0.5°C or less (Kennedy et al. 2007), while a coral reef in Florida (USA) can see swings of ~ 9°C at 20 m (Leichter et al. 2006) and a typical daily range in a South African intertidal rock pool can be as high as 10°C (Smit and Glassom 2017). As the previous chapters on this book have illustrated, the ocean is a large and complex collection of habitats so generalizations are a Sisyphean task—suffice it to say that a good understanding of the natural history of the organism under study, to identify the relevant temperature variation, is needed. Generally, a larger thermal range (i.e., amplitude of fluctuations) will lead to steeper reaction norm slopes in a population (Lande 2009, Tufto 2015).

Predictability. All modeling efforts on phenotypic and transgenerational plasticity to date (e.g., Ho and Saunders 1979, Via and Lande 1985, Sultan and Spencer 2002, Bonduriansky et al. 2012, Hoyle and Ezard 2012, Leimar and McNamara 2015) agree that the fitness benefits depend on the predictability of the environment. This is fairly intuitive: in the case of transgenerational plasticity, for instance, if you cannot predict what environment your offspring will experience (to some level of certainty), then there is no point in passing along any information (Burgess and Marshall 2014).

When referring to predictability, biologists sometimes speak in terms of "color of environmental noise": in white noise, there is no correlation between successive measurements. As the noise becomes redder, the correlation among points becomes more strongly positive. With respect to temperature, marine systems tend to be redder (i.e., have higher predictability) than terrestrial systems, which are pretty white (Vasseur and Yodzis 2004), or freshwater ones, which tend to be white or pink (Cyr and Cyr 2003). Predictability can vary across regions, too, which means that species with large distributions may experience environments that are not all equally predictable. Munch et al. (in press) empirically tested whether predictability in water temperature influenced the magnitude of transgenerational plasticity in the sheepshead minnow, *Cyprinodon variegatus*. Taking advantage of a latitudinal gradient in predictability along the near-shore western Atlantic Ocean (the east coast of the United States), they showed that the degree of TGP in these populations depends linearly on the decorrelation time (i.e., predictability) for temperature.

The same arguments made above for temporal variation apply spatially. Different levels of plasticity can evolve based on local conditions across a range, as described, for example, for an intertidal shrimp,

Betaeus emarginatus, that lives throughout the ~ 3,000 km Chilean coast (Barria et al. 2018). Coasts throughout the world, however, have different levels of spatial variation in sea surface temperature. In North America, the Atlantic Ocean experiences a ~ 25°C change (or 0.91°C lat^{-1}) in long-term mean sea surface temperature, while the Pacific Ocean only changes by ~ 10°C (0.29°C lat^{-1}; Baumann and Doherty 2013). These latitudinal gradients along coastlines, where most of the marine resources occur (Ray 1991, Costanza et al. 1997), are more heterogeneous than those found in the open ocean (Baumann and Doherty 2013).

13.3 Linking temperature to biology

The thermal performance curve

Temperature sets the pace of life in ectotherms. We know that heating up a substance causes its molecules to move faster, which leads to more collisions among them (i.e., more reactions take place; for a more detailed discussion see Chapter 7 in Clarke 2017). In a way, ectotherms are like beakers in which chemical reactions are taking place—you warm up a fish and its activity level goes up (up to a point, of course). This is all encapsulated in the thermal performance curve (TPC; Fig. 13.1), a key concept in thermal plasticity (Huey and Stevenson 1979).

The TPC, essentially a continuous reaction norm, depicts the level of performance of a trait at various temperatures (Fig. 13.1). The temperature in question (i.e., the temperatures used in the *x* axis) has been variably applied, with some biologists using chronic exposures and others choosing acute ones. The measured trait varies considerably depending on the species but is usually related to movement (e.g., maximum or sustained swimming speed), feeding (e.g., food consumption, assimilation efficiency, photosynthetic rate), or life history (e.g., growth rate, development rate, egg or offspring production, survival to a specific time point; Angilletta 2009). If the goal is to predict an organism's vulnerability to warming, however, the trait measured should be tied to fitness (Huey et al. 2012). The critical thermal maximum (CT_{max}) and minimum (CT_{min}) are the points at which performance stops, and those two points envelop what is known as the thermal tolerance range (Huey and Stevenson 1979). The optimal temperature (T_{opt}) is the point at which the curve is highest (maximal performance), although many species prefer temperatures slightly below it when presented a choice in an experimental thermal gradient, as shown elegantly by Allen et al. (2012) in sand fiddler crabs, *Uca pugilator* (see also Martin and Huey 2008). The final feature of a TPC is the thermal performance breadth, which is quantified by biologists by choosing an arbitrary performance level (e.g., 50% of T_{opt}) and then obtaining the temperature range.

The environment experienced by organisms is what shapes TPCs. Stenothermic fishes, like Antarctic or tropical ones, have a narrow thermal range, reflecting the environment to which they are adapted,

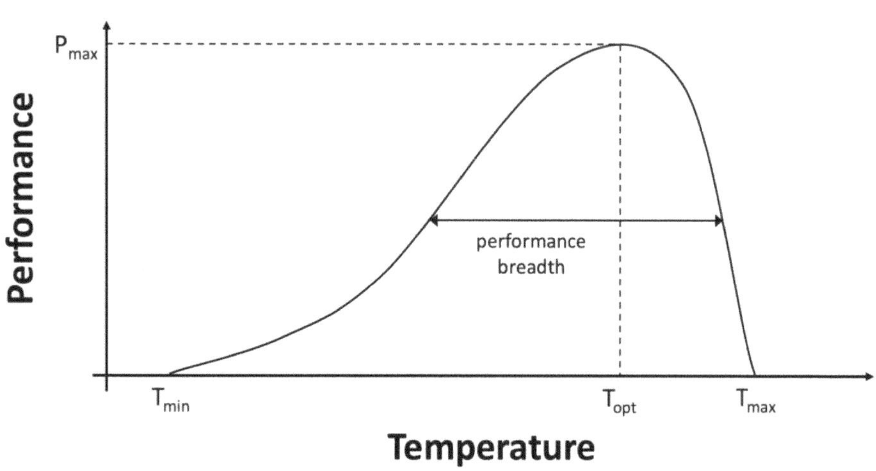

Figure 13.1. The thermal performance curve, a key concept in thermal plasticity.

resulting in narrow TPCs. Eurythermic species, many of which inhabit highly variable environments like salt marshes or the intertidal, have wide TPCs (Stillman and Somero 2000).

Worth noting is the shape of the curve on either side of T_{opt}. On the left side, the rise is often considered exponential. This rapid increase in performance with temperature forms the basis of the thermal component of the metabolic theory of ecology (MTE; Brown et al. 2004), which assumes that metabolism and other biological processes follow the Arrhenius equation,

$$M \alpha e^{\frac{-E}{kT}},$$

where M is metabolic rate, E the activation energy for the rate-limiting reactions of metabolism, k is Boltzmann's constant, and T temperature in Kelvin (Gillooly et al. 2001). The MTE has been used and extended to explain a variety of organismal and community-level processes (O'Connor et al. 2007, Munch and Salinas 2009, Regaudie-de-Gioux and Duarte 2012, Bernhardt et al. 2018a), though it also has its detractors (Clarke 2006). After T_{opt} is crossed, a steep decline in performance occurs. The mechanistic basis for this is detailed under *Physiological basis of thermal tolerance and plasticity*.

In practice, biologists measure performance under several temperatures and then perform statistical curve-fitting to estimate the entirety of the TPC. For a discussion about approaches, see Angilletta (2006) or Kingsolver et al. (2001).

13.4 A broad overview of processes governing thermal limits and plasticity

Changes in temperature have pervasive effects at all levels of organization within an organism. Although a cellular response to thermal stress is not always bad—more on this later—it is likely to manifest at the tissue, organ, and whole organismal level, impacting the overall function of an individual, and consequently the ecology of a given population (Elias et al. 2014, Clarke 2017). Here, let us focus on key molecular, biochemical, and physiological attributes to consider when examining thermal plasticity in marine ectotherms. For a more in-depth treatment of fundamentals of thermal physiology and biochemical and molecular responses to thermal stress, see Somero et al. (2017), Moyes and Schulte (2015), and Schmidt-Nielsen (1997).

As discussed in detail by Somero et al. (2017), fundamental cellular processes to consider when evaluating impacts of thermal stress are (i) rate of biochemical reactions, (ii) integrity of macromolecular structures, and (iii) solubility of small molecules. At the cellular level, metabolic reaction rates are altered in accord with the Arrhenius relationship (Q_{10} effects). Most biological systems generally follow a Q_{10} of 2–3 (Elias et al. 2014), i.e., for every 10°C increase in temperature, a given biochemical reaction rate is expected to increase by a factor of 2 to 3. In addition, changes in temperature can affect the integrity of important macromolecular structures. Proteins, nucleic acids, and lipids serve as fundamental building blocks of most cellular constituents, e.g., the phospholipid bilayer that envelops cells. These macromolecular structures rely on noncovalent bonds and weak covalent bonds for their integrity. Depending on the organism, even a few degrees of change in temperature can alter the kinetic energy within the cellular environment, perturbing these bonds and disrupting macromolecular structures. Finally, solubility of small molecules, such as dissolved gases like oxygen, and their rates of diffusion in and out of the cell are altered with changes in temperature. These fundamental shifts in cellular processes following a thermal stress can, and typically do, initiate a series of molecular reactions to compensate for these effects, with the goal of regaining homeostasis; alternatively, thermal stress can activate apoptotic pathways. As a result, the outcome of these molecular responses determine thermal limits (CT_{max} and CT_{min}) and the optimum thermal window for a given organism.

Organisms succeed in inhabiting an enormously diverse range of thermal environments, especially in the marine realm, despite the inescapable effects of thermal stress of extreme environments. For example, archaea inhabit deep sea geothermal vents at temperatures as high as ~ 125°C (Jannasch and Mottl 1985). Among eukaryotes, many intertidal species experience wide fluctuations in daily and seasonal temperatures. Small marine snail populations of the genus *Echinolittorina* are reported to live in rocky shore habitats with temperatures reaching 55°C (Marshall et al. 2011). In estuaries, teleost fishes such as the longjaw mudsucker, *Gillichthys mirabilis*, and the Atlantic killifish, *Fundulus heteroclitus*, tolerate wide temperature ranges (~ 35°C–38°C) in their habitat (Logan and Somero 2010, Healy and Schulte 2012, Jayasundara and Somero 2013). In contrast, most organisms in the Southern Ocean, including Antarctic notothenioid fish, can only withstand a few degrees of change in temperature (Hochachka and Somero 2002, Beers and Jayasundara 2015). This wide diversity of thermal habitat ranges, as well as upper and lower thermal limits, are achieved by adaptively modifying a common set of biochemical structures and processes over evolutionary time.

Acute and acclimation response to changes in temperature

The magnitude and duration of temperature change, in combination with the time available for an organism to mount a response, determine the complexity of the cellular thermal stress response. When a thermal stress persists over generations, populations may adaptively modify their biochemical constituents to thrive in the new environment. For example, genetic changes leading to shifts in amino acid sequences can produce proteins that maintain thermal stability, structural flexibility, and optimum function at the new thermal habitat; a striking example of this phenomena is protein adaptations of cytosolic malate dehydrogenase in marine mollusks as described by Dong et al. (2018). A likely consequence of such protein modifications is that the new temperature regime may no longer become 'stressful'.

Experiencing thermal stress within the lifetime of an organism results in an acute heat (or cold) shock response. When an individual first senses a change in temperature, a response is manifested through signal transduction, RNA modifications, post-translational modifications of proteins, and other rapid alterations of cellular chemistry (e.g., changes in solute composition in the intracellular *milieu*; Somero et al. 2017). If the thermal stress persists over several hours to weeks or months, some species can mount a more complex acclimation or acclimatization response (a note on definitions: "acclimation" typically refers to the adjustment of a trait in response to an environmental stimulus in the lab, while "acclimatization" encompasses the responses to the environmental variables in the field; Wilson and Franklin 2002). Acclimatization (or acclimation) shifts the critical threshold of a certain stress for a given organism (i.e., allows it to perform at a higher temperature before stress occurs) by modifying biochemical constituents and processes, which can result in molecular, cellular and organismal level reorganization. For example, at the macromolecular level, fishes alter the lipid composition of cellular membranes to account for temperature effects on membrane fluidity (Farkas et al. 2001); an elegant discussion on thermal adaptations of membranes is provided by Hazel (1995). Broadly speaking, the capacity to acclimate/acclimatize biochemical processes and constituents is likely to play a key role in determining an ectotherm's susceptibility to changes in environmental temperature.

Acclimation responses alter a series of biochemical processes by shifting gene and protein expression (altered transcription and translation; Tomanek 2011, Jayasundara et al. 2015, Logan and Buckley 2015, Madeira et al. 2016). These shifts are likely to be linked to epigenetic modifications such as changes in DNA methylation levels, histone deacetylation, and RNA silencing (e.g., through Piwi-interacting RNA; Hu and Barrett 2017). The Antarctic marine polychaete *Spiophanes tcherniai*, for instance, exhibited a marked increase in methylation in response to a transfer from –1.5°C to 4°C water, which was tied to organismal respiration rate returning to control levels in 4 weeks (Marsh and Pasqualone 2014). As discussed in detail by Somero et al. (2017), thermal stability of strands of nucleic acids (segments of DNA and RNA) may also play a thermal sensing and translation and transcription regulatory role in acute and acclimation response to temperature. Although only discovered in bacteria, a striking example of this are RNA thermometers, another set of regulatory elements that may play an important role in acute and acclimatory thermal stress responses (Weber et al. 2014). Examining the precise role of such epigenetic changes in transcriptomic and proteomic changes underpinning short- and long-term thermal acclimation and plasticity, including transgenerational thermal plasticity, is an intriguing area of future research (Hu and Barrett 2017).

The capacity to acclimate is highly species-specific and depends on the life history and the environment of the population. For example, the Antarctic notothenioid fish *Trematomus bernacchii* is a notable example of a species with limited thermal acclimation capacity that lives within a very narrow thermal niche (Beers and Jayasundara 2015). Conversely, the Atlantic killifish, *Fundulus heteroclitus*, and other fishes from estuaries, habitats whose water temperatures vary widely, show remarkable capacity to modify their biochemical and physiological processes following thermal acclimation (Logan and Somero 2010, Schulte et al. 2011, McKenzie et al. 2019). Studies show that these organisms can modify their transcriptome (gene expression—reviewed by Logan and Buckley 2015) and proteome (protein expression—reviewed by Tomanek 2011) in response to long-term thermal acclimation that is coupled to their cellular bioenergetics, cardiac function, and metabolic rates. A series of studies on the intertidal goby *Gillichthys mirabilis* demonstrate these coupled effects of acclimation from gene expression to whole animal thermal tolerance (Logan and Somero 2010, Jayasundara and Somero 2013, Jayasundara et al. 2015).

The heart as a physiological focal point of thermal tolerance

Maintaining an adequate energy supply is tightly linked to mounting an adequate stress response at the cellular level—repairing, remaking, and reorganizing macromolecular structures can be energetically costly (Kültz 2005). Furthermore, given the Q_{10} effects discussed earlier, metabolic reaction rates are expected to increase or decrease with temperature changes. Altered metabolic rates can affect the capacity of an organism to produce and allocate energy for other critical processes like growth and reproduction, especially in the presence of multiple abiotic stressors (e.g., pH and salinity change; Kültz 2005, Sokolova 2013). Accordingly, processes underlying organismal bioenergetics are considered critical in establishing thermal limits and ranges, especially for aquatic animals. In this section, we focus on cardiac function as an important physiological system of organismal energy production to discuss key considerations in defining thermal limits at the cellular level in the context of TPC.

The capacity for organisms to acquire and circulate oxygen is an important element in determining thermal limits and ranges as summarized by Pörtner et al. (2017). Temperature affects dissolved oxygen concentrations, as well as oxygen diffusion rates, through the vasculature to cells. Oxygen requirements by different tissues, it must be noted, may vary depending on their role in the thermal stress response. For instance, some organisms, such as the barred killifish, *Pseudoxiphophorus jonesii*, may increase their activity levels when exposed to thermal stress (Culumber 2020), altering the metabolic demand of peripheral muscle tissues acutely and following long-term exposure (Johnston 1993). Notably, changes in temperature impact cardiovascular performance, the key organ system in delivering oxygen and removing metabolic waste products. Because of this, cardiac function is perceived as the most important physiological system in defining thermal optima and tolerance limits in marine ectotherms (Farrell 2002). Indeed, Pacific salmon, *Oncorhynchus* spp., as highly aerobically-active fishes, demonstrate a clear link between cardiovascular activity and aerobic performance (Eliason et al. 2013, Eliason and Farrell 2016).

The interplay between oxygen availability and circulation to sustain an adequate energy supply has been widely discussed. A prominent hypothesis is the oxygen and capacity limited thermal tolerance theory (OCLTT; Pörtner 2002, Pörtner and Knust 2007, Pörtner et al. 2017). Despite many caveats (we refer interested readers to a discussion in the *Journal of Thermal Biology* special issue as highlighted by Clark and Mark 2017), the OCLTT provides a framework to examine physiological limitations imposed by thermally-induced changes in oxygen delivery among marine ectotherms. The OCLTT mechanistically links the thermal range of an aquatic animal to temperature effects on aerobic scope. Calculated as the difference between maximum and resting metabolic rates (typically presented as oxygen consumption rates), aerobic scope serves as an index for overall performance of a given organism. At high temperatures, the increase in resting metabolic rate due to Q_{10} effects reduces aerobic scope, since maximum metabolic rates are thought to be operating closer to their physiological upper limit (Pörtner et al. 2017), although this was not evident in a highly eurythermal fish (Healy and Schulte 2012). This reduced aerobic scope is compounded by the decrease in dissolved oxygen. As temperatures decrease, aerobic scope is again reduced, since maximum metabolic rates decrease rapidly. This framework helps to determine the temperature at which aerobic scope is maximized, defining an optimum thermal window for an individual organism (Pörtner et al. 2017). In the context of acclimation, the capacity to shift the window in which aerobic scope is maximized by altering temperature effects on biochemical reaction rates and macromolecular structures may lead to shifts in thermal tolerance of an organism.

Given the role of the cardiovascular system in sustaining an adequate oxygen circulation, acclimation capacity of cardiac function can play a critical role in establishing thermal plasticity under sustained thermal stress (Farrell et al. 2009, Pörtner et al. 2017). Cardiac performance and scope are strongly correlated with aerobic scope (Lillywhite et al. 1999, Farrell 2002, Farrell et al. 2009) and the capacity for heart muscles to compensate for temperature effects is critical in maintaining aerobic scope at different temperatures. Additionally, the multifaceted remodeling of biochemical systems and processes at the cellular level to optimize performance at a new temperature necessarily incur a significant energetic cost (Pörtner et al. 2012), and require a coordinated physiological response and an efficient circulation of material and humoral messages throughout the body. Therefore, the heart can be considered an important focal point of thermal tolerance.

Cardiac limitations are further compounded by its anatomical arrangement in teleost fishes (except in partially endothermic fish such as bluefin tunas) and other lower vertebrates and invertebrates in the ocean (Farrell et al. 2009, Jayasundara et al. 2013). For example, in most teleosts, cardiac tissue lacks coronary circulation and hearts rely on the oxygen in the venous return, after oxygen is distributed to the rest of the peripheral tissues (Farrell and Jones 1992, Farrell et al. 2009). Under warming conditions, when dissolved oxygen is reduced and oxygen demand is higher, the already depleted oxygen levels in venous blood may not be sufficient (Farrell et al. 2009).

Even if cardiac function is maintained, sustained thermal stress is likely to result in cellular damage to cardiac tissue. Cellular stress responses to repair damage and restore cellular homeostasis require a large fraction of cellular ATP (Kültz 2005, Hoekstra and Montooth 2013). Given that ATP production is highly dependent on oxygen availability to support the electron transport system activity in the mitochondria, reduced oxygen supply to the heart will only exacerbate the cardiac thermal damage. Thus, changes in temperature pose a complex challenge to cardiac myocytes. The thermal threshold at which myocytes do not receive sufficient oxygen to support its own function and damage repair may define cardiac thermal limits, and consequently may impact the overall organismal thermal limits. Accordingly, the capacity for cardiac myocytes to mount an adequate stress response and regain homeostasis is critical for marine organisms under changing temperatures.

The heart is a complex organ that requires coordination between multiple cell types. Rhythmic control of the cardiac pump is regulated by the extracellular matrix and the contractile and metabolic machinery of cardiac myocytes (Farrell and Jones 1992, Bers 2002). Changes in temperature remodel this regulation, and precise cardiac function through acclimation requires profound changes in cardiac energy metabolism, nervous and humoral control of cardiac contractility, and in electrical and mechanical properties of the heart (Farrell et al. 2009). Studies show that cardiac function can indeed acclimate via endocrine, paracrine, and neural control (a topic excellently reviewed by Keen et al. 2017). These changes appear to be underpinned by marked changes in a number of cellular and molecular mechanisms (e.g., cytoskeletal reorganization, iron homeostasis, oxidative stress response, energy metabolic processes) as revealed by '*omics*' studies (Vornanen et al. 2005, Castilho et al. 2009, Jayasundara et al. 2013, 2015). The complexity necessary to attain an effective level of reorganization, both in the heart and other organs, is apparent from these studies.

Among other processes, two key cellular pathways emerge from different thermal acclimation studies as highly important in thermal plasticity. First is the role of heat shock proteins and the associated downstream proteasome. Second is the role of mitochondria.

As the name suggests, heat shock proteins (HSPs) are induced immediately following a thermal stress and are molecular chaperons essential for repairing and refolding damaged proteins (Asea and Kaur 2018). There are many classes of HSPs and they play a multitude of roles in addition to responding to thermal stressors (Hofmann 1999, Clark and Peck 2009, Sen and Giri 2017). Nonetheless, inducible forms of HSPs are directly associated with coordinated cellular stress responses to fluctuations and increases in temperature. The inducible production of these proteins is believed to be highly conserved, although the Antarctic fish *Trematomus bernacchii* appears to lack one, which is thought to be linked to their limited thermal range (Hofmann et al. 2000). Proteins that cannot be repaired are ubiquitinated and degraded by forming a proteasome complex (Glickman and Ciechanover 2002). Excessive protein damage in a cell coupled with reduced protein turnover rates (i.e., making new proteins) may initiate cell death. Thus, a coordinated HSP response is a key attribute of regaining homeostasis following short- or long-term exposure to a thermal stress (Glickman and Ciechanover 2002, Kültz 2005).

Processes underlying cellular stress responses are energetically costly (Hoekstra and Montooth 2013). Hence, mitochondria have gained significant attention in this context as a key organelle responsible for setting thermal limits (Fangue et al. 2009, Iftikar et al. 2014). Mitochondria themselves show very different thermal limits within a genus, as shown in an example when examining four different congeners of abalones (genus *Haliotis*) from different temperatures and acclimation conditions (Dahlhoff and Somero 1993). Not only is the demand for energy production increased under thermal stress, but temperature changes can directly affect mitochondrial function. Changes in temperature can (i) directly alter or inhibit biochemical reactions in the mitochondria, (ii) perturb mitochondrial structures (e.g., alter lipid membrane fluidity and damage mitochondrial DNA), (iii) directly target nuclear-cytoplasmic processes that regulate mitochondrial function (e.g., oxygen availability, Ca^{2+} availability and ion regulation) and signaling (e.g., reactive oxidative species), and (iv) shift substrate (e.g., fatty acids) availability leading to compensatory mitochondrial changes (McKenzie et al. 2019). These responses may occur immediately following exposure to a thermal stress or could follow as a result of cellular attempts to regain homeostasis. Irrespective of the timing, mitochondrial dysfunction leads to a number of adverse outcomes in addition to reduced ATP supply and increased free radical formation, which can eventually result in damage and apoptosis (Schaffer and Saadeh Suleiman 2010, McKenzie et al. 2019).

Broadly, thermal limits, and TPCs more generally, are intricately shaped at various levels of biological organization, including the macromolecular, tissue, and organ.

Practical issues not to ignore when considering thermal performance curves (TPCs)

TPCs are not unvarying, fixed features of organisms. The response of one trait to temperature could be quite different to that observed in another trait within the same organism (Kellermann et al. 2019, Bozinovic et al. 2020). A response to a change in temperature, then, could be different than predictions from any one trait's TPC because of the complex trade-offs involved (Kellermann et al. 2019). TPCs also vary across ontogenetic stages. For example, the larval stages of *Littorina obtusata*, an intertidal gastropod, were much more sensitive to heat stress (i.e., had lower CT_{max} values) than adults (Truebano et al. 2018). It is not always the case that younger developmental stages are more sensitive, however. In an impressive collection of 694 published studies on fishes, Dahlke et al. (2020) showed that embryos and spawning adults have considerably smaller thermal ranges (lower T_{max} and/or higher T_{min}) than larvae and non-reproductive adults.

Of course, thermal performance is time-sensitive—past exposure matters (Schulte et al. 2011, Kingsolver and Woods 2016). Even brief exposures to extreme temperatures can cause significant changes in thermal physiology that are beneficial in subsequent stressful exposures. The reasons for this 'hardening' or hormesis, as the phenomenon is called, is probably species-specific (Costantini et al. 2010). For instance, oysters, *Isognomon nucleus*, that underwent an acute thermal stress in air (40°C), which is a common occurrence for this intertidal species, exhibited greater thermal tolerance and oxygen consumption rates when reintroduced to water (Giomi et al. 2016). Interestingly, some evidence points to the fact that the nature of the stress (temperature spike, food deprivation, toxicant exposure) may not be as relevant as experiencing one—in other words, brief exposure to heat during development may confer a higher level of resistance to other stressors (Costantini 2014). Regardless, a better understanding of thermal hormesis will, no doubt, gain importance as heat waves continue to become more frequent and severe (Oliver et al. 2018).

We also must not forget that an individual is the integration of a collection of traits in response to many environmental variables, and abiotic factors interact in non-linear ways to shape phenotypes (Kingsolver et al. 2006). For growth, a trait often tied to fitness, food quantity and/or food quality can mediate the response to temperature in fishes (Brett et al. 1969, Wurtsbaugh et al. 1983, Fonds et al. 1992, Reznick 1993, Hutchings et al. 2007, Salinas and Munch 2014). The diatom *Chaetoceros simplex* provides another example. Aranguren-Gassis and colleagues (2019) evolved this marine phytoplankton for ~ 200 generations at an above-optimal temperature (31°C) under high and low nitrogen (N) conditions.

After that period, the lines were exposed to 34°C water. Diatoms under high N conditions were the only ones able to meet the challenge and grow at the novel temperature (Aranguren-Gassis et al. 2019). Additionally, the shape of the TPC had changed significantly for the diatoms under high N, nicely demonstrating that food/nutrients can modulate the thermal stress response. Other traits, like maturation, may also vary depending on food and temperature: reducing either temperature or food tends to reduce growth in fishes, but a decrease in temperature results in delayed maturation at a larger size, while a decrease in food causes delayed maturation at a smaller size (Atkinson 1994, Berrigan and Charnov 1994, Thorpe 2004). Thus, TPCs can vary because of non-thermal factors.

Further, *how* organisms are exposed to temperature matters when estimating thermal performance. Performance under a set of constant temperatures will typically not be the same as when measured under thermal fluctuations with the same mean (Colinet et al. 2015, Salinas et al. 2019). The CT_{max} of zebrafish, *Danio rerio*, was 41.1°C under constant 28°C but was raised to 41.5°C when exposed to diel thermal cycling (28 ± 6°C; fish in both treatments were held under these conditions until adulthood). High, stressful temperatures influence the kinetic induction of heat shock proteins such as Hsp70, Hsp90, Hsp27, and Hsp22, which have differing transcriptional responses when exposed to fluctuating *vs.* constant temperature regimes (Podrabsky and Somero 2004). As stated earlier, this family of proteins plays important chaperone functions, like stabilizing proteins during folding or repairing damaged ones, and are critical in response to a number of stresses (not just thermal).

The range of fluctuating temperatures will have an effect on the resulting performance of an organism over time. As clearly explained by Denny (2017), imagine that the metabolic rate of an individual at 20°C is measured to be 3 times that of one at 10°C (Fig. 13.2a). If you tried to estimate the metabolic rate of this organism spending half its day at 10°C and half the day at 20°C, tracing a line between the two points and obtaining the metabolic rate from its midpoint would lead to an overestimate (Fig. 13.2a). Jensen's inequality provides a simple way for understanding this phenomenon. In its mathematical form, it states that $\overline{F(T)} \neq F(\overline{T})$, where F is a non-linear function of T (in our case, F is the performance trait and T is

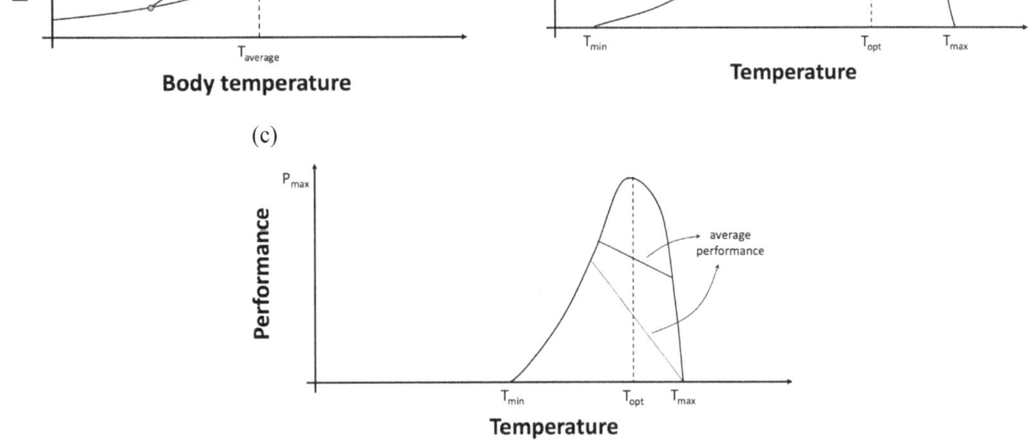

Figure 13.2. (a) Jensen's inequality illustrated. Note the overestimation when the curve is concave up (i.e., second derivative is positive). The effects of Jensen's inequality are not equal when it comes to (b) eurytherms and (c) stenotherms (after Denny 2017).

temperature). In other words, it means that the average performance across temperatures (i.e., $\overline{F(T)}$) does not equal the performance at the average temperature (i.e., $F(\bar{T})$) (Ruel and Ayres 1999). Overestimation will always happen if the curve is concave up, while underestimation will occur in the concave down part of the curve. Indeed, TPCs estimated for the green microalga *Tetraselmis tetrahele* under nine constant and nine fluctuating (same as constant temps ± 5°C) conditions were quite different. Population growth rates of the alga were greater under fluctuations over the accelerating part of the TPC and lower on the decelerating side (Bernhardt et al. 2018b).

The effect of Jensen's inequality becomes greater as variation increases and as the curve becomes narrower (Fig. 13.2b). As far as we are aware, no common patterns (i.e., thermal fluctuations lead to slower growth or higher CT_{max} than constant temperatures, for example) have been described in the literature, likely because of a lack of standardization, as studies do not normally report T_{opt} if they do not present TPCs. Regardless, it is not surprising that biased predictions of species' responses to climate change can be obtained when using only constant temperatures (Niehaus et al. 2012, Vasseur et al. 2014, Colinet et al. 2015).

All of these issues make using TPCs to predict population responses to climate change, which is a seductive proposition at face value and an approach that has been used repeatedly, problematic. Therefore, it is important to be aware of the potential pitfalls.

Developmental plasticity

Developmental plasticity, the capacity of an organism to shape its phenotype based (at least partly) on the early-life environment, is a special case of phenotypic plasticity that is worth exploring in detail (West-Eberhard 2003). As we have seen, acclimatization/acclimation and the organism's thermal history play a significant role in shaping its physiology. Developmental plasticity occurs as a result of interactions between genotype and environment during ontogeny—improvements in heat tolerance after a brief exposure to heat stress during youth is a textbook example of developmental plasticity. Developmental arrest or temperature sex determination could also be considered here (Johnston and Wilson 2006). This type of plasticity is often, but certainly not always, irreversible (Taborsky 2017). In practice, the distinction can be easily blurred as the mechanisms underlying both could very well be the same (Angilletta 2009), but it is important to highlight key features of thermal developmental plasticity.

The value of developmental plasticity is particularly obvious when thinking about hormesis, as exposure to a low level of stress during development can lead to heightened resistance later in life. These benefits can be reaped in the short-term (i.e., shortly after the phenotypic change happens) or long-term, as in the case of anticipatory effects. These predictive adaptive responses (PAR) require that the environment during the larval stage provides a reliable cue to the conditions the adult will experience. A good PAR example is provided by the spiny chromis damselfish, *Acanthochromis polyacanthus*: when raised at an elevated temperature (ambient + 3°C) for 1.5 years, they displayed lower resting oxygen consumption during warmer summer temperatures when compared to the ambient controls (Donelson et al. 2010). Examples of anticipatory effects, however, are much more common based on the food environment (in fishes, see, e.g., Lee et al. 2016, Turkmen et al. 2017).

Costs of developmental plasticity are assumed to exist, and two types of costs are shared with phenotypic plasticity (Murren et al. 2015): (i) the energy required to maintain a machinery to sense the environment and effect changes, and (ii) the costs associated with responding to a cue, like building a 'defensive helmet'. Additionally, developmental plasticity could be detrimental if the projected state of the environment is mispredicted and the plastic adjustment made results in an incomplete or inappropriate response. It is important to note, though, that developmental plasticity can interact with reversible plasticity throughout the life of an organism, as was shown for thermal tolerance limits and TPCs (e.g., Schaefer and Ryan 2006, Koumoundouros et al. 2009). Regardless, characterizing the amplitude and predictability of the thermal environment is key when studying all types of plasticity.

So far, we have discussed *intra*-generational developmental plasticity. The same concept of developmental plasticity can be extended to think about forecasting of environments *across* generations.

Transgenerational plasticity

Recently, a different form of plasticity has received considerable attention. Transgenerational plasticity (TGP) occurs when phenotypic responses to environmental variables are shaped, in part, by the environment in preceding generations, in the absence of DNA sequence changes (this concept was briefly introduced in Chapter 11). Exact definitions vary by author, so here we will explicitly define TGP as a significant interaction between parent and offspring environments affecting offspring phenotype (other terms sometimes used for TGP include environmental maternal effects, intergenerational effects, legacy or carry-over effects, cross-generational plasticity, and trans-generational acclimation). In essence, the phenotype of the offspring is a function of the environment of the offspring *and* the environment of the parents. Think about that: in many species, the phenotype is not only shaped by the temperature an organism experiences but by the interaction of its environment with that of its parents!

Let us take a look at an example. Shama et al. (2014) acclimated adult threespine stickleback, *Gasterosteus aculeatus*, to two temperatures: 17°C, a temperature they frequently experience, and 21°C, an elevated summer temperature that is associated with detrimental effects. Then, they spawned the fish and grew the offspring at the two temperatures. Interestingly, they found that offspring grown at the same temperature as the parents outperformed their mismatched counterparts (i.e., offspring at 21°C whose parents had been acclimated to 21°C had larger body sizes, and vice versa). This was true after 30 days; after 60 days, TGP was only detected under the more stressful temperature condition (Shama et al. 2014). This work illustrates nicely how parents can sense the environment and give their offspring a biological head-start of sorts (assuming the environment is predictable from parents to offspring!). The collector urchin, *Tripneustes gratilla*, also has the capacity for TGP, with larvae more tolerant to warmer waters when parents were exposed to them (Karelitz et al. 2019), as does *Artemia* (Norouzitallab et al. 2014). More and more examples continue to accumulate across the tree of life. TGP appears to be possible in a wide range of traits, too, including metabolic- (growth, aerobic scope), morphological- (vertebral number, presence of wings), and life history-related ones (age at maturation, fecundity, survival), and the environments in which offspring traits are found to be partially manipulated by the parents are quite varied as well: temperature, CO_2/acidification, food availability, predation, and so on (Jablonka and Raz 2009, Salinas et al. 2013, Munday 2014).

The common approach to test for TGP is to follow a split-brood design, exposing full-sib parents to at least two temperatures and then distributing the offspring across the same temperatures (Bonduriansky et al. 2012, Donelson et al. 2018). Other experimental designs are certainly possible, each with benefits and drawbacks (see Donelson et al. 2018 for a discussion of some of these). A very important requirement in these experiments is to quantify the mortality and fecundity of the different groups/treatments in order to account for possible selection (clonal species are, thus, the ideal model system for these types of studies). Finally, because of the great space and time needed in these experiments, a lot of the evidence comes from relatively small species with short generation times.

Another important consideration for the experimental design of TGP studies is how temperature treatments are chosen. As described above (under *Variation in water temperature*), considering predictability is key for plasticity in general. For transgenerational plasticity, it is even more so. Knowing the autocorrelation of temperature time series from the population's site, for example, can be the difference between an experiment that finds TGP and one that does not (even when it exists). If you expose sheepshead minnow, *Cyprinodon variegatus*, to high temperature for 7 days and then run a standard TGP experiment, you will find no evidence for TGP. However, if you hold parents for 30 days, the transgenerational effect will be quite striking (Salinas and Munch 2012). Thus, the timing and conditions under which TGP is expressed depends on the biology of the species.

Mechanistically, the transfer of information from parents to offspring can occur in a variety of ways. Perhaps the most common (or at least the one with the most evidence) is methylation. This requires a methyl group (CH_3) to be added or removed from cytosine, thereby altering the expression of DNA. Negative gene expression regulation can be accomplished by methylation of gene promoters, which impedes transcription, or by methyl-CpG-binding proteins, which bind to methylated DNA and change the chromatin structure (Bommarito and Fry 2019). Temperature can induce methylation changes throughout the genome, and because methylation can be passed on to offspring (Hofmeister et al. 2017), it allows for thermal information to be transferred transgenerationally (Varriale and Bernardi 2006, Marsh and Pasqualone 2014, Metzger and Schulte 2017). Methylation patterns have been linked to TGP in the spiny chromis damselfish, *Acanthochromis polyacanthus* (Veilleux et al. 2015), or the green sea urchin, *Psammechinus miliaris* (Clark et al. 2019), for example. Other mechanisms of information transfer between generations include histone modification, which involves changes in the proteins that package DNA, or non-coding RNA, which can interfere with translation, transcription, or protein structure (Ho and Burggren 2010).

Why should we care about TGP?

TGP can facilitate coping with rapid changes in the environment in a manner that is more long-lasting than phenotypic plasticity; modeling efforts of non-genetic inheritance have indeed shown that TGP can be quite important (Leimar and McNamara 2015, Greenspoon and Spencer 2018, Dury and Wade 2020). Signatures of TGP have been observed lasting more than one generation: going back to the threespine stickleback, there was an effect of grand-parent temperature, albeit diluted and weaker than the parental effect, in the grand-offspring generation (Shama and Wegner 2014). Bell and Hellmann (2019) provide many examples and a good overview of the multi-generation TGP patterns encountered so far. However, there is a need to better understand how phenotypic plasticity, transgenerational plasticity, and adaptation interact over time scales of a few generations.

TGP may, in the face of rapid environmental change, allow a population to persist until local adaptation occurs (Chevin et al. 2010, Munday et al. 2013, Nunney 2016) or may be particularly important at the edges of a population range, facilitating geographic expansions (Dyer et al. 2010, Fenesi et al. 2014). In that respect, TGP and phenotypic plasticity play a critical role in "genes as followers" or "plasticity first" conceptualizations of evolutionary change (Schwander and Leimar 2011, Forsman 2014, Levis and Pfennig 2016).

13.5 Plasticity in novel environments

As temperature continues to rise through much of the world's oceans, individuals will be exposed to rare and novel environmental conditions (Williams et al. 2007, Hobbs et al. 2009, Garcia et al. 2014). In parts of the western North Atlantic, for example, the projected departure from recent temperatures (a state continuously outside the bounds of historical variability) is predicted to occur by ~ 2050 (Mora et al. 2013). While the terms 'rare' and 'novel' are used frequently in the ecological literature, there is no clear definition of their meaning. Some define these environments on the basis of their probability of occurring; 'rare' environments are those that happen < 5% given a historical climate distribution (e.g., Hermisson and Wagner 2004, Williams et al. 2007, Snell-Rood and Papaj 2009, Smith 2011, Bailey and van de Pol 2015). Others include biological effects in their definition: a 'rare' environment is one that kills, say, 90% of the population (Kaltz and Bell 2002, Schlaepfer et al. 2002, Moreno and Møller 2011). Regardless, populations will have to adapt to these rare and novel conditions for long-term persistence, and natural selection will act based on the amount of genetic and phenotypic variation.

Under common environments, traits closely related to fitness are, in theory, expected to be similar (i.e., show low variation across individuals) because natural selection repeatedly acts on a specific range of phenotypes and can select on them (Schlichting 2008). If an organism produces a trait below the canalized trait value, it will be selected against; if it produces a higher trait value, it will be selected for, and a new canalized trait value should soon be established (Fig. 13.3). Natural selection in rare and novel environments is not as strong, simply because selection has not 'seen' the phenotypes often enough to act upon them. Theory suggests that these environments will therefore expose "hidden" phenotypes, with a concomitant increase in trait variation (Fig. 13.3; Ghalambor et al. 2007, Schlichting 2008, Paaby and Rockman 2014). This is a critical component of the adaptive capacity of a population, as the rate of evolution is proportional to the additive genetic variance (Lande 2009, Chevin et al. 2010). It is important to point out that this assumes that there is a low genetic correlation in the trait of interest between common

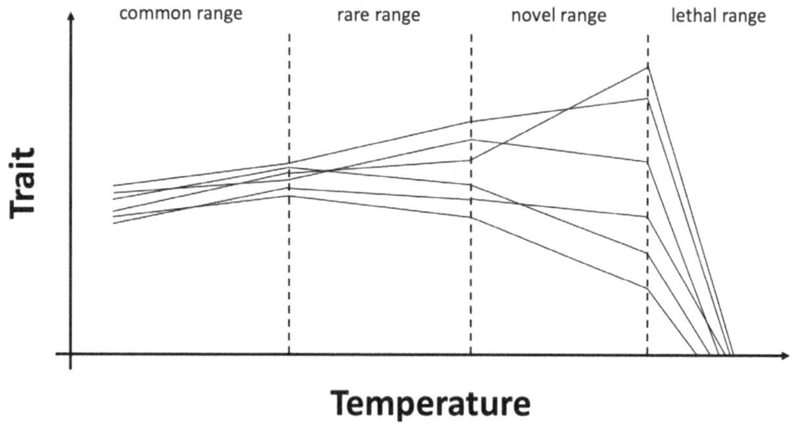

Figure 13.3. Reaction norms (each line represents a family grown at various environmental conditions) are clustered together in common environments (canalization). Selection is weak under rare and novel conditions, however, leading to higher phenotypic variance (after Ghalambor et al. 2007).

and novel environments, so that trait values at the extreme are determined in large part by the shape of the reaction norm in common environments (De Jong 2005, Chevin and Hoffmann 2017).

However, empirical evidence for these predicted changes is quite limited, and it is unclear whether variance actually increases in novel conditions (i.e., whether the fanning out in Fig. 13.3 really occurs; Hoffmann and Merilä 1999, McGuigan and Sgrò 2009, Ledon-Rettig et al. 2014). Much more work needs to be done in this relevant issue.

Genetic accommodation and assimilation

We cannot forget that phenotypic plasticity itself is a trait that can evolve (Via and Lande 1985, Scheiner 1993), although it is difficult to observe evidence of it in nature (Merilä 2012). Two processes that, in our opinion, have not received the attention they deserve are genetic accommodation and genetic assimilation (Fig. 13.4). Imagine reaction norms created by genotypes of a population, and further imagine that the population goes from one environmental state to another. Plasticity may increase in the new environment, allowing the most plastic genotypes to get closer to the phenotypic optimum—this process is known as genetic accommodation (Ehrenreich and Pfennig 2016). In a review of available evidence, Kelly (2019) found population divergence via plastic responses "extremely common" during adaptation to new climatic conditions.

Phenotypes that are currently environmentally induced via plasticity may, after some time, become genetically fixed because of costs of plasticity, thereby losing the need for the environmental stimulus. This process, known as genetic assimilation (Waddington 1942), has the potential to reduce the responsiveness of a population to changes in the environment. If, for example, directional selection on a plastic trait is imposed for multiple generations, the population reaction norm of this plastic trait may become much shallower, or even flat. After the trait is genetically fixed, or canalized, stopping the agent of directional selection will do nothing to revert the trait.

Waddington (1942) first proposed genetic assimilation with a verbal model centered on the callosities of ostriches. He noted that these callosities (on the two locations that rub against the ground when ostriches sit down) are formed during embryological development (incidentally, Darwin (1871) noticed a similar process in the thickening of the soles of human embryos). Despite genetic assimilation's similarities to Lamarckism, it should be plainly clear, as is generally accepted now (Hall 2001, Crispo 2007), that assimilation is compatible with—and in fact relies upon—neo-Darwinian evolutionary theory. Evidence for genetic assimilation in the wild comes primarily from studying the level of plasticity of populations that colonized a new ecological niche at different times. The expectation is that recently

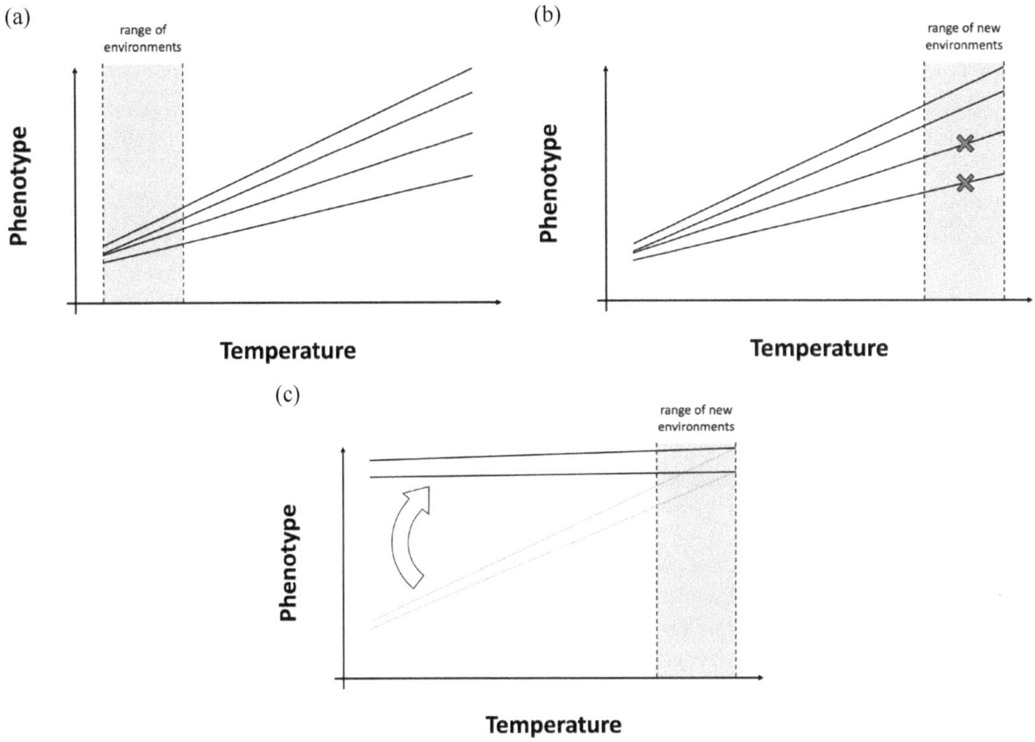

Figure 13.4. Relationship between phenotypic plasticity, loss of plasticity, and genetic assimilation. In (a), a population finds itself within a range of temperatures (shaded box). In (b), an environmental shift moves the range of temperatures experienced considerably, and some organisms die out (crosses). After some time, evolutionary changes can bring about a loss of plasticity, known as genetic assimilation (c).

separated populations will exhibit the largest degree of plasticity, while the older colonizers will show canalization with respect to an environmental variable.

Aubret and Shine (2009) exploited a natural experiment in which tiger snakes, *Notechis scutatus*, colonized five Australian islands at different times, from 40 to more than 9,000 years ago. These islands all impose selection for larger head size because of larger prey. They showed that, as predicted by genetic assimilation, snakes from "young" populations produce larger heads via phenotypic plasticity, but "older" populations do so genetically. Interestingly, snakes from the second most recently colonized island (90 years ago) had similar levels of plasticity as snakes from the island colonized 40 years ago, but they are born with larger jaws. Thus, assimilation may occur extremely rapidly. Further work suggests that the rapid assimilation is due to high costs of plasticity (Aubret and Shine 2010). Experimental manipulations made it clear that neonates with larger heads were able to achieve better swallowing performance and faster growth rates. Plasticity was also shown to be slow, taking ~ 33 days for a small-headed snake presented with larger food (easily eaten by non-plastic snakes) to be able to consume the new food. In similar fashion, the zooplankter *Daphnia melanica* rapidly lost plasticity in pigmentation when predators were introduced (Scoville and Pfrender 2010). The alpine lakes of Sierra Nevada are high UV environments, and *Daphnia* in these lakes are darkly pigmented by plastic up-regulation of the dopa decarboxylase gene (Ddc). Introduction of salmonids, visual predators of *Daphnia*, into two lakes 53 and 91 years ago caused a loss of plasticity in Ddc expression.

Although most examples involve morphological traits, genetic assimilation may occur in all trait types. In fact, canalization should be more effective in traits tightly associated with fitness (Stearns and Kawecki 1994). One such trait is soft tissue withdrawal, an anti-predator behavior by the flat periwinkle, *Littorina obtusata*. This periwinkle species has coexisted with the predatory green crab, *Carcinus maenas*, in Europe for at least 10,000 years. Green crabs, however, were introduced into the Gulf of

Maine only a century ago, setting up the perfect natural experiment to test for canalization. Periwinkles had highest behavioral flexibility in populations that had been in contact with green crabs most recently (northern Gulf of Maine, introduction ~ 60 years ago), intermediate flexibility in southern Gulf of Maine populations (introduction ~ 110 years ago), and full canalization in European populations (at least 10,000 years of co-habitation; Edgell et al. 2009). Because of the rapidity at which genetic assimilation may occur, it is entirely possible that this process may "sometimes be so fast as to pass us by" (Pigliucci and Murren 2003)—many traits that are now under genetic control may have become canalized from previously plastic forms.

Summary

1. Temperature influences the life of ectotherms in the oceans.
2. Phenotypic plasticity, the ability of a single genotype to produce different phenotypes when exposed to different environments, is ubiquitous in nature.
3. Water temperature varies spatially and temporally at various scales in the oceans, so understanding thermal plasticity is essential to understanding organisms.
4. The thermal performance curve has been used extensively to describe an organism's performance in relation to temperature.
5. Cellular processes that help determine critical thermal extremes and are involved in coping with thermal stress include (i) the rate of biochemical reactions, (ii) the integrity of macromolecular structures, and (iii) the solubility of small molecules.
6. The heart and mitochondria are two critical components of an adequate stress response.
7. There are practical issues that should not be ignored when thinking about (and using) thermal performance curves (e.g., they do change over the life of an organism, they are time-sensitive, they are only one aspect of an integrated phenotype).
8. Developmental plasticity, the capacity of an organism to shape its phenotype based (at least partly) on the early-life environment, is a special case of phenotypic plasticity that is worth exploring in detail.
9. Plasticity can occur across generations ("transgenerational plasticity"), and can be quite important in response to changes over a few generations.
10. Lastly, plasticity can play a role in adaptation to rare or novel environments, a timely consideration as climate change continues to expose populations to these uncommon conditions.

References

Allen, B.J., B. Rodgers, Y. Tuan and J.S. Levinton. 2012. Size-dependent temperature and desiccation constraints on performance capacity: Implications for sexual selection in a fiddler crab. J. Exp. Mar. Bio. Ecol. 438: 93–99.

Angilletta, M.J. 2006. Estimating and comparing thermal performance curves. J. Therm. Biol. 31: 541–545.

Angilletta, M.J. 2009. Thermal Adaptation: A Theoretical and Empirical Synthesis. Oxford University Press, Oxford, UK.

Aranguren-Gassis, M., C.T. Kremer, C.A. Klausmeier and E. Litchman. 2019. Nitrogen limitation inhibits marine diatom adaptation to high temperatures. Ecol. Lett. 22: 1860–1869.

Asea, A.A.A. and P. Kaur. 2018. Heat Shock Proteins and Stress. Springer.

Atkinson, D. 1994. Temperature and organism size—a biological law for ectotherms? Adv. Ecol. Res. 25: 1–58.

Atkinson, D. 1996. Ectotherm life-history responses to developmental temperature. pp. 183–204. *In*: Johnston, I.A. and A.F. Bennett [eds.]. Animals and Temperature: Phenotypic and Evolutionary Adaptation. Cambridge University Press, Cambridge, UK.

Aubret, F. and R. Shine. 2009. Genetic assimilation and the postcolonization erosion of phenotypic plasticity in island tiger snakes. Curr. Biol. 19: 1932–1936.

Aubret, F. and R. Shine. 2010. Fitness costs may explain the post-colonisation erosion of phenotypic plasticity. J. Exp. Biol. 213: 735–739.

Bailey, L.D. and M. van de Pol. 2015. Tackling extremes: challenges for ecological and evolutionary research on extreme climatic events. J. Anim. Ecol. 85: 85–96.
Barria, A.M., L.D. Bacigalupe, N.A. Lagos and M.A. Lardies. 2018. Thermal physiological traits and plasticity of metabolism are sensitive to biogeographic breaks in a rock-pool marine shrimp. J. Exp. Biol. 221: jeb181008.
Baumann, H. and O. Doherty. 2013. Decadal changes in the world's coastal latitudinal temperature gradients. PLoS One 8: e67596.
Beers, J.M. and N. Jayasundara. 2015. Antarctic notothenioid fish: what are the future consequences of "losses" and "gains" acquired during long-term evolution at cold and stable temperatures? J. Exp. Biol. 218: 1834–1845.
Bell, A.M. and J.K. Hellmann. 2019. An integrative framework for understanding the mechanisms and multigenerational consequences of transgenerational plasticity. Annu. Rev. Ecol. Evol. Syst. 50: 97–118.
Bernhardt, J.R., J.M. Sunday and M.I. O'Connor. 2018a. Metabolic theory and the temperature-size rule explain the temperature dependence of population carrying capacity. Am. Nat. 192: 687–697.
Bernhardt, J.R., J.M. Sunday, P.L. Thompson and M.I. O'Connor. 2018b. Nonlinear averaging of thermal experience predicts population growth rates in a thermally variable environment. Proc. R. Soc. Lond. B 285: 20181076–20181010.
Berrigan, D. and E.L. Charnov. 1994. Reaction norms for age and size at maturity in response to temperature: a puzzle for life historians. Oikos 70: 474–478.
Bers, D.M. 2002. Cardiac excitation–contraction coupling. Nature 415: 198–205.
Bommarito, P.A. and R.C. Fry. 2018. The role of DNA methylation in gene regulation. pp. 127–151. *In*: McCullough, S.D. and D.C. Dolinoy [eds.]. Toxicoepigenetics. Academic Press, Cambridge, MA.
Bonduriansky, R., A.J. Crean and T. Day. 2012. The implications of nongenetic inheritance for evolution in changing environments. Evol. Appl. 5: 192–201.
Bozinovic, F., G. Cavieres, S.I. Martel, J.M. Alruiz, A.N. Molina, H. Roschzttardtz and E.L. Rezende. 2020. Thermal effects vary predictably across levels of organization: empirical results and theoretical basis. Proc. R. Soc. B 287: 20202508.
Brett, J.R., J.E. Shelbourn and C.T. Shoop. 1969. Growth rate and body composition of fingerling sockeye salmon, *Oncorhynchus nerka*, in relation to temperature and ration size. Can. J. Fish. Aquat. Sci. 26: 2363–2394.
Brown, J.H., J.F. Gillooly, A.P. Allen, V.M. Savage and G.B. West. 2004. Toward a metabolic theory of ecology. Ecology 85: 1771–1789.
Burgess, S.C. and D.J. Marshall. 2014. Adaptive parental effects: the importance of estimating environmental predictability and offspring fitness appropriately. Oikos 123: 769–776.
Castilho, P.C., B.A. Buckley, G. Somero and B.A. Block. 2009. Heterologous hybridization to a complementary DNA microarray reveals the effect of thermal acclimation in the endothermic bluefin tuna (*Thunnus orientalis*). Mol. Ecol. 18: 2092–2102.
Chevin, L.-M., R. Lande and G.M. Mace. 2010. Adaptation, plasticity, and extinction in a changing environment: towards a predictive theory. PLoS Biol. 8: e1000357.
Chevin, L.-M. and A.A. Hoffmann. 2017. Evolution of phenotypic plasticity in extreme environments. Philos. Trans. R. Soc. Lond. B 372: 20160138–20160112.
Clark, M.S. and L.S. Peck. 2009. HSP70 heat shock proteins and environmental stress in Antarctic marine organisms: A mini-review. Mar. Genom. 2: 11–18.
Clark, M.S., C.C. Suckling, A. Cavallo, C.L. Mackenzie, M.A.S. Thorne, A.J. Davies and L.S. Peck. 2019. Molecular mechanisms underpinning transgenerational plasticity in the green sea urchin *Psammechinus miliaris*. Sci. Rep. 9: 952.
Clark, T.D. and F.C. Mark. 2017. An introduction to the special issue: "OCLTT: a universal concept?" J. Therm. Biol. 68: 147–148.
Clarke, A. 2006. Temperature and the metabolic theory of ecology. Funct. Ecol. 20: 405–412.
Clarke, A. 2017. Principles of Thermal Ecology: Temperature, Energy, and Life. Oxford University Press, Oxford.
Colinet, H., B.J. Sinclair, P. Vernon and D. Renault. 2015. Insects in fluctuating thermal environments. Annu. Rev. Entomol. 60: 123–140.
Costantini, D., N.B. Metcalfe and P. Monaghan. 2010. Ecological processes in a hormetic framework. Ecol. Lett. 13: 1435–1447.
Costantini, D. 2014. Oxidative Stress and Hormesis in Evolutionary Ecology and Physiology. Springer-Verlag, Berlin.
Costanza, R., R. d'Arge, R. de Groot, S. Farber, M. Grasso, B. Hannon, K. Limburg, S. Naeem, R.V. O'Neill, J. Paruelo, R.G. Raskin, P. Sutton and M. van den Belt. 1997. The value of the world's ecosystem services and natural capital. Nature 387: 253–260.
Crispo, E. 2007. The Baldwin effect and genetic assimilation: revisiting two mechanisms of evolutionary change mediated by phenotypic plasticity. Evolution 61: 2469–2479.
Culumber, Z.W. 2020. Thermal stress increases activity and risk-taking behavior but not anxiety in a livebearing fish. Environ. Biol. Fish. 103: 313–317.
Cyr, H. and I. Cyr. 2003. Temporal scaling of temperature variability from land to oceans. Evol. Ecol. Res. 5: 1183–1197.
Dahlhoff, E.A. and G.N. Somero. 1993. Effects of temperature on mitochondria from abalone (genus *Haliotis*): adaptive plasticity and its limits. J. Exp. Biol. 185: 151–168.

Dahlke, F.T., S. Wohlrab, M. Butzin and H.-O. Pörtner. 2020. Thermal bottlenecks in the life cycle define climate vulnerability of fish. Science 369: 65–70.

Darwin, C. 1871. The Descent of Man, and Selection in Relation to Sex. J. Murray, London.

De Jong, G. 2005. Evolution of phenotypic plasticity: patterns of plasticity and the emergence of ecotypes. New Phytol. 166: 101–118.

Denny, M. 2017. The fallacy of the average: on the ubiquity, utility and continuing novelty of Jensen's inequality. J. Exp. Biol. 220: 139–146.

Donelan, S.C., J.K. Hellmann, A.M. Bell, B. Luttbeg, J.L. Orrock, M.J. Sheriff and A. Sih. 2020. Transgenerational plasticity in human-altered environments. Trends Ecol. Evol. 35: 115–124.

Donelson, J.M., P.L. Munday, M.I. McCormick and G.E. Nilsson. 2010. Acclimation to predicted ocean warming through developmental plasticity in a tropical reef fish. Glob. Change Biol. 17: 1712–1719.

Donelson, J.M., S. Salinas, P.L. Munday and L.N.S. Shama. 2018. Transgenerational plasticity and climate change experiments: where do we go from here? Glob. Change Biol. 24: 13–34.

Dong, Y.-W., M.-L. Liao, X.-L. Meng and G.N. Somero. 2018. Structural flexibility and protein adaptation to temperature: molecular dynamics analysis of malate dehydrogenases of marine molluscs. Proc. Natl. Acad. Sci. U.S.A. 115: 1274–1279.

Dury, G.J. and M.J. Wade. 2020. When mother knows best: a population genetic model of transgenerational versus intragenerational plasticity. J. Evol. Biol. 33: 127–137.

Dyer, A.R., C.S. Brown, E.K. Espeland, J.K. McKay, H. Meimberg and K.J. Rice. 2010. The role of adaptive transgenerational plasticity in biological invasions of plants. Evol. Appl. 3: 179–192.

Edgell, T.C., B.R. Lynch, G.C. Trussell and A.R. Palmer. 2009. Experimental evidence for the Rapid evolution of behavioral canalization in natural populations. Am. Nat. 174: 434–440.

Ehrenreich, I.M. and D.W. Pfennig. 2016. Genetic assimilation: a review of its potential proximate causes and evolutionary consequences. Ann. Bot. 117: 769–779.

Elias, M., G. Wieczorek, S. Rosenne and D.S. Tawfik. 2014. The universality of enzymatic rate-temperature dependency. Trends Biochem. Sci. 39: 1–7.

Eliason, E.J., T.D. Clark, S.G. Hinch and A.P. Farrell. 2013. Cardiorespiratory performance and blood chemistry during swimming and recovery in three populations of elite swimmers: adult sockeye salmon. Comp. Biochem. Phys. A 166: 385–397.

Eliason, E.J. and A.P. Farrell. 2016. Oxygen uptake in Pacific salmon *Oncorhynchus* spp.: when ecology and physiology meet. J. Fish Biol. 88: 359–388.

Fangue, N.A., J.G. Richards and P.M. Schulte. 2009. Do mitochondrial properties explain intraspecific variation in thermal tolerance? J. Exp. Biol. 212: 514–522.

Farkas, T., E. Fodor, K. Kitajka and J.E. Halver. 2001. Response of fish membranes to environmental temperature. Aquac. Res. 32: 645–655.

Farrell, A.P. and D.R. Jones. 1992. The heart. pp. 1–73. *In*: Hoar, W.S., D.J. Randall and A.P. Farrell [eds.]. Fish Physiology, Volume XII, Part A: The cardiovascular system. Academic Press, San Diego, CA.

Farrell, A.P. 2002. Cardiorespiratory performance in salmonids during exercise at high temperature: insights into cardiovascular design limitations in fishes. Comp. Biochem. Phys. A. 132: 797–810.

Farrell, A.P., E.J. Eliason, E. Sandblom and T.D. Clark. 2009. Fish cardiorespiratory physiology in an era of climate change. Can. J. Zool. 87: 835–851.

Fenesi, A., A.R. Dyer, J. Geréd, D. Sándor and E. Ruprecht. 2014. Can transgenerational plasticity contribute to the invasion success of annual plant species? Oecologia 176: 95–106.

Flanagan, P.H., O.P. Jensen, J.W. Morley and M.L. Pinsky. 2019. Response of marine communities to local temperature changes. Ecography 42: 214–224.

Fonds, M., R. Cronie, A.D. Vethaak and P. Van der Puyl. 1992. Metabolism, food consumption and growth of plaice (*Pleuronectes platessa*) and flounder (*Platichthys flesus*) in relation to fish size and temperature. Neth. J. Sea Res. 29: 127–143.

Forsman, A. 2014. Rethinking phenotypic plasticity and its consequences for individuals, populations and species. Heredity 115: 276–284.

Garcia, R.A., M. Cabeza, C. Rahbek and M.B. Araujo. 2014. Multiple dimensions of climate change and their implications for biodiversity. Science 344: 1247579–1247579.

Ghalambor, C.K., J.K. McKay, S.P. Carroll and D.N. Reznick. 2007. Adaptive versus non-adaptive phenotypic plasticity and the potential for contemporary adaptation in new environments. Funct. Ecol. 21: 394–407.

Gillooly, J.F., J.H. Brown, G.B. West, V.M. Savage and E.L. Charnov. 2001. Effects of size and temperature on metabolic rate. Science 293: 2248–2251.

Giomi, F., C. Mandaglio, M. Ganmanee, G.-D. Han, Y.-W. Dong, G.A. Williams and G. Sarà. 2016. The importance of thermal history: costs and benefits of heat exposure in a tropical, rocky shore oyster. J. Exp. Biol. 219: 686–694.

Glickman, M.H. and A. Ciechanover. 2002. The ubiquitin-proteasome proteolytic pathway: destruction for the sake of construction. Physiol. Rev. 82: 373–428.

Gould, S.J. 1985. The median isn't the message. Discover 6: 40–42.
Greenspoon, P.B. and H.G. Spencer. 2018. The evolution of epigenetically mediated adaptive transgenerational plasticity in a subdivided population. Evolution 72: 2773–2780.
Hall, B.K. 2001. Organic selection: proximate environmental effects on the evolution of morphology and behaviour. Biol. Philos. 16: 215–237.
Hazel, J.R. 1995. Thermal adaptation in biological membranes: is homeoviscous adaptation the explanation? Annu. Rev. Physiol. 57: 19–42.
Healy, T.M. and P.M. Schulte. 2012. Thermal acclimation is not necessary to maintain a wide thermal breadth of aerobic scope in the common killifish (*Fundulus heteroclitus*). Physiol. Biochem. Zool. 85: 107–119.
Hermisson, J. and G.P. Wagner. 2004. The population genetic theory of hidden variation and genetic robustness. Genetics 168: 2271–2284.
Hobbs, R.J., E. Higgs and J.A. Harris. 2009. Novel ecosystems: implications for conservation and restoration. Trends Ecol. Evol. 24: 599–605.
Hochachka, P.W. and G.N. Somero. 2002. Biochemical Adaptation. Oxford University Press, New York.
Ho, D.H. and W.W. Burggren. 2010. Epigenetics and transgenerational transfer: a physiological perspective. J. Exp. Biol. 213: 3–16.
Ho, M.W. and P.T. Saunders. 1979. Beyond neo-Darwinism—an epigenetic approach to evolution. J. Theor. Biol. 78: 573–591.
Hoekstra, L.A. and K.L. Montooth. 2013. Inducing extra copies of the Hsp70 gene in *Drosophila melanogaster* increases energetic demand. BMC Evol. Biol. 13: 68.
Hoffmann, A.A. and J. Merilä. 1999. Heritable variation and evolution under favourable and unfavourable conditions. Trends Ecol. Evol. 14: 96–101.
Hofmann, G.E. 1999. Ecologically relevant variation in induction and function of heat shock proteins in marine organisms. Am. Zool. 39: 889–900.
Hofmann, G.E., B.A. Buckley, S. Airaksinen, J.E. Keen and G.N. Somero. 2000. Heat-shock protein expression is absent in the antarctic fish *Trematomus bernacchii* (family Nototheniidae). J. Exp. Biol. 203: 2331–2339.
Hofmeister, B.T., K. Lee, N.A. Rohr, D.W. Hall and R.J. Schmitz. 2017. Stable inheritance of DNA methylation allows creation of epigenotype maps and the study of epiallele inheritance patterns in the absence of genetic variation. Genome Biol. 18: 155.
Hoyle, R.B. and T.H.G. Ezard. 2012. The benefits of maternal effects in novel and in stable environments. J. R. Soc. Interface 9: 2403–2413.
Hu, J. and R.D.H. Barrett. 2017. Epigenetics in natural animal populations. J. Evol. Biol. 30: 1612–1632.
Huey, R.B. and R.D. Stevenson. 1979. Integrating thermal physiology and ecology of ectotherms: a discussion of approaches. Integr. Comp. Biol. 19: 357–366.
Huey, R.B., M.R. Kearney, A. Krockenberger, J.A.M. Holtum, M. Jess and S.E. Williams. 2012. Predicting organismal vulnerability to climate warming: roles of behaviour, physiology and adaptation. Philos. Trans. R. Soc. B 367: 1665–1679.
Hutchings, J.A., D.P. Swain, S. Rowe, J.D. Eddington, V. Puvanendran and J.A. Brown. 2007. Genetic variation in life-history reaction norms in a marine fish. Proc. R. Soc. B 274: 1693–1699.
Iftikar, F.I., J.R. MacDonald, D.W. Baker, G.M.C. Renshaw and A.J.R. Hickey. 2014. Could thermal sensitivity of mitochondria determine species distribution in a changing climate? J. Exp. Biol. 217: 2348–2357.
Jablonka, E. and G. Raz. 2009. Transgenerational epigenetic inheritance: prevalence, mechanisms, and implications for the study of heredity and evolution. Q. Rev. Biol. 84: 131–176.
Jannasch, H.W. and M.J. Mottl. 1985. Geomicrobiology of deep-sea hydrothermal vents. Science 229: 717–725.
Jayasundara, N., L.D. Gardner and B.A. Block. 2013. Effects of temperature acclimation on Pacific bluefin tuna (*Thunnus orientalis*) cardiac transcriptome. Am. J. Physiol. Regul. Integr. Comp. Physiol. 305: R1010–20.
Jayasundara, N. and G.N. Somero. 2013. Physiological plasticity of cardiorespiratory function in a eurythermal marine teleost, the longjaw mudsucker, *Gillichthys mirabilis*. J. Exp. Biol. 216: 2111–2121.
Jayasundara, N., L. Tomanek, W.W. Dowd and G.N. Somero. 2015. Proteomic analysis of cardiac response to thermal acclimation in the eurythermal goby fish *Gillichthys mirabilis*. J. Exp. Biol. 218: 1359–1372.
Johnston, I.A. 1993. Phenotypic plasticity of fish muscle to temperature change. pp. 322–340. *In*: Rankin, J.C. and F.B. Jensen [eds.]. Fish Ecophysiology. Chapman & Hall, London.
Johnston, I.A. and R.S. Wilson. 2006. Temperature-induced developmental plasticity in ectotherms. pp. 124–138. *In*: Warburton, S.J., W.W. Burggren, B. Pelster, C.L. Reiber and J. Spicer [eds.]. Comparative Developmental Physiology: Contributions, Tools, and Trends. Oxford University Press, Oxford.
Kaltz, O. and G. Bell. 2002. The ecology and genetics of fitness in *Chlamydomonas*. XII. Repeated sexual episodes increase rates of adaptation to novel environments. Evolution 56: 1743–1753.
Karelitz, S., M.D. Lamare, B. Mos, H. De Bari, S.A. Dworjanyn and M. Byrne. 2019. Impact of growing up in a warmer, lower pH future on offspring performance: transgenerational plasticity in a pan-tropical sea urchin. Coral Reefs 38: 1085–1095.

Keen, A.N., J.M. Klaiman, H.A. Shiels and T.E. Gillis. 2017. Temperature-induced cardiac remodelling in fish. J. Exp. Biol. 220: 147–160.
Kellermann, V., S.L. Chown, M.F. Schou, I. Aitkenhead, C. Janion-Scheepers, A. Clemson, M.T. Scott and C.M. Sgrò. 2019. Comparing thermal performance curves across traits: how consistent are they? J. Exp. Biol. 222.
Kelly, M. 2019. Adaptation to climate change through genetic accommodation and assimilation of plastic phenotypes. Philos. Trans. R. Soc. B 374: 20180176.
Kennedy, J.J., P. Brohan and S.F.B. Tett. 2007. A global climatology of the diurnal variations in sea-surface temperature and implications for MSU temperature trends. Geophys. Res. Lett. 34: L05712.
Kingsolver, J.G., R. Gomulkiewicz and P.A. Carter. 2001. Variation, selection and evolution of function-valued traits. Genetica 112-113: 87–104.
Kingsolver, J.G., J.G. Shlichta, G.J. Ragland and K.R. Massie. 2006. Thermal reaction norms for caterpillar growth depend on diet. Evol. Ecol. Res. 8: 703–715.
Kingsolver, J.G. and H.A. Woods. 2016. Beyond thermal performance curves: modeling time-dependent effects of thermal stress on ectotherm growth rates. Am. Nat. 187: 283–294.
Koumoundouros, G., C. Ashton, D.G. Sfakianakis, P. Divanach, M. Kentouri, N. Anthwal and N.C. Stickland. 2009. Thermally induced phenotypic plasticity of swimming performance in European sea bass *Dicentrarchus labrax* juveniles. J. Fish Biol. 74: 1309–1322.
Kültz, D. 2005. Molecular and evolutionary basis of the cellular stress response. Annu. Rev. Physiol. 67: 225–257.
Lande, R. 2009. Adaptation to an extraordinary environment by evolution of phenotypic plasticity and genetic assimilation. J. Evol. Biol. 22: 1435–1446.
Ledon-Rettig, C.C., D.W. Pfennig, A.J. Chunco and I. Dworkin. 2014. Cryptic genetic variation in natural populations: a predictive framework. Integr. Comp. Biol. 54: 783–793.
Lee, W.-S., P. Monaghan and N.B. Metcalfe. 2016. Perturbations in growth trajectory due to early diet affect age-related deterioration in performance. Funct. Ecol. 30: 625–635.
Leichter, J.J., B. Helmuth and A.M. Fischer. 2006. Variation beneath the surface: quantifying complex thermal environments on coral reefs in the Caribbean, Bahamas and Florida. J. Mar. Res. 64: 563–588.
Leimar, O. and J.M. McNamara. 2015. The evolution of transgenerational integration of information in heterogeneous environments. Am. Nat. 185: E55–E69.
Leroi, A.M. 2015. The Lagoon: How Aristotle Invented Science. Penguin Books, New York.
Levis, N.A. and D.W. Pfennig. 2016. Evaluating "plasticity-first" evolution in nature: key criteria and empirical approaches. Trends Ecol. Evol. 31: 563–574.
Lillywhite, H.B., K.C. Zippel and A.P. Farrell. 1999. Resting and maximal heart rates in ectothermic vertebrates. Comp. Biochem. Physiol. A Mol. Integr. Physiol. 124: 369–382.
Logan, C.A. and G.N. Somero. 2010. Transcriptional responses to thermal acclimation in the eurythermal fish *Gillichthys mirabilis* (Cooper 1864). Am. J. Physiol. Regul. Integr. Comp. Physiol. 299: R843–52.
Logan, C.A. and B.A. Buckley. 2015. Transcriptomic responses to environmental temperature in eurythermal and stenothermal fishes. J. Exp. Biol. 218: 1915–1924.
Madeira, D., J.E. Araújo, R. Vitorino, J.L. Capelo, C. Vinagre and M.S. Diniz. 2016. Ocean warming alters cellular metabolism and induces mortality in fish early life stages: a proteomic approach. Environ. Res. 148: 164–176.
Marsh, A.G. and A.A. Pasqualone. 2014. DNA methylation and temperature stress in an Antarctic polychaete, *Spiophanes tcherniai*. Front. Physiol. 5: 173.
Marshall, D.J., Y.-W. Dong, C.D. McQuaid and G.A. Williams. 2011. Thermal adaptation in the intertidal snail *Echinolittorina malaccana* contradicts current theory by revealing the crucial roles of resting metabolism. J. Exp. Biol. 214: 3649–3657.
Martin, T.L. and R.B. Huey. 2008. Why "suboptimal" is optimal: Jensen's inequality and ectotherm thermal preferences. Am. Nat. 171: E102–E118.
McGuigan, K. and C.M. Sgrò. 2009. Evolutionary consequences of cryptic genetic variation. Trends Ecol. Evol. 24: 305–311.
McKenzie, J.L., D.J. Chung, T.M. Healy, R.S. Brennan, H.J. Bryant, A. Whitehead and P.M. Schulte. 2019. Mitochondrial ecophysiology: assessing the evolutionary forces that shape mitochondrial variation. Integr. Comp. Biol. 59: 925–937.
Merilä, J. 2012. Evolution in response to climate change: in pursuit of the missing evidence. Bioessays 34: 811–818.
Metzger, D.C.H. and P.M. Schulte. 2017. Persistent and plastic effects of temperature on DNA methylation across the genome of threespine stickleback (*Gasterosteus aculeatus*). Proc. R. Soc. B 284: 20171667.
Mora, C., A.G. Frazier, R.J. Longman, R.S. Dacks, M.M. Walton, E.J. Tong, J.J. Sanchez, L.R. Kaiser, Y.O. Stender, J.M. Anderson, C.M. Ambrosino, I. Fernandez-Silva, L.M. Giuseffi and T.W. Giambelluca. 2013. The projected timing of climate departure from recent variability. Nature 502: 183–187.
Moreno, J. and A.P. Møller. 2011. Extreme climatic events in relation to global change and their impact on life histories. Curr. Zool. 57: 375–389.
Morris, M.R.J. and S.M. Rogers. 2013. Overcoming maladaptive plasticity through plastic compensation. Curr. Zool. 59: 526–536.
Moyes, C.D. and P.M. Schulte. 2015. Principles of Animal Physiology. Pearson, New York.

Munch, S.B. and S. Salinas. 2009. Latitudinal variation in lifespan within species is explained by the metabolic theory of ecology. Proc. Natl. Acad. Sci. U.S.A. 106: 13860–13864.
Munday, P.L., R.R. Warner, K. Monro, J.M. Pandolfi and D.J. Marshall. 2013. Predicting evolutionary responses to climate change in the sea. Ecol. Lett. 16: 1488–1500.
Munday, P.L. 2014. Transgenerational acclimation of fishes to climate change and ocean acidification. F1000Prime Rep. 6: 1–7.
Murren, C.J., J.R. Auld, H. Callahan, C.K. Ghalambor, C.A. Handelsman, M.A. Heskel, J.G. Kingsolver, H.J. Maclean, J. Masel, H. Maughan, D.W. Pfennig, R.A. Relyea, S. Seiter, E. Snell-Rood, U.K. Steiner and C.D. Schlichting. 2015. Constraints on the evolution of phenotypic plasticity: limits and costs of phenotype and plasticity. Heredity 115: 293–301.
Niehaus, A.C., M.J. Angilletta, M.W. Sears, C.E. Franklin and R.S. Wilson. 2012. Predicting the physiological performance of ectotherms in fluctuating thermal environments. J. Exp. Biol. 215: 694–701.
Norouzitallab, P., K. Baruah, M. Vandegehuchte, G. Van Stappen, F. Catania, J. Vanden Bussche, L. Vanhaecke, P. Sorgeloos and P. Bossier. 2014. Environmental heat stress induces epigenetic transgenerational inheritance of robustness in parthenogenetic *Artemia* model. FASEB J. 28: 3552–3563.
Nunney, L. 2016. Adapting to a changing environment: modeling the interaction of directional selection and plasticity. J. Hered. 107: 15–24.
O'Connor, M.I., J.F. Bruno, S.D. Gaines, B.S. Halpern, S.E. Lester, B.P. Kinlan and J.M. Weiss. 2007. Temperature control of larval dispersal and the implications for marine ecology, evolution, and conservation. Proc. Nat. Acad. Sci. U.S.A. 104: 1266–1271.
Oliver, E.C.J., M.G. Donat, M.T. Burrows, P.J. Moore, D.A. Smale, L.V. Alexander, J.A. Benthuysen, M. Feng, A. Sen Gupta, A.J. Hobday, N.J. Holbrook, S.E. Perkins-Kirkpatrick, H.A. Scannell, S.C. Straub and T. Wernberg. 2018. Longer and more frequent marine heatwaves over the past century. Nat. Commun. 9: 1324.
Paaby, A.B. and M.V. Rockman. 2014. Cryptic genetic variation: evolution's hidden substrate. Nat. Rev. Genet. 15: 247–258.
Peirson, B.R.E. 2012. Richard Woltereck's Concept of Reaktionsnorm. Embryo Project Encyclopedia, http://embryo.asu.edu/handle/10776/3940.
Pigliucci, M. 2001. Phenotypic Plasticity: Beyond Nature and Nurture. Johns Hopkins University Press, Baltimore, MD.
Pigliucci, M. and C.J. Murren. 2003. Genetic assimilation and a possible evolutionary paradox: can macroevolution sometimes be so fast as to pass us by? Evolution 57: 1455–1464.
Pinsky, M.L., A.M. Eikeset, D.J. McCauley, J.L. Payne and J.M. Sunday. 2019. Greater vulnerability to warming of marine versus terrestrial ectotherms. Nature 569: 108–111.
Podrabsky, J.E. and G.N. Somero. 2004. Changes in gene expression associated with acclimation to constant temperatures and fluctuating daily temperatures in an annual killifish *Austrofundulus limnaeus*. J. Exp. Biol. 207: 2237–2254.
Pörtner, H.-O. 2002. Climate variations and the physiological basis of temperature dependent biogeography: systemic to molecular hierarchy of thermal tolerance in animals. Comp. Biochem. Physiol. A Mol. Integr. Physiol. 132: 739–761.
Pörtner, H.-O. and R. Knust. 2007. Climate change affects marine fishes through the oxygen limitation of thermal tolerance. Science 315: 95–97.
Pörtner, H.-O., L.S. Peck and G.N. Somero. 2012. Mechanisms defining thermal limits and adaptation in marine ectotherms: an integrative view. pp. 379–416. *In*: Rogers, A.D., N.M. Johnston, E.J. Murphy and A. Clarke [eds.]. Antarctic Ecosystems: An Extreme Environment in a Changing World. Wiley-Blackwell, West Sussex, UK.
Pörtner, H.-O., C. Bock and F.C. Mark. 2017. Oxygen- and capacity-limited thermal tolerance: bridging ecology and physiology. J. Exp. Biol. 220: 2685–2696.
Ray, G.C. 1991. Coastal-zone biodiversity patterns. Bioscience 41: 490–498.
Regaudie-de-Gioux, A. and C.M. Duarte. 2012. Temperature dependence of planktonic metabolism in the ocean. Global Biogechem. Cy. 26: GB1015.
Reznick, D.N. 1993. Norms of reaction in fishes. pp. 72–90. *In*: Stokes, T.K., J.M. McGlade and R. Law [eds.]. The Exploitation of Evolving Resources. Springer, Berlin.
Ruel, J.J. and M.P. Ayres. 1999. Jensen's inequality predicts effects of environmental variation. Trends Ecol. Evol. 14: 361–366.
Salinas, S. and S.B. Munch. 2012. Thermal legacies: transgenerational effects of temperature on growth in a vertebrate. Ecol. Lett. 15: 159–163.
Salinas, S., S.C. Brown, M. Mangel and S.B. Munch. 2013. Non-genetic inheritance and changing environments. Non-Genetic Inheritance 1: 38–50.
Salinas, S. and S.B. Munch. 2014. Phenotypic complexity: integrated responses of life history characters to multiple environmental factors. Evol. Ecol. Res. 16: 267–284.
Salinas, S., S.E. Irvine, C.L. Schertzing, S.Q. Golden and S.B. Munch. 2019. Trait variation in extreme thermal environments under constant and fluctuating temperatures. Philos. Trans. R. Soc. Lond. B 374: 20180177.
Schaefer, J. and A. Ryan. 2006. Developmental plasticity in the thermal tolerance of zebrafish *Danio rerio*. J. Fish Biol. 69: 722–734.
Schaffer, S.W. and M.-S. Suleiman. 2010. Mitochondria: The Dynamic Organelle. Springer Science & Business Media, New York.

Scheiner, S.M. 1993. Genetics and evolution of phenotypic plasticity. Annu. Rev. Ecol. Syst. 24: 35–68.
Schlaepfer, M.A., M.C. Runge and P.W. Sherman. 2002. Ecological and evolutionary traps. Trends Ecol. Evol. 17: 474–480.
Schlichting, C.D. 2008. Hidden reaction norms, cryptic genetic variation, and evolvability. Ann. N. Y. Acad. Sci. 1133: 187–203.
Schmidt-Nielsen, K. 1997. Animal Physiology: Adaptation and Environment. Cambridge University Press, Cambridge.
Schulte, P.M., T.M. Healy and N.A. Fangue. 2011. Thermal performance curves, phenotypic plasticity, and the time scales of temperature exposure. Integr. Comp. Biol. 51: 691–702.
Schwander, T. and O. Leimar. 2011. Genes as leaders and followers in evolution. Trends Ecol. Evol. 26: 143–151.
Scoville, A.G. and M.E. Pfrender. 2010. Phenotypic plasticity facilitates recurrent rapid adaptation to introduced predators. Proc. Natl. Acad. Sci. U.S.A. 107: 4260–4263.
Sen, S.S. and S.S. Giri. 2017. Physiological role of heat shock proteins, molecular function and stress removal in fishes. pp. 215–246. *In*: Asea, A.A.A. and P. Kaur [eds.]. Heat Shock Proteins in Veterinary Medicine and Sciences. Springer International, Cham, Switzerland.
Shama, L.N.S., A. Strobel, F.C. Mark and K.M. Wegner. 2014. Transgenerational plasticity in marine sticklebacks: maternal effects mediate impacts of a warming ocean. Funct. Ecol. 28: 1482–1493.
Shama, L.N.S. and K.M. Wegner. 2014. Grandparental effects in marine sticklebacks: transgenerational plasticity across multiple generations. J. Evol. Biol. 27: 2297–2307.
Smith, M.D. 2011. The ecological role of climate extremes: current understanding and future prospects. J. Ecol. 99: 651–655.
Smit, K. and D. Glassom. 2017. Large fluctuations but constant mean temperatures allow corals to persist in intertidal rock pools on the east coast of South Africa. Helgol. Mar. Res. 71: 3.
Snell-Rood, E.C. and D.R. Papaj. 2009. Patterns of phenotypic plasticity in common and rare environments: a study of host use and color learning in the cabbage white butterfly *Pieris rapae*. Am. Nat. 173: 615–631.
Sokolova, I.M. 2013. Energy-limited tolerance to stress as a conceptual framework to integrate the effects of multiple stressors. Integr. Comp. Biol. 53: 597–608.
Somero, G.N., B.L. Lockwood and L. Tomanek. 2017. Biochemical adaptation: response to environmental challenges, from life's origins to the Anthropocene. Sinauer Associates, Sunderland, MA, USA.
Stearns, S.C. and T.J. Kawecki. 1994. Fitness sensitivity and the canalization of life-history traits. Evolution 48: 1438–1450.
Stillman, J.H. and G.N. Somero. 2000. A comparative analysis of the upper thermal tolerance limits of eastern Pacific porcelain crabs, genus *Petrolisthes*: influences of latitude, vertical zonation, acclimation, and phylogeny. Physiol. Biochem. Zool. 73: 200–208.
Sultan, S.E. 1992. Phenotypic plasticity and the Neo-Darwinian legacy. Evol. Trend. Plant. 6: 61–71.
Sultan, S.E. and H.G. Spencer. 2002. Metapopulation structure favors plasticity over local adaptation. Am. Nat. 160: 271–283.
Taborsky, B. 2017. Developmental plasticity: preparing for life in a complex world. Adv. Stud. Behav. 49: 49–99.
Thorpe, J.E. 2004. Life history responses of fishes to culture. J. Fish Biol. 65: 263–285.
Tomanek, L. 2011. Environmental proteomics: changes in the proteome of marine organisms in response to environmental stress, pollutants, infection, symbiosis, and development. Ann. Rev. Mar. Sci. 3: 373–399.
Truebano, M., P. Fenner, O. Tills, S.D. Rundle and E.L. Rezende. 2018. Thermal strategies vary with life history stage. J. Exp. Biol. 221.
Tufto, J. 2015. Genetic evolution, plasticity, and bet-hedging as adaptive responses to temporally autocorrelated fluctuating selection: a quantitative genetic model. Evolution 69: 2034–2049.
Turkmen, S., M.J. Zamorano, H. Fernández-Palacios, C.M. Hernández-Cruz, D. Montero, L. Robaina and M. Izquierdo. 2017. Parental nutritional programming and a reminder during juvenile stage affect growth, lipid metabolism and utilisation in later developmental stages of a marine teleost, the gilthead sea bream (*Sparus aurata*). Br. J. Nutr. 118: 500–512.
Varriale, A. and G. Bernardi. 2006. DNA methylation and body temperature in fishes. Gene 385: 111–121.
Vasseur, D.A., J.P. DeLong, B. Gilbert, H.S. Greig, C.D.G. Harley, K.S. McCann, V. Savage, T.D. Tunney and M.I. O'Connor. 2014. Increased temperature variation poses a greater risk to species than climate warming. Proc. R. Soc. B 281: 20132612–20132612.
Vasseur, D.A. and P. Yodzis. 2004. The color of environmental noise. Ecology 85: 1146–1152.
Veilleux, H.D., T. Ryu, J.M. Donelson, L. van Herwerden, L. Seridi, Y. Ghosheh, M.L. Berumen, W. Leggat, T. Ravasi and P.L. Munday. 2015. Molecular processes of transgenerational acclimation to a warming ocean. Nat. Clim. Chang. 5: 1074–1078.
Via, S. and R. Lande. 1985. Genotype-environment interaction and the evolution of phenotypic plasticity. Evolution 39: 505–522.
Vornanen, M., M. Hassinen, H. Koskinen and A. Krasnov. 2005. Steady-state effects of temperature acclimation on the transcriptome of the rainbow trout heart. Am. J. Physiol. Regul. Integr. Comp. Physiol. 289: R1177–1184.
Waddington, C.H. 1942. Canalization of development and the inheritance of acquired characters. Nature 150: 563–565.
Weber, G.G., J. Kortmann, F. Narberhaus and K.E. Klose. 2014. RNA thermometer controls temperature-dependent virulence factor expression in Vibrio cholerae. Proc. Natl. Acad. Sci. U.S.A. 111: 14241–14246.

West-Eberhard, M.J. 2003. Developmental Plasticity and Evolution. Oxford University Press, Oxford.
Whitman, D.W. and A.A. Agrawal. 2009. What is phenotypic plasticity and why is it important? pp. 1–63. *In*: Whitman, D.W. and T.N. Ananthakrishnan [eds.]. Phenotypic Plasticity of Insects: Mechanisms and Consequences. CRC Press, Boca Raton, FL.
Williams, J.W., S.T. Jackson and J.E. Kutzbach. 2007. Projected distributions of novel and disappearing climates by 2100 AD. Proc. Natl. Acad. Sci. U.S.A. 104: 5738–5742.
Wilson, R.S. and C.E. Franklin. 2002. Testing the beneficial acclimation hypothesis. Trends Ecol. Evol. 17: 66–70.
Wurtsbaugh, W.A. and J.J. Cech, Jr. 1983. Growth and activity of juvenile mosquitofish: temperature and ration effects. Trans. Am. Fish. Soc. 112: 653–660.

CHAPTER 14

Climate Variability and Change, their Impacts on Marine Life and Ecosystems

Jerónimo Pan[1,*] and *Reinaldo A. Maenza*[2]

14.1 Introduction—A brief history and the politics behind climate change

Throughout its 4,500 MY history, the Earth has gone through many changes. Some of these have implied such radical transformations as the coalescence of a crust and the relative movement and redistribution of tectonic plates (the Wilson cycle described in Chapter 1), the advent of life, the transition from a reducing to an oxidizing atmosphere during the Great Oxidation Event, and finally, with the creation of an atmosphere, the establishment of the climatic system as we know it today. In other words, planet Earth is used to changes and transformations. But with the advent of modern civilization, anthropic (i.e., from the Greek *ánthrōpos* = human) activities, largely related to agricultural and industrial development and based on fossil fuels as an energy source, have significantly altered biogeochemical cycles, leading to an accelerated change in many of Earth's processes.

Some scientists argue that environmental change is such that it has left an imprint on Earth's subsystems, leading to a new geologic Epoch in the Quaternary known as the Anthropocene (from the Greek *ánthrōpos* and *kainos* = new; i.e., the 'new time of man'). Despite having been proposed more than two decades ago by Crutzen and Stoermer (2000), the very existence and the onset of the Anthropocene is still reluctantly accepted and debated by a fraction of the scientific community (Lewis and Maslin 2015; Fig. 14.1). What worries many scientists is not the global scale that these climatic changes have reached—something that the planet has certainly experienced in its geologic history, and far more extreme ones if we may add—but rather the rate at which these changes are taking place and how disruptive they can be to biological systems, ranging from the thermal physiology of individual organisms (treated in Chapter 13), to the functioning of large ecosystems (Hoegh-Guldberg and Poloczanska 2017).

There is vast and overwhelming empirical evidence on how human activities have been transforming Earth's surface, and significantly altering its geomorphology and global regulatory mechanisms for the past few millennia. Come to think about it, what is somehow puzzling is how long it has taken for the concept of climate change and its derived effects on Earth's systems to become a matter of interest to society.

Climate may be considered as the resultant of a circulatory system on a planetary scale, in which the radiation that reaches the Earth from the sun warms up its surface and the atmosphere. As we will discuss in this chapter, there are constant heat fluxes between the Earth's surface, the oceans, and the atmosphere, that heat up and cool down in a dynamic equilibrium. As much as climate transversely affects all human endeavors, it probably was not until the first two decades of this century that the talk on recent climate

[1] National Scientific and Technical Research Council, Argentina (CONICET); Instituto de Geología de Costas y del Cuaternario (IGCyC); Instituto de Investigaciones Marinas y Costeras (IIMyC).
[2] National Institute for Fisheries Research and Development, Argentina (INIDEP).
* Corresponding author: jeronimopan@gmail.com

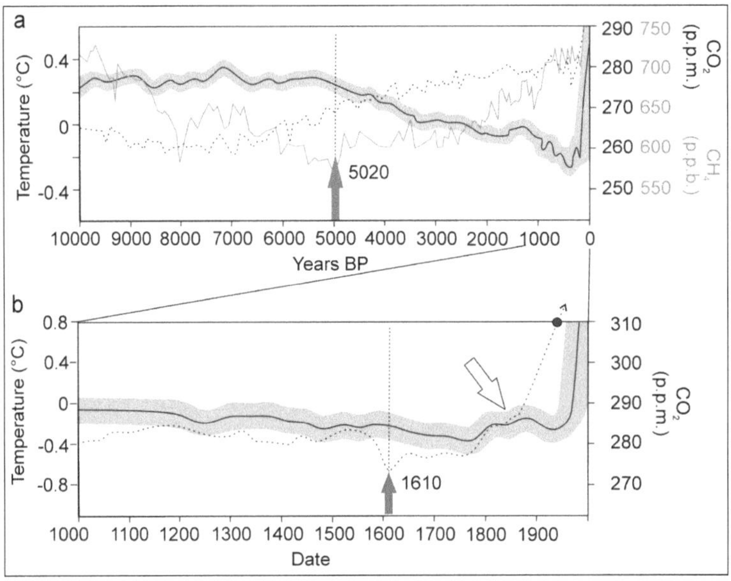

Figure 14.1. Evolution of temperature, CO_2 and methane for most of the Holocene/Anthropocene Epochs (comprising the past 10,000 yr). In the upper plot (a), global temperature anomalies (solid black line) are relative to the 1961–1990 average; the upward-pointing arrow indicates the lowest level of atmospheric methane (solid gray line), thereafter increasing due to early cultivation and ruminant cattle practices; atmospheric CO_2 data (dashed black line) compiled from various sources. On the lower plot (b), the gray arrow indicates a pronounced dip in atmospheric CO_2 (from ice core records) attributed to the collision of the Old and New World people (hence known as the 'Orbis' hypothesis), which brought along homogenization of world biotas, a large decline in human numbers, and temporary retreat in agricultural practices; this has been signaled as a putative boundary between the Holocene/Anthropocene Epochs. The white arrow on plot (b) indicates the onset of the steady increase in CO_2, attributed to the Industrial Revolution and increasing fossil fuel demand. Redrawn from Lewis and Maslin (2015).

change has percolated almost every agenda. Global and national politics, and entrepreneurial interests have certainly played a key role in this 'slow start' on climate change issues. Let us recapitulate briefly on this matter, which has been so cleverly presented in a cartoon by Monastersky and Sousanis (2015).

Scientific interest on the effects of CO_2 emissions on atmospheric temperature date back to the late 19th century, when the Swedish physicist Svante Arrhenius published the first seminal paper (1896) on the matter. Keeping up with the trend started with the Industrial Revolution, atmospheric CO_2 emissions increased steadily throughout the 20th century. This has been continuously recorded since 1958 by Scripps Institution of Oceanography at the Mauna Loa Observatory located in the middle of the North Pacific Ocean, producing what is known as the Keeling curve, a graph with cumulative daily records of atmospheric CO_2 concentration (Fig. 14.1b). On parallel environmental matters started gaining the attention of the general public and the media post World War II. In this context, the publication of Rachel Carson's *Silent Spring* in 1962 brought attention to a worldwide audience on the detrimental impacts that human activities had already caused on the air, land and water, and helped launch the environmental movement. The scenario was laid out for the Nixon administration to establish the United States Environmental Protection Agency (EPA) in 1970, with the mission to protect human health and the environment. EPA has since set up the standards for pollution thresholds worldwide. By the 1980s, matters such as the depletion of the stratospheric ozone layer were commonplace in the press and among the general public, which pushed for an international treaty (the so-called 1987 Montreal Protocol) to cut down CFCs[a] emissions and the use of aerosols.

Around this time, scientific societies organized international programs to address and tackle pressing issues of climate change and its increasingly evident effects on Earth's ecosystems. The International

[a] CFCs, or chlorofluorocarbons, are a set of synthetic organic compounds with chlorine, fluorine and carbon atoms, which were formerly and extensively used in refrigeration, as aerosol propellants and foaming agents, until their ban in 1987.

Geosphere-Biosphere Programme (IGBP), sponsored by the International Council for Science (ICSU), and running from 1986 to 2015, was perhaps the most prominent one. IGBP was largely centered on a global change scientific agenda, rather than climate issues alone (Seitzinger et al. 2015). The IGBP articulated different core projects and program-level activities (most of them focused on interphases between land, atmosphere and the oceans) such as: BAHC (Biosphere Aspects of the Hydrological Cycle); GLOBEC (Global Ocean Ecosystem dynamics); JGOFS (Joint Global Ocean Flux Study); LOICZ (Land-Ocean Interactions in the Coastal Zone); SOLAS (Surface Ocean Lower Atmosphere Study); IGAC (International Global Atmospheric Chemistry); LUCC (Land Use and Cover Change); and integrated in time and space, as exemplified by the PAGES (Past Global Changes) program. IGBP's legacy lies in the creation of international scientific networks and global databases, the emphasis on interdisciplinarity, the coining of the Anthropocene concept, and the inclusion of human actions as an integral part of the Earth system (Seitzinger et al. 2015).

Among the growing climatic concern raised by 1988, the United Nations and the World Meteorological Organization sensed a favorable scenario for the establishment of the Intergovernmental Panel on Climate Change (IPCC). Since its creation, the IPCC has pushed the agenda for reaching a scientific and political consensus on the occurrence, extent, and impacts of climate change, and seeking mitigation alternatives. The IPCC is made up of scientists, economists, social workers and lobbyists from various nations of the world that split their efforts into three work groups. Working Group I (WGI) provides the scientific basis behind the climate system and its changes; WGII provides estimations on the vulnerability of natural and socio-economic systems in the face of climate change and its consequences; and WGIII ponders the options to limit greenhouse gases (GHGs) emissions and mitigate the effects of climate change. Perhaps the most important outcome of these working groups formed by specialists from all over the world are the *assessment reports* produced periodically by IPCC's WGI (six volumes thus far since the first one published in 1990, plus other supplementary and special reports), which have become the most authoritative scientific literature on climate change.

With its consolidation throughout the 1990s and early 2000s, the IPCC introduced new perspectives into the climate change discussion, most interestingly socio-political ones. Perhaps unnoticed by the general public and disguised under the initial letter of its acronym, the role of national governments (the "I" in IPCC, stands for "Intergovernmental"), as voiced by the representatives they send to Conference of the Parties (COP) meetings and the scientists that participate in IPCC report drafting, is undoubtedly of significant weight. International climate meetings in which the wealthy nations of the world agree or disagree on a climate agenda, emissions quotas and strategies to reduce the rate of climate change are held quite regularly since the 1990s, and each new meeting is met with high expectations by environmental activists and world citizens alike. Yet, thus far the global climate agreements signed at each meeting seem to be dropped by the signatory nations shortly afterwards. This notably happened with the Kyoto Protocol signed in 1997 and later dropped by the USA in 2001, and more recently with the so-called 'Paris agreement'.

With its goal to limit global average temperature increase to $< 2°C$ by 2100, the Paris Agreement signed at the 21st COP in 2015 infused a renewed optimism into the climate discussion. On par, the United Nations set up an agenda based on Sustainable Development Goals (SDGs), which includes the oceans and coastal areas as a top priority (the targets of SDG 14 "Life below water" are specifically planned for the oceans; Reimer et al. 2020). On the one hand, unsustainable resource extraction, pollution, climate change, and habitat destruction threaten the oceans, while on the other, the ~ 600 million people currently living in low-elevation coastal zones are under serious threats caused by sea level rise, making these matters pressing. While meeting these SDGs will be critically important for growing populations, at the same time, it would be highly challenging, as there exist gaps in the suitability of some targets and the scientific capability required (Pendleton et al. 2020, Reimer et al. 2020).

In recent years, citizens and activists have taken an increasingly central role in the climate discussion. The case of Greta Thunberg is one that resonated the most. This Swedish teenage activist who, at 15 years of age so vocally advocated for immediate climate action and sustainability policies from world leaders at the 2018 COP meeting, made an international impact in the media and public opinion by showing the world that those who so reluctantly commit to climate action and sustainability issues will not be the ones facing the consequences of today's inaction, but rather the younger generations.

On the other hand, the year 2015 witnessed an unprecedented move from the Vatican with the publication of *Laudato si'*. In this encyclical, Pope Francis, one of the chief religious leaders of the world, acknowledged the science behind climate change and called for immediate action to stop its effects. The missive also represented a plea to care for the environment and find sustainable ways for development.

Sustainability, redefined for the 21st century as the capacity for Earth's biosphere and human civilization to co-exist, is a complex, multi-faceted issue, not only based on the mitigation and prevention of future climate change, but also implying radical changes in cultural practices and socio-economic paradigms. If we are to take effective climate action, societies need to start planning for cultural changes that cut the dependency on fossil fuels right away. Areas such as home building and city planning, transportation, energy generation and food production are key. There are already viable alternatives such as electric vehicles, buses and trains for human transportation, or hydrogen-powered trucks to replace shipping demands for goods and food without resorting to fossil fuels. On the other hand, renewable energies such as solar panels (to cover domestic and small industry needs) or wind turbines are becoming a reality in many territories, representing not only a source of countless new jobs but a growing industry creating new economic wealth. Now is the time to embrace these alternatives, something that will certainly demand a cultural change, often without the political and governmental support and enforcement that would be desirable.

After this introduction, let us now delve deeper into the scientific aspects of Earth's climate system, the imbalances and changes it has been subject to during the Anthropocene, and how they affect marine organisms and ecosystems at various scales. Chapter 13 in this volume has tackled some of these issues from an organismal point of view and abounds in examples of how current and projected climate changes introduced by humans have affected the physiology and evolutionary trends of marine organisms. This chapter will revisit some of the impacts described in the previous chapter and introduce others, largely from a marine-ecosystems perspective. In doing so, we will approach such task with no intention of providing either a balanced or encyclopedic survey of the topics of current interest in climate change, as the task will simply be insurmountable. Hundreds of scientific papers and policy recommendations are being written and implemented as you read these lines, so in a sense, it would be absurd to try to concentrate the breadth and depth of all climate change issues applied to the oceans in a book chapter. Instead, we here present you with an introduction to those topics which we consider relevant and aim to provide you with academic tools to build up your own critical thinking on these matters.

14.2 Components of the climate system

The climate system is made up of five subsystems: the atmosphere (gaseous layer); the hydrosphere (oceans, lakes and rivers); the cryosphere (water in solid state; such as glaciers and sea ice); the lithosphere (terrestrial surface and the topmost m of soil) and the biosphere (the organisms in land and marine ecosystems). These subsystems not only have their own dynamics, but also interact with each other through flows of matter and energy, called coupling mechanisms. As an example, the mechanisms involved in ocean-atmosphere coupling are schematized in Fig. 14.2.

The interactions involved in these coupling mechanisms can be represented by feedback processes between the components of the subsystems. Feedbacks may be positive or negative. In the former case, the response of the system generates an increase in the disturbance to it, while in the latter, the system responds by weakening an initial disturbance. In this way, positive feedbacks can generate the evolution of the system towards a new state of equilibrium, while negative feedbacks contribute to its stability. Some examples are schematized in Fig. 14.3, where the increase in air temperature produces an increment in water vapor, which translates into an increment in cloud cover. Low- and mid-altitude clouds partially block the incoming short-wave radiation, so air temperature decreases (negative feedback). In the counterpart, high altitude clouds partially absorb the outgoing long-wave radiation, causing an increase in air temperature (positive feedback).

Climate is defined as the synthesis of the meteorological conditions for a given area, characterized by long-term statistics (mean values, 30 yr) of the meteorological elements in said area. Climate has an inherent variability that arises from its many integral parts and the interactive effects and feedback

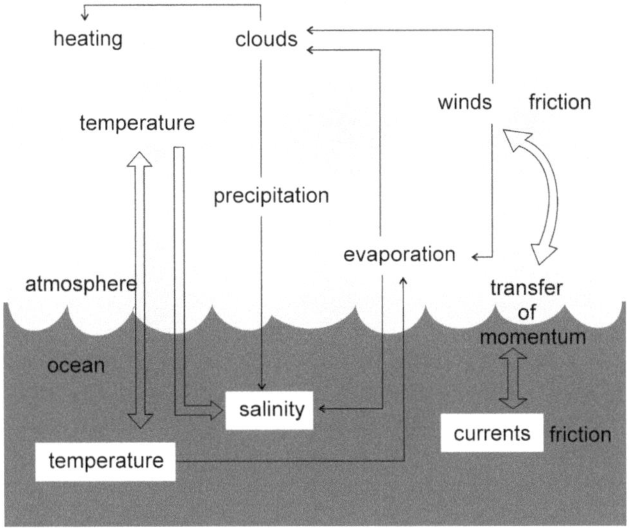

Figure 14.2. A schematic representation of ocean and atmosphere coupling mechanisms. There is a latitudinal variation in their relative importance. Modified and redrawn from McGuffie and Henderson-Sellers (2005).

processes that regulate them. Thus, a certain degree of statistical variability is to be expected. However, when the variation in any of the parameters that characterize the climate for a region (or globally) follows a steady trend for a number of years (such as the steady accumulation of GHGs in the troposphere, or pCO_2[b] in surface ocean waters), then it becomes apparent that the climate system is experiencing change. The changes in Earth's climate system are the product of internal and external *forcers*, i.e., significant alterations that induce or force the system to respond and change in a certain direction. The following section describes them in detail.

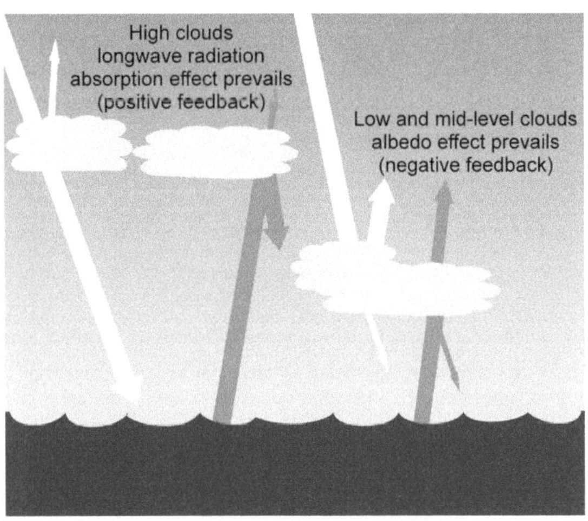

Figure 14.3. Some examples of feedbacks in the climate system related to air temperature, water vapor content, and clouds. An increase in air temperature increases water vapor content, and thus an increment in cloud cover. Low- and mid-altitude clouds partially block the incoming short-wave radiation, so air temperature decreases (negative feedback). In its counterpart, high-altitude clouds partially absorb the outgoing long-wave radiation, causing the air temperature to increase (positive feedback).

[b] pCO_2 stands for partial pressure of CO_2, i.e., the pressure that CO_2 gas would exert if it alone occupied the entire volume of an original mixture of dissolved gases, at the same temperature.

14.2.1 Internal and external forcers

The climate system is affected by various forcer factors, either internal or external. The former are the product of the links between the subsystems. For example, changes in the recurrence of a certain phenomenon (such as El Niño-Southern Oscillation, ENSO) can generate changes in the climatic conditions of a region or ocean basin on interannual to interdecadal time scales. Among the external forcers, we can cite various factors such as astronomical (e.g., variations in solar emission, or the cyclic variation in orbital parameters of the Earth known as Milankovitch cycles and introduced in Chapter 1, Fig. 14.4), geological (e.g., continental drift, volcanism and orogenic movements) and changes in atmospheric composition (for instance, in relation to GHGs and aerosols).

Let us now elaborate further on astronomical forcers. The global mean temperature of the Earth depends on the incident solar irradiance, which is modulated, on millenary scales, by cyclic variations on Earth's orbital parameters, first described by the Serbian astrophysicist Milutin Milankovitch. The three parameters in question are (i) the Eccentricity, which is related to the shape of the Earth's orbit around the sun, where a more elliptical orbit is related to a lower incident solar irradiance during the solstices, and a circular orbit receives higher incident solar irradiance during the solstices. (ii) Obliquity (or nutation), which is related to changes in the angle of the axis of rotation of the Earth with respect to the plane containing the Earth-sun system. The current incline is approximately 23.43°, and is responsible for the seasons of the year. This angle oscillates between 21.6° and 24.5° every 40,000 yr. Although the amount of radiation received by the Earth does not change, the distribution over Earth's surface does vary. (iii) Precession, which is a tilt in Earth's rotation axis which happens approximately every 26,000 yr and modifies the relative position of the solstices and equinoxes with respect to the perihelion (i.e., the point of the orbit closest to the sun) and aphelion (the farthest point). Currently, the summer solstice for the Southern Hemisphere coincides with perihelion. The position of the Earth with respect to these three orbital parameters determines, to a great extent, the global climate of the planet (Fig. 14.4a).

In turn, the combination of the above-mentioned cycles causes the alternation of glacial and interglacial periods over time scales of tens of thousands of years (Fig. 14.4b). Glacial periods are

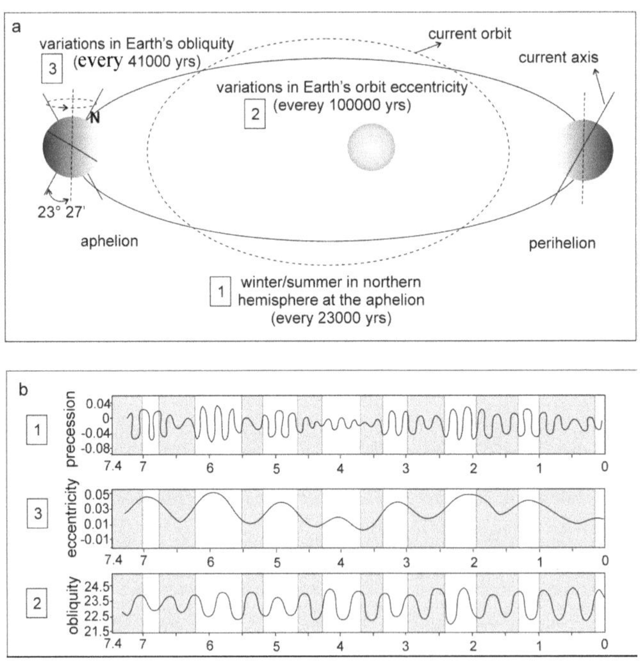

Figure 14.4. (a) Schematic representation of orbital parameters invoked in the Croll-Milankovitch Theory of ice ages. (b) Variations in the planetary orbital parameters over the past 740,000 yr (m = 100,000 yr). Grey superimposed bars represent glacial periods. Redrawn and modified from Seibold and Berger (1996).

characterized by a weak seasonal contrast as a result of the high eccentricity, low inclination and great Earth-sun distance during summer in the Northern Hemisphere. Interglacial periods are characterized by low eccentricity, large inclination, and relatively low values of Earth-sun distances in summer, resulting in contrasting seasons.

Let us now discuss El Niño-Southern Oscillation (ENSO) to exemplify the variability in atmospheric circulation and ocean-atmosphere interactions introduced in Box 1 (see also Chapter 2 in this volume). ENSO events comprise three phases: El Niño, La Niña and a Neutral (or normal) phase. El Niño episodes are linked to large-scale warming of surface waters in the central or central-east region of the equatorial Pacific; La Niña episodes, on their part, are linked to SST cooling. Fluctuations in temperatures are associated with changes in atmospheric circulation, which affect climatic conditions in various parts of the world. These far-range (thousands of km), statistically-significant climatic anomalies with impacts on remote regions of the globe are known as atmospheric teleconnections (Yeh et al. 2018).

Box 1: Global atmospheric circulation

Incident solar irradiance is uneven over the surface of the Earth. Irradiance is greater on tropical regions, where an excess of heat is produced, while, on the other hand, it affects to a lesser degree polar regions, generating a relative deficit of heat (see Fig. 2.1c in this volume). However, the tropical and polar regions are not continually warming and cooling, respectively. This is due to the fact that the atmosphere and the oceans are responsible for transporting excess heat from tropical to polar regions, through their respective circulations. This phenomenon is known as meridional heat transport.

The atmosphere is made up of different layers, called *troposphere*, *stratosphere*, *mesosphere*, *thermosphere* and *ozonosphere* (cited in ascending order according to their height). The temperature in the atmosphere varies with altitude, and in each layer its variation is different, which generates zones of atmospheric stability that prevent air mixing between the different layers.

The troposphere is the lowest layer of the atmosphere; it is located immediately above Earth's surface and extends to the *tropopause* (the boundary between the troposphere and the stratosphere). Its average height is approximately 18 km above the Equator, narrowing to ~ 8 km at the poles. The meteorological phenomena that modulate climatic conditions occur in the troposphere. Atmospheric circulation is the result of thermodynamic processes (such as differential surface heating by type of land cover) and pressure differences at a same level. Those differences create a pressure gradient, where the resulting force is opposite to the gradient.

At higher tropospheric levels, friction decreases creating what is known as a 'free atmosphere', and wind trajectories run closely parallel to isobars, describing wave-like trajectories. Low and high pressure centers developing outside of the tropical region are named extratropical cyclones and anticyclones, respectively. These cause much of the severe weather experienced in mid-latitudes, such as storm waves in coastal regions (Colle et al. 2008). Those areas where cyclone generation, propagation, and dissipation are common are known as *storm tracks* (Shaw et al. 2016), and act as guides for cyclones and anticyclones. Changes in their intensity or position can significantly affect the frequency, intensity and trajectory of meteorological phenomena such as heat- and cold waves (Lehman and Coumou 2015).

The global atmospheric circulation works as a pump that transfers heat from tropical regions to polar ones; it is made up of circulation cells (Fig. 14.5) described as follows: (i) Hadley cells are those meridional circulations located between the equatorial region and 30° latitude in both hemispheres. The high sea surface temperature (SST) over the intertropical convergence zone (ITCZ = equatorial region) warms up the air at low levels of the troposphere and loads it with humidity, reducing its density and thus forcing its ascent (sketched in Fig. 14.5 as the ascending branch of the cell). As the air rises it cools (by expansion), but its rate of cooling is diminished by the heat released in the condensation of water vapor, allowing the air to reach high levels in the troposphere to form intense storms. When reaching the tropopause (a stable layer in which the atmosphere experiences heating), the ascending branch diverges towards the poles. On its way it continues to cool, increasing its density and

Box 1 Contd. ...

...Box 1 Contd.

descending at around 30° of latitude on each hemisphere (the descending branch of the cell; Fig. 14.5). The descending air generates a belt in which high pressures predominate (i.e., the subtropical high pressure belt), in addition to originating subtropical anticyclones over the oceans. In these latitudes, little cloudiness predominates and it often coincides with the location of deserts on land.

Between 50°–60° latitude, there is a prevailing low-pressure belt in which the relatively cold air is forced to rise by the recurrent passage of storm systems (a low-pressure region called the polar front). The latter have their origin in atmospheric instability related to the collision of air masses of different characteristics, which regionally generates a strong thermal gradient between the poles and mid-latitudes, especially during winter in each hemisphere. Upon reaching the tropopause, the ascending air diverges towards lower and higher latitudes, originating (ii) Ferrel and (iii) polar cells, respectively (Fig. 14.5). In mid-latitudes, frontal zones are produced by the collision of cold (denser) and warm (less dense) air masses. The former have lower humidity since their lower temperature implies a lower capacity for water vapor content, which further promotes their contrast in density and therefore increases regional instability.

It should be noted that the above-described circulation cells result from a longitudinal global average (cross-section latitude × height). In reality, the cells are not perfectly defined across the globe, since zonal differences in temperature (west-east direction), orography and meteorological phenomena of various space-time scales can affect these cells to the point of canceling them out in certain regions. This is particularly true for Ferrel cells that depend strongly on the activity of cyclones and anticyclones.

The subtropical and polar jets are located in the upper troposphere at ~ 30° and 60° of latitude, at an average height of 13 km and 10 km, respectively, in both hemispheres (Fig. 14.5). The *jets* consist in bands of intense winds with a strong component from the west, normally several thousand km-long, several hundred km-wide and a few km-thick with strong vertical wind shear.

At low levels in the troposphere, the air in the descending branch of Hadley's cell diverges as it meets the Earth's surface, creating winds that are directed towards the Equator and towards the poles. The *trade winds* (whose name originated from 15th century transoceanic merchants who sought

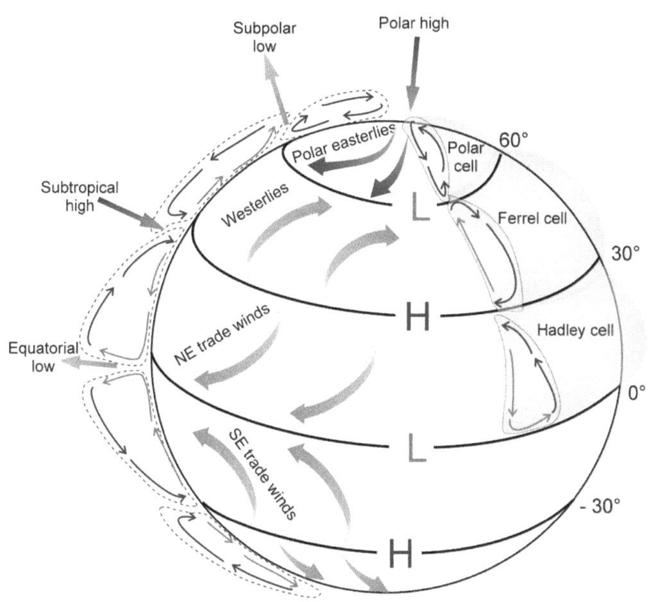

Figure 14.5. A schematic representation of global atmospheric circulations showing low and high pressure belts, prevailing wind systems and the Hadley, Ferrel and polar circulation cells (the latter only schematized for the Northern Hemisphere). Redrawn and modified from Lutgens et al. (2012).

Box 1 Contd. ...

...Box 1 Contd.

after them) are directed towards the Equator, rotating towards the west in both hemispheres by the Coriolis effect. When the trade winds from both hemispheres converge at the Equator (in the ITCZ), they contribute to the rise of a warm and humid air mass, maintaining the thermodynamic balance.

On the other hand, the winds that are directed towards the poles are responsible for transporting heat from the subtropical regions to the middle latitudes and, due to the Coriolis effect, they give rise to the *westerlies*. Finally, between the polar front and the poles is the region of the *easterly polar winds*, which arises as a product of the high pressures prevailing at the poles.

The circulation patterns described above undergo meridional (latitudinal) displacements along the different seasons of the year, following the incidence of the sun on the Earth's surface. It should be noted that the variability in the synoptic scale in mid-latitudes is very marked, since the troposphere plays a preponderant role in the transport of heat through the activity of anticyclones and cyclones, which dominate the distribution of wind, precipitation and temperature at those latitudes. For their part, the oceans play quite a relevant role in the transport at tropical regions (Fasullo and Trenberth 2008). It should also be noted that the equatorial regions show little variability throughout the year, while in mid-latitudes, and to a lesser extent the tropics, an annual cycle is observed for various climatological variables.

Warm surface currents are found on the western Pacific basin, while the cold Humboldt Current and the upwelling of deep cold waters characterize the eastern Pacific basin. During normal conditions, there is an atmospheric pressure gradient between the East and West Pacific, known as the Walker circulation or Walker cell. The Southern Oscillation Index (SOI) is a measure of the pressure difference between Tahiti (in the middle of the Pacific) and Darwin (in Western Australia). The SOI is correlated with SST anomalies in the tropical Pacific basin; in other words, anomalously-warm waters promote the formation of successive and intense storms which alter the Walker circulation. During the Neutral and La Niña phases, these storms occur mainly over the western equatorial Pacific (Fig. 14.6b). A SST pattern similar to the Neutral phase occurs during La Niña phase, but with more intense anomalies.

During an El Niño event, trade winds weaken or may even reverse, and the anomalous warming of the central-eastern region of the equatorial Pacific generates a displacement of storms towards this region (Fig. 14.6a). This shift in storm distribution patterns results in changes in the pressure systems of the Walker circulation, which are coupled with a change in the depth of the thermocline (see Chapter 2 on this volume). The deepening of the thermocline in western South America most notably weakens coastal upwelling. There are also changes in SST throughout the Pacific causing changes in the regimes of wind, precipitation and temperature, especially in the tropical and subtropical regions. It should be noted that the climatic effects of each event are never exactly the same, since they depend on the interaction with other climatic conditions, in addition to the intensity of the event and the time of year in which it occurs. Marine and lacustrine sediments accumulated in undisturbed depositional environments often keep an excellent record of this variability, and aid researchers in the reconstruction of patterns of past El Niño activity throughout the Holocene, its periodicity and the recurrence of very strong events (Rodbell et al. 1999, Rein et al. 2005).

Adding to the variability inherent to ENSO events, intra-seasonal phenomena (with a duration of 30 to 90 d) such as the Madden-Julian Oscillation (MJO) may occur simultaneously with an ENSO event. MJO consists of a stormy region that propagates eastward from the western Indian Ocean to the tropical Atlantic (Madden and Julian 1994). The event has an oscillation period of between 40 and 50 d, and although the storms tend to dissipate over the central region of the tropical Pacific, this pressure and wind anomaly continues its trail until dissipating completely. An MJO event may or may not occur during an ENSO event, being able to intensify or weaken it. It has also been found that the regional effects that an ENSO event may have depend on the state of the westerly winds (Vera and Osman 2018) since they carry the circulation anomalies generated by variations in tropical SST.

On an interdecadal time scale, the recurrence of ENSO phenomena is modulated by the Pacific Decadal Oscillation (PDO), which is associated with SST anomalies in the North Pacific (Trenberth and Hurrell 1994). The PDO has an oscillation period of between 20 and 30 yr. Its warm phase favors a

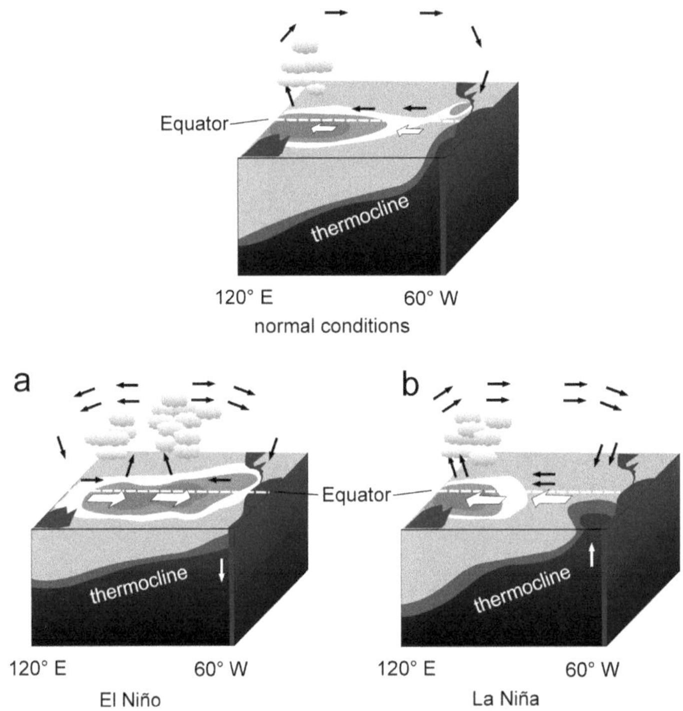

Figure 14.6. A schematic representation of atmospheric circulations and oceanic conditions during the normal conditions (on top) and El Niño (a) and La Niña (b) events. Modified and redrawn from NOAA/PMEL/TAO Project Office [M.J. McPhaden, Director].

greater recurrence of El Niño events, whereas its cold phase favors the occurrence of La Niña events. For example, in the 29 yr lapse spanning the 1947 and 1976 there was a cold phase of the PDO coinciding with a greater recurrence of La Niña events, while between 1977 and 1999 (a 22 yr-period) the warm phase of the PDO coincided with a greater recurrence of El Niño events (Mantua and Hare 2002).

In sum, from the above it can be stated that hemispheric-scale circulations in the atmosphere broadly determine the prevailing climatic conditions, as well as their interannual variability. Yet, the important role of the oceans need not be overlooked. Oceans regulate the global climate due to their high capacity to retain heat. We have described the fundamental role that equatorial Pacific SST plays in ENSO phenomena, but it should be noted that for this reason the Pacific basin responds differently than the atmosphere to external forcers. The response of the oceans usually is considerably slower than that of the atmosphere for most processes, but no less important. In addition to the surface circulation induced by the winds, the oceans present a planetary-scale circulation, the product of the variation in the density of seawater; this is the so-called meridional overturning (or *thermohaline*) circulation, discussed in Chapter 2 of this volume. It takes ~ 1,000 yr for any given cubic meter of water to complete this planetary-scale circulation along the global conveyor belt. Furthermore, thermohaline circulation not only transports and redistributes mass (water and salts) and heat across ocean basins, but it is also instrumental in regulating nutrient and CO_2 cycles globally (see Fig. 2.9b in this volume).

14.3 Stratospheric ozone depletion

The depletion of stratospheric ozone (O_3) promotes a radiative imbalance with an increase in short-wave ultraviolet (UV) radiation, and as such, it represents one of the most important global changes that human activity has introduced to the atmosphere-biosphere components of the climate system (Polvani et al. 2011). The ozone layer was created quite early as a by-product of the biological O_2 production and its accumulation in the early atmosphere (see Chapter 1, in relation to the GOE). The stratospheric ozone layer extends from 9.6 to 48 km-heights, and basically screens out most short-wave radiation from the sun (Farmer and Cook 2013). As mentioned in the introduction, synthetic, non-reactive chemicals known as CFCs and formerly used in refrigeration, accumulate in the troposphere, where upon photochemical degradation by UV radiation, release chlorine atoms that react with the O_3 molecule, in a cyclic breakdown into O_2 and O. Also, their average lifetime in the troposphere is 11–20 yr, and they have a tremendous warming potential, relative to CO_2 (Miller 2004). In large part due to CFCs accumulation, the Antarctic 'ozone hole' (AOH) opens up in Austral spring each year, and is measured both by its depth (typically a loss of about half of the total integrated column amount) and its size (often larger than 20 million km^2) (Solomon et al. 2016).

Some authors argue that the effect that ozone layer thinning has on the radiative balance has been considerably underestimated (Previdi and Polvani 2014). The AOH has generated significant surface climatic changes, not only for Antarctica and the Southern Ocean, but also on New Zealand, the southern tip of South America and Australia (Thompson et al. 2011). The influence is most pronounced during Austral summer by modulating the leading pattern of circulation anomaly variability in the Southern Hemisphere, the Southern Annular Mode. Currently, the AOH is decreasing in extent (Solomon et al. 2016) thanks to global action taken through the Montreal Protocol. However, diverse investigations suggest that regional changes observed in the Southern Hemisphere, such as the shift of winds towards the west, or an increase in summer rainfall over southeast South America (González et al. 2014), may be mainly due to the decrease in ozone layer thickness. Total column measurements of O_3 between the Earth's surface and the top of the atmosphere indicate that the ozone layer has stopped declining across the globe, but no clear increase pattern has been observed at latitudes between 60°S and 60°N (i.e., outside polar regions), where some evidence indicates that, indeed, it has continually declined since 1998 (Ball et al. 2018).

Even when stratospheric ozone depletion varies latitudinally, in areas experiencing ozone-related increase in solar UV radiation all forms of life are transversely affected, as short-wave radiation is highly reactive with the DNA molecule. UV exposure elicits different responses in marine ectotherms, such as behavioral (avoidance of high radiation exposure), photoprotective (synthesis of sunscreen compounds, e.g., mycosporine-like amino acids, MAAs), and molecular repair mechanisms. These mechanisms acting at multiple scales ultimately improve the overall fitness of UV-exposed organisms (Dahms et al. 2011).

14.4 Recent global warming

The energy source of the climate system is solar radiation. As detailed in Chapter 2, the sun emits mostly short-wave energy (with maximum intensity in the visible spectrum), given the high temperature on its surface (~ 5,500°C), and the ozone at low levels of the stratosphere absorbs much of the energy in the UV spectrum. For its part, the Earth's surface emits mostly long-wave energy (a form of heat matching the infrared portion of the electromagnetic spectrum), most of which is absorbed by GHGs and certain types of clouds, warming the atmosphere. Only a small fraction of the heat emitted by the surface reaches space (through an atmospheric window). Without this natural mechanism (called the greenhouse effect), the Earth would be too cold for the development of life. The alteration of the concentrations of GHGs results in a greater retention of the energy that Earth's surface emits into space (in other words, the atmospheric window "closes"), generating an increment in the global average temperature.

Climatic changes linked to deviations from mean conditions have been observed in various regions of the globe over the past few decades. Large-scale impacts of climate change on the oceans are expected to include increases in SST and mean elevation, decreased seasonal sea ice cover, and changes in salinity, wave conditions, and circulation, as well as ocean acidification. In the atmosphere, these changes would manifest in changes in the frequency and intensity of extreme meteorological phenomena, and changes in atmospheric circulation and/or precipitation regimes, in addition to global warming. Two hypotheses have taken on a strong relevance within the scientific community to explain the causes and origin of these climatic changes.

The first hypothesis, which links climate change to the emission of GHGs (especially CO_2, methane CH_4, and nitrous oxide NO_2), is widely known and has been supported by most researchers in recent decades (Pörtner et al. 2019a). While atmospheric CO_2 is the GHG whose concentration has increased the most, recently methane seeps have been identified in Northern- (Turner et al. 2020) and Southern (Ketzer et al. 2020) Hemisphere oceans, where massive methane leaks occur from the seabed that harbors solid methane deposits. Methane has a heating power ~ 25 times that of CO_2,[c] certainly a matter for concern. Methane seeps are linked to deep-water warming, and upon release into the atmosphere they could generate a positive feedback that would accelerate global warming.

The second hypothesis (developed over the past two decades) postulates the polar stratospheric ozone depletion as the main driver of 20th century atmospheric circulation changes in the Southern Hemisphere during the Austral summer (Polvani et al. 2011). It should be noted that the acceleration of the changes in the Southern Hemisphere circulation coincides temporarily with the appearance of the AOH. Hence, the hypothesis establishes that the AOH has induced a progressive cooling of the lower stratosphere (i.e., a negative trend in temperature) over the Antarctic region, which generates pressure anomalies in the air column, altering the wind regimes in the lower troposphere, over medium to high latitudes (Thompson and Solomon 2002). Some scientists argue that in the next half century, the recovery of the AOH could even cancel out the effect of GHGs emissions on the radiative balance (Previdi and Polvani 2014).

As a result of human activities, the average air temperature has increased globally by 1°C in relation to pre-industrial average values, and the trends and fluctuations are similar for both hemispheres (Pörtner et al. 2019a, b). There are constant heat fluxes between the ocean surface and the atmosphere that are the product of thermal gradients; i.e., both components heat up and cool down in a dynamic equilibrium. According to Pörtner et al. (2019a), the ocean has already absorbed > 90% of the heat generated by the emission of GHGs, modulating or limiting warming in other regions of the Earth. The global mean warming rate is 0.11 [0.09 to 0.13]°C per decade for the upper 75 m, decreasing to approximately 0.015°C per decade in the upper 700 m (Rhein et al. 2013). Updated data (1993–2015) of the global mean SST indicate that the global ocean has increased its temperature at a rate of $0.016 \pm 0.002°C\ yr^{-1}$ (Roquet et al. 2016, von Schuckmann et al. 2018). Furthermore, warming of the deep ocean (from 1,360 to 4,757 m deep) has recently been reported (Meinen et al. 2020), with warming trends of approximately 0.02–0.04°C per decade over the 2009–2019 period. While a large percentage of the world's oceans have increased their temperature, some high latitude areas have experienced cooling, possibly linked to the melting of glaciers, ice sheets, and sea ice (see below).

Extreme SST rise known as marine heatwaves can be defined as discrete, anomalous and intense warming events lasting for quite a prolonged time (5 d or more), extending thousands of km. They can have marked effects according to the SST rise, the duration of the event or both. They are generally a consequence of elevated air temperature, so their impacts are usually limited to the upper m of the water column (although some can also be due to eddies released from warm currents). The effects can be especially relevant for sessile organisms or those with reduced mobility that cannot move towards colder waters. It is very likely that marine heatwaves have doubled in frequency and have become longer-lasting, more intense and more extensive over the 1982–2016 period (Pörtner et al. 2019a).

The warming signal, intensified at the surface, has very likely produced an increase in the thermal stratification of the upper global ocean, reducing the mixing of water between 0 and 200 m depths (Rhein et al. 2013) and producing surface thermal stratification. As a consequence, the supply of oxygen and nutrients is reduced, which, as we will see in other sections, affects the abundance and distribution of marine organisms in coastal and open ocean areas. Additionally, the melting of glaciers, ice sheets and sea ice not only is causing global sea level rise, but also the freshening of surface waters, another factor limiting overturning and promoting stratification. Recent studies (Tzortzi et al. 2013, Boutin et al. 2016) have suggested that seawater is becoming less saline at high latitudes (due to melting of ice) and in areas dominated by rain, while in subtropical regions with high evaporation, the waters are increasingly saline.

[c] This heating power or warming potential of a greenhouse gas is measured by a parameter called radiative efficiency, quantified in $W\ m^{-2}\ ppb^{-1}$ units.

14.5 Impacts of global warming on marine ecosystems

In the previous section, we have presented the causes and trends in recent global warming. Furthermore, Chapter 13 provides a comprehensive account of temperature as one of the most important environmental variables in the life of marine organisms. So in this section, we will not delve deep on such matters again, but rather cite a few examples of how global temperature increase has affected marine organisms at various levels. Before moving on, it is important to bear in mind that marine ecosystems not only are highly responsive to alterations in the physical environment, but also highly adaptable to such changes (Steele 1998), which sometimes makes it difficult to establish baselines for comparison and evidencing regime shifts.

14.5.1 Distribution range changes in marine organisms

Shifts in the distribution range of marine species as a result of surface water warming were extensively recorded and studied in the 1990s (discussed in Pan et al. 2013). Being open-sided systems, the interconnectedness of marine ecosystems makes it possible for marine organisms to migrate seasonally and eventually expand or contrast their distributional range (Steele 1998). For example, Edwards et al. (2006) reported significant changes in the spatial patterns of HAB[d]-producing dinoflagellates in the northeast Atlantic and the North Sea, using data from the Continuous Plankton Recorder from the 1960s to the early 2000s. Also, dinoflagellate abundance increased, which is correlated to a major change in SST from the 1980s onward. As it will be discussed later, dinoflagellates not only respond physiologically to temperature increases better than other phytoplankton, but also respond to enhanced water-column stratification (Peperzak 2003).

Distribution range shifts are particularly evident in benthic organisms, which have limited motility to make rapid adjustments to their distributions. As a case study, much research has been carried out on the effects of seawater warming on seaweeds, provided many species are of commercial importance. Takao et al. (2015) evaluated the effects of seawater warming over the spatial extent of habitats for the temperate seaweed *Ecklonia cava*, using monthly mean SST records and projections spanning the period 1960–2099. Based on model projections, the trend is that continued warming may drive a poleward shift in the distribution (expansion into higher latitudes, and extinction in lower latitudes). A previous study by Tanaka et al. (2012) had observed a large increase of 0.3°C per decade in the annual mean SST influenced by the Kuroshio current over 40 years, leading to the contraction of temperate seaweed distributional ranges and expansion of tropical seaweeds.

Anomalously warm temperature events have increased by 34% in frequency and 17% in duration from 1925 to 2016 (Straub et al. 2019) and intertidal and shallow-water seaweeds are more prone to be affected. For instance, SST warming attributed to the 1997–98 ENSO was the cause of a widespread destruction of *Ecklonia* populations in Japan (Tanaka et al. 2012). The physiologic responses of seaweeds to marine heatwaves range from resistance, bleaching, changes in abundance, species invasions and local to regional extinction (Straub et al. 2019; and see below). Furthermore, experimental evidence on the effects of long-term (9-week) exposure to ambient seasonal temperatures and projected increases on cold-adapted seaweeds from Atlantic Canada, showed that the responses were not uniform, but rather species-specific (Wilson et al. 2015). While some species may show temporary acclimation, at higher temperatures growth and survival occur only for short-terms. Furthermore, while temperature stress did not affect carbon content, some species showed increased tissue nitrogen, potentially changing their nutritional quality (Wilson et al. 2015).

From a biological point of view, heatwaves may be regarded as form of acute thermal stress to marine organisms. While they are forecasted to become more frequent, gathering evidence suggests that their impacts on marine organisms vary across functional types and taxonomic groups (Leung et al. 2019), reflecting the array of physiological response strategies displayed by them. For instance, a 2017 heatwave

[d] HAB is the acronym for 'harmful algal bloom', applied to population growth of microalgal species attaining unusually-high densities in short periods, and leading to adverse ecosystemic effects such as the production and bioaccumulation of toxins, disruption of food webs, and/or generation of hypoxia (Paerl et al. 2012).

in the Gulf of Alaska negatively impacted Pacific cod (*Gadus macrocephalus*) physiology, leading to a 71% abundance decline (Barbeaux et al. 2020). Furthermore, simulated heatwaves promoted chemical defense mechanisms against bacterial biofouling and associated disease in the eelgrass *Zostera marina* (Guan et al. 2020). Along those lines, Leung et al. (2019) invoke different molecular defense mechanisms integrated with behavior, in the response of three co-occurring intertidal gastropods to heatwaves.

14.5.2 Temperature- and stratification-induced changes in phytoplankton communities

As discussed above, global temperature increase has had a double effect in surface warming and increasing the thermal stratification of the upper ocean (Rhein et al. 2013). Higher temperatures favor cyanobacteria over other phytoplankton, and they become dominant in the assemblages, due to both physiological and physical factors (O'Neal et al. 2012). In the first case, cyanobacteria exhibit optimal growth rates at relatively high temperatures (usually > 25°C, Paerl and Huisman 2008) and outcompete eukaryotic algae. In the second case, under temperature-enhanced stratification, buoyant cyanobacteria can form dense surface blooms which shade deeper non-buoyant eukaryotic phytoplankton (Paerl et al. 2011).

The evidence coming from experiments that manipulate seawater temperature (a 4°C rise) and stratification (experimentally increased) indicates that in a future scenario, the growth rates of harmful dinoflagellates and other flagellate groups (raphidophytes) will double compared to present-mixed conditions (Peperzak 2003). On the other hand, the growth rate of a non-harmful diatom may not significantly change. But it is important to bear in mind that a seawater temperature increase will also lower its viscosity, accelerating sedimentation rates for non-flagellate species such as diatoms. Altogether, it would seem that the risk of harmful dinoflagellate and raphidophyte blooms will increase rather than decrease due to climate change (Peperzak 2003).

Another widespread phenomenon of concern is the extent of coral bleaching across well-lit temperate and tropical seas. There are two main causes of coral bleaching; it either has to do with low pigment content in the symbiotic zooxanthellae (*Symbiodinium* sp.) in reef-forming corals, or a low density of zooxanthellae in the mesoglea of colonial coral polyps (Hoegh-Guldberg and Smith 1989). Regarding the latter, there is a link between SST increase and zooxanthellae density in coral tissues, as these get expelled at temperatures > 30°C (Hoegh-Guldberg et al. 2017). For the past two decades now, scientists have been pointing out the strong connection of ENSO events on SST rise and coral bleaching (Glynn et al. 2001). What is of major concern is that the frequency of coral bleaching events has been increasing in recent decades, especially following the peak of strong El Niño events (De Falco et al. 2020) and accentuating the stressful conditions that some coral reefs have been experiencing due to long-term trend in SST.

Model projections for the Southern Ocean yield a future scenario with increased warming, precipitation, and the melting and break up of glaciers, all factors which increase the freshening of buoyant surface waters, and lead up to stratification and reduced mixed layer depths (Deppeler and Davidson 2017). Increased stratification could mitigate the light limitation that phytoplankton currently experiences as the product of deep mixing and downward transport out of the eutrophic zone, by keeping cells in sunlit near-surface waters; yet, productivity would still be constrained by the availability of key nutrients (iron and silicate; see Section 14.8). On the other hand, for certain areas of the Southern Ocean, stratification may limit nutrient supply to surface waters and reduce the productivity of diatoms known to provide good nutrition to higher trophic levels, in favor of small flagellates (Deppeler and Davidson 2017). Regardless of different regional responses, there is no doubt that all regions of the Southern Ocean will experience changes in phytoplankton productivity and community composition with climate change.

14.5.3 Impacts on the phenology of marine communities

Phenology (from the Greek *phainō* = to show, make to appear) is the ecological study of the 'timing' and occurrence of cyclic processes in biological communities, that have a seasonal component (strongly linked to climate) and periodicity. Changes in phenology with respect to an established pattern in an area of the ocean (be them expressed in biomass fluctuations or in ecological rates) often lead to an ecological mismatch between otherwise coupled processes and provide particularly sensitive indicators of climate change (Hughes 2000). However, it is difficult to separate observed trends in phenology from changes in community composition caused by climate-induced biogeographical shifts (Edwards and Richardson 2004).

Edwards and Richardson (2004) studied changes in pelagic phenology in the North Sea across three trophic levels (primary, secondary and tertiary producers) using five functional groups (diatoms and dinoflagellates, copepods, non-copepod holozooplankton and meroplankton) from long-term (1958–2002) biomass surveys. Along broad lines, they concluded that many components of the pelagic assemblage are responding rapidly to temperature change (warming) through changes in the timing of their seasonal cycles, potentially leading to a mismatch between trophic levels and functional groups.

Climate-induced shifts in phenology are not limited to planktonic communities. In a recent study, Crear et al. (2020) modelled the timing of migration into Chesapeake Bay and depth-integrated habitat use by cobia (*Rachycentron canadum*), a large coastal pelagic fish that represents an important fishery, applying the projected trends in warming and increased hypoxia for the bay. They concluded that estimated arrival would occur earlier and estimated departure later when temperatures get warmer, and that by mid- and end of the century cobia may spend on average up to 30 and 65 more d in the bay, respectively.

Another example comes from kelps—a group of brown algae, forming dense canopies and forests in cold waters. Declines and losses of highly-productive kelp beds have been observed worldwide and are linked to increases in ocean temperature (Simonson et al. 2015). Their reproductive maturation is controlled at multiple levels by temperature and day length (Guiry and Dawes 1992), and it has been recently shown that seaweeds are susceptible to temperature, as evidenced by shifts in their reproductive phenology (de Bettignies et al. 2018).

14.6 Ocean acidification and its impacts on marine organisms

It is estimated that the global ocean has absorbed between 20–30% of human-induced CO_2 emissions, resulting in the acidification of the surface ocean (Le Quéré et al. 2018, Pörtner et al. 2019). Seawater has a strong capacity for CO_2 capture, but the excess in CO_2 inputs alters the dynamics of the carbonate system, with the consequent decrease in pH and the concentration of the carbonate ion, a phenomenon known as ocean acidification, OA (Feely et al. 2009, Doney et al. 2009, Gattuso et al. 2015).

Let us recapitulate briefly on how the seawater buffer systems works, and how it is being altered. In normal conditions, seawater is supersaturated with carbonate $[CO_3^{2-}]$ ions. Upon dissolution in seawater, gaseous CO_2 interacts with water and carbonate ions to form bicarbonate HCO_3^- ions in the following reaction (Doney et al. 2009):

$$CO_2 + H_2O \rightleftharpoons H_2CO_3 \rightleftharpoons H^+ + HCO_3^- \rightleftharpoons 2H^+ + CO_3^{2-}$$

In recent times, the rate of CO_2 absorption by seawater is high in relation to the supply of CO_3^{2-} ions, and therefore, bicarbonate and H^+ concentrations are on the rise (lowering pH), while carbonate ions are decreasing (Orr et al. 2005). Estimates indicate that on average, pH has decreased by 0.1 units for the surface ocean since the industrial revolution; as pH is measured on a logarithmic scale, this roughly represents an increase of ~ 26% in acidity (Gattuso et al. 2015).

14.6.1 Responses of marine organisms to ocean acidification

The intense quenching of carbonate $[CO_3^{2-}]$ ions interferes with calcification in marine organisms, but the response is variable. In general, the overall fitness of calcifying species decreases with lower pH in relation to that of non-calcifiers (Orr et al. 2005). Not all stages in the life history of marine calcifiers are equally susceptible to OA, but it can rather have more marked impacts on larval development and mortality, juvenile recruitment, and ultimately produce large fluctuations in population sizes.

For example, larval stages of bivalves show extreme sensitivity to projected levels of CO_2 (650 ppm), having marked declines in survivorship (> 50%) as well as significantly delayed metamorphosis and smaller sizes (Talmage and Gobler 2009). It has recently been found that DNA methylation (a reparative mechanism) can mediate phenotypic responses of marine calcifying species to OA. For example, Downey-

Wall et al. (2020) studied gene expression and DNA methylation in the mantle-edge tissue of adult eastern oyster (*Crassostrea virginica*) experimentally reared under three varying pCO_2 treatments over 80 d. While calcification rates were significantly lower in the high OA treatment, the authors also found that OA induced a subtle response in a large number of genes expressed in the extrapallial fluid of oysters.

Elevated CO_2 has been shown to impair temperate fish reproduction, as it experimentally increased embryonic abnormalities and egg loss in a goby, with important implications for fish recruitment (Forsgren et al. 2013). Fish calcify internal skeletal elements such as bones and otoliths, and research points that post-hatch length is significantly reduced when embryos are exposed to increased CO_2 levels (~ 1,000 ppm) (Baumann et al. 2012). The same study found fish egg stages to be significantly more vulnerable to high CO_2-induced mortality than the post-hatch larvae.

14.6.2 The role of seaweeds in the modulation of seawater pH at local scales

Seaweed stands are highly autotrophic communities that also generate habitat for an array of benthic organisms and nekton (see Chapter 10 in this volume). It has recently been pointed that due to their high autotrophic capacity (at an ecosystem level, organic matter fixed through photosynthesis exceeds respiration loses), dense stands of wild- and aquaculture-farmed seaweeds represent productivity hot-spots with associated high pH during day when photosynthesis locally drops CO_2 concentrations (reviewed in Duarte et al. 2017). This (diurnal) modulation of seawater pH has the potential to (locally) reduce the effects of OA and de-oxygenation, representing a new role for seaweed farming in OA mitigation.

In fact, there is empirical evidence that transferring rafts of farmed seaweeds to OA-threatened corals in Australia's Great Barrier Reef, and keeping them in place for 100 d increases the aragonite saturation state [Ω_a] over a km-scale, being beneficial from the point of view of local dissolved CO_2 uptake and enhancing the calcifying capacity of corals (Mongin et al. 2016). This may become a valid environmental engineering practice for OA mitigation in benthic environments with calcifiers.

14.6.3 Additive and synergistic effects of different stressors

Ecosystems are subject to a several stressors which seldom act individually, alternating or sequentially, but rather act jointly in intricate ways. Two or more stressors acting on a given ecological system may add up or combine synergistically to potentiate their effects. This certainly applies to environmental forcers related to climate change.

For instance, the dominance of cyanobacteria in planktonic assemblages projected in future scenarios with higher temperature and enhanced stratification may be further promoted by eutrophication (anthropic inputs of nitrogen and phosphorus), as many planktonic cyanobacteria have the ability to acquire organic P compounds, and they display great flexibility in the N sources they exploit leading up to blooms (Paerl et al. 2011, O'Neil et al. 2012). Managing eutrophic water bodies is difficult, as the legacy of eutrophication in the form of organically-enriched sediments can perpetuate high internal nutrient loads, even when external nutrient inputs are reduced (Paerl et al. 2011).

Gobler et al. (2014) experimentally assessed the combined effects of hypoxia and acidified water (two frequently co-occurring stressors) on early life stage bivalves. In general, they found both additive and synergistic effects of the two stressors, with more severe consequences on vital parameters (e.g., higher mortality, inhibition or reduced growth and metamorphosis) than would be expected by either stressor individually.

Cartaxana et al. (2015) experimentally assessed the effects of elevated CO_2 and temperature incubating microphytobenthic (MPB) assemblages from the Tagus estuary in microscosms. They found that elevated CO_2 had a beneficial effect on MPB biomass, only at the lower temperature tested (18°C), while elevated temperature (24°C) had a detrimental effect on MPB biomass and photosynthetic performance under both control and elevated CO_2 treatments. Furthermore, the temperature rise led to increased cyanobacterial densities and a change in the relative abundance of benthic diatoms.

Much like two or more stressors related to climate change may present additive or synergistic effects which potentiate their impacts, the biotic and abiotic components of a system may respond differently to the same stressor. For example, experimental evidence indicates that biotic processes (e.g., photosynthetic rate, aerobic and anaerobic organic matter remineralization, and growth) related to diatom MPB growing on coral-reef sediments will not be affected by projected OA conditions ($pCO_2 \sim 900$ μatm, corresponding to $\Delta pH = 0.1–0.4$). However, the same study found that OA-induced carbonatic sediment dissolution would be enhanced at the higher CO_2 concentrations (Fink et al. 2017).

14.7 Large-scale physical disruptions and their impacts on marine communities

Over the past decades, marine ecosystems have experienced large-scale physical disruptions arising from climate change. The decline in sea ice cover at polar latitudes is one example of that. For example, winter sea ice extent and duration and nutrient inputs largely regulate primary production at the Polar Front in the Southern Ocean. In turn, summer krill (*Euphausia superba*) densities correlate positively with sea-ice extent the previous winter; thus, summer food and the extent of winter sea ice are key factors explaining the high krill densities observed in the SW Atlantic sector of Antarctica (Atkinson et al. 2004). However, analyzing long-term datasets (1926–2003), these authors pointed spatial and temporal changes in krill densities over large scales, with a declining trend since the 1970s, as salps (*Salpa thompsoni*) increased in the southern part of their range. Loeb et al. (1997) link this to competition between krill and salps. Declining krill stocks are ultimately linked to winter sea ice decline and have profound effects on the Southern Ocean food web.

Likewise, model projections and satellite observations have recorded the decline in the minimum extent of sea ice in the Arctic (reviewed by Duarte 2014), often experiencing summers with abrupt sea ice melting events that go beyond projections. As a consequence of this sea ice contraction and rapid warming in the Arctic (Wang and Overland 2009), sea ice-dominated ecosystems will decline, while the solar radiation reaching coastal bottoms will favor the expansion of vegetated habitats, such as kelp beds and seagrass meadows (Krause-Jensen and Duarte 2014). These Blue Carbon ecosystems will most likely provide novel ecosystem services, some of which are known to attenuate or counter the effects of climate change, e.g., the provision of refuge for calcifiers and the local increase in pH, high rates of CO_2 sequestration, and shoreline protection from erosion. Polar zoobenthic Blue Carbon capture and secondary productivity are also bound to increase, as a consequence of the physical disruption created by climate-induced ice losses and iceberg-calving, as pointed by a study carried out in West Antarctic Peninsula shelfs (Barnes 2017).

Changes in the frequency of floods due to storm surges (which arise from differences between the observed tidal level and the astronomical tide, produced by the wind piling-up water) and the alteration of the tidal range at river mouths can be cited as notable physical disruptions to coastal (biological and human) communities (Pörtner et al. 2019). Other physical consequences of global warming include the effects of sea level rise and its associated erosion of coastal habitats. Recent eustatic sea level rise is jointly caused by changes in the cryosphere (i.e., melting of glaciers, ice sheets, and reduced sea ice extent) and, to a lesser extent, the thermal expansion of seawater. Tidal wetlands and other Blue Carbon ecosystems rank among the coastal ecosystems that will be most affected by sea level rise. For example, some conservative estimations that project the global trend in sea level rise for the 20th century (i.e., 1 to 2 mm yr^{-1} on average) into the 21st century indicate that ~ 20% of coastal wetlands would be lost by the year 2080 (Gitay et al. 2002). This has been extensively discussed in Chapter 9 so we will not delve deeper; but to close this section, it is important to point that coral reefs also stand out as one of the most vulnerable coastal ecosystems in the face of combined higher sea levels and projected more intense storms (Hoegh-Guldberg et al. 2017).

14.8 Climate models, future projections and mitigation alternatives

As much as the concept of climate is constructed upon a set of statistically-significant, long-term averages of numerous parameters, future (regional or global) climate scenarios can be projected with the aid of numerical models that also rely heavily on statistics. Broadly speaking, a model is an abstraction or representation of a system or phenomenon, comprising many integral parts and fluxes or rates connecting and regulating these. Models make it easier to understand, quantify, simulate or project current and future states of the system or phenomenon in question. To that purpose, the model is fed a set parameters (many of which come from experimental studies or short term simulations, say, the rate of calcification of a marine organism under different seawater pH), and then the model is run for a certain period (usually the next 50 or 100 yr). The outcome projections may describe more *moderate* or rather *extreme* scenarios, depending upon the different values of the parameters and other variables included in the model.

Why do we need projections of the future climate, even when, as we have learnt in the introduction to this chapter, many climatic and environmental goals set up in the agenda of international climate conferences are hardly ever met? Well, regardless of higher politics, future climate projections provide stakeholders and policymakers with scientifically-based tools so as to make educated and sound decisions about environmental processes and cultural habits that are transversely impacted by climate. So let us now go back to climate projections and how climate models are created.

Future climate projections are elaborated using Global Climate Models (GCM). These numerical models allow scientists to understand how the climate has changed in the past, the current weather, and how it may change in the future. GCMs represent the most sophisticated effort to simulate the physics, chemistry, and biology of the atmosphere, land, and oceans in great detail, and require some of the world's most powerful supercomputers to generate their climate projections.

Much like any other model, a GCM is an idealized representation of a finite number of processes (i.e., an approximation to reality, with limitations), since it cannot represent those physical processes with a spatial resolution lower than that of the model. Thus, in a numerical simulation, some sub-grid scale processes are represented using parameterizations. The latter ones are methods of accounting for the effects of sub-grid scale processes on the larger-scale variables. For example, in a numerical simulation from an atmospheric general circulation model (AGCM), low, medium, high, and convective clouds (and the corresponding derived variables, such as rainfall) are represented through model parameterizations. Also, the physical-chemical variables present a continuous spatial-temporal variability; a computer does not have the ability to treat them as such, but rather discretizes them in a mesh (Fig. 14.7). At each time step and grid point, the GCM computes the value of a variable from the values at neighboring grid points (advection of properties) and the local effects of the point itself (such as heating by radiative effect).

The impacts of increasing atmospheric concentrations of GHGs, the effects of the variability in the ozone layer on the climate system, as well as future climate projections, are all studied through GCMs. They are widely used tools in ocean and atmospheric sciences, since they allow to temporarily vary a single factor of interest (such as the concentration of GHGs, salinity, irradiance, etc.) and analyze its impacts. For example, a numerical simulation can be carried out in which the thickness of the ozone layer temporarily varies, without modifying the GHG concentration, or *vice versa*, and thus quantify the possible effect that a certain forcer has on the climate system. Projections of 21st century climate scenarios are called Shared Socioeconomic Pathways, SSPs each of which involves various scenarios to facilitate the integrated analysis of future climate impacts, vulnerabilities, adaptation, and mitigation (Riahi et al. 2017).

Despite the analytical level of abstraction reached with the use of supercomputers applied to numerical models, it should be noted that even today there are processes that have not been able to be quantified (sub-grid processes). Furthermore, it is also important to point that not all external forcers can be predicted, as for example, volcanic eruptions. Nevertheless, climate models are constantly updated, incorporating higher spatial resolution, and new parameters, such as novel physical processes and biogeochemical cycles. Modeling groups coordinate updates on climate models around the schedule of the IPCC assessment reports, publishing a set of model results, known as "runs" in the run-up to each new report.

Most projections that stem from the outcomes of future climate models indicate that the future warming of the atmosphere will depend on immediate efforts to limit GHGs emissions and actions taken towards the mitigation of the warming trends already established. If the current GHGs emission rate is continued, the global average air temperature will continue to increase. In addition, increased global temperature will result in longer, more frequent, and more intense heatwaves over most land areas (Pörtner

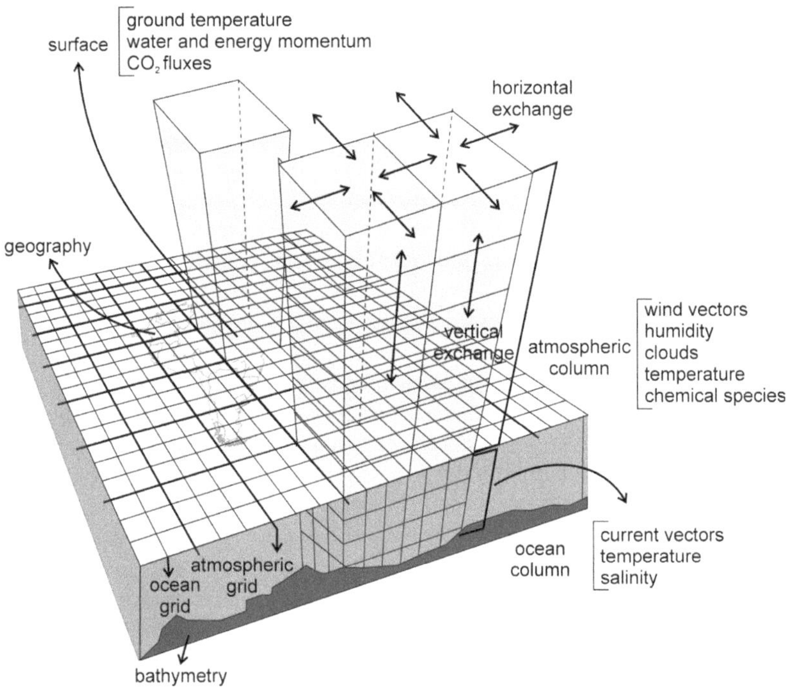

Figure 14.7. A schematic representation of Global Climatic Models (GCMs) grids and interactions between neighboring grid points. Modified and redrawn from McGuffie and Henderson-Sellers (2005).

et al. 2019). The global mean sea level will rise even if gas emissions are reduced drastically. By 2100, and if global warming is limited to 2°C, the oceans will have absorbed between 2–4 times more heat than in the present climate conditions, but if emissions are higher than those considered in such scenario, the ocean absorption of heat will be between 5–7 times greater (Pörtner et al. 2019a). In consequence, and since marine heatwaves are mainly related to air temperature, their frequency, duration, extension and intensity is expected to increase. Regarding OA, considerable decreases in pH are projected globally (albeit with regional and local variations), with values close to 7.7 and 7.5 for the years 2100 and 2300, respectively (Feely et al. 2009, Hartin et al. 2016).

Other than curbing GHGs emissions (which would entail a significant energy paradigm change, not necessarily backed by politics), mankind can appeal to natural mechanisms and biogeochemical cycles that imply the sequestration and storage of CO_2 into "inactive" compartments, for periods in the order of 100s of yr to millennia. The deep ocean (which harbors the largest carbon pools in the form of DIC and in sedimentary rocks; Le Quéré et al. 2018, see also Chapter 8) has long been considered as the most plausible candidate for CO_2 sequestration, and the biological pump has been postulated as one of the "natural" mechanisms to quench the excess atmospheric CO_2 in "active" compartments (e.g., atmosphere, surface ocean). Yet, from Chapter 4, we know that nutrients limit primary productivity in the oceans. Of particular interest in that regard are those regions of the world oceans where the concentration of dissolved nutrients above the thermocline is high, and yet present low productivity. These are known as High-Nutrient, Low Chlorophyll (HNLC) zones, and some examples are sectors of Southern Ocean, the Gulf of Alaska, or the equatorial Pacific. While inorganic nutrients are abundant in HNLC, the iron concentration in surface water is low, limiting the growth of phytoplankton > 8 µm. Iron is found in two oxidation states, Fe^{+3} (which is readily incorporated by cells) and Fe^{+2}. In alkaline seawater conditions, Fe^{+3} forms $Fe(OH)_3$, and flocculates, yielding an average concentration of $< 1 \times 10^{-9}$ moles Fe^{+3} l^{-1}. Fe^{+2} is oxidized to Fe^{+3} in seawater and also flocculates preventing direct uptake by cells. Fe is not an essential nutrient *per se*, but iron-sulfur clusters are localized in the mitochondria and plastids, where there is a high demand for these cofactors in electron transfer and catalysis (Balk and Schaedler 2014), and hence it is essential for photosynthesis.

Towards the end of the 1980s, the American oceanographer John H. Martin, director of Moss Landing Marine Laboratories, became a celebrity in the scientific community for postulating the "iron hypothesis" (Martin 1990). Martin compared Southern Ocean productivity in Fe-rich and Fe-poor areas and contrasted it with productivity during the Last Glacial Maximum, with the background knowledge that atmospheric CO_2 concentrations had increased from 200 to 280 ppm during the glacial/interglacial transition, and such variations were linked to Southern Ocean phytoplankton productivity. With evidence that dust Fe supplies were 50 times higher in the LGM than today, he posited that new productivity in the Southern Ocean (and the equatorial Pacific) was Fe-limited, and this led him to hypothesize that a large-scale Fe fertilization of such areas would trigger a pulse in the phytoplankton biological pump, significant enough to lower atmospheric CO_2 considerably (Martin 1990, Martin et al. 1991). It is rumored that Martin jokingly said among his colleagues: "*Give me a half tanker of iron, and I will give you an ice age*".

Martin went on to test this hypothesis in the equatorial Pacific off the Galápagos, in an unprecedented large-scale field experiment (IRONEX-I) that enriched an area of 64 km^2 with iron. Fe^{+3} concentration increased by two orders of magnitude from 0.06 nM to 4 nM resulting in a doubling of phytoplankton biomass, a threefold increase in chlorophyll and a fourfold increase in planktonic primary production. Similar trends were replicated in a chlorophyll-rich plume down-stream of the Galápagos Islands (Martin et al. 1994).

While vertical C transport of new production below the thermocline yielded uneven results from the IRONEX-I experiment, the iron hypothesis was to be further tested with other large-scale experiments (e.g., SOIREE, SOFEX, IRONEX-II) in following years. Under the recent conditions of extreme warming and sea ice loss in the Pacific Arctic shelf, O'Daly et al. (2020) found high export ratios of pelagic primary production, coupled with low rates of particle-associated microbial respiration (the latter adding to negligible recycling of sinking POC in the water column). Such auspicious finding highlights the extraordinary strength of the biological pump on the Arctic, and provide empirical evidence for alternatives to lower atmospheric CO_2 levels.

More recently, scientists and environmentalists alike have turned their interests to Blue Carbon ecosystems because of their role on global carbon budgets. We will not elaborate further on this subject, as it has been treated elsewhere in this volume (see Chapters 8 and 9), but let us highlight their high productivity rates and their capacity to store large amounts of organic matter in anoxic sediment layers, which prevent its oxidation and return to the atmospheric "active" carbon pool. Additionally, some projections indicate potential for Blue Carbon ecosystem expansion, such as in polar coastal ecosystems (Krause-Jensen and Duarte 2014, Barnes 2017).

Corollary

We hope to have given you an introduction to Earth's climate system and the ways in which it has been altered through anthropic influences in general, and in particular, how this has translated into the oceans, its organisms and ecosystems. We as scientists know that the planet is no stranger to drastic changes in its biotas, which are often caused by planetary-scale climatic changes. As an example, the end-Permian mass extinction affected marine ecosystems transversely during the Phanerozoic (Erwin 1994), causing the extinction of ~ 80% of marine species (Stanley and Yang 1994). However, the rate of change in climatic parameters and biological communities since the onset of the Anthropocene is a matter of concern for scientists worldwide. Year after year, new climate change impacts are discovered, or new ways in which climate forcers combine (additively or synergistically) to produce detrimental effects on marine life. As mentioned in the introduction, this chapter is by no means a thorough coverage of all the impacts that climate change has had and is currently having on marine ecosystems, but we rather intended to convey the message that the field of climate change science is budding, and so is the search for alternatives to minimize and mitigate its effects.

Besides the scientific aspects discussed here, the global climate crisis ultimately tells a story of imbalances and inequities. For example, small island nations in the Indo-Pacific, like Kiribati, Palau and the Maldives, contributed the least to climate change, yet they are the ones that are being hit the hardest by sea level rise and saltwater intrusion into aquifers. The debate on whether Third World countries should resort to 'cheap' energy sources like coal for their immediate development, or if they should invest in renewable energies can go on and on indefinitely. One thing is certain, Third World nations can do without the patronizing from developed countries which have built their wealth on fossil fuels, and are still reluctant to lead by example an energetic transition to renewable sources. On the other hand, there are auspicious initiatives such as that of the UN General Assembly calling for a Decade of Ocean Science for Sustainable Development (2021–2030) (Pendleton et al. 2020). What is perhaps more encouraging is that common people are becoming more and more engaged in climate action, exerting pressure on local and national governments and pushing for an environmental agenda that includes ocean governance, sustainable development and mitigation of climate change impacts.

If after reading this chapter you developed an interest in the science and the political and cultural background behind climate change, and would like to become more engaged in climate action, there are some fascinating books on what you can do to contribute your two cents. For the beginner, we suggest *Kick the habit* (Kirby 2008) published by the UN Environment Programme, which is not only authoritative, but also quite an entertaining guide to ways in which you can start your own path to 'climate neutrality'.

The scientific evidence is striking enough to show that the planet with its terrestrial and ocean ecosystems (and humankind) is changing at an accelerated rate and heading into a new stage. There is a normal degree of uncertainty associated with model projections on what will be the outcomes of this global-scale alteration of climatic and environmental conditions. But there is certainty in that we will witness significant change in the coming decades, as we have been witnessing since the Industrial Revolution. It lies in global politics (especially those of developed nations) to take action in order to palliate, fight or buffer the course of this unprecedented climatic change. Let us hope that mankind opts for a swift, sound, and science-based action plan that will speak for ourselves as a keystone species on this planet.

Summary

1. Human activities related to agricultural and industrial development and based on fossil fuels have altered Earth's subsystems and biogeochemical cycles, leading to a new geologic Epoch in the Quaternary, the Anthropocene.
2. Climate change became a matter of interest to society only recently, despite the disregard of its importance and impacts from certain sectors. The current environmental agenda is based on Sustainable Development Goals. Global Climate Models are used to produce projections of the future climate under different scenarios.
3. The climate system is made up of five subsystems, with coupling mechanisms represented by feedback processes between its components. Internal and external forcers produce changes in the climate system.
4. The global atmospheric circulation transfers heat from tropical to polar regions; it is made up of the Hadley, Ferrel and polar circulation cells. At low levels in the troposphere, the descending branches of each one of these cells create the trade, westerlies and easterly polar winds.
5. Variations in Earth's orbital parameters (eccentricity, obliquity and precession), known as Milankovitch cycles, are an example of external astronomical forcers responsible for the alternation of glacial and interglacial periods.
6. El Niño-Southern Oscillation (ENSO) is an example of the variability in atmospheric circulation and ocean-atmosphere interactions; it comprises three phases: El Niño (large-scale warming of surface waters in the central-east equatorial Pacific), La Niña (surface cooling) and a normal phase.
7. The depletion of stratospheric O_3 promotes a radiative imbalance with an increase in short-wave UV radiation; exposure to UV elicits different responses in marine organisms.

8. The anthropic alteration of the concentrations of greenhouse gases (GHGs: CO_2, methane, and NO_2) results in a greater retention of long-wave energy, producing the recent global warming. The average air temperature increased globally by 1°C in relation to pre-industrial values. In the oceans, this has manifested as increased surface thermal stratification, and focalized, steep SST rise (marine heatwaves).
9. In marine ecosystems, warming has impacted the distribution range of organisms, led to temperature- and stratification-induced changes in phytoplankton, and altered the phenology of marine communities.
10. The rate of CO_2 absorption by seawater is currently high in relation to the supply of carbonate ions, resulting in a decreasing pH trend (ocean acidification), which affects marine organisms differentially. Seaweed physiology might contribute to the modulation of seawater pH at local scales.
11. The biological pump and Blue Carbon ecosystems have been regarded as natural mechanisms/environments to effectively trap excess atmospheric CO_2, and palliate the effects of climate change.

Acknowledgements

Many sections of this chapter stemmed from a graduate course on *Coastal Systems and Climate Change* we taught in the spring of 2017 at National University at Mar del Plata (Argentina), after the initiative of Professors M.A. Espinosa and F.I. Isla. In this class, the community of students and instructors often found grounds for scientific debate and discussion of ideas within an interdisciplinary background. For that we are grateful, and hope to have transmitted some of the philosophy of the course into this chapter. P.D. Pratolongo designed most of the figures.

References

Arrhenius, S. 1896. On the influence of carbonic acid in the air upon the temperature of the ground. Philos. Mag. Ser. 5(41): 237–276.

Atkinson, A., V. Siegel, E. Pakhomov and P. Rothery. 2004. Long-term decline in krill stock and increase in salps within the Southern Ocean. Nature 432: 100–103.

Balk, J. and T.A. Schaedler. 2014. Iron cofactor assembly in plants. Annu. Rev. Plant Biol. 65: 125–153.

Ball, W.T., J. Alsing, D.J. Mortlock, J. Staehelin, J.D. Haigh, T. Peter, F. Tummon, R. Stübi, A. Stenke, J. Anderson, A. Bourassa, S.M. Davis, D. Degenstein, S. Frith, L. Froidevaux, C. Roth, V. Sofieva, R. Wang, J. Wild, P. Yu, J.R. Ziemke and E.V. Rozanov. 2018. Evidence for a continuous decline in lower stratospheric ozone offsetting ozone layer recovery. Atmos. Chem. Phys. 18: 1379–1394.

Barbeaux, S.J., K. Holsman and S. Zador. 2020. Marine heatwave stress test of ecosystem-based fisheries management in the Gulf of Alaska Pacific cod fishery. Front. Mar. Sci. 7: 703. doi: 10.3389/fmars.2020.00703.

Barnes, D.K.A. 2017. Polar zoobenthos Blue Carbon storage increases with sea ice losses, because across-shelf growth gains from longer algal blooms outweigh ice scour mortality in the shallows. Glob. Change Biol. 23: 5083–5091.

Baumann, H., S.C. Talmage and C.J. Gobler. 2012. Reduced early life growth and survival in a fish in direct response to increased carbon dioxide. Nat. Clim. Change 2: 38–41.

Boutin, J., Y. Chao and W.E. Asher. 2016. Satellite and *in situ* salinity: understanding near-surface stratification and subfootprint variability. Bull. Am. Meteorol. Soc. 97: 1391–1407.

Cartaxana, P., S. Vieira, L. Ribeiro, R.J.M. Rocha, S. Cruz, R. Calado and J. Marques da Silva. 2015. Effects of elevated temperature and CO_2 on intertidal microphytobenthos. BMC Ecol. 15: 10. doi:10.1186/s12898-015-0043-y.

Colle, B.A., F. Buonaiuto, M.J. Bowman, R.E. Wilson, R. Flood, R. Hunter, A. Mintz and D. Hill. 2008. New York City's vulnerability to coastal flooding. Bull. Amer. Meteorol. Soc. 89: 829–841.

Crear, D.P., B.E. Watkins, M.A.M. Friedrichs, P. St-Laurent and K.C. Weng. 2020. Estimating shifts in phenology and habitat use of cobia in Chesapeake Bay under climate change. Front. Mar. Sci. 7: 579135. doi: 10.3389/fmars.2020.579135.

Crutzen, P.J. and E.F. Stoermer. 2000. The "Anthropocene". IGBP Newslett. 41: 17–18.

Dahms, H.-U., S. Dobretsov and J.-S. Lee. 2011. Effects of UV radiation on marine ectotherms in polar regions. Comp. Biochem. Physiol. C 153: 363–371.

de Bettignies, T., T. Wernberg and C.F.D. Gurgel. 2018. Exploring the influence of temperature on aspects of the reproductive phenology of temperate seaweeds. Front. Mar. Sci. 5: 218. doi.org/10.3389/fmars.2018.00218.

De Falco, C., A. Bracco and C. Pasquero. 2020. Climatic and oceanographic controls on coral bleaching conditions in the Maldivian Region. Front. Mar. Sci. 7: 539869. doi: 10.3389/fmars.2020.539869.

Deppeler, S.L. and A.T. Davidson. 2017. Southern Ocean phytoplankton in a changing climate. Front. Mar. Sci. 4: 40. doi: 10.3389/fmars.2017.00040.

Doney, S.C., V.J. Fabry, R.A. Feely and J.A. Kleypas. 2009. Ocean acidification: the other CO_2 problem. Ann. Rev. Mar. Sci. 1: 169–192.

Downey-Wall, A.M., L.P. Cameron, B.M. Ford, E.M. McNally, Y.R. Venkataraman, S.B. Roberts, J.B. Ries and K.E. Lotterhos. 2020. Ocean acidification induces subtle shifts in gene expression and DNA methylation in mantle tissue of the Eastern oyster (*Crassostrea virginica*). Front. Mar. Sci. 7: 566419. doi: 10.3389/fmars.2020.566419.

Duarte, C.M. 2014. Global change and the future ocean: a grand challenge for marine sciences. Front. Mar. Sci. 1: 64. doi.org/10.3389/fmars.2014.00063.

Duarte, C.M., J. Wu, X. Xiao, A. Bruhn and D. Krause-Jensen. 2017. Can seaweed farming play a role in climate change mitigation and adaptation? Front. Mar. Sci. 4: 100. doi: 10.3389/fmars.2017.00100.

Edwards, M. and A.J. Richardson. 2004. Impact of climate change on marine pelagic phenology and trophic mismatch. Nature 430: 881–884.

Edwards, M., D.G. Johns, S.C. Leterme, E. Svendsen and A.J. Richardson. 2006. Regional climate change and harmful algal blooms in the northeast Atlantic. Limnol. Oceanogr. 51: 820–829.

Erwin, D.H. 1994. The Permo-Triassic extinction. Nature 367: 231–236.

Farmer, G.T. and J. Cook. 2013. Climate Change Science: A Modern Synthesis. Vol. 1-The Physical Climate. Springer, Heidelberg.

Fasullo, J.T. and K.E. Trenberth. 2008. The annual cycle of the energy budget. Part II: Meridional structures and poleward transports. J. Clim. 21: 2313–2325.

Feely, R.A., S.C. Doney and S.R. Cooley. 2009. Ocean acidification: present conditions and future changes in a high-CO_2 world. Oceanography 22: 36–47.

Fink, A., J. den Haan, A. Chennu, S. Uthicke and D. de Beer. 2017. Ocean acidification changes abiotic processes but not biotic processes in coral reef sediments. Front. Mar. Sci. 4: 73. doi: 10.3389/fmars.2017.00073.

Forsgren, E., S. Dupont, F. Jutfelt and T. Amundsen. 2013. Elevated CO_2 affects embryonic development and larval phototaxis in a temperate marine fish. Ecol. Evol. 3: 3637–3646.

Gattusso, J.P., A. Magnan, R. Billé, W.W. Cheung, E.L. Howes, F. Joos, D. Allemand, L. Bopp, S.R. Cooley, C.M. Eakin, O. Hoegh-Guldberg, R.P. Kelly, H.O. Pörtner, A.D. Rogers, J.M. Baxter, D. Laffoley, D. Osborn, A. Rankovic, J. Rochette, U.R. Sumaila, S. Treyer and C. Turley. 2015. Contrasting futures for ocean and society from different anthropogenic CO_2 emissions scenarios. Science 349: aac4722.

Gitay, H., A. Suárez, R.T. Watson and D.J. Dokken [eds.]. 2002. Climate Change and Biodiversity-IPCC Technical Paper V. Cambridge University Press, Cambridge, UK and New York, USA.

Glynn, P.W., L.J. Maté, A.C. Baker and M.O. Calderón. 2001. Coral bleaching and mortality in Panama and Ecuador during the 1997–1998 El Niño-Southern Oscillation Event: spatial/temporal patterns and comparisons with the 1982–1983 event. Bull. Mar. Sci. 69: 79–109.

Gobler, C.J., E.L. DePasquale, A.W. Griffith and H. Baumann. 2014. Hypoxia and acidification have additive and synergistic negative effects on the growth, survival, and metamorphosis of early life stage bivalves. PLoS ONE 9: e83648. doi:10.1371/journal.pone.0083648.

González, P.L., L.M. Polvani, R. Seager and G.J. Correa. 2014. Stratospheric ozone depletion: a key driver of recent precipitation trends in South Eastern South America. Clim. Dyn. 42: 1775–1792.

Guan, C., M. Saha and F. Weinberger. 2020. Simulated heatwaves lead to upregulated chemical defense of a marine foundation macrophyte against microbial colonizers. Front. Mar. Sci. 7: 463. doi: 10.3389/fmars.2020.00463.

Guiry, M.D. and C.J. Dawes. 1992. Daylength, temperature and nutrient control of tetrasporogenesis in *Asparagopsis armata* (Rhodophyta). J. Exp. Mar. Biol. Ecol. 158: 197–217.

Hartin, C.A., B. Bond-Lamberty, P. Patel and A. Mundra. 2016. Ocean acidification over the next three centuries using a simple global climate carbon-cycle model: projections and sensitivities. Biogeosciences 13: 4329–4342.

Hoegh-Guldberg, O. and G.J. Smith. 1989. The effect of sudden changes in temperature, light and salinity on the population density and export of zooxanthellae from the reef corals *Stylophora pistillata* and *Seriatopra hystrix*. J. Exp. Mar. Biol. Ecol. 129: 279–303.

Hoegh-Guldberg, O. and E.S. Poloczanska. 2017. Editorial: The effect of climate change across ocean regions. Front. Mar. Sci. 4: 361. doi: 10.3389/fmars.2017.00361.

Hoegh-Guldberg, O., E.S. Poloczanska, W. Skirving and S. Dove. 2017. Coral reef ecosystems under climate change and ocean acidification. Front. Mar. Sci. 4: 158. doi: 10.3389/fmars.2017.00158.

Hughes, L. 2000. Biological consequences of global warming: is the signal already apparent? Trends Ecol. Evol. 15: 56–61.

Ketzer, M., D. Praeg, L.F. Rodrigues, A. Augustin, M.A.G. Pivel, M. Rahmati-Abkenar, D.J. Miller, A.R. Viana and J.A. Cupertino. 2020. Gas hydrate dissociation linked to contemporary ocean warming in the southern hemisphere. Nat. Commun. 11: 3788. doi.org/10.1038/s41467-020-17289-z.

Kirby, A. 2008. CCCC Kick the habit, a UN guide to climate neutrality. United Nations Environment Programme, Nairobi. Kenya.

Krause-Jensen, D. and C.M. Duarte. 2014. Expansion of vegetated coastal ecosystems in the future Arctic. Front. Mar. Sci. 1: 77. doi.org/10.3389/fmars.2014.00077.

Le Quéré, C., R.M. Andrew, P. Friedlingstein, S. Sitch, J. Hauck et al. 2018. Global carbon budget 2018. Earth Syst. Sci. Data 10: 2141–2194.

Lehmann, J. and D. Coumou. 2015. The influence of mid-latitude storm tracks on hot, cold, dry and wet extremes. Sci. Rep. 5: 17491; doi.org/10.1038/srep17491.

Leung, J.Y.S., B.D. Russell and S.D. Connell. 2019. Adaptive responses of marine gastropods to heatwaves. One Earth 1: 374–381.

Lewis, S.L. and M.A. Maslin. 2015. Defining the Anthropocene. Nature 519: 171–180.

Loeb, V., V. Siegel, O. Holm-Hansen, R. Hewitt, W. Fraserk, W. Trivelpiecek and S. Trivelpiecek. 1997. Effects of sea-ice extent and krill or salp dominance on the Antarctic food web. Nature 387: 897–900.

Lutgens, F.K., E.J. Tarbuck and D. Tasa. 2012. Essentials of Geology, 11th Ed. Pearson/Prentice Hall, Boston. USA.

Madden, R.A. and P.R. Julian. 1994. Observations of the 40–50-day tropical oscillation—A review. Mon. Weather Rev. 122: 814–837.

Mantua, N.J. and S.R. Hare. 2002. The Pacific Decadal Oscillation. J. Oceanogr. 58: 35–44.

Martin, J.H. 1990. Glacial-interglacial CO_2 change: the iron hypothesis. Paleoceanography 5: 1–13.

Martin, J.H., M. Gordon and S.E. Fitzwater. 1991. The case for iron. Limnol. Oceanogr. 36: 1793–1802.

Martin, J.H., K.H. Coale, K.S. Johnson, S.E. Fitzwater, R.M. Gordon et al. 1994. Testing the iron hypothesis in ecosystems of the equatorial Pacific Ocean. Nature 371: 123–129.

McGuffie, K. and A. Henderson-Sellers. 2005. A Climate Modelling Primer. Third edition. John Wiley & Sons, Chichester, England.

Meinen, C.S., R.C. Perez, S. Dong, A.R. Piola and E. Campos. 2020. Observed ocean bottom temperature variability at four sites in the northwestern Argentine basin: evidence of decadal deep/abyssal warming amidst hourly to interannual variability during 2009–2019. Geophys. Res. Lett. 47: e2020GL089093.

Miller, G.T. Jr. 2004. Living in the Environment: Principles, Connections, and Solutions. Thirteenth edition. Thomson-Brooks/Cole, Toronto. Canada.

Monastersky, R. and N. Sousanis. 2015. The fragile framework. Nature 527: 427–435.

Mongin, M., M.E. Baird, S. Hadley and A. Lenton. 2016. Optimising reef-scale CO_2 removal by seaweed to buffer ocean acidification. Environ. Res. Lett. 11: 034023. doi:10.1088/1748-9326/11/3/034023.

Moss, R.H., J.A. Edmonds, K.A. Hibbard, M.R. Manning, S.K. Rose, D.P. van Vuuren, T.R. Carter, S. Emori, M. Kainuma, T. Kram, G.A. Meehl, J.F.B. Mitchell, N. Nakicenovic, K. Riahi, S. Smith, R.J. Stouffer, A.M. Thomson, J.P. Weyant and T.J. Wilbanks. 2010. The next generation of scenarios for climate change research and assessment. Nature 463: 747–756.

O'Daly, S.H., S.L. Danielson, S.M. Hardy, R.R. Hopcroft, C. Lalande, D.A. Stockwell and A.M.P. McDonnell. 2020. Extraordinary carbon fluxes on the shallow Pacific Arctic shelf during a remarkably warm and low sea ice period. Front. Mar. Sci. 7: 548931. doi: 10.3389/fmars.2020.548931.

O'Neil, J.M., T.W. Davis, M.A. Burford and C.J. Gobler. 2012. The rise of harmful cyanobacteria blooms: the potential roles of eutrophication and climate change. Harmful Algae 14: 313–334.

Orr, J.C., V.J. Fabry, O. Aumont, L. Bopp, S.C. Doney et al. 2005. Anthropogenic ocean acidification over the twenty-first century and its impact on calcifying organisms. Nature 437: 681–686.

Pan, J., M.A. Marcoval, S.M. Bazzini, M.V. Vallina, and S.G. De Marco. 2013. Chapter 2: Coastal Marine Biodiversity, challenges and threats. Pp 43-67. *In*: Arias A.H. and M.C. Menéndez [eds.] Marine Ecology in a Changing World. CRC Press, Boca Raton, USA

Paerl, H.W. and J. Huisman. 2008. Blooms like it hot. Science 320: 57–58.

Paerl, H.W., N.S. Hall and E.S. Calandrino. 2011. Controlling harmful cyanobacterial blooms in a world experiencing anthropogenic and climatic-induced change. Sci. Total Environ. 409: 1739–1745.

Pendleton, L., K. Evans and M. Visbeck. 2020. Opinion: We need a global movement to transform ocean science for a better world. Proc. Natl. Acad. Sci. USA 117: 9652–9655. doi:10.1073/pnas.2005485117.

Peperzak, L. 2003. Climate change and harmful algal blooms in the North Sea. Acta Oecol. 24: S139–S144.

Polvani, L.M., D.W. Waugh, G.J. Correa and S.W. Son. 2011. Stratospheric ozone depletion: the main driver of twentieth-century atmospheric circulation changes in the Southern Hemisphere. J. Clim. 24: 795–812.

Pörtner, H.-O., D.C. Roberts, V. Masson-Delmotte, P. Zhai, E. Poloczanska, K. Mintenbeck, M. Tignor, A. Alegría, M. Nicolai, A. Okem, J. Petzold, B. Rama and N.M. Weyer [eds.]. 2019a. Technical Summary: IPCC Special Report on the Ocean and Cryosphere in a Changing Climate.

Pörtner, H.-O., D.C. Roberts, V. Masson-Delmotte, P. Zhai, M. Tignor, E. Poloczanska, K. Mintenbeck, A. Alegría, M. Nicolai, A. Okem, J. Petzold, B. Rama and N. Weyer [eds.]. 2019b. Summary for Policymakers: IPCC Special Report on the Ocean and Cryosphere in a Changing Climate.

Previdi, M. and L.M. Polvani. 2014. Climate system response to stratospheric ozone depletion and recovery. Q. J. R. Meteorol. Soc. 140: 2401–2419.

Reimer, J.M., R. Devillers and J. Claudet. 2020. Benefits and gaps in area-based management tools for the ocean Sustainable Development Goal. Nat Sustain doi.org/10.1038/s41893-020-00659-2.

Rein, B., A. Lückge, L. Reinhardt, F. Sirocko, A. Wolf and W.-C. Dullo. 2005. El Niño variability off Peru during the last 20,000 years. Paleoceanography 20: PA4003. doi:10.1029/2004PA001099.

Rhein, M., S.R. Rintoul, S. Aoki, E. Campos, D. Chambers, R.A. Feely, S. Gulev, G.C. Johnson, S.A. Josey, A. Kostianoy, C. Mauritzen, D. Roemmich, L.D. Talley and F. Wang. 2013. Observations: Ocean. pp. 255–316. *In*: Stocker, T.F., D. Qin, G.-K. Plattner, M. Tignor, S.K. Allen, J. Boschung, A. Nauels, Y. Xia, V. Bex and P.M. Midgley [eds.]. Climate Change 2013: The Physical Science Basis. Contribution of Working Group I to the Fifth Assessment Report of the Intergovernmental Panel on Climate Change. Cambridge University Press, Cambridge, UK and New York, NY, USA.

Riahi, K., D.P. Van Vuuren, E. Kriegler, J. Edmonds, B.C. O'Neill et al. 2017. The shared socioeconomic pathways and their energy, land use, and greenhouse gas emissions implications: an overview. Glob. Environ. Change 42: 153–168.

Rodbell, D.T., G.O. Seltzer, D.M. Anderson, M.B. Abbott, D.B. Enfield and J.H. Newman. 1999. An ~ 15,000-year record of El Niño-driven alluviation in southwestern Ecuador. Science 283: 516–520.

Roquet, H., A. Pisano and O. Embury. 2016. Sea surface temperature. *In*: von Schuckmann, K. et al. [eds.]. The Copernicus Marine Environment Monitoring Service Ocean State Report. J. Oper. Oceanogr. 9(Sup. 2): s235–s320; doi:10.1080/1755876X.2016.1273446.

Seibold, E. and W.H. Berger. 1996. The Sea Floor, An Introduction to Marine Geology, Third edition. Springer-Verlag, Berlin. Germany.

Seitzinger, S.P., O. Gaffney, G. Brasseur, W. Broadgate, P. Ciais, M. Claussen, J.W. Erisman, T. Kiefer, C. Lancelot, P.S. Monks, K. Smyth, J. Syvitski and M Uematsuj. 2015. International geosphere-biosphere programme and Earth system science: three decades of co-evolution. Anthropocene 12: 3–16.

Shaw, T.A., M. Baldwin, E.A. Barnes, R. Caballero, C.I. Garfinkel, Y.T. Hwang, P.A. O'Gorman, G. Rivière, I.R. Simpson and A. Voigt. 2016. Storm track processes and the opposing influences of climate change. Nat. Geosci. 9: 656–664.

Simonson, E.J., R.E. Scheibling and A. Metaxas. 2015. Kelp in hot water: I. Warming seawater temperature induces weakening and loss of kelp tissue. Mar. Ecol. Prog. Ser. 537: 89–104.

Solomon, S., D.J. Ivy, D. Kinnison, M.J. Mills, R.R. Neely and A. Schmidt. 2016. Emergence of healing in the Antarctic ozone layer. Science 353: 269–274.

Stanley, S.M. and X. Yang. 1994. A double mass extinction at the end of the Paleozoic Era. Science 266: 1340–1344.

Steele, J.H. 1998. Regime shifts in marine ecosystems. Ecol. Appl. 8: S33–S36.

Straub, S.C., T. Wernberg, M.S. Thomsen, P.J. Moore, M.T. Burrows, B.P. Harvey and D.A. Smale. 2019. Resistance, extinction, and everything in between the diverse responses of seaweeds to marine heatwaves. Front. Mar. Sci. 6: 763. doi: 10.3389/fmars.2019.00763.

Takao, S., N.H. Kumagai, H. Yamano, M. Fujii and Y. Yamanaka. 2015. Projecting the impacts of rising seawater temperatures on the distribution of seaweeds around Japan under multiple climate change scenarios. Ecol. Evol. 5: 213–223.

Talmage, S.C. and C.J. Gobler. 2009. The effects of elevated carbon dioxide concentrations on the metamorphosis, size, and survival of larval hard clams (*Mercenaria mercenaria*), bay scallops (*Argopecten irradians*), and Eastern oysters (*Crassostrea virginica*). Limnol. Oceanogr. 54: 2072–2080.

Tanaka, K., S. Taino, H. Haraguchi, G. Prendergast and M. Hiraoka. 2012. Warming off southwestern Japan linked to distributional shifts of subtidal canopy-forming seaweeds. Ecol. Evol. 2: 2854–2865.

Thompson, D.W.J. and S. Solomon. 2002. Interpretation of recent Southern Hemisphere climate change. Science 296: 895–899.

Thompson, D.W.J., S. Solomon, P.J. Kushner, M.H. England, K.M. Grise and D.J. Karoly. 2011. Signatures of the Antarctic ozone hole in Southern Hemisphere surface climate change. Nat. Geosci. 4: 741–749.

Trenberth, K.E. and J.W. Hurrell. 1994. Decadal atmosphere-ocean variations in the Pacific. Clim. Dyn. 9: 303–319.

Turner, P.J., B. Ball, Z. Diana, A. Fariñas-Bermejo, I. Grace, D. McVeigh, M.M. Powers, L. Van Audenhaege, S. Maslakova, C.M. Young and C.L. Van Dover. 2020. Methane seeps on the US Atlantic margin and their potential importance to populations of the commercially valuable deep-sea red crab, *Chaceon quinquedens*. Front. Mar. Sci. 7: 75. doi: 10.3389/fmars.2020.00075.

Tzortzi, E., S.A. Josey, M. Srokosz and C. Gommenginger. 2013. Tropical Atlantic salinity variability: new insights from SMOS. Geophys. Res. Lett. 40: 2143–2147.

Vera, C.S. and M. Osman. 2018. Activity of the Southern Annular Mode during 2015–2016 El Niño event and its impact on Southern Hemisphere climate anomalies. Int. J. Climatol. 38: e1288–e1295.

von Schuckmann, K., P.Y. Le Traon, N. Smith, A. Pascual, P. Brausser, K. Fennel and S. Djavidnia. 2018. Copernicus marine service ocean state report. J. Oper. Oceanogr. 11(Sup. 1): S1–S142. doi:10.1080/1755876X.2018.1489208.

Wang, M. and J.E. Overland. 2009. A sea ice-free summer Arctic within 30 years? Geophys. Res. Lett. 36: L07502. doi:10.1029/2009GL037820.

Wilson, K.L., L.M. Kay, A.L. Schmidt and H.K. Lotze. 2015. Effects of increasing water temperatures on survival and growth of ecologically and economically important seaweeds in Atlantic Canada: implications for climate change. Mar. Biol. 162: 2431–2444.

Yeh, S.-W., W. Cai, S.-K. Min, M.J. McPhaden, D. Dommenget, B. Dewitte, M. Collins, K. Ashok, S.-I. An, B.-Y. Yim and J.-S. Kug. 2018. ENSO atmospheric teleconnections and their response to greenhouse gas forcing. Rev. Geophys. 56: 185–206.

Index

C

D

H

I

J

K

L

S

T

About the Editors

Jerónimo Pan

Jerónimo Pan is Adjunct Research Scientist at the National Research and Technical Council of Argentina (CONICET). He received his double B.Sc. degree in Ecology and Botany from the National University at La Plata (UNLP) in 2000. He then obtained a M.A. degree in Higher Education, also from UNLP. In 2004, he was awarded the prestigious Fulbright fellowship to pursue graduate studies in the United States; he went on to earn his *Ph.D.* in Marine and Atmospheric Sciences in 2010, from the State University of New York at Stony Brook. Dr. Pan was the 2000 recipient of the 'J.V. González' Distinction as class Valedictorian, and in 2011 he was presented the Antarctica Service Medal of the USA from NSF, among other honors.

Dr. Pan's research focus is in coastal and estuarine ecosystems. Throughout his career, he has studied the ecosystemic role of bivalve suspension-feeders, microphytobenthos and phytoplankton dynamics, and the ecological role of modern microbial mats in coastal environments, on which he has dozens of publications. He is currently on the editorial board for the journal *Frontiers in Marine Science*. Dr. Pan has participated in oceanographic campaigns in the northwest- and southwest Atlantic Ocean, and the Ross Sea in Antarctica. In addition to carrying out research in Marine Biology, he has taught several undergraduate and graduate courses in the USA and Argentina, led educational field trips and scientific campaigns to coastal areas, and mentored students.

Paula Pratolongo

Paula Pratolongo is an Independent Researcher at the National Research and Technical Council of Argentina (CONICET), and Professor of Marine Ecology at Universidad Nacional del Sur, Bahía Blanca. She graduated in Biology in 2000 and obtained her *Ph.D.* in Biological Sciences from Universidad de Buenos Aires in 2005. In 2009, Dr. Pratolongo was awarded a Fulbright fellowship to pursue postdoctoral studies at East Carolina University, in the United States. In 2002, she was one of the six young scientists worldwide to receive the UNESCO-MAB Young Scientists Awards, endorsed by the International Coordinating Council of the Man and the Biosphere Program, in support of her research on Ecosystems and Water. Current research of Dr. Pratolongo is dedicated to wetlands' ecology and coastal oceanography.